HOW TO SUCCEED IN ORGANIC CHEMISTRY

JOHN E. GORDON, Ph.D.

Department of Chemistry
Kent State University

John Wiley & Sons, Inc.
New York · Chichester · Brisbane · Toronto

Editors: Judy Wilson and Irene Brownstone
Editorial/Production Supervision: Service to Publishers, Inc.

Copyright © 1979, by John Wiley & Sons, Inc.

All rights reserved. Published simultaneously in Canada.

Reproduction or translation of any part of this work beyond that permitted by Sections 107 or 108 of the 1976 United States Copyright Act without the permission of the copyright owner is unlawful. Requests for permission or further information should be addressed to the Permissions Department, John Wiley & Sons, Inc.

Library of Congress Cataloging in Publication Data

Gordon, John E.
    How to succeed in organic chemistry.
    (Wiley self-teaching guides)
    Includes index.
    1. Chemistry, Organic—Programmed instruction.
I. Title.
QD251.2.G67   547'.007'7   78-21496
ISBN 0-471-03010-4

Printed in the United States of America

79   80   10 9 8 7 6 5 4 3 2 1

Acknowledgments

In a very real sense this is a book by students, for students. I listened, reacted, and organized, but the material itself originated in student response to organic chemistry and was repeatedly plowed back into successive courses, where many generations of students made their contributions in turn. Irene Brownstone, the series editor, helped improve not only the language and organization, but the logic and the science as well.

To the Reader

Today's textbooks are better than ever, but they leave out one thing—they don't tell you *what to do* to learn organic chemistry successfully. Neither do the study guides that accompany the textbooks. This guide tells you what to do and how to do it.

You can use this book as a supplement to your organic course; as a guide for independent study; or as a self-instructional review for achievement tests (GRE, MCAT). Whatever your purpose, look first at Unit 1, where you will find practical and detailed plans for using this book and learning organic chemistry.

Several hundred students have already contributed to the improvement of this guide by commenting on their experience with it. If you would like to make suggestions for its future development, please address them to:

>Editor, Self-Teaching Guides
>John Wiley & Sons, Inc.
>605 Third Avenue
>New York, NY 10016

To the Instructor

Using this guide as a textbook supplement in the organic class can help you in several ways. By taking over repetition of worked-out examples and answering many questions not answered by the text, it should free some class time for other purposes. This guide provides a concrete and individualized self-study method that you can recommend to students who need more help than you can provide during office hours. It will be especially helpful in pulling the poorly prepared student through the first part of the course. By improving students' notational skills, it makes them better able to concentrate on content. Unit 1 gives students a detailed, practical plan for learning organic chemistry. The Cross-Reference Chart on pages xii to xv keys this guide to whatever textbook you are using in your organic chemistry course.

Suggestions for further improvements in this guide are welcome. Please write to:

Editor, Self-Teaching Guides
John Wiley & Sons, Inc.
605 Third Avenue
New York, NY 10016

Contents

To the Reader — iv

To the Instructor — v

Cross-Reference Chart — viii

Important Charts and Tables — xii

Guide to Standard Problem Types and Solution Techniques — xiv

Unit 1	Introduction	1
Unit 2	Structures	12
Unit 3	Function and Skeleton	32
Unit 4	Geometry of Organic Molecules	47
Unit 5	Seeing Structures Analytically	71
Unit 6	Structural Isomerism	77
Unit 7	Systematic Nomenclature	96
Unit 8	Stable Versus Unstable Molecules	133
Unit 9	Resonance	142
Unit 10	Mechanism	165
Unit 11	Acids and Bases	178
Unit 12	Reactions of the Aliphatic Hydrocarbons	198
Unit 13	Stereochemistry	247
Unit 14	Halides	305
Unit 15	Alcohols	335
Unit 16	Aromatic Hydrocarbons and Their Derivatives	355
Unit 17	Aldehydes and Ketones (Carbonyl Compounds)	391
Unit 18	Carboxylic Acids and Their Derivatives	439
Unit 19	Amines	481
Unit 20	Structure Problems	502
Unit 21	Synthetic Problems	540

Index — 593

Cross-Reference Chart

To locate the units of this guide that you may want to consult in studying the various chapters of the following textbooks, consult the chart on pages xiv and xv.

Allinger, N. J., et al., *Organic Chemistry*, 2nd ed., Worth Publishers, New York, 1976.
Baumgarten, R. L., *Organic Chemistry, a Brief Survey*, Ronald Press, New York, 1978.
Brown, R. F., *Organic Chemistry*, Wadsworth, Belmont, Calif., 1975.
DePuy, C. H., and K. L. Rinehart, Jr., *Introduction to Organic Chemistry*, 2nd ed., Wiley, New York, 1975.
English, J., Jr., H. G. Cassidy, and R. L. Baird, *Principles of Organic Chemistry*, 4th ed., McGraw-Hill, New York, 1971.
Fessenden, R. J., and J. S. Fessenden, *Organic Chemistry*, Willard Grant Press, Boston, 1979.
Finar, I. L., *Organic Chemistry*, 6th ed., Longman, London, 1973.
Geissman, T. A., *Principles of Organic Chemistry*, 4th ed., W. H. Freeman, San Francisco, 1977.
Gerig, J. T., *Introduction to Organic Chemistry*, Academic Press, New York, 1974.
Griffin, R. W., Jr., *Modern Organic Chemistry*, McGraw-Hill, New York, 1969.
Gutsche, C. D., and D. J. Pasto, *Fundamentals of Organic Chemistry*, Prentice-Hall, New York, 1975.
Hart, H., and R. D. Schuetz, *Organic Chemistry, a Short Course*, 4th ed., Houghton Mifflin, Boston, 1972.
Hendrickson, J. B., D. J. Cram, and G. S. Hammond, *Organic Chemistry*, 3rd ed., McGraw-Hill, New York, 1970.
Holum, J. R., *Organic Chemistry, a Brief Course*, Wiley, New York, 1975.
Kice, J. L., and E. N. Marvell, *Modern Principles of Organic Chemistry*, 2nd ed., Macmillan, New York, 1974.
Leffler, J. E., *A Short Course in Modern Organic Chemistry*, Macmillan, New York, 1973.
Menger, F. M., D. J. Goldsmith, and L. Mandell, *Organic Chemistry, a Concise Approach*, Benjamin, Menlo Park, Calif., 1972.
Moore, J. A., *Elements of Organic Chemistry*, W. B. Saunders, Philadelphia, 1974.
Moore, J. A., and T. J. Barton, *Organic Chemistry, an Overview*, W. B. Saunders, Philadelphia, 1978.
Morrison, R. T., and R. N. Boyd, *Organic Chemistry*, 3rd ed., Allyn and Bacon, Boston, 1973.
Neckers, D. C., and M. P. Doyle, *Organic Chemistry*, Wiley, New York, 1977.
O'Leary, M. H., *Contemporary Organic Chemistry*, McGraw-Hill, New York, 1976.
Reusch, W. H., *An Introduction to Organic Chemistry*, Holden-Day, San Francisco, 1977.

Roberts, J. D., R. Stewart, and M. C. Caserio, *Organic Chemistry, Methane to Macromolecules*, Benjamin, Menlo Park, Calif., 1971.

Solomons, T. W. G., *Organic Chemistry*, Wiley, New York, 1976.

Streitwieser, A., Jr., and C. H. Heathcock, *Introduction to Organic Chemistry*, Macmillan, New York, 1976.

Ternay, A. L., Jr., *Contemporary Organic Chemistry*, W. B. Saunders, Philadelphia, 1976.

Zimmerman, H., and I. Zimmerman, *Elements of Organic Chemistry*, Glencoe, Encino, Calif., 1977.

Zlatkis, A., E. Breitmaier, and G. Jung, *A Concise Introduction to Organic Chemistry*, McGraw-Hill, New York, 1973.

UNITS OF THIS SUPPLEMENT FOR USE

Text \ Chapter	1	2	3	4	5	6	7	8	9	10	11	12	13	14
Allinger et al.	—	1, 2, 4	4–7	3–5, 7, 11	—	14	7, 8, 9, 13	7, 11, 17, 18	—	11, 18	8, 9	11	—	10, *12*, 13
Baumgarten	1, 2, 4	3, 5, 6	7	11	8, 9	10	12	12	12	16	15	15	14	17
Brown	1, 3	7, 10, 11	7, 12, 14	3, 5, 6, 7	2, 4, 8, 9	4, 6, *13*	10, *14*	—	7, *15*, 20, 21	—	—	4, *12*	*12*, 21	8, 9
DePuy and Rinehart	1, 2, 4, 6, 7	—	4, 7, 9, 10, *12*	12	7, 9, *16*	7, *15*	13, 14	7, *17*	13	16, *19*	18			
English, Cassidy, and Baird	1, 2, 4, 9, 10	3, 5–7, *12*	7, *12*	6, 12	*12*	9, *16*	13	7, *15*	*14*, 16	15	7, *17*	11, *18*	18	11, *19*, 20, 21
Fessenden and Fessenden	1–3, 8, 11	4, 5, 9	6, 7	13	10, *14*	14	15	—	12	16	17	18	18	17, 18, 21
Finar	1–3, 5, 6	4, 10, 11	6, 7, *12*	7, 9, *12*, 13	14	15	15	7	—	11, *17*	7, *17*, 21	15	21	18, 19
Geissman	1, 2, 6	2, *3*, 5	2, 4	11	10	13	13, 14	—	8, 9	7, *12*, 13	7, 13, *15*	15	—	13
Gerig	1, 4, 11	2–4, 6, 7, *12*	7, *12*, 13	9, 12	—	15	7, 11, 19	10, 14	7, *15*	13	11, 18			
Griffin	1, 2, 4, 9	6, 7, *12*	12	7, *12*	11, *12*	—	7, 9, *16*	7, 14, 16	7, 10, *15*, 20	—	15	13	17	11, *18*, 21
Gutsche and Pasto	1	2, 4	3, 5–7, 13	7–9, *13*	—	—	9, 11	*12*, 16	7, *14*	7, 11, *15*, 21	7, *15*	7, 11, *19*	7, *17*, 21	7, 11, *18*
Hart and Schuetz	1–6	7, *12*	7, *12*, 13	9, 10, *12*	7, 9, *16*	11, *15*	10, 15	*14*, 16	*17*, 21	11, *18*	—	*19*, 21	—	21
Hendrickson, Cram, and Hammond	1	2, 4, 6	3, 5, 7, 20	5, 8	9	4, 13	—	11	10, 12	13, 14	14, 21	*17*, 21	*18*, 21	14
Holum	1–6	7, *12*	4, 7, 9, 10, *12*	7, 9, *16*	—	7, *15*, 21	14	11, *19*, 21	7, *17*	11, *18*	—	17, 18, 21	13	
Kice and Marvell	1	2–6	7, *12*	7, *14*, 15, 19	7, *17*, 18	21	13	—	8, 9	11	10	14, 15	12	
Leffler	1, 2, 3, 8	11	4, 6	4, 7, *12*	4, *12*	9, *16*	7, *15*	7, *17*	13	7, *18*	19			
Menger, Goldsmith, and Mandell	1, 2, 4	8, 9	3, 5	11	13	4, 6, 20	10, 14, 21	14	14	16	17, 18, 21			
Moore	1–6, 10	7, *12*	12	9, *16*	13	14	11, *15*, 19	—	*17*, 21	11, *18*	19, 21			
Moore and Barton	1–6, 11	4, 7, *12*	4, 7, 10, *12*, 13	7–9, *16*	13	13, 14	15	17	20	18	13	11, 16, 19	19	
Morrison and Boyd	1–6	12	7, *12*	13	10, 14	9, *12*	13	11, *12*, 21	12	7, 9	*16*	16	—	14
Neckers and Doyle	1–3, 11	2, 4, 6, 7	3, 11, 12, 14	20	7, *12*, 13	8, 10, 22	9, 12	21	13	—	8, 15, 17	7, *16*	10, 14, 22	11, *18*
O'Leary	1	2–5	6, 7, 9, 10, 12	13	7, 11, *14*, *15*	19	7, *17*	21	7, *18*	18	7, *16*			
Reusch	1–6, 8, 10, 20	7, *12*	7, 14	7, *12*, 13, 20	*12*, 13	12	13	15	7, 9, *16*	16	7, *19*	7, *17*	7, *18*	18
Roberts, Stewart, and Caserio	1–4, 11	4, 12	5–7, 12	7, *12*	7, 11, *12*	8, 9	—	7, 10, *14*	*14*	11, *15*	7, *17*	17	11, *18*	13
Solomons	1, 2, 4, 6	3–5	4, 7	10, *12*	4, 7, 13	*12*	13	—	7, *12*	8, 9	—	9	16	
Streitwieser and Heathcock	1	2–5, 8, 9	—	6, 7	12	7	13	10, 11, *14*	14	—	7, 11, *15*	7, *12*	7, *12*	
Ternay	1	2–5, 9	6, 7, *12*	13	10, *14*	14	—	7, *12*, 13	7, 11, *12*	7, *15*, 21	15	8, 9, *12*	—	9
Zimmerman and Zimmerman	1–6	7, *12*	7, *12*	7–9, 11, *12*	7–9, *16*	16, 21	15	11, *16*	7, *15*	7, *17*	17	13	13	11, *18*
Zlatkis, Breitmaier, and Jung	1–4	5–7, *12*	7, 12	12	14	13	12	8, 9	16	16	7, *15*	7, 11, *15*	7, *18*	7, *17*

Unit numbers in italic type indicate major treatment of the topic in question.

WITH THE VARIOUS TEXTBOOK CHAPTERS

15	16	17	18	19	20	21	22	23	24	25	26	27	28	29	30	31	32	Other
9, 10, 16	*14*, 16	15	*17*	11, 18	17, 18, *19*, 21	12	21	21	–	–	–	–	–	–	–	–	20	Ch.33: 21
18	19	13	13,15, 17	18	11,18, 19	–	20	–	20									
7, *16*	*17*, 21	*18*, 21	*18*, 21	–	18, *19*	–	16,19, 21	14										
17, 18																		
–	–	18, 21	21	–	21													
19	16	13,15, 17	11,18, 19	18														
14, 21	–	13	–	–	7, 9, *16*	14	19	19, 21										
–	11, 19	20	17	17	17	–	11, 17	7	18	–	20	–	16	16	16	16, 19	16, 19	Ch.40: 20
18	11,*19*, 21																	
7, *18*	18,*19*, 21	–	7, 20, 21	13	10, 14	13,*14*, 16	*15*, 21	15	19	21	16	–	21	17, 18	17, 18	*16*, 20	–	Ch.39: 21
12	16	–	18	–	–	–	19	–	–	20								
12, 13	16	–	15	–	–	–	21	–	–	20								
16, 21	17	18																
7, 21	*15*, 21	15	18	17	18	17	19	11,19, 20,21	15,16, 20	14, 16	21							
13	11, 19	8, 12	12	16	16, 21	14	14,17, 21	–	–	17	18							
16	8, 12, 13	–	13	21														
–	11,18, *19*	–	–	–	16	14, *16*	16, 21	–	16									
11, 14	7, 11, 15,21	14, 16	7,*17*, 21	7, *18*	18, 21	7, 11, *19*,21												
7, *17*	–	7, 11, *18*	7, 18	21	9	9, *16*	–	–	21	–	21	7, 11, *19*	–	16	16	–	21	
16	7, *17*	17	7, *18*	7, 21	16, 19	16												
18	–	7, 19																
7, 11, 19	*16*, 21	–	14															

Important Charts and Tables

Molecular Structure and Isomerism
   Common bonding units (part structures)   13
   Names of the functions (generic family names)   35
   Unsaturation number, structural possibilities for common values of   88
   How to estimate the stability of any structure   137
   Electron-donor and -acceptor functions   154
   How to deduce the relationship between structures A and B   248, 265

Nomenclature
   Family, substituent, and aromatic parent names and priorities   97, 117
   Alkane names   102
   Flowchart for compound naming   128

Reaction Products and Reagents
   Relationships between functions, graphic representations
      Alkanes, alkenes, alkynes, halides   220
      Alkanes, alkenes, alkynes, alcohols, halides, and carbonyl compounds   317
      Alkanes, alcohols, carbonyl compounds, carboxylic acids   342
      Alkanes, alkenes, alcohols, carbonyl compounds, carboxylic acids   410
      Interconversion of carboxylic acids and derivatives   441
      Alkanes, alcohols, carbonyl compounds, and carboxylic acid derivatives   453, 454
      Nitrogen analogs of oxygenated functions   482
   Structural feature → basic mode of reactivity → standard reactions charts
      Aliphatic hydrocarbons   209
      Alkyl halides   306
      Alcohols   336
      Aromatic rings   364
      Aldehydes and ketones   392
      Carboxylic acids and derivatives   442
      Amines   484
   Slots and slot fillers (see pp. 208, 308)
      Reagents for electrophilic addition   213
      Reagents for radical addition   214
      Alkene oxidation products   218
      Reactants/reagents for nucleophilic substitution   308, 338
      Reactants/reagents for E2 elimination   309
      Reactants/reagents for aliphatic electrophilic substitution   310
      Reactants/reagents for electrophilic aromatic substitution   366
      Reactants/reagents for nucleophilic aromatic substitution   367, 368
      Reactants/reagents for nucleophilic addition   416
   Other types
      Chemical analogies   236
      Possible stereochemical outcomes of reactions   289, 290

Common oxidizing, reducing, dehydrating, alkylating, etc., agents	456
Reactions that make C–C or C=C	267, 268
Reactions that break C–C or C=C	524
Reactions that form difunctional compounds	569

Mechanism

The nine standard mechanisms	175
Chemical analogies	236
Recognition clues for the nine standard mechanisms	239
Flowchart for writing nucleophilic addition mechanisms	420

Synthesis

Reactions that make C–C and C=C	567, 568
Groups introducible via electrophilic aromatic substitution	546
Groups introducible via aromatic nucleophilic substitution	547
Reactions that form difunctional compounds	569
Flowchart for constructing carbon skeletons	572
Flowchart for functional-group adjustment	579
Reactions that lengthen/shorten carbon chains	586

Acids/Bases

Common bases (acids) in organic chemistry	186 (187)
Orders of acid and base strengths	192
Acid–base species present at various pH values	487

Diagnostic/Degradative Reactions

Alkene oxidation products	218
Diagnostic reactions	510
Degradative reactions	524

Reactivity

Orders of acid and base strengths	192
Activating/deactivating groups in electrophilic aromatic substitution	379

Guide to Standard Problem Types and Solution Techniques

Molecular Structure and Isomerism
    The "Draw a correct Lewis (dot) structure for . . ." Problem   17
    The "Translate this condensed structure" Problem   28
    The "Show the shape of this molecule" Problem   51ff.
    The "Deduce the relationship of structure A to structure B" Problem   62, 78, 248, 266, 284
    The "Draw all isomers of . . ." Problem   80, 83, 91, 94
    The "Draw a complete resonance-hybrid structure for . . ." Problem   152–164
    The "Specify the configuration of this compound" Problem   272
    The "Given these chemical data on compound X, deduce the structure of X" Problem   502

Nomenclature
    The "Write the systematic (IUPAC) name for . . ." Problem   96, 116
    The "Draw the structure corresponding to this systematic name" Problem   130
    The "Specify the configuration of this compound" Problem   272

Reaction Products and Reagents
    The "Write (predict) all (the major) products of these reactions" Problem   224ff., 234, 329, 344, 384, 409, 456
    The "Supply the missing reagent" Problem   456
    The "Predict the stereochemistry of the products of these reactions" Problem   289–304, 319–322

Mechanism
    The "Write all steps in the mechanisms of the following reactions" Problem   239ff., 318, 344, 418, 470

Synthesis
    The "Write reactants in a Grignard synthesis of this alcohol" Problem   427
    The "Write reactants in a Wittig synthesis of this alkene" Problem   430
    The "Write the carbonyl compound from which this compound can be made by aldol condensation (plus further reactions)" Problem   431
    The "Synthesize compound A from any starting materials in one step" Problem   543
    The "Synthesize compound A starting from . . ." Problem
        A = carboxylic acid or derivative   479
        Aromatic A   545
        Aliphatic A   566

Acids/Bases
    The "Draw the structure of the conjugate acid (base) of this compound" Problem   182
    The "To which side will the following acid-base equilibria lie?" Problem   192, 196
    The "Predict the state of ionization of this acid (base) at pH x" Problem   486

UNIT ONE

Introduction

WHY IS THIS BOOK NECESSARY?

Your regular textbook and your instructor's lectures are composed of (a) a large body of experimental facts, qualitative and quantitative, (b) a network of theory that forms the basis for understanding and interrelating the facts, and (c) a notational system capable of communicating this material. The textbook (if it is well written) and the lectures (if your instructor is worth his/her salt) will give you a thoroughly explained, balanced, cohesive, and rational account of organic chemistry. This generally leads to *understanding* most individual topics of the subject, but it does not necessarily lead to success in the course. Why do many students feel that they "understand the material" but cannot answer questions and solve problems? Mainly because, although hearing lectures and reading books are reasonable ways to learn theoretical concepts, concepts are only *one third* of the subject. The other two parts—organic chemical facts and organic chemical notation—*are poorly learned by the traditional approach of lectures and reading alone.* This book teaches you how to get these facts and notation under control and how to analyze, solve, and respond to all the standard types of organic chemistry problems.

This book also aims to fill another set of gaps in the lecture/reading approach. Because time is short, your lecturer can hit only the high points, leaving out many of the small steps in the development of each topic that he/she thinks (or hopes) you can make by yourselves. However, some students attending any lecture fail to make one of these steps somewhere along the way and are still puzzling over that step when the lecturer is winding up the argument. Even in the most thoughtfully developed textbooks, some students invariably find that a link in the chain of understanding is missing because the author thought it was too basic to include. This guide provides many of the small helps, hints, and gap fillers that speed learning.

Our remarks above apply equally well to strong students. A sizable number of A students in general chemistry run into difficulty in repeating their performance in the organic course. Even if you encounter no real difficulties in learning organic chemistry, you will do it faster and more easily using an efficient method of organizing and learning the factual material.

WHAT ARE THE OBJECTIVES
OF THE ORGANIC COURSE AND THIS GUIDE?

At the end of the organic chemistry course you should be able to:

1. Given the structural formula or name of any of a large assortment of organic compounds, predict (a) the detailed three-dimensional *structure* of the molecule; (b) the *reaction products* arising from treatment with a variety of reagents; (c) the way in which bonds are made and broken to bring about product formation in these reactions (that is, the *reaction mechanism*); and (d) the effect of structural variations on *reactivity* (rate or position of equilibrium) in these reactions.

2. Show sufficient facility with the *notation* of organic chemistry to communicate this knowledge (understand and answer questions, look things up).

3. Design reaction paths by which a great variety of moderately complex organic compounds can be prepared from simple, readily available compounds (*synthetic* capability).

4. Deduce the structural formula of an unknown organic compound from spectroscopic or chemical data.

5. Predict very roughly the *physical* and *physiological* properties of an organic compound given its structural formula.

6. Demonstrate some knowledge of the *sources* of and *uses* for organic compounds in the practical world.

I have arranged the goals roughly according to my own opinion of their decreasing importance for chemistry majors. For preprofessional students, 5 and 6 may be more important than 3 and 4. The order of these goals is not perfectly arbitrary because the highly interwoven and cumulative nature of organic chemistry requires that basic facts of structure and reactivity be learned before later material can make any sense.

This guide focuses on goals 1 to 4, in the following way:

(a) It gives you an effective organizational network of visual and verbal cues which, if practiced as directed, guarantees efficient learning and recall of the large body of facts on which goal 1 is based. And it develops the rules and analogies from this body of facts to allow you to predict information about new compounds in new situations.

(b) It will help you achieve complete notational competence.

(c) It develops detailed analyses and solution techniques for each of the common problem types.

Goals 5 and 6 are better covered in textbooks and laboratory work and will not be discussed here. Specific objectives are stated at the beginning of each unit.

HOW IS LEARNING ORGANIC CHEMISTRY LIKE LEARNING A SECOND LANGUAGE?

Many people have observed the similarity of processes and problems between learning organic chemistry and learning a second language. Some similarities are: (a) the new notation that is required for communicating chemical ideas is like a new language; (b) a large volume of new information must be stored in readily accessible form; (c) the study methods that are successful in second-language learning also work in the study of organic chemistry. Actually, the analogy holds to a surprising extent through a detailed technical analysis of the two subjects. Two points are especially important to us:

1. A limited number of "basic sentence patterns" of a language are general frameworks that we use to generate an unlimited number of meaningful, specific constructions. In language learning it has proved to be more important to master and be able to manipulate easily the forms and patterns within a limited vocabulary than it is to acquire a large vocabulary. Similarly, a limited number of "allowed reactions" and simple rules of analogy generate the unlimited number of chemical transformations actually observed to occur in nature. Much of this guide's method is based on:

- Thorough learning of the basic patterns of organic chemical behavior and extensive drill on turning these rapidly into large numbers of specific equations, mechanisms, structures, and so on.
- Developing rules that, like grammatical rules, summarize large areas of experience and can be rapidly applied to new situations.
- Adapting study techniques that have proved effective in second-language learning.

2. All learning depends upon motivation, perception, and exercise. The language learner's most important task is to internalize the basic patterns and to acquire a new system of language habits so that he/she can react automatically to the structural signals of the second language. This can be accomplished only by drill. Theoretical study of a language does not improve your ability to speak. All this applies equally well to organic chemistry learning. Structure recognition and structure drawing must be automatic and accurate. The eye must learn to assemble all cues and size up just what has gone on in a given reaction—quickly, as a matter of habit. Obviously, achieving this proficiency will also require drill. The exercises in this guide are designed to give you this essential practice.

WHAT ARE THE MOST COMMON STUDENT DIFFICULTIES AND HOW CAN YOU AVOID THEM?

Most students perform at a fraction of their potential in organic chemistry because they are troubled by one or more of the following problems:

1. Lack of organization.
2. Notational problems.
3. Too little drill.
4. Falling behind.
5. Poor problem analysis.
6. Difficulties in seeing and manipulating three-dimensional objects in two dimensions.
7. Poor study habits.

Let's discuss these one by one.

1. *Lack of organization.* This can affect every aspect of your work, from allocation of time and the preparation of factual material to be learned, to the acquisition of notational skill and the analysis and solution of problems. The basic set of organic chemical facts consists of perhaps 150 characteristic reactions. To internalize this list in readily usable form, you must have help from the logical relationships within the material, from an underlying theory with predictive power, or from mnemonic or other devices. We will develop a fourfold organization that very much simplifies acquisition, recall, and use. Thus the approximately 150 reactions can be classified in four different ways (don't worry now about what each means—we take that up later):

   (a) By mechanism; the approximately 150 reactions occur by means of only nine basic mechanisms (standard sequences of bond-breaking/bond-making steps).
   (b) By functional group.
   (c) By change in oxidation or alkylation state.
   (d) By visual charting of structural relationships.

In recalling most organic chemical facts you have these four classifiers to use as cues—it is rare when all four cues fail.

In Units 14 to 19, where we discuss material that constitutes the bulk of the "memory work" of organic chemistry, we lay out specific instructions for organizing the material to be learned.

2. *Notational problems.* Difficulty in keeping up with the lecturer while taking notes, failure to finish exams, loss of points due to illegible exam answers, inability to manipulate three-dimensional structures on paper—these are the common symptoms of inadequacy in using organic chemical notation. Instructors often lack patience with students who, after several weeks or months, insist that they "know the material" but cannot demon-

strate it because mistakes in structural formulas or poorly drawn reaction mechanisms make it impossible to judge conceptual ability. Suppose that a problem consists of writing the structural formulas of the products of a number of reactions (reactants given). If you write an incorrect structural formula for a product, your instructor will not usually be able to tell whether you understood the reaction well enough to predict the product; you cannot expect much credit in such a case.

However, notational proficiency is important for a still more basic reason. Structural formulas, accurate equations, and reaction mechanisms are your main tools in organizing the material, in internalizing it, and in checking and preserving this learning by self-quizzing. In general, your understanding of facts or concepts will be no more accurate than your notation. Introducing and propagating errors or fuzziness in memory results in confusion rather than in learning.

Unit 2 will give you a good start in interpreting structural formulas and drawing correct ones, and Unit 5 backs this up. However, like any skill, drawing correct structures requires mainly continuous practice. At first this will be just plain notational practice. As time goes on, you can get enough notational practice as you study reactions, mechanisms, and so on—if you stick to the prime rule of doing all your studying with pencil and paper in hand. The spoken language is a very poor way to communicate chemical ideas. Do not get into the habit of asking or answering oral questions by loose verbal descriptions of structure; saying "that thing with the CO ion hanging off it" and the like is a waste of time and an invitation to misunderstanding. When you have structural information to convey or ask about, *draw what you mean* and ask others to do the same.

3. *Too little drill.* Because learning organic chemistry *is* so much like learning a foreign language, drill sessions are necessary and vital for success. Other things being equal, the scientist with the greater number of facts at his/her disposal is the more able scientist. Drill is also the only way to acquire skill in handling organic chemical notation, as already discussed.

Why do students fail because of lack of repetitive practice when it is not difficult to do and does not really consume much time? In part because textbooks do not have the space to include sufficient drill exercises. In part because often no one tells students that drill is necessary. This guide provides drill exercises in almost every frame. You should work these and check your answers before proceeding to the next frame. We also show you how to make up your own drills.

4. *Falling behind.* This is deadly. Students underestimate this danger for three reasons. First, the course starts off with material that is a review of freshman chemistry, and it seems easy. The first new conceptual material that comes up also seems easy. (It's true—none of the concepts of organic chemistry is very difficult.) So students can easily drop into the pitfall of thinking that everything is under control as long as they read the book and collect lecture notes.

Second, the material is very cumulative. Factual material is best handled in small batches; if allowed to pile up, it becomes much more tedious and difficult to internalize. Concepts build on one another, so that the new ones soon cease to make sense if the previous concepts haven't been fully learned. Students often either put off clarifying questions or else do not study enough to locate their unanswered questions and trouble areas until the day before an exam. That's too late. For one thing, after such questions are answered, a second round of questions usually comes to light on the night before the exam, when the instructor is not available. In any event, such late clarification of problem areas does not leave time for this information to serve as the basis on which later concepts are built. The notational skill that you acquire in the earlier stages is also a prerequisite to handling subsequent material.

Third, a traditional arrangement of material in the organic course causes the chemistry of the carbonyl compounds and of the carboxylic acid derivatives to be studied close together somewhere near the end of the second or the beginning of the third quarter (in a year course). In this arrangement, the number of new reactions to be learned in just a few weeks nearly equals the total number of reactions studied up to that point. A number of students who were already somewhat behind go under at that point.

So the word on falling behind is—*don't*. In our later discussion of study habits, you'll get some good pointers on how to avoid this pitfall.

5. *Poor problem analysis.* Methods of approaching many specific types of problems are taken up in detail throughout this book (see Guide to Standard Problem Types, p. xviii). Right now we want to look very generally at how you should attack problems in order to avoid some common difficulties. First we describe the symptoms of difficulties, then identify some of the causes and their remedies, and finally summarize with a suggested stepwise routine for analyzing problems.

The commonly observed symptoms are (a) the blank response ("I just couldn't see how to begin."), and (b) solving a problem *other* than the one given, in some cases a fictitious problem in which everything is the reverse of the original. Let's examine the various causes for each of these fruitless responses.

The simplest cause of the blank response is just lack of practice due to lack of study. With conscientious students, however, the cause is more likely to be the habit of trying to "see the solution" or a path to the solution in one's head before putting pencil to paper. This is a bad habit. Many problems involve several stages in their solution; the actual answer or key to the answer cannot be "seen" until all but the last step has been worked out on paper. You may not be able to crack even single-stage problems in your head if they involve not verbal connections but chemical (especially three-dimensional structural) relationships.

The cure for this difficulty is the use of several pump-priming techniques, which must be applied during practice sessions as well as in examination situations.

(a) After you are satisfied that you understand what is being asked, mentally scan the problem types that you have encountered in the past to see if the one in hand corresponds to any of them or to any of them played backward.

(b) To increase the effectiveness of the mental scanning, keep a growing list of problem types as you encounter new ones and study all the problems you can get your hands on, especially solved problems. This book contains many solved problems; other sources are textbooks besides your own or your instructor's old exams. (Ask for an exam and the answer sheet; if the answer sheet is not available, work the exam and ask the instructor to check your solutions.)

(c) Translate the problem, which is usually stated in words, into notation closer to the real content of the problem. In organic chemistry this usually means turning names of compounds into structures, statements that A reacts (or fails to react) with B into equations, and so on (on paper, of course). Segregate the *given* information in a certain section of the page, and write a description of what is required as an answer (using chemical notation where possible) in another section. Connect the two with an arrow. If you now see a way to get from the givens to the answer, get on with it. It often happens that either the notation or the visual layout of the problem will cue you to see the method of solution.

(d) If you do not, ask yourself what given information you *would need* to deduce the answer. Then try to see a way of getting *this* information from the givens; this transforms the problem from

$$A \xrightarrow{\text{(connection not yet seen)}} Z$$
$$\text{(given)} \qquad \qquad \text{(required)}$$

into

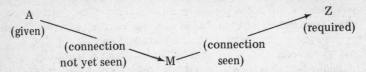

M acts as a bridge; hopefully the conceptual distance between A and M is smaller than that between A and Z, so that a connection between A and M can be found. If not, you must either find another bridge in place of M that *can* be related to A, or else ask yourself what information you *would need* to get to M. Use this information as a second bridge, making it a three-stage problem: A $\xrightarrow{?}$ L $\longrightarrow$ M $\longrightarrow$ Z. We will illustrate this technique of problem solving in later units.

Although learning to solve one type of chemical problem ordinarily makes new problem types easier for you to handle, prior training sometimes has a *negative* effect. In one such case prior training conditions you to perceive a class of objects in a certain way that has proved useful in solving problems in the past, but which is counterproductive when applied to the new problem. For example, habitually seeing formic acid (H–C(=O)OH) as a carboxylic acid (H–C(=O)OH) may make it difficult for you to recognize it as an aldehyde (H–C(=O)OH), causing you to draw a blank on a problem that requires the latter view. Sometimes just the act of translating a name or condensed structure into a fully detailed structure will cue your perception of aspects of the molecule that were at first unseen. If translating the problem into structures, equations, and so on, as discussed above, doesn't help, you can try to identify all the categories of structure or behavior that might be used to describe the elements given in the problem.

A related difficulty can be caused by fixation on habitual techniques of problem solution. If you work a number of problems that conform to a single pattern and are then confronted with a new type that does not conform (even though it *looks* similar), you may suddenly find yourself unable to proceed. Or else you may unconsciously redefine the problem into the kind you are in the habit of solving, and answer that instead. This is the situation in which our response is likely to be "trick question!" The spell can often be broken by warning from the outside (for example, telling subjects in psychological experiments to wake up). When chemists make up problems for students, if they are aware that they are setting up a potentially fixating situation, they sometimes provide the warning in the statement of the problem. Phrases such as "Be careful to consider all possibilities.", "Don't answer with . . . ." and "Hint: . . . ." can be signals that the method you are most likely to use will not work on this problem. It is more difficult to snap *yourself* out of this type of fix by deliberately trying to tell yourself to wake up and change thought patterns, but you can try. In a sense this difficulty is a case of stubborn misreading. Fight back by concentrating, rereading, and separating what is given from what is required.

Now we can incorporate some of these suggestions and some common sense steps into a routine to follow in analyzing problems:

- *Read* the problem.
- Translate words into symbols where possible, and lay out the known and unknown portions separately on paper.
- Reread the problem to verify your analysis of what is given and what solution is required.
- Scan memory to see if the problem fits a problem pattern previously solved. Be

INTRODUCTION 7

careful with problems only slightly different from ones you have done previously; don't solve the old problem instead of the required one.
- If you don't see the connection between the given material and the required answer, try listing the alternative ways of representing the given material, or its different properties, to see if one of these other aspects is the one that shows you the path to the answer.
- If you still can't see how to proceed, try to break the solution into two stages, as described on page 6.

6. *Seeing and manipulating three-dimensional objects in two dimensions.* Students vary widely in aptitude for this aspect of organic chemistry. However, do not panic, even if you find the spatial problems difficult at first. An adequate ability can usually be acquired by relatively little practice, such as that provided in Units 4 and 13. In addition, the amount of this type of work in the organic course is not large.

Test

You can get an idea of your three-dimensional perception by taking the following test.

Directions: Answer the following questions; then check the correct answers given below the line of dashes. (a) The curved arrow goes clockwise. Now imagine you are behind the paper looking at the arrow. Which way does it go? (b) The clockwise order of the letters is d-b-c-a (or b-c-a-d or c-a-d-b or a-d-b-c). What is it for an observer behind the paper? (c) The tetrahedron shown has four flat, triangular faces; call them front, bottom, left rear, and right rear. (The dashed lines should project behind the paper. If you are seeing rear, bottom, left front, and right front, you are letting the dashed lines come *forward* instead.) Each face has three numbered vertices. What is the clockwise order of the vertex numbers for each face of the tetrahedron when the face is viewed from the outside? What is it for each when viewed from *inside* the tetrahedron?

— — — — — — — — —

(a) Counterclockwise.

(b) d-a-c-b or a-c-b-d or c-b-d-a or b-d-a-c.

Face	From without	From within
(c)		
Front	2-3-1	2-1-3
Bottom	1-3-4	4-3-1
Left rear	2-1-4	2-4-1
Right rear	2-4-3	2-3-4

If you had no difficulty with this exercise, you will probably have no space-perception problems in your organic course, and you probably will not need a set of molecular models of your own. You should arrange to handle some borrowed models, at least brief-

8 HOW TO SUCCEED IN ORGANIC CHEMISTRY

ly, however, during your study of molecular geometry and stereochemistry.

If you had difficulty with the exercise, you should (1) budget some extra time for practice on Units 4 and 13 and related problems. (2) Start now to practice seeing, in your mind's eye, various objects from any vantage point. (3) Buy an inexpensive set of student molecular models. Do the model exercises in Unit 4. Throughout your study of isomerism and the geometry of molecules, make models of the molecules under discussion to help you see the correspondence between the three-dimensional molecule and its two-dimensional representations on paper.

HOW SHOULD YOU STUDY?

To succeed with organic chemistry you need (in addition to a modest amount of talent and a reasonable general chemistry preparation) a well-organized approach and sufficient time and commitment that you can stick to a fairly rigorous study schedule. Being psychologically "up" for the course also helps. We discuss these plans next, beginning with directions on how to study.

1. *Allocate your time and set study goals in advance.* You will require *no less than* 10 hours of study time a week. You get much more value per hour spent if you plan in advance what you intend to accomplish in each study session. As nearly as possible, set up your main daily study session in the same time and place. Choose your time and place to minimize interruption and distraction. Adapt an all-business attitude from the moment you sit down—do your relaxing during breaks.

Your plan for each new chapter or week should contain these activities, roughly in this order:

- Scan your text and this guide.
- Plan the week.
- Read text and guide, and mesh with lecture notes.
- Make summaries.
- Work exercises.
- List ambiguities, difficulties, inconsistencies, and unsolvable exercises, and get the instructor to clear them up.
- Drill on old and new material.

2. *Practice daily.* Just as in studying a foreign language, several daily sessions yield better results than the same time spent in a single weekly session.

The key to efficiency is using your large time slots for large jobs and using your small scraps of time for short drills instead of wasting them. Five minutes is enough for a drill that uses a previously prepared summary. You will find specific instructions on preparing summaries and conducting drills in Unit 12, frames 15 and 16, and in Units 14 to 19.

3. *Study actively*, with pencil in hand. As you read, draw structures of the compounds you are reading about. When you look at a summary, speak or copy the items. Experimental psychologists have proved the effectiveness of recitation, silent rehearsal, and self-quizzing many times over. If you learn material by passive methods, you can usually recognize it as correct if you encounter it later, but often you can't recall the material at will, which is the aim of learning.

Underlining textbook passages with a transparent marker is *too passive*. Time spent in "reading the notes and the chapter through three times" would usually produce vastly better results if spent on active study. Transcribing notes, making summaries, drill, and problem solving all qualify as active study.

4. *Intersperse study periods with breaks.* Stick strictly to business during each period and consider the following break the reward for completing the scheduled tasks.

5. *Make summaries and drill on them:*

- Include in summaries mainly items that
    (a) Have predictive power for newly encountered problems (e.g., rules for predicting reaction product structure, reactivity and stability orders).
    (b) Summarize many specific cases (e.g., generic equations that hold for whole families of compounds or standard mechanisms that apply to many different reactions).
- Practice reproducing each entry in your summary.
- Make a supersummary that is just a list of cues (name, word, single structure) to the items in your summary.
- Practice recalling at will each of the items in your supersummary.
- Experiment with visual summaries (oxidation-state charts, flowcharts, etc.; examples are included in most units of this book).

This study plan and the specific learning and problem-solving methods suggested in this guide are based on long-term observations of (and by) many students and on proven psychological principles. The total method works best if you support it in these ways:

1. *Put your detailed plan of study into operation at the earliest possible moment,* preferably *before* the first class meeting. Start this program with the maximum possible momentum. Anticipate temptations to break your study schedule and fight them actively. Skip a study session only when you can manage to complete the scheduled material in advance.

2. *Make a list of the reasons why you need good grades* in the organic course. Then if later you feel yourself losing momentum or are tempted to give up your study plan, go back and reread the list.

3. *Use all the resources available to you.* (Many students don't.) Some of these resources are:

(a) *Your instructor's office hours.* In addition to getting answers to questions, you will get insight into what your instructor thinks is important. You also get the opportunity to show that you are taking the work seriously, and you extend your chance to show your abilities beyond the examination periods. The conversation gives the instructor the opportunity to identify and correct some of your misconceptions, and perhaps to suggest some tailor-made aids to learning. How welcome will you and your questions be? That depends largely on how you have prepared beforehand. Questions such as "Would you go over aldehydes?" imply that you have not done much for yourself, while "Would you check this summary for errors?" shows that you have progressed to the point of narrowing down specific problem areas. The instructor will accomplish more with a limited amount of time in responding to questions such as the second one above.

(b) *Help sessions.* The benefit to you again depends on your preparation. Watching someone else work a problem that you have not yet tried is a passive experience with limited learning value for you. On the other hand, confirmation of a solution which you have reached yourself, or seeing an *alternative* solution, is very valuable. So is seeing a successful analysis of a problem that you have gotten stuck on.

(c) *Exam postmortems.* Detailed study of your returned exam may be painful, but it's the best way to learn all you can from the exam. Don't let the matter rest until you feel you know exactly how you made each mistake and have learned how to avoid repeating it.

(d) *Old exams.* If your instructor doesn't make them available, ask to see some. They make extremely practical problem sets, especially if you can check your answers against your instructor's key.

4. *Adopt a positive attitude.* Forget what you may have heard about the difficulty of the course; remember that most of these comments originate from people not *prepared*

and organized to handle the material. Using the method of this guide gives you a pronounced advantage.

5. *Try to make other aspects of your life easy on yourself* as you move into the organic course. You are putting together a major effort against a substantial hurdle in your career, and you deserve whatever help you can get from reducing other demands on your time and energy. Try to arrange some substantial aid, like a relatively light course load, or at least a couple of extra hours a week free for the organic course. Good habits (including the good study habits discussed above) are most easily established, and bad ones broken, when other hassles in your life are minimized. This is not the time to decide to stop smoking or lose 20 pounds.

6. *Make a resolution and keep repeating it* to yourself: "I am going to stay on schedule in organic."

HOW SHOULD YOU USE THIS BOOK?

You can use this guide in two ways: as a troubleshooting handbook to which you turn for help with specific subjects or problem types as the need arises, or as an integrated, individualized method and workbook. In either case the examples and exercises will very substantially expand those available in your textbook.

If you use this guide for troubleshooting, the following lists (located at the front of the book) will help you find what you need:

- Cross-Reference Chart (p. xii)
- Important Charts and Tables (p. xvi)
- Guide to Standard Problem Types and Solution Techniques (p. xviii)
- Table of Contents (p. xi)

Using these lists, you can turn directly to the section you want.

If you use this guide as a self-paced method of learning organic chemistry, in either a classroom or self-teaching situation, you will get best results by working through the units in order.

In either case you will find that Units 1 to 11 are a rather cohesive block arranged for progressive mastering of basic skills needed for learning organic chemistry. This block is designed to be studied with the first few chapters of the standard textbooks even though it does not correspond to the usual arrangement of topics. Units 12 to 19 correspond more-or-less exactly to specific chapters in most textbooks. Units 20 and 21 pull together subjects that are usually scattered through several textbook chapters.

As implied above, you will need a textbook as well as this guide. If you are not in a classroom situation, you can use any of the textbooks listed in the Cross-Reference Chart. The best of these for self-study purposes would be those by Morrison and Boyd, Neckers and Doyle, Solomons, or Streitwieser and Heathcock.

As noted earlier, this guide concentrates on the "language" of organic chemistry, on learning the many facts of the subject, and on using these things to solve problems. We will rely on your textbook to supply theoretical concepts and descriptive material, such as physical properties, origins, and uses of organic compounds.

Whatever your purpose, you should generally work through a given unit from beginning to end, following all directions, working all exercises as they come up, and carefully checking the solutions before proceeding. Each unit opens with a statement of objectives (which you can use to help you decide if the unit will provide what you need), followed by a sequence of numbered sections called *frames*. Each frame marks roughly one step in acquiring the skills listed as objectives. Each frame consists of a teaching section, including examples, and exercises followed by a dashed line and the answers. The examples usually take the form of solved problems. When you study the examples, correlate the steps in the solution with the preceding discussion or outlined solution method. Keep plenty of scratch paper and pencils handy; avoid pens, since it is difficult to draw structural formulas without making frequent corrections. As you approach the exercises, cover the answers with a sheet of paper; you will benefit most from working the exercises if you do not see the solutions prematurely. *Save your solutions to the exercises* because later exercises often refer back to them.

Ready? Then on to Unit 2.

UNIT TWO

Structures

Organic molecules always contain one or more *carbon* atoms. They usually contain hydrogen, and often contain other elements. The *structural formula* of an organic compound is a *map* showing how the various atoms are bonded together in each molecule. In everyday speech we often say just "structure" instead of "structural formula." Structural formulas are the words in the language of organic chemistry. They are basic to understanding and communicating everything else in organic chemistry.

OBJECTIVES

When you complete this unit, you should be able to:

- Draw correct structural formulas quickly (frames 1 to 6, 8, 9).
- Detect errors in incorrect structures (frames 7, 8).
- Write structures representing whole families of compounds (frame 10).
- Read condensed structural formulas (frame 11).

If you think you have already mastered these objectives and might skip all or part of this unit, take a self-test, consisting of the problems identified by a star in the margin. If you answer all these questions correctly, you are probably ready to move on to the next unit. If you miss any questions, study the frames in question and rework the problem before proceeding.

If you are not yet ready for a self-test, proceed to frame 1.

RECOGNIZING AND COMBINING PART STRUCTURES OR BONDING UNITS

1. Most organic compounds contain only elements drawn from the boxed portion of the periodic table:

IA							0
1 H	IIA	IIIA	IVA	VA	VIA	VIIA	2 He
3 Li	4 Be	5 B	6 C	7 N	8 O	9 F	10 Ne
11 Na	12 Mg	13 Al	14 Si	15 P	16 S	17 Cl	18 Ar
19 K	20 Ca	31 Ga	32 Ge	33 As	34 Se	35 Br	36 Kr
37 Rb	38 Sr					53 I	

Practically all of these organic compounds have structural formulas that are simple combinations of the part structures or bonding units listed in Table 2.1.

Table 2.1 *Bonding Units in Organic Molecules*

Part structure	Number of covalent bonds to rest of molecule	Number of unshared pairs	Charge
Monovalent nonmetals: H–	1	0	0
:F:⊖, :Cl:⊖, :Br:⊖, :I:⊖	0	4	–1
:F̈–, :C̈l–, :B̈r–, :Ï–	1	3	0
Divalent elements: ⊖:Ö–, ⊖:S̈–	1	3	–1
–Ö–, :Ö=, –S̈–, :S̈=	2	2	0
–Ö⊕, –Ö⊕=, –S̈⊕, –S̈⊕=	3	1	+1
–S²⁺–	4	0	+2
Trivalent elements: –N̈⊖–, :N̈⊖=	2	2	–1
\N̈–, =N̈–, ≡N:, \P̈–	3	1	0
\N⊕, =N⊕, ≡N⊕–, \P⊕	4	0	+1
Tetravalent element: \C/, \C=, –C≡	4	0	0
Metals: Li⊕, Na⊕, K⊕, Mg²⁺, Ca²⁺	0	0	+1 or +2

Note in Table 2.1 that:

1. ⊕ and ⊖ indicate one unit of positive or negative electrical charge on an atom. The circles help to keep ⊕ *charges* from getting confused with algebraic + *signs* in chemical equations. Example: H⊕ + :NH₃ ⇌ NH₄⊕.

2. The symbol : indicates an <u>unshared</u> (nonbonding) pair of electrons on an atom.

3. – indicates one end of a covalent bond to the rest of the molecule; = and ≡ are one end of a double and triple bond, respectively. Part structures are combined to give complete structural formulas by coalescing the bonds: H– plus –F̈: ⟶ H–F̈:. No bonds in the part structures may remain uncoalesced. For example, the process –Ö– plus

–H ⟶ –Ö–H does not produce the structural formula for a complete molecule because of the leftover bond (pointed out by the heavy vertical arrow); <u>–Ö–H</u> is just a new *part* structure. In the following structures, circles have been drawn around the three types of bonding units of the carbon atom to show how these look as they exist in completed structural formulas:

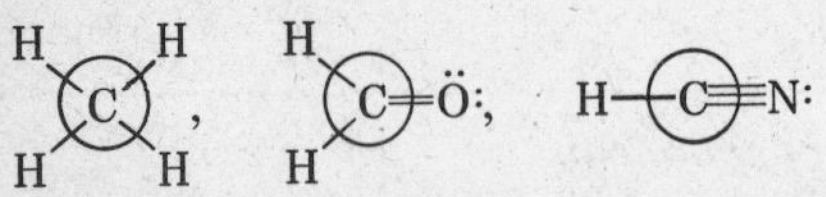

$$H_2C, \quad H_2C=\ddot{O}:, \quad H-C\equiv N:$$

To make the first of these structures, four H— units were combined with one $\diagup C \diagdown$ unit to get $H_2C H_2$. To make the second, two H—, one $\diagup C=$, and one $=\ddot{O}:$ unit were combined.

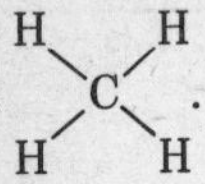

Exercises

1. From what bonding units was this structure made? $:\ddot{Br}-C\equiv C-H$.
2. What structure can you make from two $\diagup C=$ units and four $:\ddot{F}-$ units?

— — — — — — — — —

1. From one $:\ddot{Br}-$, two $-C\equiv$, and one $-H$

2. 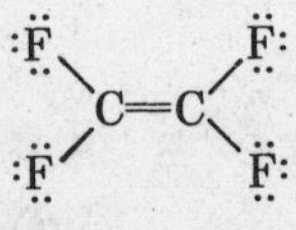 $$\ddot{F}\diagdown_{\ddot{F}}C=C\diagup^{\ddot{F}}_{\ddot{F}}$$

2. Let's continue our notes on Table 2.1.

   4. Each bonding unit listed has the inert gas structure, that is, 2 electrons, or a duet, for H; 8 electrons, or an octet, for all other atoms. You can verify this by adding the number of bonds to the number of unshared pairs, getting four electron pairs, or 8 electrons, in each case (one pair, or 2 electrons, for H—). The structures that you write by combining these bonding units will automatically be correct ones that represent stable molecules.

   5. Except in the case of metals and the sulfur unit $\overset{\oplus}{\underset{\diagdown}{S}}\diagup$, bonding units involving doubly charged atoms, such as $-\overset{..}{\overset{\textcircled{2+}}{N}}:$, are too unstable to exist under ordinary conditions.

   6. S behaves like O because it is in the same family of the periodic chart, which means that the two have the same number of valence electrons. P behaves like N for the same reason. P and S, from the second row of the periodic chart, rarely form multiple bonds, however.

   7. The directions in which the bonds point in part structures and complete structural formulas mean nothing; $H\underset{H}{\overset{H}{\diagdown}}C-$ is equivalent to $H-\overset{H}{\underset{H}{C}}-$. These two-dimensional struc-

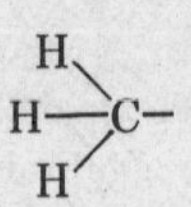

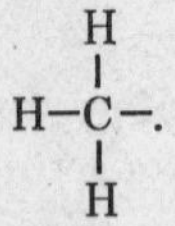

tures represent real molecules that have their own definite shapes in three dimensions. The ordinary structural formula that we are working on here does not try to depict the three-dimensional geometry. We consider ways to do so in Unit 4.

You must learn to recognize and draw without hesitation the bonding units of Table 2.1. These skills are basic to everything else in organic chemistry. As with all memorization tasks, use the systematic nature of the material. Note that trading one bond for one unshared pair decreases a positive electrical charge by one unit ($-\overset{\oplus}{O}\diagup \longrightarrow -\ddot{O}-$) or

STRUCTURES 15

increases a negative charge by one unit ($-\ddot{\text{O}}- \longrightarrow -\ddot{\text{O}}:^{\ominus}$); it amounts to a net gain of one electron. Observe how, as the number of covalent bonds increases from 0 to 4, the number of unshared pairs decreases from 4 to 0; enough unshared pairs are present to fill out the octet exactly. Note that in their uncharged (electrically neutral) bonding units, the monovalent atoms (H, F, Cl, Br, I) form one bond, the divalent atoms (O, S) form two, the trivalent (N, P) three, and the tetravalent carbon forms four. Note that neutral nitrogen exists in just three kinds of bonding units in Table 2.1 because there are just three ways to arrange three bonds—as three single, one double plus one single, or one triple bond. Remember, your memorization will go faster if you study actively—writing the part structures on scratch paper from memory, using as few cues as possible, and checking your result for correctness—than it will if you study passively (reading, looking). Practice these ideas on your own; then proceed to the following exercises.

Exercises

1. Some of these part structures are correctly written, some are not. Correct the incorrect ones by adjusting *unshared pairs* and/or electrical *charges* on the atoms (but do not change any *bonds*). Refer to Table 2.1 where necessary.

   (a) $\diagup\text{C}\diagdown$ (b) $=\ddot{\text{O}}$
   (c) $-\text{Cl}$ (d) $-\ddot{\text{N}}\diagdown$
   (e) $-\ddot{\text{O}}:^{\ominus}$ (f) $\text{F}^{\ominus}$
   (g) Na^{+2} (h) $\diagdown\text{N}=^{\ominus}$

2. Do as in exercise 1 but without using Table 2.1.

   (a) $-\text{H}$ (b) $-\ddot{\text{O}}-$
   (c) $=\ddot{\text{O}}-^{\oplus}$ (d) $-\text{C}\equiv$
   (e) $-\ddot{\text{N}}=$ (f) $-\ddot{\text{S}}^{\ominus}$
   (g) $=\text{C}\diagup$ (h) $\diagdown\text{N}^{\oplus}\diagup$

3. Circle each <u>bonding unit</u> in the following structures, and verify that each is correctly written.

   (a) (H$-\ddot{\text{O}}-$H) (b) H$-\ddot{\text{O}}:^{\ominus}$ Na$^{\oplus}$

   (c) H$\diagdown$N$^{\oplus}\diagup$H :Br:$^{\ominus}$ (d) H$-$C$-$C$-\ddot{\text{Cl}}:$ with H's and =$\ddot{\text{O}}:$
       H$\diagup$ $\diagdown$H

4. The number of bonds on each atom is correct in the following structures, but there are some other errors. Circle each of the *incorrectly written* bonding units in the following molecules, and rewrite them correctly:

   (a) H$-\text{O}^{\oplus}\diagup$H I$^{\ominus}$ I (b) H$\diagdown$C$-\ddot{\text{O}}-\ddot{\text{O}}-$H
        $\diagdown$H   H$\diagup$

   (c) H$\diagdown$C=$\ddot{\text{O}}:$ (d) H$-$N$-\ddot{\text{O}}:^{\ominus}$
       H$\diagup$ |
                                             H

(e) H−C̈≡C−F̈: (f) H−B̈r:⁺⁻

(g) H−Ö−C̈l:⁺ (h) Na⁺ :Ö−N̈=Ö:⁻

5. Although they are not covered by Table 2.1, and are rare, the following species do exist. See if you can extend the patterns in Table 2.1 to predict the unshared pairs and charges on the unusual bonding units in these molecules:
   (a) H−F−H⁺   (b) H−C≡C⁻ Na⁺

- - - - - - - - -

1. (a) 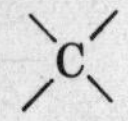 \>C<   (b) =Ö:
   (c) −C̈l: (d) 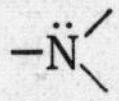 −N̈<
   (e) −Ö:⁻   (f) :F̈:⁻
   (g) Na⁺   (h) 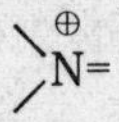 \>N=⁺
2. (a) −H   (b) −Ö−
   (c) =Ö−⁺   (d) −C≡
   (e) −N̈=   (f) −S̈:⁻
   (g) 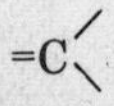 =C<   (h) \>N<⁺
3. (a) H−Ö−H   (b) H−Ö:⁻ Na⁺
   (c) 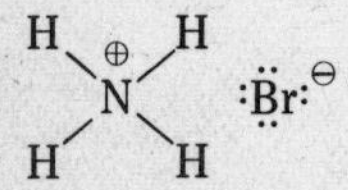 H\>N⁺<H / H H   :B̈r:⁻   (d) H\>C−C̈(=O:)−C̈l: / H

4. (a) 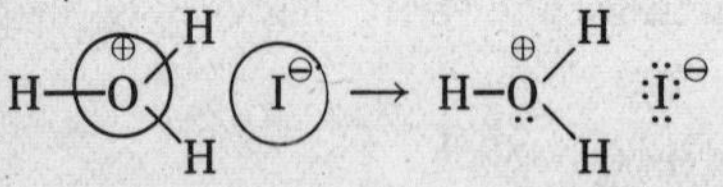 H−(O<H/H)⁺ (I⁻) → H−O<H/H + :Ï:⁻

   (b) 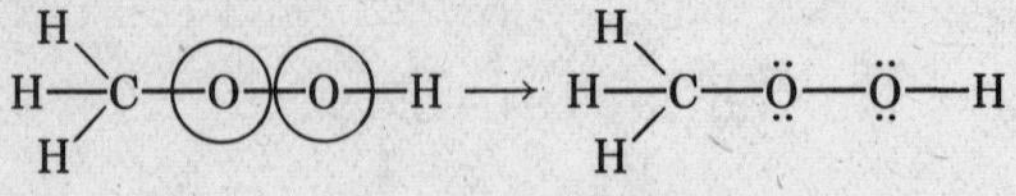 H\>C−(O−O)−H / H → H\>C−Ö−Ö−H / H

   (c) H\>C=(O:)⁺ / H → H\>C=Ö: / H

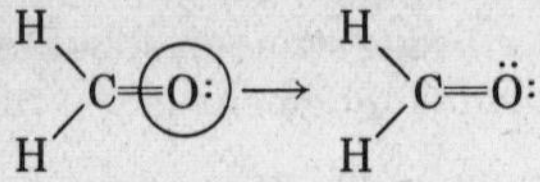

   (d) 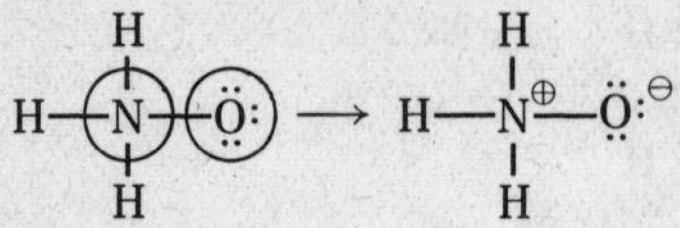 H−(N<H/H)−(Ö:) → H−N⁺<H/H−Ö:⁻

   (e) 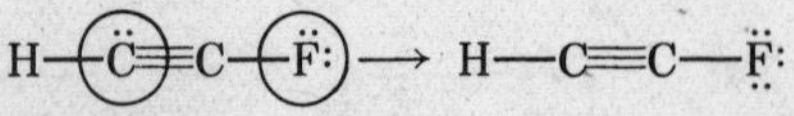 H−(C̈≡C−F̈:) → H−C≡C−F̈:

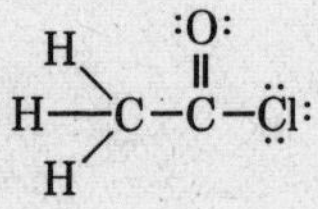

   (f) (H−Br:)⁺⁻ → H−B̈r:

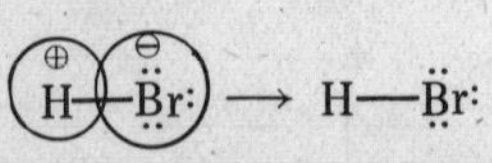

(g) H—(O⊕—Cl) ⟶ H—Ö—Cl̈:

(h) Na⊕ (O—N≠O) ⟶ Na⊕ ⊖:Ö—N̈=Ö:

5. (a) H—F—H has *two* bonds on F. Since increasing the number of bonds by one in the other cases of Table 2.1 decreases the unshared pairs by one and adds one unit of positive charge, it should do the same here: —F̈: ⟶ —F̈⊕, ∴ H—F̈⊕—H. (Throughout this book, we will use the symbol ∴ to mean "therefore.")

(b) In H—C≡C Na, the right-hand C has three bonds, one less than usual, so it should have one more unshared pair and one more unit of negative charge than usual: ≡C— ⟶ ≡C:⊖, ∴ H—C≡C:⊖ Na⊕.

DRAWING A CORRECT STRUCTURAL FORMULA

3. The *molecular formula* of a compound is an inventory of the exact numbers of each kind of atom present in a molecule: for example, CH_4O (shorthand for $C_1H_4O_1$). Given the molecular formula, you need to be able to draw a correct *structural* formula for the compound. (Ordinarily, one can draw more than one structure for a given molecular formula; we work on this complication in Unit 6.)

It is useful to mentally divide a structural formula into two parts: a central framework made from di-, tri-, and tetravalent atoms, and a skin made from the monovalent atoms. For example,

$$-\overset{|}{\underset{|}{C}}-\overset{|}{\underset{|}{C}}-\ddot{N}\diagup \quad + \quad 7\,H- \quad \longrightarrow \quad H-\overset{H}{\underset{H}{\overset{|}{C}}}-\overset{H}{\underset{H}{\overset{|}{C}}}-\ddot{N}\diagup\overset{H}{\diagdown H}$$

framework   monovalent atoms   whole molecule

The systematic method of generating structural formulas from molecular formulas is:

- Mentally set aside the atoms with <u>one bond only</u> (monovalent atoms).
- Make a framework by combining suitable bonding units for the C, O, S, N, and P atoms present.
- Add the monovalent atoms to the framework.

Example

Draw a structural formula for molecular formula C_2H_6.

Solution:

Set aside: six —H units.

Make skeleton from two —C— units: —C— + —C— ⟶ —C—C—.

Add H— units: 6 H— + —C—C— ⟶ H—C—C—H.

18 HOW TO SUCCEED IN ORGANIC CHEMISTRY

If you try to use the $\diagup\!\!\!\diagdown\!\!\mathrm{C}=$ or $-\mathrm{C}\equiv$ bonding units instead, you end up with H— units left over, so these possibilities must be abandoned:

$$\diagup\!\!\!\mathrm{C}= \; + \; =\mathrm{C}\diagdown\!\! \longrightarrow \; \diagup\!\!\!\mathrm{C}=\mathrm{C}\diagdown\!\!$$

$$\diagup\!\!\!\mathrm{C}=\mathrm{C}\diagdown\!\! \; + \; 6\,\mathrm{H-} \longrightarrow \; \underset{\substack{\uparrow\\ \text{complete}\\ \text{molecule}}}{\overset{\mathrm{H}\quad\mathrm{H}}{\underset{\mathrm{H}\quad\mathrm{H}}{\mathrm{C}=\mathrm{C}}}} \; + \; \underset{\substack{\uparrow\\ \text{left}\\ \text{over}}}{2\,\mathrm{H-}} \quad \therefore \text{ This skeleton is impossible for } \mathrm{C_2H_6}.$$

Example

Draw a structural formula for CH_4O.

Solution:

Set aside: 4 H—.

Make skeleton: $-\overset{|}{\underset{|}{\mathrm{C}}}-\; + \; -\ddot{\mathrm{O}}-\; \longrightarrow \;-\overset{|}{\underset{|}{\mathrm{C}}}-\ddot{\mathrm{O}}-.$

Add H— units: $\mathrm{H}-\overset{\mathrm{H}}{\underset{\mathrm{H}}{\overset{|}{\underset{|}{\mathrm{C}}}}}-\ddot{\mathrm{O}}-\mathrm{H}.$

★ *Exercises*

Draw a correct structural formula for molecules possessing each molecular formula.

1. CH_4
2. CH_3I
3. C_2HCl_5
4. $C_2H_6O_2$

— — — — — — — — —

1. $\mathrm{H}-\overset{\mathrm{H}}{\underset{\mathrm{H}}{\overset{|}{\underset{|}{\mathrm{C}}}}}-\mathrm{H}$

2. $\mathrm{H}-\overset{\mathrm{H}}{\underset{\mathrm{H}}{\overset{|}{\underset{|}{\mathrm{C}}}}}-\ddot{\underset{..}{\mathrm{I}}}:$

3. $\mathrm{H}-\overset{:\ddot{\mathrm{C}}\mathrm{l}:\;:\ddot{\mathrm{C}}\mathrm{l}:}{\underset{:\ddot{\mathrm{C}}\mathrm{l}:\;:\ddot{\mathrm{C}}\mathrm{l}:}{\overset{|\quad|}{\underset{|\quad|}{\mathrm{C}-\mathrm{C}}}}}-\ddot{\mathrm{C}}\mathrm{l}:$

4. $\mathrm{H}-\ddot{\mathrm{O}}-\overset{\mathrm{H}\;\mathrm{H}}{\underset{\mathrm{H}\;\mathrm{H}}{\overset{|\;\;|}{\underset{|\;\;|}{\mathrm{C}-\mathrm{C}}}}}-\ddot{\mathrm{O}}-\mathrm{H},\quad \mathrm{H}-\overset{\mathrm{H}\;:\ddot{\mathrm{O}}-\mathrm{H}}{\underset{\mathrm{H}\;:\ddot{\mathrm{O}}-\mathrm{H}}{\overset{|\quad|}{\underset{|\quad|}{\mathrm{C}-\mathrm{C}}}}}-\mathrm{H},\quad \mathrm{H}-\ddot{\mathrm{O}}-\overset{\mathrm{H}\;\;\mathrm{H}}{\underset{\mathrm{H}\;\;\mathrm{H}}{\overset{|\;\;\;|}{\underset{|\;\;\;|}{\mathrm{C}-\ddot{\mathrm{O}}-\mathrm{C}}}}}-\mathrm{H},\quad \text{or}\quad \mathrm{H}-\overset{\mathrm{H}\;\;\;\;\;\mathrm{H}}{\underset{\mathrm{H}\;\;\;\;\;\mathrm{H}}{\overset{|\quad\quad|}{\underset{|\quad\quad|}{\mathrm{C}-\ddot{\mathrm{O}}-\ddot{\mathrm{O}}-\mathrm{C}}}}}-\mathrm{H}$

4. If a metal is present in the molecular formula, set the metal atom aside immediately as the appropriate positive ion (cation). Then combine the remaining atoms to make a negative ion (anion) using the methods described above.

Example

Draw the structural formula for CH₃ONa.

Solution:

Set aside: Na as Na⊕. It will be ionically bonded rather than covalently.

Set aside: 3 H—.

Make a skeleton: —C— + { —Ö:⁻, —Ö—, —Ö⊕ } which to choose?

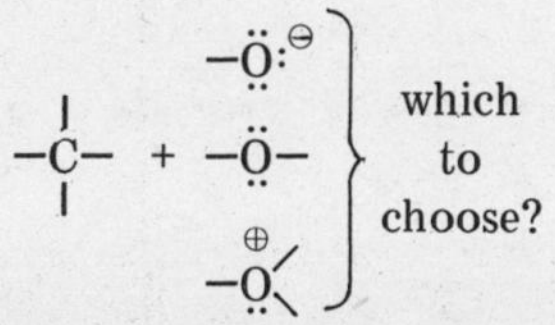

Two ways to decide: (1) The molecule as a whole is electrically neutral because the molecular formula shows no charge; therefore, O must be —Ö:⁻ to balance off the Na⊕.

(2) The skeleton from —Ö— + —C— is —C—Ö—, which requires four monovalent atoms for completion (heavy arrows). The skeleton from —Ö:⁻ and —C— is —C—Ö:⁻, which requires three monovalent atoms. We have only three (3 H—); therefore, use —Ö:⁻. Next, add the H— units:

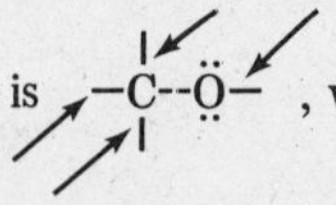

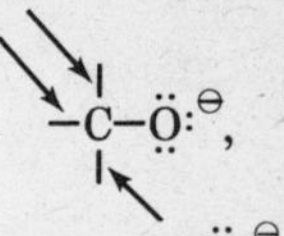

—C—Ö:⁻ + 3 H— → H—C(H)(H)—Ö:⁻

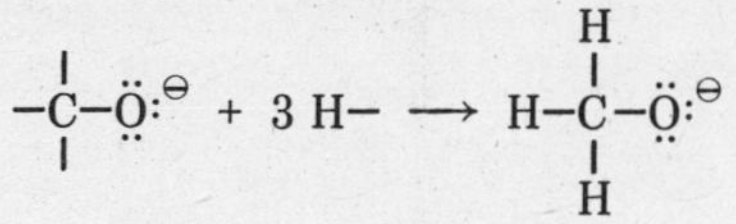

Finally, bring up Na⊕ close to the anion to get the structural formula:

H—C(H)(H)—Ö:⁻ Na⊕   *Note:* No covalent bond here.
    ionic bond

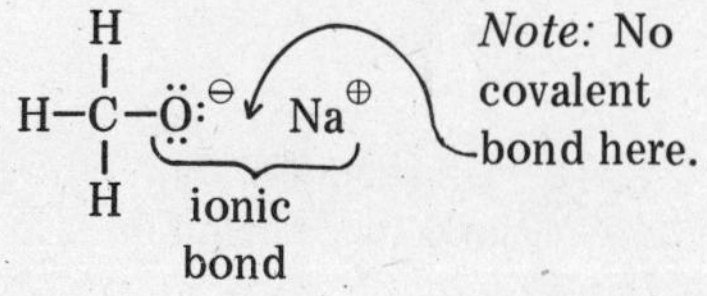

★ *Exercise*

Draw a structural formula for C₂H₅KS.

— — — — — — — —

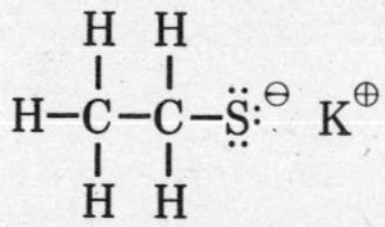

H—C(H)(H)—C(H)(H)—S̈:⁻ K⊕

5. If your skeleton requires more monovalent atoms than you have available in the molecular formula, change a pair of singly bonded part structures to doubly or triply bonded ones.

Example

Draw a structural formula for C_2F_3Cl.

Solution:

Set aside: 3 $:\ddot{F}-$, 1 $:\ddot{Cl}-$.

Try to make a skeleton: 2 $-\overset{|}{\underset{|}{C}}-$ → $-\overset{|}{\underset{|}{C}}-\overset{|}{\underset{|}{C}}-$ this skeleton requires exactly 6 monovalent atoms; but 4 are available.

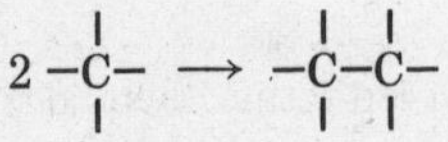

Therefore, try: 2 $>C=$ → $>C=C<$.

And finish: $>C=C< + 3 :\ddot{F}- + :\ddot{Cl}-$ → $\overset{:\ddot{F}:}{\underset{:\ddot{F}:}{>}}C=C\overset{\ddot{F}:}{\underset{\ddot{Cl}:}{<}}$.

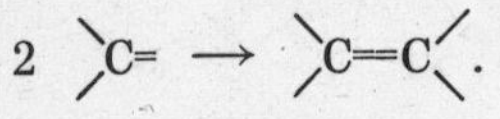

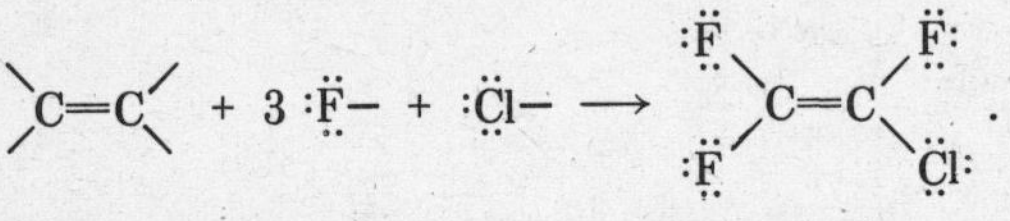

★ *Exercises*

Draw a structural formula for each compound.

1. N_2
2. CH_2O
3. CHN
4. CNNa
5. CH_3ON

— — — — — — — — — —

1. $:N\equiv N:$
2. 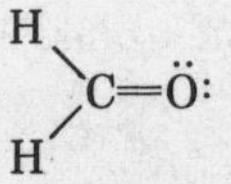 $\overset{H}{\underset{H}{>}}C=\ddot{O}:$
3. $H-C\equiv N:$
4. $Na^{\oplus} \ :C\equiv N:^{\ominus}$
5. 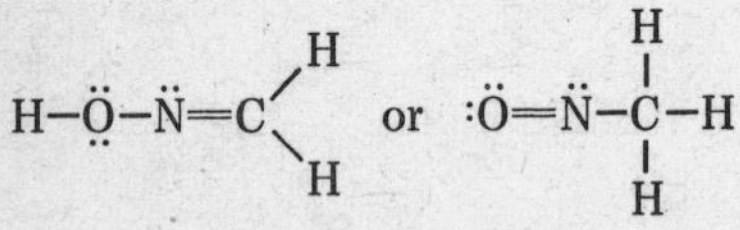 $H-\ddot{O}-\ddot{N}=C\overset{H}{\underset{H}{<}}$ or $:\ddot{O}=\ddot{N}-\overset{H}{\underset{H}{C}}-H$

6. If you have more monovalent atoms than your skeleton can accommodate, you have two possibilities. The first is to change a triple or a double bond to a single bond. In the next example, an initial bad guess is corrected by this procedure.

Example

Draw a structural formula for CH_3N.

Solution:

Set aside: 3 H–.

Try to make a skeleton: $-C\equiv + \equiv N:$ → $-C\equiv N:$ this skeleton requires exactly 1 monovalent atom; but 3 are available.

Therefore, try $>C= + =\ddot{N}-$ → $>C=\ddot{N}-$.

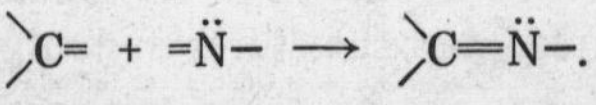

STRUCTURES 21

And finish: $\ce{>C=\overset{..}{N}-} + 3\text{ H}- \longrightarrow \ce{>C=\overset{H}{\underset{H}{N}}-H}$.

But suppose that you already have only single bonds in your framework and still have too many monovalent atoms; then the molecule must contain an atom capable of bearing a negative charge, usually a *halogen* or *oxygen* atom (occasionally, S or N). Make an *anion* out of this atom by choosing one of its anionic part structures from Table 2.1 and combining it with other of the available atoms if necessary. Make a *cation* out of the rest. Combine.

Example

Draw a structural formula for NH_4Br.

Solution:

Set monovalent atoms aside: $-H, -H, -H, -H, -\ddot{\underset{..}{B}}r:$.

Try to make a skeleton: it consists of a single atom: $>\!\ddot{N}\!-$ this skeleton requires exactly 3 monovalent atoms; but 5 are available.

Can't change multiple to single bonds.

Therefore, set aside Br as $:\ddot{\underset{..}{B}}r:^{\ominus}$.

Make a cation of the rest (choose a cationic nitrogen bonding unit from Table 2.1):

$$-\overset{|}{\underset{|}{N}}{\oplus}- + 4\text{ H}- \longrightarrow \text{H}-\overset{H}{\underset{H}{\overset{|}{\underset{|}{N}{\oplus}}}}-\text{H}.$$

Combine to get: $\text{H}-\overset{H}{\underset{H}{\overset{|}{\underset{|}{N}{\oplus}}}}-\text{H} \quad :\ddot{\underset{..}{B}}r:^{\ominus}$.

　　　　　　　　　　　　　　　　↖ no covalent bond here

Example

Draw a structural formula for CH_5OCl.

Solution:

Set aside: 5 $-H$ and the one $-\ddot{\underset{..}{C}}l:$.

Make a skeleton: $-\overset{|}{\underset{|}{C}}- + -\ddot{\underset{..}{O}}- \longrightarrow -\overset{|}{\underset{|}{C}}-\ddot{\underset{..}{O}}-$ this skeleton requires exactly 4 monovalent atoms; but 6 (5 H— + Cl—) are available.

No multiple bonds are available to turn into single bonds.

Therefore, make Cl ionic $\longrightarrow :\ddot{\underset{..}{C}}l:^{\ominus}$.

Make the rest a cation (choose a ⊕ oxygen bonding unit):

22 HOW TO SUCCEED IN ORGANIC CHEMISTRY

$-\overset{|}{\underset{|}{C}}- + -\overset{\oplus}{\underset{\cdot\cdot}{O}}\!\!<\; \longrightarrow\; -\overset{|}{\underset{|}{C}}-\overset{\oplus}{\underset{\cdot\cdot}{O}}\!\!<$ this skeleton requires 5 monovalent atoms and exactly 5 are available.

Complete the cation by adding H— units: $H-\overset{H}{\underset{H}{\overset{|}{C}}}-\overset{\oplus}{\underset{H}{O}}\!\!<^{H}$.

Combine: $H-\overset{H}{\underset{H}{\overset{|}{C}}}-\overset{\oplus}{\underset{H}{O}}\!\!<^{H}\quad :\overset{..}{\underset{..}{Cl}}\!:^{\ominus}$.

★ *Exercises*

Draw a structural formula for each compound.

1. N_2H_4
2. N_2H_6O
3. H_3ClO
4. CH_3NaO_3S (S in this oxidation state uses a $-\overset{|}{\underset{|}{\overset{2\oplus}{S}}}-$ part structure)

— — — — — — — — —

1. $\overset{H}{\underset{H}{>}}\!\overset{..}{N}-\overset{..}{N}\!\!<^{H}_{H}$

2. $\overset{H}{\underset{H}{>}}\!\overset{..}{N}-\overset{\oplus}{\underset{H}{N}}\!-H \;\; :\overset{..}{\underset{..}{O}}-H$ or $\overset{H}{\underset{H}{>}}\!\overset{..}{N}-\overset{..}{\underset{..}{O}}\!:^{\ominus}\; H-\overset{\oplus}{\underset{H}{N}}\!-H$ or $H-\overset{H}{\underset{H}{\overset{\oplus}{N}}}-\overset{..}{\underset{..}{O}}-H\;\; H-\overset{..}{\underset{H}{N}}\!-H^{\ominus}$

3. $\overset{H}{\underset{H}{>}}\!\overset{\oplus}{O}-H\;\; :\overset{..}{\underset{..}{Cl}}\!:^{\ominus}$

4. $H-\overset{H}{\underset{H}{\overset{|}{C}}}-\overset{:\overset{..}{O}:^{\ominus}}{\underset{:\overset{..}{O}:_{\ominus}}{\overset{|}{\overset{2\oplus}{S}}}}-\overset{..}{\underset{..}{O}}\!:^{\ominus}\;\; Na^{\oplus}$

CHECKING STRUCTURES FOR ERRORS

7. Three checks should be run on all structures drawn. At first this must be done deliberately and systematically; with sufficient practice it becomes automatic and rapid.

1. *Octet check:* verify that each atom is surrounded by 8 electrons (exception: 2 for H), either in the form of covalent bonds or as unshared pairs.
2. *Atom check:* verify that your structure contains the exact number of atoms of each element specified by the molecular formula.
3. *Electron check:* the following relation must be satisfied:

STRUCTURES 23

sum of electrons in structural formula = sum of electrons in the neutral, isolated atoms specified by the molecular formula − net charge on the structure (if any)

↑ [count bonds + unshared pairs]

↑ [count valence electrons only:
1 for each H, Na, or K
2 for each Mg or Ca
3 for each B or Al
4 for each C
5 for each N or P
6 for each O or S
7 for each halogen]

↑ [positive number for cations; negative number for anions]

Example

Check 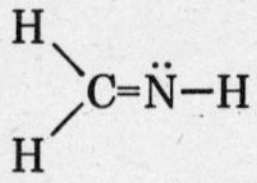
$$\begin{array}{c} H \\ \diagdown \\ C=\ddot{N}-H \\ \diagup \\ H \end{array}$$
for correctness.

Solution:

Valence electrons in atoms: 3 H, 3 × 1 = 3
                                 1 C, 1 × 4 = 4
                                 1 N, 1 × 5 = 5

                                  total =   12 electrons

Net charge on molecule: 0.

Therefore, total electrons in structural formula should = 12 − 0 = 12 electrons.

Verify: 5 bonds × 2 electrons per bond = 10
          1 unshared pair × 2 electrons   = 2

                                             12 electrons. Checks.

Example

Check 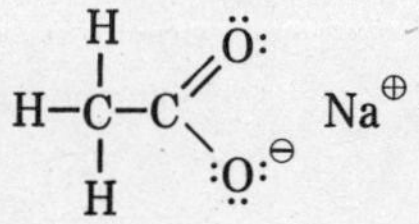
$$H-\underset{\underset{H}{|}}{\overset{\overset{H}{|}}{C}}-C\overset{\ddot{O}:}{\underset{:\ddot{O}:^{\ominus}}{\diagup\!\!\!\diagdown}}\;\;Na^{\oplus}$$
for correctness versus the molecular formula $C_2H_3O_2Na$.

Solution:

Atoms: H, 3 × 1 =   3
             C, 2 × 4 =   8
             O, 2 × 6 = 12
            Na, 1 × 1 =   1

                            24 electrons

Structure should have 24 − (0) = 24 electrons.
                                    ↖ zero net charge on compound (= cation + anion)

Verify: 7 covalent bonds × 2 electrons per bond = 14
5 unshared pairs × 2 electrons per pair = 10
24 electrons. Checks.

The structural formula for any individual ion must also pass the three tests.

Example

Check just the anion in the preceding example, $C_2H_3O_2^\ominus$.

Solution:

Valence electrons in atoms: 3 H, 3 × 1 = 3
2 C, 2 × 4 = 8
2 O, 2 × 6 = 12
23 electrons

Structure should have: 23 − (−1) = 23 + 1 = 24 electrons.
↖ net charge on the anion

Verify: 7 covalent bonds × 2 electrons per bond = 14 electrons
5 unshared pairs × 2 electrons per pair = 10
24 electrons. Checks.

Exercise

Perform the checks for correctness on any two structures drawn in the exercises of frame 2.

Suppose that you check each individual bonding unit in a structure against Table 2.1 or your memory of it, verifying that each has the proper number of bonds and unshared pairs and the correct charge. This accomplishes the same thing as the electron and octet checks (but not the atom check) described above. Once the table is firmly in mind, this is usually the easier method.

HANDLING FORMAL CHARGES

8. A charge borne by an atom in a molecule is called a "formal charge." The net (overall) charge on any molecule or ion is the sum of the formal charges on all its atoms. If you have the part structures from Table 2.1 clearly in mind, you automatically *know by sight* all the common situations that require formal charges. If you ever need to *figure out* the formal charge on a given atom of a molecule or ion, here is the method:

1. Count up the valence electrons *effectively owned* by the atom. Each unshared pair counts as two (owned outright) and each bond counts as one (half a "shared pair").
2. Compare this number with the number of valence electrons this atom would have if it were alone and uncharged:

   1 for H, Li, Na, K
   2 for Mg, Ca
   3 for B, Al
   4 for C
   5 for N, P
   6 for O, S
   7 for halogen

3. If the atom in the molecule has the *same* number of electrons as the isolated, neutral atom, it is neutral in the molecule also (*zero* formal charge). If it has *one less* electron in the molecule, it has a formal charge of +1 in the molecule. If it has *one more* electron in the molecule than in the neutral atom, it has a formal charge of −1 in the molecule. Zero formal charges are not written; all nonzero formal charges must be written as superscripts on the atom involved.

Example

What are the formal charges on the atoms in
$$\begin{array}{c} H\ \ H \\ |\ \ \ | \\ H-C-N-H? \\ |\ \ \ | \\ H\ \ H \end{array}$$

Solution:

C: has 1 electron per bond × 4 bonds = 4, versus 4 in the isolated atom, therefore zero formal charge.

Each H: has 1 electron per bond × 1 bond = 1, versus 1 in the isolated atom, therefore zero formal charge.

N: has 1 electron per bond × 4 bonds = 4, versus 5 in the isolated atom, therefore +1 formal charge, and the structure must show it:
$$\begin{array}{c} H\ \ H \\ |\ \ \ | \\ H-C-N^{\oplus}H. \\ |\ \ \ | \\ H\ \ H \end{array}$$

★ *Exercises*

1. Supply the missing formal charges (if any) in each structure. Try both the by-sight method and the calculation method.

    (a) $:\!\ddot{C}l-\overset{\overset{\displaystyle :\!\ddot{O}:}{\|}}{C}-\ddot{C}l\!:$

    (b) $\begin{array}{c} H \\ | \\ H-C: \\ | \\ H \end{array}$

    (c) $\begin{array}{c} H\ \ H\ \ H \\ |\ \ \ |\ \ \ | \\ H-C-O-C-H \\ |\ \ \ \ \ \ \ | \\ H\ \ \ \ \ \ H \end{array}$

    (d) $\begin{array}{c} CH_3 \\ | \\ H_3C-P-CH_3 \\ | \\ CH_3 \end{array}\ \ :\!\ddot{\underline{I}}\!:$

    (e) $H-\ddot{\underline{S}}:$

    (f) $\begin{array}{c} H \\ | \\ H-\ddot{N}-\ddot{O}-H \end{array}$

    (g) $\begin{array}{c} H \\ | \\ H-N-\ddot{O}: \\ | \\ H \end{array}$

    (h) $\begin{array}{c} H\ \ :\!\ddot{O}:\ \ H \\ |\ \ \ |\ \ \ | \\ H-C-S-C-H \\ |\ \ \ \ \ \ \ | \\ H\ \ \ \ \ \ H \end{array}$

    (i) $\begin{array}{c} H \\ | \\ H-C-\dot{N}=\ddot{O}: \\ | \\ H \end{array}$

    (j) $\begin{array}{c} H \\ | \\ H-C-N \\ | \\ H \end{array}\!\!\!\begin{array}{c} \nearrow\ddot{O}: \\ \\ \searrow\ddot{\underline{O}}: \end{array}$

2. Which method is easier?

1. (a) None.

(b) $H-\overset{\overset{H}{|}}{\underset{\underset{H}{|}}{C}}:^{\ominus}$

(c) $H-\overset{\overset{H}{|}}{\underset{\underset{H}{|}}{C}}-\overset{\oplus}{\underset{..}{O}}-\overset{\overset{H}{|}}{\underset{\underset{H}{|}}{C}}-H.$

(d) $H_3C-\overset{\overset{CH_3}{|}}{\underset{\underset{CH_3}{|}}{\overset{\oplus}{P}}}-CH_3 \quad :\overset{..}{\underset{..}{I}}:^{\ominus}.$

(e) $H-\overset{..}{\underset{..}{S}}:^{\ominus}.$

(f) None.

(g) $H-\overset{\overset{H}{|}}{\underset{\underset{H}{|}}{\overset{\oplus}{N}}}=\overset{..}{\underset{..}{O}}:^{\ominus}.$

(h) $H-\overset{\overset{H}{|}}{\underset{\underset{H}{|}}{C}}-\overset{\overset{:\overset{..}{O}:^{\ominus}}{|}}{\underset{\underset{}{}}{\overset{\oplus}{S}}}-\overset{\overset{H}{|}}{\underset{\underset{H}{|}}{C}}-H.$

(i) None.

(j) $H-\overset{\overset{H}{|}}{\underset{\underset{H}{|}}{C}}-\overset{\oplus}{N}\overset{\nearrow \overset{..}{\underset{..}{O}}:^{\ominus}}{\searrow \overset{..}{\underset{..}{O}}:}$

2. In the case of structures containing only the familiar bonding units, the by-sight method is much faster. But unfamiliar cases (e.g., the correction of $H-\overset{\overset{H}{|}}{\underset{\underset{H}{|}}{B}}-H$ to $H-\overset{\overset{H}{|}}{\underset{\underset{H}{|}}{\overset{\ominus}{B}}}-H$) can only be done the long way.

9. The structural formulas you have been drawing can be called *complete* structural formulas—or *Lewis structures*. However, writers of structural formulas often omit unshared pairs to save time, expense, or clutter. When should you take the trouble to put in the unshared pairs? Beginning students should probably *always* do so. The unshared pairs make it easier to spot errors in structures, and since they are one of the important reactive sites on a molecule, their visibility makes it easier to recall or predict the molecule's chemical behavior.

Formal charges must *never* be omitted from a structural formula that has been fully drawn out. The species $H-C\overset{\nearrow \overset{..}{\underset{..}{O}}:}{\searrow \overset{..}{\underset{..}{O}}:^{\ominus}}$ must not be written $H-C\overset{\nearrow \overset{..}{\underset{..}{O}}:}{\searrow \overset{..}{\underset{..}{O}}:}$. However, if the structure is condensed to a single line, as is usually done in printing, it becomes $H-CO_2^{\ominus}$, or $HCO_2^{\ominus}$. Since neither oxygen is shown individually, the location of the charge within the ion is not expressed; the $\ominus$ is just referring to the net, overall charge on the ion. But you know from frame 4 how to locate the charge on specific atoms in such a case. Other facets of condensed structures, however, require attention; we look into these in frame 11.

WRITING GROUP STRUCTURES AND GENERIC STRUCTURES

10. The *generic* name for a plant or animal identifies a whole group of related species, which are then distinguished by their specific names. The generic name of a drug applies to a group of closely related pharmaceutical preparations based on that drug, which are distinguished from one another by their trade names. Similarly, a family of closely related organic compounds is given a generic name to emphasize the similarities among the compounds. Each individual compound in the family receives its specific name. We will study the generic names in Unit 3 and the specific names in Unit 7. In this frame you should learn to read and write *generic structural formulas*; this requires, first, familiarity with the notion of a group.

A *group* is a part structure that is complete except for one bond: for instance,

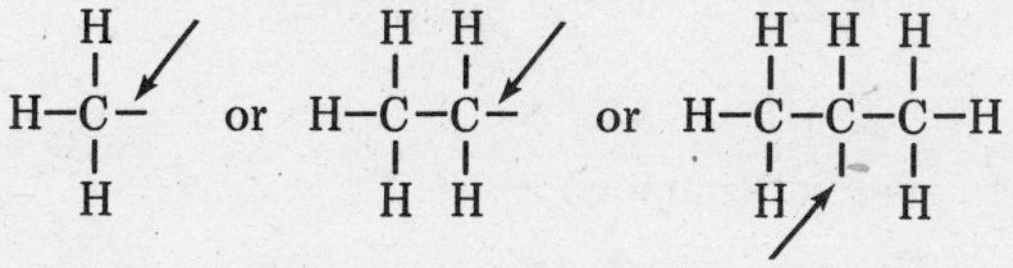

$$\begin{array}{c} H \\ | \\ H-C- \\ | \\ H \end{array} \quad \text{or} \quad \begin{array}{cc} H & H \\ | & | \\ H-C-C- \\ | & | \\ H & H \end{array} \quad \text{or} \quad \begin{array}{ccc} H & H & H \\ | & | & | \\ H-C-C-C-H \\ | & | & | \\ H & & H \end{array}$$

The unfinished bond in the structure (arrows in the examples) is called an open bond or valence. The groups drawn above, and all the many other possible ones that are made up solely of $-\overset{|}{\underset{|}{C}}-$ and $H-$ bonding units, are called *alkyl* groups. Since a chain of carbon atoms can be of any length, there are a great many alkyl groups. The standard symbol used to represent any and all of them is $R-$.

Another concept that we will find useful is the perfectly general group, composed of any conceivable combination of bonding units, and having any size, subject only to the restriction of obeying all the rules developed above for the construction of complete molecules, but having one open valence. There is no standard symbol for such a general group. Some people use $R-$ in this sense, instead of in its strict, original meaning. In this book we will use the symbol $G-$ for the general group.

Examples

$R-$ 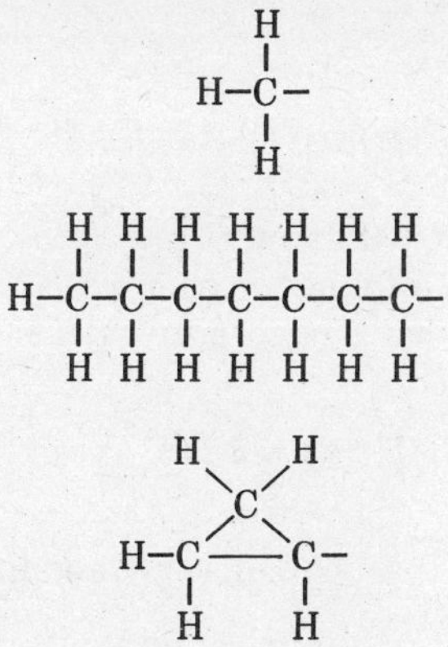	$G-$ 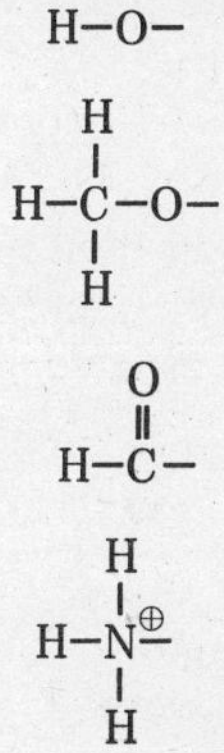
$\begin{array}{c} H \\ | \\ H-C- \\ | \\ H \end{array}$	$H-O-$
$\begin{array}{ccccccc} H & H & H & H & H & H & H \\ | & | & | & | & | & | & | \\ H-C-C-C-C-C-C-C- \\ | & | & | & | & | & | & | \\ H & H & H & H & H & H & H \end{array}$	$\begin{array}{c} H \\ | \\ H-C-O- \\ | \\ H \end{array}$
$\begin{array}{c} H \quad H \\ \diagdown \diagup \\ C \\ \diagup \diagdown \\ H-C-\!\!-\!\!-C- \\ | \quad\quad | \\ H \quad\quad H \end{array}$	$\begin{array}{c} O \\ \| \\ H-C- \end{array}$
	$\begin{array}{c} H \\ | \\ H-N^{\oplus}\!\!- \\ | \\ H \end{array}$

$R-$ and $G-$ are used for writing generic structural formulas. If we want to talk about all the possible organic compounds whose structures contain a $Cl-\overset{\overset{\displaystyle Cl}{|}}{\underset{\underset{\displaystyle Cl}{|}}{C}} -$ group, we can write the generic structure for this set or *family* (chemists prefer "family") of compounds as G–C(Cl)(Cl)–Cl. If we want to single out the family of compounds containing an H–O– group bound to a framework composed of –C– and H– bonding units, we write R–O–H.

★ *Exercises*

1. Draw the structural formula for a *group* having the molecular formula
   (a) HO_2
   (b) C_2F_3

2. Write a generic structure for each family of compounds.

   (a) H–C(H)(H)–C(H)=N–H, H–C(H)(H)–C(H)(H)–C(H)=N–H, H–C(H)(H)–C(H)(H)–C(H)(H)–C(H)=N–H, etc.

   (b) H–C(=O)–O–H, H–O–C(=O)–O–H, H(H)N–C(=O)–O–H, H–C(H)(H)–C(=O)–O–H, etc.

– – – – – – – – – –

1. (a) H–Ö–Ö–
   (b) (:F:)(:F:)C=C(:F:)(F:)

2. (a) R–C(H)=N̈–H
   (b) G–C(=O:)–Ö–H

READING CONDENSED STRUCTURAL FORMULAS

11. A condensed structure is one that has been boiled down, to save time and space, so that it fits on a single line of the printed page. The information eliminated in the condensation is: unshared pairs, some explicit details of what is bonded to what, and sometimes formal charges. Thus H–C(H)(H)–C(=Ö:)(H) becomes CH_3CHO. You will rarely need to write condensed structures, but you will often have to read them. A simple method for doing this consists of these steps:

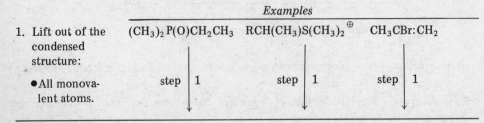

STRUCTURES

- All atoms or groups enclosed in ( ).
- All abbreviations for groups (R—, G—).

2. Connect in a chain the remaining atoms (mainly C, some O, N, S, P) with single bonds except where : or ⦂ indicates that double or triple bonds should be used.

3. Working from the left, attach the lifted groups and atoms to the chain so as to give each atom in the chain its correct valence before moving on to the one on its right.

4. Add any formal charges required and verify that they sum to the net charge shown on the condensed structure. Add unshared pairs.

$$CH_3- \quad H- \quad H-$$
$$O- \quad H- \quad H-$$
$$CH_3- \quad H- \quad H-$$

P   C   C

$$R- \quad CH_3- \quad CH_3-$$
$$H- \quad CH_3-$$

C   S

$$H- \quad H-$$
$$H- \quad Br- \quad H-$$
$$H- \quad H-$$

C   C   :   C

↓ step 2

$$CH_3- \quad H- \quad H-$$
$$O- \quad H- \quad H-$$
$$CH_3- \quad H- \quad H-$$

P—C—C

$$R- \quad CH_3-CH_3-$$
$$H- \quad CH_3-$$

C—S

$$H- \quad H-$$
$$H- \quad Br- \quad H-$$
$$H- \quad H-$$

C—C=C

↓ step 3

$$CH_3\diagdown O H H$$
$$ | | |$$
$$ P-C-C-H$$
$$CH_3\diagup | |$$
$$ H H$$

$$ H CH_3$$
$$ | \diagup$$
$$R-C-S$$
$$ | \diagdown$$
$$ H CH_3$$

$$ H Br$$
$$ | | H$$
$$H-C-C= \diagup$$
$$ | \diagdown H$$
$$ H$$

↓ step 4

$$CH_3\diagdown :\ddot{O}:^{\ominus} H H$$
$$ | | |$$
$$ P^{\oplus}-C-C-H$$
$$CH_3\diagup | |$$
$$ H H$$

(+1) + (−1) = 0

$$ H CH_3$$
$$ | \diagup$$
$$R-C-\overset{\oplus}{S}$$
$$ | \ddot{} \diagdown$$
$$ H CH_3$$

$$ H :\ddot{Br}:$$
$$ | | H$$
$$H-C-C=\diagup$$
$$ | \diagdown H$$
$$ H$$

Special cases:

(a) If insufficient groups are available to fill out all valences on the chain, it usually means that an O or S atom presently *in* the chain should instead be attached *to* the chain atom on its left by a double bond. For instance,
$$-\underset{|}{\overset{|}{C}}-O- \longrightarrow -\underset{|}{\overset{\overset{O}{\|}}{C}}-$$

$(CH_3)(C_2H_5)CHCOOCH_3$
↓

$$CH_3- H- H-$$
$$C_2H_5- H-$$
$$ H-$$

C   C   O   O   C
↓

$$CH_3- H- H-$$
$$C_2H_5- H-$$
$$ H-$$

C—C—O—O—C
↓

30 HOW TO SUCCEED IN ORGANIC CHEMISTRY

$$\text{CH}_3-\underset{\underset{H}{|}}{\overset{\overset{C_2H_5}{|}}{C}}-\underset{}{\overset{\overset{O}{\|}}{C}}-O-C \xleftarrow{} \text{CH}_3-\underset{\underset{H}{|}}{\overset{\overset{C_2H_5}{|}}{C}}-\underset{}{\overset{\overset{H-\ H-\ H-}{}}{C}}-O-O-C \xleftarrow{\text{back up}} \text{CH}_3-\underset{\underset{H}{|}}{\overset{\overset{C_2H_5}{|}}{C}}-\underset{\underset{H}{|}}{\overset{\overset{\downarrow}{H}}{C}}-O-O-\underset{\underset{?}{|}}{\overset{\overset{H}{|}}{C}}-?$$

$$\downarrow$$

$$\text{CH}_3-\underset{\underset{H}{|}}{\overset{\overset{C_2H_5}{|}}{C}}-\overset{\overset{\ddot{O}:}{\|}}{C}-\ddot{\text{O}}-\underset{\underset{H}{|}}{\overset{\overset{H}{|}}{C}}-H$$

(b) If the whole condensed structure has this form, (   )₂, then the contents of the (   ) must be a group with one open bond, G—, and the whole molecule is G—G.

$$(\text{ClCH}_2\text{CH}_2\text{CCl})_2$$
$$\downarrow$$
$$\text{Cl}-,\ \text{H}-,\ \text{H}-,\ \text{H}-,\ \text{H}-,\ \text{Cl}-,\ \text{Cl}-$$
$$(\text{C}-\text{C}-\text{C})_2$$
$$\downarrow$$

$$:\ddot{\text{Cl}}-\underset{\underset{H}{|}}{\overset{\overset{H}{|}}{C}}-\underset{\underset{H}{|}}{\overset{\overset{H}{|}}{C}}-\underset{\underset{:\ddot{\text{Cl}}:}{|}}{\overset{\overset{:\ddot{\text{Cl}}:}{|}}{C}}-\underset{\underset{:\ddot{\text{Cl}}:}{|}}{\overset{\overset{:\ddot{\text{Cl}}:}{|}}{C}}-\underset{\underset{H}{|}}{\overset{\overset{H}{|}}{C}}-\underset{\underset{H}{|}}{\overset{\overset{H}{|}}{C}}-\ddot{\text{Cl}}: \quad \xleftarrow{} \quad (\text{Cl}-\underset{\underset{H}{|}}{\overset{\overset{H}{|}}{C}}-\underset{\underset{H}{|}}{\overset{\overset{H}{|}}{C}}-\underset{\underset{\text{Cl}}{|}}{\overset{\overset{\text{Cl}}{|}}{C}}-)_2$$

(c) If a (   ) contains a part structure with *two* rather than *one* open bond, this group is *part of* the chain, not attached to it. A subscript on the (   ) tells how many times to repeat the contents of the (   ) in the chain. For example, (CH₂)₃ indicates that the chain contains the group —CH₂—CH₂—CH₂— at this point.

$$\text{CCl}_3(\text{CH}_2)_3\text{Cl} \diagdown\!\!\!\diagup$$
$$\begin{array}{l}\text{Cl}-\ -\text{CH}_2-\\ \text{Cl}-\ -\text{CH}_2-\ \text{Cl}\\ \text{Cl}-\ -\text{CH}_2-\end{array}$$

$$\text{Cl}-,\ \text{Cl}-,\ \text{Cl}-,\ \text{Cl}-$$
$$\text{C}-\text{CH}_2-\text{CH}_2-\text{CH}_2-$$
$$\downarrow$$
$$:\ddot{\text{Cl}}-\underset{\underset{:\ddot{\text{Cl}}:}{|}}{\overset{\overset{:\ddot{\text{Cl}}:}{|}}{C}}-\text{CH}_2-\text{CH}_2-\text{CH}_2-\ddot{\text{Cl}}:$$

(d) Sometimes you will have to puzzle out the construction of the group in (   ).

$$(\text{CH}_3\text{CO})_3\text{CH} \rightarrow \begin{array}{l}\text{CH}_3\text{CO}-,\ \text{CH}_3\text{CO}-\\ \text{CH}_3\text{CO}-,\ \text{H}-\\ \text{C}\end{array}$$

Four groups on four open bonds, but what does CH₃CO— mean?

$$\begin{array}{c}\text{H}-,\ \text{H}-,\ \text{H}-\\ \text{C}-\text{C}-\text{O}\end{array} \xleftarrow{\text{apply steps 1-3 to CH}_3\text{CO}-}$$

obviously insufficient groups, so $\searrow$

$$\begin{array}{c}\text{H}-,\ \text{H}-,\ \text{H}-\\ \underset{\text{C}-\text{C}}{\overset{\overset{O}{\|}}{}}\end{array} \rightarrow \text{H}-\underset{\underset{H}{|}}{\overset{\overset{H}{|}}{C}}-\overset{\overset{O}{\|}}{C}- \curvearrowleft \text{open bond}$$

Therefore,

$$\text{CH}_3-\underset{\underset{\underset{:O:}{\|}}{\underset{C-\text{CH}_3}{|}}}{\overset{\overset{\overset{:O:}{\|}}{\overset{:O:\ \ C-\text{CH}_3}{}}}{\overset{}{C}}}-\underset{}{\overset{}{C}}-H$$

STRUCTURES 31

★ *Exercises*

Translate each condensed structure into a full one (omit unshared pairs).

1. $CH_3SCH_2CH_2CH_2OH$
2. $CH_2{:}CHCHO$
3. CH_3COCHO
4. $CH_3OCH{:}CHC{:}CCOCH_3$
5. $R(OH)CHCOOCH_3$
6. $(CH_3)_2NCH{:}CHCOR$
7. $H_2N(CH_2)_5NH_2$
8. $(CH_3CO)_2$
9. $R_2C(OBr)CO_2H$ [CO_2H should be written $C(O)OH$, or at least $COOH$, but CO_2H has become a widespread habit]

— — — — — — — — —

1. 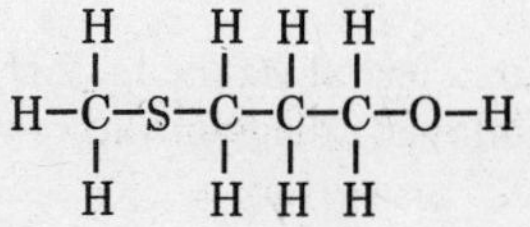
```
    H     H H H
    |     | | |
H—C—S—C—C—C—O—H
    |     | | |
    H     H H H
```
or

$CH_3-S-CH_2-CH_2-CH_2-OH$

(this half-condensed style is a useful compromise—compact but readable)

2. 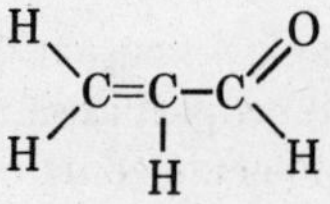
```
 H           O
  \         //
   C=C—C
  /   |    \
 H    H     H
```

3. 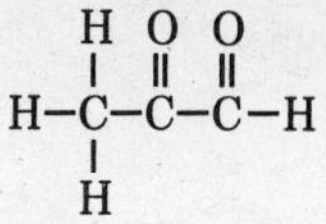
```
    H O O
    | ‖ ‖
H—C—C—C—H
    |
    H
```

4. 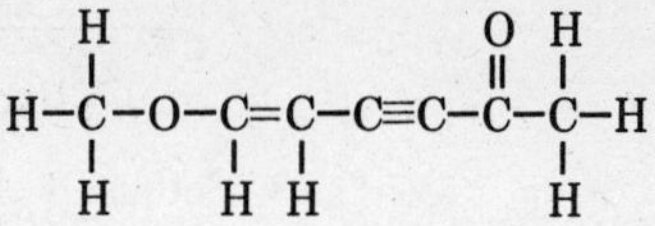
```
    H         O H
    |         ‖ |
H—C—O—C=C—C≡C—C—C—H
    |     | |       |
    H     H H       H
```

5. 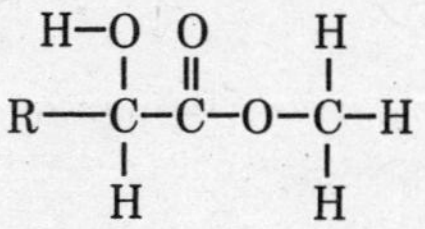
```
   H—O O     H
      |  ‖   |
R—C—C—O—C—H
      |     |
      H     H
```

6. 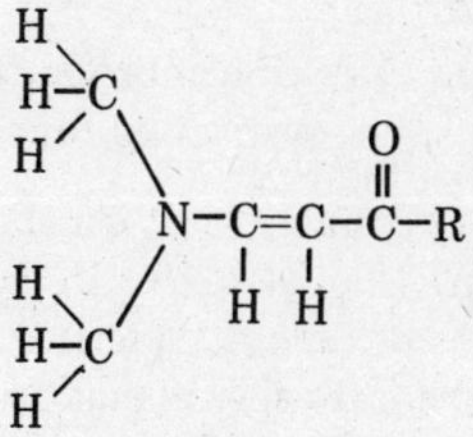
```
    H
   H—C
  H/        O
   \        ‖
    N—C=C—C—R
   /    | |
  H     H H
   \
   H—C
    H/
```

7. 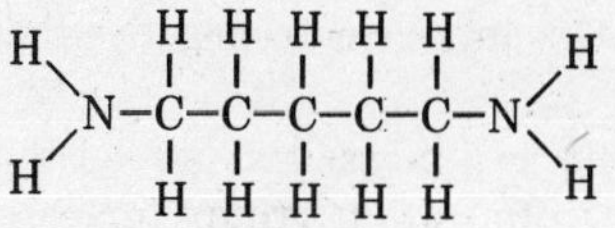
```
 H\    H H H H H    /H
   N—C—C—C—C—C—N
 H/    | | | | |    \H
       H H H H H
```

8. 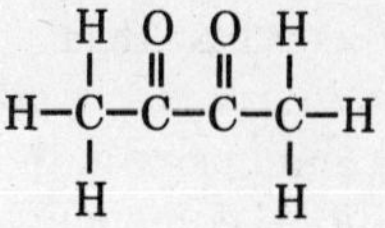
```
    H O O H
    | ‖ ‖ |
H—C—C—C—C—H
    |     |
    H     H
```

9. 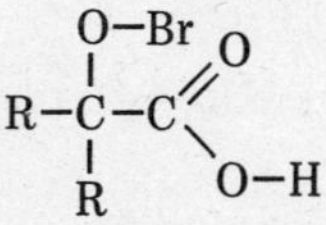
```
     O—Br O
     |    ‖
R—C—C
     |    \
     R     O—H
```

UNIT THREE

Function and Skeleton

In Unit 2 you learned that we can write generic structures representing whole familes of compounds, for example R—Cl, R—OH, R—C(=O)OH . Now we put this classification to practical use. Writing R—Cl, etc., divides a compound into two parts, the *skeleton*, R, and the nonskeletal part, called the *functional group* (—Cl, —OH, —C(=O)OH in the cases above). This separation is important, because it divides the molecule in a useful way—one that matches well the observed properties of the molecules. The following diagram shows what the various parts of the molecule contribute to the compound's properties.

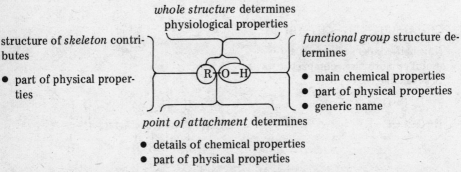

Since a compound's chemical properties are localized in its functional group, people usually learn organic chemistry one functional group at a time by:

1. Learning a set of facts (e.g., observable chemical reactions) and correlating them with structure ("all alcohols, ROH, react with Na metal to give R—O$^\ominus$ Na$^\oplus$ + ½H$_2$; ethers, R—O—R, do not," etc.).
2. Learning a theoretical explanation of the correlation whenever possible (e.g., "R—O—R is a weaker base than R$_2$N—R because O is more electronegative than N").
3. Putting the facts and theory in the form of rules that can be used to predict the behavior of additional compounds (e.g., "Electron-withdrawing groups increase acidity").

This pattern is repeated for each family of compounds, for the R—Cl, the R—OH, the R—C(=O)OH , and so on. To make the correlations and rules accurate and useful, we must distinguish not only between the various common functional groups (−Cl, −OH, −C(=O)OH, etc.) but also between several types of skeleton and between several ways of connecting the functional group to the skeleton. These classifications are shown below.

Classifications of functions, skeletons, and modes of connection discussed in this unit:

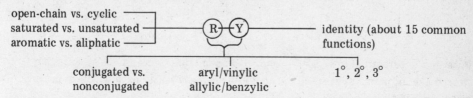

In this unit we take up classification of functional groups, skeletons, and points of attachment, in that order.

OBJECTIVES

When you complete this unit, you should be able, given a structural formula, to:

- Distinguish its skeleton and functional group (frame 1).
- Identify the functional group by name (frame 1).
- Recognize mono-, di-, tri-, and tetrasubstituted varieties of the C=C function and terminal vs. nonterminal varieties of the C≡C function (frame 2).
- Distinguish saturated, unsaturated, and aromatic skeletons, and open-chain (acyclic) from cyclic skeletons (frame 3).
- Read line-segment abbreviations for skeletal structures (frame 4).
- Identify primary, secondary, and tertiary sites for placement of the functional group on saturated skeletons (frame 5).
- Identify vinylic and allylic sites on aliphatic skeletons and aryl and benzylic sites on aromatic skeletons (frame 6).
- Identify conjugated and nonconjugated functions (frame 7).

If you think you have already mastered these objectives and might skip all or part of this unit, take a self-test consisting of the problems identified by a star in the margin. If you answer all these questions correctly, you are probably ready to move on to the next unit. If you miss any questions, study the frame(s) in question and rework the problem before proceeding.

If you are not yet ready for a self-test, proceed to frame 1.

FUNCTIONAL GROUPS

1. In everyday speech the functional group is often just "function." The identity of the function determines the chemical behavior of an organic compound almost completely. As a result, all the compounds in the family R−OH (or R−Cl or R−C(=O)OH, etc.) have approximately the same chemical properties. The type of site on the skeleton that the functional group occupies *modifies* this chemical behavior to a certain extent. The chemical role of the skeleton lies not in its size or shape but in the type of sites it offers for the attachment of functional groups.

Ordinarily, the functional group is what is left if you set aside the carbon atoms and the hydrogens attached to them.

Examples

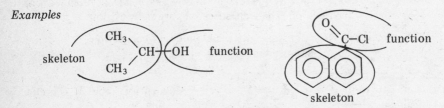

However, the part structures >C=C< and $-\text{C}\equiv\text{C}-$ are functional groups because they undergo a set of predictable, specific, and useful chemical reactions in whatever molecules they happen to be located.

Examples

There is one family of compounds that lacks a functional group: the *alkanes*, composed of only $-\overset{|}{\underset{|}{\text{C}}}-$ and $-\text{H}$ bonding units, for instance $\text{CH}_3(\text{CH}_2)_n\text{CH}_3$ (n = 1, 2, 3, ...). An alkane is 100% skeleton. It is not 100% lacking in chemical reactivity, but it comes close. The reactions that alkanes undergo (1) require potent rather than mild reaction conditions and (2) take place unselectively at almost all the bonds in the molecule. As a result, these reactions are less useful than the reactions more easily brought about at a functional group. The skeletal part of a molecule that contains a functional group (e.g., R in R—OH) behaves chemically just like an alkane. So the functional group generally reacts more readily, predictably, and usefully, and we do not often carry out reactions of the skeleton.

A molecule can have two or more functional groups, and these can be identical or different. We learn the chemistry of molecules containing one function first (*monofunctional* compounds), then go on to *polyfunctional* compounds.

Thousands of monofunctional compounds are known, and an unlimited number could be made. But the vast majority belong to only about 15 families derived from the 15 most common functional groups. This means that one must learn only 15 sets of chemical properties, not several thousand, in order to understand and predict the behavior of monofunctional compounds. The 15 standard functional groups and their names are listed in Table 3.1.

Note that some of the functional "groups" are part structures with more than one uncompleted bond, which means that the skeleton is actually in two or more parts. To complete the generic structure of an ester, $-\text{C}\begin{smallmatrix}\nearrow\text{O}\\\searrow\text{O}-\end{smallmatrix}$, for example, we need *two* alkyl groups. If we write this generic structure as $\text{R}-\text{C}\begin{smallmatrix}\nearrow\text{O}\\\searrow\text{O}-\text{R}\end{smallmatrix}$, we are limiting ourselves to

FUNCTION AND SKELETON

Table 3.1 *Common Functional Groups*

Functional group	Name of function = generic name of family of compounds containing it
$-F$, $-Cl$, $-Br$, $-I$ $\}$ $-X$	alkyl fluoride, alkyl chloride, alkyl bromide, alkyl iodide $\}$ alkyl halide
$-OH$	alcohol
$-O-$	ether
$\mathrm{C=C}$	alkene or olefin
$-C\equiv C-$	alkyne or acetylene
$-\underset{\parallel}{\overset{O}{C}}-$	ketone $\}$ carbonyl compounds, carbonyl function
$-\underset{\parallel}{\overset{O}{C}}-H$	aldehyde
$-C(=O)O-H$	carboxylic acid
$-C(=O)O-$	ester
$-C(=O)Cl$	acyl chloride
$-C(=O)NH_2$, $-C(=O)NH-$, $-C(=O)N\!<$	amide
$-C\equiv N$	nitrile $\}$ carboxylic acid derivatives
$-NO_2$	nitro
$-SO_3H$	sulfonic acid
$-NH_2$, $-NH-$, $-N\!<$	amine

esters with the two skeletal parts identical; this excludes many members of the family. To get around this difficulty, we often use R, R′, and R″, or R^1, R^2, and R^3 as different symbols for any alkyl group. Then the ester family can be written perfectly generally as

$$R-C\underset{O-R'}{\overset{\parallel O}{}}$$

, which includes cases with R = R′ as well as R ≠ R′. The symbol —X represents any alkyl halide function.

Don't memorize the 15 functions and their names now. Instead, refer back to the list while you work through the next few units. Whenever you encounter a functional group, check to see if it is in the list and look up its name. By the end of Unit 11 you should know most of them.

★ *Exercises*

1. Write the generic structural formula for each family of compounds.

   (a) nitriles
   (b) ketones
   (c) alkyl halides
   (d) alkanes

2. Draw complete part structures for the 15 common functions listed in Table 3.1 by supplying missing bonds, unshared pairs, and formal charges.

3. Do the same for these less common functions:

   (a) —NO (nitroso)
   (b) >NOH (hydroxylamine)
   (c) >C=NOH (oxime)
   (d) >C=N— (nitrone)
                   |
                   O
   (e) >NO (amine oxide)

— — — — — — — — —

1. (a) R—C≡N
   (b) R—C(=O)—R'
   (c) R—X
   (d) R—H

2. 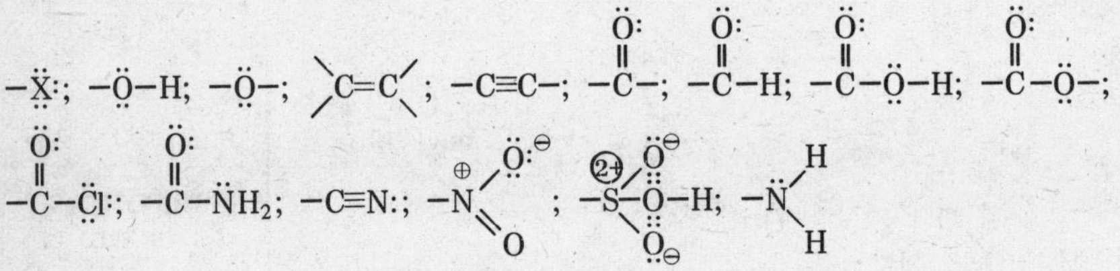 —$\ddot{\text{X}}$:; —$\ddot{\text{O}}$—H; —$\ddot{\text{O}}$—; >C=C< ; —C≡C—; —C(=$\ddot{\text{O}}$:)—; —C(=$\ddot{\text{O}}$:)—H; —C(=$\ddot{\text{O}}$:)—$\ddot{\text{O}}$—H; —C(=$\ddot{\text{O}}$:)—$\ddot{\text{O}}$—;

   —C(=$\ddot{\text{O}}$:)—$\ddot{\text{Cl}}$:; —C(=$\ddot{\text{O}}$:)—$\ddot{\text{N}}$H₂; —C≡$\ddot{\text{N}}$:; —N⁺(=O)($\ddot{\text{O}}$:⁻); —$\overset{2+}{\text{S}}$(=$\ddot{\text{O}}$)(—$\ddot{\text{O}}$—H)($\ddot{\text{O}}$:⁻); —$\ddot{\text{N}}$(H)(H)

3. 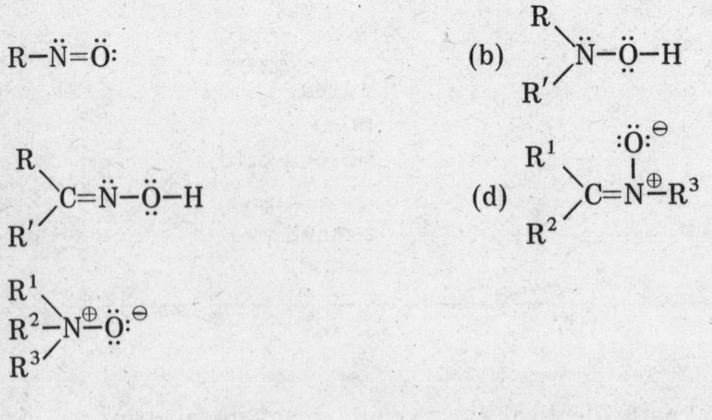 (a) R—$\ddot{\text{N}}$=$\ddot{\text{O}}$:
   (b) R\\$\ddot{\text{N}}$—$\ddot{\text{O}}$—H / R'
   (c) R\\C=$\ddot{\text{N}}$—$\ddot{\text{O}}$—H / R'
   (d) R¹\\C=N⁺(—R³)(:$\ddot{\text{O}}$:⁻) / R²
   (e) R²—N⁺(R¹)(R³)—$\ddot{\text{O}}$:⁻

SPECIAL CASES—THE >C=C< AND —C≡C— FUNCTIONS

2. Each of the 15 common functional groups in our list could be made into a complete compound by adding a "skeleton" consisting of H— rather than R—. In some cases this gives an organic compound, for instance —C(=O)(O—H) → H—C(=O)(O—H), in some cases an inorganic one (H—F, H—O—H, H—N(H)(H)). In some cases it generates a family already listed

$\left(-O- \rightarrow -O-H, -C{\overset{O}{\underset{}{\diagdown}}} \rightarrow -C{\overset{O}{\underset{H}{\diagdown}}} \right)$. In the case of $>C=C<$ and $-C\equiv C-$, the families resulting from the various possible combinations of R and H must be discussed in more detail.

Alkenes are classified according to the degree of "alkyl substitution" (sometimes "alkylation" or "substitution") of the $>C=C<$ function. The word *substitution* refers to the conceptual act of replacing H– by R–. Consequently, all the varieties of $>C=C<$ can be generated by starting with $\overset{H}{\underset{H}{>}}C=C\overset{H}{\underset{H}{<}}$ and "substituting alkyl":

$$\left.\begin{array}{l} CH_2=CH_2 \quad \text{unsubstituted} \\ R-CH=CH_2 \quad \text{monosubstituted} \\ R-CH=CH-R' \\ \quad \text{and} \\ \overset{R}{\underset{R'}{>}}C=CH_2 \end{array}\right\} \text{disubstituted}$$

$\overset{R}{\underset{R'}{>}}C=C\overset{R''}{\underset{H}{<}}$ trisubstituted

$\overset{R}{\underset{R'}{>}}C=C\overset{R''}{\underset{R'''}{<}}$ tetrasubstituted

⇩ direction of increasing stability ⇩

This categorization is important because the relative stabilities of the alkenes depend on the extent of alkyl substitution; the order is indicated above. The main difficulty here lies in confusing number of alkyl groups with number of carbon atoms.

Example

This is a monosubstituted alkene $\boxed{CH_3-CH_2-CH_2-CH_2-CH_2}\!\!\nearrow\!\!CH=CH_2$ because only one of the four original H– units on $>C=C<$ has been replaced by R. Or look at it this way: R (the boxed part of the structure) contains five carbon atoms, but it is attached to $>C=C<$ through only one bond (arrow), so it is just one alkyl group.

This is a trisubstituted alkene:

$\boxed{CH_3-CH_2}\diagdown \quad H$
$\qquad\qquad C=C$
$\boxed{CH_3-CH_2}\diagup \quad \diagup\boxed{CH_3}$

It contains the same number of carbon atoms as the first example, but they are arranged in three alkyl groups, attached to $>C=C<$ by three bonds.

The alkyne function, $-C\equiv C-$, can take either of these forms:

$R-C\equiv C-H$ $\qquad$ $R-C\equiv C-R'$
terminal alkyne $\quad$ nonterminal alkyne

The distinction is important because H attached directly to $-C\equiv C-$ is weakly acidic, so terminal alkynes can form salts with strong bases, whereas nonterminal ones cannot.

★ Exercises

1. Determine the degree of substitution of each alkene.

   (a) 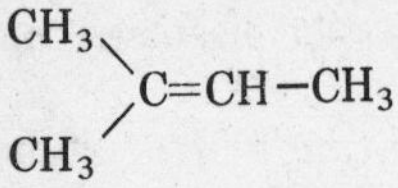 $\begin{array}{c}CH_3\\ \diagdown\\ C=CH-CH_3\\ \diagup\\ CH_3\end{array}$

   (b) $CH_3-CH=CH_2$

   (c) $CH_3-CH_2-CH_2-CH=CH_2$

   (d)  (cyclohexene)

   (e) 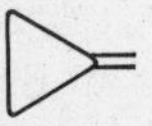 (methylenecyclopropane)

2. Classify the $-C\equiv C-$ in each molecule as terminal or nonterminal.

   (a) $CH_3-CH_2-C\equiv C-CH_3$

   (b) 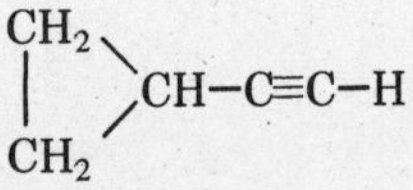 $\begin{array}{c}CH_2\\ |\diagdown\\ CH-C\equiv C-H\\ |\diagup\\ CH_2\end{array}$

   (c) (cyclic structure with $C\equiv C$)

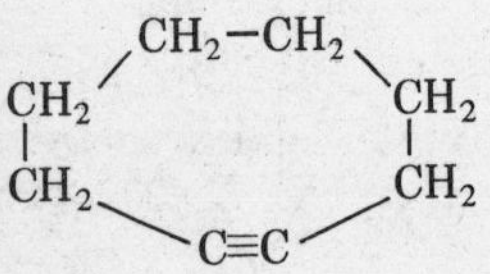

   (d) $H-C\equiv C-CH=CH-CH_3$

---

1. (a) 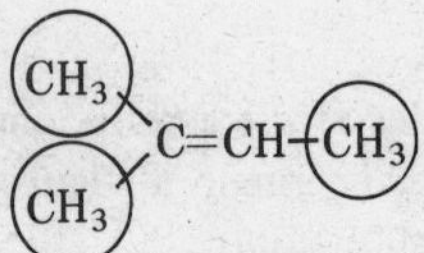 $\begin{array}{c}\boxed{CH_3}\\ \diagdown\\ C=CH-\boxed{CH_3}\\ \diagup\\ \boxed{CH_3}\end{array}$ trisubstituted

   (b) 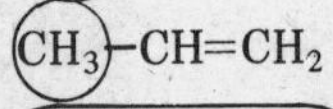 $\boxed{CH_3}-CH=CH_2$ monosubstituted

   (c) 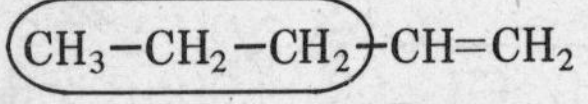 $\boxed{CH_3-CH_2-CH_2}-CH=CH_2$ monosubstituted

   (d) 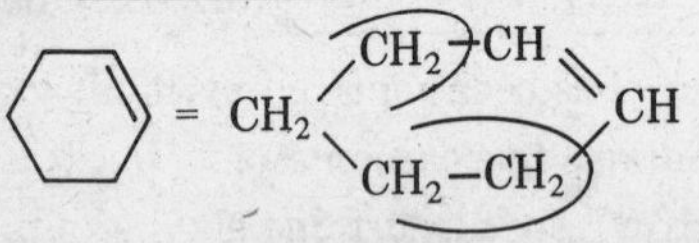 ⬡ = $CH_2\begin{array}{c}CH_2-CH\\ \diagdown\diagdown\\ CH\\ \diagup\diagup\\ CH_2-CH_2\end{array}$ disubstituted

The circled part of the molecule counts as two alkyl groups because it is attached to $\diagup C=C \diagdown$ by two bonds. $\diagup C=C \diagdown$ feels the effect of two alkyl groups; the fact that they are tied together to make a ring is irrelevant.

   (e) 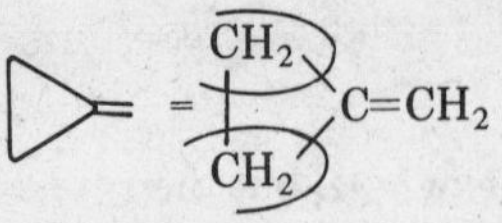 ▷= $\begin{array}{c}CH_2\\ |\diagdown\\ C=CH_2\\ |\diagup\\ CH_2\end{array}$ disubstituted

2. (a) nonterminal   (b) terminal

   (c) nonterminal   (d) terminal

CLASSES OF SKELETONS

3. A skeleton, carbon atom, group, or compound having one or more multiple bonds

FUNCTION AND SKELETON 39

(any $\diagup\!\!\!\!C\!=$, $-C\!\equiv$, or $=\!C\!=$ bonding unit) is called *unsaturated* because it could accommodate more monovalent atoms than it presently has. If multiply bonded carbon is absent, the skeleton is *saturated*, built entirely of *saturated* carbon atoms, $-\overset{|}{\underset{|}{C}}-$.

Both saturated and unsaturated skeletons can have one or more rings (*cyclic* skeletons, for instance CH₂⟨CH₂–CH₂⟩CH–) or exist strictly as straight or branched chains (*acyclic* or *open-chain* skeletons). A completely unsaturated ring (no $-\overset{|}{\underset{|}{C}}-$) usually has entirely different properties from those of ordinary unsaturated skeletons. Such rings are called aromatic; examples are

$$\text{CH}\diagup\overset{\text{CH=CH}}{\diagdown}\text{CH} \quad \text{and} \quad \text{(naphthalene-like structure)}$$

Aromatic skeletons contain one or more of these rings. (There are some special cases of aromatic rings with more or fewer than six carbons; your textbook will discuss these at the appropriate point.) Skeletons that are not aromatic are called *aliphatic*. The following diagram shows the relationships:

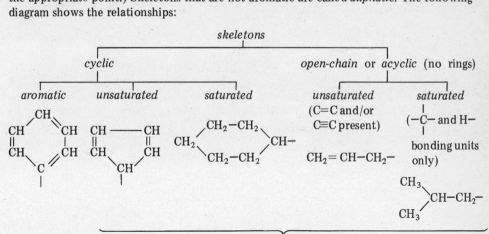

★ *Exercises*

1. Identify each skeleton as aromatic or aliphatic. Identify the aliphatic cases further as cyclic or acyclic and as saturated or unsaturated.

   (a) $CH_2=CH-\underset{|}{CH}-CH_3$

   (b) $H-C\equiv C-\underset{|}{C}=CH_2$

   (c) $\begin{array}{c} CH_2-CH- \\ |\quad\quad | \\ CH_2-CH_2 \end{array}$

   (d) $\begin{array}{c} CH_2-C- \\ |\quad\quad \| \\ CH_2-CH \end{array}$

   (e) $CH\diagup\overset{CH-CH}{\diagdown_{CH=CH}}C-C\diagup\overset{CH=CH}{\diagdown_{CH-CH}}C-$

   (f) $-CH_2-CH_3$

40   HOW TO SUCCEED IN ORGANIC CHEMISTRY

2. Circle the unsaturated carbon atoms in each skeleton.

(a) H–C≡C–CH$_2$–

(b) 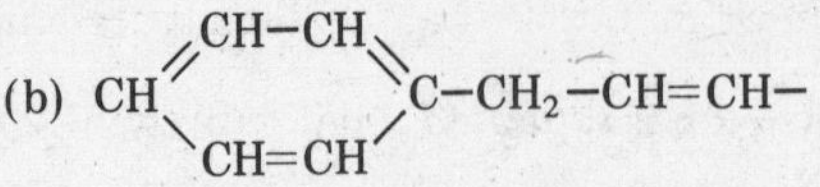 CH=CH–CH=CH–C(–CH–CH=)–CH$_2$–CH=CH–

- - - - - - - - - -

1. (a) and (b) are aliphatic, acyclic, unsaturated. (b) contains no –C–, but is not aromatic because it is acyclic. (c) is aliphatic, cyclic, and saturated. (d) is aliphatic, cyclic, and unsaturated. (e) is aromatic. (f) is aliphatic, acyclic, and saturated.

2. 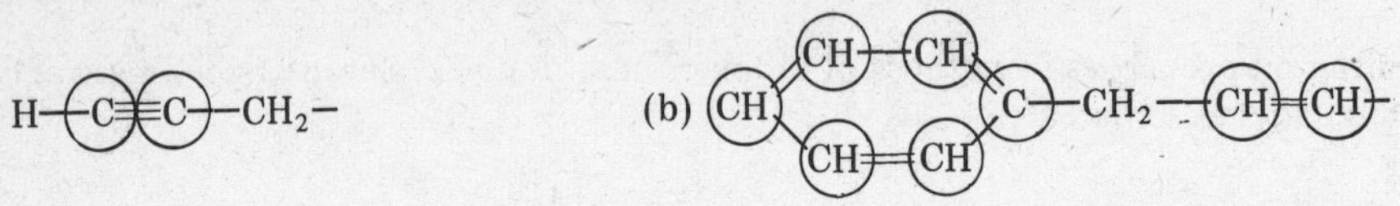 (a) H–(C)≡(C)–CH$_2$–   (b) (CH)=(CH)–(CH)=(CH)–(C)–CH$_2$–(CH)=(CH)–

READING SHORTHAND (LINE-SEGMENT) SKELETONS

4. A shorthand notation for skeletons has been developed to speed up the writing of structures and to simplify typesetting. To translate a standard skeleton into shorthand, (a) remove the C and H symbols, and (b) extend the bonds until they meet.

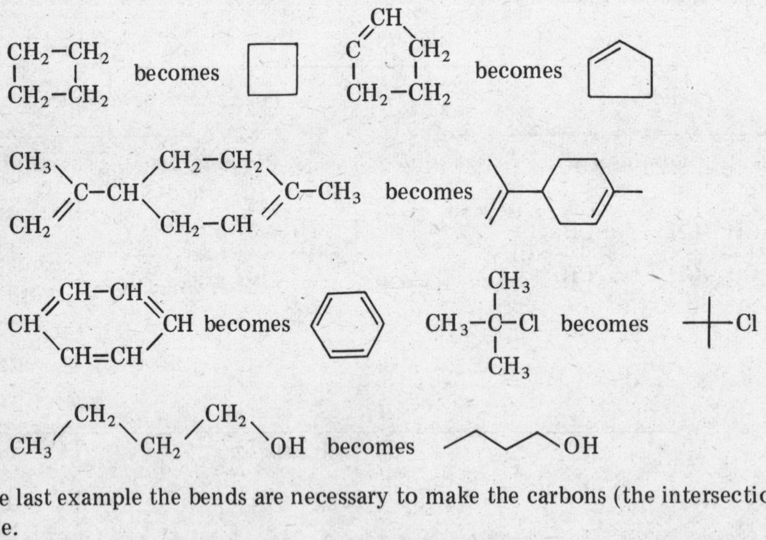

CH$_2$–CH$_2$ / CH$_2$–CH$_2$ becomes   □

CH=C(–CH$_2$)–CH$_2$–CH$_2$ becomes   ⬠ (cyclopentene)

CH$_3$–C(=CH$_2$)–CH–CH$_2$–CH$_2$–C(–CH$_2$–CH)=CH$_3$ becomes   (substituted cyclohexadiene shorthand)

CH=CH–CH=CH–CH=CH   becomes   ⬡ (benzene)

CH$_3$–C(CH$_3$)(CH$_3$)–Cl   becomes   ⊢Cl

CH$_3$–CH$_2$–CH$_2$–CH$_2$–OH   becomes   ∿∿OH

In the last example the bends are necessary to make the carbons (the intersections) visible.

It is a little harder to reconstruct the full structure from the shorthand, because one must decide how many hydrogens to put on each carbon. The answer is simple—just enough to give carbon a total of four bonds—but some practice is required to gain facility in this notation.

For reasons developed in Unit 9, aromatic rings are often written with a circle replacing the double bonds:

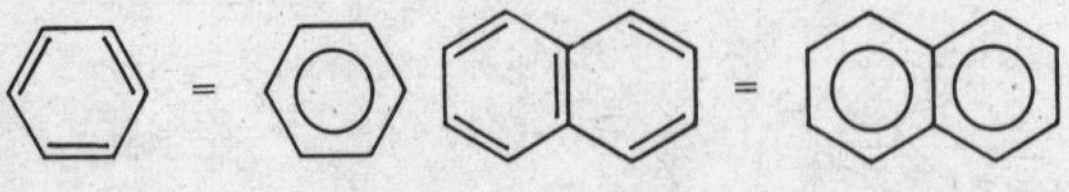

⬡ = ⬡(○)   ⬡⬡ = ⬡⬡(○○)

★ *Exercises*

1. Write a shorthand (line-segment) structure for each.

(a) 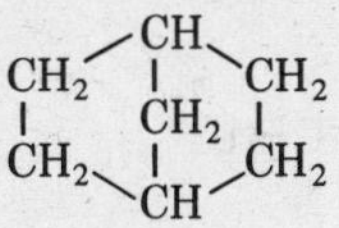

(b) CH⁝CCH:CHCH$_2$CHO

2. Write a standard structure for each.

(a) 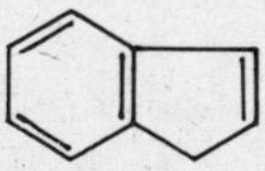

(b) 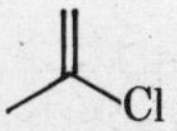

(c) 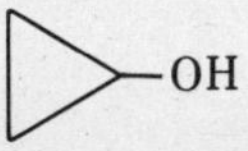

(d) 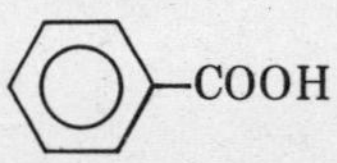

(e) 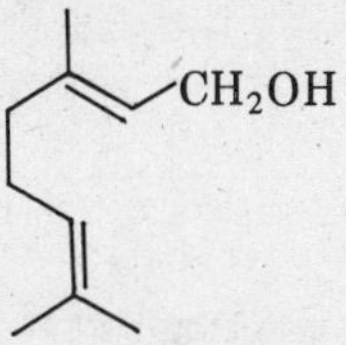

---

1. (a) 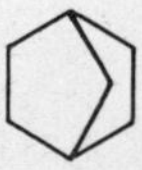

(b) 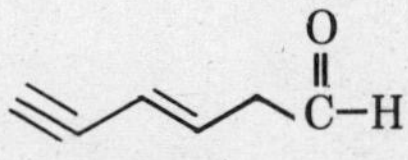

2. (a) 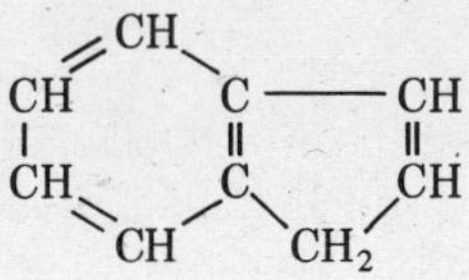

(b) 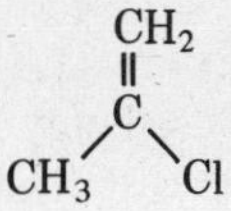

(c) 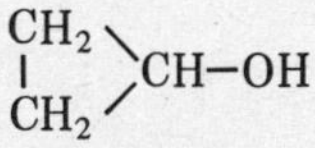

(d) 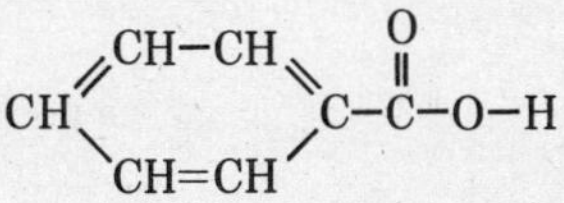

(e) 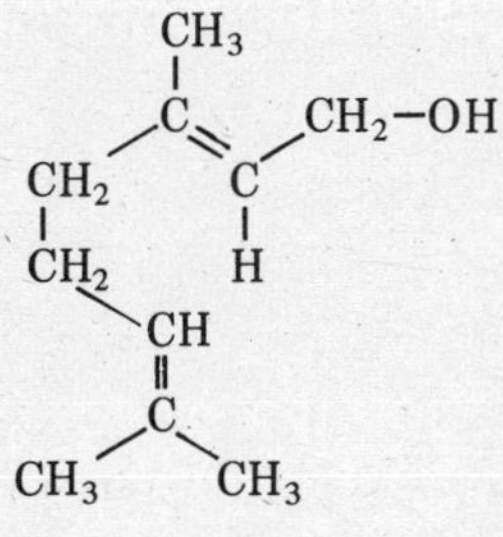

POINTS OF ATTACHMENT ON SATURATED SKELETONS

5. We distinguish four types of carbon atoms in saturated skeletons:

$$\begin{array}{cccc}
\text{C} & \text{C} & \text{C} & \text{H} \\
| & | & | & | \\
\text{C}-\text{C}-\text{C} & \text{C}-\text{C}-\text{H} & \text{C}-\text{C}-\text{H} & \text{C}-\text{C}-\text{H} \\
| & | & | & | \\
\text{C} & \text{C} & \text{H} & \text{H}
\end{array}$$

name:	quaternary	tertiary	secondary	primary
symbol:	4°	3°	2°	1°

To determine if a carbon atom is 1°, 2°, 3°, or 4°, count the number of other carbon atoms directly bonded to it. Do not count carbon atoms farther away.

Example

$$\text{CH}_3-\overset{\overset{\displaystyle \text{CH}_3}{|}}{\underset{\underset{\displaystyle \text{CH}_3}{|}}{\text{C}}}-\boxed{\text{C}}\text{H}_2-\text{CH}_2-\text{CH}_3$$

The boxed carbon atom is 2° because it is attached to *two* carbons (arrows); the remaining carbons are farther removed and are not counted.

Hydrogen atoms and certain functional groups are also labeled as primary, secondary, and tertiary. They take on the label of the carbon they are attached to:

$$\text{CH}_3-\text{CH}_2-\text{OH} \qquad \text{CH}_3-\overset{\overset{\displaystyle \text{CH}_3}{|}}{\underset{\underset{\displaystyle \text{CH}_3}{|}}{\text{C}}}-\text{H} \leftarrow 3° \text{ hydrogen}$$

1° carbon ↗ ↖ 1° alcohol function   3° carbon ↗

The 1°, 2°, and 3° labels are used only for the halide (−F, −Cl, −Br, −I), alcohol (−OH), nitro (−NO$_2$), and ester (−O−CO−R) functions. The 1°, 2°, and 3° distinctions are important because reaction rates and equilibrium positions often vary drastically for the same function attached to these various types of sites.

The 1°, 2°, and 3° labels can be used *only on saturated carbon atoms*, $-\overset{|}{\underset{|}{\text{C}}}-$. They can be used on the saturated carbon atoms in skeletons that have unsaturation elsewhere.

Example

$$\bigcirc-\text{CH}_2-\overset{\overset{\displaystyle \text{CH}_3}{|}}{\underset{\underset{\displaystyle \text{Br}}{|}}{\text{C}}}-\text{CH}_2-\text{CH}_2-\text{CH}=\text{CH}_2$$

Br ↘ 3°

Functional groups attached at sites closer to C=C or C≡C than in this example get a special classification (frame 6).

★ *Exercises*

Label each underscored atom or group as 1°, 2°, 3°, 4°.

FUNCTION AND SKELETON   43

1. 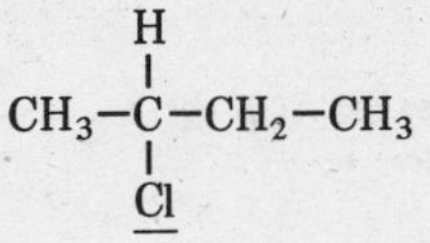 $CH_3-\underset{\underset{Cl}{|}}{\overset{\overset{H}{|}}{C}}-CH_2-CH_3$

2. $\begin{array}{c} CH_2-CH_2 \\ |\qquad\quad | \\ CH_2-CH-CH_2-\underline{NO_2} \end{array}$

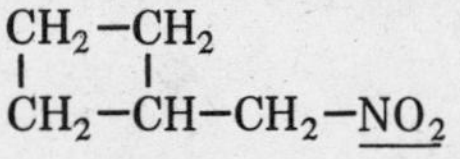

3. $(CH_3)_3\underline{C}CHClCH_2CN$

4. 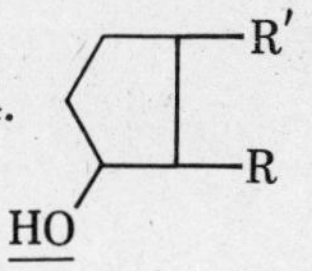 cyclopentane with HO– and substituents R′ and R

5. 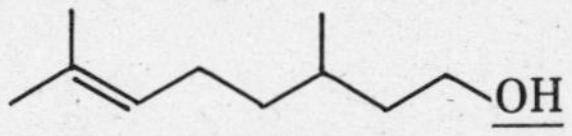 (structure with OH)

- - - - - - - - - -

1. 2°
2. 1°
3. 4°
4. 2°
5. 1°

ATTACHING SATURATED GROUPS
TO UNSATURATED SKELETONS

6. A saturated group is labeled according to its distance from the unsaturation in the skeleton. Points of attachment close to C=C and to aromatic rings get different labels:

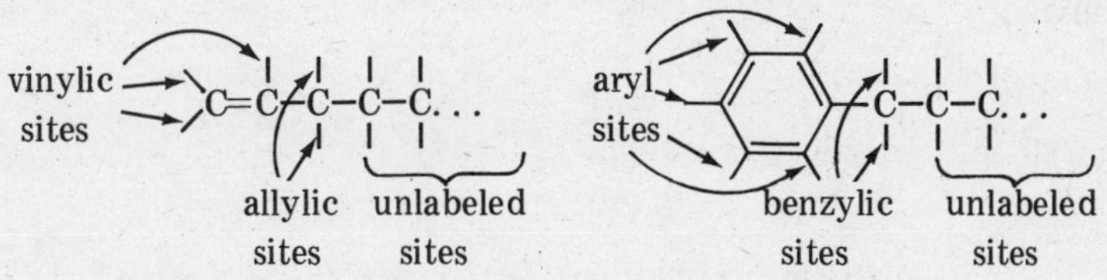

vinylic sites → C=C–C–C–C... allylic sites, unlabeled sites

aryl sites → (ring)–C–C–C... benzylic sites, unlabeled sites

Groups take the label of the site they are attached to.

Examples

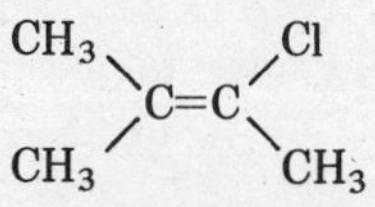

$\begin{array}{c} CH_3 \\ \diagdown \\ C=C \\ CH_3\diagup\diagdown CH_3 \\ \text{a vinylic halide} \end{array}$ with Cl

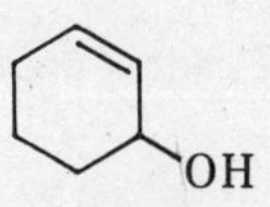

an allylic alcohol (cyclohexenol)

$CH_3-\bigcirc-Br$ an aryl halide

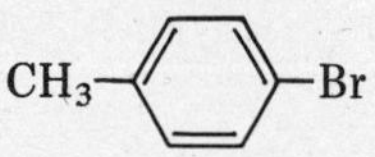

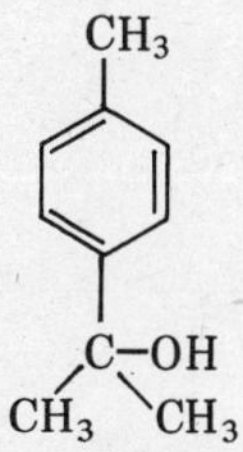

a benzylic alcohol (p-tolyl C(CH_3)_2OH)

Vinylic and aryl groups are attached to unsaturated carbon, so the 1°, 2°, and 3° labels cannot be applied to them. Allylic and benzylic groups are attached to saturated carbon; they can be identified *both* as allylic or benzylic *and* 1°, 2°, or 3°.

Examples

a 3° benzylic alcohol   a 2° allylic chloride

★ *Exercises*

Label the location of the circled groups as completely as possible.

1. Cl–CH₂–CH₂–C(CH₃)=C(CH₃)(Br)

2. (cyclopentene with CH₃ and H on sp³ carbon)

3. O₂N–C₆H₄–CH(Cl)–CH₃

4. (dihydronaphthalene with Br, Cl, OH)

----

1. Cl is 1°, Br is vinylic.
2. H is 3° and allylic.
3. NO₂ is aryl, Cl is 2° and benzylic.
4. Br is aryl; Cl is 2° and benzylic; OH is 2° and allylic.

ATTACHING UNSATURATED FUNCTIONAL GROUPS TO UNSATURATED SKELETONS

7. A multiple bond in the functional group has one of three locations relative to unsaturation in the skeleton. The following examples use $\rangle$C=O as the function. The number of single bonds separating the multiple bonds is the criterion.

no single bonds	one single bond	more than one single bond
$\rangle$C=C=O	$\rangle$C=C–C=O	$\rangle$C=C–C–C=O, $\rangle$C=C–C–C–C=O,
cumulated unsaturation	conjugated unsaturation	and greater separations
		isolated unsaturation

More than two multiple bonds can be in conjugation:

$\rangle$C=C–C=C–C=O

An aromatic ring is not only a completely unsaturated system (all carbons are $\rangle$C=),

it is a completely conjugated system as well: 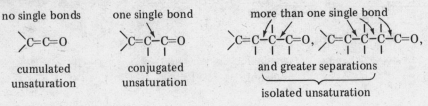 When a multiple bond outside the ring is attached to the ring by one single bond, it forms a lengthened conjugated system and is usually said to be conjugated "with the ring." In the following structure the perfect alternation of multiple and single bonds is visible when one reads in the order a, b, c, d:

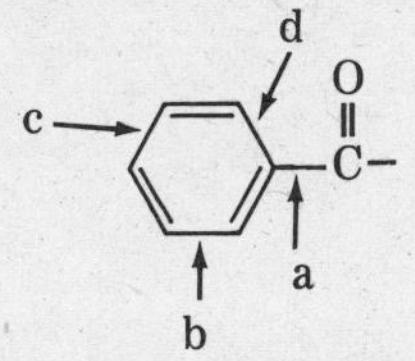

The notion of conjugation still holds when the alternating multiple and single bonds are all within the skeleton or all drawn from functional groups. These are all conjugated systems:

$$CH_2=CH-CH=CH-CH_2-Cl \qquad CH_3-C{\equiv}C-CH=CH-\text{[phenyl]}$$

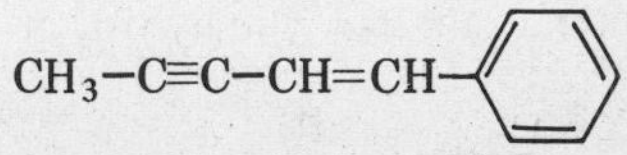

$$\underset{H}{\overset{O}{\|}}C-\underset{}{\overset{O}{\|}}C-H \qquad N{\equiv}C-CH=CH-\overset{O}{\overset{\|}{C}}-NH_2$$

In these examples the conjugation is *complete*—all the multiple bonds form a single, unbroken sequence of alternating multiple and single bonds. In the following case the conjugation is *broken* by an extra single bond:

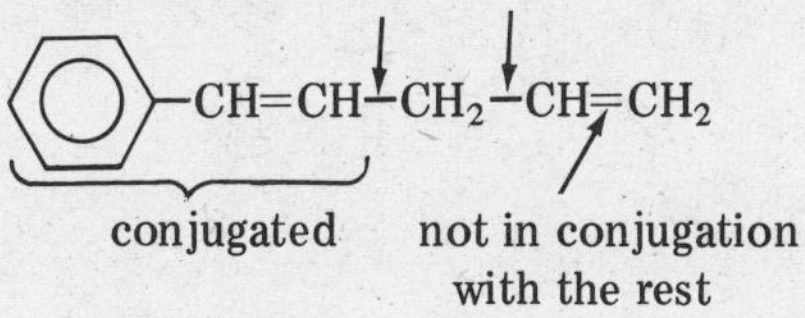

[phenyl]—CH=CH—CH$_2$—CH=CH$_2$

 conjugated   not in conjugation
               with the rest

★ *Exercises*

1. Label each molecule as conjugated or nonconjugated.

   (a) $CH_2=CH-CH=CH_2$ (b) $CH_2=CH-CH_2-C{\equiv}C-H$

   (c) $CH_3-\overset{O}{\overset{\|}{C}}-CH=CH_2$ (d)  [cyclopentene]

   (e) 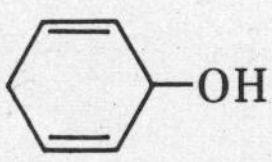 [cyclohexadiene]—OH   (f) 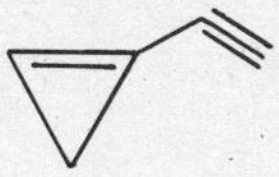 [cyclopropane with ethynyl group]

2. How long (measured in number of connected atoms) is the longest conjugated system in each molecule?

   (a) 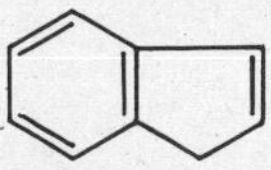 [indene]   (b) 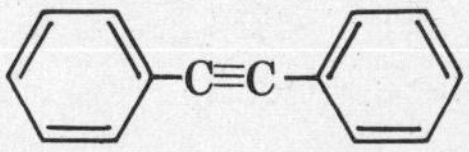 [phenyl]—C≡C—[phenyl]

   (c) [phenyl]—CH$_2$—[phenyl]

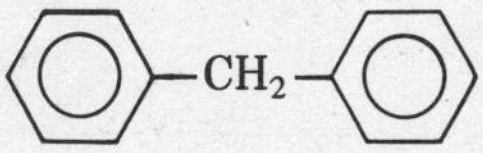

46 HOW TO SUCCEED IN ORGANIC CHEMISTRY

1. Molecules (a), (c), and (d) have one C—C intervening between C=C and C=C or C=C and C=O; they are conjugated. Molecules (b) and (e) have two C—C bonds intervening and are nonconjugated. Redraw (f) to see that it is conjugated:

$$\begin{array}{c} CH_2 \\ | \\ CH_2 \end{array} C-C\equiv C-H$$

2. (a) 8 atoms (b) 14 atoms
   (c) 6 atoms

★   **REVIEW EXERCISE**

8. Use the symbols 1°, 2°, 3°, V, Ar, Al, and B to label the primary, secondary, tertiary, vinylic, aryl, allylic, and benzylic hydrogens in

UNIT FOUR
Geometry of Organic Molecules

In the structural formulas we have been drawing, we have pointed the bonds arbitrarily in whatever directions were convenient. However, we also called a structure a kind of map, which implies definite spatial relations. Experimental results show that the network of bonds in a molecule is indeed laid out on a precise and predictable geometric pattern. In most molecules this pattern is three-dimensional, so the structural formula must be a relief map of sorts. Since the structural formula does not show the geometric information explicitly, the reader must translate the structure into three dimensions, by eye or hand, using principles to be learned in this unit. When necessary, we switch to a more explicit notation, perspective structures, which we begin to use in this unit and develop more completely in Unit 13.

In this unit you will start to sharpen space-perceptual skills that will be used throughout the study of organic and biochemistry. You will progress most rapidly by doing a lot of drawing and manipulation of structures on paper rather than by just looking at structures drawn by others. This unit also includes exercises in manipulation of molecular models. If you do not have a model kit, try to borrow one for use in frame 6.

OBJECTIVES

When you have completed this unit, you should be able to:

- Predict the three-dimensional shape of a molecule by inspection of the structural formula (frames 2 to 5).
- Judge whether two structures represent the same or different molecules (frames 6 to 8).
- Read and write the standard notation for representing the three-dimensional structure of molecules on paper (frames 5 and 9).

If you think that you might have already mastered these objectives and might skip all or part of this unit, take a self-test composed of those exercises marked with a star in the margin. If you solve these problems correctly, you are probably ready for Unit 5. If you miss any problems, study the frames in question and rework the problems before proceeding. If you are not ready for a self-test, proceed to frame 1.

DISTRIBUTION OF ELECTRONS IN SPACE AND THE GEOMETRY OF MOLECULES

1. The regions in space occupied by the electrons on an atom are the atomic orbitals. If you have not studied the shapes, energies, and names of the atomic orbitals recently, you should read the appropriate sections of your textbook before proceeding with this unit. The geometric shapes of organic molecules are, in fact, predictable without knowledge of the identities or shapes of the orbitals used in bond formation. However, understanding some of the properties of organic molecules does require manipulation of atomic orbitals.

For our purposes at present we can visualize the formation of covalent bonds as being the overlapping of an atomic orbital on one atom with an atomic orbital on another atom:

$$A\supset \; + \; 2 \text{ electrons} \; + \; \subset B \; \longrightarrow \; A\supset\!\cdot\!\cdot\!\subset B \; = \; A-B$$

The new orbital is a molecular orbital, which contains the shared electron pair of the covalent bond. Any unshared electron pairs on A or B are housed in atomic orbitals on that atom.

Now visualize the covalent bond between an atom A and a hydrogen atom. Visualize the bond first, then add the H and A atoms and consider the geometric possibilities. The molecular orbital defines a line, which is the bond direction. The H atom will be located on this line, and the only decision to be made about its location is how far it should be from A.

$$A \; + \; - \; + \; H \; \longrightarrow \; A-H$$

There is nothing more to specify about H because it has no other bonds and no unshared electron pairs. Suppose, however, that A is an atom other than H. It will be surrounded by an octet of electrons. Two are in the A–H bond. The other six must be either in atomic orbitals on A (unshared pairs) or in other covalent bonds formed by A (shared pairs in molecular orbitals). In either case these orbitals on A each have a definite direction in space. To complete the description of A, then, we not only must specify the distance from H but also the angles that the A–H bond makes with the other three orbitals on A:

When the second orbital on A is a molecular orbital, the angle in question is the angle between two covalent bonds, for instance in

When the second orbital is an atomic orbital, the angle is between an unshared pair and the A–H bond, for instance in

GEOMETRY OF ORGANIC MOLECULES

To complete the description of C or O in these examples, one would also have to specify angles for two remaining orbitals on C or O:

Reduced to its simplest terms, this problem is just the problem of describing the directions of several lines (the bonds or unshared-pair orbitals) radiating from a central point (the atom). In general, one needs to specify as many angles as there are lines:

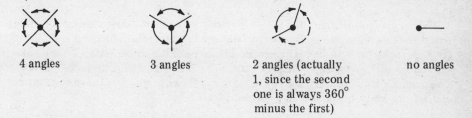

4 angles　　　3 angles　　　2 angles (actually 1, since the second one is always 360° minus the first)　　　no angles

In each of these cases there is an unlimited number of possible ways (combinations of angles) in which the lines can be oriented about the point. For real molecules the situation is simple, however. Nature selects a *single* orientation of bonds/unshared pairs about each atom—the orientation that makes the angles about that atom equal to one another (or nearly equal). These angles are most easily visualized in terms of the geometric figures drawn in frame 2.

BASIC SHAPES

2. If the angles are made equal in the four-angle, three-angle, and two-angle diagrams above, they become

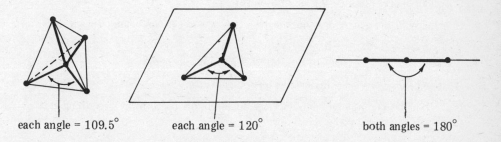

each angle = 109.5°　　　each angle = 120°　　　both angles = 180°

These arrangements are most easily visualized in terms of the basic geometric shapes they define: the *tetrahedron*, the *equilateral triangle*, and the *line*. The tetrahedron is a pyramid whose four faces are identical equilateral triangles.

Next visualize some atom (other than hydrogen) at the center of each figure, and the heavy lines as the bond and unshared pair directions. You now have a picture of the three-dimensional shapes of the bonding units or part structures for that atom. Which shape corresponds to which bonding unit is discussed in frame 3. The names given to atoms or part structures in these arrangements come from the names of the figures:

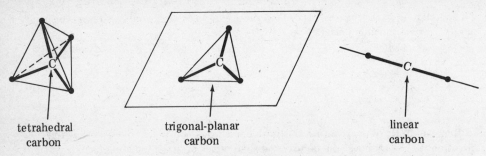

tetrahedral carbon   trigonal-planar carbon   linear carbon

When we speak of the orientation of bonds/unshared pairs about atoms, or the shapes of molecules, we often say the "geometry" of the atom or molecule, with these figures in mind.

We can predict the shape of a whole molecule by deducing the shapes of the individual bonding units (frame 3), and then putting them together (frames 2 and 5). In fact, only the angles between *bonds* determine the shape of the whole surface to the molecule. This means that we focus on the geometry of the bonding units with more than one bond. These are the bonding units of B, C, N, O, P, and S—what might be called the *central atoms* from which the molecular skeleton is built.

★ *Exercises*

Consider some basic properties of the tetrahedron shown resting on one of its faces at the right. Like all tetrahedra, its center lies on a line connecting the center of any face with the opposite vertex. The center is one-fourth of the way from face to vertex. This tetrahedron is drawn with atoms at the center and each vertex (dots), and with each center-to-vertex distance equal to 1.54 Å, the normal C–C single bond distance. Each edge then turns out to be 2.52 Å long, and the height of the top vertex above the base is 2.05 Å.

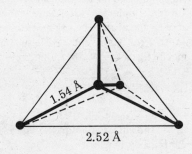

1. How many edges does the tetrahedron have?

2. How many of the other edges does a given edge intersect? How many does it not intersect?

3. A tetrahedron has ___ pairs of nonintersecting (opposite) edges.

4. What angle does any edge make with its opposite edge?

5. Is a tetrahedron completely defined (size, location, orientation) by describing any single pair of opposite edges?

— — — — — — — — —

1. Six edges.

2. Each edge intersects four of the other five; it does not intersect the remaining (opposite) edge.

3. Three.

4. An edge and its opposite edge lie at right angles to one another. This is most easily seen from this drawing:

where the (solid) horizontal and (dashed) vertical lines are opposite edges. (In reading perspective drawings, make sure that your eye pushes dashed lines *back* behind the paper plane.)

5. Yes; given ─┼─, all vertices are fixed, and the four missing edges can unambiguously be drawn in.

PREDICTING THE THREE-DIMENSIONAL SHAPES OF PART STRUCTURES AND SMALL MOLECULES

3. To predict the geometry of a given bonding unit in a molecule, follow these three steps:

1. Draw the complete structural formula and focus on one central atom (B, C, N, O, P, or S). Count the number of other groups to which the central atom is directly attached by covalent bonds.

Examples

(The central atom is indicated by ↑. In the first two examples the groups attached to the central atom are circled.)

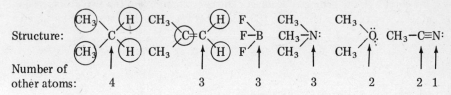

2. To this number add the number of unshared electron pairs on the central atom.

Examples

Structure: CH₃\\ /H   CH₃\\   /H   F\\   CH₃\\   CH₃\\
               C   C=C   F—B   CH₃—N: Ö: CH₃—C≡N:
              CH₃/ \\H   CH₃/ \\H   F/   CH₃/   CH₃/
                   ↑   ↑   ↑   ↑   ↑   ↑↑

Number of other atoms:	4	3	3	3	2	2	1
Number of unshared pairs:	0	0	0	1	2	0	1
Total	4	3	3	4	4	2	2

3. Use the total of the two numbers as follows.

(a) If the total is 4, the geometry is *tetrahedral* or close to it; that is, with the central atom at the center of a tetrahedron, the four directly bonded atoms or unshared pairs will be oriented toward the corners of the tetrahedron. The angles between these bonds or unshared pairs on the central atom are about 109.5°.

Example

The first and fifth molecules from the preceding example look like this:

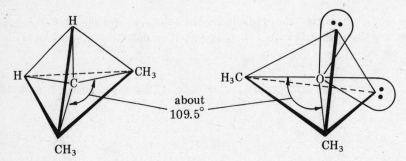

about 109.5°

(b) If the total is 3, the geometry is *trigonal-planar*. The central atom and the three atoms or unshared pairs attached to it all lie in the same plane. The three groups or unshared pairs are oriented at about 120° angles around the central atom.

Example

The second and third molecules from the previous examples look like this:

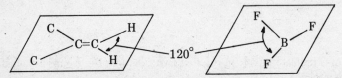

(c) If the total is 2, the geometry is *linear*. The central atom and the two atoms or unshared pairs attached to it lie on a line (180° angle between attached atoms or unshared pairs).

Example

The last molecule in the preceding example is linear:

$$CH_3-C\equiv N: \quad \text{orbital containing unshared pair}$$

180°, 180°

The physical principle behind these rules is simply that electron pairs repel one another. Equal angles and resulting tetrahedral, trigonal-planar, and linear geometries produce maximum angular separation of 4, 3, or 2 pairs of electrons ranged about an atom.

Since an unshared electron pair occupies more space than an electron in a covalent bond (shared pair), tetrahedral and trigonal geometries involving unshared pairs are found to be somewhat distorted in the direction of forcing the shared pairs (bonds) closer together.

Example

The bond angles predicted for H_2O and NH_3 are 109.5°. The angles actually observed are:

GEOMETRY OF ORGANIC MOLECULES 53

104.5°—Ö: H–N–H
H   H   H H 106.6°

In addition, larger groups attached to the central atom may interfere spatially with one another and force the shared pairs farther apart, distorting the tetrahedron slightly in the opposite direction, as in dimethyl ether:

:Ö:
CH₃   CH₃
  112°

Exercises

1 and 2. Refer to the set of 16 correct bonding units that make up your answers to exercises 1 and 2 from Unit 2, frame 2. Predict the approximate angle between the covalent bonds in each of these bonding units.

★ 3. Refer to the set of complete Lewis structures of the 15 common functional groups that you generated in exercise 2, Unit 3, frame 1. Predict the geometric arrangement (tetrahedral, trigonal, linear) of bonded atoms and unshared pairs about each C, N, and O atom in these functions.

★ 4. Predict the approximate geometry (tetrahedral, trigonal, or linear) of each of the following molecules. Answer only for the underlined central atom.

(a) $(CH_3)_3\underline{B}$ (b) $(CH_3)_4\underline{N}^{\oplus}$
(c) $H_3\underline{C}:^{\ominus}$ (d) $H_3\underline{O}:^{\oplus}$
(e) $CH_3-\underline{\ddot{O}}-H$ (f) $:\underline{\ddot{N}}H_2^{\ominus}$
(g) $:\underline{\ddot{P}}H_2$

— — — — — — — — — —

1. (a) 109° (b) no geometry; only one group attached to O (c) no geometry (d) 109° (e), (f), (g) no geometry (h) 120°

2. (a) no geometry (b) 109° (c) 120° (d) 180° (e) 120° (f) 109° (g) 120° (h) 109°

3. $-\ddot{O}-H$ and $-\ddot{O}-R$, tetrahedral $\diagdown C=C \diagup$, trigonal

   $-C\equiv C-$, linear

   :Ö: :Ö:
    ‖   ‖   all
   $-C-R$ and $-C-H$, trigonal

   :Ö: :Ö: :Ö: :Ö:
    ‖   ‖   ‖   ‖
   $-C-\ddot{O}-H$, $-C-\ddot{O}-R$, $-C-\ddot{C}l:$, and $-C-\ddot{N}H_2$, all trigonal

   O   O   O
   ‖   ‖   ‖
   $-C-\ddot{O}-H$, $-C-\ddot{O}-R$, and $-C-\ddot{N}H_2$, all tetrahedral

−C≡N:, both linear

−N⁺(=O)(O⁻), trigonal

⁻O−S(=O)(O⁻)−O−H, tetrahedral

−ṄH₂, tetrahedral

4. (a) trigonal (b)-(f) tetrahedral (g) trigonal

CONNECTING THREE-DIMENSIONAL PART STRUCTURES BY MULTIPLE BONDS

4. When we build up larger molecules by combining part structures, a new geometric possibility appears. This is the state of alignment of bonds in one part structure with bonds in the other part structure. Consider the molecule C_2H_4, made by combining two H₂C= part structures to give H₂C=CH₂. The geometry about each C is trigonal-planar according to frame 3, but how do the two ends line up? Perpendicular as in (a), or parallel as in (b), or somewhere in between?

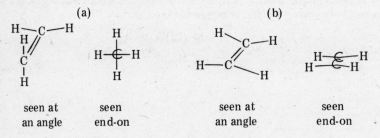

Another way to put the question is seen better in figure (c): are the two H₂C= planes rotating in the real molecule, and if not, what is the angle between them?

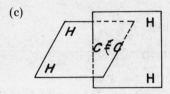

The experimental results show that the two H₂C= planes *coincide* and that the rotation of one end versus the other to produce (a) from (b), after 90° rotation, and finally (b) back again, after 180° rotation, *does not occur* at ordinary temperatures. So C_2H_6 is locked into the geometry shown in figure (d).

(d) All six atoms lie in the same plane. All bond angles are 120°.

This is true of all double bonds.

Rule: When two trigonal atoms are linked to make a double bond, these two atoms and the four atoms directly bonded to them all lie in the same plane, and this part of the molecule is rigid. If one of the two trigonal atoms has unshared pair(s) in place of electron-pair bonds, the orbitals containing the unshared pair(s) lie in the plane.

Example

Show the geometry of $CH_2=NH$.

Solution:

trigonal

The physical reason for the overall planarity and absence of internal rotation in doubly bonded molecules lies in the spatial distribution of electrons in the double bond, which is different from that in the single bond. Consult your textbook for a description of the electronic structure of the $>C=C<$ bond.

Combination of two $-C\equiv$ part structures to make a triple bond is simpler, for instance in C_2H_2:

$$-C\equiv\ +\ \equiv C-\ +\ 2H-\ \rightarrow\ H-C\equiv C-H$$

with 180° angles at each end.

Here the two ends, $H-C\equiv C$ and $C\equiv C-H$, both have linear geometry. In the real C_2H_2 molecule, the two lines coincide so that the whole molecule is linear:

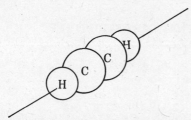

Rule: Two atoms joined by a triple bond, and the two atoms directly connected to them, all lie on a line. If one of the triply bonded atoms has an unshared pair instead of a bond to another atom, the orbital containing the unshared pair lies on the line through the triply bonded atoms.

Example

What is the three-dimensional structure of $H-C\equiv N$?

Solution:

Write the complete structural formula: H−C≡N:. C is linear, N is linear. Combine:

H−C≡N: All three atoms and the unshared pair lie on the same line.
(180°, 180°)

★ *Exercises*

Show the three-dimensional shape of each molecule or ion.

1. $CF_2=CF_2$

2. $F-C\equiv C-F$

3. $H-\overset{\overset{O}{\|}}{C}-H$

4. $H-C\equiv \overset{\oplus}{O}:$

5. $H-\overset{\oplus}{C}=N-H$ with H, H bonded to C and N

6. $H-C\equiv C:^{\ominus}$

— — — — — — — — —

1. All six atoms in the same plane, all bond angles about 120°.

2. All four atoms lie on a line; both bond angles 180°.

3. The four atoms and two unshared pairs lie in a plane:

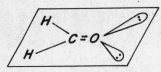

4. H, C, O, and the unshared pair lie on a line.

5. All six atoms in the same plane.

6. H, C, C, and the unshared pair lie on a line.

CONNECTING THREE-DIMENSIONAL PART STRUCTURES BY SINGLE BONDS

Single Bonds of Type ⟩C—C⟨

5. Suppose that we combine the two part structures H⟩C− and −C⟨H to make
H⟩C−C⟨H (with H's). The individual part structures are tetrahedral, and they remain tetrahedral in the finished molecule. But one decision remains—how are the two tetrahedra oriented relative to one another? It is hard to discuss this problem without a picture that shows the alignment of the two ends clearly. This is obtained by sighting directly down the new C−C bond: The carbon atoms are not drawn (to reduce clutter), the near carbon atom is

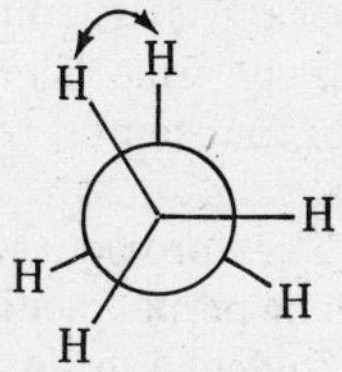

the intersection of the Y, and the far carbon atom hides behind the circle. The state of rotation about the C—C bond is easily measured by the dihedral angle between one of the near and one of the far bonds (in formula a). Let us now start at 0° in formula b and increase the dihedral angle:

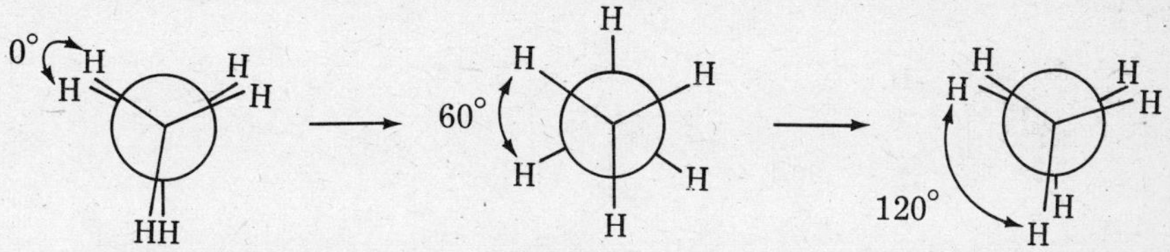

These various stages of *internal rotation* are called *conformations* of the molecule. Conformation b is called the *eclipsed* conformation, because it has the near C—H bonds exactly covering the far C—H bonds. Rotation by 60° puts the near C—H bonds precisely between the far C—H bonds (formula c); this is the *staggered* conformation. Continuing to 120° gives an eclipsed conformation (d) that cannot be distinguished from the original one (b). With this molecule it is clear that further rotation produces nothing new—just b, c, d and stages in between. However, in other molecules (e.g., in CH_2Cl-CH_2Cl) several distinguishably different staggered and eclipsed conformations may exist. We will return to the question of different versus indistinguishable conformations in frame 7.

Which conformation represents the structure of real CH_3-CH_3 molecules? The experimental results show that:

1. Almost all the molecules are in the staggered conformation.
2. At room temperature each molecule is carrying out internal rotation at a rapid rate, so the lifetime of any molecule in a particular conformation (called a *conformational isomer* or *conformer*) is only about 10^{-10} to 10^{-11} second.

The C—C bond is rather rigid with respect to resistance to stretching and bending, but not with respect to the internal rotation just described. This combination of properties is best understood by manipulating a molecular model of CH_3-CH_3. An exercise for this purpose will be found in frame 6.

Study of more complex molecules, such as $CH_3-CH_2-CH_2-CH_3$, adds one more general result:

3. If there are two or more distinguishably different conformations, most of the molecules will adopt the conformation that puts the largest groups farthest apart. Thus $CH_3-CH_2-CH_2-CH_3$ exists mainly in conformation f, partly in conformation e. At any given instant, a few molecules will be in eclipsed conformations, in transit between the

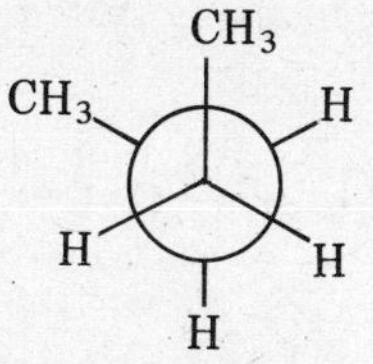

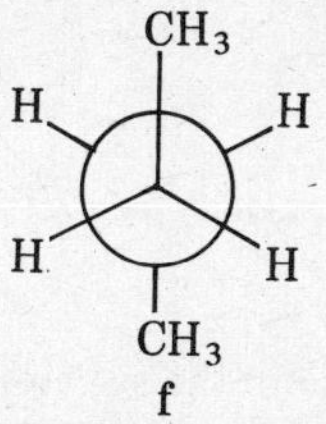

stable staggered conformations. Conformation e is called *gauche* (large groups, CH_3, 60° apart); f is called *anti* (large groups, 180° apart).

The staggered conformers are more stable than the eclipsed ones because staggering gets the electrons in the near and far C—H or C—C bonds as far apart as possible. This is just an extension of the principle behind the tetrahedral orientation of four bonds, trigonal-planar orientation of three bonds, and so on, on the same atom.

The formulas used above are called *Newman projection formulas*. They project the whole three-dimensional molecule (or sometimes part of it) onto a plane perpendicular to the C—C bond undergoing internal rotation. There is another way to depict internal rotation—by perspective rather than projection formulas. These represent conformations e and f as

e   and   f

These are sometimes called *sawhorse structures*.

Single Bonds of Type =C–C=

Here a single bond connects two trigonal-planar atoms rather than two tetrahedral ones. Consider the molecule $CH_2=CH-CH=CH_2$. In this case we can use projection formulas again, but they look different because each $\begin{array}{c}CH_2\\ \diagdown\\ H\end{array}\!\!C-$ unit is planar:

g
perpendicular

h
s-cis (planar)

i
s-trans (planar)

The experimental results show that internal rotation about the central C—C bond occurs, but not nearly as freely as in the $\rangle C-C\langle$ case. Individual conformers have much longer lifetimes. There is a second difference: the "eclipsed" conformations are the more stable, although in this case they are called planar conformations. This preference arises from an interaction between the C=C which stabilizes the molecule only when the planes of the two $\rangle C=C\langle$ units coincide. We consider this interaction further in Unit 9. Note that these systems connecting C=C to C=C by C—C are by definition *conjugated* systems (Unit 3, frame 7).

Rule: When two part structures of type $-C=C-$ are combined to give $-C=C-C=C-$:

1. The $-C=C-C=C-$ system and the six atoms directly bonded to it all lie in a single plane.
2. Most of the molecules exist in the s-trans conformation.

GEOMETRY OF ORGANIC MOLECULES

Example

Draw the three-dimensional structure for the most abundant conformer of $CH_2=CH-CH=CH_2$.

Solution:

An easily drawn perspective structure is

The Newman projection formula was drawn in formula i.

★ *Exercises*

1. Draw the three-dimensional structure of the most abundant conformer of each molecule or ion.

   (a) CF_3-CH_3 (b) CH_3-O-H

   (c) $\underset{\|}{\overset{O}{H-C}}-\underset{\|}{\overset{O}{C}}-H$ (C=O behaves like C=C)

2. We did not consider the possible existence of more than one conformation when combining H–C≡ + ≡C–H to give H–C≡C–H.

   (a) Do various distinguishable conformations of this molecule exist?

   (b) Would you expect internal rotation to be possible around –C≡C– in H–C≡C–H?

   (c) Would various conformations exist for H–C≡C–C≡C–H?

3. Draw sawhorse structures and Newman projection formulas for gauche and anti CH_2Cl-CH_2Cl.

1.

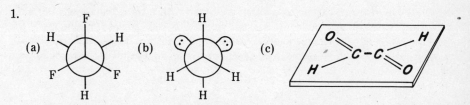

2. (a) No. Since the four atoms lie on a line and each is a sphere, the view down the C≡C bond is perfectly symmetrical about the bond axis: Ⓗ . (This is called *cylindrical symmetry*.)

   (b) No. Since rotation does not occur about C=C, it should not occur about C≡C either. (Electronically, the triple bond is two perpendicular double bonds; see your text.)

(c) One would expect rotation about the C—C bond to be restricted, like that between two double bonds. However, since the whole molecule is linear with cylindrical symmetry (end view = (H)), different stages of rotation cannot be distinguished from one another, and conformers do not exist.

3.

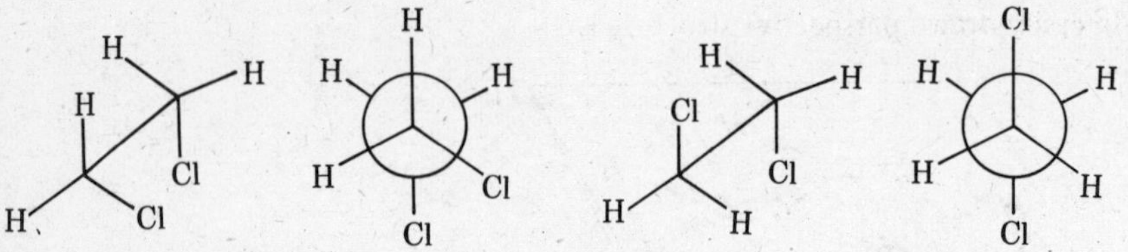

gauche   anti

USING MOLECULAR MODELS

6. At times (especially while first learning) it is worthwhile to use actual three-dimensional models of organic molecules. Kits for constructing such models are a familiar part of most organic courses, and you should make an effort to handle one, at least briefly. If you do not buy a set of student models (see p. 7) for continuing use, borrow one long enough to run through the exercises below.

Return to the models whenever you have difficulty with molecules in three dimensions. It helps to have a model available for comparison when learning the notation for representing the three-dimensional structure of molecules in two dimensions, with pencil and paper. We take up this problem in frame 8.

Exercises

1. Build a model of CH_3-CH_3. Most kits are true to nature in showing that CH_3-CH_3 has resistance to compression of the C—C bond ($\vec{C}-\overleftarrow{C}$) and to stretching of it ($\overleftarrow{C}-\vec{C}$) but little resistance to rotation about it (C↻C). Observe the appearance of this internal rotation from all angles by holding one CH_3 group fixed and twisting the other. The real molecule is executing this internal rotation rapidly at room temperature. It is also tumbling rapidly in space as a whole.

2. Many kits include some carbon atoms that are bored with tetrahedral angles (109.5°), some with trigonal-planar angles (120°), and some with 180° angles (linear). The smaller kits generally use all tetrahedral carbons and construct double and triple bonds by using curved or flexible connectors. In this case, ethene, $CH_2=CH_2$, is made as follows:

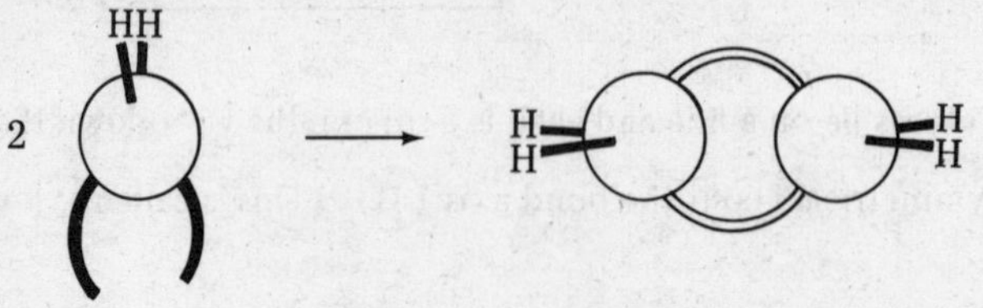

Build the ethene model. Verify the planarity. Note that the model is true to life in not being internally rotatable about the C=C bond. The direction of the C=C bond has to be inferred as the straight line connecting C and C. This type of model makes the

=C$\langle$ bond angles 109.5° instead of the actual 120°, but it gives a fair picture of the shape of the molecule.

3. Build and inspect models of $CH_2=CH-CH_3$ (partly planar) and $H-C\equiv C-H$ (linear).

4. Build CH_4. Get acquainted with the properties of the tetrahedron. Perhaps the easiest way to picture tetrahedral geometry without a model is to note that two of the H atoms and the central C of the model before you form one plane, and C taken together with the two other H's define a second plane perpendicular to the first. Look at the model and visualize these planes.

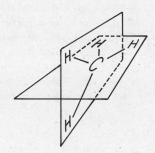

SUPERIMPOSABILITY

7. Before you attack the problems of representing three-dimensional structures in two dimensions in frame 8 and structural isomerism in Unit 6, you need practice with one more concept, that involving the criteria of superimposability and chemical equivalence.

To test the superimposability of two objects, perform the following thought experiment: (1) put them in the same orientation in space; (2) set one upon the other; and (3) imagine the top one settling upon the other so that they occupy the same space jointly. (4) Now, if each and every feature of the one has an exact counterpart in the other, the two are *superimposable*. If there is any failure in correspondence or any point of mismatch, they are *nonsuperimposable*.

Example

Imagine a 1974 and a 1975 penny. Put heads up on both and point them in the same direction. Put one on top of the other and conceptually squeeze them into sharing the same region in space. Now the figure, the building, all of the lettering, and the numbers 197 on the top and bottom coins match precisely. But 4 overlays 5, which is not a precise match. The two coins are nonsuperimposable.

Superimposability is a definitive proof of identity in organic chemistry. If two structures or models are superimposable, they represent the same molecule. However, they may represent the same molecule even if they do not superimpose. First, when dealing with structures drawn on paper, the structures may not superimpose because they represent different views of the same molecule. The way around this difficulty is to visualize both structures as three-dimensional models before trying to superimpose them. Second, two structures may not superimpose because they represent different conformers of the same compound. In these cases we say that the structures are chemically *equivalent* because they represent different momentary aspects of the same compound, even though they do not superimpose.

★ *Exercises*

1. Draw sawhorse structures for all the staggered conformations of CH_3-CFCl_2. Which ones are superimposable on one another?

2. Repeat exercise 1 for $CH_2Br-CFCl_2$. Which structures are chemically equivalent?

3. Repeat exercise 2 using Newman projections.

----

1. All are superimposable. To see this, pick up structure **A** and give the whole molecule a 120° turn clockwise. Labeling the H atoms helps you see where they end up:

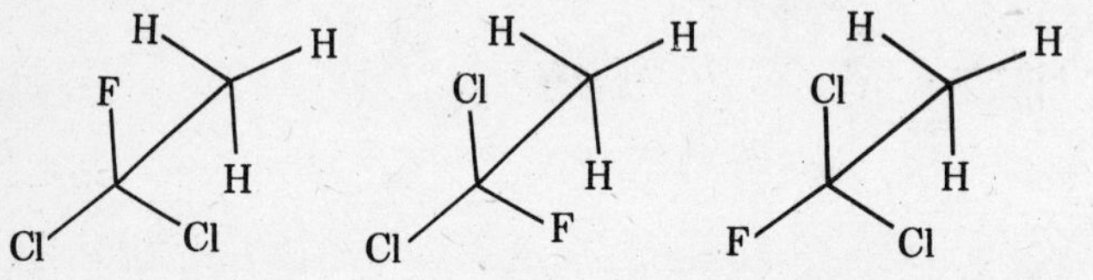

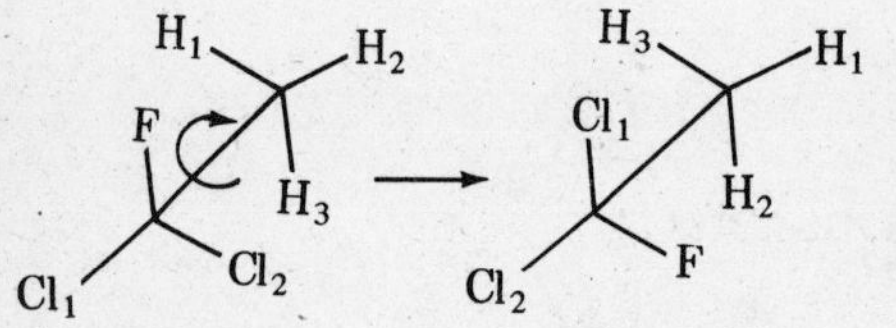

This new structure is superimposable on **B**, so the original **A** and **B** were superimposable. The same can be done for **A** and **C** and for **B** and **C**.

2. None of these superimposes on any other. If you try lifting and turning these as in exercise 1, you get the Br on the rear carbon out of position for superimposing on the rest.

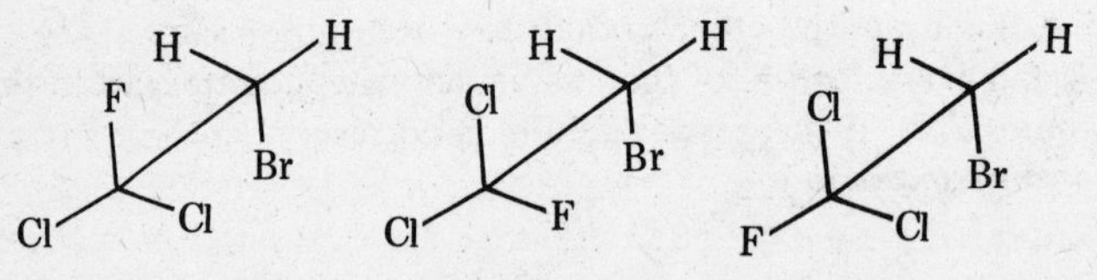

These three structures are chemically equivalent. Since the molecules they represent are interconverting rapidly at room temperature, they correspond to a single compound.

3. These projections are planar structures that you can slide and rotate in the plane of the paper to check for superimposability. The result is the same as in exercise 2.

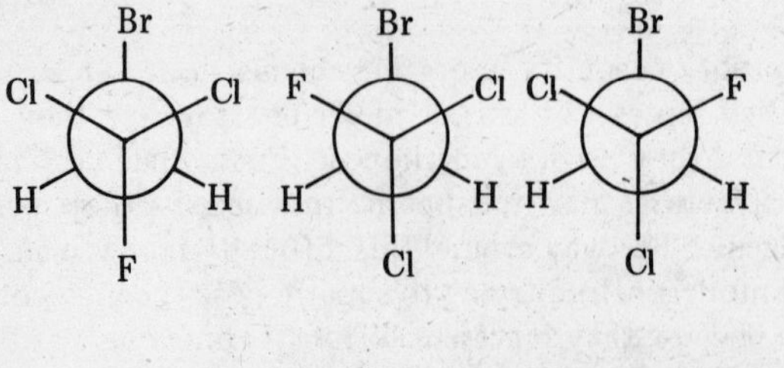

REPRESENTING THREE-DIMENSIONAL MOLECULES BY MEANS OF ORDINARY STRUCTURAL FORMULAS

8. The standard structural formulas that we have been using since Unit 2 do not attempt

to show the molecular geometry explicitly. They show it implicitly in the sense that we can apply the rules of frames 3 to 5 to figure out the shape of the molecule. Or, with practice, you can visualize it directly when looking at the structural formula.

For beginning students the principal difficulty in using this scheme lies in distinguishing whether two given structures represent the same or different three-dimensional molecules. For example, the following structures all represent the same compound, and they are equally correct ways of drawing it.

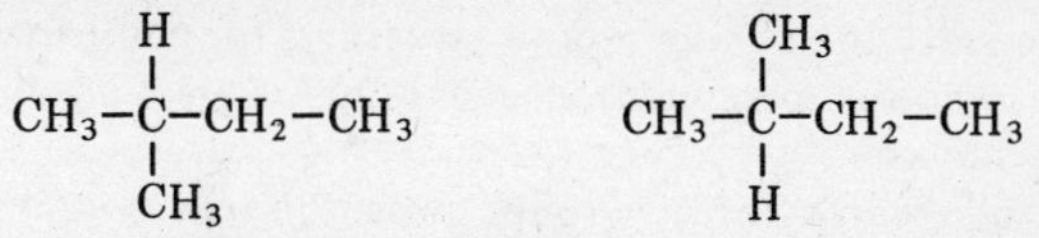

$$\begin{array}{cc} H & CH_3 \\ | & | \\ CH_3-C-CH_2-CH_3 & CH_3-C-CH_2-CH_3 \\ | & | \\ CH_3 & H \\ a & b \end{array}$$

$$\begin{array}{ccc} CH_2-CH_3 & CH_3-CH_2 & CH_2-CH_3 \\ | & | & | \\ CH_3-C-CH_3 & CH_3-C-CH_3 & CH_3-C-H \\ | & | & | \\ H & H & CH_3 \\ c & d & e \end{array}$$

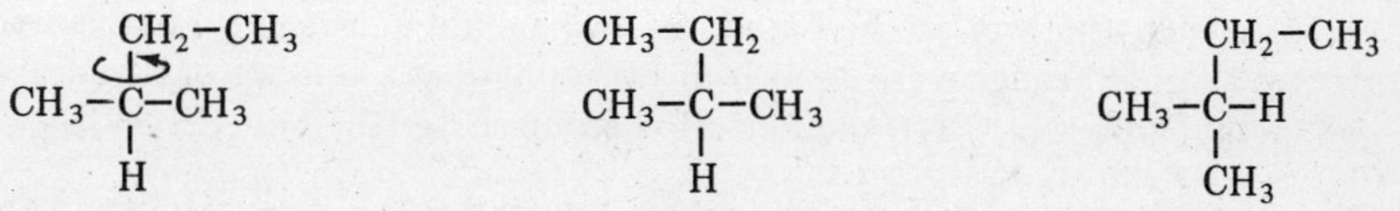

The easiest way to see their equivalence is to imagine the molecular model for this compound, and to look at it from all sides. Then you will see that some of the structures are just different views of the model, for instance a versus b. Others represent different conformations of the same model, made by rotation around a single bond. For example, the 180° rotation indicated by the small arrow in c converts c into d, so the two are equivalent structures. Rotation around single bonds occurs rapidly in the real molecule, and of course rotation of the whole molecule in space does, too. You must get used to visualizing the real molecule or its model executing these twists as shown above, and learn to recognize any of these views of it. In the first exercise at the end of this frame you will use models to practice this recognition.

Now let's set down two practical methods for deciding, without using models, whether two structures are equivalent (represent the same compound) or different (represent different compounds).

Method 1: Actually carry out, in the mind's eye, the act of trying to superimpose one two-dimensional structure, or a model of it, upon the other (or its model). Here you usually need to visualize some manipulations of one or the other structure (or their models) to get the two into the same shape (conformation) and to get the same view of that shape. For example, to compare e with the other structures, proceed in this way:

- Decide to compare with structure b, which has a CH_3 branch on a four-carbon chain.
- Find a bent chain of four carbon atoms in e: $\boxed{\begin{array}{c} CH_2-CH_3 \\ | \\ CH_3-C{+}H \\ | \\ CH_3 \end{array}}$
- Straighten it out: $\begin{array}{c} C-C \\ | \\ C-C \end{array} \rightarrow$ C–C–C–C. (If we are visualizing the model, this amounts to doing some internal rotations about C–C bonds.)
- Write the four-carbon chain down in the same way it appears in the comparison structure, b, and add the remaining groups:

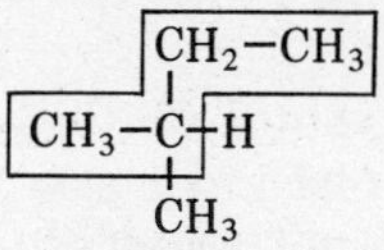

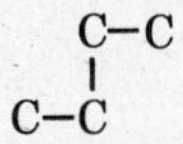

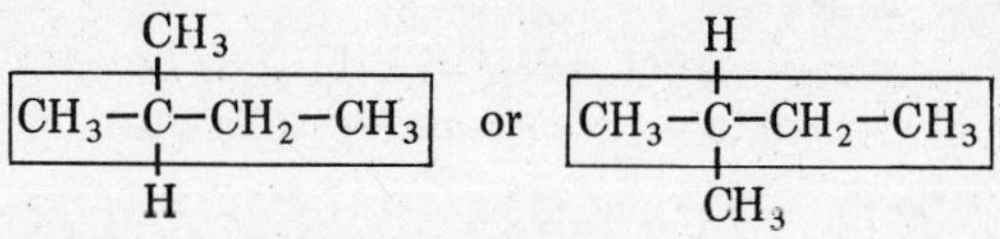

$$\boxed{\begin{array}{c} CH_3 \\ | \\ CH_3-C-CH_2-CH_3 \\ | \\ H \end{array}} \quad \text{or} \quad \boxed{\begin{array}{c} H \\ | \\ CH_3-C-CH_2-CH_3 \\ | \\ CH_3 \end{array}}$$

(If we are visualizing the model, this amounts to viewing it from a useful direction.)
- These flat, ink-on-paper structures can now be tested for superimposability on b. The left one superimposes perfectly on b, which proves that e and b are equivalent representations of the same molecule, since all we did to e was equivalent to twisting and turning the model or the molecule. If we had been visualizing the model instead of written structures, we would test the superimposability of the model on a model of b in this step.

At first it may be necessary to actually redraw structures for comparison with others. However, the ability to do the manipulations mentally is a great advantage, and you should cultivate it by practice.

Method 2: Take an inventory of the groups present in one structure and compare it with that of another structure. This turns the visual comparison into a verbal one. If one finds that the group inventories match precisely, the two structures are equivalent. If any structural difference whatever can be found, the two are not equivalent. For example, we can show that all of a through e are equivalent by noting that a consists of a hydrogen ($-H$), two methyl groups ($-CH_3$), and one ethyl group (CH_3-CH_2-, $-CH_2-CH_3$, $\overset{|}{CH_2-CH_3}$, etc.), all attached to a central carbon. By reciting this inventory while looking in turn at each of the other structures, we find that each has the same inventory and so must be equivalent to a. The following example is a negative case.

Example

Do the following structures represent the same molecule?

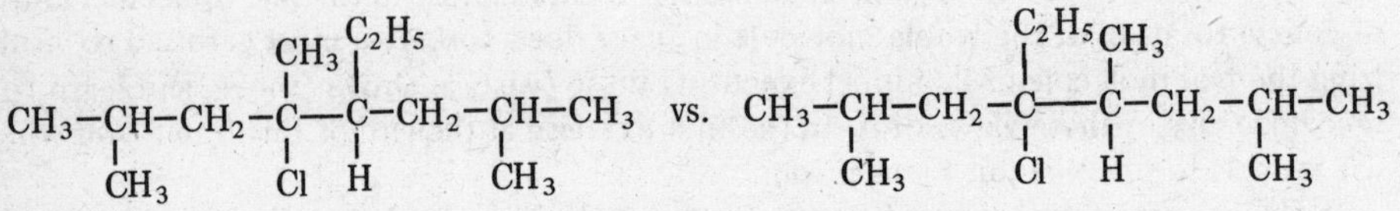

$$CH_3-CH(CH_3)-CH_2-\underset{\underset{Cl}{|}}{\overset{\overset{CH_3}{|}}{C}}-\underset{\underset{H}{|}}{\overset{\overset{C_2H_5}{|}}{C}}-CH_2-CH(CH_3)-CH_3 \quad \text{vs.} \quad CH_3-CH(CH_3)-CH_2-\underset{\underset{Cl}{|}}{\overset{\overset{C_2H_5}{|}}{C}}-\underset{\underset{H}{|}}{\overset{\overset{CH_3}{|}}{C}}-CH_2-CH(CH_3)-CH_3$$

Solution:

Both have the Cl on a 3° carbon atom, but in one case this atom bears a CH_3 group, while in the other it bears a C_2H_5. Consequently, the two structures represent different molecules.

There is a third way, but we are not yet ready to use it. This still-more-verbal method consists of building the systematic name for each structure. If the appropriate names are identical, the structures are equivalent.

Exercises

1. Make two models of $CH_3CH(CH_3)_2$. Lower one onto the other to see if the corresponding atoms of the two models are all in contact. If they are not, make some adjustments by rotating internally about C—C bonds until the shapes correspond exactly. This is the superimposability test, using models; we did the same thing mentally in the comparison of b and e.

2. Use one of your models to verify that the changes a ⟶ b ⟶ c ⟶ d ⟶ e can be accomplished just by rotating the whole model before the eye, or by performing internal rotations about single bonds (or both).

3. Again using both models, locate the tertiary carbon, $CH_3\overset{*}{C}H(CH_3)_2$, on one of them. Interchange any of the four groups attached to it (by disconnecting, exchanging, and

reconnecting them). Use a superimposability test with the other (unchanged) model to find out if interchange of two groups on the same carbon alters a structure or leaves it unchanged.

★ 4. Indicate which of the structures in each of the following groups are equivalent representations of the same molecule.

(A) 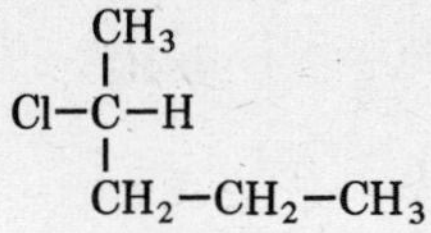
$$Cl-\underset{\underset{CH_2-CH_2-CH_3}{|}}{\overset{\overset{CH_3}{|}}{C}}-H$$
a

$CH_3-CH_2-CH_2-CHCl-CH_3$
b

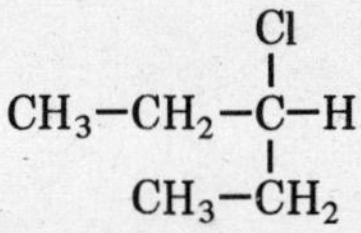

$$CH_3-CH_2-\underset{\underset{CH_3-CH_2}{|}}{\overset{\overset{Cl}{|}}{C}}-H$$
c

$$Cl-\underset{\underset{CH_3}{|}}{\overset{\overset{CH_3-CH_2}{|}}{CH}}-CH_2$$
d

(B) 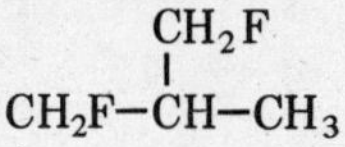 $CH_2F-\underset{\underset{a}{}}{\overset{\overset{CH_2F}{|}}{CH}}-CH_3$ $CH_2F-\underset{\underset{b}{}}{\overset{\overset{CH_3}{|}}{CF}}-CH_3$ $CH_2F-\underset{\underset{c}{}}{\overset{\overset{CH_3}{|}}{CH}}-CH_2F$ $CH_3-\underset{\underset{d}{}}{\overset{\overset{CH_2F}{|}}{CH}}-CH_2F$

(C) 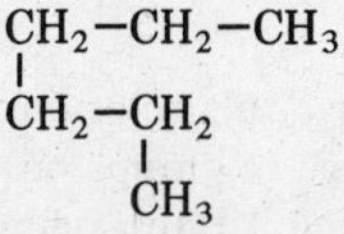
$$\begin{array}{l} CH_2-CH_2-CH_3 \\ | \\ CH_2-CH_2 \\ | \\ CH_3 \end{array}$$
a

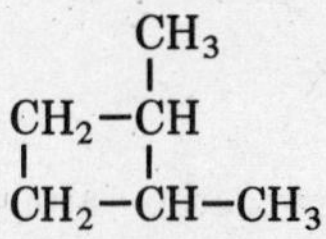

$$\begin{array}{l} \quad\; CH_3 \\ \quad\; | \\ CH_2-CH \\ |\quad\quad | \\ CH_2-CH-CH_3 \end{array}$$
b

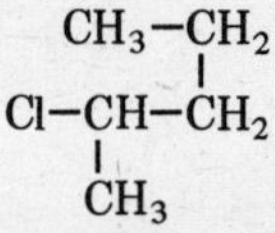

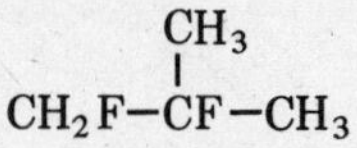

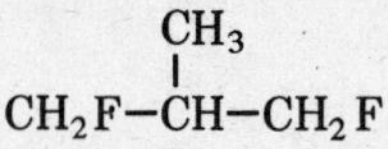

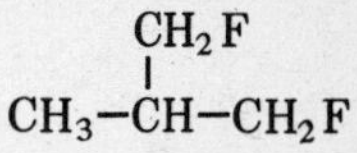

$CH_3(CH_2)_4CH_3$
c

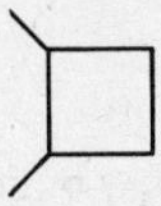

[square with two methyl substituents]
d

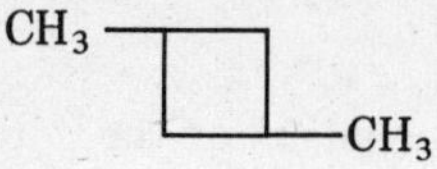

[square with CH₃ and CH₃ substituents]
e

(D) $CH_3-CHCl-CHF-CH_2-CH_3$
a

$CH_3-CH_2-CHF-CHCl-CH_3$
b

$CH_3-CH_2-CHCl-CHF-CH_3$
c

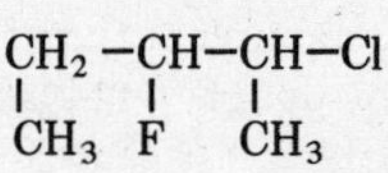

$\underset{\underset{CH_3\;\; F\;\; CH_3}{|\quad|\quad|}}{CH_2-CH-CH-Cl}$
d

(E) $CH_3-CH_2-CCl=CH-CH_2-CH_2-CH_3$
a

$CH_3-CH_2-CH_2-CCl=CH-CH_2-CH_3$
b

$CH_3-CH_2-CH_2-CH=CCl-CH_2-CH_3$
c

$CH_3-CH_2-CHCl-CH=CH-CH_2-CH_3$
d

$CH_3-CH_2-CH_2-CH_2-CCl=CH-CH_3$
e

(F)

a:
$$CH_3-\underset{\underset{CH_3}{|}}{\overset{\overset{CH_3}{|}}{C}}-\underset{\underset{CH_2}{|}}{\overset{\overset{H}{|}}{C}}-\underset{\underset{CH_3}{|}}{\overset{\overset{CH_3}{|}}{C}}-CH_2-CH_3$$
with CH_2 branching to $CH(CH_3)_2$

b:
$$CH_3-CH-CH_2-CH_2-CH-\underset{\underset{CH_2-CH_3}{|}}{\overset{\overset{C(CH_3)_3}{|}}{C}}-CH_3$$
with CH_3 on first CH

c:
$$CH_3-CH-CH_2-CH_2-\underset{\underset{(CH_2)_2-CH(CH_3)_2}{|}}{\overset{\overset{C(CH_3)_3}{|}}{C}}-H$$
with CH_3 on first CH

d:
$$CH_3-CH_2-\underset{\underset{CH_3}{|}}{\overset{\overset{CH_3}{|}}{C}}-CH-CH_2-CH_2-CH-CH_3$$
with $C(CH_3)_3$ and CH_3 branches

---

3. Interchange(s) of groups attached to the same carbon atom leave this compound (and most compounds) unchanged. The only exception is interchange of groups on a carbon that bears four different groups; this problem is discussed in Unit 13. Interchange of two groups attached to *different* carbons usually does change the compound into a new one.

4. (A) a, b, and d are equivalent. c puts Cl on the third carbon in the chain, the rest put Cl on the second carbon. (B) a, c, and d are equivalent. Easiest way: in these three, both F's are primary; in b, one is primary, one tertiary. (C) a and c are equivalent. Since they are C_6H_{14}, they cannot be equivalent to the rest, which are C_6H_{12}. b and d are equivalent; e differs from them in having the two methyls on opposite, not adjacent, carbons. (D) a, b, and d are equivalent. c has Cl rather than F on the center carbon. (E) a and c are equivalent (just flipped left for right). The rest are different molecules: e because the double bond is in a new place; d because the Cl is no longer on a doubly bonded carbon; b because Cl is on the center C, not the third one in. (F) a, b, and d are equivalent. Easiest way: all have a central carbon bearing an H, a tertiary butyl group $\left(-C\begin{smallmatrix}CH_3\\-CH_3\\CH_3\end{smallmatrix}\right)$, an isopentyl group $\left(-CH_2-CH_2-CH\begin{smallmatrix}CH_3\\CH_3\end{smallmatrix}\right)$, and a tertiary pentyl group $\left(-C\begin{smallmatrix}CH_3\\-CH_3\\CH_2-CH_3\end{smallmatrix}\right)$; c does not. Or c can be recognized as differing by lack of an ethyl group, $-CH_2-CH_3$, anywhere in the molecule, which the others have.

REPRESENTING THREE-DIMENSIONAL MOLECULES BY MEANS OF PERSPECTIVE STRUCTURES

9. Four different conventions are available for representing tetrahedral geometry:

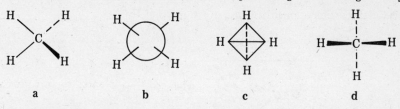

In a the normal bonds, —, and the atoms they connect are in the plane of the paper. The dashed bond (— — —) and its H recede below the paper and the heavy (◄■■■) bond (and its H) project above the paper.

The following drawing emphasizes how one should read structure a:

The same effect is obtained in b by letting one bond and its H peek out from behind an imagined sphere, whose center is the carbon atom, which is not written. c actually draws the tetrahedron; the carbon is imagined at its center, and a bond extends to each vertex. It is important to read structure c correctly in space from the beginning; the top and bottom H's are at the rear, "pulled back" by the dashed edge of the tetrahedron, which must recede. The left and right H's come forward with their connecting solid edge. Do not misread the tetrahedron *as if* the dashed edge came forward and the solid one receded. Seen correctly it has a flat top and a pointed bottom. A tetrahedron without a dashed edge cannot be read unambiguously: ◇ . d is c with the tetrahedron erased and the central carbon shown; it has replaced c in many current textbooks.

Each of the perspective structures above can be used to advantage in certain types of problems. Each can be used to represent the shape of molecules containing more than one carbon atom. Style a is the easiest—one puts the carbon chain in extended zigzag form as shown below. Style c is good when the order of groups must be read, as we will see in Unit 13. Thus $CH_3-CF_2-CH_2-CCl_3$ can be drawn as follows:

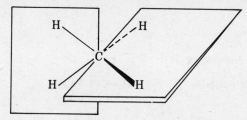

These perspective formulas can be used in combination with the notation used previously for unsaturated systems.

Example

What is the three-dimensional structure of $CH_3-CH=CH_2$?

Solution:

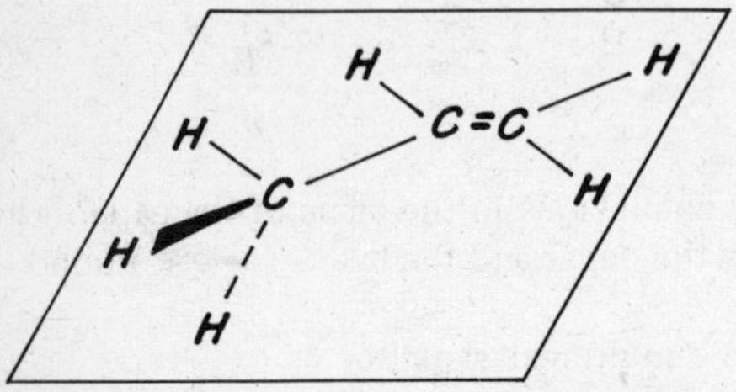

The $\begin{array}{c} H \\ \diagdown \\ C=C \\ \diagup \\ C \end{array} \begin{array}{c} H \\ \diagup \\ \\ \diagdown \\ H \end{array}$ part structure lies entirely in a plane. But no more than one of the hydrogens of the CH_3- group can lie in this plane because the CH_3- group is tetrahedral.

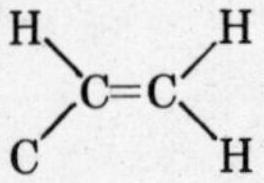

★ *Exercises*

1. Draw perspective formulas for each compound.

    (a) CF_2Cl_2 (do this one in all four notations)   (b) $CH_3-CCl_2-CH_3$

    (c) $CH_3-\underset{\underset{CH_3}{|}}{CH}-C{\equiv}C-H$

2. Which structures of each group represent the same molecule?

    (A) a, 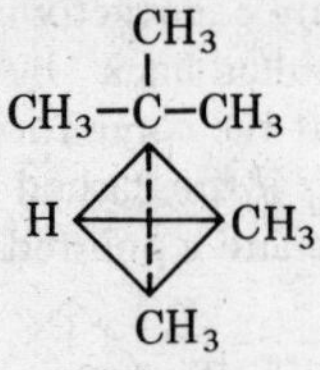 b, 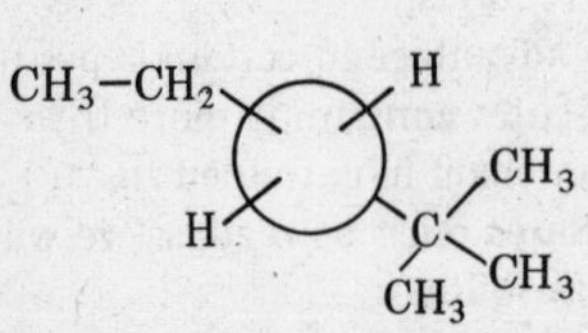 c, 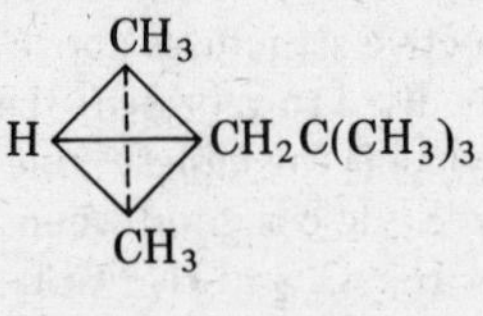 d

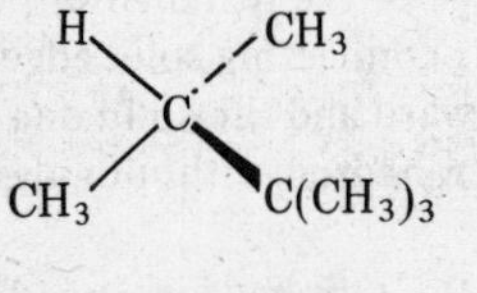

    (B) a, 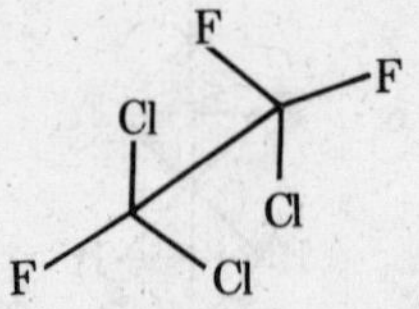 b, 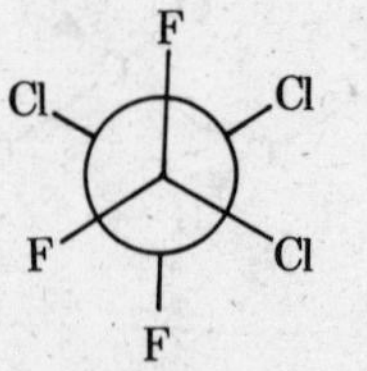 c, 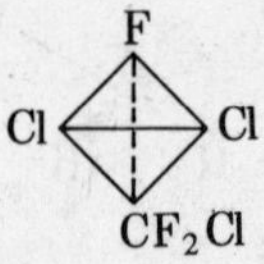 d

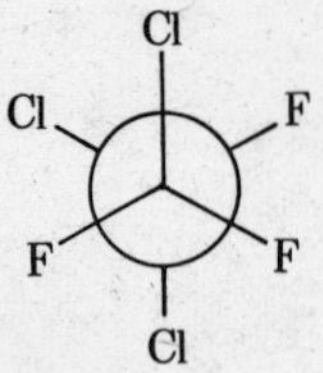

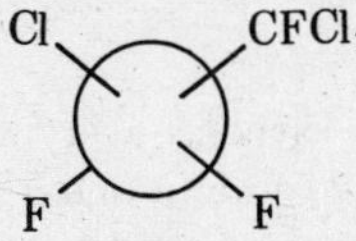

GEOMETRY OF ORGANIC MOLECULES 69

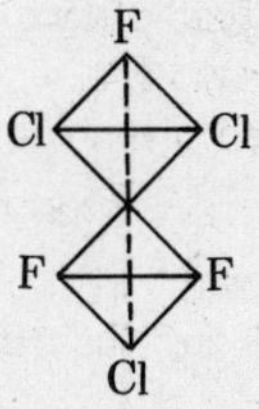

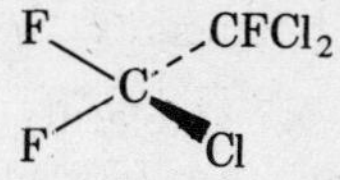

(C) 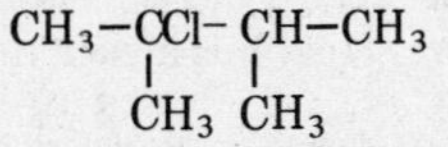 $CH_3-CCl-CH-CH_3$
      $||$
      CH_3CH_3

a

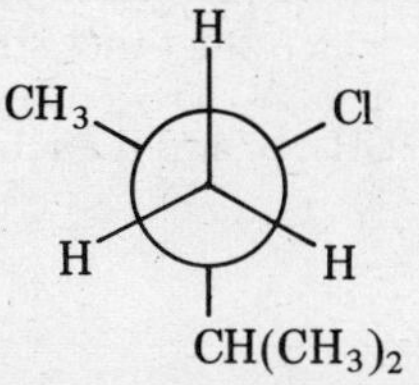

b

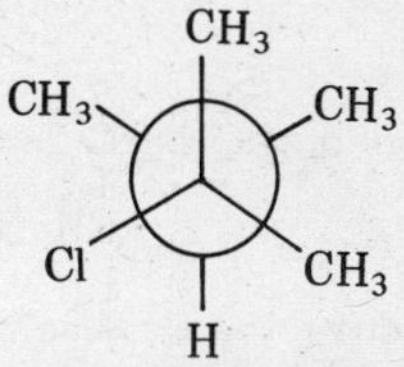

c

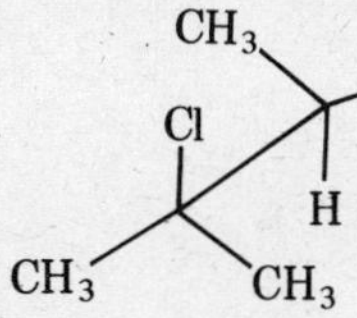

d

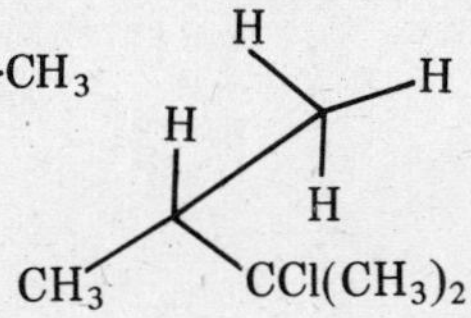

e

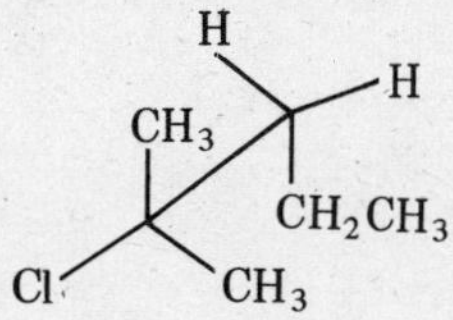

f

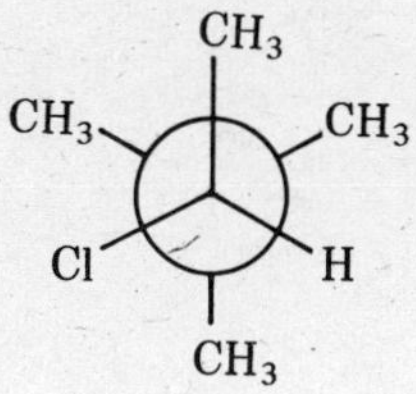

g

---

1. (a)

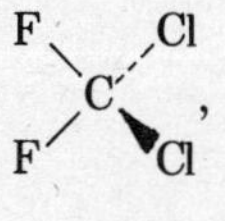

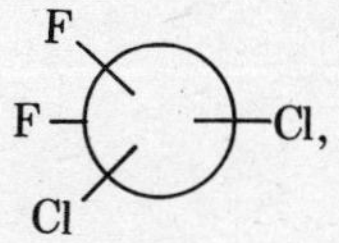

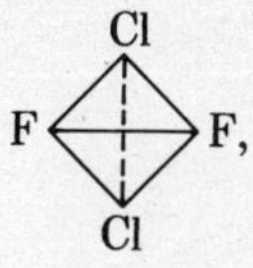

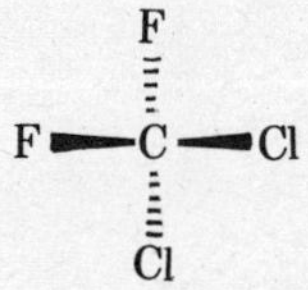

All the alternative ways of arranging the F's and Cl's on these formulas are equally correct.

(b)

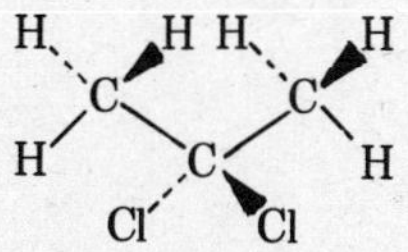

(c)

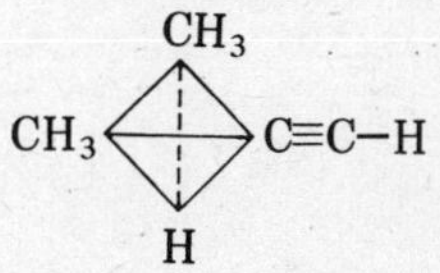

2. (A) a and b represent the same molecule. Instead of two CH_3- groups, an $H-$ and a $-C(CH_3)_3$ group, the central carbon of c has two H's, a $-CH_2CH_3$, and a $-C(CH_3)_3$ group. d actually has eight carbons rather than seven (do not forget the carbons hidden in a sphere or tetrahedron).

(B) All represent the same molecule. Easiest way: with only two carbons, every molecule in which one carbon has 2 F's and one Cl and the other has 2 Cl's and one F has to be the same.

(C) Only f and g differ from the rest (and from one another). g has a $-C(CH_3)_3$ group, and f has a CH_3-CH_2- group, which the other structures lack. b is a view down the $\overset{1}{C}-\overset{2}{C}$ bond of $\overset{1}{C}H_3-\overset{2}{C}Cl-\overset{3}{C}H-\overset{4}{C}H_3$. c and d are both views down the $\overset{2}{C}-\overset{3}{C}$ bond. e is
$\qquad\qquad\qquad\quad\ \ |\quad\ |$
$\qquad\qquad\qquad CH_3\ CH_3$

a view down the $\overset{3}{C}-\overset{4}{C}$ bond.

UNIT FIVE
Seeing Structures Analytically

Looking at a structural formula *analytically* means looking at

$$HO-CH_2-\langle\bigcirc\rangle-CHCH_3-CHCl-CH_2-CH=CH-CH_3$$

for example, and seeing

[structural diagram with labels: benzylic, planar, benzylic, disubstituted planar, aryl, 2°, vinylic, H←allylic, H–Ö]

OBJECTIVES

When you have completed this unit, you should be on your way toward:
- Forming the habit of analyzing each structural formula encountered into functional group(s) and skeleton, noting how they are connected, and visualizing the shape of the molecule (frame 1).
- Being able to summarize the important features of the molecule in a short verbal description (frame 2).

If you think you may have already achieved these objectives and might skip this short unit, take a self-test consisting of the exercises marked with a star in the margin. If you miss any of these, review the frames in question and rework the problems before proceeding. If you are not ready for a self-test, proceed to frame 1.

1. The trained eye looks at a new structural formula *actively*. Active looking *tests* the

structure. The eye verifies that the valence of each atom is satisfied, verifies or adds unshared pairs where needed, and verifies the presence or absence of charge on each atom. Active looking analyzes the structure and puts life into it, as in the example shown above. The eye senses a knobby tetrahedral texture over most of an aliphatic molecule but is alert to spot planar and linear localities. It identifies the degree of alkyl substitution of double bonds and identifies the site of attachment of each functional group as a 1°, 2°, 3°, vinylic, aryl, allylic, or benzylic carbon. Aromatic molecules are recognized as planar, rigid discs of hydrocarbon material with functional groups attached to the edge. The eye fastens on the identity of the functional group(s) as the prime determinant of the molecule's chemical properties.

The beginner does all of this slowly and consciously, but with experience the process becomes rapid and automatic. We have already done each of these things separately in previous units. In training your eye to look at structures analytically, you only need to put them together and to practice. You can improve your perceptual skills markedly by proper exercise.

If you train yourself now in the art of seeing all the signals that a structural formula offers, you will find later learning very much easier. A good eye allows you to avoid errors; it speeds up both study and exam response.

The following exercises are a good beginning, but only everyday practice in both seeing and drawing structures will give you real proficiency in working with them. A few minutes spent in reviewing your solutions to the exercises in Unit 3 will make the exercises in this unit easier.

Exercises

1. Cover the questions on the right-hand side below, then study the first structural formula at the left for 15 seconds. Next, cover the structure and answer the questions. Do the same for the second structural formula.

   (A) 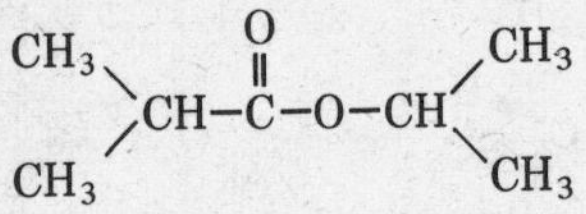 $(CH_3)_2CH-C(=O)-O-CH(CH_3)_2$

   (a) Is the skeleton branched or unbranched?
   (b) What is the functional group?
   (c) How many isopropyl groups, $-CH(CH_3)_2$, are present?

   (B) 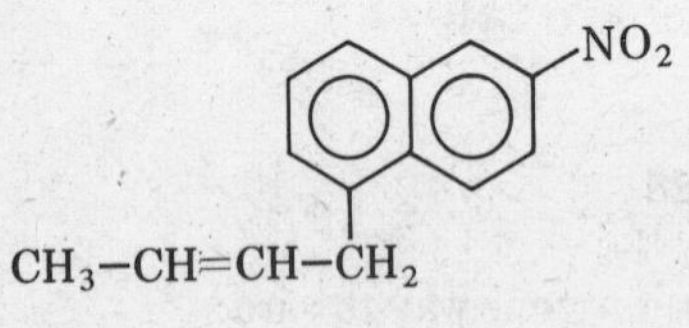 naphthalene with NO_2 substituent and $CH_3-CH=CH-CH_2-$ side chain

   (a) What functional group(s) are present?
   (b) How many rings are present?
   (c) Is the double bond conjugated?
   (d) How substituted is the double bond?
   (e) Is the carbon side chain branched?
   (f) Are there any benzylic hydrogens?

2. Study the first structural formula below for 15 seconds; then cover it and draw the structure from memory. Do the same for each of the other structural formulas.

   (a) 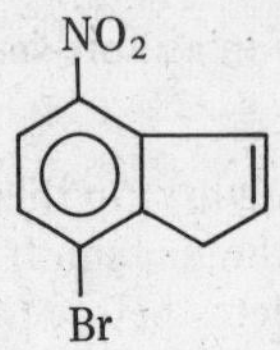 indene with NO_2 and Br substituents

   (b) 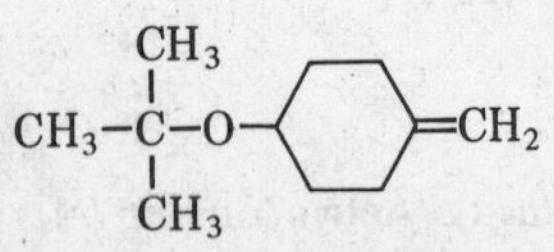 $CH_3-C(CH_3)(CH_3)-O-$ cyclohexyl $=CH_2$

(c) 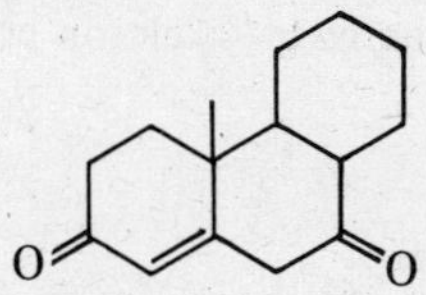 [structure: decalin-like bicyclic system with two ketone groups (=O) and a methyl group]

(d) $CCl_3-CH_2-CH=C-CBr_3$
              |
              CH_3

(e) 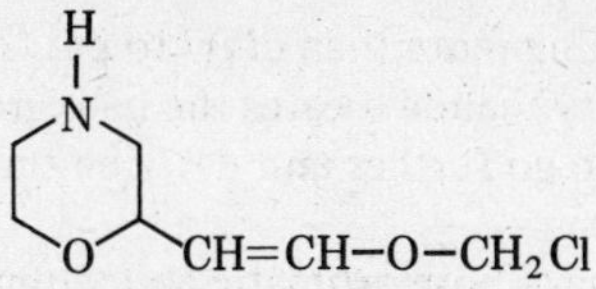 [morpholine-like ring with NH, with substituent $CH=CH-O-CH_2Cl$]

- - - - - - - - - -

1. (A) (a) branched (b) ester (c) two

   (B) (a) alkene, nitro (b) two (c) no (d) di (e) no (f) two

2. Summarizing the results of your inspection of a structural formula in a short verbal description is useful both for fixing in mind what you have seen and also for talking with others about the structure. Such a description is informal and generic; it bears little relation to the formal, systematic name of the compound (which we consider in Unit 7). The same molecule may be given different informal descriptions, depending on the amount of detail and the structural features emphasized for an immediate purpose.

   As an example of this type of informal description, we might call the molecule

   $CH_3-CH-CH_2-CH_2-CH_2-CH_2-NO_2$
         |
         CH_3

   an open-chain, saturated, branched, primary nitro compound—or just an aliphatic nitro compound. It may help you at this point to go back and review briefly the classes of carbon skeletons in Unit 3, frame 3. The terms "aliphatic" and "aromatic" are useful as broad classifiers; between them, they cover all carbon skeletons. The compound above is an *aliphatic* nitro compound because it has an aliphatic, not an aromatic, skeleton.

Examples

Aromatic compounds	Aliphatic compounds
$CH_3-\bigcirc-CHO$	$CH_3-CH_2-O-CH_3$
an aromatic aldehyde	an aliphatic ether
$\bigcirc-SO_3H$	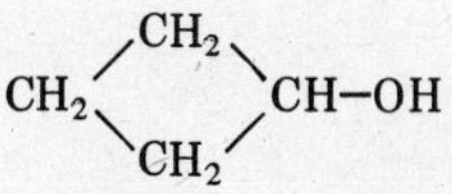 cyclopentane-CH-OH (an aliphatic alcohol)
an aromatic sulfonic acid	cyclohexene-C(=O)-CH_3
	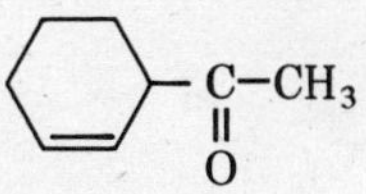 an aliphatic ketone

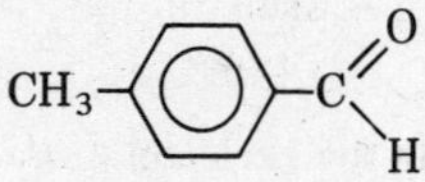

Sometimes "alkyl" takes the place of "aliphatic." The meaning is "skeleton composed of alkyl groups."

Example

$CH_3-CH_2-CH(CH_3)-CH_2-Cl$ is an alkyl halide. If only an alkyl group (or groups) is present, the skeleton *must* be aliphatic, and you need say nothing more than *alkyl* to get the notion of *aliphatic* across. Similarly, the description *aryl* halide locates the halo group on an aromatic ring automatically, and there is no need to go further and describe the compound as aromatic.

What do we do with compounds whose skeletons have both aromatic and aliphatic parts? The term "aralkyl" is handy for these compounds.

Example

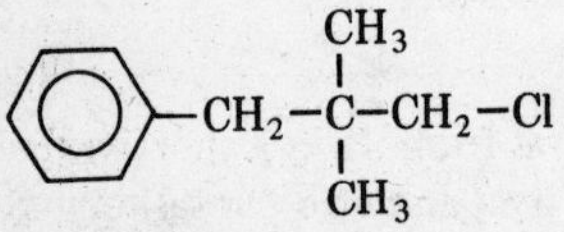

Ph–CH_2–$C(CH_3)_2$–CH_2–Cl   is an aralkyl chloride.

If the skeleton is in two parts, as with ethers and ketones, the parts can be identified separately as aryl or alkyl.

Examples

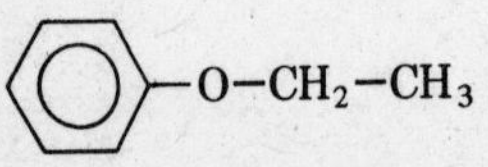

Ph–O–CH_2–CH_3

an aryl alkyl ether

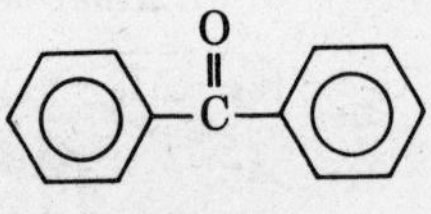

Ph–CO–Ph

a diaryl ketone

The short verbal description usually communicates three things: functional group, skeleton type, and point of attachment.

Example

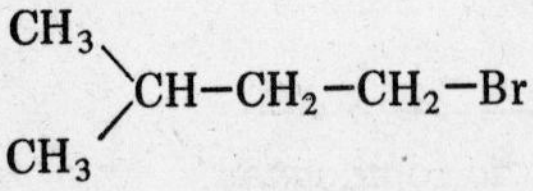

$(CH_3)_2CH-CH_2-CH_2-Br$ is a primary alkyl halide

point of attachment ↑ skeleton type ↑ functional group

To this description, we can add modifiers describing the skeleton more completely, if desired.

Example

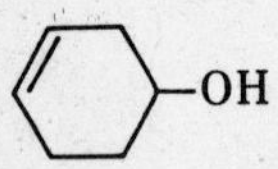

(cyclohexenyl)–OH   is a cyclic, unsaturated, secondary alcohol.

Exercises

1. Draw a specific example of each type of compound.

   (a) a cyclic secondary alcohol   (b) a tetrasubstituted alkene

(c) an aryl carboxylic acid
(d) a conjugated, nonterminal alkyne
(e) a vinylic ether
(f) an aryl dinitrile
(g) an unsaturated, open-chain ketone
(h) a cyclic, trisubstituted alkene
(i) a tertiary amine
(j) a benzylic *and* tertiary iodide
(k) a cyclic, allylic nitro compound
(l) a tertiary alkyl ester

2. Study the following structures and for each write a concise, descriptive name that sums up the functional group, skeleton type, and point of attachment as accurately as possible.

(a) 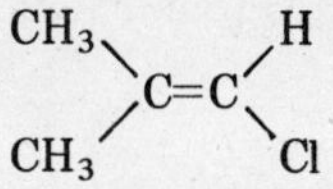 $(CH_3)_2C=CHCl$ (with H on same carbon as Cl)

(b) $CH_3-CH_2-CH_2-CH(CH_3)-CO_2H$

(c) 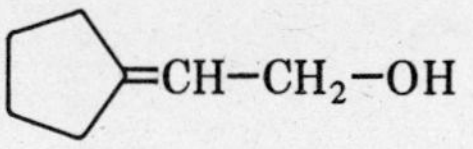 cyclopentylidene=CH-CH$_2$-OH

(d) 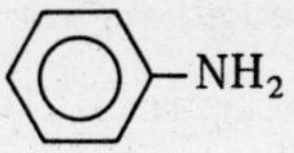 phenyl-NH$_2$

(e) 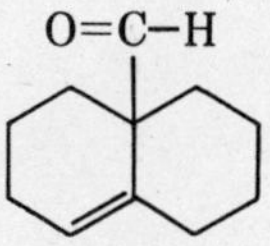 decalin-type bicyclic with O=C-H (aldehyde) at ring junction and a ring double bond

(f) phenyl-CH$_2$-CH(CN)-CH$_3$

(g) 1,3-cyclohexadiene

(h) $CH_3-\overset{O}{\underset{\|}{C}}-O-CH(CH_3)_2$

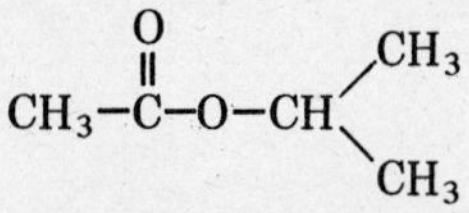

(i) 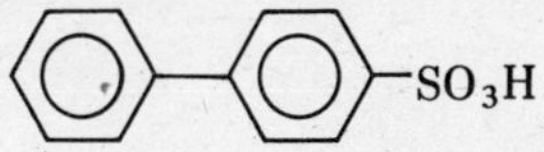 biphenyl-SO$_3$H

(j)  tetrahydropyran with =C(CH$_3$)$_2$ exocyclic

(k) 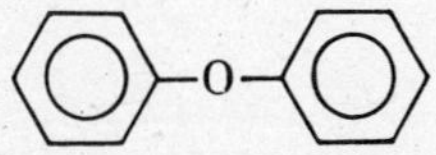 diphenyl ether (Ph-O-Ph)

(l) $CH_3-CH_2-CH_2-C\equiv C-H$

---

1. (illustrative answers) (a) 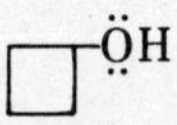 cyclobutyl-ÖH   (b) 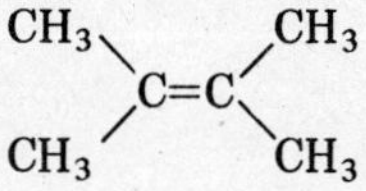 $(CH_3)_2C=C(CH_3)_2$ (c) naphthalene-CO$_2$H

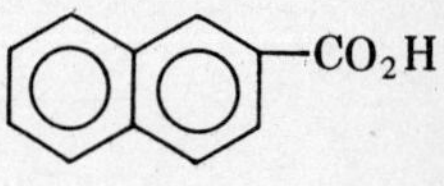

(d) $CH_3-C\equiv C-CH=CH_2$ (e) 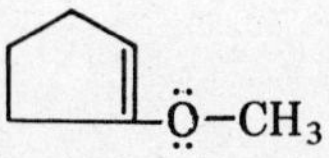 cyclopentyl-Ö-CH$_3$ (f) 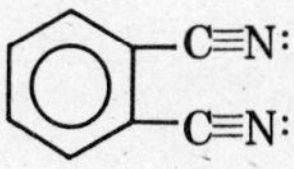 benzene-1,2-di(C≡N)

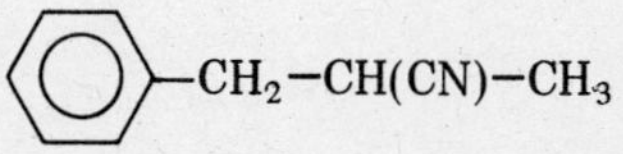

(g) 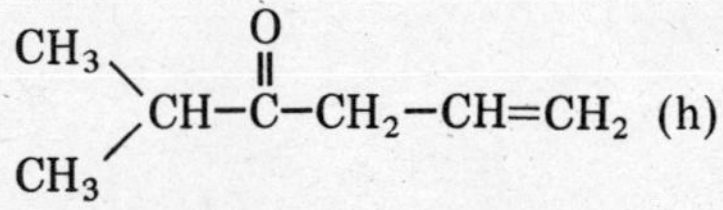 $(CH_3)_2CH-\overset{O}{\underset{\|}{C}}-CH_2-CH=CH_2$ (h) 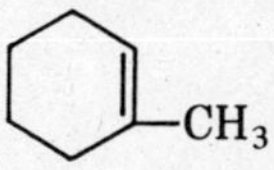 1-methylcyclohexene   (i) 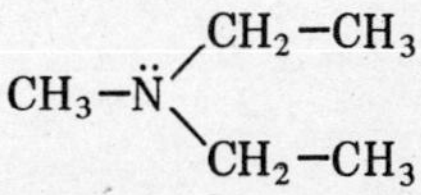 $CH_3-\ddot{N}(CH_2-CH_3)_2$

(j) 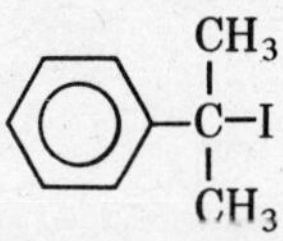 phenyl-C(CH$_3$)$_2$-I   (k) 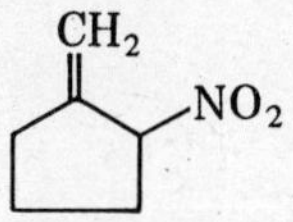 cyclopentane with =CH$_2$ and NO$_2$ (l) 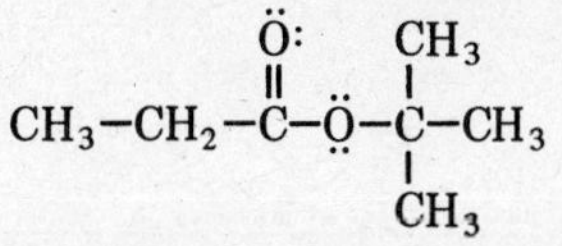 $CH_3-CH_2-\overset{\ddot{O}:}{\underset{\|}{C}}-\ddot{\underset{}{O}}-\overset{CH_3}{\underset{CH_3}{C}}-CH_3$

2. (All frame references below are to Unit 3 unless otherwise noted.)

   (a) a vinylic chloride (frames 1, 6)

   (b) an open-chain, saturated carboxylic acid (frames 1, 3)

   (c) a cyclic allylic alcohol (frames 1, 3, 7)

   (d) a primary aromatic amine (frames 1, 6)

   (e) a cyclic, unsaturated aldehyde (frames 1, 3, 4)

   (f) an aralkyl nitrile (frames 1,6)

   (g) a cyclic, conjugated diene (standard contraction for di-alkene; frames 1, 3, 4, 7)

   (h) a secondary aliphatic (or alkyl) ester (frames 1, 3, 5)

   (i) an aryl sulfonic acid (frames 1, 6)

   (j) a cyclic, unsaturated ether (frames 1, 3)

   (k) a diaryl ether (Unit 5, frame 2)

   (l) an open-chain, terminal alkyne (frames 1, 2, 3)

UNIT SIX
Structural Isomerism

In any natural language, a small number of sounds combine to produce an unlimited number of words. This is possible because, in principle, the length of a word has no limit, and because a chosen set of sounds can combine in more than one way to give more than one word: for instance, $a + b + t$ in English $\longrightarrow$ *bat* and *tab*. The language of organic chemistry works in the same way. We can write an unlimited number of structures using only half a dozen elements. In part this is because we can make each structure contain as many atoms as we please (CH_4, CH_3-CH_3, $CH_3-CH_2-CH_3$, and so on without limit). In part it is because a given set of atoms can usually be combined in more than one way to give more than one compound. A "given set of atoms," say C_2H_6O, is a *molecular formula*. Most molecular formulas correspond to more than one *structural formula* because the atoms can be bonded to one another in more than one arrangement:

$$C_2H_6O \begin{cases} CH_3-CH_2-O-H \\ CH_3-O-CH_3 \end{cases}$$

This phenomenon is called *structural isomerism*.

In both English and organic chemistry, not all sets of sounds (atoms) can combine to give genuine words (structures). We require rules to avoid generating nonsense words or nonsense structures, for instance $p + b + t \longrightarrow pbt$ (not a word in English) and $2C + H \longrightarrow C-C-H$ (a molecule not capable of existence). Much of your work in Units 2 and 4 was invested in learning rules for writing meaningful structures representing real, stable molecules. In this unit we study the problem of finding all the possible structural formulas corresponding to a given molecular formula. The solution to this problem is a basic part of many later problems.

OBJECTIVES

When you have completed this unit, you should be able to:

- Write the structural formulas of all the possible compounds having a given molecular formula (frames 2, 3, 5 to 7).
- Identify equivalent sites in part structures and equivalent atoms or groups in complete structures (frame 1).
- Calculate and interpret the unsaturation number, Δ (frame 4).

If you think you may be able to skip all or part of this unit, take a self-test consisting of the exercises marked with a star in the margin. If you miss some problems, review the frames in question and rework the problems. If you are not ready for a self-test, proceed to frame 1.

78 HOW TO SUCCEED IN ORGANIC CHEMISTRY

NATURE OF THE PROBLEM. EQUIVALENT SITES

1. Two compounds that have the same molecular formula but different structural formulas are *isomers*. Their molecules or structures are *isomeric*. The number of possible isomers cannot be predicted from the molecular formula in any simple way; it increases rapidly as the number of atoms in the molecule increases. Students experience two main difficulties with the "Draw the structures of all possible isomers of ___" problem: (1) how to be certain that all structures have been generated—the systematic approach developed in frames 2 to 7 solves this difficulty; (2) how to weed out or avoid duplicate structures (different representations of the same isomer). You have already learned to recognize equivalent structures in Unit 4, frame 8. In this unit we practice examples more closely related to the problem of generating isomeric structures.

First, you should be able to recognize two *part* structures as equivalent or different using the methods of Unit 4, frame 8.

Examples

The following sets of drawings are *equivalent* representations of a single part structure.

1. $-CH_2-CH_2-Br$, $Br-CH_2-CH_2-$, and CH_2-CH_2-
   $|$
   Br

2. $-CH_2-CH_2-CF_2-$, $-CH_2\diagup^{CH_2}\diagdown CF_2-$, and $-CF_2-CH_2-CH_2-$

3. $-\underset{|}{\overset{|}{C}}-\underset{|}{\overset{|}{C}}-\underset{|}{\overset{|}{C}}-$, $-\underset{|}{\overset{|}{C}}-\underset{|}{\overset{|}{C}}-\underset{|}{\overset{|}{C}}-$, $\underset{|}{\overset{-C-}{}}\overset{|}{\underset{\diagup|}{C}}-\underset{|}{\overset{|}{C}}-$, $-\underset{|}{\overset{-C-}{\underset{|}{C}}}-\underset{|}{\overset{|}{C}}-$, and $-\underset{|}{\overset{-C-}{\underset{|}{C}}}-\underset{|}{\overset{-C-}{\underset{|}{C}}}-$

4. $-\underset{|}{\overset{|}{C}}-\underset{|}{\overset{|}{C}}-O-\underset{|}{\overset{|}{C}}-$ and $-\underset{|}{\overset{|}{C}}-O-\underset{|}{\overset{|}{C}}-\underset{|}{\overset{|}{C}}-$

Examples

The following sets of drawings represent *different* part structures.

1. $-CH_2-CH_2Br$ (open bond is on a C bearing only H and C atoms) and $CH_3-CHBr-$ (open bond on C bearing Br)

2. $CH_3-CH_2-CH_2-$ and $CH_3-CH-CH_3$
   $\qquad\qquad\quad|$
   open bond $\qquad\qquad$ open bond
   on 1° C $\qquad\qquad\;$ on 2° C

3. $-CH_2-CH_2-O-CH_3$ and $-CH_2-O-CH_2-CH_3$
   oxygen is two atoms $\quad$ oxygen is one atom
   away from the open $\quad\;$ away from the open
   bond $\qquad\qquad\qquad\quad\;$ bond

4. $-\underset{|}{\overset{|}{C}}-\underset{|}{\overset{|}{C}}-O-\underset{|}{\overset{|}{C}}-\underset{|}{\overset{|}{C}}-$ and $-\underset{|}{\overset{|}{C}}-\underset{|}{\overset{|}{C}}-\underset{|}{\overset{|}{C}}-O-\underset{|}{\overset{|}{C}}-$

   O splits the molecule $\qquad$ O splits the molecule
   into $C_2 + C_2$ $\qquad\qquad\quad\;$ into $C_3 + C_1$

STRUCTURAL ISOMERISM 79

Second, atoms of the same element in part structures or complete structures, and open bonds in part structures, are either equivalent or different. The atoms or open bonds can be called *sites* on the structures. We sometimes label equivalent sites with the same lowercase letter, different sites with different letters.

Example

In 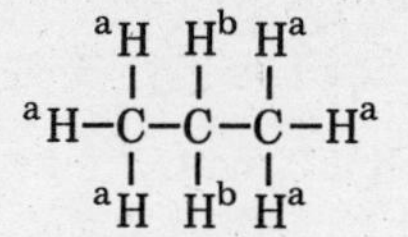
^{a}H−C−C−C−H^a with ^{a}H H^b H^a above and ^{a}H H^b H^a below, the a sites are 1° hydrogens and the b sites are 2° hydrogens.

The easiest test for equivalent sites is this: put two different atoms, say Cl and F, at the two sites. Then interchange the two. If the structures before and after the interchange are equivalent, the sites are equivalent. In the example above, F−C−C−C−H (with Cl H H on top and H H H on bottom)

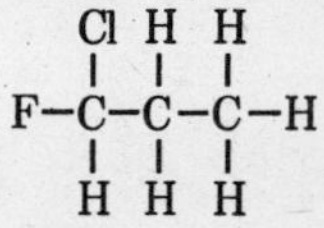

is equivalent to Cl−C−C−C−H (with F H H on top and H H H on bottom) because internal rotation about the left-hand C−C bond interconverts them. Therefore, the circled *sites* are equivalent in

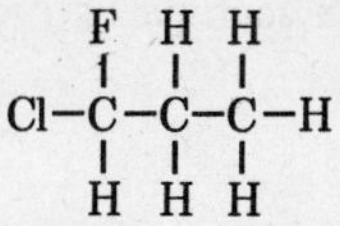

(H)−C−C−C−H with (H) H H on top and H H H on bottom

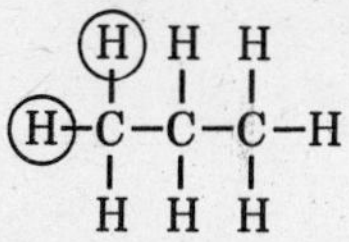

Similarly, the open bonds in CH$_3$−CH−CH$_2$− are *different* sites because CH$_3$−CH−CH$_2$ (with F, Cl substituents)

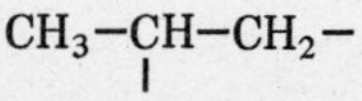

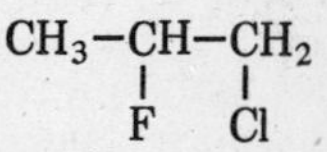

and CH$_3$−CH−CH$_2$ (with Cl, F substituents) are not equivalent structures—no internal rotation about C−C bonds or reorientation of the whole molecule can make them superimpose.

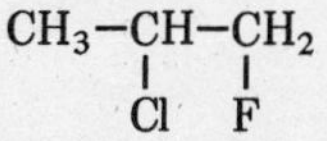

Examples

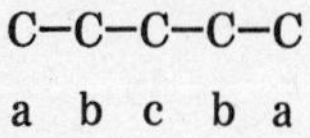

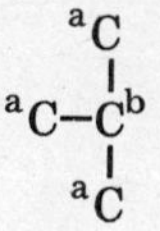

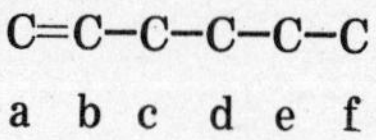

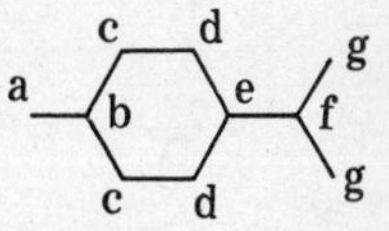

C−C−C−C−C
a b c b a

^{a}C−C^b with ^{a}C on top and ^{a}C on bottom

C=C−C−C−C−C
a b c d e f

[benzene ring with substituents labeled a, b, c, d, c, d, e, f, g, g]

★ *Exercises*

1. Which are equivalent part structures? (Open bonds have been omitted to reduce clutter.)

   (a) 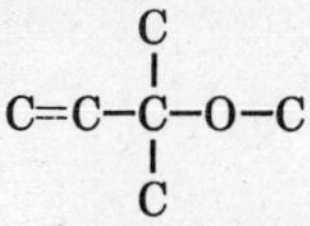 C=C−C−O−C with C above and C below the third C

   (b) C=C−O−C−C with C above and C below the fourth C

   (c) C−O−C−C with C=C above and C below the third C

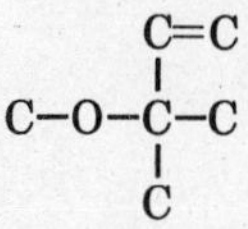

   (d) 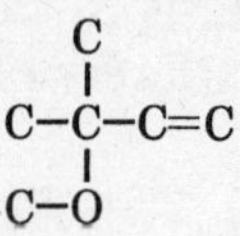 C−C−C=C with C above and C−O below the second C

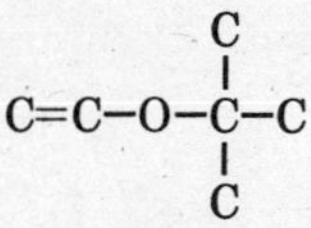

80 HOW TO SUCCEED IN ORGANIC CHEMISTRY

2. Which are equivalent part structures?

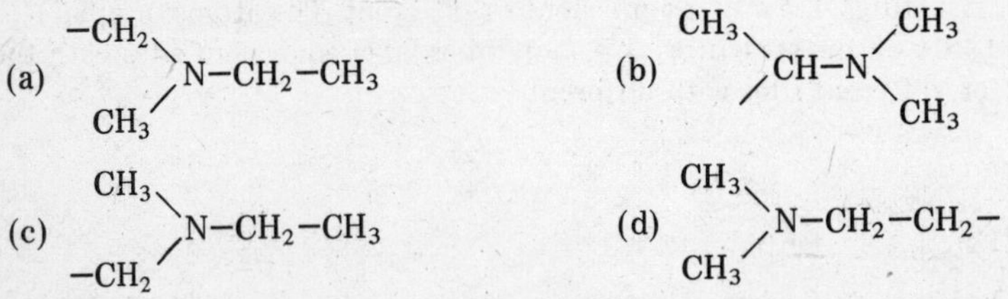

(a) $\begin{array}{c}-CH_2\\ \searrow\\ CH_3\nearrow\end{array}N-CH_2-CH_3$

(b) $\begin{array}{c}CH_3\searrow\\ CH-N\\ \nearrow\end{array}\begin{array}{c}\nearrow CH_3\\ \searrow CH_3\end{array}$

(c) $\begin{array}{c}CH_3\searrow\\ N-CH_2-CH_3\\ -CH_2\nearrow\end{array}$

(d) $\begin{array}{c}CH_3\searrow\\ N-CH_2-CH_2-\\ CH_3\nearrow\end{array}$

3. Label the hydrogen atoms to show which are equivalent, which different.

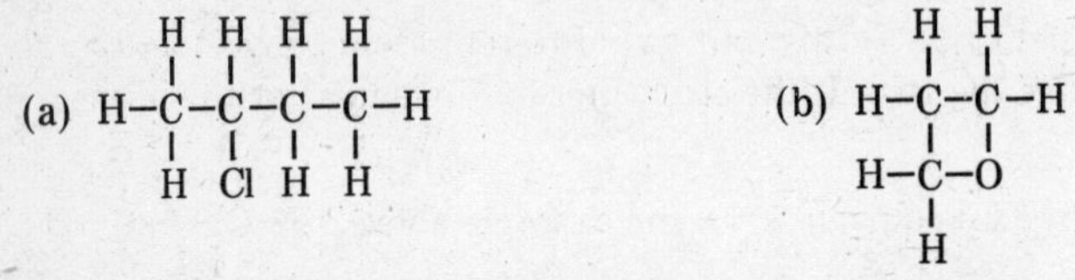

(a)
$$\begin{array}{ccccc}&H&H&H&H\\&|&|&|&|\\H-&C-&C-&C-&C-H\\&|&|&|&|\\&H&Cl&H&H\end{array}$$

(b)
$$\begin{array}{ccc}&H&H\\&|&|\\H-&C-&C-H\\&|&|\\H-&C-&O\\&|&\\&H&\end{array}$$

4. How many *different types* of site do the open bonds in each part structure provide?

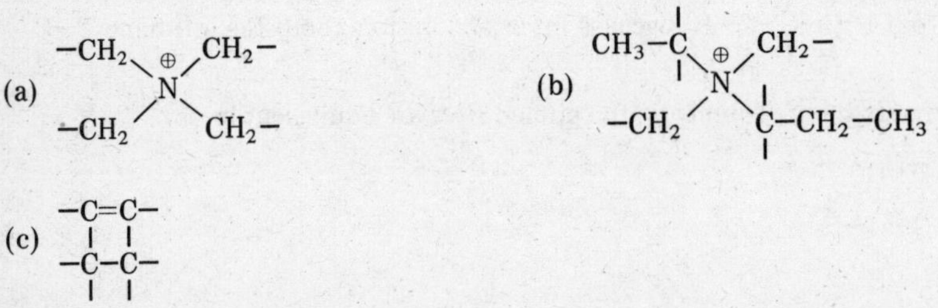

(a) $\begin{array}{c}-CH_2\searrow\nearrow CH_2-\\ \overset{\oplus}{N}\\ -CH_2\nearrow\searrow CH_2-\end{array}$

(b) $\begin{array}{c}CH_3-\overset{|}{C}\searrow\nearrow CH_2-\\ \overset{\oplus}{N}\\ -CH_2\nearrow\searrow\overset{|}{C}-CH_2-CH_3\\ |\end{array}$

(c)
$$\begin{array}{c}-C=C-\\||\\-C-C-\\||\end{array}$$

_ _ _ _ _ _ _ _ _

1. (a), (c), and (d) are equivalent. (b) has a $-C(CH_3)_3$ group that the others lack.

2. (a) = (c)
   (b) ≠ (d)

3. 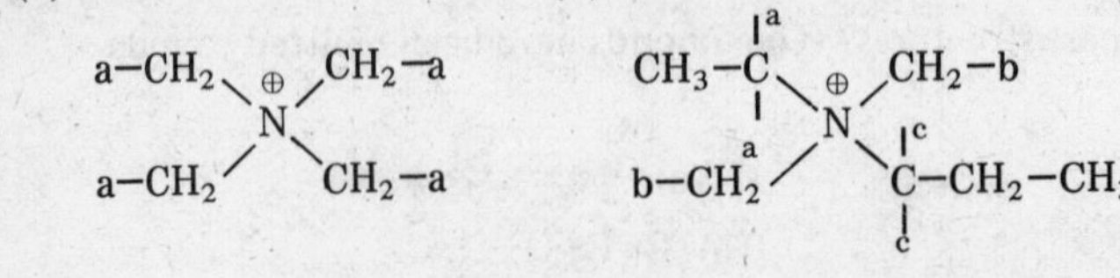 (a)
$$\begin{array}{ccccc}&^aH&^bH&^cH&^dH\\&|&|&|&|\\^aH-&C-&C-&C-&C-^dH\\&|&|&|&|\\&^aH&Cl&^cH&^dH\end{array}$$

(b)
$$\begin{array}{ccc}&^aH&^bH\\&|&|\\^aH-&C-&C-^bH\\&|&|\\^bH-&C-&O\\&|&\\&^bH&\end{array}$$

4. (a) one:

$\begin{array}{c}a-CH_2\searrow\nearrow CH_2-a\\ \overset{\oplus}{N}\\ a-CH_2\nearrow\searrow CH_2-a\end{array}$

(b) three:

$\begin{array}{c}CH_3-\overset{|\,a}{C}\searrow\nearrow CH_2-b\\ \overset{\oplus}{N}{}^c\\ b-CH_2\nearrow\searrow\overset{|}{C}-CH_2-CH_3\\ |\,c\end{array}$

(c) two:

$$\begin{array}{c}a-C=C-a\\||\\b-C-C-b\\||\\bb\end{array}$$

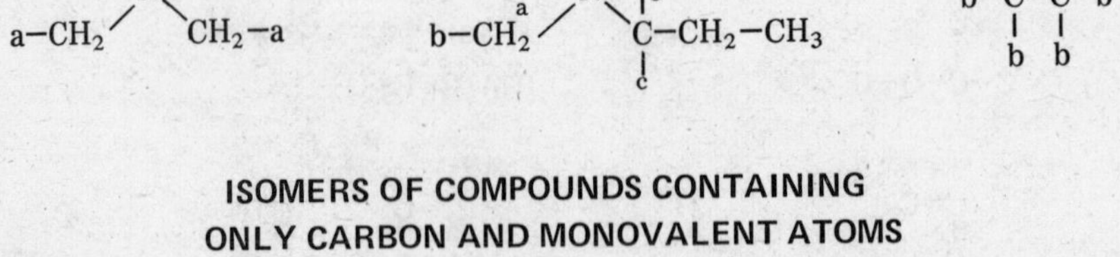

ISOMERS OF COMPOUNDS CONTAINING ONLY CARBON AND MONOVALENT ATOMS

2. If the molecular formula contains only carbon and monovalent elements (H, halogen), the complete set of isomeric structural formulas can be found in three steps:

- Form all possible carbon skeletons.
- Attach the monovalent atoms in all possible arrangements.
- Weed out any duplicate structures.

To generate all possible skeletons, proceed systematically: first, put all the carbons in a straight chain, then go to a straight chain one carbon shorter, using the extra carbon as a branch. Then use two carbons as branches, and so on. Locate branches at all possible carbons of the chain. For example, with five carbons, the possible skeletons are obtained as follows:

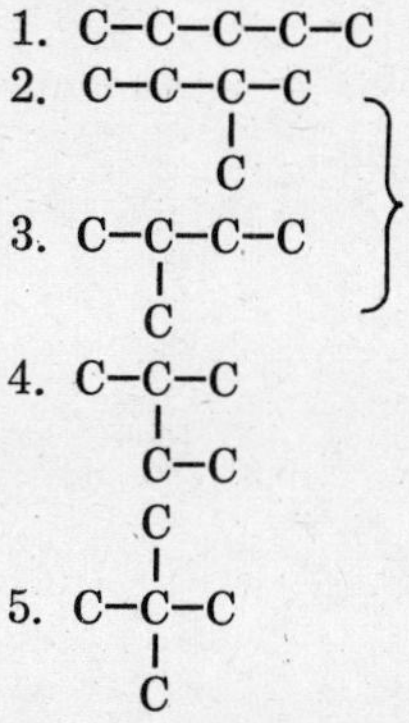

1. C–C–C–C–C   all five in a straight chain
2. C–C–C–C
       |
       C   ⎫
3. C–C–C–C   ⎬ C_4 chain + one C_1 branch
       | ⎭
       C
4. C–C–C
       |
       C–C
       |
       C
5. C–C–C C_3 chain + two C_1 branches
       |
       C

(Note that a branch can't be put at the end of a chain—it would just lengthen the straight chain.)

Are any of these skeletons equivalent? Yes, 2, 3, and 4 are equivalent. So there are just three different five-carbon skeletons:

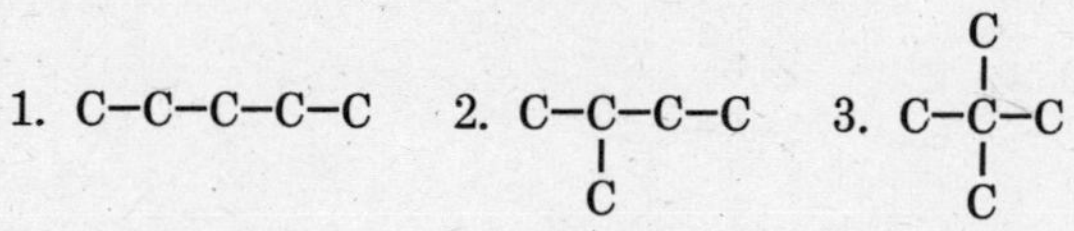

1. C–C–C–C–C   2. C–C–C–C   3. C–C–C
                       | |
                       C   C
                                      |
                                      C (top)

You must also proceed systematically in finding all possible arrangements of the monovalent atoms on a skeleton. The simplest case occurs when all the monovalent atoms are identical. Then there is only one possible arrangement on each skeleton.

Example

Write structures for all isomers of C_5H_{12}.

Solution:

The three possible skeletons for C_5 have already been generated above. Each has 12 open bonds, and all one can do is put the 12 H– on them, giving

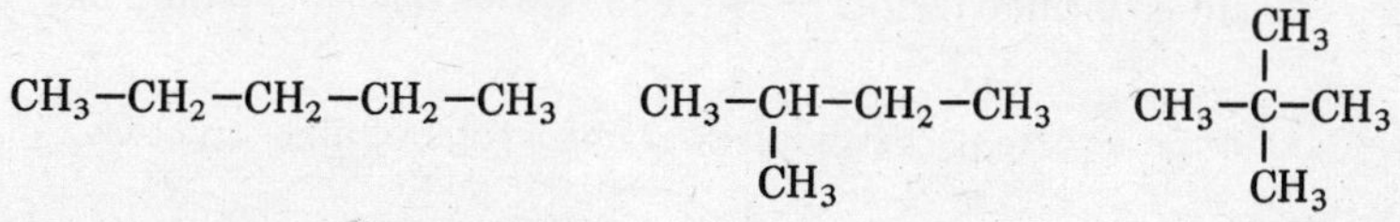

$$CH_3-CH_2-CH_2-CH_2-CH_3 \qquad CH_3-CH-CH_2-CH_3 \qquad CH_3-\underset{\underset{CH_3}{|}}{\overset{\overset{CH_3}{|}}{C}}-CH_3$$
$$\underset{CH_3}{|}$$

as the only possible C_5H_{12} structures.

When one monovalent atom is different from the rest, examine each skeleton and label the carbons to identify all the different types of sites, as was done in frame 1. Then put the unique monovalent atom on each different site in turn.

Example

For C_3H_7Br the only possible skeleton is C–C–C. Draw all the possible isomers of C_3H_7Br.

Solution:

Label: $\overset{a\ b\ a}{C-C-C}$, two different sites, therefore two isomers:

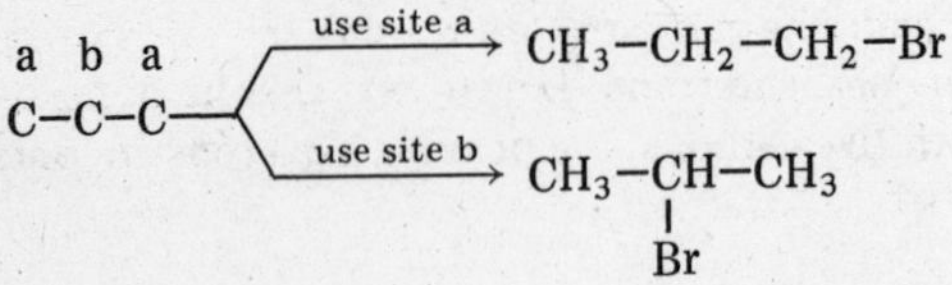

$\overset{a\ b\ a}{C-C-C}$ $\underset{\text{use site b}}{\overset{\text{use site a}}{\Big\langle}}$ $CH_3-CH_2-CH_2-Br$
$CH_3-\underset{Br}{CH}-CH_3$

When two of the monovalent atoms are different from the rest, use the same technique, but relabel the carbon atoms after attaching the first monovalent atom.

Example

Repeat the preceding example for C_3H_6FCl.

Solution:

Label and attach F:

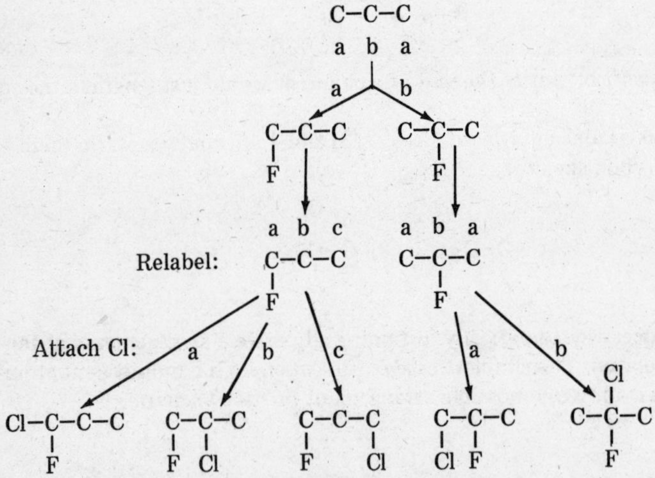

$C-C-C$
$a\ b\ a$

$\overset{a}{\swarrow}\quad \overset{b}{\searrow}$

$\underset{F}{C-C-C}\qquad \underset{F}{C-C-C}$

$\downarrow\qquad\qquad \downarrow$

Relabel: $\overset{a\ b\ c}{\underset{F}{C-C-C}}\qquad \overset{a\ b\ a}{\underset{F}{C-C-C}}$

Attach Cl: $\overset{a}{\swarrow}\ \overset{b}{\downarrow}\ \overset{c}{\searrow}\qquad \overset{a}{\swarrow}\ \overset{b}{\searrow}$

$\underset{F}{Cl-C-C-C}\quad \underset{F\ Cl}{C-C-C}\quad \underset{F\ \ Cl}{C-C-C}\quad \underset{Cl\ F}{C-C-C}\quad \underset{F}{\overset{Cl}{C-C-C}}$

Attachment of 6 H gives five structures, all different.

Identifying the sites on the skeletons is a way of looking ahead to avoid drawing duplicate structures that then just have to be eliminated. The long way of doing the preceding example would put the first F on *every* site of C-C-C; not just every *different* site. So $\underset{F}{C-C-C}$ would be included with $\underset{\ \ F}{C-C-C}$ and $\underset{\ \ \ \ F}{C-C-C}$. Then Cl would have been added at *every* site in these three part structures, giving a total of nine structures:

A $\underset{F}{Cl-C-C-C}$ B $\underset{F\ Cl}{C-C-C}$

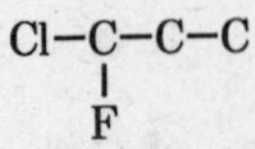

C $\underset{F\ \ \ Cl}{C-C-C}$ D $\underset{Cl\ F}{C-C-C}$

E $\underset{F}{\overset{Cl}{C-C-C}}$ F $\underset{F\ Cl}{C-C-C}$

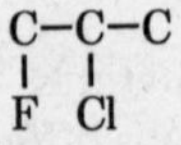

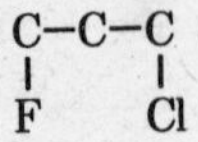

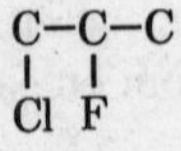

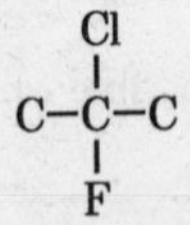

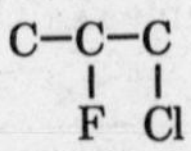

STRUCTURAL ISOMERISM 83

```
G   C–C–C              H   C–C–C
    |   |                  |   |
    Cl  F                  Cl  F

        Cl
        |
I   C–C–C
        |
        F
```

Four of these (**F-I**) duplicate others and would have to be eliminated. This not only takes time, it is an additional opportunity for error.

Exercises

Draw the structures of all possible isomers of each of the following.

1. C_6H_{14} 2. $C_4H_8F_2$
3. C_4H_8FCl

— — — — — — — — —

1. $CH_3CH_2CH_2CH_2CH_2CH_3$ $CH_3CH_2CH_2CHCH_3$ $CH_3CH_2CHCH_2CH_3$
                                                 | |
                                                 CH_3 CH_3

$$CH_3CH-CHCH_3 \quad CH_3\overset{\overset{CH_3}{|}}{\underset{\underset{CH_3}{|}}{C}}CH_2CH_3$$
   | |
   CH_3 CH_3

2. $CH_3-CH_2-CH_2-CHF_2$ $CH_3-CH-CHF_2$
                                            |
   $CH_3-CH_2-CF_2-CH_3$ CH_3

   $CH_3-CH_2-CHF-CH_2F$ $CH_3-CH-CH_2F$
                                            |
   $CH_3-CHF-CH_2-CH_2F$ CH_2F

   $CH_2F-CH_2-CH_2-CH_2F$ $CH_3-CF-CH_2F$
                                            |
   $CH_3-CHF-CHF-CH_3$ CH_3

3. $CH_3-CH_2-CH_2-CHFCl$ $CH_3-CHF-CH_2-CH_2Cl$ $CH_3-CH-CHFCl$
                                                             |
   $CH_3-CH_2-CFCl-CH_3$ $CH_3-CHCl-CH_2-CH_2F$ CH_3

   $CH_3-CH_2-CHF-CH_2Cl$ $CH_2F-CH_2-CH_2-CH_2Cl$ $CH_3-CH-CH_2Cl$
                                                              |
   $CH_3-CH_2-CHCl-CH_2F$ $CH_3-CHF-CHCl-CH_3$ CH_2F

                                                     $CH_3-CCl-CH_2F$
                                                            |
                                                            CH_3

   $CH_3-CF-CH_2Cl$
         |
         CH_3

ISOMERS OF COMPOUNDS CONTAINING C ATOMS, MONOVALENT ATOMS, AND DI- OR TRIVALENT ATOMS

3. When one of the divalent or trivalent elements (O, S, N, or P) is present, the following procedure is most efficient:

- Set aside monovalent atoms.
- Divide the carbon atoms into two (one O or S present) or three (one N or P present) piles. Repeat the division in all possible ways.
- From each pile of carbons, form all possible skeletons.
- Attach a skeleton from each pile to the O, S, N, or P atom. Do this for each set of piles.
- Distribute monovalent atoms in all possible ways over every C + O, C + S, C + N, or C + P skeleton generated above.

Example

Draw structural formulas for all the isomers of $C_5H_{12}O$.

Solution:

Set aside: 12 H

Divide 5 C into two piles: (a) 3 C + 2 C
(b) 4 C + 1 C
(c) 5 C + 0 C
} only three possible divisions

Form skeletons: scheme (a): C–C–C only and C–C only

scheme (b): C–C–C–C
or
C–C–C
|
C
} and C only

scheme (c): C–C–C–C–C
or
C–C–C–C
|
C
or
C
|
C–C–C
|
C

Attach to O: scheme (a): two different sites

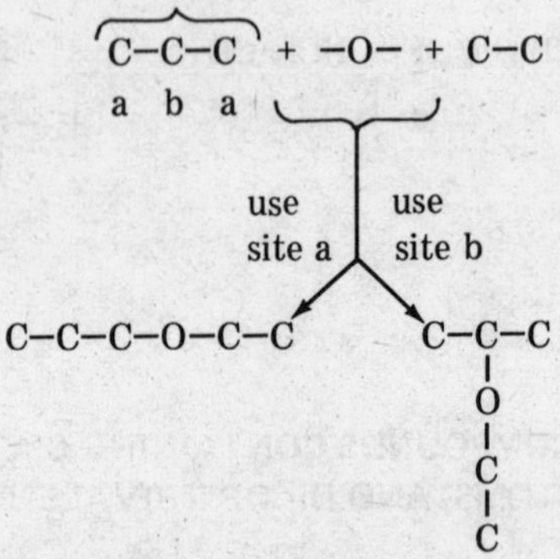

C–C–C + –O– + C–C
a b a

use site a | use site b

C–C–C–O–C–C   C–C–C
             |
             O
             |
             C
             |
             C

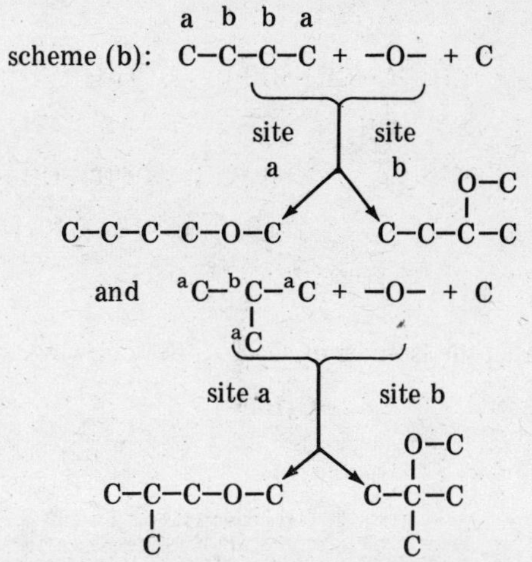

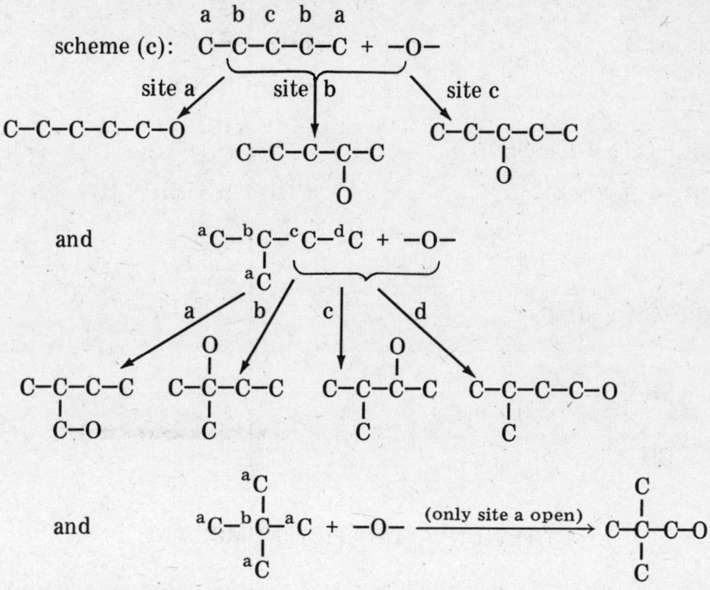

STRUCTURAL ISOMERISM

scheme (b):
$$\overset{a\ b\ b\ a}{C-C-C-C} + -O- + C$$

site a / site b

C–C–C–C–O–C   C–C–C–C with O–C on third C

and $^aC-^bC-^aC + -O- + C$ with ^aC branch

site a / site b

C–C–C–O–C with C branch   C–C–C with O–C branch and C branch

scheme (c):
$$\overset{a\ b\ c\ b\ a}{C-C-C-C-C} + -O-$$

site a: C–C–C–C–C–O
site b: C–C–C–C–C with O on middle
site c: C–C–C–C–C with O on third

and $^aC-^bC-^cC-^dC + -O-$ with ^aC branch

a: C–C–C–C with C–O branch
b: C–C–C–C with =O and C branches (O on 2nd, C on 2nd)
c: C–C–C–C with O and C branches
d: C–C–C–C–O with C branch

and $^aC-^bC-^aC + -O-$ with ^aC and ^aC branches $\xrightarrow{\text{(only site a open)}}$ C–C–C–O with C and C branches

Add 12 H atoms and collect structures:

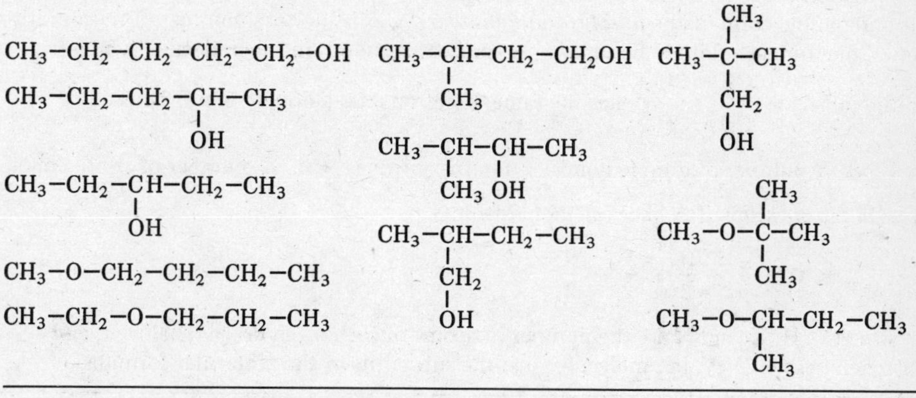

CH$_3$–CH$_2$–CH$_2$–CH$_2$–CH$_2$–OH

CH$_3$–CH$_2$–CH$_2$–CH–CH$_3$
                              |
                              OH

CH$_3$–CH$_2$–CH–CH$_2$–CH$_3$
              |
              OH

CH$_3$–O–CH$_2$–CH$_2$–CH$_2$–CH$_3$

CH$_3$–CH$_2$–O–CH$_2$–CH$_2$–CH$_3$

CH$_3$–CH–CH$_2$–CH$_2$OH
      |
      CH$_3$

CH$_3$–CH–CH–CH$_3$
      | |
      CH$_3$ OH

CH$_3$–CH–CH$_2$–CH$_3$
      |
      CH$_2$
      |
      OH

CH$_3$
 |
CH$_3$–C–CH$_3$
 |
CH$_2$
 |
OH

       CH$_3$
        |
CH$_3$–O–C–CH$_3$
        |
        CH$_3$

CH$_3$–O–CH–CH$_2$–CH$_3$
          |
          CH$_3$

$$\text{CH}_3-\underset{\underset{\text{CH}_3}{|}}{\overset{\overset{\text{OH}}{|}}{\text{C}}}-\text{CH}_2-\text{CH}_3 \qquad \text{CH}_3-\underset{\underset{\text{CH}_3}{|}}{\text{CH}}-\text{O}-\text{CH}_2-\text{CH}_3$$

$$\text{CH}_3-\underset{\underset{\text{CH}_3}{|}}{\text{CH}}-\text{CH}_2-\text{O}-\text{CH}_3$$

Exercises

Draw structural formulas for all the isomers of each of the following.

1. $C_4H_{10}S$

2. C_3H_9N

— — — — — — — — —

1. $CH_3-CH_2-S-CH_2-CH_3$

   $CH_3-CH_2-CH_2-S-CH_3$

   $\underset{CH_3}{\overset{CH_3}{>}}CH-S-CH_3$

   $CH_3-CH_2-CH_2-CH_2-SH$

   $CH_3-CH_2-\underset{\underset{SH}{|}}{CH}-CH_3$

   $\underset{CH_3}{\overset{CH_3}{>}}CH-CH_2-SH$

   $\underset{CH_3}{\overset{CH_3}{>}}C\underset{CH_3}{\overset{SH}{<}}$

2. $CH_3-\underset{\underset{CH_3}{|}}{N}-CH_3$

   $CH_3-CH_2-CH_2-NH_2$

   $CH_3-\underset{\underset{NH_2}{|}}{CH}-CH_3$

   $CH_3-CH_2-NH-CH_3$

THE UNSATURATION NUMBER, Δ

4. In generating all isomeric structural formulas for a given molecular formula, it is very helpful to know if there are any multiple bonds in the molecule, and if so, how many. This information is contained in the molecular formula; and it is obtained from the molecular formula in the form of the *unsaturation number*, Δ. Δ is the total number of unsaturation units (one for each double bond, two for each triple bond) in the molecule. A ring (e.g., in the molecule $\begin{pmatrix} CH_2-CH_2 \\ | \quad | \\ CH_2-CH_2 \end{pmatrix}$ has the same effect on Δ as a double bond. Thus

Δ = number of double bonds + number of rings + 2 × number of triple bonds

This is the *meaning* of Δ. Δ is *calculated* from

$$\Delta = C + 1 - \frac{H}{2} - \frac{X}{2} + \frac{N}{2}$$

in which C, H, X, and N are the number of atoms of carbon, hydrogen, halogen, and nitrogen, respectively, per molecule—just the subscripts in the molecular formula.

STRUCTURAL ISOMERISM

Example

Calculate Δ for $C_6H_7N_3OBr_2$.

Solution:

$$\Delta = 6 + 1 - 7/2 - 2/2 + 3/2 = 4$$

The most common values of the unsaturation number for aliphatic compounds are $\Delta = 0$ to 3. Higher values of Δ usually indicate an aromatic or aralkyl skeleton. Each benzene ring, ⬡ , contributes 4 units to Δ. A fractional value of Δ indicates that the molecular formula is an impossible one, leading to no stable structures. Negative values of Δ do have meaning; this is discussed in frame 6.

Exercises

Calculate Δ for each compound.

1. C_8H_5Cl
2. $C_{20}H_{40}$
3. $C_6H_{12}O_6$
4. NH_4Cl

— — — — — — — — —

1. 6
2. 1
3. 1
4. -1

THE USE OF Δ IN GENERATING ISOMERIC STRUCTURES

5. If we begin to draw the possible isomeric structures for an unsaturated compound, we soon discover by trial and error that the compound is unsaturated. For example, C_3H_6 ought to be C—C—C + 6 H atoms. However, one finds that 6 H atoms are too few to finish off all eight bonds in $-\overset{|}{C}-\overset{|}{C}-\overset{|}{C}-$ and make a complete structure. One then sees a solution using a multiple bond:

$$\text{>C=C-C- + 6 H} \longrightarrow \text{H}_2\text{C=CH-CH}_3$$

The difficulty with the trial-and-error method is that some structures are often overlooked. For example, C_3H_6 (above) has another isomer: cyclopropane (CH₂-CH₂-CH₂ ring). A better approach would be to have an overview, in advance, of all the possible ways a given degree of unsaturation can show up in isomeric structures. This is provided by Table 6.1, which logs all the structural combinations that can produce Δ values of 0, 1, 2, or 3. For example, there are just four ways a molecule can have $\Delta = 2$; it can have one triple bond, two double bonds, two rings, or one double bond and one ring.

Now let's see how to use Table 6.1 in a specific example.

88 HOW TO SUCCEED IN ORGANIC CHEMISTRY

Table 6.1 *Structural Elements Making Up Common Values of* Δ

Δ	Number of double bonds	Number of rings	Number of triple bonds	Examples	
0	0	0	0	$CH_3-CH_2-CH_2-CH_2-CH_2-CH_3$	C_6H_{14}
1	1	0	0	$CH_2=CH-CH_2-CH_2-CH_2-CH_3$	C_6H_{12}
1	0	1	0	$CH_2CH_2CH-CH_2-CH_2-CH_3$	
2	2	0	0	$CH_2=CH-CH=CH-CH_2-CH_3$	
2	0	2	0	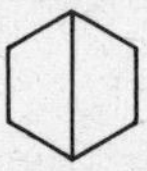 (bicyclic ring structure)	
2	0	0	1	$CH_3-CH_2-CH_2-CH_2-C{\equiv}C-H$	C_6H_{10}
2	1	1	0	$\begin{array}{c}CH_2\\|\\CH_2\end{array}\!\!\!\!\!\!C=CH-CH_2-CH_3$	
3	3	0	0	$CH_2=CH-CH=CH-CH=CH_2$	
3	0	3	0	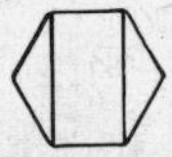 (tricyclic ring structure)	
3	0	1	1	$\begin{array}{c}CH_2\\|\\CH_2\end{array}\!\!\!\!CH-CH_2-C{\equiv}C-H$	C_6H_8
3	1	0	1	$CH_2=CH-C{\equiv}C-CH_2-CH_3$	
3	1	2	0	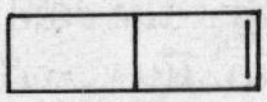 (bicyclic structure)	
3	2	1	0	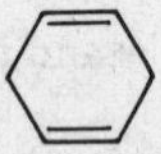 (benzene ring)	

Example

Write all possible structural formulas for C_3H_6O.

Solution:

Calculate: $\Delta = 3 + 1 - 6/2 = 1$

Interpret Δ: one double bond or one ring present

Set aside: 6 H; 1 O (*divalent*)

Divide 3 C into *two* piles in all possible ways:

  scheme 1: 2 C + 1 C
  scheme 2: 3 C + 0 C

STRUCTURAL ISOMERISM 89

Form skeletons: scheme 1: C—C and C
 scheme 2: C—C—C

Attach O: scheme 1: C—C + —O— + C ⟶ C—C—O—C

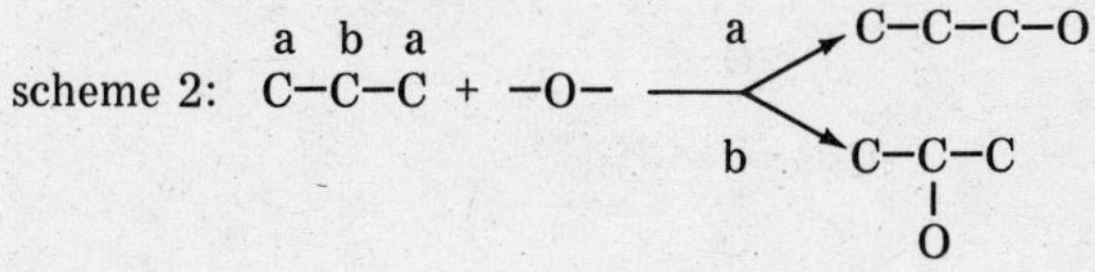

scheme 2:
a b a
C—C—C + —O— $\xrightarrow{}$ ^{a}⟶ C—C—C—O
 $_b$↘ C—C—C
|
O

Now add the information from the unsaturation number to these three carbon + oxygen skeletons:

First in the form of one double bond:

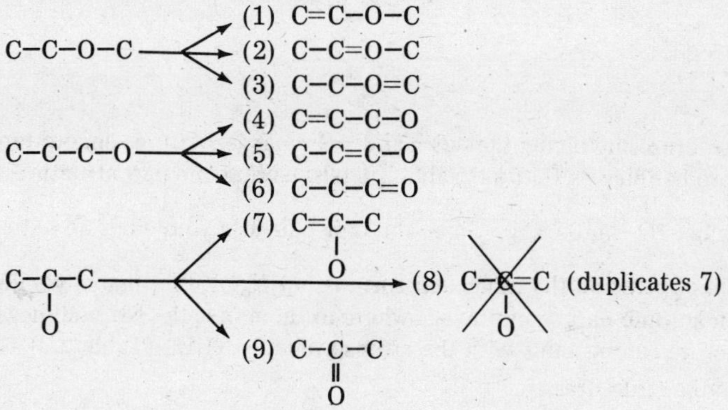

C—C—O—C ⟶
(1) C=C—O—C
(2) C—C=O—C
(3) C—C—O=C

C—C—C—O ⟶
(4) C=C—C—O
(5) C—C=C—O
(6) C—C—C=O

C—C—C ⟶
|
O
(7) C=C—C
|
O
(8) C—C̶=̶C̶ (duplicates 7)
|
O
(9) C—C—C
‖
O

Second in the form of one ring:

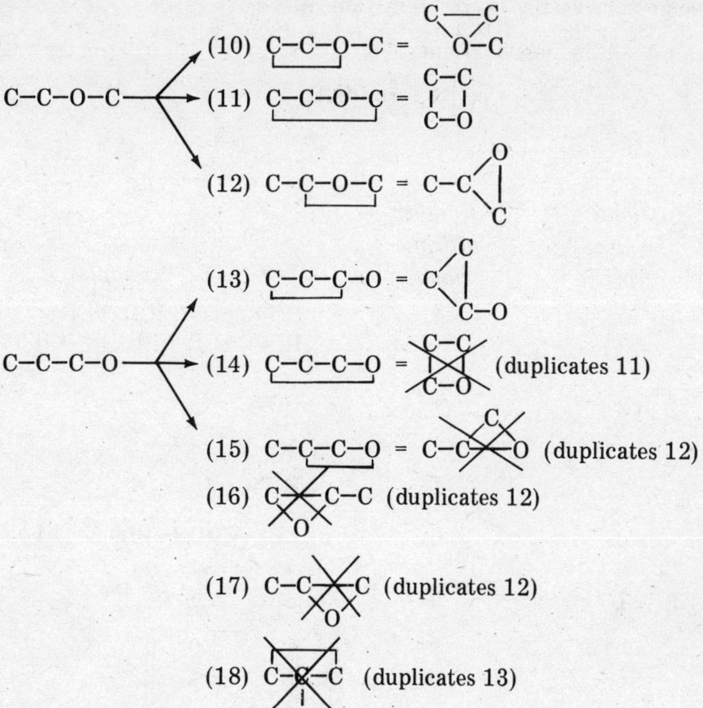

C—C—O—C ⟶
(10) C—C—O—C = C—C
 ＼ ／
 O—C

(11) C—C—O—C = C—C
| |
C—O

(12) C—C—O—C = C—C⟨^O_C

C—C—C—O ⟶
(13) C—C—C—O = C⟨$^C_{C—O}$

(14) C—C—C—O = ⨉ (duplicates 11)

(15) C—C—C—O = C—C⟨^C_O (duplicates 12)

(16) C̶—̶C—C (duplicates 12)
⨉
O

(17) C—C̶—̶C̶ (duplicates 12)
⨉
O

(18) C̶—̶C̶—̶C̶ (duplicates 13)
|
O

90 HOW TO SUCCEED IN ORGANIC CHEMISTRY

Note: A ring must be at least three-membered; a two-membered "ring" is a double bond.

Add peripheral H's and rewrite rings in normal shapes:

$CH_2=CH-CH_2-O-H$ $H_2C=C-CH_3$ $CH_3-O-CH=CH_2$
                                          |
$CH_3-CH=CH-O-H$ $O-H$

$CH_3-CH_2-CH=O$ CH_3-C-CH_3
                                          ‖
                                          O

```
        CH2
       /   \
   CH2------CH-O-H
```

```
   CH2-CH2
   |    |
   CH2-O
```

```
           CH2
          /   \
   CH3-CH-----O
```

Those possible structures involving the $=O-$ and $-\overset{|}{O}-$ part structures encountered in the example above are handled as follows. Table 2.1 tells us that the part structures would really have to be $=\overset{\oplus}{\underset{..}{O}}-$ and $-\overset{\oplus}{\underset{|}{\ddot{O}}}-$. Since the molecule as a whole has no net charge (no + or − superscript appears on the molecular formula, C_3H_6O), this charge would have to be balanced by an anionic part structure elsewhere in the molecule. No such negatively charged part structure can be built with the atoms available, C_3H_6 (Table 2.1). These are therefore impossible structures.

Exercises

1. Fill in the nine possible entries for $\Delta = 4$ in Table 6.1.

★ 2. Draw structures of all the possible isomers of

    (a) C_4H_8 (b) C_4H_6

    (c) C_4H_7F

― ― ― ― ― ― ― ―

1. Number of double bonds	Number of rings	Number of triple bonds	Examples	
0	0	2	$HC\equiv C-C\equiv C-CH_2-CH_3$	
2	0	1	$HC\equiv C-CH=CH-CH=CH_2$	
0	2	1	$HC\equiv C-\diamondsuit$ 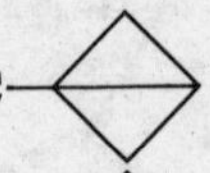	
1	1	1	$HC\equiv C-\square$ 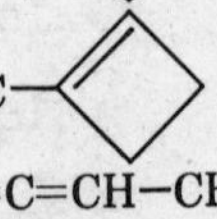	$\Big\}C_6H_6$
4	0	0	$CH_2=C=CH-CH=C=CH_2$	
3	1	0	⬡	
2	2	0	⋈	

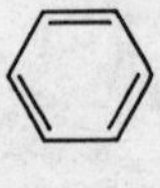

STRUCTURAL ISOMERISM

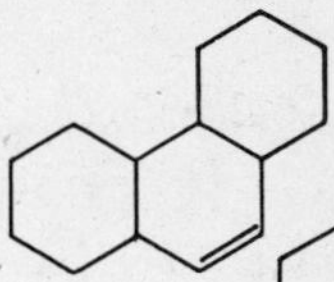

1	3	0	(fused ring structure)	$C_{13}H_{20}$
0	4	0	(fused ring structure)	$C_{19}H_{32}$

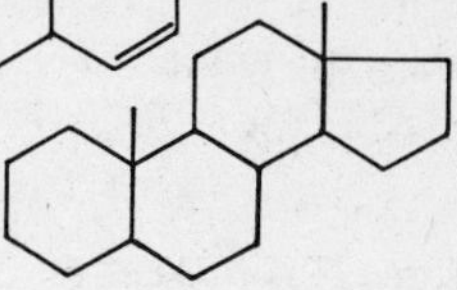

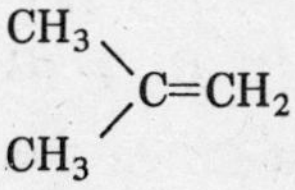

2. (a) $CH_3-CH_2-CH=CH_2$ □ $(CH_3)_2C=CH_2$

   $CH_3-CH=CH-CH_3$ △—CH_3

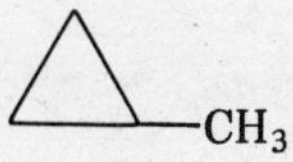

   (b) $CH_3-CH_2-C\equiv C-H$ ⊡

   $CH_3-C\equiv C-CH_3$ 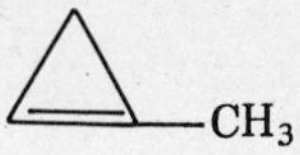 △—CH_3

   $CH_2=CH-CH=CH_2$ 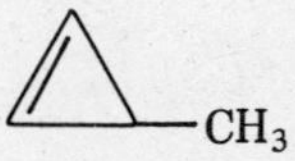 △⃫—CH_3

   $CH_2=C=CH-CH_3$ △=CH_2

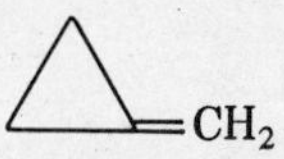

   (c) $CH_3-CH_2-CH=CHF$ $CH_3-CH=CF-CH_3$ $CH_2F\diagdown C=CH_2 / CH_3$

   $CH_3-CH_2-CF=CH_2$ □—F

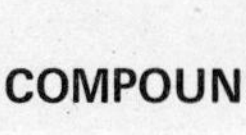

   $CH_3-CHF-CH=CH_2$ 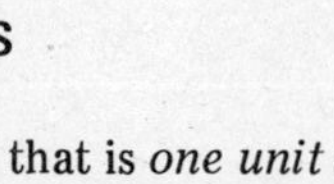   △—CH_2F $CH_3\diagdown C=CHF / CH_3$

   $CH_2F-CH_2-CH=CH_2$ ⋈—F, CH_3

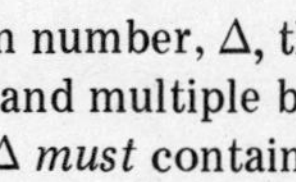

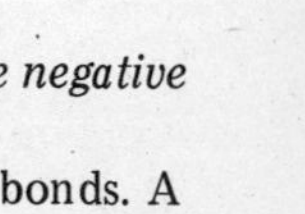

   $CH_3-CH=CH-CH_2F$ F—△—CH_3

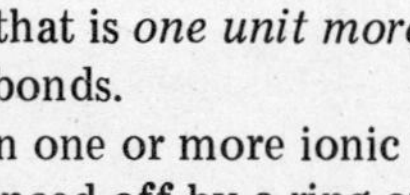

IONIC COMPOUNDS

6. An ionic structure has an unsaturation number, Δ, that is *one unit more negative* than one would expect by counting rings and multiple bonds.

A compound with a negative value of Δ *must* contain one or more ionic bonds. A compound with $\Delta = 0$ may contain an ionic bond balanced off by a ring or by a double bond's contribution of +1 to Δ. In either case, some atom capable of bearing a negative charge must be present; this is usually X or O (rarely S or N). The positive charge is usually borne on a metal atom, N, P, O, or S. The ⊕ bonding units for the latter atoms are $\overset{\oplus}{\diagup}N\diagdown$, $\overset{\oplus}{\diagup}N=$, $-\overset{\oplus}{N}\equiv$, $\overset{\oplus}{\diagup}P\diagdown$, $\overset{\oplus}{\diagup}O-$, $-\overset{\oplus}{O}=$, and $\overset{\oplus}{\diagup}S-$. The following examples illustrate the systematic method of generating isomeric structures for compounds of this type.

Example

Write all possible structural formulas for C_2H_8NI.

Solution:

Calculate: $\Delta = 2 + 1 - 8/2 - 1/2 + 1/2 = -1$; therefore, the compound must be ionic.

Identify: probably, the anion is $I^\ominus$; if so, the cation must be $C_2H_8N^\oplus$.

Separate: $2C + N + 8H$

Divide up 2 C (since N is present as $>\!\!\overset{\oplus}{N}\!\!<$, tetravalent, need four piles): the only possibilities are scheme 1: $2C + 0C + 0C + 0C$
scheme 2: $1C + 1C + 0C + 0C$

Form skeletons: scheme 1: C—C only; scheme 2: C and C

Attach $>\!\!\overset{\oplus}{N}\!\!<$: scheme 1 $\rightarrow$ C—C—$\overset{\oplus}{N}$, scheme 2 $\rightarrow$ C—$\overset{\oplus}{N}$—C

Add monovalent H's: $CH_3-CH_2-\overset{\oplus}{N}H_3$ and $CH_3-\overset{\oplus}{N}H_2-CH_3$

Add anion: $CH_3-CH_2-\overset{\oplus}{N}H_3\ I^\ominus$ and $CH_3-\overset{\oplus}{N}H_2-CH_3\ I^\ominus$

Example

Draw structures of all possible isomers of C_2H_6NI.

Solution:

Calculate: $\Delta = 2 + 1 - 6/2 - 1/2 + 1/2 = 0$

Identify possibilities for $\Delta = 0$: (1) no rings or multiple bonds; (2) since I is present to play the role of anion, $\Delta = 0$ could also mean one double bond + one ionic bond, or (3) one ring + one ionic bond.

Divide up 2 C, form skeletons, attach to N: C—C—N and C—N—C

Pursue option (1)—no rings or multiple bonds: (add monovalent atoms, 6 H + 1 I, in all possible ways) $\rightarrow$

(a) $I-CH_2-CH_2-NH_2$

(b) $CH_3-CHI-NH_2$

(c) CH_3-CH_2-NHI

(d) $I-CH_2-NH-CH_3$

(e) $CH_3-\underset{I}{N}-CH_3$

Pursue option (2)—1 double bond + 1 ionic bond—anion will be $I^\ominus$, cation will be $C_2H_6N^\oplus$:

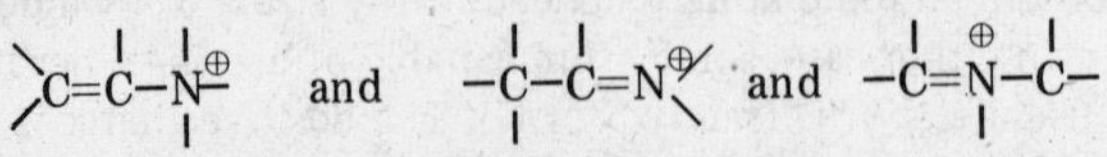

$>\!\!C=\overset{|}{C}-\overset{|}{\underset{|}{N}}\!\!\overset{\oplus}{{}}$ and $-\overset{|}{C}-\overset{|}{C}=\overset{\oplus}{N}\!\!<$ and $-\overset{|}{C}=\overset{\oplus}{\underset{|}{N}}-\overset{|}{C}-$

Add monovalent atoms = 6 H, and bring up $I^\ominus$:

(f) $CH_2=CH-\overset{\oplus}{N}H_3\ I^\ominus$

(g) $CH_3-CH=\overset{\oplus}{N}H_2\ I^\ominus$

(h) $CH_2=\overset{\oplus}{N}H-CH_3\ I^\ominus$

Pursue option (3)—1 ring + 1 ionic bond:

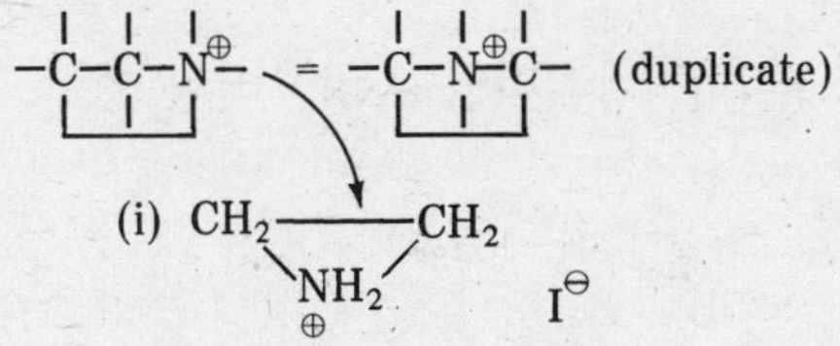

$$-\overset{|}{\underset{|}{C}}-\overset{|}{\underset{|}{C}}-\overset{|}{\underset{}{N}}\overset{\oplus}{-}\quad = \quad -\overset{|}{\underset{|}{C}}-\overset{|}{\underset{}{N}}\overset{\oplus}{=}\overset{|}{\underset{|}{C}}-\quad \text{(duplicate)}$$

(i) $CH_2\text{———}CH_2$
    $\diagdown NH_2 \diagup$
    $\underset{\oplus}{} \qquad I^{\ominus}$

Exercises

1. Under what circumstances could a compound with $\Delta = 1$ contain an ionic bond?

2. Draw structures of all the isomers of (a) C_3H_9BrO (check Table 2.1 for possible ionic structures); (b) $C_3H_{11}NO$ (consider only structures with N in the cation).

— — — — — — — — —

1. In addition to the ionic bond, two double bonds, one triple bond, two rings, or one ring and one double bond would also have to be present.

2. (a) $\Delta = 3 + 1 - 9/2 - 1/2 = -1$; therefore, probably one ionic bond and no double bonds or rings present.

$CH_3-CH_2-CH_2-\overset{\oplus}{O}H_2 \ Br^{\ominus}$ $\qquad CH_3-CH_2-\overset{\oplus}{O}H-CH_3 \ Br^{\ominus}$

$\begin{array}{c}CH_3\diagdown\\ \diagup\end{array}\overset{\oplus}{CH}-\overset{\oplus}{O}H_2 \ Br^{\ominus}$ $\qquad CH_3-\overset{\oplus}{\underset{|}{O}}-CH_3 \ Br^{\ominus}$
$CH_3\diagup \qquad\qquad\qquad\qquad\qquad\quad CH_3$

(b) $\Delta = 3 + 1 - 11/2 + 1/2 = -1$; therefore, an ionic bond is present.

Table 2.1 shows that the cationic part structure could use a $-\overset{|}{\underset{|}{N}}{}^{\oplus}, =\overset{\oplus}{N}\!\!<, -\overset{\oplus}{\underset{\ddot{}}{O}}-$, or $=\overset{\oplus}{\underset{\ddot{}}{O}}-$ bonding unit and that the anionic part structure could use a $-\overset{\ddot{}}{\underset{\ddot{}}{O}}{:}^{\ominus}$ or $-\overset{\ddot{}}{\underset{\ddot{}}{N}}{}^{\ominus}$. Since $\Delta = -1$ means one ionic bond + no double bonds, we rule out $=\overset{\oplus}{N}\!\!<$ and $-\overset{\oplus}{O}=$. The restriction given with the problem limits us to N in the cation. The bonding units in the two ions must thus be $-\overset{|}{\underset{|}{N}}{}^{\oplus}$ and $-\overset{\ddot{}}{\underset{\ddot{}}{O}}{:}^{\ominus}$. The possible distribution of the three carbon atoms between cation and anion is as follows:

Anion	Cation	
3	0	→ structures *a, b*
2	1	→ structure *c*
1	2	→ structures *d, e*
0	3	→ structures *f* to *i*

Application of the general method leads to these structures:

a $NH_4{}^{\oplus} \ CH_3-CH_2-CH_2-\overset{\ddot{}}{\underset{\ddot{}}{O}}{:}^{\ominus}$
b $NH_4{}^{\oplus} \ \begin{array}{c}CH_3\diagdown\\ \diagup\end{array}CH-\overset{\ddot{}}{\underset{\ddot{}}{O}}{:}^{\ominus}$
$\qquad\qquad\qquad\qquad\qquad\qquad\qquad\qquad CH_3\diagup$

c $CH_3\overset{\oplus}{N}H_3$ $CH_3CH_2\overset{..}{\underset{..}{O}}{:}^{\ominus}$

d $CH_3-CH_2-\overset{\oplus}{N}H_3$ $CH_3-\overset{..}{\underset{..}{O}}{:}^{\ominus}$

e $CH_3-\overset{\oplus}{N}H_2-CH_3$ $CH_3-\overset{..}{\underset{..}{O}}{:}^{\ominus}$

f $CH_3-CH_2-CH_2-\overset{\oplus}{N}H_3$ $H-\overset{..}{\underset{..}{O}}{:}^{\ominus}$

g $CH_3-CH_2-\overset{\oplus}{N}H_2-CH_3$ $H-\overset{..}{\underset{..}{O}}{:}^{\ominus}$

h $\begin{matrix}CH_3\\ \\ CH_3\end{matrix}\!\!\!>\!\!CH-\overset{\oplus}{N}H_3$ $H-\overset{..}{\underset{..}{O}}{:}^{\ominus}$

i $\begin{matrix}CH_3\\ \\ CH_3\end{matrix}\!\!\!>\!\!\overset{\oplus}{N}H-CH_3$ $H-\overset{..}{\underset{..}{O}}{:}^{\ominus}$

AROMATIC COMPOUNDS

7. Aromatic compounds do not have so many isomers as aliphatic compounds. In substituted benzenes, the only variable is the relative positions of the substituents. In the disubstituted case, there are just these possibilities:

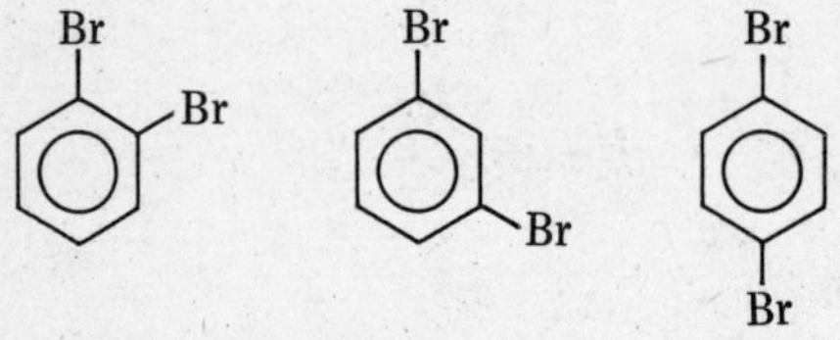

The number of isomeric trisubstituted benzenes depends on whether the substituents are alike or different:

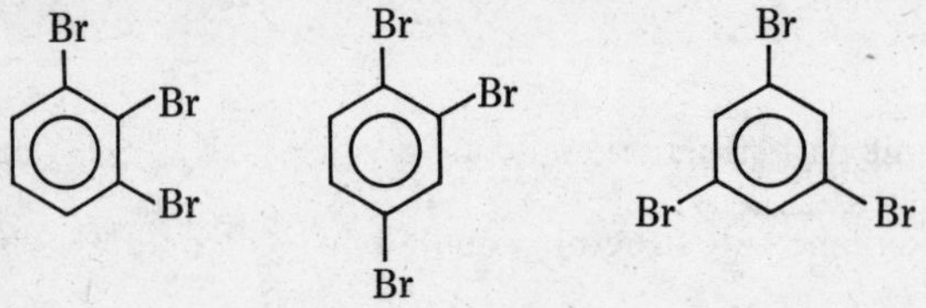

but:

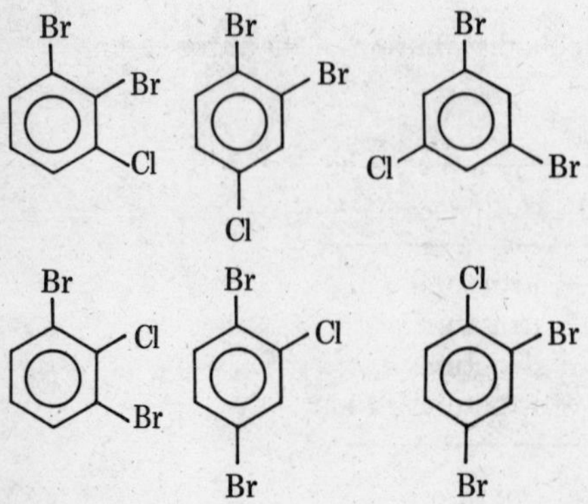

Note that

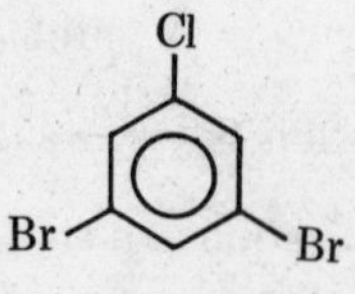

is the same as

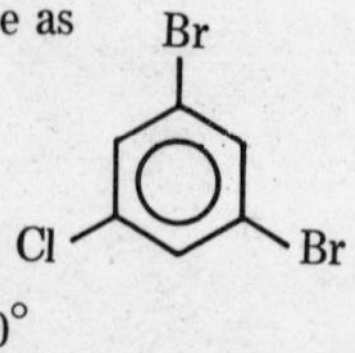

rotation by 120° turns one into the other

Naphthalene and the other aromatic rings have more possibilities for isomerism because not all positions on the rings are equivalent. Thus there are two monobromonaphthalenes in contrast to one monobromobenzene:

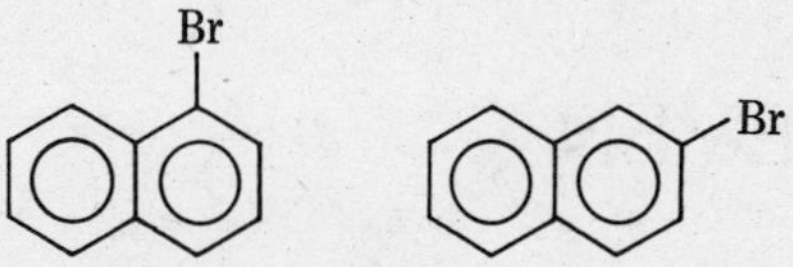

And there are 10 dibromonaphthalenes versus 3 dibromobenzenes.

★ *Exercises*

Draw structural formulas for all isomers of each compound.

1. Fluorochlorobromobenzene, C_6H_3FClBr
2. Dibromonaphthalene, $C_{10}H_6Br_2$

1.

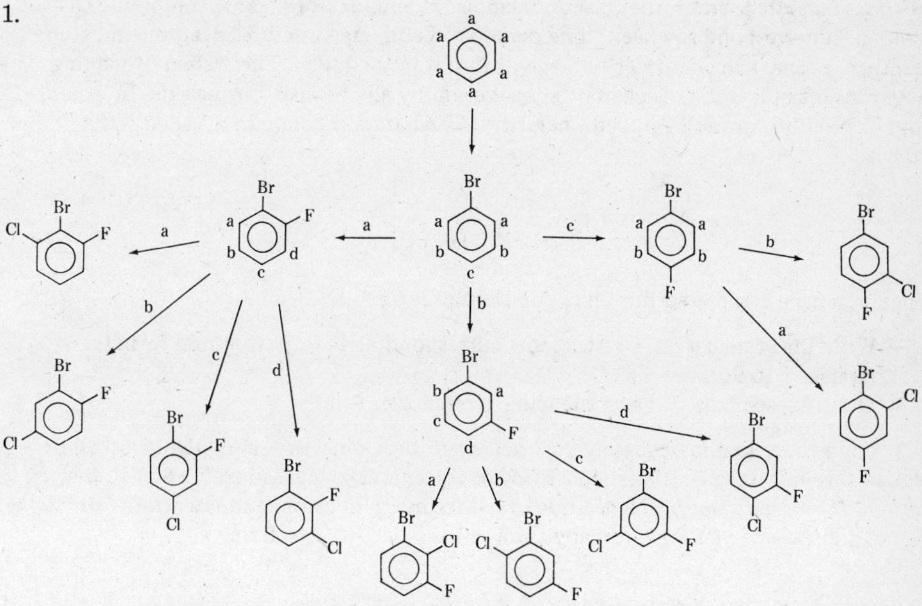

The same 10 isomers are obtained by introducing Br, F, and Cl in any other order.

2.

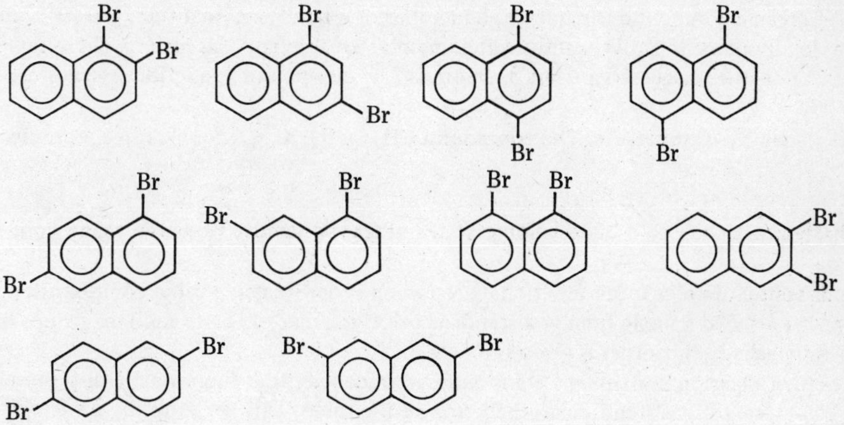

UNIT SEVEN
Systematic Nomenclature

The informal descriptions or generic names that we worked with in Unit 5 are most useful in conversation or whenever we have the structural formulas in front of us to eliminate ambiguities. The formal, systematic names presented in this unit are designed mainly for information storage and retrieval. For example, in an index of organic compounds, each name must correspond to exactly one compound and one structure. Everyone must use exactly the same name in order to place or find it in the index. The system of naming taught in this unit is the official one agreed upon by all chemists through the International Union of Pure and Applied Chemistry. The names generated are called *IUPAC names*.

OBJECTIVES

When you have completed this unit, you should be able to:

- Write the systematic (IUPAC) name corresponding to any structural formula (frames 1 to 20).
- Draw the structure if given the name (frame 22).

If you believe you may already have achieved these objectives and might skip all or part of this unit, take a self-test consisting of the exercises marked with a star in the margin. If you miss any problems, reread the frames in question and rework the problems before going on. If you are not ready for a self-test, proceed to frame 1.

GENERAL PRINCIPLES AND INITIAL DECISIONS

1. An organic compound is named as a member of a family. A qualifying phrase is added to the family name to show exactly which member of the family is meant. These families are the same ones discussed in Unit 3, frame 1. The compound's functional group determines its family membership. The compound $CH_3-CH_2-C{\overset{O}{\underset{H}{\diagdown}}}$, for example, contains an aldehyde function, $-C{\overset{O}{\underset{H}{\diagdown}}}$, and belongs to the aldehyde family. However, many compounds contain two or more functional groups. In order to assign these compounds unambiguously to a single family, a standard priority order of the functional groups has been established. This order is given in Table 7.1.

The first decision in naming a compound is to identify that functional group present in the compound that stands nearest the top of the table. This becomes the *principal*

Table 7.1 *Family and Substituent Names*

Priority	Group[a]	Family name	Substituent name
1	$-(C)\begin{smallmatrix}\diagup O\\ \diagdown OH\end{smallmatrix}$	alkanoic acid	carboxy
2	$-SO_3H$	alkanesulfonic acid	sulfo
3	$-(C)\begin{smallmatrix}\diagup O\\ \diagdown O-\end{smallmatrix}$	alkyl alkanoate	alkoxycarbonyl = $-\overset{\overset{O}{\|}}{C}-O-R$
4	$-(C)\begin{smallmatrix}\diagup O\\ \diagdown X\end{smallmatrix}$	alkanoyl halide	haloformyl
5	$-(C)\begin{smallmatrix}\diagup O\\ \diagdown NH_2\end{smallmatrix}$	alkanamide	carbamoyl
6	$-(C)\begin{smallmatrix}\diagup O\\ \diagdown H\end{smallmatrix}$	alkanal	formyl
7	$-(C)\equiv N$	alkanonitrile	cyano
8	$=O$	alkanone	oxo
9	$-OH$	alkanol	hydroxy
10	$-NH_2$	alkylamine	amino
11	$-O-$	alkyl alkyl ether	alkoxy = $-O-R$
12	$-(C)\equiv(C)-$	alkyne	yne
13	$>(C)=(C)<$	alkene	ene
14	None, $-X$, or $-NO_2$	alkane	[b]

[a] Carbon atoms in parentheses are counted as part of the skeleton when this group is the principal group, but as part of the substituent group when this group is not the top-priority one.

[b] The groups $-X$ ($-F$, $-Cl$, $-Br$, and $-I$) and $-NO_2$ never play the role of principal group. Compounds containing these as the only functional groups are named as derivatives of the alkanes. These groups, then, are substituent groups, for which we use the substituent group names halo (fluoro, chloro, bromo, and iodo) and nitro, respectively.

group. All the other groups present besides the principal group are treated as *subsidiary groups* (C=C and C≡C) or as *substituents*, that is, groups that have replaced an $-H$ group somewhere in the skeleton. This makes each name consist of a noun element (identifying the family) and various modifiers (identifying the other parts of the molecule). The noun element comes last, and the name has the form

adjective string adjective noun
    // || ||
other groups skeleton family
  present   identifier name

The symbol ⌣ indicates that two parts of the name are to be put together without a space or punctuation. The process of constructing the name is best carried out in reverse order: get the family name, add the skeleton, add the substituent groups.

Second, one must make the decision between two basic schemes of nomenclature—the aliphatic and aromatic schemes. The criterion is simple: *if the principal group is located on an aromatic ring, the compound is named by the aromatic scheme;* otherwise, it is named by the aliphatic scheme. If you are in doubt about the distinction between aromatic and aliphatic skeletons or parts of a skeleton, review Unit 3, frame 3. All the common aromatic skeletons contain one or more benzene rings, ⌬ or ⌬.

Note that this criterion produces an unambiguous decision in the case of structures that have both aliphatic and aromatic portions because "located on an aromatic ring" has a clear-cut meaning. In the compound Br—⌬—CH—CH$_3$, the Br— group is
                                                    |
                                                    Cl

located on the aromatic ring; the Cl— group is not.

We study the aromatic naming scheme in frames 13 to 18. The aliphatic scheme (frames 2 to 12) will be taken up as soon as you gain some practice in the decisions discussed above. Finally, in frame 20, we return to an overview and a general strategy of naming.

Learn Table 7.1 gradually. It is worthwhile to memorize the family names now. After repeated use of the table in working exercises, you will probably find the priority order and the group names falling into place automatically.

Exercises

Identify the principal group in each compound, and determine if it will be named by the aliphatic or the aromatic scheme.

1. $CH_3-\underset{\underset{CH_3}{|}}{\overset{\overset{Cl}{|}}{C}}-\overset{\overset{O}{||}}{C}-CH=CH_2$

2. [benzene ring with CO$_2$H, Br, OH substituents]

3. N≡C—⌬—CH$_2$—C≡C—O—CH$_3$

4. [naphthalene with CH$_3$—CH—CH$_2$—NH$_2$, F, NO$_2$ substituents]

----

1. $-\overset{\overset{O}{||}}{C}-$; aliphatic

2. $-\overset{\overset{O}{||}}{C}-OH$; aromatic

3. —C≡N; aromatic   4. —NH$_2$; aliphatic

ALIPHATIC NAMING

Step 1: Writing the Family Name and Identifying Subsidiary Groups

2. The first part of the name to write down is the noun part—the family name. It is found in the third column of Table 7.1 opposite the principal group identified according to frame 1. Several comments on the table are necessary.

(a) Each function has two names. The family name is used for the principal group only. Groups playing the role of substituents use their substituent names (the fourth column in Table 7.1).

(b) Since every carbon atom present in the structure must be accounted for in the systematic name, a convention is required for those functional groups that contain one or two carbon atoms. These carbons are shown in ( ) in the table. They are counted as part of the skeleton when the group is the principal group or subsidiary group, but as part of the substituent when the group is playing the role of substituent. Thus the word *cyano* contributes one carbon atom to a name of which it is part, but the suffix *onitrile* contributes no carbon atoms to a name. So the skeleton identifier will have to be chosen to contain one more carbon atom in the latter case than in the former.

(c) The last entry in the table is peculiar. An aliphatic compound having no functional group is an alkane, composed of $-\overset{|}{\underset{|}{C}}-$ and —H bonding units only. You can think of —H as the "functional group" in this case, for example

$$CH_3-CH_2-CH_2- \;+\; -H \longrightarrow CH_3CH_2-CH_2-H \text{ or } CH_3-CH_2-CH_3$$

   (skeleton)   (function)   (alkane)

(d) The halo and nitro groups never play the role of principal group. If they are the only functions present, then (1) there is no principal group, (2) —NO$_2$ and —X play the role of substituents, and (3) the compound is named as a member of the alkane family.

(e) Each of the family names in the table contains the syllable *alk*, which represents the skeleton. Two of the family names contain *alk* twice, once in the separate word "alkyl." This *alk* is a slot or space maintainer or dummy syllable. *Alk* is converted in the finished name to the name of the specific skeleton present in the compound named. "Alkyl" is always written as a separate word unconnected to the rest of the name.

(f) The groups C=C and C≡C are treated differently from all the rest. When either one is the only function in the molecule, it plays the role of principal group in normal fashion. However, when either is present but is not the principal group, it plays the special role of subsidiary group. The names *ene* and *yne* in the substituent name column of the table are really the subsidiary group names; these functions are never treated as substituent groups.

The following examples illustrate the process of assigning the family name. These same compounds will be carried from example to example through frames 3 to 8, while we gradually build up complete names for them.

Examples

Write the family name to be used in naming each compound.

1. $CH_3-\underset{\underset{OH}{|}}{CH}-CH_2-\underset{\underset{O}{\|}}{C}-H$ 2. $F_3C-CH_2-CH=CH-CO_2H$

3. HO−CH$_2$−CH$_2$−O−CH$_2$−CH$_2$−Cl

4. 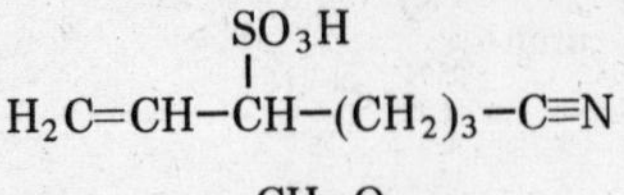 H$_2$C=CH−CH(SO$_3$H)−(CH$_2$)$_3$−C≡N

5. 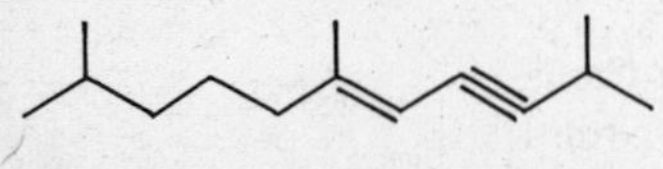 [structure: branched chain with double bond and triple bond]

6. 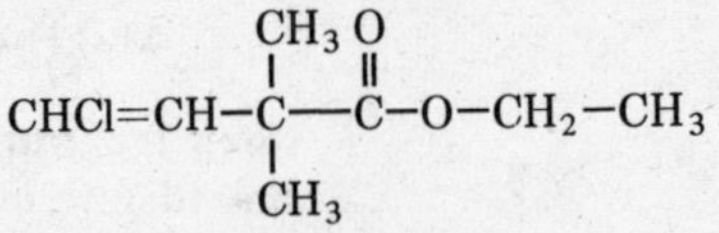 CHCl=CH−C(CH$_3$)$_2$−C(=O)−O−CH$_2$−CH$_3$ (with CH$_3$ and CH$_3$ on the central C)

Solutions:

1. alkanal ($-\overset{\underset{\|}{O}}{C}-$H takes precedence over −OH)
2. alkanoic acid
3. alkanol
4. alkanesulfonic acid
5. alkyne
6. alkyl alkanoate

Exercises

For each compound, write the family name and the name of the subsidiary group, if any.

1. H−C(=O)−CH=CH−C(=O)−CH$_3$

2. 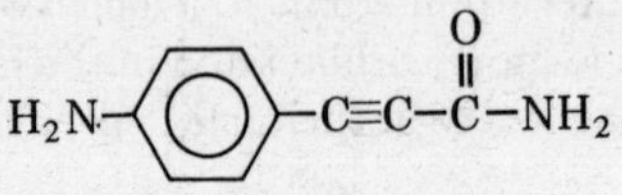 H$_2$N−⟨C$_6$H$_4$⟩−C≡C−C(=O)−NH$_2$

3. (CH$_3$)$_2$-cyclohexyl−CF$_3$

— — — — — — — — —

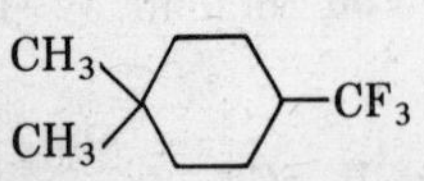

1. alkanal; subsidiary group = ene
2. alkanamide; subsidiary group = yne
3. alkane; no subsidiary group

Step 2: Identifying the Main Chain

3. The main chain consists of carbon atoms only, and must be continuous, as in

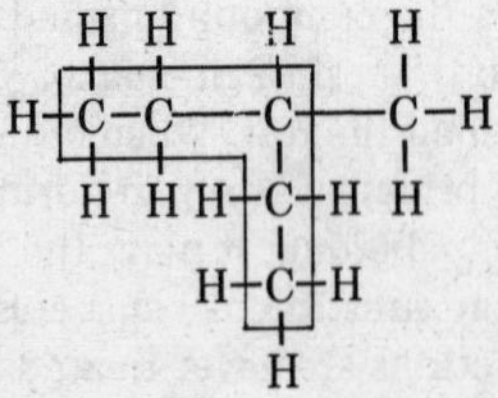

[structure showing continuous carbon chain]

not broken, as in

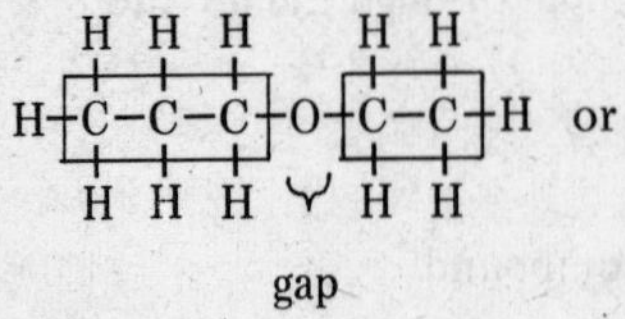

[structure with gap through O] or [structure with gap between CH$_2$ groups]

gap

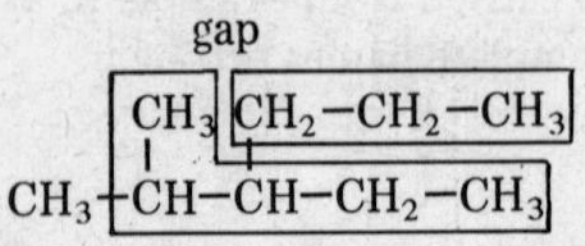

To choose it, apply the following rules. You *must* follow rule a. You use as many of rules b to d (*in order*) as you find necessary to narrow the possibilities down to a single chain:

(a) The main chain must contain or bear the principal group.
(b) The main chain must contain the maximum possible number of subsidiary groups.
(c) The main chain must contain the maximum possible number of carbon atoms.
(d) The main chain has the maximum number of substituents attached to it.

Examples

Identify the main chain in the examples from frame 2.

1. $\boxed{CH_3-CH-CH_2-C}-H$ The main chain is boxed. Rules a and c were used. Rule b
            OH   O
doesn't apply.

2. $F_3\boxed{C-CH_2-CH=CH-C}O_2H$ Rules a, b, and c were used. Note that the carbon atom of $-CO_2H$, and that of $-CHO$ in example 1, are included in the main chain (skeleton) because these are the principal groups. The carbon atoms of the $-CH=CH-$ part structure are included because this is the subsidiary group.

3. $HO\boxed{-CH_2-CH_2-}O-CH_2-CH_2-Cl$ Rules a and c were used.

4. $H_2\boxed{C=CH-CH-CH_2-CH_2-}C\equiv N$ Rules a, b, and c used. Note that the C of the
                    SO_3H
$-C\equiv N$ group is not included in the main chain. Because $-C\equiv N$ plays the role of substituent, not principal group, this C is not counted as part of the skeleton (frame 2).

5. [structure] Rules a, b, and c used.

6. $Cl-\boxed{CH=CH-C-C}\begin{smallmatrix}=O\\O-CH_2-CH_3\end{smallmatrix}$ with CH_3 substituents on the middle carbon. Rules a and b used.

Exercises

Identify the main chain in each of the following.

1. $CH_3-CH_2-CH-CH_2-CH_3$
              |
              CH_2-OH

2. $\begin{smallmatrix}CH_2\\CH_3-CH_2\end{smallmatrix}C-CH_2-CH_2-NH_2$

3. $HO_3S-CH_2-CH-CH_2-CH_2-\overset{O}{\overset{\|}{C}}-CH_3$
                 |
                 $CH=CH_2$

4. $CH_3-CH=CH-CH-CH=CH_2$
                  |
                  CO_2H

5. $Cl-CH_2-CH=CH-CH-CH=CH-CH_2-CH_3$
                      |
                      $Cl-C=O$

- - - - - - - - -

1. $\boxed{CH_3-CH_2-CH}\dashbox{-CH_2-CH_3}$ Via rules a and c. The dashed chain is longer, but it
              |
              $\boxed{CH_2}-OH$ doesn't contain the principal group.

2. [structure with CH₂, C–CH₂–CH₂–NH₂, CH₃–CH₂]   Via rules a and b. The dashed chain is longer, but it doesn't contain the subsidiary group.

3. HO₃S–CH₂–CH–CH₂–CH₂–C–CH₃ ; CH=CH₂ ; O   Via rules a and b. Rule b takes precedence over rules c and d, which would give the dashed chain.

4. CH₃–CH=CH–CH–CH=CH₂ ; CO₂H   The two marked chains both satisfy rules a and b. Rule c decides.

5. Cl–CH₂–CH=CH–CH–CH=CH–CH₂–CH₃ ; Cl–C=O   Both marked chains satisfy rules a and b. The right-hand alternative satisfies rule c; the left-hand one satisfies rule d. Rule c takes precedence

Step 3: Writing the Skeleton Identifier

4. The middle part of the name, which describes the main chain, is completed next. Since the reader knows the chain to be continuous and made up only of carbon atoms, only the length is in question. The length is not written as a number, but as the root of the name of the alkane of the same length. The alkane names are given in Table 7.2. The root is obtained by dropping the final *ane*.

Table 7.2 *Names of the Common Normal Alkanes*, $H-(CH_2)_n-H$

n	Name	n	Name	n	Name
1	methane	8	octane	15	pentadecane
2	ethane	9	nonane	16	hexadecane
3	propane	10	decane	17	heptadecane
4	butane	11	undecane	18	octadecane
5	pentane	12	dodecane	19	nonadecane
6	hexane	13	tridecane	20	eicosane
7	heptane	14	tetradecane		

For the generic term *alk* in the family name, substitute the root that describes the length of the main chain in the compound being named. For those family names that contain *alk* twice, once in the word *alkyl*, this *alk* must also be converted to denote the specific chain length of the second carbon chain in the molecule, the one attached to –O–.

Examples

For each compound in the examples in frame 3, add the skeleton identifier to the family name.

1. CH₃–CH–CH₂–C–H ; OH ; O C₄   ∴ alkanal → butanal (partial name)

2. 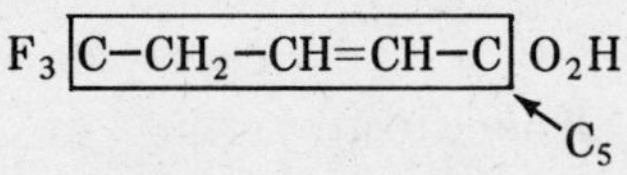 F$_3$ |C–CH$_2$–CH=CH–C| O$_2$H ∴ alkanoic acid → pentanoic acid (partial name)
   ↖C$_5$

3. 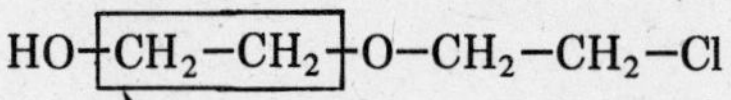 HO–|CH$_2$–CH$_2$|–O–CH$_2$–CH$_2$–Cl
   ↖C$_2$, ∴ alkanol → ethanol (partial name)

4. 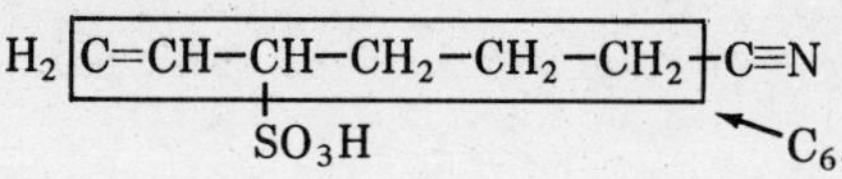 H$_2$ |C=CH–CH–CH$_2$–CH$_2$–CH$_2$|–C≡N
             |
             SO$_3$H   ↖C$_6$, ∴ alkanesulfonic acid → hexanesulfonic acid (partial name)

5. 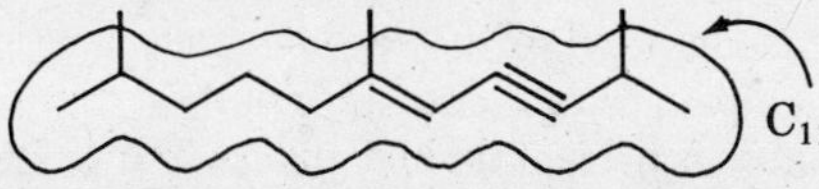 (squiggle structure) C$_{11}$, ∴ alkyne → undecyne (partial name)

6. 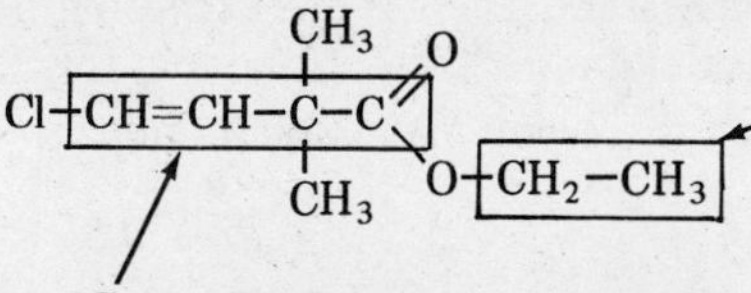 Cl–|CH=CH–C–C| ←C$_2$, ∴ alkyl → ethyl (partial name)
              CH$_3$ O
              | ∥
              C
              | O–|CH$_2$–CH$_3$|
              CH$_3$
   ↗
   C$_4$, ∴ alkanoate → butanoate

So alkyl alkanoate → ethyl butanoate (partial name).

Exercises

Write the family name for each compound and convert the skeleton identifier to its proper, specific form.

1. 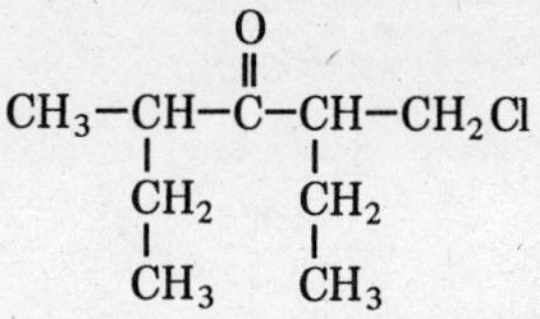 CH$_3$–CH–C–CH–CH$_2$Cl
          | ∥ |
         CH$_2$ O CH$_2$
          | |
         CH$_3$ CH$_3$

2. CH$_3$\
              C=CH–C–Cl
   CH$_3$/   ∥
                O

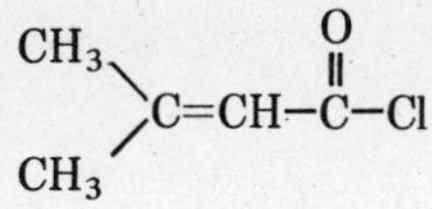

3. ⟨○⟩–CH$_2$–C–O–CH$_3$
              ∥
              O

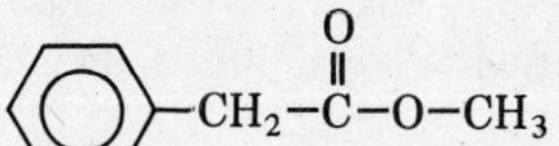

(In naming aliphatic compounds, an aromatic ring present cannot be part of the main chain—it must be a substituent.)

- - - - - - - - - -

1. 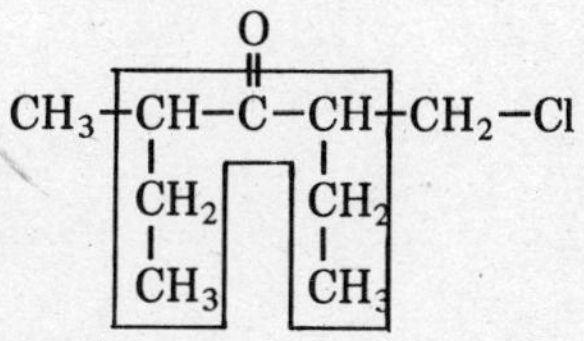 CH$_3$–|CH–C–CH|–CH$_2$–Cl   alkanone → heptanone (partial name)
          | ∥ |
         CH$_2$ O CH$_2$
          | |
         CH$_3$ CH$_3$

104   HOW TO SUCCEED IN ORGANIC CHEMISTRY

2. 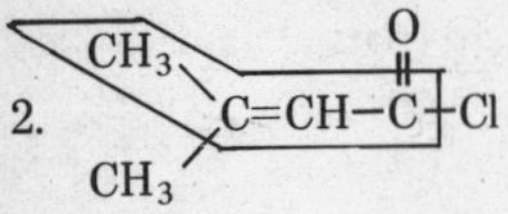 $(CH_3)_2C=CH-\overset{O}{\overset{\|}{C}}{\mid}Cl$ alkanoyl halide $\rightarrow$ butanoyl chloride (partial name)
(The generic term *halide* should be converted to the specific form at this point also.)

3. $C_6H_5-CH_2-\overset{O}{\overset{\|}{C}}-O-CH_3$ 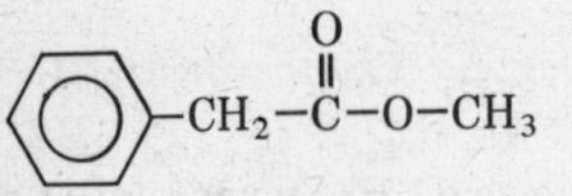   alkyl alkanoate $\rightarrow$ methyl ethanoate (partial name)

Step 4: Numbering the Main Chain

5. Starting at one end, number each successive carbon atom of the carbon chain identified in step 2. Start at whichever end gives the principal group the lowest number. If the only function present is halo or nitro, or if no function is present (as in naming a branched alkane), the longest carbon chain is numbered in whichever direction gives the lowest numbers to the alkyl, nitro, and/or halo substituents present. There is one exception to the procedure we have described. If an alkyne also contains a >C=C< function, the chain is numbered in the direction that gives the >C=C< group the lowest number.

Examples

Number the main chain in each example from frame 4.

1. 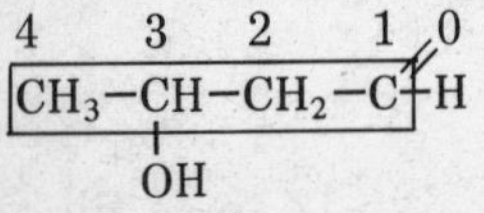 $\overset{4}{C}H_3-\overset{3}{C}H-\overset{2}{C}H_2-\overset{1}{C}\overset{0}{\mid}H$ 2. $F_3\overset{5}{C}-\overset{4}{C}H_2-\overset{3}{C}H=\overset{2}{C}H-\overset{1}{C}\mid O_2H$
      $\quad\ \ \ |$
      $\quad\ \ \ OH$

3. $HO\mid\overset{1}{C}H_2-\overset{2}{C}H_2\mid O-CH_2-CH_2-Cl$ 4. 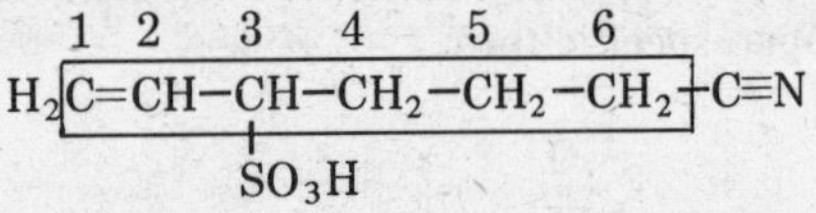 $H_2\overset{1}{C}=\overset{2}{C}H-\overset{3}{C}H-\overset{4}{C}H_2-\overset{5}{C}H_2-\overset{6}{C}H_2\mid C\equiv N$
                                                                                                        $|$
                                                                                                       SO_3H

5. (chain with numbering 11, 10, 9, 8, 7, 6, 5, 4, 3, 2, 1)   6. $\mid\overset{4}{CHCl}=\overset{3}{C}H-\overset{2}{\underset{|}{C}}-\overset{CH_3}{\underset{|}{\overset{1}{C}}}\mid\overset{0}{O}-CH_2-CH_3$
                                                                                                        $\quad\quad\quad CH_3\ \ CH_3$

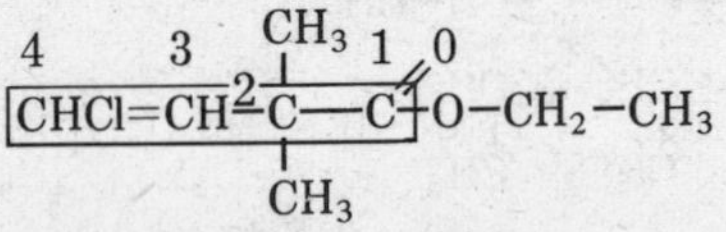

When the choice between the two possible directions of numbering the main chain has to be made on the criterion of lowest numbers for the substituents, the comparison is done as shown below for

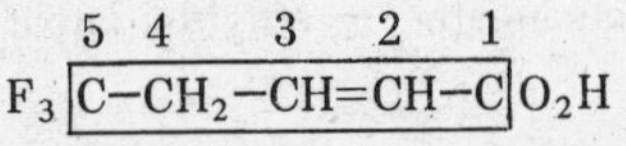

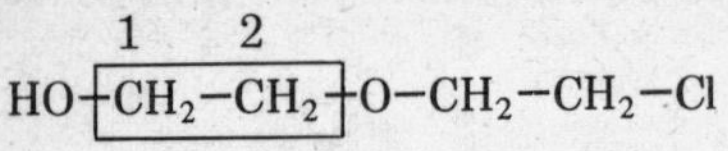

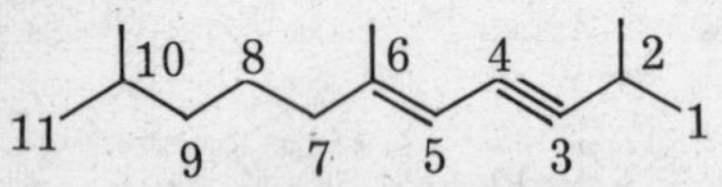

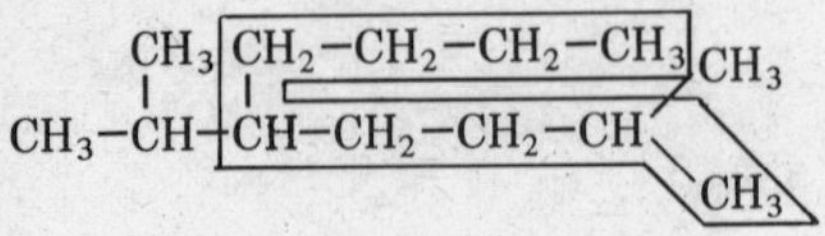

$$CH_3-\underset{|}{CH}\mid\overset{CH_3}{\underset{|}{CH}}-\overset{CH_2-CH_2-CH_2-CH_3}{\overset{|}{CH_2-CH_2-CH}}\diagdown\overset{CH_3}{\underset{CH_3}{}}$$

Locate substituents (everything that is not the principal group or part of the main chain is a substituent) and number the main chain:

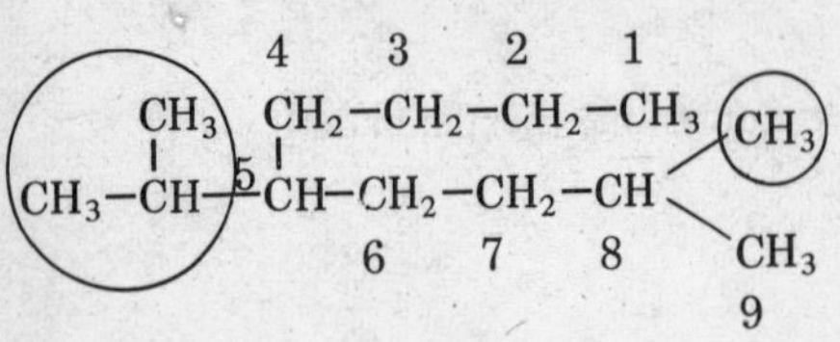

(a) $CH_3-\overset{5}{CH}-\overset{4}{\underset{|}{CH_3}}\overset{3}{\underset{|}{CH_2}}-\overset{2}{CH_2}-\overset{1}{CH_2}-CH_3\diagdown\overset{CH_3}{\underset{CH_3}{}}$ chain numbered 6,7,8,9

or

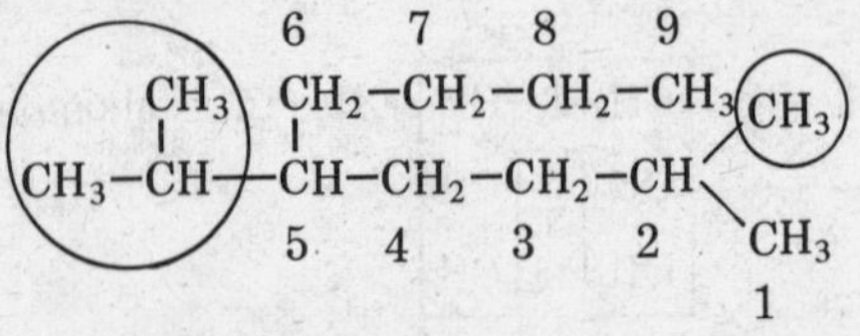

(b) $CH_3-\overset{5}{CH}-\overset{4}{\underset{|}{CH}}-\overset{3}{CH_2}-\overset{2}{CH_2}-\overset{1}{CH}\diagdown\overset{CH_3}{\underset{CH_3}{}}$ with 6,7,8,9 on other branch

(a)   (b)

Write the increasing orders of substituent locations: scheme (a): 5, 8
scheme (b): 2, 5

Choose the scheme that gives the lower number at the first point of difference.

In the present case: scheme (a) [5], 8
scheme (b) [2], 5

differ in the first comparison: adopt scheme (b)

If the choice were between: scheme (a): [2], [2], [3], 10
scheme (b): [2], [2], [4], 6

no   no   differ, choose scheme (a)
diff. diff.

Note that this is not the scheme that would be chosen on the basis of sum, which is smaller (14) for (b) than for (a) (17).

Suppose that both schemes give identical strings of substituent locations, as in

```
     4                              1
    CH₃  3   2   1                CH₃  2   3   4
      \CH–CHCl–CH₃                   \CH–CHCl–CH₃
   CH₃      (a)                  CH₃       (b)
```

2,3 ←————— substituents at —————→ 2,3

Then the scheme is chosen that gives the lower number to the substituent group whose name (Table 7.1, last column) comes earlier in alphabetical order. In this case it is <u>ch</u>loro versus <u>m</u>ethyl, and we choose scheme (a), which puts chloro in the 2 position.

Exercises

Number the main chain in each compound.

1. $\begin{array}{c} O \\ \| \\ CH_3-CH-C-CH_2-CH_3 \\ | \\ CH_2 \\ | \\ CH_3 \end{array}$

2. $\begin{array}{c} ClO \\ |\| \\ CH_3-C-C-CH-CH_3 \\ || \\ CH_2CH_2 \\ || \\ CH_3CH_3 \end{array}$

3. $\begin{array}{c} IOBr \\ |\|| \\ CH_3-C-C-C-CH_3 \\ || \\ CH_2CH_2 \\ || \\ CH_3CH_3 \end{array}$

– – – – – – – – – –

1. $\begin{array}{c} 43\overset{O}{\underset{\|}{}}21 \\ CH_3-CH-C-CH_2-CH_3 \\ | \\ 5\,CH_2 \\ | \\ 6\,CH_3 \end{array}$ The reverse numbering puts the principal group at 4 rather than 3.

2. 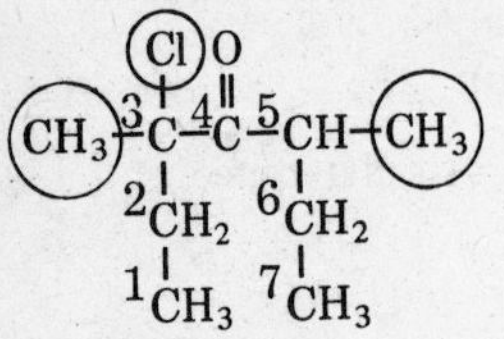
$$\text{(CH}_3\text{)}\overset{3}{\text{C}}\overset{4}{-}\overset{\overset{\text{Cl}}{|}\overset{\text{O}}{\|}}{\text{C}}\overset{5}{-}\text{CH}-\text{(CH}_3\text{)}$$
with $^2\text{CH}_2$, $^6\text{CH}_2$, $^1\text{CH}_3$, $^7\text{CH}_3$

Both directions put the principal group at 4, but the reverse direction puts the substituents at 3,5,5 rather than at 3,3,5.

3. 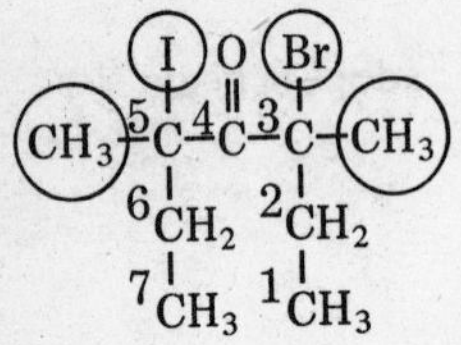
$$\text{(CH}_3\text{)}\overset{5}{\text{C}}\overset{4}{-}\overset{\overset{\text{I}}{|}\overset{\text{O}}{\|}\overset{\text{Br}}{|}}{\text{C}}\overset{3}{-}\overset{2}{\text{C}}\text{-(CH}_3\text{)}$$
with $^6\text{CH}_2$, $^2\text{CH}_2$, $^7\text{CH}_3$, $^1\text{CH}_3$

Both directions give substituent locations 3,3,5,5. The one shown gives <u>b</u>romo a lower number (3) than <u>i</u>odo (5).

Step 5: Write the Location of the Principal Group

6. Add a number, flanked with hyphens, to the beginning of the skeleton identifier; this number identifies the position of the principal function on the chain. However, omit this number if the principal function is $-\text{C}(\text{=O})\text{OH}$, $-\text{C}(\text{=O})\text{OR}$, $-\text{C}(\text{=O})\text{X}$, $-\text{C}(\text{=O})\text{NH}_2$, $-\text{C}(\text{=O})\text{H}$, or $-\text{C}\equiv\text{N}$, all of which can only occur at the end of the chain.

Examples

Do step 5 for the compounds of the previous examples.

1. $\overset{4}{\text{CH}_3}-\overset{3}{\text{CH}}-\overset{2}{\text{CH}_2}-\overset{1}{\text{C}}(\text{=O})\text{H}$, with OH on C3

    partial name: butanal, not 1-butanal—the 1 is redundant for the functions that can only be located at the end of the chain.

2. $\overset{5}{\text{F}_3\text{C}}-\overset{4}{\text{CH}_2}-\overset{3}{\text{CH}}=\overset{2}{\text{CH}}-\overset{1}{\text{CO}_2\text{H}}$

    partial name: pentanoic acid, needs no number

3. $\text{HO}-\overset{1}{\text{CH}_2}-\overset{2}{\text{CH}_2}-\text{O}-\text{CH}_2-\text{CH}_2-\text{Cl}$

    partial name: -1-ethanol; in this case the number 1 is necessary because OH can occur anywhere on the chain

4. $\overset{1}{\text{H}_2\text{C}}=\overset{2}{\text{CH}}-\overset{3}{\text{CH}}-\overset{4}{\text{CH}_2}-\overset{5}{\text{CH}_2}-\overset{6}{\text{CH}_2}-\text{C}\equiv\text{N}$, with SO_3H on C3

    partial name: -3-hexanesulfonic acid

5. 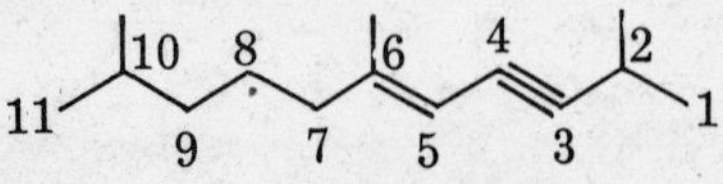 (chain numbered 11-10-9-8-7-6-5-4-3-2-1 with C≡C between 3 and 4, C=C between 5 and 6)

    partial name: -3-undecyne; in locating >C=C< or $-\text{C}\equiv\text{C}-$ functions on the chain, use the lower of the numbers of the two unsaturated carbons

The numbering has been done so as to give >C=C< the lowest possible number (see frame 5); in this case, it also happens to give $-\text{C}\equiv\text{C}-$ its lowest possible number, but this will not generally happen.

6. $\text{CHCl}=\text{CH}-\underset{\underset{\text{CH}_3}{|}}{\overset{\overset{\text{CH}_3}{|}}{\text{C}}}-\text{C}(\text{=O})\text{O}-\text{CH}_2-\text{CH}_3$

    partial name: ethyl butanoate, needs no number

Exercises

Go through steps 1 to 5 for these compounds:

1. $\begin{array}{c}CF_3\\CF_3\end{array}C=C\begin{array}{c}CF_2-CF_3\\CF_3\end{array}$

2. $CH_3-CH_2-\underset{\underset{NH_2}{|}}{\overset{\overset{CH_3}{|}}{C}}-\underset{\underset{NO_2}{|}}{\overset{\overset{CH_3}{|}}{C}}-CH_2-CH_3$

3. ⌬—$CH_2-\overset{\overset{O}{\|}}{C}-O-CH_3$

- - - - - - - - - -

1. $\begin{array}{c}CF_3\\CF_3\end{array}\overset{2}{C}=\overset{3}{C}\begin{array}{c}\overset{4}{CF_2}-\overset{5}{CF_3}\\CF_3\end{array}$ alkene → pentene → 2-pentene

2. $\boxed{\overset{1}{CH_3}-\overset{2}{CH_2}-\overset{3}{\underset{\underset{NH_2}{|}}{C}}-\overset{\overset{CH_3\;CH_3}{|\;\;\;\;|}}{\underset{\underset{NO_2}{|}}{\overset{4}{C}}}-\overset{5}{CH_2}-\overset{6}{CH_3}}$ alkylamine → hexylamine → 3-hexylamine

3. ⌬—$\boxed{\overset{C_2}{\overset{↙}{\overset{2}{CH_2}}}\overset{O}{\overset{↓}{\overset{1}{\|}}}}\overset{1}{C}$—O—$\boxed{\overset{C_1}{\overset{↘}{CH_3}}}$ alkyl   alkanoate
                                                                 ↓   ↓
                                                              methyl   ethanoate (no number needed)

Step 6: Including the Subsidiary Groups in the Name

7. Check for the presence of subsidiary groups, $>C=C<$ and $-C\equiv C-$. If there are none, go on to step 7. If one or more subsidiary groups are present, the way you include them in the name depends upon what the name already consists of.

- If there is no number already prefixed to the skeleton identifier (because the principal group is $-CO_2H$, $-CHO$, etc.), you now prefix the location of the subsidiary group there and change *an* in the skeleton identifier to *en* or *yn* (e.g., alkanal → *x*-alkenal).
- If there is already a number prefixed to the skeleton identifier, bump this number to the rear of the skeleton identifier before prefixing the location of the subsidiary group and changing *an* to *en* or *yn* (e.g., *y*-alkanone → *x*-alken-*y*-one).
- If the principal group is $C\equiv C$, there is no *an* in the name. In this case, you bump the prefixed number back to the end of the skeleton identifier, prefix in its place the location of the subsidiary $C=C$ group, and insert *ene* at the end of the skeleton identifier: *y*-alkyne → *x*-alkene-*y*-yne.

In a name containing more than one number, each group is located by the number preceding it:

4-buten-2-one

$\boxed{\overset{5}{CH_2}=\overset{4}{CH}-\overset{3}{CH_2}}-\boxed{\overset{2}{\underset{\underset{O}{\|}}{C}}-\overset{1}{CH_3}}$

If two subsidiary groups are present, follow the procedures above, but change *an* to *adien* or *adiyn* and use *two* numbers in the prefix:

$^6\text{CH}_2\overset{5}{=}\text{CH}\overset{4}{-}\text{CH}\overset{3}{=}\text{CH}\overset{2}{-}\overset{\overset{O}{\|}}{\text{C}}\overset{1}{-}\text{CH}_3$ = 3,5-hexadien-2-one

Examples

Add the principal group's location to the names in the previous examples.

1. No $\text{\textgreater}C=C\text{\textless}$ or $-C\equiv C-$; skip to step 7.

2. $\overset{5}{\text{F}_3\text{C}}-\overset{4}{\text{CH}_2}-\overset{3}{\text{CH}}=\overset{2}{\text{CH}}-\overset{1}{\text{CO}_2\text{H}}$ partial name: pentanoic acid. Include the $\text{\textgreater}C=C\text{\textless}$ function by changing *an* to *en*. As the location of the $\text{\textgreater}C=C\text{\textless}$, choose the lower of the two possible numbers—in this case, 2 rather than 3:

   pentanoic acid ⟶ -2-pentenoic acid

3. No $\text{\textgreater}C=C\text{\textless}$ or $-C\equiv C-$

4. $\overset{1}{\text{H}_2\text{C}}=\overset{2}{\text{CH}}-\underset{\underset{\text{SO}_3\text{H}}{|}}{\overset{3}{\text{CH}}}-\overset{4}{\text{CH}_2}-\overset{5}{\text{CH}_2}-\overset{6}{\text{CH}_2}-\text{C}\equiv\text{N}$ partial name: -3-hexanesulfonic acid. In this case, the prefix -3- already at the front must be bumped to the rear: -3-hexane-sulfonic acid ⟶ -1-hexene-3-sulfonic acid

5. [structure with numbers 11, 10, 9, 8, 7, 6, 5, 4, 3, 2, 1] partial name: -3-undecyne

   This alkyne has a subsidiary group ($\text{\textgreater}C=C\text{\textless}$) at position 5. We bump the number -3-, insert *-ene* for the subsidiary function, and prefix its location:

   -3-undecyne ⟶ -5-undecene-3-yne

6. $\overset{4}{\text{CHCl}}=\overset{3}{\text{CH}}-\underset{\underset{\text{CH}_3}{|}}{\overset{2}{\overset{|}{\underset{}{\text{C}}}}}\overset{\text{CH}_3}{}-\overset{1}{\overset{\overset{O}{\|}}{\text{C}}}-\text{O}-\text{CH}_2-\text{CH}_3$ partial name: ethyl butanoate. $\text{\textgreater}C=C\text{\textless}$ is present at position 3. Therefore, add -3- and change *an* to *en*:

   ethyl butanoate ⟶ ethyl 3-butenoate

Exercises

Include the identity and location of the subsidiary group(s) in each name.

1. $\text{CF}_3-\text{CF}=\text{CF}-\text{CF}_2-\overset{\overset{O}{\|}}{\text{C}}-\text{NH}_2$ partial name: pentanamide

2. $\text{CH}_3-\text{CH}_2-\text{CH}_2-\text{C}\equiv\text{C}-\text{CH}_2-\text{OH}$ partial name: 1-hexanol

3. $\underset{\text{CH}_3}{\overset{\text{CH}_3}{\text{\textgreater}}}\text{CH}-\underset{\underset{\text{Cl}}{|}}{\text{C}}=\text{CH}-\text{C}\underset{\text{CH}_3}{\overset{\text{CH}_2}{\text{\textless}}}$ partial name: pentanoyl chloride
   (with C=O branch)

4. $\text{CH}_3-\text{CCl}_2-\text{CH}=\text{CHCl}-\text{CH}_2-\text{C}\equiv\text{C}-\text{H}$ partial name: 6-heptyne

SYSTEMATIC NOMENCLATURE   109

1. pentanamide ⟶ 3-pentenamide
2. 1-hexanol ⟶ 2-hexyn-1-ol
3. pentanoyl chloride ⟶ 2,4-pentadienoyl chloride
4. 6-heptyne ⟶ 3-heptene-6-yne

Step 7: Adding the Substituents

8. Now list all parts of the molecule that have not yet been put into the name. This list will make up the adjective string at the beginning of the name. Give each substituent group its proper group name from the last column of Table 7.1. Give each of these group names a numerical prefix, which is the number of the main-chain carbon atom to which it is attached, and flank the number by two hyphens (e.g., -2-bromo). If two or three of the substituent groups are the same, use the prefix di or tri before prefixing the numbers (e.g., dimethyl or trichloro). In these cases, use two or three numbers in the prefix, one for each substituent group. Separate the numbers by comma(s) and flank the numbers with hyphens (e.g., -2,3-dimethyl and -1,1,1-trichloro-). Finally, prefix the numbered substituent group names to the skeleton identifier-plus-family name, arranging the substituents in alphabetical order. Ignore di, tri, sec-, and tert- while alphabetizing, but alphabetize isopropyl, isobutyl, isopentyl, and isohexyl under i.

Examples

Add the substituents to the partial names developed in the previous examples.

1. $\overset{4}{C}H_3-\overset{3}{C}H-\overset{2}{C}H_2-\overset{1}{C}\overset{O}{\underset{}{-}}H$ One substituent, therefore write -3-hydroxy for the adjective string.

   with (OH) on C3; butanal; -3-|hydroxy ⟶ 3-hydroxybutanal (the front hyphen is always deleted)

2. $\overset{5}{(F_3C)}-\overset{4}{C}H_2-\overset{3}{C}H=\overset{2}{C}H-\overset{1}{C}O_2H$ Three substituents, all fluoro; therefore, write 5,5,5-trifluoro for the adjective string.

   -2-pentenoic acid ⟶ 5,5,5-trifluoro-2-pentenoic acid (full name)
   (partial name)

3. $\overset{1}{HO}-\overset{2}{CH_2}-CH_2-(O-CH_2-CH_2-Cl)$ -1-ethanol (partial name)

   There is only one substituent. Its name is not in Table 7.1, therefore it must be named. We will work out the name of this group in frame 10, meanwhile call it "group."
   So -1-ethanol ⟶ 2-group-1-ethanol (temporary name).

4. $\overset{1}{H_2C}=\overset{2}{CH}-\overset{3}{CH}-\overset{4}{CH_2}-\overset{5}{CH_2}-\overset{6}{CH_2}-(C\equiv N)$ -1-hexene-3-sulfonic acid (partial name)
   $\quad\quad\quad\quad\;\;|$
   $\quad\quad\quad\quad SO_3H$

   The only substituent is —C≡N (cyano). Therefore, 6-cyano-1-hexene-3-sulfonic acid (full name).

5. (structure with numbering 11-10-9-8-7-6-5-4-3-2-1)   -5-undecene-3-yne (partial name)

110 HOW TO SUCCEED IN ORGANIC CHEMISTRY

There are three substituents, all methyl. Therefore,

2,6,10-trimethyl-5-undecene-3-yne (full name)

6.
```
       H   4   3  CH₃  O
        \           ‖
         C=CH—²C—C—O—CH₂—CH₃     ethyl 3-butenoate (partial name)
        /         | 1
       Cl        CH₃
```

There are three substituents: two methyl groups and one chloro group. With locations, they are:

-2-methyl
-2-methyl
-4-chloro

Transform to: -2,2-dimethyl and -4-chloro

Alphabetize: -4-chloro-2,2-dimethyl

Add to the partial name and drop the front hyphen:

ethyl 4-chloro-2,2-dimethyl-3-butenoate

Exercises

Include the substituents in these partial names.

1.
$$CH_3-\overset{\overset{O}{\|}}{C}-CF_2-CF_3$$ partial name: 2-butanone

2.
$$CH_3-\underset{\underset{Cl}{|}}{\overset{\overset{Cl}{|}}{C}}-\underset{\underset{CH_3}{|}}{\overset{\overset{CH_3}{|}}{C}}-C\equiv N$$ partial name: butanonitrile

3.
$$Br-\underset{\underset{Cl}{|}}{\overset{\overset{CH_3}{|}}{C}}-\underset{\underset{Cl}{|}}{\overset{\overset{CH_3}{|}}{C}}-CH_2-CCl=CH-CH_2OH$$ partial name: 2-hepten-1-ol

— — — — — — —

1. 2-butanone ⟶ 3,3,4,4,4-pentafluoro-2-butanone

2. butanonitrile ⟶ 3,3-dichloro-2,2-dimethylbutanonitrile

3. 2-hepten-1-ol ⟶ 6-bromo-3,5,6-trichloro-5-methyl-2-hepten-1-ol

EXERCISES IN NAMING ALIPHATIC COMPOUNDS

9. The following exercises provide practice in putting the whole naming procedure together.

★ *Exercises*

Name these compound by the IUPAC system. (The group name for is phenyl.)

SYSTEMATIC NOMENCLATURE 111

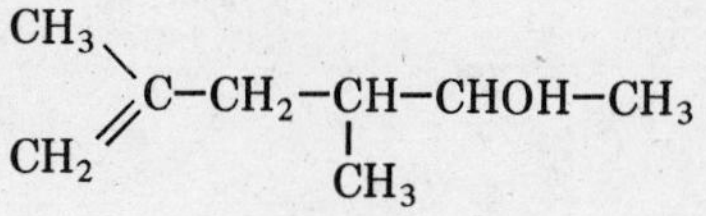

1. 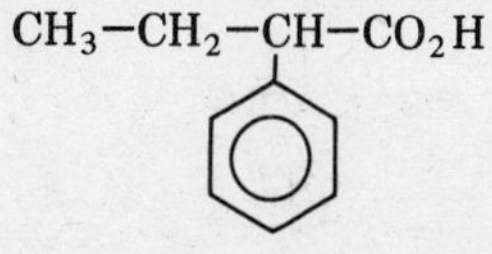 CH$_3$−CH$_2$−CH−CO$_2$H
          |
      (phenyl)

2. CH$_3$\
   $\quad$ C−CH$_2$−CH−CHOH−CH$_3$
   CH$_2$/ |
              CH$_3$

3. (structure with Cl and C=O)

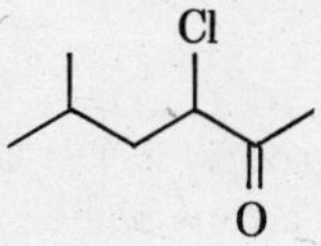

4. 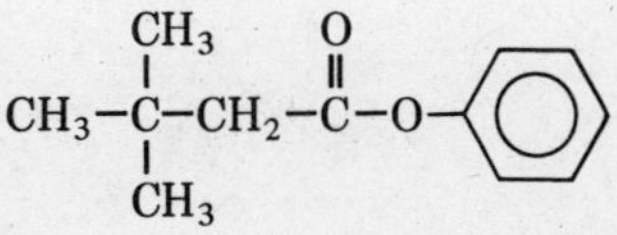 $\quad$ CH$_3$ $\quad$ O
   $\quad\quad$ | $\quad\quad\quad$ ||
   CH$_3$−C−CH$_2$−C−O−(phenyl)
   $\quad\quad$ |
   $\quad\quad$ CH$_3$

5. H−C≡C−(CH$_2$)$_{11}$−CH$_2$−SO$_3$H

6. 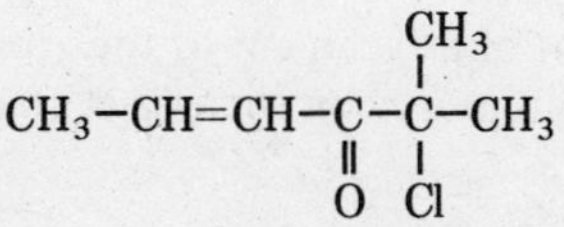 $\quad\quad\quad\quad\quad$ CH$_3$
   $\quad\quad\quad\quad\quad$ |
   CH$_3$−CH=CH−C−C−CH$_3$
   $\quad\quad\quad\quad$ || |
   $\quad\quad\quad\quad$ O Cl

7. 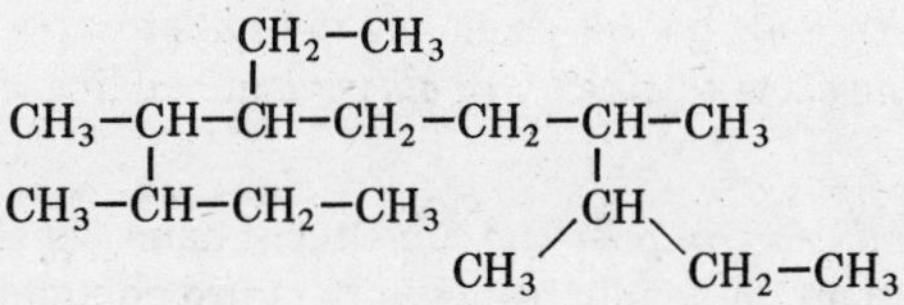 $\quad\quad\quad$ CH$_2$−CH$_3$
   $\quad\quad\quad$ |
   CH$_3$−CH−CH−CH$_2$−CH$_2$−CH−CH$_3$
   $\quad$ | $\quad\quad\quad\quad\quad$ |
   CH$_3$−CH−CH$_2$−CH$_3$ $\quad$ CH
   $\quad\quad\quad\quad$ CH$_3$/ $\quad$ \CH$_2$−CH$_3$

─ ─ ─ ─ ─ ─ ─ ─ ─

1. 2-phenylbutanoic $\quad\quad\quad\quad$ 2. 3,5-dimethyl-5-hexen-2-ol
3. 3-chloro-5-methyl-2-hexanone $\quad\quad$ 4. phenyl 3,3-dimethylbutanoate
5. 13-tetradecyne-1-sulfonic acid $\quad\quad$ 6. 2-chloro-2-methyl-4-hexen-3-one
7. 5-ethyl-3,4,8,9-tetramethylundecane

ALIPHATIC GROUP NAMES

10. In step 7 of the scheme we have been discussing, you require names for all groups that are substituents on the main chain. Some of these group names are found in Table 7.1. Others are simple alkyl or aryl groups with the following names:

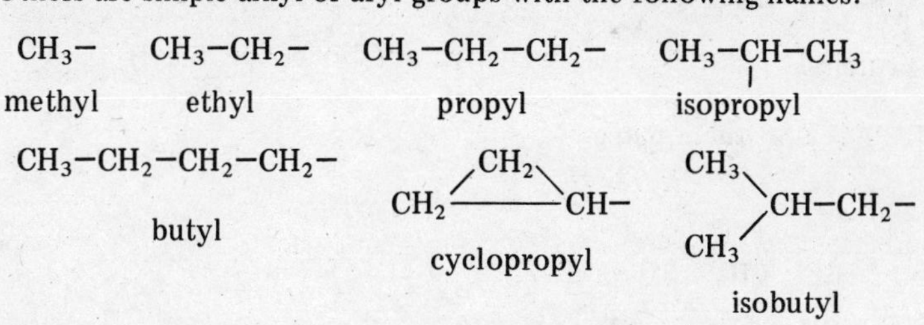

$\quad\quad$ CH$_3$− $\quad\quad$ CH$_3$−CH$_2$− $\quad\quad$ CH$_3$−CH$_2$−CH$_2$− $\quad\quad$ CH$_3$−CH−CH$_3$
$\quad\quad\quad\quad\quad\quad\quad\quad\quad\quad\quad\quad\quad\quad\quad\quad\quad\quad\quad\quad\quad\quad\quad\quad\quad$ |
$\quad\quad$ methyl $\quad\quad\quad$ ethyl $\quad\quad\quad\quad$ propyl $\quad\quad\quad\quad$ isopropyl

$\quad\quad$ CH$_3$−CH$_2$−CH$_2$−CH$_2$− $\quad\quad\quad$ CH$_2$\ $\quad\quad\quad\quad$ CH$_3$\
$\quad\quad\quad\quad\quad\quad\quad\quad\quad\quad\quad\quad\quad$ CH$_2$───CH− $\quad\quad\quad$ $\quad\quad$CH−CH$_2$−
$\quad\quad\quad\quad$ butyl $\quad\quad\quad\quad\quad\quad\quad\quad\quad\quad\quad\quad\quad\quad\quad\quad$ CH$_3$/
$\quad\quad\quad\quad\quad\quad\quad\quad\quad\quad\quad\quad\quad$ cyclopropyl $\quad\quad\quad\quad$ isobutyl

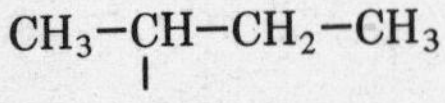

CH₃−CH−CH₂−CH₃ *sec*-butyl

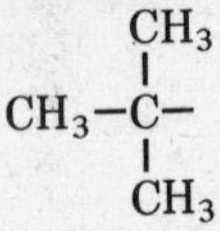

$$CH_3-\underset{CH_3}{\overset{CH_3}{\underset{|}{\overset{|}{C}}}}-$$
tert-butyl

CH₂=CH−CH₂− allyl

CH₂=CH− vinyl

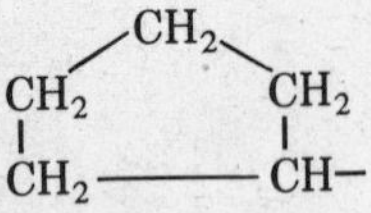

cyclopentyl

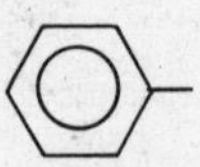

phenyl

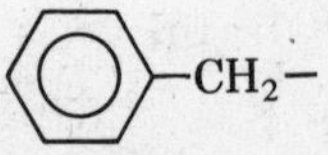

⟨⟩−CH₂− benzyl

You make the names of longer straight-chain alkyl groups (*n*-alkyl groups) in just the same way you make the word "alkyl" from the word "alkane." Take the alkane name (Table 7.2), substract "ane," add "yl." For example:

$$\text{decane } (C_{10}H_{22}) \xrightarrow[+yl]{-ane} \text{decyl}$$

The names of more complex groups must be built. The process is simpler than whole-compound naming and follows these steps:

Step A. Number the carbon with the open valence (point of attachment of the group to some molecule) 1. Locate the longest continuous carbon chain that starts with this carbon, and number it to its end.

Step B. Write the name of the *n*-alkyl group of this length.

Step C. Add the group names of all substituents on this chain and their numbers. Indicate unsaturation by changing *alkyl* to *alkenyl* or *alkynyl* and adding the position number.

There is never a principal function in naming a group, so the substituent names in the fourth column of Table 7.1 are used rather than the family names in the third column.

Example

Name the group $-CH_2-CH=CH-\overset{\overset{O}{\|}}{C}-CH_3$.

Solution:

Number: $-\overset{1}{C}H_2-\overset{2}{C}H=\overset{3}{C}H-\overset{\overset{O}{\|}}{\underset{4}{C}}-\overset{5}{C}H_3$

Write: pentyl
↓
2-pentenyl
↓
4-oxo-2-pentenyl   (*not* -4-keto-2-pentenyl, *not* 4-one-2-pentenyl)
          ↖ IUPAC substituent name

informal generic name   IUPAC *family* name

Example

Name the group $-\underset{\underset{CH_3}{\overset{|}{CH_2}}}{CCl}-CH_2-CH_2-CH_3$.

Solution:

$$\text{Number: } \underset{\underset{CH_2-CH_3}{|}}{-\overset{1}{C}Cl}-\overset{2}{C}H_2-\overset{3}{C}H_2-\overset{4}{C}H_3$$

Write:

        butyl
         ↓
1-chloro-1-ethylbutyl (*not* 3-chloro-3-hexyl)

The groups above all have their open bond on carbon. Two groups in Table 7.1 attach through O instead of C:

$$-O-R \qquad\qquad -\overset{\overset{O}{\|}}{C}-O-R$$

   alkoxy   alkoxycarbonyl

These are named just by converting "alk" to the root for the specific R group involved:

$$-O-CH_3 \quad C_1, \therefore \text{meth}yl \qquad\qquad -\overset{\overset{O}{\|}}{C}-O-CH_2-\!\!\bigcirc$$

                  ↓   benzyl ⟶ benz
           meth

so *alk*oxy ⟶ *meth*oxy   so *alk*oxycarbonyl ⟶ *benz*oxycarbonyl

If substituents are present on R, they are handled as in step C above.

Example

    Name the group $-O-CH_2-CH_2-Cl$.

Solution:

Identify type: alkoxy

Find longest carbon chain and number it: $-O-\boxed{\overset{1}{C}H_2-\overset{2}{C}H_2}-Cl$ (the carbon nearest the open bond is always 1)

Get root: $C_2, \therefore$ eth(yl) ⟶ eth

Insert root in "alkoxy": *alk*oxy ⟶ *eth*oxy

Add substituent: 2-chloroethoxy

    If the name of a group built using the procedure illustrated in this frame contains numbers, the numbers refer to the own internal numbering system of the group and bear no relation to the numbering system of the main chain of any compound in which the group plays the role of substituent. To keep the two sets of numbers separate, the group name must be enclosed in parentheses before it is used.

Example

    Use the $-O-CH_2-CH_2-Cl$ group name deduced in the preceding example to complete example 3 from frame 9.

Solution:

$$\underset{1}{HO-CH_2}-\underset{2}{CH_2}-\boxed{O-CH_2-CH_2-Cl}$$ temporary name: 2-group-1-ethanol

group = $-\boxed{O-CH_2-CH_2-Cl}$ = 2-chloroethoxy

combine:

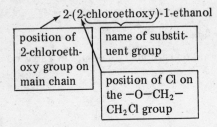

2-(2-chloroethoxy)-1-ethanol
- position of 2-chloroethoxy group on main chain
- name of substituent group
- position of Cl on the $-O-CH_2-CH_2Cl$ group

★ *Exercises*

Name the groups.

1. $CH_3-\underset{\underset{CH_3}{|}}{CH}-CH_2-CH_2-$

2. $CH_3-CH=CH-CH_2-$

3. $CH_3-\underset{\underset{|}{}}{CH}-CH_2-CH_2-CH_3$

4. $-CH_2-CH_2-C\equiv N$

5. $-O-\underset{\underset{CH_3}{|}}{CCl}-CH_2-CH_2-CH_3$

6. $-\underset{\underset{O}{\|}}{C}-O-CH_2-CH_2-CH_2-CH=CH_2$

- - - - - - - - -

1. 3-methylbutyl
2. 2-butenyl
3. 1-methylbutyl
4. 2-cyanoethyl
5. 1-chloro-1-methylbutoxy
6. 4-pentenoxycarbonyl

NAMING ETHERS

11. Of the families in Table 7.1, the ethers and amines are not named quite like the rest, in that *group* names are used to designate the carbon skeleton.

If the two alkyl groups in an ether, R—O—R, are different, take the family name *alkyl alkyl ether* and for alkyl substitute the names of the specific alkyl (or aryl) groups involved:

$$\underset{CH_3}{\overset{CH_3}{>}}CH-CH_2-O-CH_3 \text{ is isobutyl methyl ether}$$

- isobutyl
- methyl
- separate words and alphabetical order

If the two alkyl (aryl) groups are identical, do the same but drop one "alkyl" (redundant):

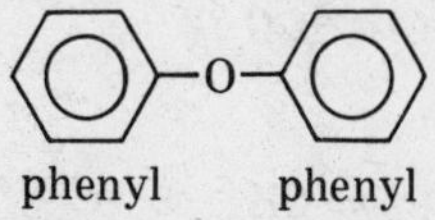

⬡—O—⬡   is   phenyl ether
phenyl phenyl

★ *Exercises*

Name the ethers.

1. $CH_3-CH_2-CH_2-O-CH_2-CH_2-CH_3$ 2. $CH_2=CH-CH_2-O-CH_2-CH_3$

3. ⬡—O—CH—CH$_2$Cl
         |
         CH$_3$

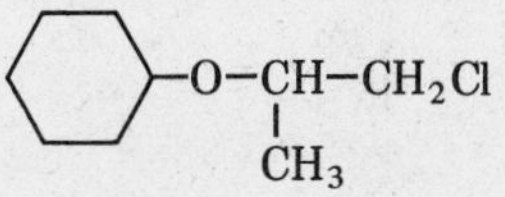

— — — — — — — — —

1. propyl ether

2. allyl ethyl ether

3. 2-chloro-1-methylethyl cyclohexyl ether

NAMING AMINES

12. The amines, $R-NH_2$, $R-NH-R'$, and $R-N(R')-R''$, require the following procedure.

    1. Choose the largest (most C atoms) of the groups on N. Write its name and add the suffix amine. If two or three groups of this same kind are present on N, add the prefix di or tri, respectively.
    2. If any of the groups on nitrogen have not yet been included in the name, add their names onto the front end of the amine name (no space or punctuation). Each should be preceded by -*N*-, which shows that the group is attached to nitrogen, not to a carbon in the original alkyl group. If two identical groups are added in this step, write *N,N*-dialkyl (e.g., *N,N*-diethyl) instead of *N*-alkyl-*N*-alkyl.

Example

Name the amines.

1. 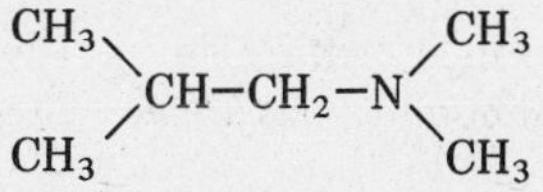 (CH$_3$)$_2$CH—CH$_2$—N(CH$_3$)$_2$ Write: isobutylamine → *N,N*-dimethylisobutylamine

2. 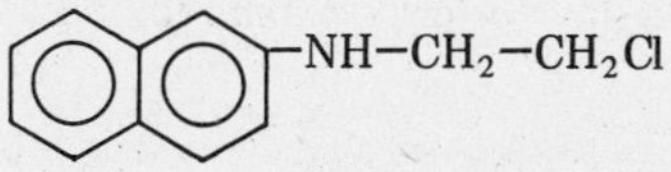 2-naphthyl—NH—CH$_2$—CH$_2$Cl   Write: 2-naphthylamine → *N*-(2-chloroethyl)-2-naphthylamine

3. 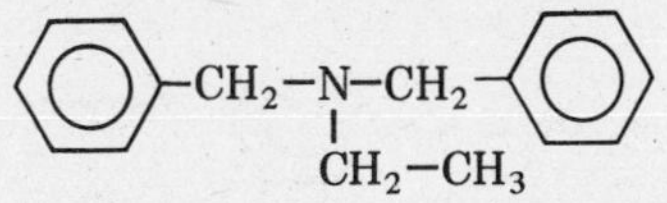 ⬡—CH$_2$—N(CH$_2$CH$_3$)—CH$_2$—⬡   Write: dibenzylamine → *N*-ethyldibenzylamine

★ *Exercises*

Name the amines.

116 HOW TO SUCCEED IN ORGANIC CHEMISTRY

1. $CH_3-CH_2-NH-CH_2-CH_3$

2. $CH_3-CH_2-\underset{\underset{CH_3}{|}}{N}-CH_2-CH_3$

3. $CH_3-CH_2-NH-CH_2-CH_2-CH_3$

4. $CH_3-CH_2-\underset{\underset{CH_2-CH_2-CH_2-CH_3}{|}}{N}-CH_2-CH_3$

5. $CH_3-\underset{\underset{CH_2-CH_3}{|}}{N}-CH_2-CH_2-CH_3$

— — — — — — — — —

1. diethylamine
2. diethylamine ⟶ N-methyldiethylamine
3. propylamine ⟶ N-ethylpropylamine
4. butylamine ⟶ N,N-diethylbutylamine
5. propylamine ⟶ N-ethyl-N-methylpropylamine
                         ↖   ↗
                     alphabetical

AROMATIC NAMING

Deciding Which Parent Compound to Use

13. Here we are naming compounds whose principal function is attached to an aromatic ring. Since there are not nearly so many different aromatic rings as there are aliphatic skeletons, the principal group + aromatic ring is treated as a unit called a *parent compound*. Aromatic compounds are then named as derivatives of the appropriate parent compound. Since more than 90% of the aromatic compounds named in beginning organic chemistry have a simple benzene skeleton, we limit the work here to the parent compounds based on benzene. The method of naming is easily extended to other skeletons.

The parent compounds are arranged in order, from the highest to the lowest priority, in Table 7.3.

This order parallels the order of functional group priorities in aliphatic naming. The halobenzenes and nitrobenzenes are not treated as parent compounds, but are named as substituted benzenes.

To begin an aromatic name, write the name of the highest-priority parent compound that can be found within the structure to be named.

Example

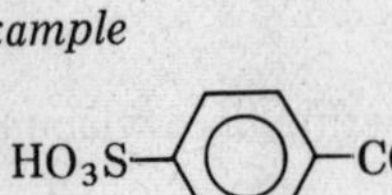

$HO_3S-\bigcirc-CO_2H$ is named as a derivative of benzoic acid, $\bigcirc-CO_2H$,

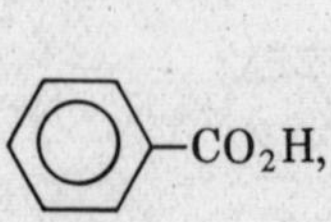

since this parent lies higher in the table than $\bigcirc-SO_3H$.

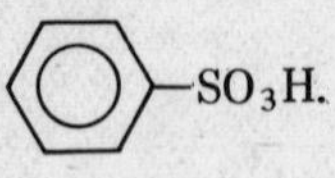

Table 7.3 *Parent Compounds, Benzene Series*

Priority	Parent	Name
1	Ph–CO$_2$H	benzoic acid
2	Ph–C(=O)–O–R	alkyl benzoate
3	Ph–SO$_3$H	benzenesulfonic acid
4	Ph–C(=O)–X	benzoyl halide
5	Ph–NH–C(=O)–CH$_3$	acetanilide
6	Ph–C(=O)–NH$_2$	benzamide
7	Ph–C(=O)–H	benzaldehyde
8	Ph–C≡N	benzonitrile
9	CH$_3$-C$_6$H$_4$-OH (three isomers)	p-cresol, m-cresol, o-cresol
10	Ph–OH	phenol
11	CH$_3$-C$_6$H$_4$-NH$_2$, etc.	p-toluidine, m-toluidine, o-toluidine
12	Ph–NH$_2$	aniline
13	CH$_3$–C$_6$H$_4$–CH$_3$, etc.	p-xylene, m-xylene, o-xylene
14	Ph–CH$_3$	toluene
15	C$_6$H$_6$	benzene

118 HOW TO SUCCEED IN ORGANIC CHEMISTRY

Exercises

Write the parent compound name for each structure.

1. 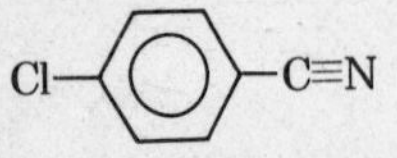 Cl—⟨ ⟩—C≡N

2. 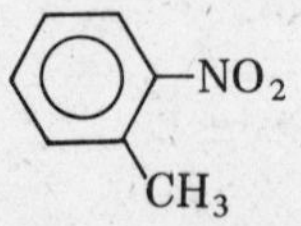 ⟨ ⟩—NO$_2$, CH$_3$

3. Br—⟨ ⟩—OH, with Br (top), CH$_3$ (bottom left), Br (bottom right)

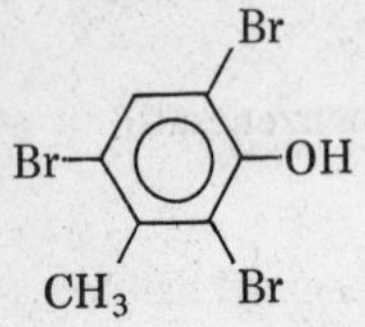

4. 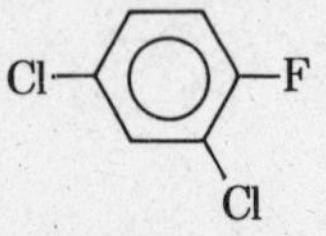 Cl—⟨ ⟩—F, with Cl

- - - - - - - - -

1. benzonitrile
2. toluene
3. *m*-cresol
4. benzene

Numbering the Ring

14. Some aromatic ring systems have their own numbering systems that are permanently in force:

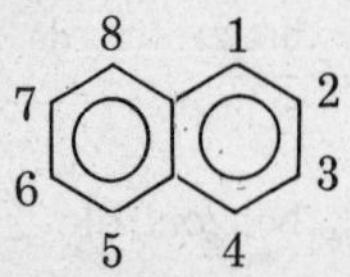

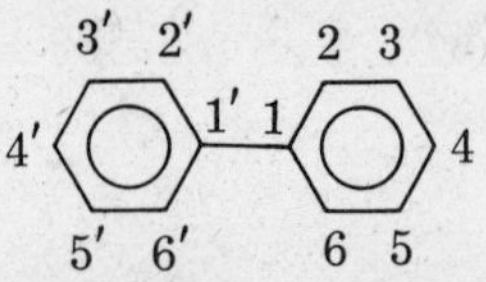

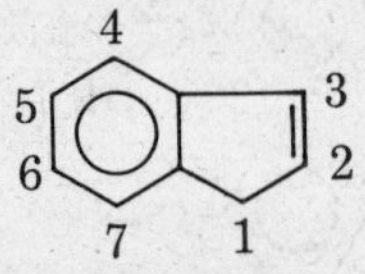

(Naphthalene numbered 1–8; biphenyl numbered 1–6 and 1′–6′; indene numbered 1–7)

The benzene ring, however, has no landmarks that distinguish the carbons from one another. The ring must be numbered, after the parent compound is identified, using the following scheme. To number a given structure, go step by step, and go to the next step only if the current one fails to decide the numbering completely.

Step 1. Carbon no. 1, and in some cases other ring carbons, are fixed by the location of the group(s) in the parent compound. These ring carbon numbers are shown on the parent compounds in Table 7.3; they are used as the first step in numbering the ring.

Examples

The following compounds *are* numbered uniquely in step 1:

1. 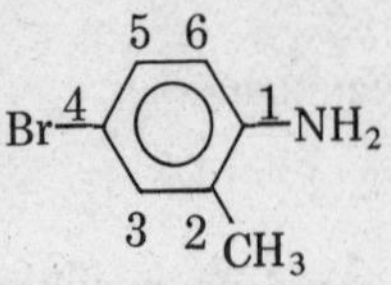 Br—⟨ ⟩—NH$_2$ (positions 5,6 top; 3,2 bottom with CH$_3$; Br at 4, NH$_2$ at 1) — The numbering is completely spelled out in the parent compound structure.

2. 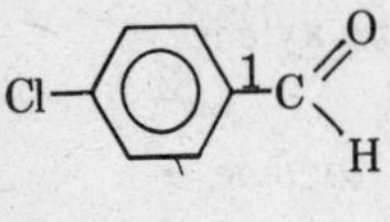 Cl—⟨ ⟩—CHO — Here the numbering is not completely fixed in the parent, but whether one proceeds clockwise or counterclockwise to number the rest of the ring starting from carbon no. 1, the result is the same: Cl is on carbon no. 4:

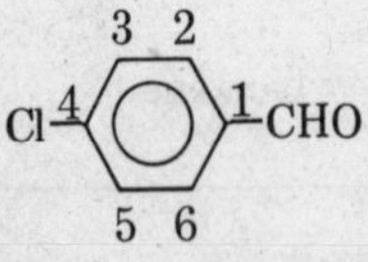

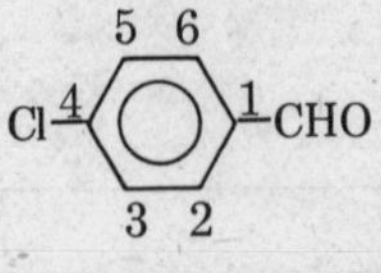

Cl—⟨ ⟩—CHO (3,2 top; 5,6 bottom; Cl at 4, CHO at 1)   Cl—⟨ ⟩—CHO (5,6 top; 3,2 bottom; Cl at 4, CHO at 1)

SYSTEMATIC NOMENCLATURE

Examples

The following compounds are *not* numbered uniquely in step 1:

1. 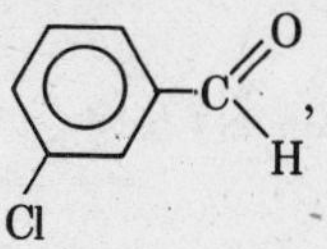 [3-chlorobenzaldehyde structure], which could be [structure numbered 1,2,3,4,5,6 with Cl at 5] or [structure numbered with Cl at 3]

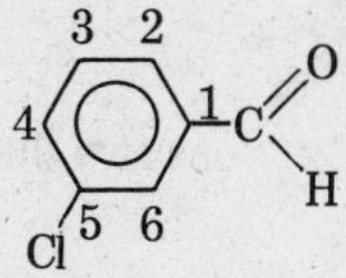

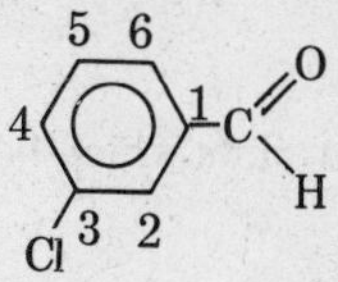

   5-chlorobenzaldehyde   3-chlorobenzaldehyde

2. 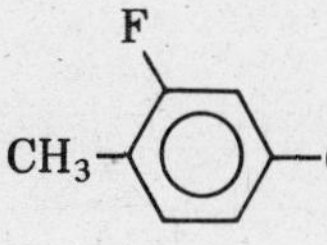 CH₃–[F-substituted ring]–CH₃, which could be CH₃–[numbered 1,2,3,4,5,6 with F at 3]–CH₃ or CH₃–[numbered with F at 5]–CH₃

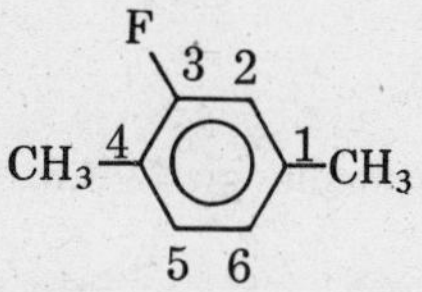

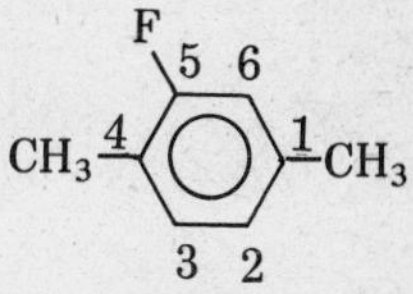

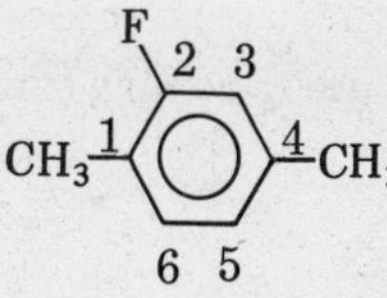

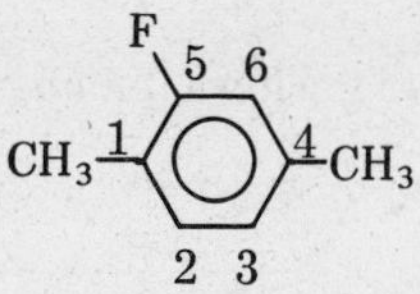

   or CH₃–[numbered 1,2,3,4,5,6 with F at 2]–CH₃ or CH₃–[numbered with F at 6]–CH₃

3. F–[ring with NO₂]–Cl, which could be numbered in six different ways.

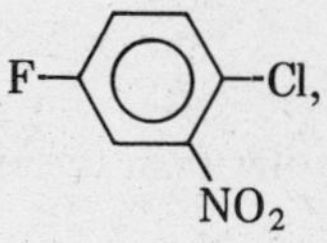

Step 2. Start from carbon no. 1 and number in the direction that gives the lowest set of numbers (judged by 1:1 comparison, as in aliphatic naming).

Examples

The following compounds are numbered uniquely in step 2:

1. 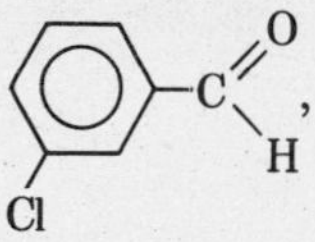 [structure], which must be [numbered with Cl at 3] (1,3 vs. 1,5 for [alternative numbering with Cl at 5])

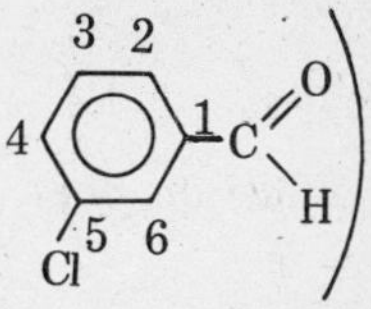

2. 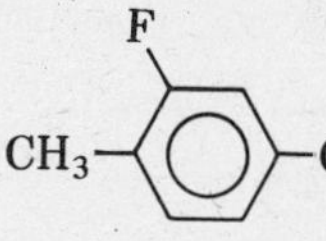 CH₃–[F ring]–CH₃, which must be CH₃–[numbered 1,2,3,4,5,6 with F at 2]–CH₃ (1,2,4 vs. 1,3,4 or 1,4,5 or 1,4,6)

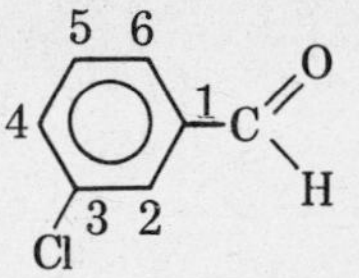

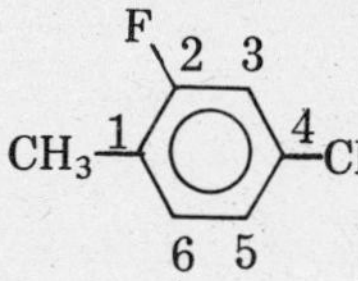

3. 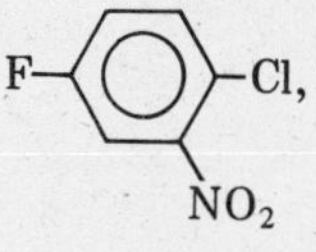 F–[ring with NO₂]–Cl, which must be F–[numbered with NO₂ at 2]–Cl (1,2,4) rather than

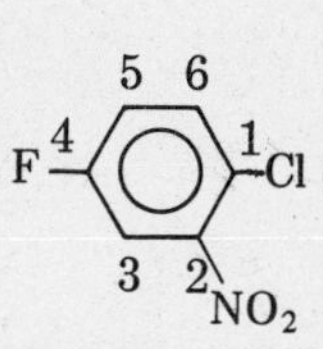

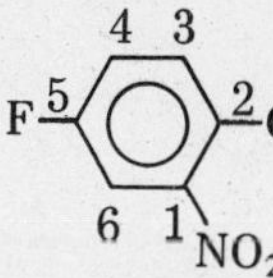

   F–[numbered 1,2,3,4,5,6]–Cl (1,2,5), F–[numbered]–Cl (1,3,4), or any other

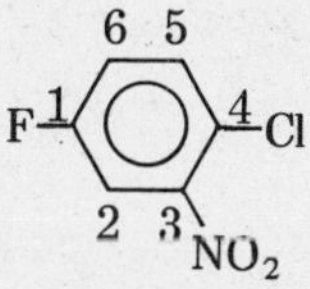

Examples

The following compounds are *not* numbered uniquely in step 2:

1. 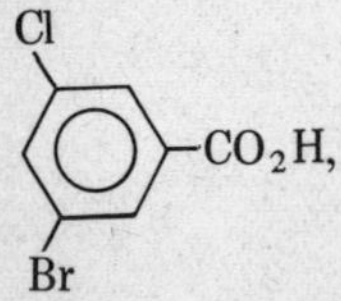 (benzene ring with Cl, Br, CO_2H)—CO_2H, in which both directions give the same set of numbers for the substituents, namely 1,3,5.

2. 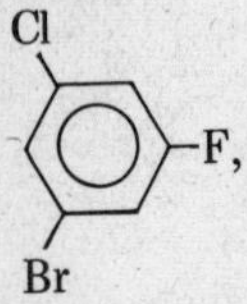 (benzene ring with Cl, Br, F)—F, for which all schemes give 1,3,5.

Step 3. Use the direction that gives the lower number to the group coming earlier in alphabetical order.

Examples

This step settles (benzene with Cl, Br, CO_2H)—CO_2H as (numbered ring: Cl at 4, CO_2H at 1, Br at... with positions 5,6,4,3,2)—CO_2H (3-bromo rather than the alternative 3-chloro) and (benzene with Cl, Br, F)—F as (numbered ring 3,4,2,5,1,6)—F (1-bromo, 3-chloro, 5-fluoro).

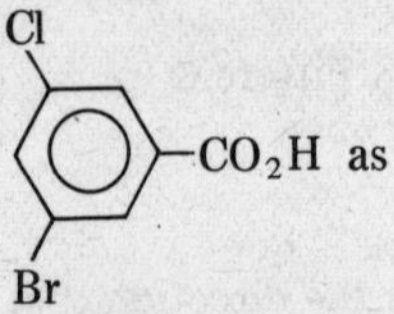

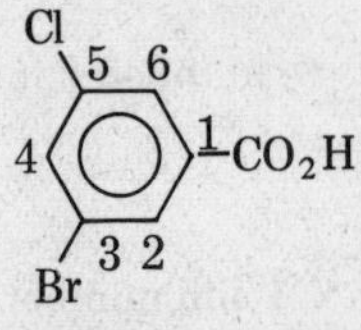

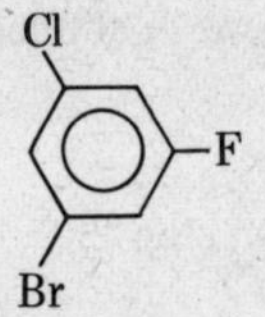

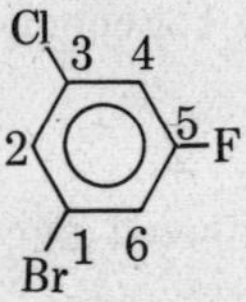

Exercises

Number the rings.

1. 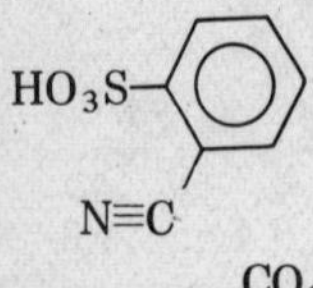 HO_3S—(benzene ring with N≡C)

2. 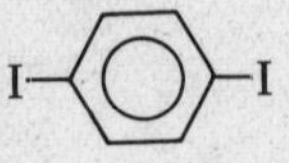 I—(benzene ring)—I

3. 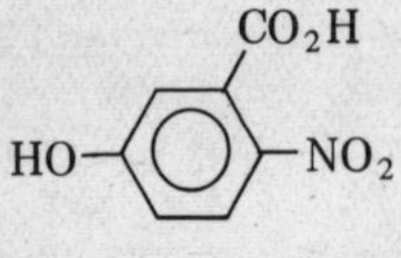 HO—(benzene ring with CO_2H and NO_2)

4. 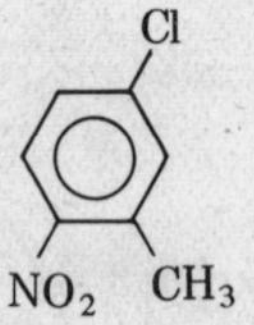 (benzene ring with Cl, NO_2, CH_3)

5. 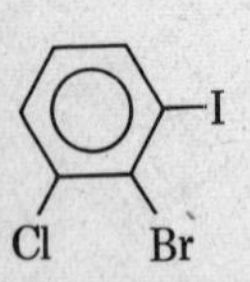 (benzene ring with I, Cl, Br)

6. 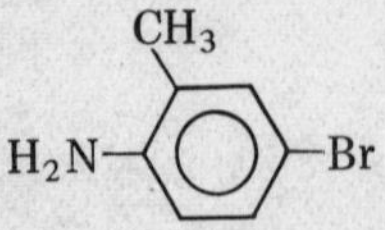 H_2N—(benzene ring with CH_3, Br)

1. HO₃S–[ring: 1, N≡C at 2, 3, 4, 5, 6]–4, via steps 1 and 2; parent = [ring]–¹SO₃H

2. I–[ring: 1, 2, 3, 4, 5, 6]–4 I, via step 2, parent = benzene

3. HO–[ring: 5, 6, 1-CO₂H, 2-NO₂, 3, 4]–, via steps 1 and 2, parent = [ring]–¹CO₂H

4. 3–[ring: 4, 5, 6, Cl at top, NO₂ at 2, CH₃ at 1]–6, via steps 1 and 2, parent = [ring]–¹CH₃

5. 6–[ring: 5, 4, 3-I, 2, 1, Cl at 1, Br at 2]–, via steps 1, 2, and 3, parent = [ring]

6. H₂N–[ring: 1, CH₃ at 2, 3, 4-Br, 5, 6]–, via step 1, parent = H₂N–[ring: 1, CH₃ at 2, 3, 4, 5, 6]

Adding the Substituent Names

15. How the completed name is assembled depends upon the total number of groups attached to the ring.

(a) *One group on the ring.* No numbers are necessary. If the group is present in the parent compound, the name is just the parent compound name, for instance *benzoic acid* for [ring]–CO₂H. If the group is a substituent on benzene-acting-as-a-parent, the substituent name is added to benzene, for instance in *chlorobenzene* for [ring]–Cl.

(b) *Two groups on the ring* (whether substituents or part of the parent compound does not matter). The name has the form

{ o- or m- or p- } { first substituent's group name (if present) } { second substituent's group name (if present) } { parent compound name }

The first and second substituents are placed in alphabetical order. Choose the prefix depending on whether the groups are in a 1,2 or 1,3 or 1,4 relationship:

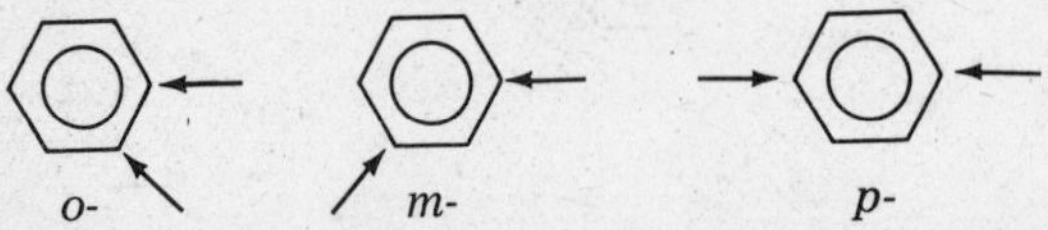

o-   m-   p-

Examples

Finish naming the compounds in exercises 1 and 2, frame 14, and $CH_3-\bigcirc-CH_3$.

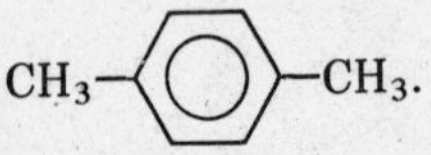

1. 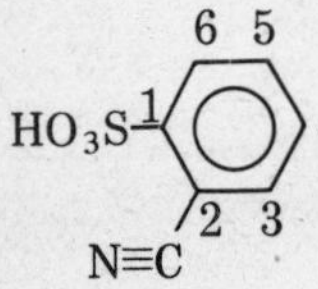 $HO_3S-\bigcirc-$ (1,2) $N\equiv C$ = o-cyanobenzenesulfonic acid

2. 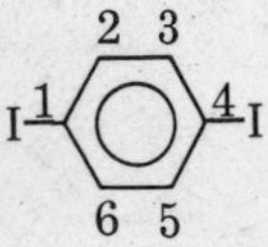 $I-\bigcirc-I$ = p-diiodobenzene (as in aliphatic naming, write diiodo instead of iodoiodo)

3. $CH_3-\bigcirc-CH_3$ 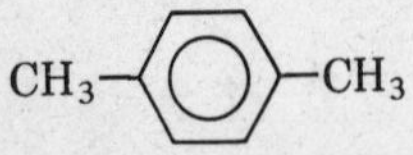 = p-xylene (no substituent names added—both are included in the parent compound)

(c) *Three or more groups on the ring.* Prefix each substituent's group name with its location number and a hyphen. String these together in front of the parent compound name.

Examples

Finish naming the compounds in examples 3 and 6 from frame 14.

1. 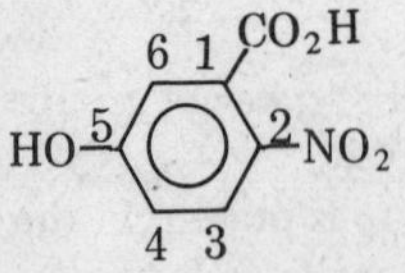 $HO-\bigcirc(CO_2H)(NO_2)$ = 5-hydroxy-2-nitrobenzoic acid

   alphabetical order ↑   ↑

   not numerical order

2. 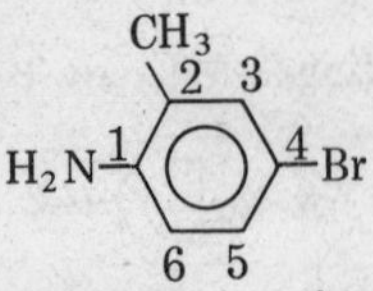 $H_2N-\bigcirc(CH_3)(Br)$ = 4-bromo-o-toluidine (a mixture of numbers and letters to locate the groups on the ring is legitimate—the letters come from the parent compound name, the numbers locate the substituents).

Exercises

1. Finish naming the compounds in exercises 4 and 5 from frame 14.

2. Finish naming each compound.

(a) 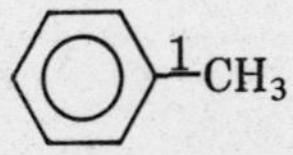 [benzene ring]—¹CH₃

(b) 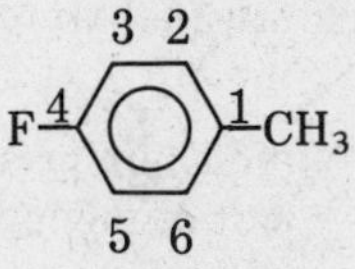 F—⁴[benzene ring ³,²]—¹CH₃
         5 6

(c) 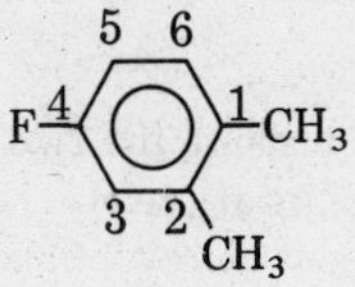 F—⁴[benzene ring ⁵,⁶ / ³,²]—¹CH₃
                                CH₃

(d) 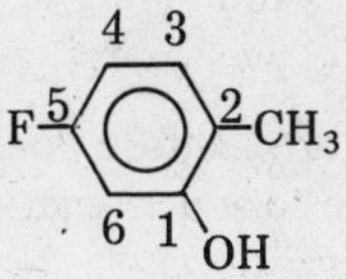 F—⁵[benzene ring ⁴,³ / ⁶,¹]—²CH₃
                                OH

- - - - - - - - -

1. 5-chloro-2-nitrotoluene and 2-bromo-1-chloro-3-iodobenzene
2. (a) toluene (b) *p*-fluorotoluene
   (c) 4-fluoro-*o*-xylene (d) 5-fluoro-*o*-cresol

Complete Example of Aromatic Compound Naming

16. The steps in naming

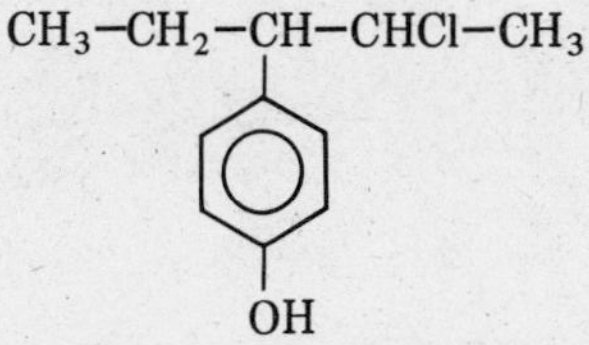

$$CH_3-CH_2-CH-CHCl-CH_3$$
          |
     [benzene ring]
          |
         OH

are:

1. Principal group = OH (Table 7.1)
2. Principal group is on the ring; ∴ use aromatic naming scheme
3. Parent compound: [benzene]—OH, phenol

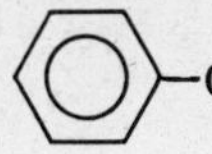

4. Number ring: 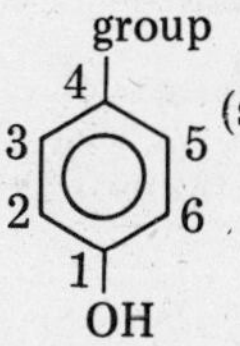

   group
    4
   3   5
   2   6
    1
    OH

   (step 1 of frame 14 suffices)

5. Get group name for $CH_3-CH_2-CH-CHCl-CH_3$
                                    |

   a. Number the point of attachment: $CH_3-CH_2-\underset{|}{\overset{1}{CH}}-CHCl-CH_3$

   b. Find longest chain starting with carbon no. 1 (C_1) and number it:

   $$CH_3-CH_2-\overset{1}{CH}-\overset{2}{CHCl}-\overset{3}{CH_3}$$
              |

   c. Substituents: CH_3-CH_2- = ethyl; $-Cl$ = chloro
   d. Group name: 2-chloro-1-ethylpropyl
6. Add group name to aromatic parent name:

2-chloro-4-(2-chloro-1-ethylpropyl)phenol

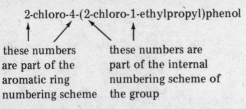

these numbers are part of the aromatic ring numbering scheme

these numbers are part of the internal numbering scheme of the group

Note that in this case chloro...chloro does not collapse to dichloro because the two chlorines are in the aliphatic and aromatic portions of the molecule, respectively. Note also the ordering of the substituents: 2-*chloro*-1-*ethylpropyl* comes later in alphabetical order than *chloro*.

Exercises in Naming Aromatic Compounds

*17. Write a systematic name for each compound.

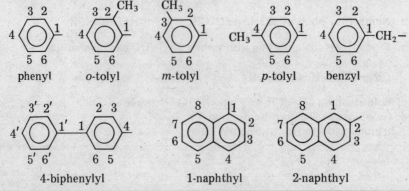

- - - - - - - - - -

1. *p*-hydroxyacetanilide
2. methyl *o*-nitrobenzoate
3. 1-chloro-2,4-dinitrobenzene
4. *m*-cyanobenzaldehyde
5. 2-chloro-6-fluoro-*p*-xylene
6. 2,3,5,6-tetrafluoro-*p*-cresol

Naming Aromatic Groups

18. The common unsubstituted aromatic groups and their permanent numbering systems are:

SYSTEMATIC NOMENCLATURE

Substituted group names are made from these by fusing the substituent's group name to the front end. If the resulting ring has one substituent, prefix *o-*, *m-*, or *p-* to show the relationship of substituent to the open bond.

Example

Name Cl—⟨phenyl ring⟩—.

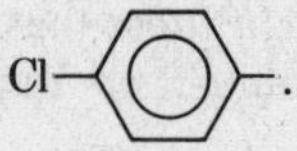

Parent = phenyl   Substituent = chloro   ∴ *p*-chlorophenyl

If the ring has two or more substituents, fuse on the group names using numbers for locations. If one of the substituents is CH$_3$, use *o*-tolyl, *m*-tolyl, or *p*-tolyl as the parent group name and attach substituent group names and numbers.

Examples

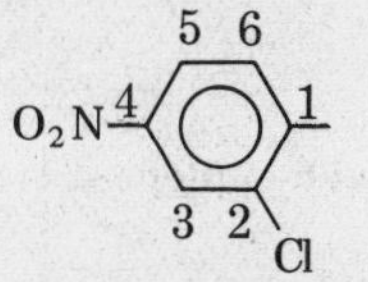

2-chloro-4-nitrophenyl 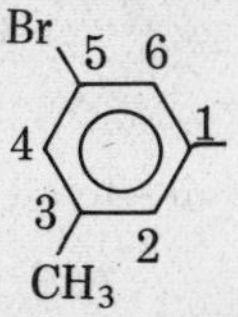   5-bromo-*m*-tolyl

★ *Exercises*

Name each group.

1. 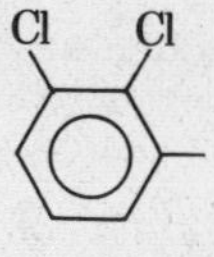 (dichlorophenyl structure)

2. 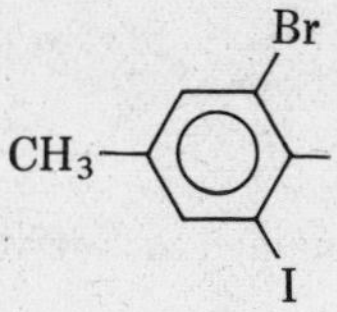 (CH$_3$, Br, I substituted ring)

3. (O$_2$N-naphthyl structure)

-- -- -- -- -- -- --

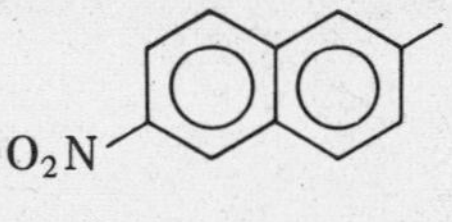

1. 2,3-dichlorophenyl   2. 2-bromo-6-iodo-*p*-tolyl

3. 6-nitro-2-naphthyl

NAMING COMPOUNDS WITH TWO IDENTICAL PRINCIPAL GROUPS

19. There are some exceptions, but most aliphatic compounds in which the principal group occurs twice are named by this slight variation on the usual scheme:

1. The top-priority criterion for choice of the main chain is that it should contain *both* principal groups.
2. The family names are modified as follows (most common cases only):

2 CO$_2$H	alkanedioic acid	2 CHO	alkanedial
2 SO$_3$H	alkanedisulfonic acid	2 C=O	alkanedione

2 CO_2R	dialkyl (or alkyl alkyl') alkanedioate	2 OH	alkanediol
2 CO_2NH_2	alkanediamide	2 C≡C	alkadiyne
2 C≡N	alkanedinitrile	2 C=C	alkadiene

3. Provide two numbers to locate the two principal groups (except those that can only occur at the ends of the main chain, in which case no numbers are **required**).

Examples

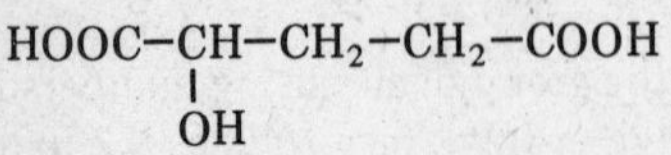

HOOC–CH–CH$_2$–CH$_2$–COOH   2-hydroxypentanedioic acid 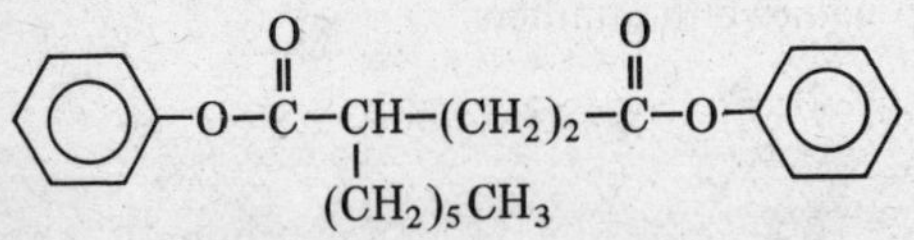
     |
    OH

C$_6$H$_5$–O–C(=O)–CH–(CH$_2$)$_2$–C(=O)–O–C$_6$H$_5$ diphenyl 2-hexylpentanedioate 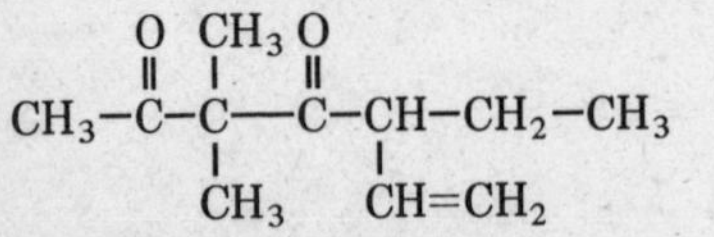
             |
            (CH$_2$)$_5$CH$_3$

CH$_3$–C(=O)–C(CH$_3$)$_2$–C(=O)–CH(CH=CH$_2$)–CH$_2$–CH$_3$ 5-ethyl-3,3-dimethyl-6-heptene-2,4-dione

Cl$_3$C–CH=CH–CH=CH–CF$_3$ 1,1,1-trichloro-6,6,6-trifluoro-2,4-hexadiene

Aromatic compounds with two identical principal groups on the aromatic ring are named exactly as all other aromatic compounds are named. The disubstituted ring acts as parent compound. Look up these parent compound names as you need them.

Examples

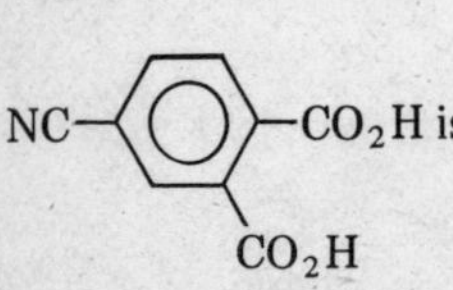

NC–C$_6$H$_4$–CO$_2$H is named as a derivative of the parent: 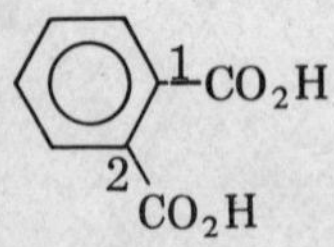
    CO$_2$H

C$_6$H$_4$(CO$_2$H)$_2$ (1,2-) = phthalic acid

Therefore, it is 4-cyanophthalic acid.

★ *Exercises*

Name each compound.

1. O=CH–(CH$_2$)$_8$–CHO

2. 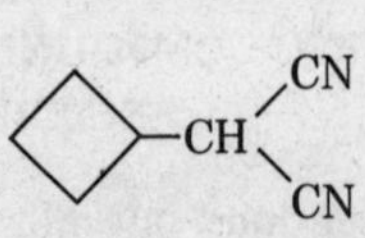 cyclobutyl–CH(CN)$_2$

3. CH$_3$–CHOH–CH=CH–CH$_2$OH

4. Br–CH$_2$–(CH$_2$)$_4$–CH$_2$–Br

– – – – – – – – –

1. decanedial
2. cyclobutylpropanedinitrile (numbers would be redundant)

3. 2-pentene-1,4-diol

4. 1,6-dibromohexane (Br never a principal group)

FLOWCHART FOR NAMING COMPOUNDS

20. Now, having studied the details of the aliphatic and aromatic naming schemes, let us focus on the big picture. The flowchart on the following pages gives the best overall view of the logical flow and necessary order of operations.

The chart uses rectangles for operations performed and diamonds for decisions made. What you write as part of the growing name is given in capital letters, everything else in lowercase.

The flowchart shows clearly how the initial decision between the aliphatic and aromatic branches and the presence of ether or amine as the principal group determine the course of the naming process. These early decisions break the chart into four tracks, each terminating in a finished name. I suggest you trace each of the four tracks (aromatics, ethers, amines, other aliphatics) now and relate the track as well as you can to what you have studied in previous frames. You can use the flowchart as a guide while naming compounds, but because the directions for carrying out each operation are abbreviated, it may be necessary for you to look up details in frames 2 to 19.

GENERAL EXERCISES

21. In these exercises, carry out the entire naming process for each compound. The exercises require putting together things you have in some cases practiced only separately, such as making the aliphatic/aromatic choice and generating group names for use in compound naming.

1. $CCl_3-CH=CH-\overset{O}{\underset{\|}{C}}-CH_3$

2. $Ph-\underset{\underset{CH_3}{|}}{\overset{\overset{CH_3}{|}}{C}}-\underset{\underset{OH}{|}}{CH}-\overset{O}{\underset{\|}{C}}-H$

3. $Cl-C_6H_4-C{\equiv}C-\overset{O}{\underset{\|}{C}}-O-Ph$

4. $\underset{CH_3}{\overset{CH_3}{\diagdown}}CH-NH-\underset{\underset{CH_3}{|}}{CH}-CH_2-CH=CH_2$

5. $HOOC-C_6H_4-SO_3H$

6. $\underset{CH_3}{\overset{CH_3}{\diagdown}}CH-O-CH_2-C_6H_4-NO_2$ (O$_2$N on ring)

7. $F-C_6H_4-\overset{O}{\underset{\|}{C}}-O-CH_2-CH_2-OH$

8. $CH_3-(CH_2)_4-\underset{\underset{CH_2-CH_2-C{\equiv}N}{|}}{CCl}-CH_2-CH_2-\overset{O}{\underset{\|}{C}}-Cl$

9. $F-C_6H_3(NO_2)-CH=CH-\overset{O}{\underset{\diagdown}{C}}{-}NH_2$

10. $HO-C_6H_3(CH_3)-CH_2-CH_2-CH_2-NH_2$

- - - - - - - - - -

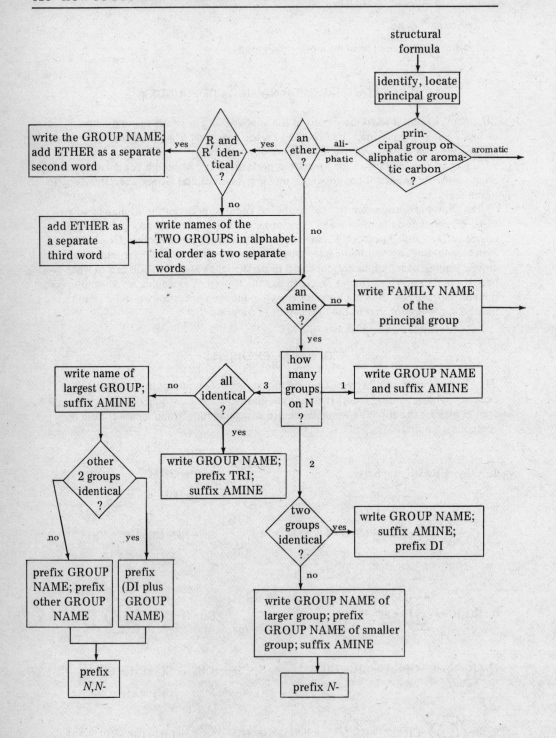

SYSTEMATIC NOMENCLATURE

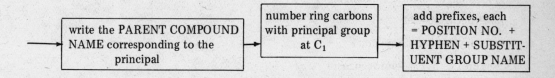

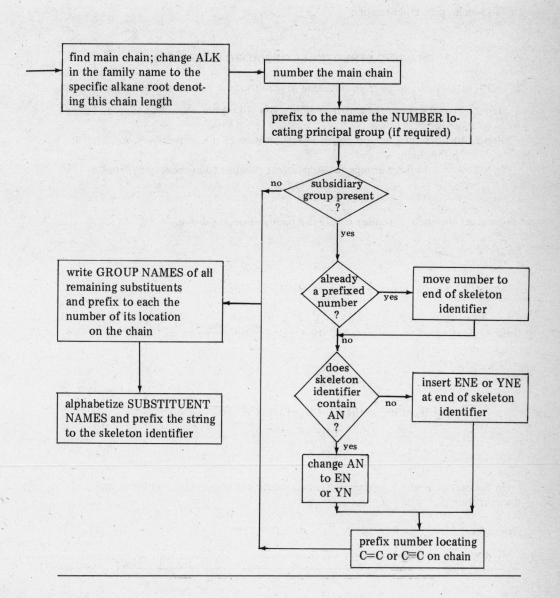

1. 5,5,5-trichloro-3-penten-2-one
2. 2-hydroxy-3-methyl-3-phenylbutanal
3. phenyl 3-(*p*-chlorophenyl)propynoate (the 2 locating the triple bond is omitted—on a C—C—COO— chain it cannot be located anywhere else)
4. *N*-isopropyl-1-methyl-3-butenylamine
5. *p*-sulfobenzoic acid
6. isopropyl *o*-nitrobenzyl ether
7. 2-hydroxyethyl *m*-fluorobenzoate
8. 4-chloro-4-(2-cyanoethyl)nonanoyl chloride
9. 3-(4-fluoro-2-nitrophenyl)propenamide
10. 4-(3-aminopropyl)-*m*-cresol

DRAWING STRUCTURES OF NAMED COMPOUNDS

22. It is much easier to go from the name to the structure than to name the compound.

Step 1. Find the skeleton identifier, write a straight carbon chain that long, and number it.

Step 2. Identify the principal group and its location (1 if not specified), and place it on the chain.

Step 3. Add subsidiary groups and substituent groups. Add necessary hydrogens.

Example

Draw the structural formula of 5-ethyl-2-phenyl-4-octene-3-one.

Solution:

Skeleton identifier is oct $\xrightarrow{\text{step 1}}$
$\overset{1}{C}-\overset{2}{C}-\overset{3}{C}-\overset{4}{C}-\overset{5}{C}-\overset{6}{C}-\overset{7}{C}-\overset{8}{C}$

$\overset{1}{C}-\overset{2}{C}-\overset{3}{C}-\overset{4}{C}-\overset{5}{C}-\overset{6}{C}-\overset{7}{C}-\overset{8}{C}$ $\xrightarrow{\text{step 2}}$ $\overset{1}{C}-\overset{2}{C}-\underset{3}{\overset{\overset{O}{\|}}{\overset{3}{C}}}-\overset{4}{C}-\overset{5}{C}-\overset{6}{C}-\overset{7}{C}-\overset{8}{C}$ $\xrightarrow{\text{step 3}}$ $\overset{1}{C}-\overset{2}{C}-\underset{3}{\overset{\overset{O}{\|}}{C}}-\overset{4}{C}=\overset{5}{C}-\overset{6}{C}-\overset{7}{C}-\overset{8}{C}$

$\downarrow$ step 3

$\text{CH}_3-\underset{\underset{\underset{\bigcirc}{|}}{|}}{\text{CH}}-\overset{\overset{O}{\|}}{\text{C}}-\text{CH}=\text{C}-\text{CH}_2-\text{CH}_2-\text{CH}_3$
$\qquad\qquad\qquad\qquad\quad|$
$\qquad\qquad\qquad\qquad\text{CH}_2$
$\qquad\qquad\qquad\qquad\quad|$
$\qquad\qquad\qquad\qquad\text{CH}_3$

$\xleftarrow{\text{step 3}}$ $\overset{1}{C}-\overset{2}{C}-\underset{3}{\overset{\overset{O}{\|}}{C}}-\overset{4}{C}=\overset{5}{\underset{\underset{\text{CH}_3}{\underset{|}{\text{CH}_2}}}{C}}-\overset{6}{C}-\overset{7}{C}-\overset{8}{C}$

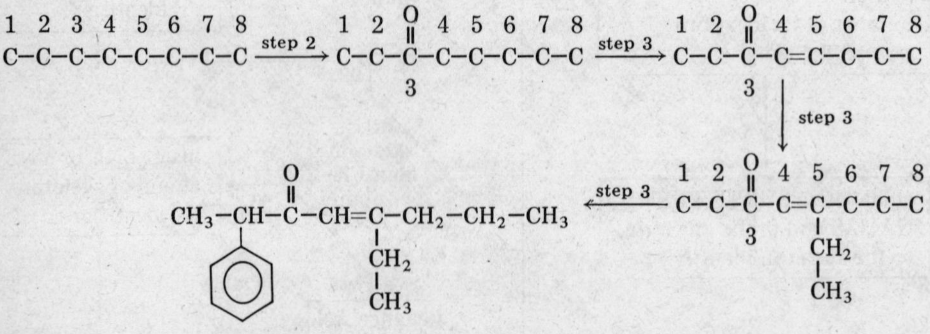

In the case of aromatic compounds, draw the parent compound, number it, and add the substituents.

Example

3-bromo-4-hydroxybenzonitrile

Solution:

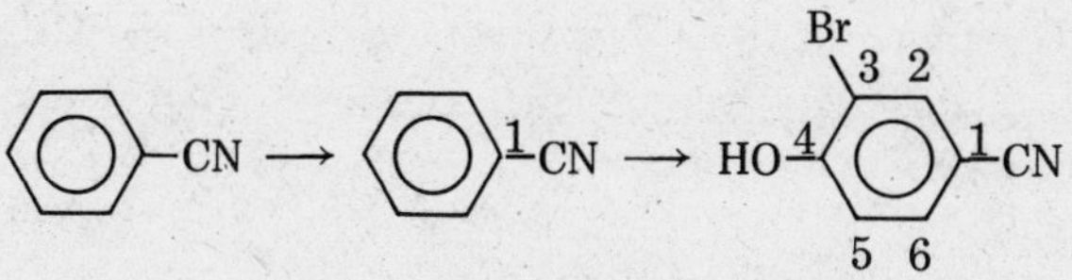

Ph—CN → (4-HO-C₆H₄)—CN → 3-Br-4-HO-C₆H₃—CN (positions 1-CN, 2, 3-Br, 4-OH, 5, 6)

★ Exercises

1. Draw the structural formula corresponding to each name.
   (a) 2-butanone
   (b) 4-chloro-2-nitroheptane
   (c) *m*-bromotoluene
   (d) *o*-aminobenzoic acid
   (e) 2,2,4-trimethylpentane
   (f) 5,5-difluoro-5-iodo-1-pentanol
   (g) 3-pentenal
   (h) 3-pentyn-2-one
   (i) 2,6-dichloro-*p*-cresol
   (j) phenyl 2-chloro-2-pentenoate

2. Draw structural formulas for each group.
   (a) pentyl
   (b) 5,5,5-trifluoropentyl
   (c) 1,2-dimethylpropyl
   (d) 3,3,3-trifluoro-1,2-dimethylpropyl

3. Draw the structural formula corresponding to each name.
   (a) 5-(3,3,3-trichloropropyl)decane
   (b) *p*-(2-chloro-3-cyanobutyl)benzenesulfonic acid
   (c) heptafluoroisopropyl ether
   (d) *N*-methyldihexylamine

– – – – – – – – –

1. (a) $CH_3-\underset{\underset{O}{\|}}{C}-CH_2-CH_3$

   (b) $CH_3-\underset{NO_2}{CH}-CH_2-CHCl-CH_2-CH_2-CH_3$

   (c) 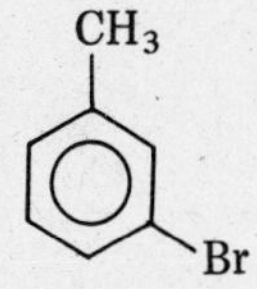 3-bromotoluene (benzene ring with CH₃ and Br in meta positions)

   (d) 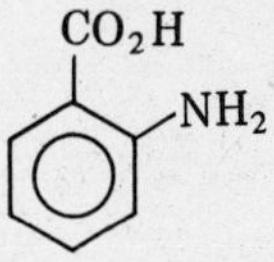 2-aminobenzoic acid (benzene ring with CO₂H and NH₂ in ortho positions)

(e) $CH_3-\underset{\underset{CH_3}{|}}{\overset{\overset{CH_3}{|}}{C}}-CH_2-\underset{\underset{}{}}{\overset{\overset{CH_3}{|}}{CH}}-CH_3$

(f) $CF_2I-CH_2-CH_2-CH_2-CH_2-OH$

(g) $CH_3-CH=CH-CH_2-\overset{\overset{O}{\nearrow}}{C}-H$

(h) $CH_3-C\equiv C-\overset{\overset{O}{\|}}{C}-CH_3$

(i) HO—[2,6-dichloro-4-methylphenyl ring]—CH_3 (with Cl at 2 and 6 positions)

(j) $CH_3-CH_2-CH=CCl-\overset{\overset{O}{\|}}{C}-O-$[phenyl]

2. (a) $CH_3-CH_2-CH_2-CH_2-CH_2-$

(b) $CF_3-CH_2-CH_2-CH_2-CH_2-$

(c) $CH_3-\underset{\underset{CH_3}{|}}{CH}-\underset{\underset{CH_3}{|}}{CH}-$

(d) $CF_3-\underset{\underset{CH_3}{|}}{CH}-\underset{\underset{CH_3}{|}}{CH}-$

3. (a) $CH_3-CH_2-CH_2-CH_2-\underset{\underset{CH_2-CH_2-CCl_3}{|}}{CH}-(CH_2)_4-CH_3$

(b) [benzene ring with SO_3H at top and $CH_2-CHCl-\underset{\underset{CN}{|||}}{CH}-CH_3$ at bottom, where the middle CH has C≡N substituent]

(c) $\underset{CF_3}{\overset{CF_3}{\diagdown}}CF-O-CF\underset{CF_3}{\overset{CF_3}{\diagup}}$

(d) $CH_3-N\underset{(CH_2)_5-CH_3}{\overset{(CH_2)_5-CH_3}{\diagup}}$

UNIT EIGHT
Stable Versus Unstable Molecules

As we have seen, structural formulas are the words in the language of organic chemistry. Just as language learning develops ability to write words in correct form and connect sounds with them, your work in Units 2 to 7 focused on how to draw and name structures accurately and visualize the shapes of the molecules they represent. This is the notational aspect of organic chemistry. Most of the following units deal with something more analogous to the *meaning* of the words—the *behavior* of the real molecules, which is only hinted at by their structural formulas. The first question of behavior is: "Does the molecule exist?" It may not exist just because it has not yet been made, or because it is not stable enough to exist. We deal with this question here, and other aspects of behavior (chemical reactivity) in later units.

OBJECTIVES

When you have completed this unit, you should be able to:

- Explain how stability is related to energy content and equilibrium reactivity (frame 1).
- Predict the stability of the molecules represented by a given structural formula in such terms as probably nonexistent, existing as a short-lived intermediate only, or indefinitely stable (frames 2, 3).
- Classify the common reactive intermediates in organic reactions as free radicals, carbenes, carbonium ions, or carbanions and write equations illustrating reactions by which they frequently stabilize themselves (frames 4, 5).

If you think you may be able to skip all or part of this unit, take a self-test consisting of the exercises marked with a star in the margin. If you miss some problems, review the frames in question and rework the problems. If you are not ready for a self-test, proceed to frame 1.

THE MEANING OF STABILITY

1. The *stability* of a compound *is the inverse of the energy content* of its molecules. This is a general principle. The more energy a molecule possesses, the greater the driving force for it to shed some energy by changing into something more stable via chemical reactions, either with itself or with any other compounds present. This means that there is a general connection between stability and chemical equilibrium: stability is the inverse of equilibrium reactivity.

Greater "equilibrium reactivity" means a larger equilibrium constant (equilibrium lying farther toward the product side). However, this principle can only be easily applied to reactions that form the same product. Suppose that the energy change accompanying the reaction A→C looks like this:

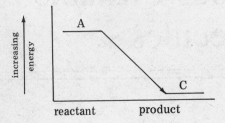

And the change for the B→C reaction looks like this:

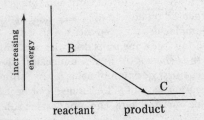

A→C is energetically more downhill than B→C, so the A⇌C equilibrium constant is bigger than the B⇌C equilibrium constant. So A has greater equilibrium reactivity than B in this reaction. Now since the product is the same in the two reactions, they can be plotted together like this:

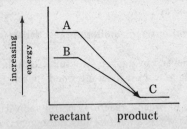

One can read from the graph that B must lie lower on the energy scale than A, so its energy is less than that of A. That means that B is more stable than A and has smaller equilibrium reactivity than A. But you can see that this would not hold necessarily if the two reactions did not give the same product. Furthermore, the reactivity-stability relationship holds only for equilibria—the rate of the A→C reaction can be either larger or smaller than the rate of the B→C reaction.

The same rules apply to formation of alternative sets of products in a reaction; if the alternative products are formed in equilibrium with each other under the reaction conditions, the most stable will always predominate: X ⇌ Y or ⇌ Z. But if the alternative products are not in equilibrium, X → Y or → Z, then the product formed fastest will predominate and this may or may not be the more stable one.

★ *Exercises*

1. If the energies of compounds X and Y are as shown, which is more stable?

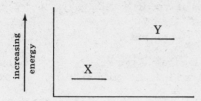

2. Suppose that X and Y in exercise 1 are alternative products of some reaction. If they are formed in equal amounts, is the product composition being determined by the rates of formation of X and Y (called kinetic control) or by their relative stabilities (called thermodynamic control)?

3. Suppose that compound L reacts partly to give compound M and partly to give N, together with unchanged L, all in equilibrium with one another. If the composition of this product mixture is 10% M, 30% N, and 60% L, put the three on an energy diagram and decide which is more stable, L, M, or N.

— — — — — — — — —

1. X, which has the lower energy content, is more stable.

2. Since X is more stable than Y, X would predominate at equilibrium. Since X does not predominate in this product mixture, the products are not in equilibrium with one another. Therefore, the product composition must have been determined by the rates of formation of X and Y.

3. L→M and L→N are both uphill energetically, the former more uphill than the latter. Therefore,

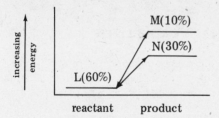 and L is the most stable compound.

THE CONNECTION BETWEEN STRUCTURE AND STABILITY

2. The stability of a molecule is *increased* if every H atom in it is surrounded by a duet and every first- or second-row element by an octet of electrons; if any positive charge present is borne on an electropositive atom; if any negative charge present is borne on an electronegative atom; if atoms bearing *like* charges (⊕ and ⊕ or ⊖ and ⊖) are as far apart as possible; if atoms bearing opposite charges are as close together as possible; if all the bond angles have the optimum values described in Unit 4.

Most molecules stable enough to store on the laboratory shelf satisfy all these requirements. Most of the structural formulas that we write by combining the part structures of Table 2.1 automatically satisfy these requirements and represent such storable compounds. However, there are exceptions—reasonably stable compounds that exist despite the fact that their structures violate one of these principles. For example, BF_3 is stuck with a sextet on B. $(CH_3)_3O^{\oplus}$ puts a positive charge on oxygen. $AlCl_4^{\ominus}$ and $AlH_4^{\ominus}$ have a negative charge on electropositive Al. Some ions of type $\overset{\oplus}{R_3}N-\overset{\oplus}{N}R_3$ exist, and structures

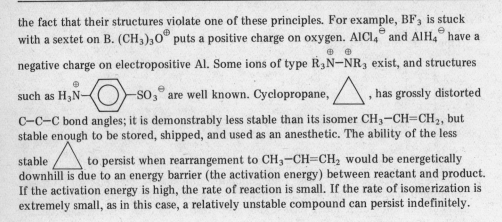

such as $H_3\overset{\oplus}{N}-\!\!\bigcirc\!\!-SO_3^{\ominus}$ are well known. Cyclopropane, △, has grossly distorted C–C–C bond angles; it is demonstrably less stable than its isomer $CH_3-CH=CH_2$, but stable enough to be stored, shipped, and used as an anesthetic. The ability of the less stable △ to persist when rearrangement to $CH_3-CH=CH_2$ would be energetically downhill is due to an energy barrier (the activation energy) between reactant and product. If the activation energy is high, the rate of reaction is small. If the rate of isomerization is extremely small, as in this case, a relatively unstable compound can persist indefinitely.

Exercises

Inspect the following structures and identify which of the principles listed at the beginning of this frame each one violates.

1. H–C≡N:

2. $\begin{matrix} H & & H \\ & \searrow\!\!\!N\!\!\!\swarrow & \\ H & \overset{|}{:\!\ddot{Cl}\!:} & H \end{matrix}$

3. $H-\overset{\oplus}{\underset{..}{N}}-H$

----

1. none
2. decet (10 electrons) on N
3. sextet on N and ⊕ on N

PREDICTION OF RELATIVE STABILITIES

3. Now let's make the principles in frame 2 into a very rough and arbitrary but useful scheme for classifying and predicting the stability of a molecule or ion from its structural formula. We arrange the structural defects in order of decreasing seriousness:

1. Exceeding the octet in first-row atoms (exceeding the duet on H).
2. Falling short of the octet on first-row atoms (fewer than duet on H).
3. Putting ⊕ on an electronegative, or ⊖ on an electropositive atom.
4. Putting ⊕ and ⊕ (or ⊖ and ⊖) into the same molecule.
5. Putting ⊕ and ⊖ onto different atoms in the same molecule.
6. Putting ⊕ or ⊖ onto atoms of intermediate electronegativity, such as C and H.

Exceeding the octet destabilizes first-row atoms so badly that we do not need to consider structures containing such atoms. This restriction does not apply to the second-row elements; for example, S and P can tolerate 10 electrons. To deal with the other defects, let us define a quantity called the *stability deficit* by summing all the defects present in a structure; let type 2 count double:

stability deficit = number of defects of + 2 × number of defects of
 types (3) to (6) type (2)

STABLE VERSUS UNSTABLE MOLECULES

Using this quantity and Table 8.1, we can make a rough estimate of the stability of the chemical species represented by any structure.

Table 8.1 *Stability Predictions*

Stability deficit	Rough description of stability
0-1	Storable in bottles for long periods
1-3	Reactive intermediates with short lifetimes
4 or more	Unlikely to exist

Examples

Predict the qualitative stability of each species.

1. $CH_3-\ddot{O}:^{\ominus}$

2. $:\ddot{Cl}^{\oplus}$

3. 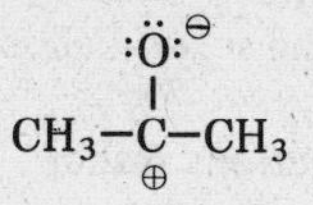 $CH_3-\underset{\oplus}{\overset{\overset{\displaystyle :\ddot{O}:^{\ominus}}{|}}{C}}-CH_3$

Solutions:

1. Zero stability deficit, therefore stable in bottles.

2. Stability deficit = 3 (defects 2, 3), therefore transient intermediate.

3. Stability deficit = 4 (defects 2, 5, 6), therefore nonexistent compound.

★ *Exercises*

1. Consider the stability defects in these species, and rank them in order of decreasing stability. Assume the anions (cations) to be balanced electrically by the proper number of stable cations (anions).

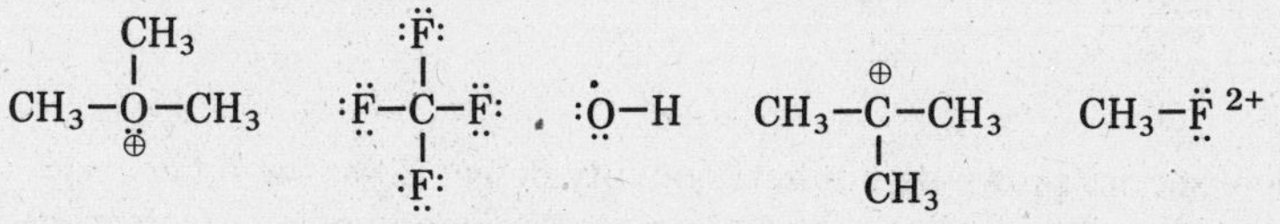

    $CH_3-\underset{\oplus}{\overset{\overset{\displaystyle CH_3}{|}}{O}}-CH_3$ $:\ddot{F}-\underset{\overset{\displaystyle :\ddot{F}:}{|}}{\overset{\overset{\displaystyle :\ddot{F}:}{|}}{C}}-\ddot{F}:$ $:\dot{O}-H$ $CH_3-\underset{\overset{\displaystyle CH_3}{|}}{\overset{\oplus}{C}}-CH_3$ $CH_3-\ddot{F}:^{2+}$

2. Predict the qualitative stability class (Table 8.1) of the following molecules:

    (a) 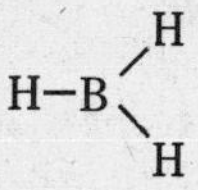 $H-B\begin{smallmatrix}\nearrow H \\ \searrow H\end{smallmatrix}$

    (b) $H-\dot{N}-H$

    (c) 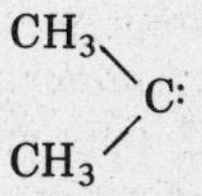 $\begin{smallmatrix}CH_3 \searrow \\ CH_3 \nearrow\end{smallmatrix}C:$

1. Stability order: $CF_4 > (CH_3)_3\ddot{O}:^{\oplus} > \cdot\ddot{O}-H > CH_3-\underset{\oplus}{\overset{\overset{\displaystyle CH_3}{|}}{C}}-CH_3 > CH_3-\ddot{\ddot{F}}\,^{2+}$

   Stability deficit: 0   1   2   3   4

   Defects present: none   (3)   (2)   (2), (6)   (2), (3) twice

2. (a) Sextet on B, deficit = 2; therefore, probably not stable enough to store. BH_3 actually exists mainly as the dimer B_2H_6, in which the electron deficiency on B is relieved by use of one electron pair to make two "3-center bonds":

   $\begin{array}{c}H\diagdown\quad\diagup H\diagdown\quad\diagup H\\ \quad B\cdots\cdots B\\ H\diagup\quad\diagdown H\diagup\quad\diagdown H\end{array}$

   (b) and (c) Sextets on N and C, deficit = 2, reactive intermediates.

TRANSIENT INTERMEDIATES

4. We have seen many examples of the first of the three stability classes in Table 8.1. Almost all the structural formulas we have been working with in earlier units represent stable, storable molecules. The third class, nonexistent compounds, you will not encounter in real life, but we will make use of many structures of this type in Unit 9. In the rest of the present unit we will examine the most commonly met species in the second class—those having 1 to 3 stability defects.

   These species are important as *transient intermediates* in chemical reactions rather than as reactants or reaction products. The most important ones are classified below according to the number of valence electrons present.

   species 2 electrons short of octet or duet: $\underset{R}{\overset{R}{\diagdown}}\overset{\oplus}{C}\underset{|}{\diagup}R \qquad \underset{R}{\overset{R}{\diagdown}}C: \qquad :\ddot{\ddot{Br}}^{\oplus} \qquad H^{\oplus}$

   species 1 electron short of octet or duet: $\underset{R}{\overset{R}{\diagdown}}\underset{|}{C\cdot}\underset{R}{\diagup} \qquad R-\ddot{O}\cdot \qquad :\ddot{\ddot{Br}}\cdot \qquad H\cdot$

   octet/duet species: $\underset{R}{\overset{R}{\diagdown}}\overset{\ominus}{C}\underset{|}{\diagup}R \qquad H:^{\ominus}$

   The octet/duet species shown do not have stability defect (2) as the other species listed do, but they do put a negative charge on an atom of intermediate electronegativity (C, H). They could almost be classed as stable ions; their metal salts can indeed be isolated, but they generally involve some stabilizing covalent interaction between cation and anion.

   The molecules that fall one electron short of an octet/duet are easy to recognize because each one has a single, "unpaired" valence electron (single dot, as in H·). When these species consist of one atom only, they are called "atoms" (H· = hydrogen atom). When they contain more than one atom, they are called *free radicals* or just *radicals*. The 2-electrons-short species not only fall short of the octet/duet, most of them put $\oplus$ on an electronegative (Br) or intermediate (C, H) atom. More abbreviated generic structures and generic names can be written for those species having C as the central atom:

STABLE VERSUS UNSTABLE MOLECULES

$R:^{\ominus}$ $R\cdot$ $R^{\oplus}$ $\begin{matrix}R\\R\end{matrix}\!\!>\!\!C:$

carbanion   alkyl radical   carbonium ion   carbene

The most important points to remember about all of these reactive or transient intermediates are that these are molecules or ions of borderline stability, which means that they have a relatively high energy content, and that they can be formed by chemical reactions of ordinary, stable molecules only if sufficient energy is available—from heat in the surroundings, from light, or from chemical energy stored in relatively weak bonds in the reactants. The energy profile for reactions involving these species generally looks like this:

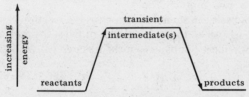

Energy is put into formation of the intermediate. The intermediate has a short lifetime (often milliseconds or less). It is destroyed by chemical reactions that drain off the excess energy, re-forming stable molecules, and the system returns to a lower energy level in the product state. These intermediates cannot be isolated, so *they should never be written as final products of any reaction.*

Exercises

1. Label each species as carbonium ions, carbanions, free radicals, carbenes, or atoms.

   (a) $\begin{matrix}CH_3\\CH_3\end{matrix}\!\!>\!\!\dot{C}\!-\!CH_3$ (b) $:\underline{\ddot{I}}\cdot$

   (c) $\begin{matrix}Cl\\Cl\end{matrix}\!\!>\!\!\overset{\ominus}{C}\!-\!Cl$ (d) ⌬$-\overset{\oplus}{C}H\!-\!CH_3$

   (e) ⌬$=CH\!-\!CH\!:$ (f) [cyclopentadienyl anion with H]

2. Find the stability deficit of the structure $\begin{matrix}H\\H\end{matrix}\!\!>\!\!\overset{\oplus}{C}\!-\!\overset{\ominus}{C}\!\!<\!\!\begin{matrix}H\\H\end{matrix}$. Does a compound with this structure probably exist? Find a different arrangement of the 12 electrons that provides six bonds instead of five. What is its stability deficit?

--- --- --- --- --- --- ---

1. (a) radical   (b) atom

   (c) carbanion   (d) carbonium ion

   (e) carbene   (f) carbanion

2. Defects: 2, 5, 6 (twice). Stability deficit = 5. Nonexistent. The structure

$\mathrm{H}_2\mathrm{C}=\mathrm{CH}_2$ uses the 12 electrons to make six covalent bonds. The new structure does represent a stable molecule (no defects, stability deficit = 0). It's just an alkene. What was the original structure? An electronically deformed alkene. Such unstable, high-energy species can sometimes be made by causing the molecule to absorb high-energy light; their lifetimes are generally exceedingly short.

THE FATES OF UNSTABLE, TRANSIENT INTERMEDIATES

5. We are not yet in a position to discuss all the chemical reactions that can destroy transient intermediates. We will look at two important ones.

A very general way in which free radicals can stabilize themselves is a reaction called *dimerization* (a reaction of two identical species to give a product of twice the molecular weight). An unpaired electron on each radical combines to make a shared pair; this fills out the octet on each species:

$$\mathrm{H_3C\cdot} + \mathrm{\cdot CH_3} \longrightarrow \mathrm{H_3C:CH_3} \text{ or } \mathrm{H_3C-CH_3}$$

or generally,

$$\mathrm{R\cdot} + \mathrm{\cdot R} \longrightarrow \mathrm{R-R}$$

This process has such a high likelihood of occurring that essentially every collision between two radicals causes them to react and be destroyed (by this or other processes). Then how is it that radicals survive long enough to collide and react with other molecules to give products more useful than the dimerization product? The answer is that at any given instant the concentration of radicals is minute, so that collisions of radicals with radicals is an improbable event. Most other reactive intermediates also exist only in very low concentrations. This is largely a result of the shape of the energy profile. If formation of the intermediate is energetically unfavorable and destruction is energetically quite favorable, as is typical (frame 4), destruction will tend to be faster than formation, a situation guaranteed to keep the intermediate from building up an appreciable concentration.

Those unstable intermediates that are two electrons short of an octet or duet have a strong tendency to fill out the octet/duet by accepting an electron pair from some molecule with an unshared electron pair. The reaction is a lot like radical dimerization, but both electrons come from the same molecule:

$$\mathrm{H}^{\oplus} + \mathrm{:\ddot{O}H_2} \longrightarrow \mathrm{H:\ddot{O}H_2}^{\oplus} = \mathrm{H-\ddot{O}H_2}^{\oplus}$$

↑ ↑ ↑ ↑
2 electrons octet duet octet
short of duet

Since most solvents have unshared pairs, reactive intermediates of this type often undergo this reaction with solvent, as in the above case of $\mathrm{H}^{\oplus}$ dissolved in water, which exists solely as $\mathrm{H_3O}^{\oplus}$. This reaction is a very fundamental one called a *coordination* or generalized acid-base reaction. The electron-deficient molecule is the acid and the electron-donor molecule is the base. These processes are treated in detail in Unit 11. For our present purposes, note just two things. Coordination with the solvent sometimes stabilizes the

STABLE VERSUS UNSTABLE MOLECULES 141

intermediate without destroying it. This happens when the coordination reaction is a reversible equilibrium so that it amounts to temporary storage of the intermediate, which can be rapidly liberated again. This is the case with the $H^\oplus + H_2O \rightarrow H_3O^\oplus$ coordination. On the other hand, if the base is too strong, it may never give up the intermediate once it is captured; coordination is effectively irreversible and just destroys the intermediate. This is the case with carbonium ions generated in the presence of ammonia:

$$
\begin{array}{c}
R' \quad H \\
| \quad | \\
R-C^\oplus \curvearrowleft :N-H \\
| \quad | \\
R'' \quad H
\end{array}
\rightarrow
\begin{array}{c}
R' \quad H \\
| \oplus / \\
R-C-N-H \\
| \quad | \\
R'' \quad H
\end{array}
$$

sextet octet   octets

Since the experimental evidence speaks against the existence of free $H^\oplus$ in solutions, we try to avoid ever writing it in the free condition, $H^\oplus$, and show it always coordinated to some base, often the solvent. In water as solvent, this turns it into $H_3O^\oplus$. In CH_3-OH solvent, it is $CH_3-\overset{\oplus}{\underset{H}{O}}\diagdown^{H}$; in $:NH_3$, it is $H-\overset{\oplus}{N}H_3$; and so on.

★ Exercises

Write an equation to show one way in which each unstable intermediate might stabilize itself under the conditions indicated.

1. $H-\ddot{N}-H$ in the gas phase (no solvent).

2. $\begin{array}{c}:\ddot{F}:\\ | \\ :\ddot{F}-B\\ | \\ :\ddot{F}:\end{array}$ in solution in $CH_3-C\equiv N:$ solvent

3. $\begin{array}{c}CH_3\\ \diagdown\\ C:\\ \diagup\\ CH_3\end{array}$ in the gas phase

4. $:\ddot{Br}^\oplus$ in solution in $CH_3-\ddot{O}-H$ solvent

— — — — — — — — —

1. $\begin{array}{c}H\\ \diagdown\\ \ddot{N}\cdot\\ \diagup\\ H\end{array} + \begin{array}{c}H\\ \diagdown\\ \cdot\ddot{N}\\ \diagup\\ H\end{array} \rightarrow \begin{array}{c}H\\ \diagdown\\ \ddot{N}-\ddot{N}\\ \diagup\diagdown\\ HH\end{array}$

2. $\begin{array}{c}F\\ |\\ F-B\\ \uparrow\;|\\ F\end{array} \quad :N\equiv C-CH_3 \rightarrow \begin{array}{c}F^\ominus\\ |\;\;\oplus\\ F-B-N\equiv C-CH_3\\ |\\ F\end{array}$

   acid   base   octets

3. $\begin{array}{c}CH_3\\ \diagdown\\ C:\\ \diagup\\ CH_3\end{array} \begin{array}{c}CH_3\\ \diagup\\ :C\\ \diagdown\\ CH_3\end{array} \rightarrow \begin{array}{c}CH_3\\ \diagdown\diagup CH_3\\ C=C\\ \diagup\diagdown CH_3\\ CH_3\end{array}$

   sextets   octets

4. $:\overset{\oplus}{\ddot{Br}} \quad :\ddot{O}\diagdown^{CH_3}_H \rightarrow :\ddot{Br}-\overset{\oplus}{\ddot{O}}\diagdown^{CH_3}_H$

UNIT NINE

Resonance

In Unit 3 you learned these basic principles: (1) the properties of an organic compound are largely predictable from its structural formula; (2) the physical properties are determined by the identity of the functional group, the skeleton, and the manner of their connection; and (3) the chemical properties are principally determined by the identity of the functional group and to a smaller degree by the manner of its connection to the skeleton. In this unit we will look in detail at how one manner of connection—conjugation—modifies the properties. The second purpose of this unit is to modify the basic principles in the following way. Study of a great variety of compounds has shown that there is a class of compounds whose properties are *not* very well accounted for by the ordinary structural formula. We must see which compounds these are and understand in what ways and why their properties are not the expected ones.

OBJECTIVES

When you have completed this unit, you should be able to:

- Recognize those molecules that cannot be adequately represented by a single structural formula (frames 2, 8, 11, 12).
- Accurately describe their structures using a standard notation (frames 3, 5, 9, 11, 13 to 16).
- Predict how their properties differ from those expected from any single structural formula (frames 4 to 7).

If you think you have already accomplished some of these objectives and may be able to skip part or all of this unit, take a self-test consisting of the exercises marked with a star in the margin. If you miss any questions, study the frame(s) involved and rework the questions before proceeding. If you are not ready for a self-test, proceed to frame 1.

THE STRUCTURE OF RESONANCE-HYBRID MOLECULES

1. For some compounds we cannot write a single structure that accounts for the observed properties (energy content, bond lengths, dipole moments, chemical behavior). We are forced to substitute for the usual single structural formula an imaginary composite or average of two or more structural formulas. The averaging process is called *resonance*. The composite structure is called a *resonance-hybrid* structure. The individual structural formulas used in the averaging process are called the *limiting structures* or *contributing structures*.

A resonance hybrid is not a compound whose molecules spend part of their time in one limiting structure and part in the other. It is not a compound, half of whose molecules are of one limiting structure and half of the other. In a resonance-hybrid compound, all molecules possess the same structure all the time; that structure is intermediate between the limiting structures.

This is probably the most abstract concept in beginning organic chemistry, but it is not really a difficult one. Most student difficulties lie in (a) recognizing those compounds that require a resonance-hybrid structure, (b) finding all the limiting structures needed for description of the hybrid, and (c) confusing resonance with chemical equilibrium. We concentrate on methods of solving or avoiding these problems in succeeding frames.

Exercise

A certain compound is a resonance hybrid of limiting structures A and B. Which of the following statements about the compound is most accurate?

(a) The compound can be separated into molecules of structure A and molecules of structure B provided that extremely efficient separation methods are available.

(b) Structures A and B interconvert at the speed of light.

(c) A and B exist only in the imagination.

— — — — — — — — — .

The last statement is the only correct one. The limiting structures do not correspond to any real molecules; they are mental props for visualizing and manipulating real resonance-hybrid molecules.

2. The compounds that require a resonance-hybrid description are the ones for which you can write two or more structural formulas that *differ only in the position of electrons*. Looking for alternative electron arrangements is not the easiest way to recognize these compounds, but it is important in understanding their properties.

Structures that differ only in the way the electrons are distributed over the framework of atomic nuclei are not capable of independent existence. Only one compound corresponding to all such structures exists. Its molecules have the resonance-hybrid structure resulting from conceptually averaging the limiting structures.

Example

What is the relationship between structures

$$:N{\equiv}C-\ddot{\underset{..}{O}}:^{\ominus} \quad \text{and} \quad :\overset{\ominus}{\underset{..}{N}}{=}C{=}\ddot{O}: \quad ?$$
$$\phantom{:N{\equiv}C-\ddot{O}:}a \phantom{\text{and}\quad\quad\quad} b$$

Solution:

They differ in the locations of two electron pairs. One can be obtained from the other by moving two electron pairs:

$$:N{\equiv}C-\ddot{\underset{..}{O}}:^{\ominus} \quad \text{turns into} \quad :\overset{\ominus}{\underset{..}{N}}{=}C{=}\ddot{O}:$$

$$:\overset{\ominus}{\underset{..}{N}}{=}C{=}\ddot{O}: \quad \text{turns into} \quad :N{\equiv}C-\ddot{\underset{..}{O}}:^{\ominus}$$

These structures, together with some cation, say $Na^{\oplus}$, should represent a single compound, and in fact only one is known that has this number of electrons on this arrangement of nuclei—sodium cyanate.

By comparing these two structures, we can identify the key feature of resonance-hybrid molecules. Structure a implies that the three unshared pairs on oxygen are *localized* there. Structure b localizes two unshared pairs on N. But the interconversion of a and b by electron shifts shows that one pair on O is actually partly on O, partly shared between C and O as part of a double bond. Similarly, one pair on N in b is not localized there but is able to occupy the region between N and C as part of a triple bond. In real molecules electrons are *delocalized* in this way whenever possible. The real molecule and the resonance-hybrid structure we use to describe it differ from the limiting structures in having some electrons delocalized that are localized in the limiting structures. The molecules that require a resonance-hybrid description are those that are capable of delocalization.

The delocalization doesn't have to involve unshared pairs versus multiple bonds. Electrons can be delocalized between multiple bonds in different locations.

Example

$CH_2=CH-CH_2^{\oplus}$ has the alternative electronic arrangement $^{\oplus}CH_2-CH=CH_2$. One electron pair in the double bond is delocalized over the regions in space between both pairs of carbon atoms.

In both of the examples above we can also say that the charge, $\oplus$ or $\ominus$, is delocalized over two atoms. Since the charges are just local excesses or deficiencies of electrons, charge delocalization is just one aspect of electron delocalization.

In molecules that are capable of it, delocalization is a spontaneous natural phenomenon. If, by experimental means as yet unknown, one could make molecules with structure a or with structure b above, they would instantaneously change their electron distribution pattern to that of the hybrid.

Exercises

Which of the following pairs of structures differ only in the position of electrons?

1. $:N\overset{\oplus}{=}\underset{..}{O}:$ and $:N\overset{\oplus}{\equiv}O:$
2. $H-C\equiv\overset{\ominus}{N}:$ and $:\overset{\ominus}{C}\equiv\overset{\oplus}{N}-H$
3. $H-C\begin{smallmatrix}\nearrow\overset{..}{\underset{..}{S}}:\\ \searrow\underset{..}{O}:^{\ominus}\end{smallmatrix}$ and $H-C\begin{smallmatrix}\nearrow\overset{..}{\underset{..}{S}}:^{\ominus}\\ \searrow\underset{..}{O}:\end{smallmatrix}$
4. $:\overset{\ominus}{\underset{..}{O}}-\overset{\oplus}{O}=\underset{..}{O}:$ and $:\underset{..}{O}=\overset{\oplus}{O}-\overset{\ominus}{\underset{..}{O}}:$
5. $H-\underset{..}{O}-\overset{\ominus}{\underset{..}{S}}:$ and $H-\underset{..}{S}-\underset{..}{O}:^{\ominus}$

Pairs 1, 3, and 4 differ only in the position of electrons. Pairs 2 and 5 differ in the position of both electrons and nuclei (H moves). Pair 4 could be read as representing different views of the same molecule; pancake-flipping interconverts the two. But it can also represent the molecule held fixed in place with two different electron-distribution schemes; this is the interpretation that is necessary in order to recognize pair 4 as a resonance-hybrid molecule. This interpretation can be made clear by labeling the end oxygens as a and b so that the reader can tell them apart: $:\overset{\ominus}{\underset{..}{O}}\underset{a}{}-\overset{\oplus}{O}\underset{b}{}=\underset{..}{O}:$ and $:\underset{..}{O}\underset{a}{}=\overset{\oplus}{O}-\overset{\ominus}{\underset{..}{O}}\underset{b}{}:$.
$\qquad\qquad\qquad\qquad\qquad\qquad\qquad\qquad\qquad\qquad\qquad$ I $\qquad\qquad\quad$ II

So labeled, they can only be interpreted as alternative electron arrangements. Since such structures are not generally labeled for you, you need to watch for sets of structures that can be interpreted as differing in position of electrons, even if there is also a trivial interpretation in terms of repositioning of the whole structure. Structures I and II describe two genuinely different ways of distributing the valence electrons; consequently, they are limiting structures in a resonance-hybrid description of the molecule. You would weed one out if you were trying to write all the *isomers* of O_3 (Unit 6), but you must keep both when trying to write a resonance-hybrid structure for the molecule. On the other hand, structures that differ only in the *identity* of the electrons that occupy certain locations do not qualify as different electron distributions and are not limiting structures of a resonance-hybrid molecule. For example, we might argue that

$$\underset{\text{III}}{\overset{H}{\underset{H}{>}}C\overset{H}{\underset{H}{<}}} \quad \text{and} \quad \underset{\text{IV}}{\overset{H}{\underset{H}{>}}C\overset{H}{\underset{H}{<}}} \quad \text{are interconverted by the electron shifts shown, but the differ-}$$

ence produced by these shifts is only a difference in electron *identity*. To qualify as limiting structures of a resonance-hybrid molecule, two structures must have different electron *density* in the neighborhood of at least one atom in the molecule; this is what is meant by "different electron distributions." Structure I has a lower electron density on atom b and structure II has a higher electron density on b; these are limiting structures because they represent different electron distributions. Structures III and IV have identical electron density on whatever atom we might test; they are not different electron distributions and are not limiting structures of a resonance-hybrid molecule.

3. If no one of the limiting structures is adequate, what does one use for a written description of a resonance-hybrid molecule? The standard notation connects the limiting structures by double-headed arrows:

$$:\overset{\ominus}{\underset{..}{O}}-C\equiv N: \longleftrightarrow :\overset{..}{\underset{..}{O}}=C=\overset{\ominus}{\underset{..}{N}}: \qquad \overset{\ominus O}{\underset{\ominus O}{>}}C=O \longleftrightarrow \overset{O}{\underset{\ominus O}{>}}C-O^{\ominus} \longleftrightarrow \overset{\ominus O}{\underset{O}{>}}C-O^{\ominus}$$

Sometimes we add curved arrows that indicate the shifts of electrons required to turn each limiting structure into the next one:

$$:\overset{\ominus}{\underset{..}{O}}-C\equiv N: \longleftrightarrow :\overset{..}{\underset{..}{O}}=C=\overset{\ominus}{\underset{..}{N}}: \qquad \overset{\ominus:\ddot{O}:}{\underset{\ominus:\ddot{O}:}{>}}C=\ddot{O}: \longleftrightarrow \overset{:\ddot{O}:}{\underset{\ominus:\ddot{O}:}{>}}C-\ddot{O}:^{\ominus} \longleftrightarrow \overset{\ominus:\ddot{O}:}{\underset{:\ddot{O}:}{>}}C-\ddot{O}:^{\ominus}$$

These curved, full-headed arrows are standard notation; they are also used for charting the movement of electrons in chemical reactions (Unit 10). The arrows are an optional part of the resonance-hybrid structure.

An unpaired electron in a molecule can often be delocalized just as electron pairs are. Standard notation uses a half-headed, curved arrow to move a single electron:

$$\dot{C}H_2=CH-\dot{C}H_2 \longleftrightarrow \dot{C}H_2-CH=\dot{C}H_2$$

A rule of physics states that every limiting structure combined in a resonance-hybrid structure must have the same number of unpaired electrons. The hybrid above has one unpaired electron on each limiting structure. All the hybrids written previously in this unit have zero unpaired electrons on each limiting structure. According to this rule, the electron shifts carried out by curved arrows cannot create or destroy unpaired electrons, only move them about. The example above maintains the number of unpaired electrons, but the following one is illegal because it does not:

$$CH_2=CH_2 \ \ \cancel{\longleftrightarrow} \ \ \dot{C}H_2-\dot{C}H_2$$

The curved-arrow notation must be kept accurate by placing the end of the arrow tail precisely at the electron(s) to be moved, and placing the tip of the arrowhead precisely at the new bond or unshared-pair location they are to occupy.

★ *Exercises*

1. Supply the second limiting structure that is generated by the electron shifts shown.

   (a) $CH_3-CH=CH_2$ (with arrow)

   (b) $\overset{\oplus}{CH_2}-CH=CH_2$ (with arrow)

   (c) $:\overset{\ominus}{CH_2}-CH=CH_2$ (with arrow)

   (d) cyclohexadienyl cation with H⊕ (with arrows)

   (e) cyclohexadienyl cation (with arrows)

   (f) $CH_3-C\overset{NH_2}{\underset{\overset{\oplus}{NH_2}}{\diagup\diagdown}}$ (with arrow)

   (g) $\underset{CH_3}{\overset{CH_3}{\diagdown\diagup}}C=\ddot{O}:$ (with arrow)

   (h) $CH_3-C\overset{O:}{\underset{\overset{\ominus}{\ddot{O}:}}{\diagup\diagdown}}$ (with arrow)

   (i) $CH_3-C\overset{=\ddot{O}:}{\underset{\ddot{N}H_2}{|}}$ (with arrow)

   (j) $CH_3-C\equiv\ddot{N}:$ (with arrow)

2. For each pair of structures, supply the curved arrows necessary to turn the first structure into the second and the second into the first.

   (a) $CH_3-CH=CH-\overset{\cdot}{C}H_2 \longleftrightarrow CH_3-\overset{\cdot}{C}H-CH=CH_2$

   (b) $\underset{CH_3}{\overset{CH_3}{\diagdown\diagup}}\overset{\oplus}{C}-\overset{\ominus}{\underset{\cdot\cdot}{N}}-H \longleftrightarrow \underset{CH_3}{\overset{CH_3}{\diagdown\diagup}}C=\ddot{N}-H$

   (c) $CH_3-\overset{\cdot}{C}H-\ddot{\underset{\cdot\cdot}{S}}:^{\ominus} \longleftrightarrow CH_3-CH=\ddot{\underset{\cdot\cdot}{S}}:^{\ominus}$

   (d) $CH_2=CH-\overset{\oplus}{CH}-CH=CH_2 \longleftrightarrow \overset{\oplus}{CH_2}-CH=CH-CH=CH_2$

   (e) $CH_2=CH-CH=CH_2 \longleftrightarrow {}^{\ominus}\ddot{C}H_2-CH=CH-\overset{\oplus}{C}H_2$

   (f) phenoxide ↔ cyclohexadienone anion

   (g) phenyl acetate ↔ cyclohexadienone oxocarbenium

   (h) $\underset{Br}{\overset{Cl}{\diagdown}}\!\!\diagup\!\!-\overset{\ominus}{\underset{O}{\overset{\cdot}{N}\!\!\rightarrow\!\!O}} \longleftrightarrow \underset{Br}{\overset{Cl}{\diagdown}}\!\!=\!\!N\!\!\rightarrow\!\!O \atop :\ddot{O}:^{\ominus}$

1. (a) $CH_3-\overset{\curvearrowleft}{CH}=CH_2 \longleftrightarrow CH_3-\overset{\ominus}{CH}-\overset{\oplus}{CH_2}$

   (b) $\overset{\oplus}{CH_2}-\overset{\curvearrowleft}{CH}=CH_2 \longleftrightarrow CH_2=CH-\overset{\oplus}{CH_2}$

   (c) $:\overset{\ominus}{CH_2}-CH=CH_2 \longleftrightarrow CH_2=CH-\overset{\ominus}{\underset{..}{CH_2}}$

   (d) [cyclohexadienyl cation resonance structure] ⟷ [cyclohexadienyl cation resonance structure]

   (e) [cyclohexadienyl cation resonance structure] ⟷ [cyclohexadienyl cation resonance structure]

   (f) $CH_3-C\begin{matrix}\nearrow NH_2 \\ \searrow \underset{\oplus}{NH_2}\end{matrix} \longleftrightarrow CH_3-C\begin{matrix}\nearrow \overset{\oplus}{NH_2} \\ \searrow NH_2\end{matrix}$

   (g) $\begin{matrix}CH_3 \\ CH_3\end{matrix}\!\!>\!\!C=\underset{..}{\overset{..}{O}}: \longleftrightarrow \begin{matrix}CH_3 \\ CH_3\end{matrix}\!\!>\!\!\overset{\oplus}{C}-\overset{..}{\underset{..}{O}}:^{\ominus}$

   (h) $CH_3-C\begin{matrix}\nearrow \overset{..}{O}: \\ \searrow \underset{..}{\overset{..}{O}}:_{\ominus}\end{matrix} \longleftrightarrow CH_3-C\begin{matrix}\nearrow \overset{..}{\underset{..}{O}}:^{\ominus} \\ \searrow \overset{..}{\underset{..}{O}}:\end{matrix}$

   (i) $CH_3-C\begin{matrix}\nearrow \overset{..}{O}: \\ \searrow NH_2\end{matrix} \longleftrightarrow CH_3-C\begin{matrix}\nearrow \underset{..}{\overset{..}{O}}:^{\ominus} \\ \searrow \underset{\oplus}{NH_2}\end{matrix}$

   (j) $CH_3-\overset{\curvearrowleft}{C}\equiv N: \longleftrightarrow CH_3-\overset{\oplus}{C}=\overset{\ominus}{N}:$

2. (a) $CH_3-\overset{\curvearrowleft}{CH}=CH-\overset{\curvearrowleft}{CH_2} \longleftrightarrow CH_3-\dot{C}H-CH=CH_2$

   (b) $\begin{matrix}CH_3 \\ CH_3\end{matrix}\!\!>\!\!\overset{\oplus}{C}-\overset{\ominus}{N}-H \longleftrightarrow \begin{matrix}CH_3 \\ CH_3\end{matrix}\!\!>\!\!C=N-H$

   (c) $CH_3-\dot{C}H-\overset{..}{\underset{..}{S}}:^{\ominus} \longleftrightarrow CH_3-CH=\dot{S}:^{\ominus}$

   (d) $CH_2=CH-CH=CH-\overset{\oplus}{CH_2} \longleftrightarrow \overset{\oplus}{CH_2}-CH=CH-CH=CH_2$

   (e) $\overset{\curvearrowleft}{CH_2}=CH-CH=CH_2 \longleftrightarrow \overset{..}{CH_2}-CH=CH-\overset{\oplus}{CH_2}$
   $\underset{\ominus}{}$

(f) [resonance structures of phenoxide]

(g) [resonance structures of acetylbenzene cation]

(h) [resonance structures of nitro compound with Cl and Br substituents]

4. How should the molecule represented by :Ö—C≡N: ⟷ :O=C=N̈: be visualized? The O—C bond is intermediate between single and double. The C—N bond is somewhere between double and triple. The one unit of negative charge is borne partly by O, partly by N.

The implications about positions of the nuclei are less obvious, but important. You are to visualize the nuclear positions as those that are optimum for the real, hybrid molecule, not those that would be optimum for either limiting structure. The limiting structures above would have the following approximate internuclear distances, since bond lengths generally increase in the order triple < double < single:

The real, hybrid molecule (with bonds intermediate between single and double, or double and triple) will have intermediate distances, approximately

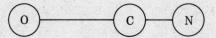

The rule that limiting structures can differ only in the position of electrons, not in position of nuclei, means that when we look at the limiting structures, we should see them not with their own internuclear distances, but with the single, intermediate set of distances that is appropriate for the hybrid molecule.

★ *Exercises*

Describe the bonds and charge distribution in these resonance-hybrid molecules (ions):

1. $\text{H}_2\text{C}=\ddot{\text{O}}:$ ⟷ $\text{H}_2\overset{\oplus}{\text{C}}-\ddot{\text{O}}:^{\ominus}$

2. $^{\ominus}\text{O}_2\text{C}=\text{O}$ ⟷ (resonance structures of carbonate ion)

----

1. The C—O bond is intermediate between single and double; it is longer (C and O farther apart) than a normal double bond, shorter (C and O closer together) than a normal

single bond. C bears a fractional positive charge, O bears a fractional negative charge.

2. The three oxygens are equivalent in every way. To figure the charge on each O, consider that each one must bear one-third of the total charge, or $1/3 \times -2 = -2/3$ on each oxygen. Each C—O bond is intermediate between single and double (closer to single since a given bond is single in two limiting structures, double in one limiting structure).

5. The more stable a limiting structure, the more it contributes to a resonance-hybrid structure. You will not be able to apply this principal very quantitatively, but by using the stability criteria from Unit 8, you can make rough judgments.

If one limiting structure is much more stable than all the rest, that one most stable structure by itself is a reasonable description of the molecule for most purposes. The contribution of the other structures is small, and for many purposes we will not need to bother to write the molecule as a resonance-hybrid structure. An arbitrary rule of thumb is this:

- Structures with a stability deficit (p. 137) of 4 or more can generally be disregarded altogether.
- Structures with a stability deficit 3 or 4 units larger than that of the most stable structure are usually omitted for most purposes (writing equations for reactions, for example), but must be included in a resonance-hybrid description if one wants to predict the structure or stability of the compound.
- Structures that differ in stability deficit by 2 units or less should always be included, and the hybrid structure written out.

Example

The molecule $CH_3-C\overset{\displaystyle O}{\underset{\displaystyle H}{\diagup}}$ has this resonance-hybrid structure:

$$CH_3-C\overset{\displaystyle \ddot{O}:}{\underset{\displaystyle H}{\diagup}} \longleftrightarrow CH_3-\overset{\oplus}{C}\overset{\displaystyle :\ddot{O}:^{\ominus}}{\underset{\displaystyle H}{\diagup}}$$

a   b

Limiting structure a has no stability deficit; structure b has a deficit of 4. Write just $CH_3-C\overset{\displaystyle O}{\underset{\displaystyle H}{\diagup}}$ for most purposes, but write the full hybrid structure for interpretation of stability, reactivity, and so on. The real, hybrid molecule can be described as having (1) a carbon-oxygen double bond, with a small amount of single-bond character and (2) small fractional positive and negative charges on C and O, respectively.

Example

In the resonance-hybrid ion

$$CH_3-\overset{\oplus}{\underset{\displaystyle \underset{\displaystyle CH_3}{|}}{C}}-\ddot{O}-CH_3 \longleftrightarrow CH_3-\underset{\displaystyle \underset{\displaystyle CH_3}{|}}{C}=\overset{\oplus}{\underset{\displaystyle ..}{O}}-CH_3$$

a   b

the stability deficit of structure a is 3, that of structure b is 1. They differ by only 2 units, so both are important contributors, and the full hybrid structure should be written for practically all purposes. The ⊕ is borne partly on C but mostly on O.

Example

In the description of $CH_2=CH-CH=CH_2$, one possible limiting structure is $\overset{\oplus}{C}H_2-\overset{\ominus}{\underset{..}{C}}H-\overset{\ominus}{\underset{..}{C}}H-\overset{\oplus}{C}H_2$. It has at least 8 defects. Exclude it for all purposes.

One other case requires special handling. When two or more structures are *identical* in stability, they contribute equally and the hybrid molecule is structurally precisely midway between them. The only time structures are identical in energy is when they are entirely equivalent except for the identity of one or more nuclei.

Example

In $CH_2=CH-\overset{\oplus}{C}H_2 \longleftrightarrow \overset{\oplus}{C}H_2-CH=CH_2$, the two limiting structures are equivalent, are identical in stability, and contribute precisely equal weights to the hybrid structure. Exactly half the charge is on each of the end carbons, none on the central carbon. Each C—C bond is a "1½ bond." In cases like this, all the equivalent structures must be shown for essentially all purposes since the real molecule differs so much from any limiting structure. Some abbreviations used to reduce the writing involved are taken up in frame 15.

Exercise

Consider the molecule

$CH_3-C\equiv N: \longleftrightarrow CH_3-\overset{\oplus}{C}=\overset{\ominus}{\underset{..}{N}}: \longleftrightarrow CH_3-\overset{\ominus}{\underset{..}{C}}=\overset{\oplus}{N}: \longleftrightarrow CH_3-\overset{(2+)}{C}-\overset{(2-)}{\underset{..}{N}}:$

(a) Which of these limiting structures should really be included in the resonance-hybrid structure?
(b) Give a description of the hybrid molecule.

— — — — — — — — — —

(a) From left to right, the limiting structures have stability deficits of 0, 4, 5, and >8 defects (the last one is so bad the rules can only be approximately applied). Eliminate the last two. Write $CH_3-C\equiv N:$ for everyday use, but $CH_3-C\equiv N: \longleftrightarrow CH_3-\overset{\oplus}{C}=\overset{\ominus}{\underset{..}{N}}:$ for careful investigation of its properties.

(b) The C—N bond is triple with a little double-bond character. Small fractional positive and negative charges are located on C and N, respectively.

EFFECT OF RESONANCE ON ENERGY CONTENT

6. The real, hybrid molecule has an energy content that is lower, or a stability that is greater, than that of the most stable limiting structure. The difference in energy between the real compound and the best of the limiting structures is called the *resonance energy*. It is energy that you would expect the molecule to have if you looked just at limiting structures, but which the real molecule in fact does not have. Consequently, many people find it more useful to call this energy quantity the *resonance stabilization*.

Example

The resonance stabilization of $CH_3-C\equiv N$ (example from frame 5) can be visualized on an energy diagram as follows:

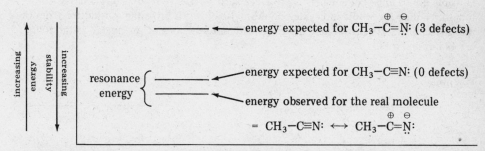

The resonance stabilization results from the delocalization of electrons in the hybrid relative to the limiting structures. Other things being equal, the greater the number of limiting structures, the more extensive the delocalization and the greater the resonance stabilization.

The more similar to one another in structure and energy the limiting structures are, the greater the resonance stabilization. If there is only one reasonably stable limiting structure, the resonance stabilization is small.

Exercise

Predict the order of resonance stabilization in $CH_3-C\overset{O}{\underset{H}{\diagdown}}$, $CH_3-\overset{\oplus}{\underset{CH_3}{C}}-O-CH_3$, and $CH_2=CH-\overset{\oplus}{C}H_2$, all examples from frame 5.

— — — — — — — — —

The difference in stability deficit between the best and next-best limiting structures for these three compounds (ions) was 4, 2, and 0 units, respectively. So $CH=CH-\overset{\oplus}{C}H_2$ has the most similar limiting structures and the largest resonance stabilization; CH_3-CHO has the least similar limiting structures and the smallest resonance stabilization; $CH_3-\overset{\oplus}{\underset{CH_3}{C}}-O-CH_3$ is intermediate.

EFFECT OF RESONANCE ON MOLECULAR GEOMETRY

7. Resonance alters bond lengths and induces planarity in conjugated systems. If resonance introduces some double-bond character into single bonds, and so on, one result should be changes in bond length, since the order of lengths is triple < double < single.

Example

$$CH_3-C\overset{O}{\underset{O^\ominus}{\diagdown}} \longleftrightarrow CH_3-C\overset{O^\ominus}{\underset{O}{\diagdown}}$$

Predicted for either limiting structure: one C—O single bond (1.43 Å long) and one C—O double bond (1.20 Å). Observed for the real ion: both C—O bonds identical in length at 1.26 Å, a value intermediate between the single- and double-bond values.

In most cases, conversion of one limiting structure into another destroys one multiple bond and creates another (see the examples above). For the resonance phenomenon to occur and the resonance stabilization to be collected, the planes of the old and new multiple bonds must coincide. This produces more extensive regions of planarity, sometimes including the whole molecule. You can visualize this principle as resulting from the rule that limiting structures may not have different positions of the nuclei, as the following example shows.

Example

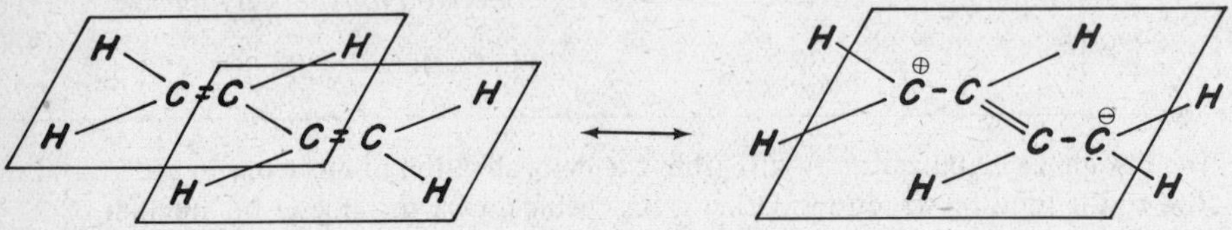

In the structure at the left, the two sets of atoms must be in the planes shown (geometry of the double bond, Unit 4), but the two planes need not coincide. In the structure at the right a different set of atoms must lie in a plane, as shown in the drawing. The only way all the atoms can occupy the same positions in the left-hand *and* right-hand structures is if the three planes drawn coincide. The molecule can rotate internally about the center C—C bond only by destroying the planarity. In order to collect the resonance stabilization, the molecule has to give up some freedom and restrict rotation about this bond. Rotation can occur, but only when the surroundings provide sufficient energy to offset the lost resonance stabilization, in this case about 3.5 kcal/mole.

Exercise

What effect(s) does resonance have on the geometry of $H-\overset{\overset{O}{\|}}{C}-NH_2 \longleftrightarrow H-\overset{\overset{O^{\ominus}}{|}}{C}=\overset{\oplus}{N}H_2$?

— — — — — — — — —

The right-hand limiting structure has every atom in the same plane:

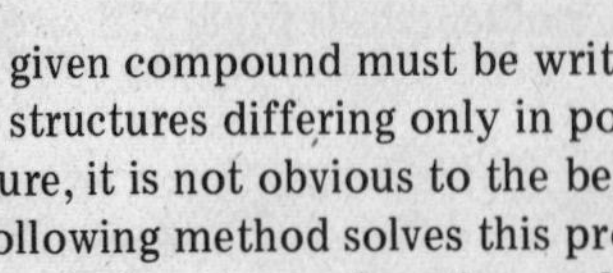

For resonance to occur, this must be the geometry of the real molecule. The C—N bond will be somewhat shorter than normal C—N single bonds; the C—O bond will be somewhat longer than normal C—O double bonds.

RECOGNIZING RESONANCE-HYBRID MOLECULES

8. A given compound must be written as a resonance hybrid if two or more reasonably stable structures differing only in position of electrons can be drawn. But, given a single structure, it is not obvious to the beginner whether another such structure can be found. The following method solves this problem. We substitute the following criterion: Whenever an electron-donor function and an electron-acceptor function are connected by a single bond, a resonance interaction exists between them. If the donor donates a pair of electrons to the acceptor, a new limiting structure results that, together with the original, makes up a resonance-hybrid structure.

This method involves learning which functions are donors and which are acceptors (see Table 9.1); this has to be done later, for other purposes, anyhow. If you use these ideas now, their later applications will make more sense. This way of looking at resonance also has the advantage of emphasizing that resonance is an *interaction* between two parts of a molecule. Electron-donor and electron-acceptor functions are normal when alone; resonance and the changes that go with it result from putting one next to the other in the same molecule.

Table 9.1 *Electron-Donor and -Acceptor Functions*

Donor functions	Acceptor functions
$^\ominus\!:\!\ddot{O}\!-$	$-C\!\!\stackrel{\oplus}{\diagup}$
$>\!\!\stackrel{\ominus}{\ddot{C}}\!-$	$>\!C\!=\!O$
$-\ddot{O}\!-$	$>\!C\!=\!N\!-$
$>\!\ddot{N}\!-$	$-C\!\equiv\!N$
$-\ddot{S}\!-$	$-N\!\!\stackrel{\diagup O}{\diagdown O}$
$>\!C\!=\!C\!<$	$>\!C\!=\!C\!<$
(benzene ring)	(benzene ring)

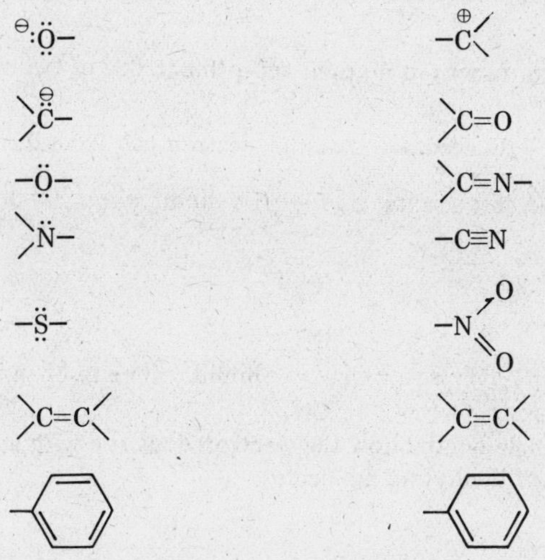

Note in the table that two functions, $>\!C\!=\!C\!<$ and (benzene ring), can play the role of either electron donor or electron acceptor.

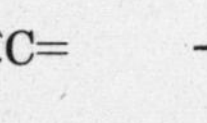

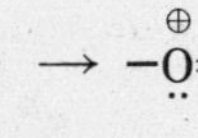

The shift of electrons from donor to acceptor is carried out by the curved arrows that frame 3 introduced. Each of the different donors operates in the same way; it donates an unshared pair (or a shared pair in a multiple bond—that is, a π pair) to the adjacent single bond to make the latter double:

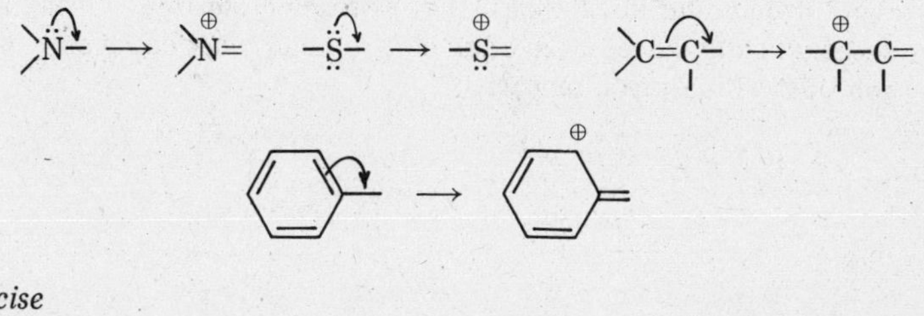

$^\ominus\!:\!\ddot{O}\!- \;\longrightarrow\; :O\!=$ $>\!\ddot{C}^\ominus\!- \;\longrightarrow\; >\!C\!=$ $-\ddot{O}\!- \;\longrightarrow\; -\overset{\oplus}{O}\!=$

$>\!\ddot{N}\!- \;\longrightarrow\; >\!\overset{\oplus}{N}\!=$ $-\ddot{S}\!- \;\longrightarrow\; -\overset{\oplus}{S}\!=$ $>\!C\!=\!\overset{\ominus}{C}\!- \;\longrightarrow\; -\overset{\oplus}{C}\!-\!C\!=$

(benzene with arrow) → (benzene cation)

Exercise

Refer back to exercise 1 in frame 3. For parts b to d, f, h, and i, circle the electron-donor function.

- - - - - - - -

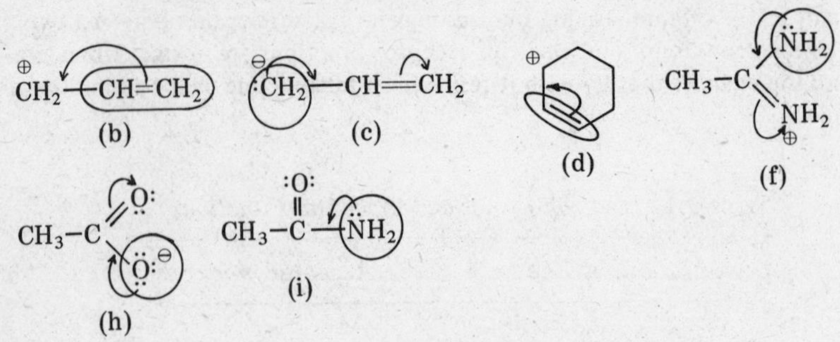

154 HOW TO SUCCEED IN ORGANIC CHEMISTRY

(b) $CH_2\!-\!CH\!=\!CH_2$ (cation, arrow from double bond to $CH_2^{\oplus}$)

(c) $\overset{\ominus}{C}H_2\!-\!CH\!=\!CH_2$ (arrow from carbanion lone pair)

(d) benzene ring with $\oplus$ substituent

(f) $CH_3\!-\!C(\ddot{N}H_2)(\overset{\oplus}{N}H_2)$

(h) $CH_3\!-\!C(\!=\!\ddot{O})(\ddot{\ddot{O}}{:}^{\ominus})$

(i) $CH_3\!-\!\overset{:\ddot{O}:}{\underset{\|}{C}}\!-\!\ddot{N}H_2$

9. The electron-acceptor functions do their accepting in one of two ways:

Only $-\overset{\oplus}{C}\!\!<$ can directly accommodate the electron pair from the donor because it is the only one with just a sextet on C: $>\!\!\overset{\oplus}{C}\!\!\curvearrowleft\!\text{donor} \longleftrightarrow >\!\!\overset{\oplus}{C}\!=\!\text{donor}$

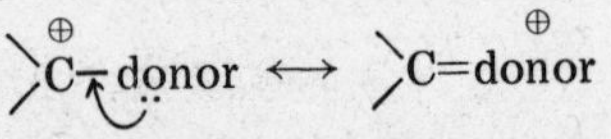

Example

According to Table 9.1, $:\!\overset{\ominus}{\ddot{O}}\!-$ and $-\overset{\oplus}{C}\!\!<\!\!\overset{CH_3}{\underset{CH_3}{}}$ should form a resonance-hybrid molecule when connected by a single bond. Show the electron donation with appropriate arrows, and write the structure of the hybrid molecule.

Solution:

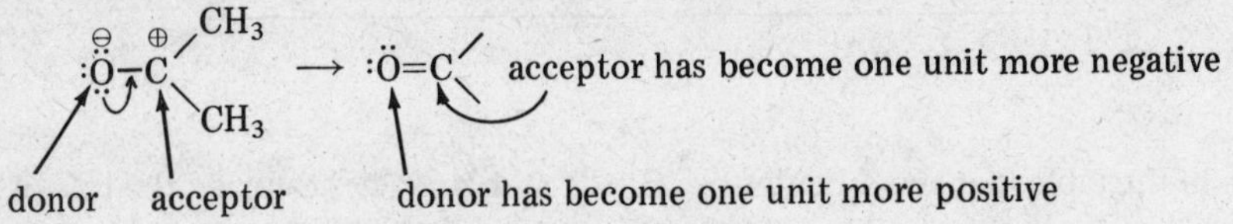

$:\!\overset{\ominus}{\ddot{O}}\!-\!\overset{\oplus}{C}\!\!<\!\!\overset{CH_3}{\underset{CH_3}{}} \longrightarrow :\!\ddot{O}\!=\!C\!\!<\!\!\overset{CH_3}{\underset{CH_3}{}}$ acceptor has become one unit more negative

donor   acceptor   donor has become one unit more positive

Hybrid structure: $:\!\overset{\ominus}{\ddot{O}}\!-\!\overset{\oplus}{C}\!\!<\!\!\overset{CH_3}{\underset{CH_3}{}} \longleftrightarrow :\!\ddot{O}\!=\!C\!\!<\!\!\overset{CH_3}{\underset{CH_3}{}}$

           minor contribu-   major contributor
           tor (violates rules
           2, 5, 6 of Unit
           8)

The remaining acceptor functions involve carbon or nitrogen atoms that already have four bonds and/or unshared pairs. Acceptance of an electron pair must be accompanied by an internal electron shift within the acceptor:

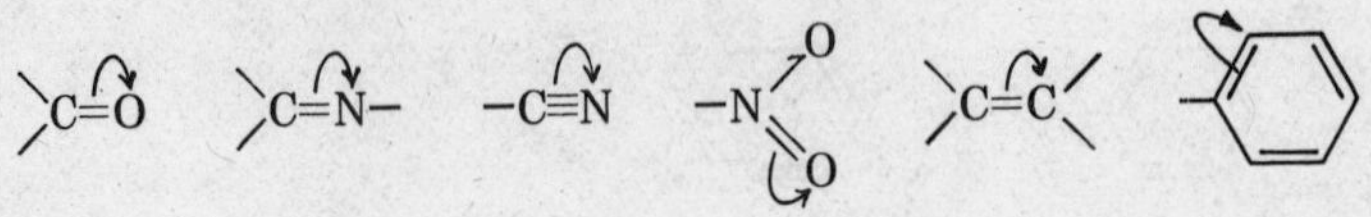

$>\!C\!=\!\ddot{O}$ $>\!C\!=\!\ddot{N}\!-$ $-C\!\equiv\!N$ $-N\!\!<\!\!\overset{O}{\underset{O}{}}$ $>\!C\!=\!C\!\!<$ (benzene ring)

Example

Combine the functions $\overset{CH_3}{\underset{CH_3}{}}\!\!>\!\overset{\ominus}{C}\!-$ and $-\overset{O}{\underset{\|}{C}}\!-\!CH_3$ to make a resonance-hybrid ion.

RESONANCE

Solution:

donor-acceptor shift →

$$\underset{\text{donor}}{\underset{CH_3}{\overset{CH_3}{>}}C}-\underset{\text{acceptor}}{\overset{\overset{\ddot{O}:}{\|}}{C}}-CH_3 \xrightarrow[\text{shift}]{\text{acceptor's internal}} \underset{CH_3}{\overset{CH_3}{>}}C=\underset{}{\overset{:\overset{\ominus}{\ddot{O}:}}{C}}-CH_3$$

Hybrid: $\underset{\ominus}{\overset{:\ddot{O}:}{>C}}\overset{\|}{-}\overset{}{C}-\leftrightarrow \overset{}{>}C=\overset{:\ddot{O}:^\ominus}{\underset{|}{C}}-$ (This is why adjacent $-\overset{O}{\overset{\|}{C}}-$ stabilizes carbanions.)

Example

Combine ⟨phenyl⟩— and ⟨phenyl⟩— to make a resonance-hybrid molecule.

Solution:

donor-acceptor shift ↘ ↙ acceptor's internal shift

[biphenyl structure with donor and acceptor labels]

Hybrid: [biphenyl] ⟷ [quinoid cation-anion form] ⟷ other analogous structures (see frame 13)

(This is one explanation of why the central C—C bond in [biphenyl] is unusually short for a single bond.)

Exercises

Identify the acceptor and donor functions in the following resonance-hybrid molecules and ions, and supply the curved arrow(s) that carry out the electron shift.

1. $CH_2=CH-\ddot{\underset{}{O}}:^\ominus \leftrightarrow {}^\ominus:CH_2-CH=\ddot{O}:$ 2. $CH_3-\ddot{\underset{}{O}}-\overset{\oplus}{C}H_2 \leftrightarrow CH_3-\overset{\oplus}{\underset{}{O}}=CH_2$

3. ${}^\ominus:CH_2-CH=\ddot{N}-H \leftrightarrow CH_2=CH-\overset{\ominus}{\ddot{N}}-H$ 4. $CH_3-\underset{\overset{\|}{:O:}}{C}-\ddot{\underset{}{S}}-CH_3 \leftrightarrow CH_3-\underset{\underset{:\ddot{O}:^\ominus}{|}}{C}=\overset{\oplus}{\underset{}{S}}-CH_3$

5. $\underset{:\ddot{O}:}{\overset{O}{>}}N-\overset{\ominus}{\ddot{C}}H_2 \leftrightarrow \underset{{}^\ominus:\ddot{O}:}{\overset{O}{>}}N=CH_2$

- - - - - - - - -

1. $\overset{\frown}{C}H_2=CH\overset{\frown}{-}\ddot{\underset{}{O}}:^\ominus \leftrightarrow {}^\ominus:CH_2-CH=\ddot{O}:$ 2. $CH_3-\overset{\frown}{\ddot{\underset{}{O}}}\overset{\frown}{-}\overset{\oplus}{C}H_2 \leftrightarrow CH_3-\overset{\oplus}{\underset{}{O}}=CH_2$

acceptor donor   donor   acceptor

3. 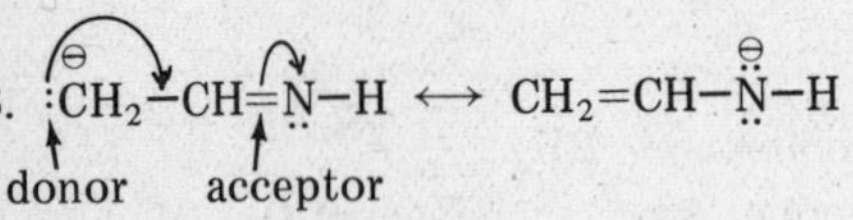 :CH₂⁻⁻CH=N−H ⟷ CH₂=CH−N̈−H   4. 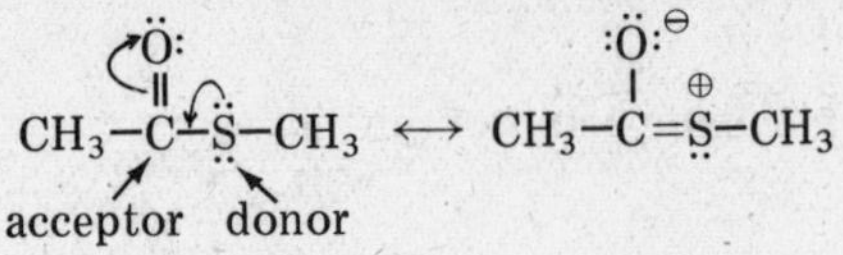 CH₃−C(=Ö:)−S̈−CH₃ ⟷ CH₃−C(−Ö:⁻)=S̈−CH₃
   donor acceptor   acceptor donor

5. 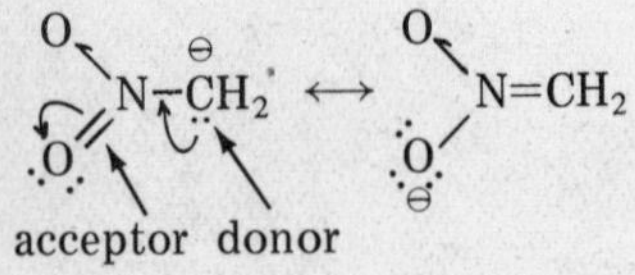 (O)(Ö:)N−CH₂⁻ ⟷ (O)(Ö:⁻)N=CH₂
   acceptor donor

10. Aromatic rings are a special case with three C=C in a loop, each playing the role of donor *and* acceptor:

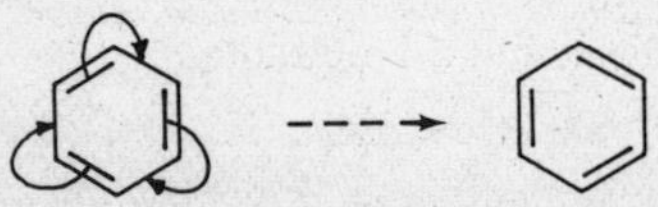

These shifts always just "realternate" the single and double bonds around the ring. Since only two arrangements are possible in a single aromatic ring, when you see one arrangement, you can always immediately write down the other by inspection and connect the two:

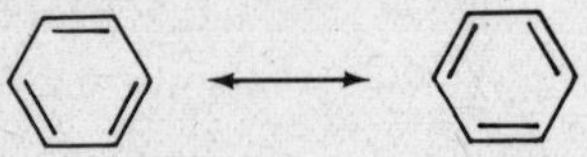

Since these two structures are perfectly equivalent, they contribute equally, and the resonance stabilization is large (36 kcal/mole). This resonance is called the *Kekulé resonance* of aromatic rings. It is in effect no matter what the environment of the aromatic ring:

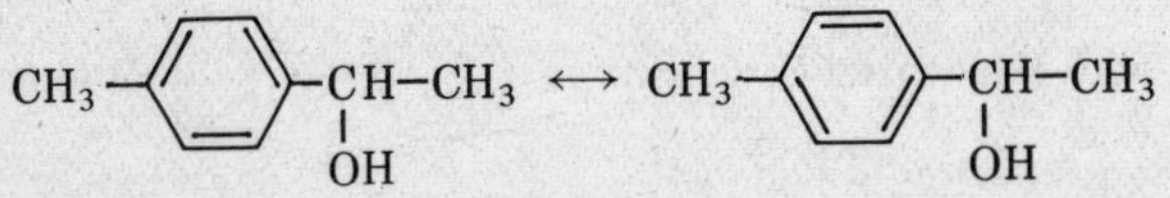

CH₃−⟨ ⟩−CH(OH)−CH₃ ⟷ CH₃−⟨ ⟩−CH(OH)−CH₃

Because of the Kekulé resonance, the six C−C bonds in the  ⟨ ⟩ ring are all identical, "1½ bonds." This resonance is a major factor in determining the chemical behavior of aromatic rings. The ⟨O⟩ notation illustrated in Unit 3 is partly motivated by convenience, but it is also a good reminder of the equivalence of all the C−C bonds and the absence of real double bonds. It represents the *hybrid* structure, not either limiting structure:

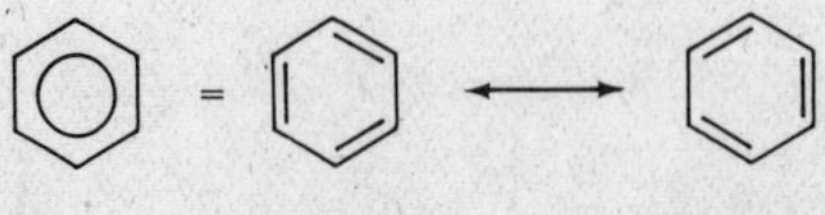

⟨O⟩ = ⟨ ⟩ ⟷ ⟨ ⟩

Exercise

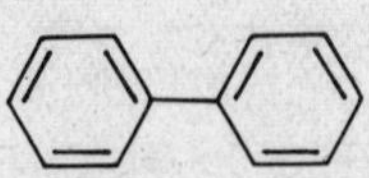

How many ways can you realternate the double bonds in ⟨ ⟩−⟨ ⟩ to make new limiting structures? How many in [naphthalene] ?

- - - - - - - - - -

[Diagram showing resonance structures of biphenyl and naphthalene with labels: "realternate right ring only", "left only", "left and right", "realternate left ring", "right ring", "left and right", "pentavalent C", "nonsense structure"]

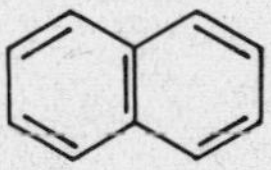

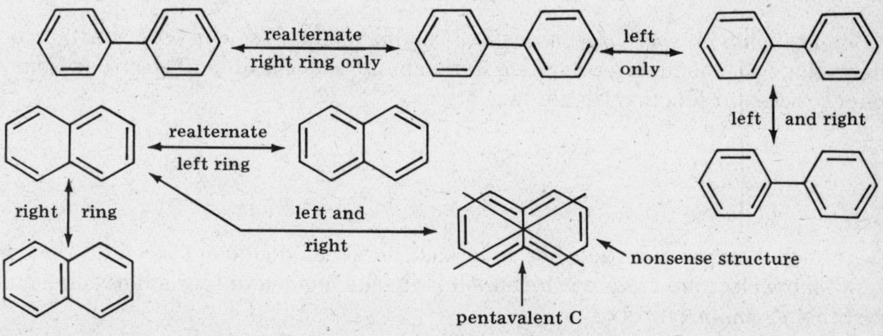

11. The last way of generating new limiting structures is to carry out the acceptor function's internal electron shift without doing anything else.

Example

Draw resonance-hybrid part structures for the acceptor functions >C=O and $-NO_2$.

Solution:

$\text{>C=Ö:} \rightarrow \text{>C-Ö:}$, so the function is really $\text{>C=O:} \leftrightarrow \text{>C-O:}^{\oplus\ \ominus}$. Note that we came to the same conclusion from the opposite direction in an example in frame 9. (This is one reason for the polarity and electron-withdrawing power of the >C=O group.)

All molecules containing an acceptor function *alone* are resonance-hybrid molecules.

Exercise

Draw resonance-hybrid part structures for the acceptor functions >C=Ṅ-, $-C\equiv N:$, and >C=C<.

- - - - - - - - - -

$\text{>C=Ṅ-} \leftrightarrow \text{>C-Ṅ-}^{\oplus\ \ominus}$ $-C\equiv N: \leftrightarrow -C=\ddot{N}:^{\oplus\ \ominus}$ $\text{>C=C<} \leftrightarrow \text{>C-C<}^{\oplus\ \ominus}$

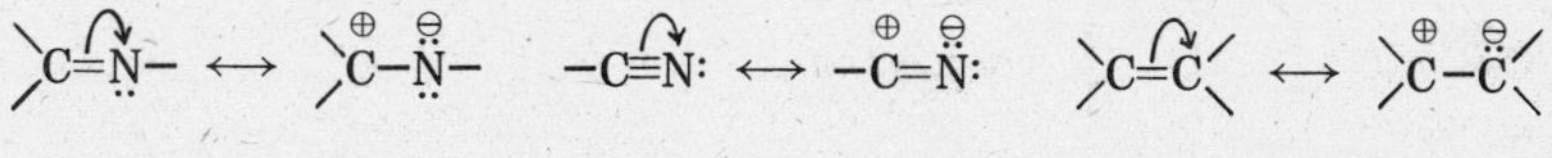

CONJUGATION

12. In Unit 3 we used the label *conjugated* for systems having two double bonds separated by a single bond. Now we can see why conjugated systems are important. Resonance is possible when two multiple bonds are placed in conjugation. Choosing an unsaturated donor and an unsaturated acceptor function from Table 9.1, for instance >C=C< and >C=O, and connecting them by a single bond gives a conjugated molecule that must have a resonance-hybrid description:

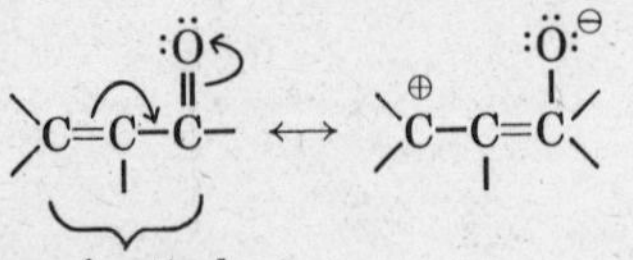

$$\text{>C=C-C(=Ö:)-} \longleftrightarrow \text{>C(⊕)-C=C(-Ö:⁻)<}$$

conjugated

Conjugation is necessary for resonance. Combining these same functions with two intervening single bonds gives a system in which the delocalization of electrons from donor to acceptor function cannot take place:

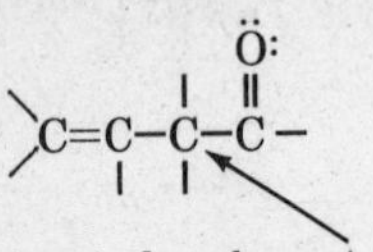

$$\text{>C=C-C-C(=Ö:)-}$$

The saturated carbon atom acts like an insulator between donor and acceptor; the donor cannot move electrons toward it because it is already bonded to four atoms and has an octet that it cannot get rid of.

The other donor functions are also said to be conjugated when they are spaced properly to interact via resonance with an unsaturated acceptor:

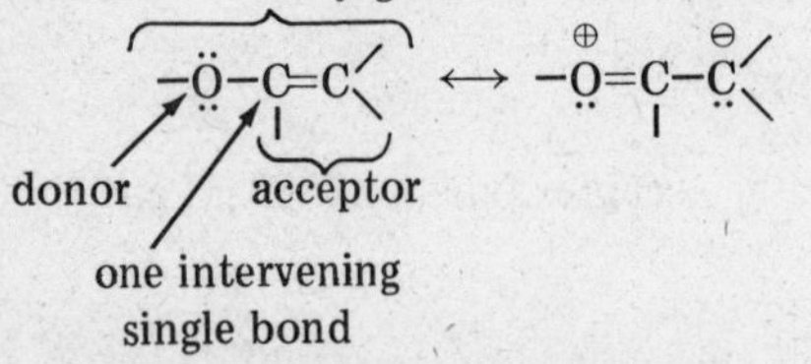

conjugated

$$\text{-Ö-C=C<} \longleftrightarrow \text{-O(⊕)=C-C(⁻)<}$$

donor / acceptor

one intervening single bond

Conjugated systems can be longer:

$$\text{>C=C-C=C-C=O} \qquad \text{-Ö-C=C-C=C<}$$

conjugated   conjugated

The only requirement is that the pattern of *alternating* multiple and single bonds must not be broken.

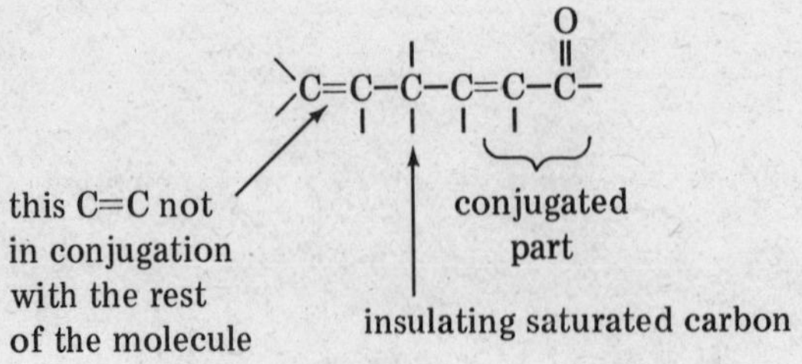

$$\text{>C=C-C-C=C-C(=O)-}$$

this C=C not in conjugation with the rest of the molecule

conjugated part

insulating saturated carbon

Extended conjugated systems like the two above can be looked at as having donor and acceptor functions conjugated "through a double bond." In such cases resonance is possible because the transfer of electrons can be done through the intervening double bond:

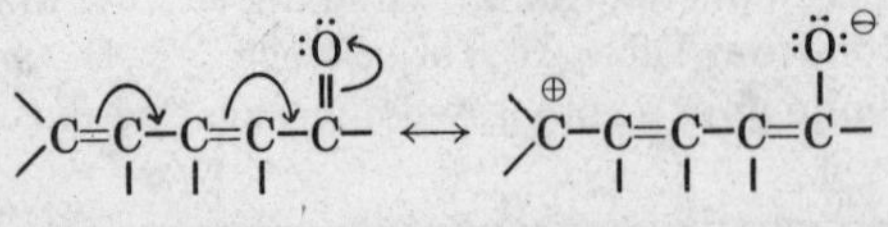

$$\text{>C=C-C=C-C(=Ö:)-} \longleftrightarrow \text{>C(⊕)-C=C-C=C(-Ö:⁻)-}$$

The more extended the conjugation, the greater the number of limiting structures, and the greater the resonance stabilization.

An aromatic ring also transmits conjugation:

$$\text{donor} \quad \underset{\text{fully conjugated connecting link}}{\text{ring}} \quad \text{acceptor}$$

Exercises

1. Combine the following donor-acceptor pairs by a single bond to get a simple conjugated system. Write the resonance-hybrid structure and show the electron shifts necessary to produce each limiting structure.

   (a) $CH_2=CH-$ and $-C\equiv N:$

   (b) $^\ominus:CH_2-$ and $-N{\overset{\oplus}{\underset{O^\ominus}{\diagup\hspace{-2pt}=\hspace{-2pt}O}}}$

2. Connect the following pair by means of an intervening $-CH=CH-$ group and write the hybrid structure for the conjugated system as in (1):

   $CH_3-\overset{..}{\underset{}{S}}-$ and $-\overset{\oplus}{C}\diagdown_{CH_3}^{CH_3}$

3. Connect the following pair via a —⬡— group and write the resonance-hybrid structure of the conjugated molecule: $CH_2=CH-$ and $-\overset{..}{N}H_2$.

- - - - - - - - -

1. (a) $CH_2\overset{\frown}{=}CH\overset{\frown}{-}C\equiv N: \longleftrightarrow \overset{\oplus}{C}H_2-CH=C=\overset{\ominus}{\underset{..}{N}}:$

   (b) $^\ominus\overset{\frown}{C}H_2\overset{\oplus}{-}N\diagdown^{\overset{\frown}{O}:}_{\underset{..}{O}:^\ominus} \longleftrightarrow CH_2=N\diagdown^{\overset{..}{O}:^\ominus}_{\underset{..}{O}:^\ominus}$

2. $CH_3-\overset{..}{\underset{}{S}}\overset{\frown}{-}CH\overset{\frown}{=}CH\overset{\oplus}{-}C\diagdown^{CH_3}_{CH_3} \longleftrightarrow CH_3-\overset{\oplus}{\underset{..}{S}}=CH-CH=C\diagdown^{CH_3}_{CH_3}$ (see frame 13)

3. $CH_2=CH-$ could be donor or acceptor, but $\overset{..}{N}H_2-$ can only be the donor; therefore,

   $H_2\overset{\frown}{N}-⬡-CH\overset{\frown}{=}CH_2 \longleftrightarrow H_2\overset{\oplus}{N}=⬡=CH-\overset{\ominus}{C}H_2$ (see frame 13)

FINDING ALL THE LIMITING STRUCTURES

13. No simple rule can tell you how many limiting structures will be important contributors to a given resonance-hybrid molecule's structure. Consequently, after you have tried the internal shift in acceptor functions and the donor-acceptor shift, you must reinspect the new limiting structure obtained to see if it now has a new juxtaposition of acceptor and donor functions capable of yielding a third limiting structure, and so on. You must repeat this until the new limiting structure has no more capability for electron shifts, or until you get the original structure back again. As always, the secret is to make the shifts systematically.

Example

Write the complete resonance-hybrid structure for benzyl cation, Ph–CH$_2^\oplus$.

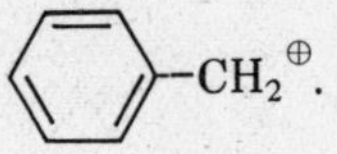

Solution:

Identify functions: $>\!\!-CH_2^\oplus$ is $>\!\!C\!-\!CH_2^\oplus$ ← acceptor; C ← donor

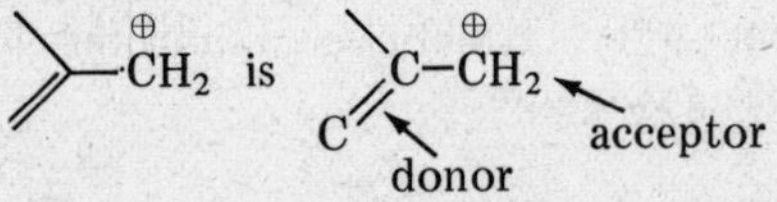

Shift electrons and formulate hybrid: [benzyl cation] ↔ [cyclohexadienyl cation with CH$_2$=]

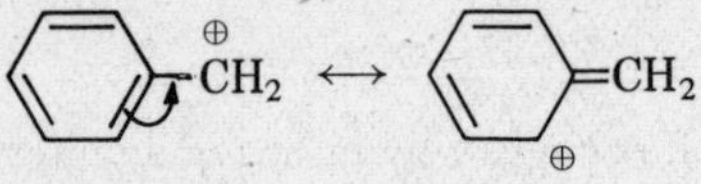

Examine second structure: donor → [ring] =CH$_2$, acceptor → Therefore, it is capable of further shift.

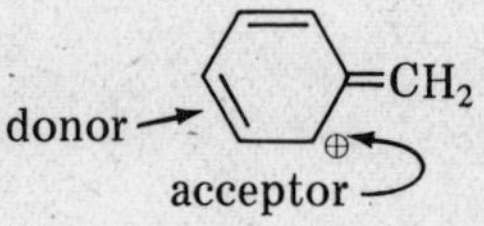

Second shift: [structure with ⊕ on ring] =CH$_2$ ↔ ⊕[ring]=CH$_2$

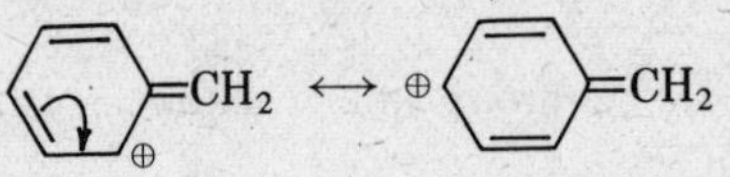

Examine third structure: ⊕[ring]=CH$_2$ ← donor; acceptor → Therefore, repeat.

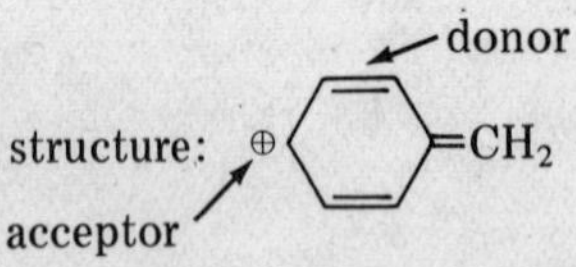

Third shift: ⊕[ring]=CH$_2$ ↔ [ring with ⊕]=CH$_2$

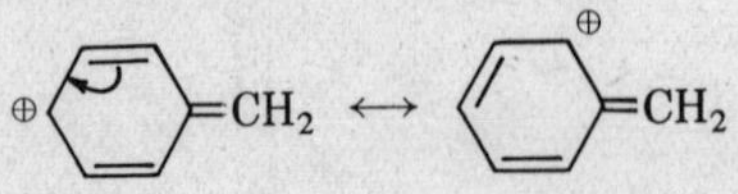

Examine: acceptor ⊕[ring]=CH$_2$ ← donor   Therefore, repeat.

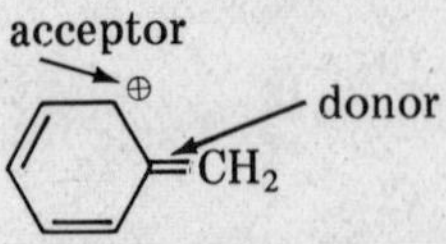

Fourth shift: ⊕[ring]=CH$_2$ ↔ [ring]–CH$_2^\oplus$ Original structure; therefore, stop.

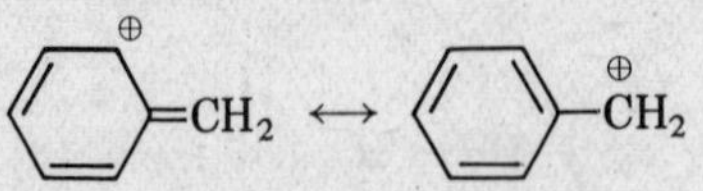

Collect all limiting structures in hybrid notation:

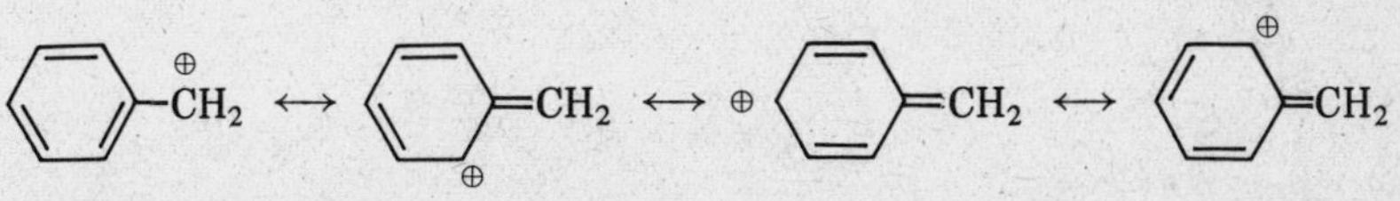

RESONANCE

[Resonance structures of benzyl cation showing four limiting structures with CH₂⁺ and ring charges]

Remember that the longer the conjugated system, the more limiting structures there are. Locate the donor function and move the electrons from it along the conjugated system one step at a time, getting a new limiting structure after each step.

Example

Exercises 2 and 3 in frame 12 have more limiting structures than are indicated there, for example:

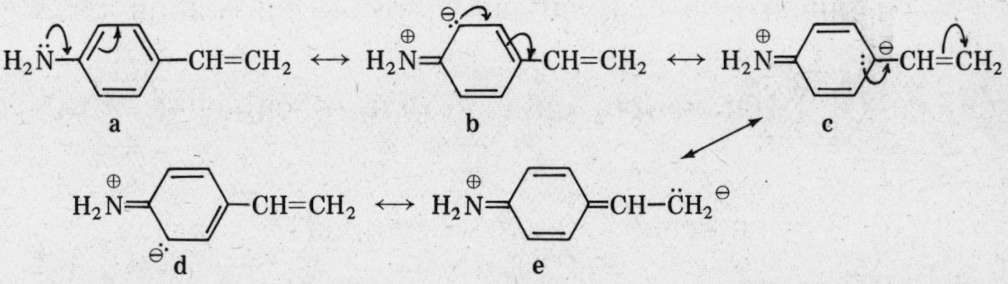

[Resonance structures a, b, c of H₂N-C₆H₄-CH=CH₂ showing electron movement]

a   b   c

[Resonance structures d and e]

d   e

Whenever you have a limiting structure with a charge on one of the carbons of an aromatic ring (for example, b above), there will be two more, similar structures that put that charge on alternate carbons around the ring (c and d in the case above).

★ *Exercises*

Write a complete resonance-hybrid structure for each molecule.

1. 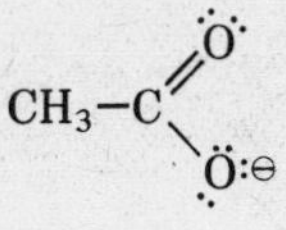 $CH_3-C(\ddot{\ddot{O}})(\ddot{\ddot{O}}{:}^{\ominus})$

2. $C_6H_5-\ddot{\ddot{O}}{:}^{\ominus}$

3. [cyclopentadienyl anion] $:^{\ominus}$

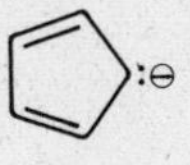

4. $CH_2=CH-\overset{O}{\overset{\|}{C}}-CH=CH_2$

5. $:CH_2^{\ominus}-CH=CH-CH=O$

6. 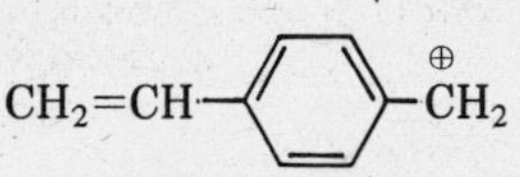 $CH_2=CH-C_6H_4-\overset{\oplus}{C}H_2$

7. $CH_3-O-C_6H_4-\overset{\oplus}{C}H_2$

---

1. $CH_3-C(\overset{\ddot{O}}{\|})(\ddot{\ddot{O}}{:}^{\ominus}) \longleftrightarrow CH_3-C(\ddot{\ddot{O}}{:}^{\ominus})(\ddot{\ddot{O}})$

2. [Resonance structures of phenoxide: O⁻ on oxygen ↔ negative charge on ortho, para, ortho carbons of ring with C=O, and back to phenoxide]

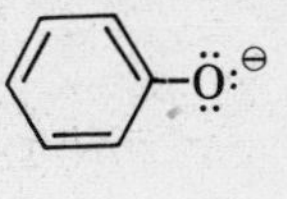

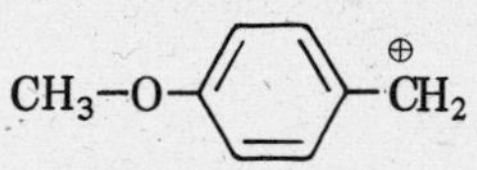

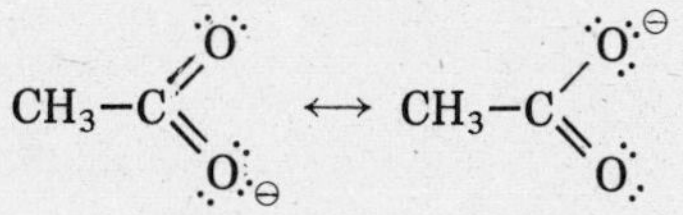

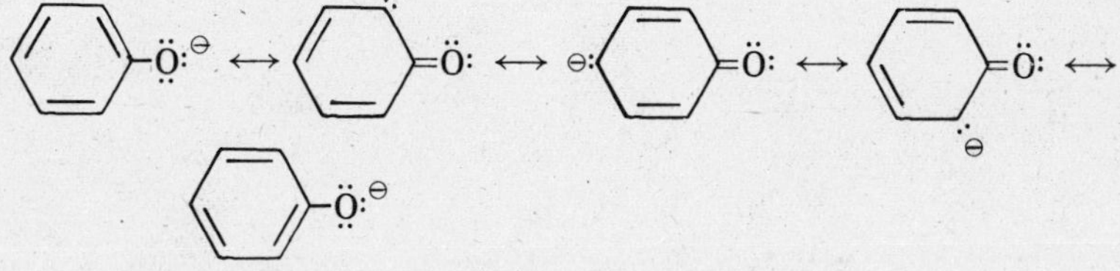

3. [resonance structures of cyclopentadienyl anion with curved arrows showing electron movement]

4. $CH_2=CH-\overset{\overset{\ddot{O}}{\|}}{C}-CH=CH_2 \longleftrightarrow CH_2-CH=\overset{\overset{:\ddot{O}:^\ominus}{|}}{C}-CH=CH_2 \longleftrightarrow CH_2=CH-\overset{\overset{:\ddot{O}:^\ominus}{|}}{C}=CH-\overset{\oplus}{CH_2} \longleftrightarrow CH_2=CH-\overset{\overset{:\ddot{O}:^\ominus}{|}}{\underset{\oplus}{C}}-CH=CH_2$

5. $^\ominus\ddot{C}H_2-CH=CH-CH=\ddot{O}: \longleftrightarrow CH_2=CH-\overset{\ominus}{C}H-CH=\ddot{O}: \longleftrightarrow CH_2=CH-\overset{\ominus}{C}H-CH=O:$

6. $CH_2=CH-\underset{}{\bigcirc}-\overset{\oplus}{CH_2} \longleftrightarrow CH_2=CH-\bigcirc=CH_2 \longleftrightarrow CH_2=CH-\bigcirc=CH_2$
$\longleftrightarrow \overset{\oplus}{CH_2}-CH=\bigcirc=CH_2 \longleftrightarrow CH_2=\overset{\oplus}{CH}-\bigcirc=CH_2$

7. $CH_3-O-\bigcirc-\overset{\oplus}{CH_2} \longleftrightarrow CH_3-O-\overset{\oplus}{\bigcirc}=CH_2 \longleftrightarrow CH_3-O-\bigcirc=CH_2$
$\longleftrightarrow CH_3-O-\underset{\oplus}{\bigcirc}=CH_2 \longleftrightarrow CH_3-\overset{\oplus}{O}=\bigcirc=CH_2$

SOME NOTATIONAL PROBLEMS

14. The symbol $\longleftrightarrow$ denotes *resonance hybrid*. It connects two structures that are so similar that their only difference is in the position of *electrons*.

The symbol $\rightleftharpoons$ denotes *chemical equilibrium*. The structures it connects differ in the positions of electrons and at least some *nuclei*, and they represent different compounds.

Example

Supply the proper symbol between a and b.

1. $CH_3-C\overset{\ddot{O}:}{\underset{\ddot{O}:^\ominus}{\diagup}}$ $CH_3-C\overset{\ddot{O}:^\ominus}{\underset{\ddot{O}:}{\diagup}}$
   a   b

2. $CH_3-\overset{O}{\overset{\|}{C}}-\overset{H}{\underset{}{C}}\overset{H}{\diagdown H}$ $CH_3-\overset{O-H}{\underset{}{C}}=C\overset{H}{\diagdown H}$
   a   b

Solutions:

1. The arrows do turn a into b. Only electrons moved. Therefore, ⟷.

2. Nothing is wrong with the arrows, but O and H are not within bonding distance in a, and consequently the H has moved drastically in structure b. Therefore, ⇌.

★ *Exercises*

Supply the missing symbol, ⟷ or ⇌, connecting the structures.

1. 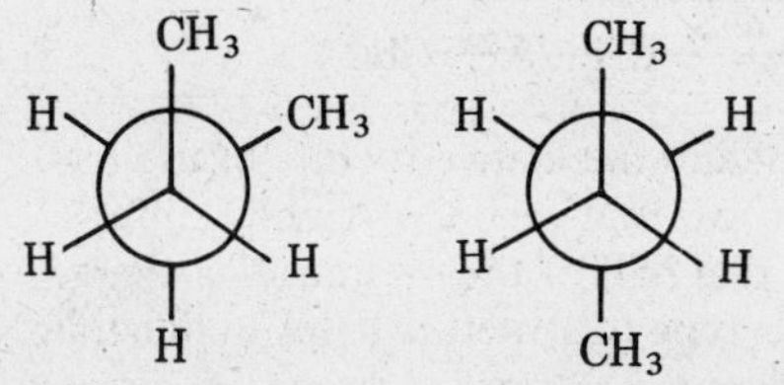 [Newman projection with CH$_3$, CH$_3$, H, H, H, H]   [Newman projection with CH$_3$, H, H, H, H, CH$_3$]

2. $\underset{H}{\overset{H}{>}}\overset{\ominus}{C}-C\equiv N:$ $\underset{H}{\overset{H}{>}}C=C=\overset{\ominus}{\underset{..}{N}}:$

3. $\underset{H}{\overset{H}{>}}H-C-C\equiv N:$ $\underset{H}{\overset{H}{>}}C=C=\overset{..}{N}-H$

4. $CH_3-\overset{\oplus}{\underset{\underset{CH_3}{|}}{C}}-\overset{CH_3}{\underset{}{|}}C=CH_2$ $CH_3-\overset{CH_3}{\underset{\underset{CH_3}{|}}{C}}=C-\overset{\oplus}{C}H_2$

5. $CH_3-\overset{\oplus}{C}-\overset{CH_3}{\underset{\underset{CH_3}{|}}{C}}-CH_3$ $CH_3-\overset{CH_3}{\underset{\underset{CH_3}{|}}{C}}-\overset{\oplus}{C}-CH_3$

- - - - - - - - - - - -

1. 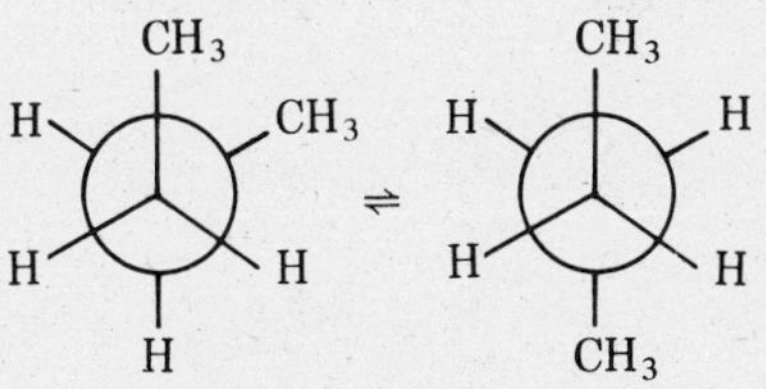 [Newman projection with CH$_3$, CH$_3$, H, H, H, H] ⇌ [Newman projection with CH$_3$, H, H, H, H, CH$_3$]   Five nuclei move.

2. 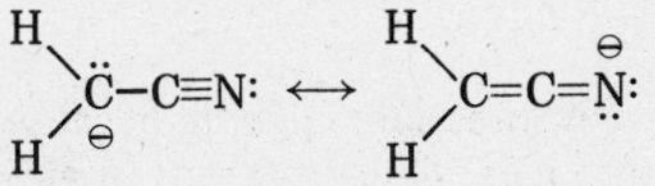 $\underset{H}{\overset{H}{>}}\overset{\ominus}{\underset{..}{C}}-C\equiv N:$ ⟷ $\underset{H}{\overset{H}{>}}C=C=\overset{\ominus}{\underset{..}{N}}:$

3. 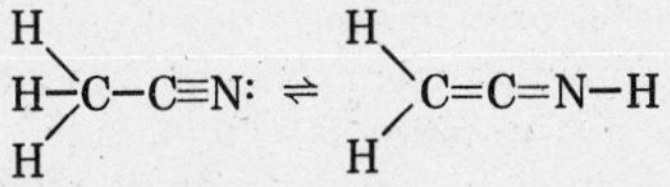 $\underset{H}{\overset{H}{>}}H-C-C\equiv N:$ ⇌ $\underset{H}{\overset{H}{>}}C=C=N-H$ H nucleus moves.

4. 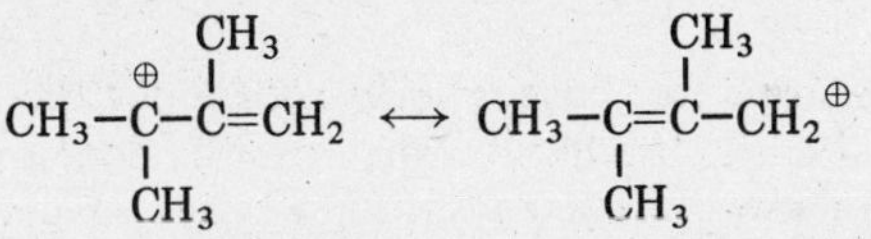 $CH_3-\overset{\oplus}{\underset{\underset{CH_3}{|}}{C}}-\overset{CH_3}{\underset{}{|}}C=CH_2$ ⟷ $CH_3-\overset{CH_3}{\underset{\underset{CH_3}{|}}{C}}=C-\overset{\oplus}{C}H_2$

164 HOW TO SUCCEED IN ORGANIC CHEMISTRY

5. 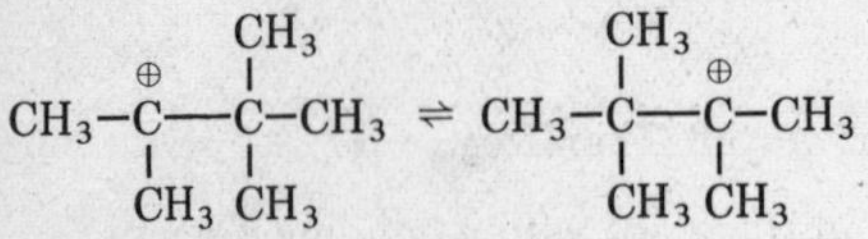 $CH_3-\overset{\oplus}{\underset{\underset{CH_3}{|}}{\overset{\overset{CH_3}{|}}{C}}}-\underset{\underset{CH_3}{|}}{\overset{}{C}}-CH_3 \rightleftharpoons CH_3-\underset{\underset{CH_3}{|}}{\overset{\overset{CH_3}{|}}{C}}-\overset{\oplus}{\underset{\underset{CH_3}{|}}{C}}-CH_3$ CH_3 moves.

15. The resonance-hybrid notation is clumsy. Sometimes we try to streamline it by drawing a single, average structure using dashed lines for fractional bonds. For instance, in place of $CH_2=CH-CH_2^{\oplus} \longleftrightarrow {}^{\oplus}CH_2-CH=CH_2$ we might substitute

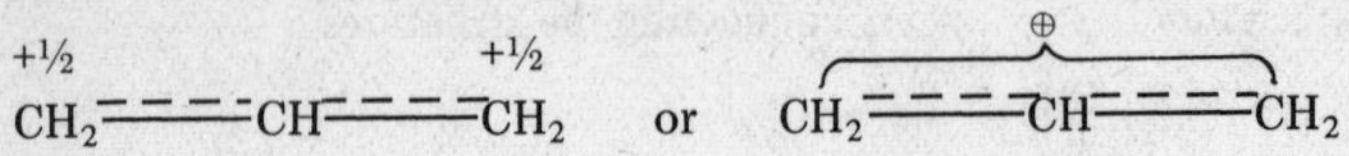

$$\overset{+\frac{1}{2}}{CH_2}=\!=\!=\!=CH=\!=\!=\!=\overset{+\frac{1}{2}}{CH_2} \quad \text{or} \quad CH_2=\!=\!=\!=\overset{\overset{\oplus}{\frown}}{CH}=\!=\!=\!=CH_2$$

These average structures give a direct picture of the intermediate structure of the real molecule. In the cation above, for example, they show that each C—C bond is intermediate between a single and a double bond, and that half of the unit positive charge is borne on each of the end carbons. Although this type of structure helps to illustrate the meaning of the resonance phenomenon and saves pencil work, I discourage you from relying on it in preference to the traditional $\longleftrightarrow$ notation for several reasons. (a) Such structures may be less accurate in complex molecules. For example, the equivalent of the $\longleftrightarrow$ structure for the ⌬—$CH_2^{\oplus}$ cation is $+\delta$ ⌬$=\!=\!=\!=CH_2$. This implies that all the C—C bonds have the same mixture of single- and double-bond character, which is false. The $\longleftrightarrow$ notation, on the other hand, is capable of expressing every possible subtlety of electronic structure by including a sufficient number of limiting structures and giving them proper weights. (b) There is no easy way for you to generate an accurate $=\!=\!=\!=\!=\!=$ structure without generating the $\longleftrightarrow$ structure first. (c) Most instructors do not consider a $=\!=\!=\!=\!=\!=$ structure an adequate answer to the "Draw a complete resonance-hybrid structure for compound X" problem.

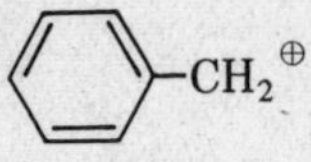

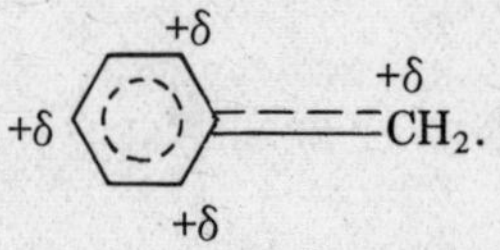

Exercise

Draw a single structure that summarizes some important features of the resonance-hybrid ion $CH_3-CO_2^{\ominus}$.

- - - - - - - - - -

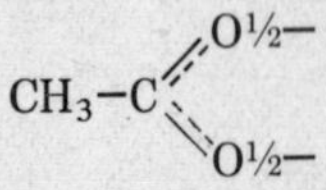

$$CH_3-C\underset{\diagdown O\frac{1}{2}-}{\overset{\diagup O\frac{1}{2}-}{}}$$

16. The notation $CH_2\overset{\frown}{=}CH\overset{\oplus}{-}CH_2 \longleftrightarrow \overset{\oplus}{CH_2}-CH=CH_2$ confuses some people because it combines curved and doubleheaded arrows in a single statement. This is actually a way of combining in one step (a) the operation of getting the second limiting structure, (b) the writing of the second structure, and (c) the act of combining them conceptually into the resonance hybrid. This saves writing. When the same $\frown$ notation is applied to writing reaction mechanisms (Unit 10), the saving of rewriting intermediate stages becomes even greater.

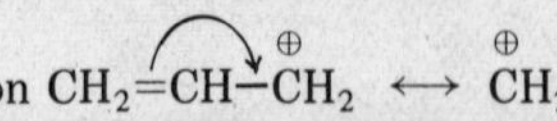

The curved arrow on the left-hand limiting structure above *is not part of that structure.* You should read it *after* you read the structure. It tells you what must be done to that structure to turn it into the following structure.

UNIT TEN

Mechanism

The mechanism of a reaction is an account of *how* the reaction occurs. To understand why the observed products and not others are formed, and to be able to predict the products of related reactions, you must understand the mechanism of a reaction. This knowledge is necessary for you also to understand how the rate of reaction depends on the structure of the reactants and how the spatial arrangement of groups in reactants is affected by the reaction. The mechanism also gives you an additional way of remembering a list of reactant-product combinations.

The vast majority of organic reactions take place by just nine standard mechanisms. In this unit, after considering the general characteristics of all mechanisms and practicing the standard notation for writing them, we survey these nine standard mechanisms. Each of the nine is treated in detail in later chapters, where reactions employing it first come up. You can use Table 10.1 as an index for locating these detailed discussions.

Other aspects of writing and understanding mechanisms are treated in:

- Unit 12, frames 24 to 26 (the "Write the mechanism for the following reaction" problem).
- Unit 13, frames 12 to 14 (stereochemical consequences of the standard mechanisms).

OBJECTIVES

When you have completed this unit, you should be able to:

- Read and write standard mechanistic notation (frames 1 to 5).
- Distinguish plausible from implausible mechanisms (frames 1, 4).

If you think you may already have achieved these objectives and may be able to skip all or part of this unit, take a self-test consisting of the exercises marked with a star in the margin. If you miss any questions, study the frames involved and rework the questions before proceeding.

If you are not ready for a self-test, proceed to frame 1.

GENERAL CHARACTERISTICS OF MECHANISMS

1. A reaction mechanism is a concise history of the detailed way in which reactant molecules are transformed into product molecules. It specifies the order in which bond breaking and bond making proceed, and identifies the structures of all reaction intermediates formed between reactants and products. It does this by breaking up the overall reaction into discrete steps—one step for the formation of each intermediate involved and one for the formation of the final product. Each step written down is thought to represent a real chemical event in the reaction mixture.

Each written step involves as reactant(s) either a single molecule or ion (unimolecular step) or two molecules or ions (bimolecular step). This corresponds to decomposition of a single molecule (ion) or a collision between two molecules (ions) that results in reaction. Since simultaneous collisions of three or more molecules or ions are rare events, we ordinarily do not write mechanistic steps involving three (termolecular step) or more molecules or ions. However, since solvent is always in contact with all molecules in solution, a solvent molecule can participate in a mechanistic step without regard to this rule.

Each single step in a reaction mechanism usually breaks and/or makes one or two bonds. The sum of all the steps in a reaction mechanism is the equation for the overall reaction.

The mechanism of each organic reaction is deduced from experimental results. It cannot be deduced from theory, although similar reactions have similar mechanisms, and the reaction mechanism for the ethyl compound, for example, can usually be successfully carried over to the propyl compound by analogy. With practice you can learn to extend such analogies to the prediction of mechanisms for unfamiliar reactions with considerable success.

Exercises

Given the mechanism

$$CH_3-\ddot{O}-H \quad H-\overset{\oplus}{O}{\scriptstyle\begin{array}{c}H\\H\end{array}} \longrightarrow CH_3-\overset{\oplus}{O}{\scriptstyle\begin{array}{c}H\\H\end{array}} + \ddot{O}{\scriptstyle\begin{array}{c}H\\H\end{array}}$$

$$:\overset{\ominus}{Br}: \quad CH_3-\overset{\oplus}{O}{\scriptstyle\begin{array}{c}H\\H\end{array}} \longrightarrow :\ddot{Br}-CH_3 + \ddot{O}{\scriptstyle\begin{array}{c}H\\H\end{array}}$$

1. Write the balanced equation for the net, overall reaction.
2. Label each step as unimolecular, bimolecular, or termolecular.
3. Determine the number of bonds made and broken in each step.

— — — — — — — — —

1. Add the two sides separately to get

$$CH_3OH + H_3O^\oplus + Br^\ominus + CH_3OH_2^\oplus \longrightarrow CH_3OH_2^\oplus + H_2O + Br-CH_3 + H_2O$$

Cancel species that occur on both sides ($CH_3OH_2^\oplus$), giving

$$CH_3OH + H_3O^\oplus + Br^\ominus \longrightarrow H_2O + Br-CH_3 + H_2O$$

Collect species occurring more than once on the same side (H_2O):

$$CH_3OH + H_3O^\oplus + Br^\ominus \longrightarrow 2\, H_2O + Br-CH_3$$

2. 1st step and 2nd step = bimolecular.

3. 1st step: 1 bond broken = $H-\overset{\oplus}{O}{\scriptstyle\begin{array}{c}H\\H\end{array}}$, 1 bond made = $CH_3-\overset{\oplus}{O}{\scriptstyle\begin{array}{c}H\\H\end{array}}$

   2nd step: 1 bond broken = $CH_3-\overset{\oplus}{O}{\scriptstyle\begin{array}{c}H\\H\end{array}}$, 1 bond made = $Br-CH_3$

MECHANISTIC NOTATION

2. We use a standard notation for writing reaction mechanisms. It describes the individual events of bond making, bond breaking, charge creation, and charge neutralization by mapping out the *movements of valence electrons* in and between molecules. The movements of the nuclei and uninvolved electrons are not mapped; they are assumed to take up their equilibrium positions automatically following any valence electron shift. A full-headed curved arrow (⤻) indicates movement of an electron pair; a half-headed one (⇁) moves a single (unpaired) electron. The locations between which the electrons are moved are the atomic orbitals that house unshared electron pairs (and unpaired single electrons) and the molecular orbitals that house shared (bonding) electron pairs.

If you have studied Unit 9, you are already acquainted with this notation, because it is exactly the same as the one used to convert one limiting structure into another. The only difference is that in the case of limiting structures there is no accompanying movement of the nuclei—only the electrons move.

Let's illustrate the notation using some simple acid-base (proton transfer) reactions.

Consider first the reactions that take place when HCl gas is dissolved in aqueous ammonia. The acid's proton can be transferred to either of two bases by these one-step mechanisms:

$$:\ddot{Cl}-H \quad :\ddot{O}\!\!\begin{array}{c}H\\ \\H\end{array} \longrightarrow :\ddot{Cl}:^{\ominus} + H-\overset{\oplus}{O}\!\!\begin{array}{c}H\\ \\H\end{array} \tag{1}$$

$$:\ddot{Cl}-H \quad :N\!\!\begin{array}{c}H\\-H\\H\end{array} \longrightarrow :\ddot{Cl}:^{\ominus} + H-\overset{\oplus}{N}\!\!\begin{array}{c}H\\-H\\H\end{array} \tag{2}$$

These reactions use two curved arrows each because two bonds are altered; the right-hand arrow *makes* the new H—base bond; the left-hand arrow *breaks* the old Cl—H bond. Verify for yourself that a new bond or unshared pair appears in the products where the head of each arrow pointed. Verify that each old bond or unshared pair located at the tail of an arrow has disappeared in the products. The arrows move electrons only. The plus sign connecting the two reactant molecules is missing; it is often omitted to make room for the arrows.

Just as in the problem of generating one limiting structure from another (Unit 9), the curved arrow on or between reactant molecules is not a part of their structure(s). It is the *operation* which, if carried out in the mind's eye, changes reactants into products. The eye should read the structure first (e.g., Cl—H), then let the arrow operate on it (C͡l—H→).

Reactions (1) and (2) are bimolecular; they require collision of two reactant molecules. Here is an example of a termolecular reaction:

$$:\ddot{Cl}-H \quad :\ddot{O}-H \quad :N\!\!\begin{array}{c}H\\-H\\H\end{array} \longrightarrow :\ddot{Cl}:^{\ominus} + H-\ddot{O}: + H-\overset{\oplus}{N}\!\!\begin{array}{c}H\\-H\\H\end{array} \tag{3}$$

This is one of those feasible termolecular steps, mentioned in frame 1, in which the solvent is one of the three colliding reactant molecules. Verify that the products contain exactly the bonds and unshared pairs dictated by the arrows. Note that the center arrow destroys one bond and makes another.

The mechanism of a unimolecular process looks like this:

$$:\ddot{Cl}-H \longrightarrow :\ddot{Cl}:^{\ominus} + H^{\oplus} \tag{4}$$

This particular one does *not* occur because $H^{\oplus}$ never exists in the free state in solutions

168 HOW TO SUCCEED IN ORGANIC CHEMISTRY

(Units 8, 11). [Reactions (1) to (3) *are* observed in aqueous ammonia solutions.]

A two-step mechanism that accomplishes the same thing as (3) is this one:

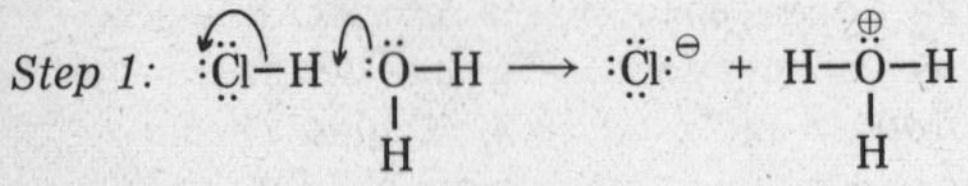

Step 1: $\ddot{Cl}-H + \ddot{O}(H)-H \longrightarrow :\!\ddot{Cl}:^{\ominus} + H-\overset{\oplus}{O}(H)-H$ (5)

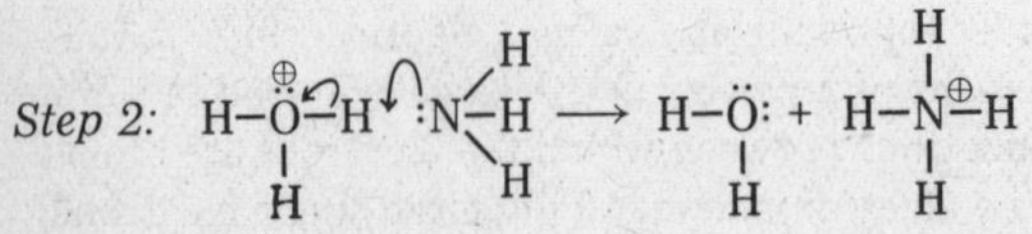

Step 2: $H-\overset{\oplus}{O}(H)-H + :N(H)(H)H \longrightarrow H-\ddot{O}(H) + H-\overset{\oplus}{N}(H)(H)H$ (6)

Verify that the balanced equation for the net, overall reaction is the same as that for reaction (3). In contrast to mechanism (3), this mechanism does not imply a three-way collision of reactants. It uses two successive two-way collisions instead. The species $H-\overset{\oplus}{O}(H)-H$, which is formed in the first step and destroyed in the second step, is called an *intermediate* in the reaction.

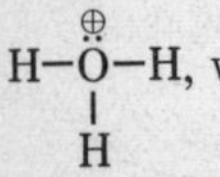

Now consider a mechanism that involves a resonance-hybrid molecule. In this mechanism the base (H_2O) removes a proton from the acid and then puts it back in a new location.

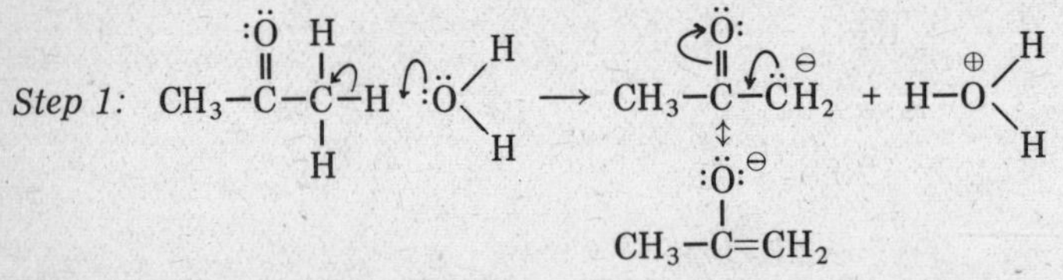

Step 1: $CH_3-\overset{\ddot{O}}{\overset{\|}{C}}-\overset{H}{\underset{H}{C}}-H + :\ddot{O}(H)(H) \longrightarrow CH_3-\overset{\ddot{O}}{\overset{\|}{C}}-\ddot{C}H_2^{\ominus} + H-\overset{\oplus}{O}(H)(H)$ (7)

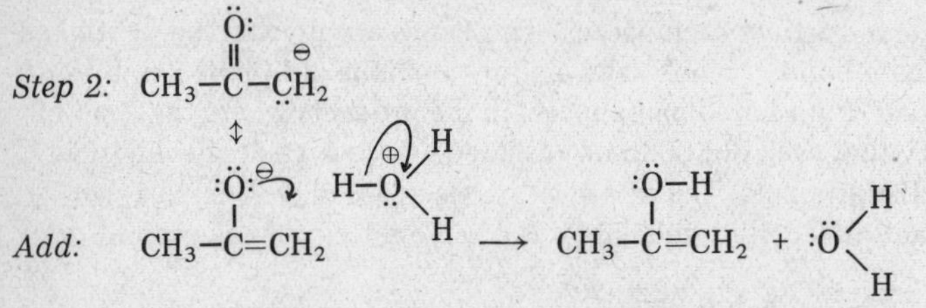

$\updownarrow$

$CH_3-\overset{:\ddot{O}:^{\ominus}}{\underset{}{C}}=CH_2$

Step 2: $CH_3-\overset{\overset{\ddot{O}:}{\|}}{C}-\overset{\ominus}{C}H_2$

$\updownarrow$

Add: $CH_3-\overset{:\ddot{O}:^{\ominus}}{\underset{}{C}}=CH_2 + H-\overset{\oplus}{O}(H)(H) \longrightarrow CH_3-\overset{:\ddot{O}-H}{\underset{}{C}}=CH_2 + :\ddot{O}(H)(H)$ (8)

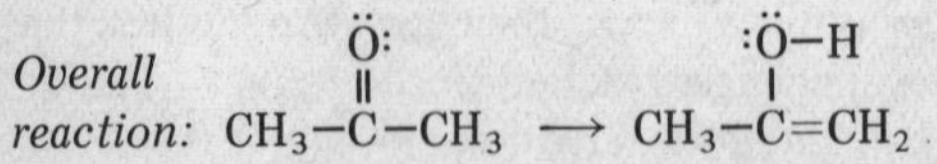

Overall reaction: $CH_3-\overset{\ddot{O}}{\overset{\|}{C}}-CH_3 \longrightarrow CH_3-\overset{:\ddot{O}-H}{\underset{}{C}}=CH_2$ (H_2O, $H_3O^{\oplus}$ cancel out in addition)   (9)

The first new item here is seen on the right side of step 1. The arrows there convert the top limiting structure of the hybrid into the bottom limiting structure. Verify that they do. This mixing of mechanistic and resonance arrows is perfectly valid because the two notations are identical. The second point leads to an important general principle. Equation (8) portrays step 2 as proceeding through the bottom limiting structure. But *the mechanism of any reaction of a resonance-hybrid molecule can be written using any of its limiting structures*. So we could have written step 2 using the top structure. A different set of arrows is required:

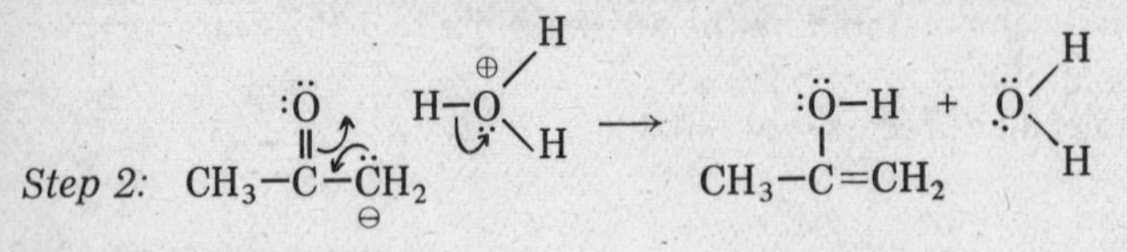

Step 2: $CH_3-\overset{\overset{:\ddot{O}}{\|}}{C}-\overset{\ominus}{C}H_2 + H-\overset{\oplus}{O}(H)(H) \longrightarrow CH_3-\overset{:\ddot{O}-H}{\underset{}{C}}=CH_2 + :\ddot{O}(H)(H)$ (10)

The result of (10) is precisely the same as that of (8).

Some mechanisms involve movement of single, unpaired electrons. Again the notation employed is exactly that used for interconverting limiting structures. A half-headed arrow moves a single electron, as in step (11):

$$H-H \quad \cdot \ddot{B}r: \longrightarrow H\cdot + H-\ddot{B}r: \tag{11}$$

The H—H symbol portrays an electron-pair bond breaking homolytically, that is, sending one electron off with each fragment: $H\cdot\cdot H \longrightarrow H\cdot + \cdot H$.

★ Exercises

1. Supply the products that result from the electron shift(s).

   (a) Br—Br

   (b) Br—Br

   (c) Li—H

   (d) $CH_3-\ddot{O}-\ddot{C}l:$

   (e) $CH_3-\overset{\oplus}{\underset{CH_3}{O}}-H \quad :\ddot{O}\underset{H}{\overset{H}{<}}$

   (f) $H-\overset{H}{\underset{:\ddot{C}l:}{C}}-\overset{\oplus}{\underset{H}{C}}\overset{H}{<}_H$

   (g) $CH_3-CH=CH-CH_3$
       $\oplus CH_2-CH_3$

   (h) $F-\overset{F}{\underset{F}{\overset{|}{B}}}{}^{\ominus}-F$

   (i) $CH_3-\overset{\oplus}{\underset{CH_3}{O}}-CH_3 \quad :\ddot{I}:$

   (j) $H\ddot{O}:\quad H-\overset{H}{\underset{}{\diagup}}\text{(cyclic ketone)} \quad \ddot{O}:$

   (k) $H\ddot{O}:\quad H-\text{(cyclic ketone with }:\ddot{O}:)$

   (l) $H\ddot{O}:\quad H-\text{(cyclic ketone)} \quad H-\ddot{C}l:$

   (m) $CH_3-CH_2 \quad Br-Br$

   (n) $CH_3-\ddot{O}\cdot \quad H-CH_3$

   (o) $:N\equiv C-\ddot{O}:^{\ominus} \quad CH_3-I$
       $\updownarrow$
       $^{\ominus}:\ddot{N}=C=\ddot{O}:$

   (p) $:N\equiv C-\ddot{O}:^{\ominus}$
       $\updownarrow$
       $^{\ominus}:\ddot{N}=C=\ddot{O}: \quad CH_3-I$

   (q) $:N\equiv C-\ddot{O}:^{\ominus}$
       $\updownarrow$
       $^{\ominus}:\ddot{N}=C=\ddot{O}: \quad CH_3-I$

2. Supply the curved arrows that accomplish each chemical change.

   (a) $:\ddot{B}r\cdot\ddot{I}: \longrightarrow :\ddot{B}r:^{\ominus} + \ddot{I}:^{\oplus}$

   (b) $H-\ddot{\underset{\ominus}{O}}: \quad H-\overset{\oplus}{\underset{H}{O}}\overset{H}{<} \longrightarrow H-\ddot{O}-H + :\ddot{O}\overset{H}{\underset{H}{<}}$

170   HOW TO SUCCEED IN ORGANIC CHEMISTRY

(c) $\begin{array}{c}CH_3\\ \\ CH_3\end{array}\!\!C=\ddot{O}:\ \ H \longrightarrow \begin{array}{c}CH_3\\ \\ CH_3\end{array}\!\!\overset{\oplus}{C}-\ddot{O}-H$
$|$
$:\ddot{Cl}::\ddot{Cl}:^{\ominus}$

(d) $CH_3-\overset{\oplus}{\underset{H}{\overset{H}{O}}}:\overset{H}{\underset{\underset{\ominus}{H}}{N}}-H \longrightarrow CH_3-\ddot{O}-H + H-\overset{H}{\underset{H}{\overset{|}{N}}}-H$

(e) $\overset{R}{\underset{R}{R}}\!\!C-\overset{\oplus}{\underset{H}{\overset{H}{O}}} \longrightarrow \overset{R}{\underset{R}{R}}\!\!\overset{\oplus}{C}:\ddot{O}\!\!\overset{H}{\underset{H}{}}$

(f) $CH_3-CH=CH_2\ \ \overset{\oplus}{Br} \longrightarrow CH_3-\overset{\oplus}{C}H-CH_2-Br$

(g) $CH_3-CH=CH_2\ \ :\ddot{Br}-\ddot{Br}: \longrightarrow \ CH_3-CH-CH_2$
$\overset{\oplus}{}$
$:\ddot{Br}:\ :\ddot{Br}:^{\ominus}$

(h) $CH_3-CH_2\cdot\ \cdot Br \longrightarrow CH_3-CH_2-Br$

(i) $CH_3-CH=CH_2 \longrightarrow CH_3-\overset{\cdot}{C}H-CH_2$
$\cdot CH_3 |$
CH_3

(j) $CH_3-C-O-N=O \longrightarrow CH_3-C=\ddot{O}:\ +\ :\overset{\ominus}{\ddot{O}}-N=\ddot{O}:$
$\|\updownarrow\updownarrow$
$OCH_3-C\equiv O:^{\oplus}:\ddot{O}=N-\ddot{O}:^{\ominus}$

(k) $:\overset{\ominus}{\ddot{O}}-N=\ddot{O}:\ \ \overset{\overset{\oplus\ddot{O}}{\|}}{\underset{\underset{CH_3}{|}}{C}} \longrightarrow :\ddot{O}=N-\ddot{O}-\overset{\overset{:\ddot{O}:}{\|}}{C}\!\!\diagdown_{CH_3}$

- - - - - - - - -

1. (a) $:\overset{\curvearrowleft}{\ddot{Br}-\ddot{Br}:} \longrightarrow :\ddot{Br}:^{\ominus}\ \ \overset{\oplus}{\ddot{Br}}:$

(b) $:\overset{\curvearrowleft\ \curvearrowright}{\ddot{Br}-\ddot{Br}:} \longrightarrow :\ddot{Br}\cdot\ \ \cdot\ddot{Br}:$

(c) $Li\overset{\curvearrowright}{-}H \longrightarrow \overset{\oplus}{Li}\ +\ :H^{\ominus}$

(d) $CH_3-\overset{\curvearrowleft}{\ddot{O}}-\ddot{Cl}: \longrightarrow CH_3-\overset{\ominus}{\ddot{O}}:\ +\ \overset{\oplus}{\ddot{Cl}}:$

(e) $CH_3-\overset{\oplus}{\underset{CH_3}{\overset{|}{O}}}H\overset{\curvearrowleft}{}:\overset{H}{\underset{H}{\overset{|}{\ddot{O}}}} \longrightarrow CH_3-\overset{\ominus}{\underset{CH_3}{\overset{|}{\ddot{O}}}}:\ +\ H-\overset{\oplus}{\underset{H}{\overset{H}{O}}}$

(f) $\overset{H}{\underset{\overset{\ominus}{:}\ddot{Cl}:}{\overset{|}{H-C}}}\!\!\overset{\curvearrowleft}{\underset{\curvearrowright}{}}\!\!\overset{\oplus}{\underset{H}{\overset{H}{C}}}\!\!\diagdown_{H} \longrightarrow \overset{H}{\underset{H}{}}\!\!C=C\!\!\overset{H}{\underset{H}{}}\ +\ :\ddot{Cl}-H$

(g) $CH_3-CH=CH-CH_3CH_3-CH-CH-CH_3$
$\overset{\longrightarrow}{}\overset{\oplus}{}$
$\oplus CH_2-CH_3CH_2-CH_3$

(h) $F-\overset{F}{\underset{F}{\overset{|}{B}}}-F \rightarrow F-\overset{F}{\underset{F}{B}} + F^{\ominus}$

(i) $CH_3-\overset{\oplus}{\underset{CH_3}{O}}-CH_3 \quad :\overset{\ominus}{\underset{..}{I}}: \rightarrow CH_3-\overset{..}{\underset{CH_3}{O}}: + CH_3-\overset{..}{\underset{..}{I}}:$

(j) [cyclopentanone with α-H being abstracted by hydroxide, giving enolate + $H-\overset{..}{\underset{..}{O}}-H$]

(k) [cyclopentanone enolization to enol/enolate with hydroxide]

(l) [cyclopentanone with water, giving enol + $H-\overset{..}{\underset{..}{O}}-H$]

(m) $CH_3-\overset{.}{C}H_2 \quad Br-Br \rightarrow CH_3-CH_2-Br + \cdot Br$

(n) $CH_3-\overset{..}{\underset{..}{O}}\cdot \quad H-CH_3 \rightarrow CH_3-\overset{..}{\underset{..}{O}}-H + \cdot CH_3$

(o), (p), and (q) $:N\equiv C-\overset{..}{\underset{..}{O}}-CH_3 + :\overset{..}{\underset{..}{I}}:^{\ominus}$

2. (a) $:\overset{..}{\underset{..}{Br}}:\overset{..}{\underset{..}{I}}: \rightarrow :\overset{..}{\underset{..}{Br}}:^{\ominus} + \overset{..}{\underset{..}{I}}:^{\oplus}$

(b) $H-\overset{\ominus}{\underset{..}{O}}: \quad H-\overset{\oplus}{\underset{H}{O}}\overset{H}{\diagdown} \rightarrow H-\overset{..}{\underset{..}{O}}-H + :\overset{..}{\underset{H}{O}}\overset{H}{\diagdown}$

(c) $\overset{CH_3}{\underset{CH_3}{\diagup}}C=\overset{..}{\underset{..}{O}}: \quad H \atop :\underset{..}{\overset{..}{Cl}}: \rightarrow \overset{CH_3}{\underset{CH_3}{\diagup}}\overset{\oplus}{C}-\overset{..}{O}-H \quad :\overset{..}{\underset{..}{Cl}}:^{\ominus}$

(d) $CH_3-\overset{..}{\underset{\oplus}{O}}\overset{H}{\diagdown} \quad :\overset{\ominus}{\underset{H}{N}}\overset{H}{\diagdown} \rightarrow CH_3-\overset{..}{\underset{..}{O}}-H + H-\overset{..}{\underset{H}{N}}\overset{H}{\diagdown}$

(e) $\overset{R}{\underset{R}{\diagup}}\overset{R}{C}-\overset{\oplus}{\underset{H}{O}}\overset{H}{\diagdown} \rightarrow \overset{R}{\underset{R}{\diagup}}\overset{\oplus}{C} + :\overset{..}{\underset{H}{O}}\overset{H}{\diagdown}$

(f) $CH_3-CH=CH_2 \quad Br^{\oplus} \rightarrow CH_3-\overset{\oplus}{C}H-CH_2-Br$

(g) $CH_3-CH=CH_2 \atop \qquad :\overset{..}{\underset{..}{Br}}-\overset{..}{\underset{..}{Br}}: \rightarrow CH_3-\overset{\oplus}{C}H-CH_2 \atop \qquad\qquad\qquad\quad :\overset{..}{\underset{..}{Br}}: + :\overset{..}{\underset{..}{Br}}:^{\ominus}$

(h) $CH_3-CH_2\cdot \,\frown\, \cdot Br \longrightarrow CH_3-CH_2-Br$

(i) $CH_3-CH=CH_2 \;+\; \cdot CH_3 \longrightarrow CH_3-\overset{\cdot}{C}H-CH_2-CH_3$

(j) $CH_3-\underset{\underset{O}{\|}}{C}-O-N=O \longrightarrow CH_3-\overset{\oplus}{C}=\ddot{O}: \;+\; :\overset{\ominus}{\ddot{O}}-N=\ddot{O}:$

$\qquad\qquad\qquad\qquad\qquad \updownarrow \qquad\qquad\quad \updownarrow$

$\qquad\qquad\qquad\qquad CH_3-C\equiv O:^{\oplus} \quad :\ddot{O}=N-\ddot{O}:^{\ominus}$

(k) $:\overset{\ominus}{\ddot{O}}-N=\ddot{O}: \;+\; \underset{\underset{CH_3}{|}}{\overset{\overset{:\overset{\oplus}{\ddot{O}}:}{\|}}{C}} \longrightarrow :\ddot{O}=N-\ddot{O}-\underset{CH_3}{\overset{\overset{:\ddot{O}:}{\|}}{C}}$

STRINGS OF STEPS

3. To save space and the time required for rewriting the products of step 1 as the reactants of step 2, we often write the successive mechanistic steps in a string. For example, equations (5) and (6) can be written

$$:\ddot{C}l-H \;\frown\; :\underset{H}{\overset{|}{\ddot{O}}}-H \longrightarrow :\ddot{C}l:^{\ominus} \;+\; H-\underset{H}{\overset{|}{\overset{\oplus}{\ddot{O}}}}-H \xrightarrow{:NH_3} H-\underset{H}{\overset{|}{\ddot{O}}}: \;+\; H-\underset{H}{\overset{|}{\ddot{O}}}: \;+\; H-\underset{H}{\overset{\overset{H}{|}}{\overset{\oplus}{N}}}-H \qquad (12)$$

In such notation the original reactant(s) and each intermediate have curved arrows, which always refer to the *following* step in the mechanism. In such a scheme, where each intermediate is both the product of the preceding step and the reactant for the following step, there is no real room for the second molecule (reactant, solvent, etc.) involved in the second step (e.g., NH_3 in the mechanism above). We generally put it *on the arrow*, as shown, and position it conveniently so that it can be connected appropriately to the intermetiate with curved arrows. The general form of our mechisms is thus

```
reactant(s)        product(s) of 1st step                products of 2nd step
of 1st step    ⟶          ‖            reagent                    ‖
                    reactant(s) of 2nd step   for 2nd    reactant(s) of 3rd step
                                               step
                                                                      |
                                                                      ↓
              etc.     product(s) of 3rd step     reagent for
                 ⟵            ‖                   3rd step
                       reactant for 4th step
```

Sometimes the positioning of reagent for the next step is more off the arrow than on it, but you can always identify it because it is connected to the intermediate by a curved arrow.

Exercise

Translate this string mechanism into a sequence of steps:

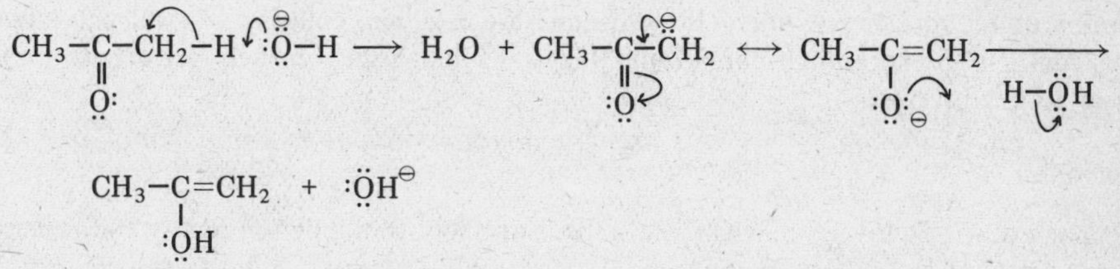

$$CH_3-\underset{\underset{\ddot{O}:}{\|}}{C}-\overset{\frown}{CH_2-H} \overset{\curvearrowleft}{:}\overset{\ominus}{\ddot{O}}-H \longrightarrow H_2O + CH_3-\underset{\underset{:\ddot{O}:^\ominus}{\|}}{C}\overset{\ominus}{-}CH_2 \longleftrightarrow CH_3-\underset{\underset{:\ddot{O}:^\ominus}{|}}{C}=CH_2 \xrightarrow{H-\ddot{O}H}$$

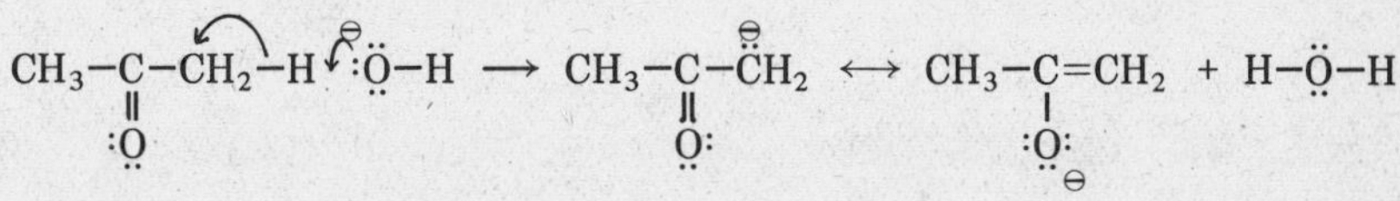

$$CH_3-\underset{\underset{:\ddot{O}H}{|}}{C}=CH_2 + :\ddot{O}H^\ominus$$

- - - - - - - - -

Step 1: $CH_3-\underset{\underset{:\ddot{O}}{\|}}{C}-\overset{\frown}{CH_2-H} \overset{\curvearrowleft}{:}\overset{\ominus}{\ddot{O}}-H \longrightarrow CH_3-\underset{\underset{\ddot{O}:}{\|}}{C}-\overset{\ominus}{\ddot{C}H_2} \longleftrightarrow CH_3-\underset{\underset{:\ddot{O}:^\ominus}{|}}{C}=CH_2 + H-\ddot{O}-H$

Step 2: $CH_3-\underset{\underset{:\ddot{O}}{\|}}{C}-\overset{\ominus}{\ddot{C}H_2} \longleftrightarrow CH_3-\underset{\underset{^\ominus:\ddot{O}:}{|}}{C}=CH_2 \xrightarrow{H-\ddot{O}-H} CH_3-\underset{\underset{:\ddot{O}-H}{|}}{C}=CH_2 + H-\ddot{O}:^\ominus$

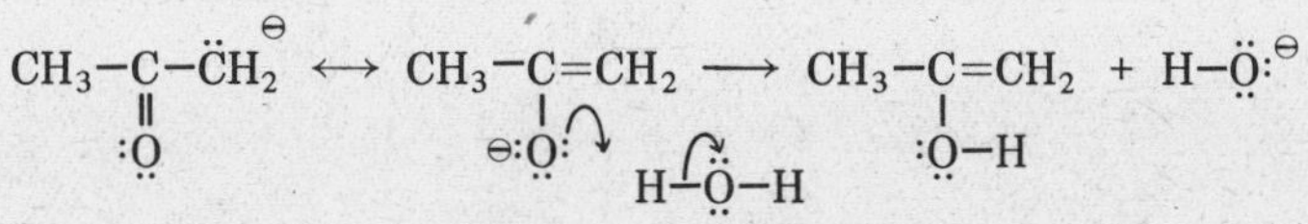

CHECKS ON ACCURACY

4. You can avoid mistakes in writing mechanisms by performing these checks:
   1. Make sure that each step in your mechanism involves a reaction between no more than two molecules or ions.
   2. Make sure that each step breaks and makes a total of no more than four bonds.
   3. Make sure that each step balances—in atoms, in electrons, in charge. Even with mechanisms written in string style you can readily check by eye because there are rarely more than two reactant molecules plus two product molecules per step. A very common student error is writing a mechanistic step of the sort

   neutral molecule + cation $\longrightarrow$ neutral molecule + neutral molecule

This cannot be correct because charge must be conserved in each step—in this case the products would have to have a net positive charge somewhere.
   4. Make sure that every reactant species in each step you have written is either a reactant given in the problem or a product of an earlier step in the mechanism. In other words, do not introduce reagents that are not part of the given reaction mixture.
   5. Make sure that your drawing of reactant molecules and curved arrows is careful enough and complete enough that a reader could supply the product side of each step using only the information you have written on the left.
   6. Make sure that the source of electrons to be shifted by a curved arrow is shown explicitly in the form of an unshared pair, unpaired electron, or covalent bond at the base of the arrow. If you write $CH_3-CH_3 \overset{\frown}{} \overset{\frown}{}Cl$, the reader cannot really tell that you mean $CH_3-CH_2\overset{\frown}{}H \overset{\frown}{}\cdot Cl$ because your arrow doesn't show exactly where you are getting the electron from.
   7. Make sure that all intermediates written are plausible. Using the criteria of Unit 8, that means they must be either stable molecules (0-1 stability deficit) or transient intermediates (1-3 stability deficits). Grossly unstable intermediates (such as

$\underset{H}{\overset{H}{>}}C^{2\oplus}$ are impossible because their formation would require a very large input of

energy. The energy would have to be possessed by the reactant species. At ordinary

174 HOW TO SUCCEED IN ORGANIC CHEMISTRY

temperatures such highly energetic molecules are rare, and collision of two such rare molecules is an exceedingly rare event. Reactions depending on such rare events are too slow to be observed.

Exercises

Each of the following mechanistic steps is unlikely to happen in reality for at least one of the reasons discussed above, or it is written incorrectly. Identify the error in each case.

1. 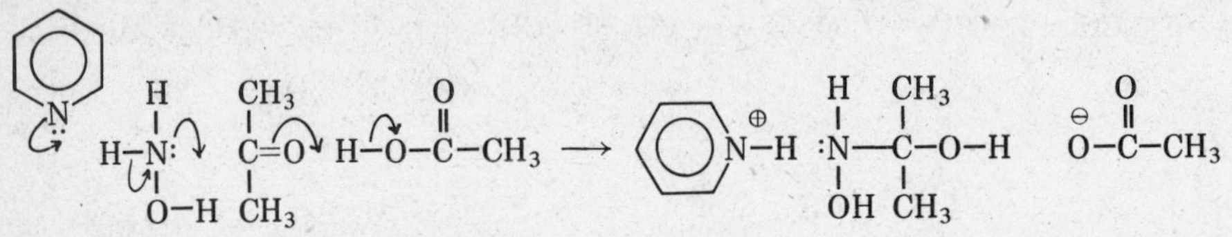 [pyridine + H–N(H)(O–H) + CH$_3$–C(=O)–... + H–O–C(=O)–CH$_3$ → pyridinium–N–H + :N(H)(CH$_3$)(OH)–C–O–H + $^{\ominus}$O–C(=O)–CH$_3$]

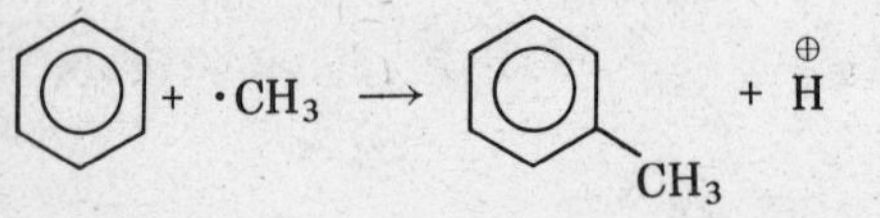

2. [benzene] + ·CH$_3$ → [toluene] + $^{\oplus}$H

3. [cyclopentane-1,1-diol with OH$^{\ominus}$] → [cyclopentanone] + H$_2$O

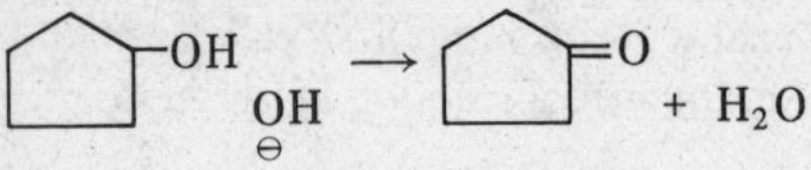

4. 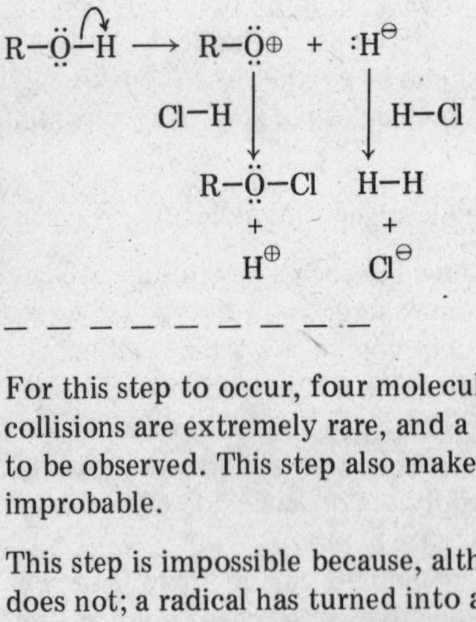 R–Ö–H → R–Ö$^{\oplus}$ + :H$^{\ominus}$

   Cl–H ↓   ↓ H–Cl

   R–Ö–Cl   H–H
   +   +
   H$^{\oplus}$ Cl$^{\ominus}$

- - - - - - -

1. For this step to occur, four molecules would have to collide simultaneously. Such collisions are extremely rare, and a reaction depending upon them will be too slow to be observed. This step also makes/breaks five bonds, another indication that it is improbable.

2. This step is impossible because, although the atom count balances, the charge count does not; a radical has turned into a cation. Since such a change implies destruction of one electron, it is impossible. (Don't be fooled by the apparent absence of H in [benzene]—remember, it is really [benzene with all H's shown].)

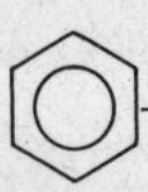

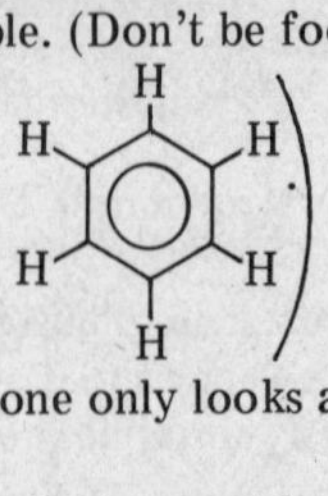

3. This step seems to balance if one only looks at the atoms that are written out, but in fact one H in [cyclopentane with H and OH] has disappeared in the product, [cyclopentanone]. Consequently, the atom count does not balance. Neither do the charges: left, net 1−; right, neutral.

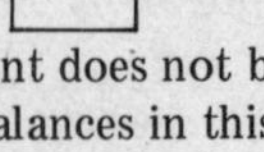

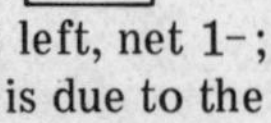

4. Everything balances in this mechanism. Its implausibility is due to the formation

of two unstable species in the first step. In R—O$^\oplus$, oxygen bears a positive charge and has only a sextet of electrons.

According to the rules of Unit 8, both R—O$^\oplus$ and H:$^\ominus$ are classified as unstable species capable of only a transient existence. With very few exceptions, the decomposition of a stable molecule into *two* such species is a step so energetically unfeasible that it cannot be observed. The most common exceptions involve either photochemical processes (in which the extra energy required is supplied by absorption of light, for instance in Br—Br ⟶ Br· + ·Br) or high-temperature reactions, or a reactant molecule with a very weak bond (for instance, R—O—O—R $\xrightarrow{\Delta}$ R—O· + ·O—R).

THE NINE STANDARD MECHANISMS

5. Of the approximately 100 to 150 basic reactions studied in beginning organic chemistry, about 80% take place according to one of the nine standard mechanisms summarized in Table 10.1. Of the remaining 20%, part occur by simple variations on one of these nine, part occur by uncommon mechanisms of much less general applicability, and part by as-yet-unknown mechanisms.

Table 10.1 *The Nine Standard Mechanisms*

			Page
I	Radical substitution	A—B $\xrightarrow[\text{or light}]{\text{ROOR}}$ A· + ·B initiation R—H ·A ⟶ R· + H—A ⎫ R· B—A ⟶ R—B + A· ⎬ propagation R· ·R ⟶ R—R, etc. termination ⎭		211

II	Electrophilic addition	C=C E—Nu ⟶ —C—C— + Nu:$^\ominus$
		                                    E
		
		$^\ominus$Nu: —C—C—$^\oplus$ ⟶ —C—C—
		                 E Nu E
				212

III	Radical addition	A—B $\xrightarrow[\text{or light}]{\text{ROOR}}$ A· + ·B initiation
		
		C=C ·A ⟶ —C—C—A ⎫
		                                          ⎬ propagation
		—C—C—A ⟶ —C—C— + A· ⎭
		    B—A B A
				214

(table continues)

Table 10.1 (continued)

			Page
		$\overset{\frown}{>}C\cdot\overset{\frown}{\frown}\cdot C\overset{\frown}{<} \longrightarrow -\overset{|}{\underset{|}{C}}-\overset{|}{\underset{|}{C}}-$, etc. termination	
IV	$S_N1/E1$	$-\overset{|}{CH}-\overset{|}{\underset{|}{C}}\overset{\frown}{-X} \longrightarrow -\overset{|}{CH}-\overset{\oplus}{\underset{|}{C}}- + X^{\ominus}$	320
		$-\overset{|}{CH}-\overset{\oplus}{C}\overset{\frown}{\cdot}\text{:Nu}^{\ominus} \longrightarrow -\overset{|}{CH}-\overset{|}{\underset{|}{C}}-\text{Nu} \quad S_N1$	
	and/or	$-\overset{|}{\underset{H\overset{\curvearrowright}{:}B}{C}}-\overset{\oplus}{C} \longrightarrow \;>C=C< + HB^{\oplus} \quad E1$	
V	S_N2	$^{\ominus}\text{Nu:}\overset{\frown}{\downarrow}R\overset{\frown}{-}X \longrightarrow \text{Nu-R} + X^{\ominus}$	319
VI	$E2$	$-\overset{\overset{\curvearrowright X}{|}}{\underset{\overset{|}{H}}{C}}-\overset{|}{\underset{|}{C}}- \longrightarrow \;>C=C<\overset{X^{\ominus}}{\;}$	321
		$^{\ominus}B:\overset{\curvearrowright}{\quad} \qquad\qquad H-B$	
VII	Electrophilic aromatic substitution	[benzene + $E^{\oplus}$] $\longrightarrow$ [arenium ion with E, H] $\longleftrightarrow$ etc.	369
		etc. $\longleftrightarrow$ [arenium with E, H:B] $\longrightarrow$ [Ar–E] + $BH^{\oplus}$	
VIII	Nucleophilic addition	$\text{Nu:}^{\ominus}\overset{\frown}{\downarrow}\;>C\overset{\frown}{=}O \longrightarrow \text{Nu}-\overset{|}{\underset{|}{C}}-O^{\ominus}$	396
		$\text{Nu}-\overset{|}{\underset{|}{C}}-O^{\ominus}\overset{\frown}{\downarrow}H\overset{\frown}{-}\text{Nu} \longrightarrow \text{Nu}-\overset{|}{\underset{|}{C}}-O-H + \text{Nu:}^{\ominus}$	
IX	(a) acyl S_N2	$^{\ominus}\text{Nu:}\overset{\frown}{\downarrow}\overset{O\overset{\frown}{\;}}{\underset{}{\overset{\|}{C}}}-X \longrightarrow \text{Nu}-\overset{O^{\ominus}}{\underset{|}{\overset{|}{C}}}\overset{\frown}{-}X \longrightarrow \text{Nu}-\overset{O}{\overset{\|}{C}}- + X^{\ominus}$	445
	(b) aryl S_N2	[Ar(W)–X with :Nu] $\longrightarrow$ [Meisenheimer with X, Nu, W$^{\ominus}$] $\longleftrightarrow$ [with X, Nu, W$^{\ominus}$] $\longrightarrow$ [Ar(W)–Nu] + $X^{\ominus}$	371
		etc.	

MECHANISM 177

You should not sit down and learn the standard mechanisms now. Learn each one when it first comes up as you study specific reactions of the functional groups. Your ultimate goal is to be able to write the mechanism of any given reaction (at least of the 80% that go by standard mechanisms). To do this you will (1) learn the general form of each mechanism, (2) learn which mechanism applies at the time you learn each new reaction, and (3) rewrite the *generic* form of the mechanism, as given in Table 10.1, in *specific* form using the structures of the specific reactants in the problem.

Note that the standard mechanisms fall into two classes—those with *ionic* intermediates (II, IV to IX) and those with *radical* intermediates (I, III). These types never mix; if a mechanism starts out by forming radical intermediates, it never switches over to ionic intermediates, and vice versa.

Every step in each of the *ionic* mechanisms can be looked at as a variation on the same simple process. This process is donation of a pair of electrons from some electron donor (a species with an unshared electron pair or a multiple bond) to form a new single bond to some electron-deficient species. The electron-acceptor species is called a Lewis acid (A) or an electrophile (E). The electron donor species is called a base (B:) or a nucleophile (Nu:). Thus the fundamental process is

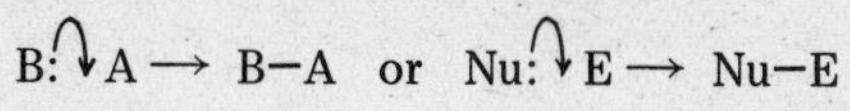

$$B:\curvearrowright A \longrightarrow B-A \quad \text{or} \quad Nu:\curvearrowright E \longrightarrow Nu-E$$

Some steps from Table 10.1 correspond exactly to this pattern:

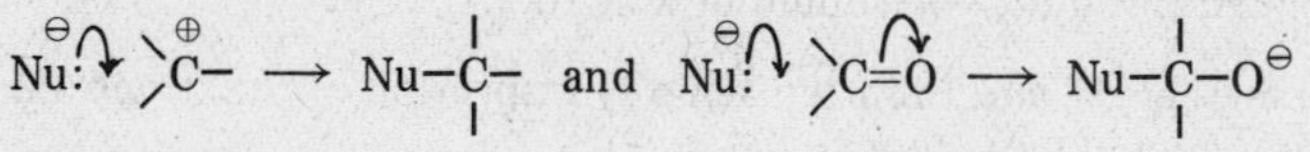

$$Nu:^{\ominus}\curvearrowright \overset{\oplus}{>}\!\!\overset{|}{C}\!- \longrightarrow Nu-\overset{|}{\underset{|}{C}}- \quad \text{and} \quad Nu:^{\ominus}\curvearrowright >\!\!C\!\!=\!\!O \longrightarrow Nu-\overset{|}{\underset{|}{C}}-O^{\ominus}$$

Others are examples of the reverse process:

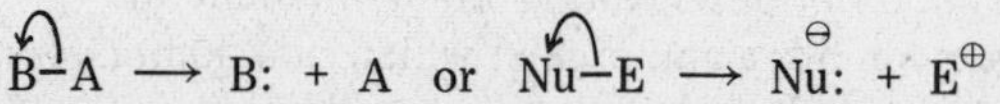

$$B\overset{\curvearrowleft}{-}A \longrightarrow B: + A \quad \text{or} \quad Nu\overset{\curvearrowleft}{-}E \longrightarrow Nu:^{\ominus} + E^{\oplus}$$

For example,

$$R\overset{\curvearrowright}{-}X: \longrightarrow R^{\oplus} + :\ddot{X}:^{\ominus}$$

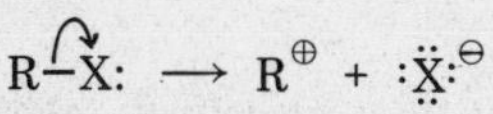

These generalized acid-base reactions are discussed in Unit 11.

Exercise

Verify that each of the 23 individual steps of the nine mechanisms in Table 10.1 (a) obeys the charge-balance rule; (b) requires collision of no more than two molecules or ions; and (c) breaks/makes a total of no more than four bonds. What is the most common number of bonds made and broken per step?

— — — — — — — — —

Of the 23 steps represented, 7 break/make one bond, 11 break/make two bonds, 4 break/make three bonds, and only one breaks/makes four bonds.

UNIT ELEVEN

Acids and Bases

Many reactants, catalysts, and products in the organic reactions you will be studying are acids and bases. Furthermore, we saw in Unit 10 that, if we use a suitably broad concept of acids and bases, all the steps in the mechanisms can be visualized as simple variations on reaction of an acid with a base. So it is advantageous to you, at this point, to learn to identify acids and bases and manipulate acid-base equilibria.

OBJECTIVES

When you have completed this unit, you should be able to:

- Identify bases, Lewis acids, and Brønsted acids by inspection of the structural formulas of molecules and ions (frames 1, 5 to 7).
- Draw the structure of the conjugate acid (base) of any given base (acid) (frame 3).
- Arrange any set of acids (bases) in order of increasing strength (frame 8).
- Write the equations for acid-base equilibria and predict which side dominates at equilibrium (frames 2, 4, 9, 11).
- Use these skills to predict the conditions for the existence of acidic and basic species (frame 10).

If you think you may have already achieved some or all of these objectives, take a self-test consisting of the exercises marked with a star in the margin. If you miss any problems, study the frames in question and rework the exercises before proceeding. If you are not ready for a self-test, proceed to frame 1.

IDENTIFYING ACIDIC AND BASIC SPECIES

1. Acids are of two types. A *Brønsted acid* contains an O—H, N—H, S—H, or X—H bond. A *Lewis acid* contains an atom that is two electrons short of an octet (duet for H), for instance $\overset{R}{\underset{R}{>}}\overset{\oplus}{C}-R$, $\overset{X}{\underset{X}{>}}B-X$, $H^\oplus$, $:\ddot{B}r^\oplus$, or $Ag^\oplus$. Brønsted acids are much more numerous than Lewis acids.

A *base* is a compound that possesses either:

(a) An unshared pair on $-\ddot{O}-$, $=\ddot{O}:$, $-\ddot{O}:^\ominus$, $>\!\ddot{N}-$, $=\ddot{N}-$, $\equiv N:$, $-\ddot{N}-^\ominus$, $=\ddot{N}:^\ominus$, $-\ddot{S}-$, $-\ddot{S}:^\ominus$, $>\!\ddot{P}-$, $>\!C:^\ominus$, or $:\ddot{X}:^\ominus$; *or*

(b) A shared pair made available by reorganizing a π bond in $>\!C=C\!<$, $-C\equiv C-$, or

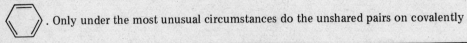. Only under the most unusual circumstances do the unshared pairs on covalently

bound halogen, $-\ddot{\underset{..}{X}}:$, or those in σ bonds (e.g., $\underset{}{\overset{\downarrow}{\underset{}{>}C-H}}$) act as centers of basicity. The same molecule can possess both basic and acidic sites.

★ *Exercises*

For each molecule or ion, label all acidic sites a and all basic sites b.

1. $F^{\ominus}$
2. $H-O-H$
3. $H-Br$
4. CH_4
5. $HSO_4^{\ominus}$
6. H_2SO_4
7. NH_3
8. $CH_2=CH_2$
9. $NH_4^{\oplus}$
10. CH_3-O-H
11. $H_3O^{\oplus}$
12. $C_6H_5-NH_3^{\oplus}$

- - - - - - - - -

1. $:\ddot{\underset{..}{F}}:^{\ominus} \leftarrow b$
2. $H\underset{a}{\overset{b}{\rightarrow}}\ddot{O}\underset{}{\overset{b}{\leftarrow}}H$
3. $H\underset{a}{\overset{\uparrow}{-}}\ddot{\underset{..}{Br}}:$
4. $\underset{H}{\overset{H}{>}}C\underset{H}{\overset{H}{<}}$ neither
5. $H\underset{a}{\overset{\uparrow}{-}}\ddot{O}-\underset{\underset{a\ :\ddot{O}:^{\ominus}\ b}{|}}{\overset{\overset{:\ddot{O}:^{\ominus}\ b}{|}}{S^{2\oplus}}}-\ddot{O}:^{\ominus}\ b$
6. $H\underset{a}{\overset{\uparrow}{-}}\ddot{O}-\underset{\underset{a\ :\ddot{O}:^{\ominus}\ b}{|}}{\overset{\overset{:\ddot{O}:^{\ominus}\ b}{|}}{S^{2\oplus}}}-\ddot{O}-H\ a$
7. $a\overset{H}{\underset{H}{<}}N\underset{a}{\overset{b}{-}}H$
8. $CH_2\overset{b}{=}CH_2$
9. $a\overset{H}{\underset{H}{<}}\overset{\oplus}{N}\overset{H}{\underset{H}{>}}a$
10. $CH_3-\underset{b\nearrow\ \ \ \ a}{\ddot{O}-H}$
11. $a\overset{H}{\underset{H}{<}}\overset{\oplus}{O}\underset{b}{\overset{-H}{\ \ \ \ \ a}}$
12. $C_6H_5-\overset{\oplus}{N}\underset{H}{\overset{H\ \ a}{\underset{\ \ \ \ a}{-H}}}$ (b on ring)

REACTIONS OF ACIDS WITH BASES

2. Acids react rapidly with bases. This reaction is the most fundamental one in organic chemistry. It is important by itself and as one step in many other reactions. Bases always donate their unshared pairs or π electron pairs in this reaction. What the acid does depends on whether it is a Lewis or a Brønsted acid.

First, let's look at the reaction of *Lewis* acids with bases, called *coordination*. The base donates its unshared pair to the region in space between the basic and acidic atoms,

---

*However, see frame 6.

where it becomes a new σ bond. The product is rather loosely called a *complex*.

Example

Write the equation for the acid-base reaction of BF_3 with CH_3-O-CH_3.

Solution:

$$\underset{\substack{\text{Lewis acid} \\ \text{(sextet on} \\ \text{B)}}}{\ddot{\underset{|}{\overset{|}{\underset{:\ddot{F}:}{\overset{:\ddot{F}:}{F}-B}}}}} \quad \underset{\substack{\text{base} \\ (\ddot{:}\text{ on O})}}{\overset{CH_3}{\underset{CH_3}{\ddot{O}}}} \longrightarrow \underset{\substack{\text{electron acceptor} \\ \text{atom now one unit} \\ \text{more negative}}}{\ddot{\underset{|}{\overset{|}{\underset{:\ddot{F}:}{\overset{:\ddot{F}:}{F}-B^{\ominus}}}}}-\overset{\oplus}{\underset{CH_3}{\overset{CH_3}{O}}}} \quad \text{electron-donor atom now one unit more positive}$$

When the basic site is a π bond, as in the next example, the resulting π complex is less easy to describe. X-ray diffraction results show that the Lewis acid sits in one of the π-electron clouds of the molecule. The π electrons are only partly donated.

Example

The equations for the Lewis acid-base reaction of $Ag^{\oplus}$ with ethylene and benzene are written using an arrow to symbolize the electron donation:

$$H_2C=CH_2 + Ag^{\oplus} \longrightarrow \overset{CH_2}{\underset{CH_2}{\|}} \longrightarrow Ag^{\oplus}$$

The most important Lewis acid is $H^{\oplus}$.

Example

Write the equation for the acid-base reaction of $H^{\oplus}$ with CH_3-O-H.

Solution:

$$\underset{\substack{\text{Lewis acid} \\ \text{(zero elec-} \\ \text{trons)}}}{H^{\oplus}} \quad \underset{\substack{\text{base} \\ (\ddot{:}\text{ on O})}}{\overset{CH_3}{\underset{CH_3}{\ddot{O}}}} \longrightarrow \underset{\substack{\text{electron} \\ \text{acceptor now} \\ \text{one unit} \\ \text{more neg-} \\ \text{ative}}}{H} - \overset{\text{new σ bond}}{\underset{CH_3}{\overset{\oplus}{O}}} \overset{CH_3}{\underset{}{}} \quad \substack{\text{electron-donor} \\ \text{atom now one} \\ \text{unit more} \\ \text{positive}}$$

However, $H^{\oplus}$ is so strong that it, and its coordination reaction, can only be observed in vacuum at rather low pressures. In solutions (where more than 99.9% of our chemistry is carried out) *free* $H^{\oplus}$ has never been observed—it is always coordinated to some base. The coordination complex formed is held together by a new σ bond formed from the original unshared pair of the base.

Apparently because of its great strength, $H^{\oplus}$ does not seem to form π complexes with the >C=C< of ⬡ functions, but produces the electrophilic addition reaction (see

ACIDS AND BASES 181

Unit 12, frame 11) instead, destroying the π bond. This produces a carbonium ion, for instance $CH_2{=}CH_2 \;\; H^\oplus \longrightarrow \overset{\oplus}{C}H_2{-}CH_2{-}H$.

All Lewis acid-base coordination reactions are reversible, equilibrium processes.

★ Exercises

Write the equation for each Lewis acid-base reaction.

1. $CH_3{-}\underset{CH_3}{\overset{CH_3}{\underset{|}{\overset{|}{C}}}}{\oplus} \; + \; :NH_3$

2. $:\ddot{B}r^\ominus \; + \; CH_3{-}O{-}H$

3. $H_2S \; + \; \underset{CH_3}{\overset{CH_3}{{\diagdown}}}B{-}CH_3$

4. $H^\oplus \; + \; :\ddot{F}:^\ominus$

5. ⌬ + $Li^\oplus$

6. $Ag^\oplus \; + \; H_2O$

7. ⬡ + $H^\oplus$

- - - - - - - - - -

1. $CH_3{-}\underset{CH_3}{\overset{CH_3}{\underset{|}{\overset{|}{C}}}}{\oplus} \; + \; :NH_3 \longrightarrow CH_3{-}\underset{CH_3}{\overset{CH_3}{\underset{|}{\overset{|}{C}}}}{-}\underset{H}{\overset{H}{\underset{|}{\overset{|}{\overset{\oplus}{N}}}}}{-}H$

2. $:\ddot{B}r^\ominus \; + \; \underset{H}{\overset{CH_3}{\diagup}}\ddot{O} \longrightarrow :\ddot{B}r{-}\underset{H}{\overset{CH_3}{\underset{\diagdown}{\overset{\diagup}{\ddot{O}}}}}{}^\oplus$

3. $CH_3{-}\underset{CH_3}{\overset{CH_3}{\underset{|}{\overset{|}{B}}}} \; + \; \underset{CH_3}{\overset{CH_3}{\diagup}}\ddot{S} \longrightarrow CH_3{-}\underset{CH_3}{\overset{CH_3}{\underset{|}{\overset{|}{\overset{\ominus}{B}}}}}{-}\underset{CH_3}{\overset{CH_3}{\underset{\diagdown}{\overset{\diagup}{\overset{\oplus}{\ddot{S}}}}}}$

4. $H^\oplus \; + \; :\ddot{F}:^\ominus \longrightarrow H{-}\ddot{F}:$

5. ⌬ + $Li^\oplus \longrightarrow$ ⌬⁻→$Li^\oplus$

6. $Ag^\oplus \; + \; \underset{H}{\overset{H}{\diagup}}\ddot{O} \longrightarrow Ag{-}\underset{H}{\overset{H}{\underset{\diagdown}{\overset{\diagup}{\overset{\oplus}{\ddot{O}}}}}}$

7. (cyclohexene with H, attacked by $H^\oplus$) → (cyclohexyl cation with $\oplus$, H)

3. Now let's look at the reactions of Brønsted acids with bases. The fact that $H^\oplus$ exists in solution only in coordinated form means that all its acid-base chemistry in solution consists of transfer from one base to another. Now, note that coordination of $H^\oplus$ by a base produces a Brønsted acid, for instance in

$$CH_3{-}\ddot{O}{:} + H^\oplus \rightarrow CH_3{-}\overset{\oplus}{O}{-}H,$$ or in general terms:

$$H^\oplus + {:}B \rightarrow H{-}B^\oplus$$

This means that *the characteristic acid-base reaction in solution is reversible proton transfer from a Brønsted acid to a base:*

$$\overset{\oplus}{B_1}{-}H + {:}B_2 \rightleftharpoons B_1{:} + H{-}\overset{\oplus}{B_2}$$
protonated base = Brønsted acid   base

Transfer of $H^\oplus$ *to* base B_2 (protonation of B_2) converts it into $H{-}\overset{\oplus}{B_2}$, a new Brønsted acid. Transfer of $H^\oplus$ *from* acid $\overset{\oplus}{B_1}{-}H$ (deprotonation of $\overset{\oplus}{B_1}{-}H$) restores its unshared pair and converts it into $:B_1$, a base. As a result, the Brønsted-acid-base reaction always has the symmetry: acid + base = base + acid. We can label the species in the equilibrium above:

$$\overset{\oplus}{B_1}{-}H + {:}B_2 \rightleftharpoons B_1{:} + H{-}\overset{\oplus}{B_2}$$
acid   base   base   acid

Bases are generally either neutral molecules or anions, Brønsted acids either neutral molecules or cations. The proton-transfer equilibrium can consequently occur in the following four charge patterns:

$$\overset{\oplus}{B_1}{-}H + {:}B_2 \rightleftharpoons B_1{:} + H{-}\overset{\oplus}{B_2} \qquad (1)$$
acid   base   base   acid

$$\overset{\oplus}{B_1}{-}H + {:}\overset{\ominus}{B_2} \rightleftharpoons B_1{:} + H{-}B_2 \qquad (2)$$
acid   base   base   acid

$$B_1{-}H + {:}B_2 \rightleftharpoons \overset{\ominus}{B_1}{:} + H{-}\overset{\oplus}{B_2} \qquad (3)$$
acid   base   base   acid

$$B_1{-}H + {:}\overset{\ominus}{B_2} \rightleftharpoons \overset{\ominus}{B_1}{:} + H{-}B_2 \qquad (4)$$
acid   base   base   acid

The base formed by removing $H^\oplus$ from acid $B_1{-}H$ or $\overset{\oplus}{B_1}{-}H$ is called the *conjugate base* of that acid. $B_1{-}H$ and $\overset{\ominus}{B_1}{:}$ are a *conjugate* acid-base pair, as are $\overset{\oplus}{B_1}H$ and $B_1{:}$. $B_1{-}H$ is the *conjugate acid* of $\overset{\ominus}{B_1}{:}$ and $\overset{\oplus}{B_1}{-}H$ is the conjugate acid of $B_1{:}$. To generate the conjugate acid of a base, coordinate $H^\oplus$ with it ($B{:} + H^\oplus \rightarrow \overset{\oplus}{B}{-}H$). To generate the conjugate base of any Brønsted acid, remove $H^\oplus$, leaving an unshared pair behind ($\overset{\oplus}{B}{-}H \rightarrow B{:} + H^\oplus$).

ACIDS AND BASES 183

Example

Draw the structural formula of the conjugate acid of $Br^\ominus$. Draw the conjugate base of CH_3-O-H.

Solution:

$$:\overset{\ominus}{\underset{..}{Br}}: + H^\oplus \rightarrow :\underset{..}{Br}-H \qquad CH_3-\underset{..}{\overset{..}{O}}-H \rightarrow CH_3-\underset{..}{\overset{..}{O}}:^\ominus + H^\oplus$$

  base   conjugate   acid   conjugate
             acid   base

★ *Exercises*

1. Draw the structure of the conjugate acid of each base.

   (a) $:\underset{..}{\overset{..}{F}}:^\ominus$
   (b) $H_2\underset{..}{\overset{..}{O}}:$

   (c) $:\underset{..}{\overset{..}{O}}H^\ominus$
   (d) $CH_3-\underset{..}{\overset{..}{N}}H_2$

   (e) $CH_3-\underset{..}{\overset{\ominus}{N}}H$
   (f) NH_2-OH (two answers)

   (g) $CH_3-\underset{\underset{\overset{\|}{\underset{..}{O}:}}{}}{C}-CH_3$

2. Draw the structure of the conjugate base of each Brønsted acid.

   (a) H_2O
   (b) $\overset{\oplus}{NH_4}$

   (c) NH_2-OH (two answers)
   (d) $CH_3-C\underset{O-H}{\overset{\diagup\!\!\!\!\!\!O}{}}$

   (e) $CH_3-O-O-H$
   (f) $CH_3-S\underset{\underset{O}{O-H}}{\overset{\diagup\!\!\!\!\!\!O}{}}$

---

1. (a) $:\underset{..}{\overset{..}{F}}:^\ominus \rightarrow :\underset{..}{\overset{..}{F}}-H$

   (b) $\underset{H}{\overset{H}{\diagdown}}\underset{..}{\overset{..}{O}}: \rightarrow \underset{H}{\overset{H}{\diagdown}}\overset{\oplus}{\underset{..}{O}}-H$

   (c) $H-\underset{..}{\overset{\ominus}{\underset{..}{O}}}: \rightarrow H-\underset{..}{\overset{..}{O}}-H$

   (d) $CH_3-\underset{\underset{H}{|}}{\overset{\overset{H}{|}}{N}}: \rightarrow CH_3-\underset{\underset{H}{|}}{\overset{\overset{H}{|}}{\overset{\oplus}{N}}}H$

   (e) $CH_3-\underset{\underset{\ominus}{}}{\overset{\overset{H}{|}}{\underset{..}{N}}}: \rightarrow CH_3-\overset{\overset{H}{|}}{\underset{..}{N}}-H$

(f) $\text{H}_2\ddot{\text{N}}-\ddot{\text{O}}-\text{H} \longrightarrow \text{H}-\overset{\oplus}{\text{N}}\text{H}_2-\ddot{\text{O}}-\text{H}$ or $\text{H}_2\text{N}-\overset{\oplus}{\text{O}}\text{H}_2$

(g) $\text{CH}_3-\overset{:\ddot{\text{O}}:}{\underset{\|}{\text{C}}}-\text{CH}_3 \longrightarrow \text{CH}_3-\overset{:\overset{\oplus}{\text{O}}-\text{H}}{\underset{\|}{\text{C}}}-\text{CH}_3$

2. (a) $\text{H}-\ddot{\text{O}}-\text{H} \longrightarrow \text{H}-\ddot{\text{O}}:^{\ominus}$

(b) $\text{H}_3\overset{\oplus}{\text{N}}\text{H} \longrightarrow \text{H}_2\ddot{\text{N}}-\text{H}$

(c) $\text{H}_2\ddot{\text{N}}-\ddot{\text{O}}-\text{H} \longrightarrow \text{H}_2\ddot{\text{N}}-\ddot{\text{O}}:^{\ominus}$ or $\text{H}-\overset{\ominus}{\ddot{\text{N}}}-\ddot{\text{O}}-\text{H}$

(d) $\text{CH}_3-\text{C}(\ddot{\text{O}})-\ddot{\text{O}}-\text{H} \longrightarrow \text{CH}_3-\text{C}(\ddot{\text{O}})-\ddot{\text{O}}:^{\ominus}$

(e) $\text{CH}_3-\ddot{\text{O}}-\ddot{\text{O}}-\text{H} \longrightarrow \text{CH}_3-\ddot{\text{O}}-\ddot{\text{O}}:^{\ominus}$

(f) $\text{CH}_3-\text{S}(\text{O})(\text{O})-\ddot{\text{O}}-\text{H} \longrightarrow \text{CH}_3-\text{S}(\text{O})(\text{O})-\ddot{\text{O}}:^{\ominus}$

4. Each of the standard types of Brønsted acid-base equilibrium, eqs. (1) to (4) in frame 3, is constructed from two conjugate acid-base pairs, for instance

conjugate acid-base pair

$$B_1\text{-}H^{\oplus} + B_2 \rightleftharpoons B_1 + B_2\text{-}H^{\oplus} \quad (1)$$

conjugate acid-base pair

Example

Write the equation for the acid-base equilibrium established between $NH_4^{\oplus}$ and $CH_3-CH_2-\ddot{O}-CH_2-CH_3$, and label the species.

Solution:

Identify: $H-\overset{\oplus}{N}H_3 =$ acid only   its conjugate base $= :NH_3$

(no :, has N—H)

$CH_3CH_2-\ddot{O}-CH_2CH_3 =$ base only   its conjugate acid

(no O—H, etc., has :)

$= CH_3CH_2-\overset{\oplus}{O}H-CH_2CH_3$

Write:

$$\overset{\oplus}{NH_4} + CH_3CH_2-\overset{..}{\underset{..}{O}}-CH_2CH_3 \rightleftharpoons :NH_3 + CH_3CH_2-\overset{H}{\underset{|}{\overset{\oplus}{O}}}-CH_2CH_3$$

acid — base — base — acid

(conjugate a/b pair brackets connect NH$_4^\oplus$/:NH$_3$ and the ether/oxonium)

★ *Exercises*

Write the equation for the acid-base equilibrium established between each pair and label the species.

1. H—I + $:\overset{..}{\underset{..}{Cl}}:^\ominus$

2. $CH_3-\overset{..}{\underset{..}{O}}-CH_3$ + H—Br

3. $NH_4^\oplus$ + $:H^\ominus$

— — — — — — — — —

1. H—I + $:\overset{..}{\underset{..}{Cl}}:^\ominus$ ⇌ $:\overset{..}{\underset{..}{I}}:^\ominus$ + H—Cl

   acid   base   base   acid

2. $CH_3-\overset{..}{\underset{..}{O}}-CH_3$ + H—Br ⇌ $CH_3-\overset{\oplus}{\underset{|}{O}}-CH_3$ + $:\overset{..}{\underset{..}{Br}}:^\ominus$
   $\phantom{CH_3-\overset{..}{\underset{..}{O}}-CH_3 + H—Br \rightleftharpoons CH_3-}H$

   base   acid   acid   base

3. $NH_4^\oplus$ + $:H^\ominus$ ⇌ $\overset{..}{N}H_3$ + H—H

   acid   base   base   acid

COMMON ACIDS AND BASES

5. Tables 11.1A and B list the common bases and Brønsted acids of organic chemistry.

Note that only the electrically neutral acids are listed in Table 11.1B and only the electrically neutral bases in Table 11.1A. There are some *cationic* acids also; you will find these in the conjugate-acid column of Table 11.1A. Similarly, there are some anionic bases; these are found in the conjugate-base column of Table 11.1B.

Note that the position taken up by $H^\oplus$ in coordinating with some of the neutral bases in Table 11.1A represents a choice between two types of available sites (unshared pairs) in the base. In the following example, arrows indicate two available sites:

$$R-C\begin{smallmatrix}\nearrow\ddot{O}:\leftarrow\\ \searrow\ddot{N}H_2\leftarrow\end{smallmatrix}$$

So two conjugate acids are possible:

$$R-C\begin{smallmatrix}\nearrow\ddot{O}:\\ \searrow NH_3^\oplus\end{smallmatrix} \qquad R-C\begin{smallmatrix}\nearrow\overset{\oplus}{O}-H\\ \searrow\ddot{N}H_2\end{smallmatrix}$$

   a   b

Table 11.1A The Common Bases of Organic Chemistry

Base	$\xrightarrow{H^\oplus}$	its conjugate acid

Base	its conjugate acid
:NH₃ *	NH₄⊕
R–ṄH₂ *	R–NH₃⊕
R₂ṄH *	R₂NH₂⊕
R₃N:	R₃N–H⊕
R₂C=Ṅ–R	R₂C=N⊕(H)(R)
R–C≡N:	R–C≡N–H⊕
H₂Ö: *	H₃O:⊕
R–Ö–H *	R–O⊕(H)(H)
R–Ö–R	R–O⊕(H)–R
R₂C=Ö:	R₂C=O⊕–H
R–C(=Ö)Ö–H *	R–C(=O⊕–H)Ö–H
R–C(=Ö)Ö–R	R–C(=O⊕–H)Ö–R
R–C(=Ö)ṄH₂ *	R–C(=O⊕–H)ṄH₂
R–N⊕(=Ö)(O⊖)	R–N⊕(=Ö)Ö–H

*These compounds are both acids and bases.

ACIDS AND BASES 187

Table 11.1B *The Common Acids of Organic Chemistry*

Acid	$\xleftarrow{-H^{\oplus}}$	its conjugate base
:NH$_3$ *		:ṄH$_2^{\ominus}$
R–ṄH$_2$ *		R–ṄH$^{\ominus}$
R$_2$Ṅ–H *		R$_2$Ṅ$^{\ominus}$
H–Ö–H *		:Ö–H$^{\ominus}$
R–Ö–H *		R–Ö:$^{\ominus}$
R–C(=O)–Ö–H *		R–C(=O)–Ö:$^{\ominus}$
R–C(=O)–NH$_2$ *		R–C(=O)–ṄH$^{\ominus}$
R–S(=O)$_2$–Ö–H		R–S(=O)$_2$–Ö:$^{\ominus}$

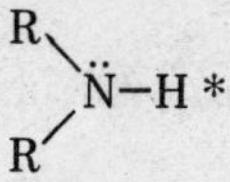

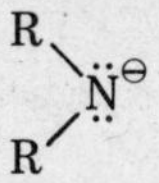

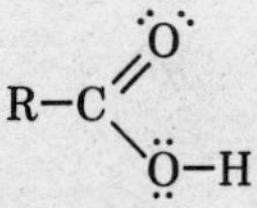

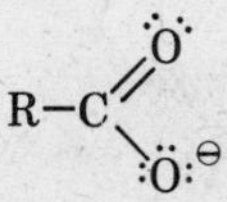

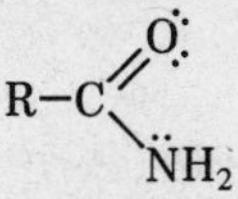

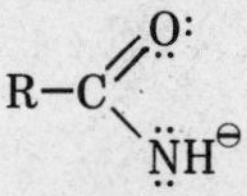

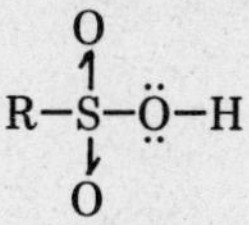

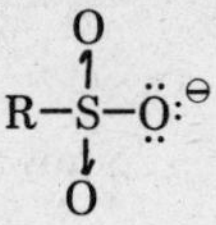

* These compounds are both acids and bases.

Experiment shows that **b** is the species actually formed. This means it is more stable than **a**. The reason for its greater stability is resonance:

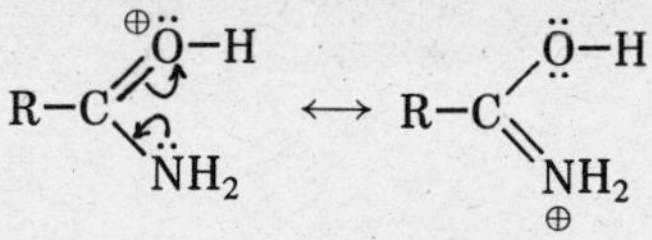

$$R-C\begin{smallmatrix}\oplus\ddot{O}-H\\ \\NH_2\end{smallmatrix} \longleftrightarrow R-C\begin{smallmatrix}\ddot{O}-H\\ \\NH_2\\ \oplus\end{smallmatrix}$$

The positive charge is delocalized over two atoms.

Exercises

The species $R_2C=\overset{\oplus}{\underset{..}{O}}-H$, $R-C(=\overset{..}{\underset{..}{O}})-\overset{..}{\underset{..}{O}}:^{\ominus}$, $R-C\begin{smallmatrix}\overset{\oplus}{\ddot{O}}-H\\ \\\ddot{O}-R\end{smallmatrix}$, and $R-C\begin{smallmatrix}\ddot{O}:\\ \\N-H\\ \oplus\end{smallmatrix}$ in Table 11.1 all have electron-donor functions adjacent to electron-acceptor functions, and thus represent only one limiting structure of resonance-hybrid molecules. Write the hybrid structures for these molecules.

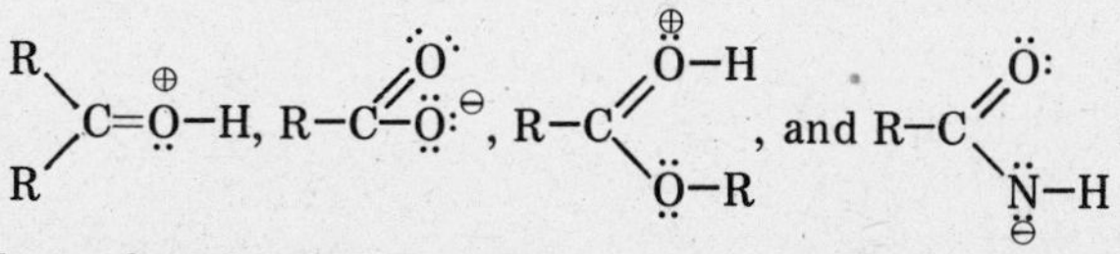

– – – – – – – –

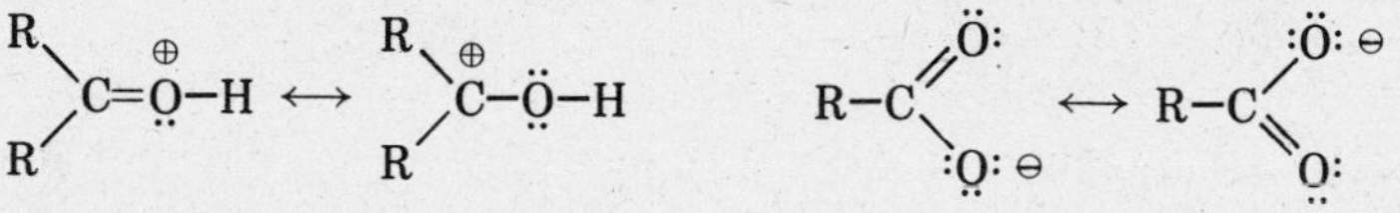

$$R_2\overset{\oplus}{C}=\ddot{O}-H \longleftrightarrow R_2\overset{\oplus}{C}-\ddot{O}-H \qquad R-C\begin{smallmatrix}\ddot{O}:\\ \\\ddot{O}:^{\ominus}\end{smallmatrix} \longleftrightarrow R-C\begin{smallmatrix}\ddot{O}:^{\ominus}\\ \\\ddot{O}:\end{smallmatrix}$$

$$R-C\overset{\overset{\oplus}{O}-H}{\underset{\ddot{O}-R}{}} \leftrightarrow R-C\overset{\ddot{O}-H}{\underset{\underset{\oplus}{\ddot{O}-R}}{}} \qquad R-C\overset{\ddot{O}:}{\underset{\underset{\ominus}{N}-H}{}} \leftrightarrow R-C\overset{:\ddot{O}:^{\ominus}}{\underset{\ddot{N}-H}{}}$$

6. According to the structural requirements we have been using, the species $\overset{H}{\underset{H}{}}\overset{\oplus}{O}-H$ should be both an acid and a base. Actually, its basic properties are not detectable. Consequently, we must revise our criteria as follows: bases whose conjugate acids would have a double electrical charge, 2+, on the basic atom have negligible strength. The same is true of acids whose conjugate bases would have a 2− charge on one atom.

Examples

1. The following ions do *not* have appreciable acidic properties.

    (a) $:\overset{\ominus}{\ddot{O}}-H \quad (\rightarrow :\ddot{O}:^{2\ominus} + H^{\oplus})$

    (b) $\overset{CH_3}{\underset{H}{}}\overset{\ominus}{\ddot{N}}: \quad (\rightarrow CH_3-\ddot{N}:^{2\ominus} + H^{\oplus})$

    too unstable to exist under ordinary conditions

2. The following ions do not have appreciable basic properties:

    (a) $H-\overset{\oplus}{\underset{H}{\ddot{O}}}\overset{H}{\underset{}{}} \quad \left(\xrightarrow{H^{\oplus}} \overset{H}{\underset{H}{}}\overset{2\oplus}{O}\overset{H}{\underset{H}{}}\right)$

    (b) $CH_3-\overset{\oplus}{\underset{H}{\ddot{O}}}\overset{H}{\underset{}{}} \quad \left(\xrightarrow{H^{\oplus}} CH-O\overset{2\oplus}{\underset{H}{}}\overset{H}{\underset{}{}}\right)$

    unstable

★ *Exercises*

1. Practically speaking, is $\overset{H}{\underset{H}{}}C=\overset{\oplus}{\underset{}{\ddot{O}}}-H$ an acid? Is it a base?

2. Is $:\overset{\ominus}{\ddot{N}}H_2$ an acid? A base?

3. Is $^{\ominus}O-\!\!\bigcirc\!\!-O-H$ an acid? A base?

- - - - - - - - - -

1. An acid, not a base $\left(\cancel{\leftrightarrow} \overset{H}{\underset{H}{}}C=\overset{2\oplus}{O}\overset{H}{\underset{H}{}}\right)$.

2. A base, not an acid ($\cancel{\leftrightarrow} :\ddot{N}H^{2\ominus}$).

3. Acid and base. Its conjugate base, $^{\ominus}O-\!\!\bigcirc\!\!-O^{\ominus}$, puts the two $^{\ominus}$ on different atoms.

AMPHIPROTIC COMPOUNDS

7. One of the most puzzling aspects of acid-base chemistry for beginners is the fact that several of the common organic functional groups are simultaneously acid and base. These compounds are identified by the asterisks in Table 11.1. The dual character is caused by the simultaneous presence of an unshared pair and an O—H or N—H bond, as in $CH_3-\ddot{O}-H$.

Such compounds are called amphiprotic; they have *both* a conjugate acid (e.g., $CH_3-\overset{\oplus}{\underset{H}{O}}\overset{H}{\diagdown}_H$)

and a conjugate base (e.g., $CH_3-\ddot{\underset{..}{O}}:^\ominus$). We must always first decide whether the amphiprotic compound is playing the role of an acid or a base in the reaction at hand, judging by the available clues, as in the next examples.

Examples

Write the equations for the acid-base equilibrium (if any) established by reaction of
(a) $CH_3OH + :\overset{\ominus}{N}(CH_3)_2$; (b) $HF + CH_3-CO_2H$; (c) $H_2O + CH_3OH$; (d) $CH_3OH + NH_2^\ominus$; (e) $H_3O^\oplus + CH_3NH_3^\oplus$.

Solutions

(a) Write reactants' Lewis structures: $CH_3-\ddot{O}-H + :\overset{\ominus}{N}\diagdown_{CH_3}^{CH_3}$

Identify which is to play the role of acid and which the base:

$CH_3-\ddot{O}-H$ $^\ominus :N\diagdown_{CH_3}^{CH_3}$

could be base or acid   base only

Since $\overset{\ominus}{N}(CH_3)_2$ must act as base, CH_3OH must act as an acid. Write the conjugate base of CH_3OH and the conjugate acid of $\overset{\ominus}{N}(CH_3)_2$ on the right-hand side:

$CH_3-\ddot{O}-H + ^\ominus\!:\!\ddot{N}\diagdown_{CH_3}^{CH_3} \rightleftharpoons CH_3-\overset{\ominus}{\underset{..}{O}}: + H-\ddot{N}\diagdown_{CH_3}^{CH_3}$

Check for charge balance: $0 + (-1) = -1 + 0$.

Check for atomic balance: $3 C + 10 H + 1 N + 1 O = 3 C + 10 H + 1 N + 1 O$.

Label acids and bases:

$CH_3-\ddot{O}-H + :\overset{\ominus}{N}\diagdown_{CH_3}^{CH_3} \rightleftharpoons CH_3-\overset{\ominus}{\underset{..}{O}}: + H-\ddot{N}\diagdown_{CH_3}^{CH_3}$

     acid   base   base   acid

Confirm that the pattern is a + b = b + a.

(b) $:\ddot{F}-H + CH_3-\overset{\overset{\ddot{O}:}{\|}}{C}-O-H$ Therefore, HF = acid, CH_3CO_2H = base.

    acid only   acid and base

$\ddot{\text{F}}$–H + CH$_3$–C(=Ö:)–O–H ⇌ :$\ddot{\text{F}}$:$^{\ominus}$ + CH$_3$–C(=$\overset{\oplus}{\text{O}}$–H)–O–H

acid   base   base   acid

0 + 0 = –1 + 1

2 C, 5 H, F, O = 2 C, 5 H, F, O

(c)   H$_2$Ö: CH$_3$–Ö–H

     acid and base   acid and base   In this case, two acid-base equilibria coexist.

H$_2$Ö: + CH$_3$–Ö–H ⇌ H$_2$$\overset{\oplus}{\text{O}}$–H + CH$_3$–Ö:$^{\ominus}$

base   acid   acid   base

H$_2$Ö: + CH$_3$–Ö–H ⇌ H–Ö:$^{\ominus}$ + CH$_3$–$\overset{\oplus}{\text{O}}H_2$

acid   base   base   acid

When this happens, both equilibria are actually set up in solution, but they may have very different equilibrium constants.

(d)   CH$_3$–Ö–H   :$\overset{\ominus}{\text{N}}H_2$ In principle, acid and base; but acidic property is negligible (frame 6).

     acid and base

Therefore,

CH$_3$–O–H + :NH$_2^{\ominus}$ ⇌ CH$_3$–O:$^{\ominus}$ + :NH$_3$

acid   base   base   acid

(e)   H$_2$$\overset{\oplus}{\text{O}}$–H   CH$_3$–$\overset{\oplus}{\text{N}}H_3$ Therefore, no acid-base reaction.

     acid only   acid only

★ *Exercises*

Write equations for the acid-base equilibria established by each pair of molecules, and label the acid and base species.

1. NH$_4^{\oplus}$ + H$_2$O

2. RCO$_2$H + C$_6$H$_5$–O$^{\ominus}$

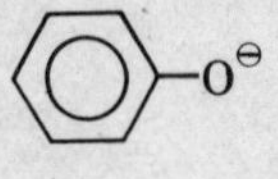

3. R–C≡C$^{\ominus}$ + NH$_3$

4. HSO$_4^{\ominus}$ + NH$_4^{\oplus}$

5. CH$_3$SO$_3$H + H–C(=O)–CH$_2^{\ominus}$

1. $NH_4^{\oplus}$ + H–Ö–H ⇌ :NH$_3$ + H–Ö–H with extra H (H$_3$O$^{\oplus}$)

   acid   base   base   acid

2. $R-C(=O)-O-H$ + $C_6H_5-\ddot{O}:^{\ominus}$ ⇌ $R-C(=O)-O:^{\ominus}$ + $C_6H_5-\ddot{O}-H$

   acid   base   base   acid

3. $R-C\equiv C:^{\ominus}$ + :NH$_3$ ⇌ $R-C\equiv C-H$ + :$\ddot{N}H_2^{\ominus}$

   base   acid   acid   base

4. $H-O-S(=O)_2-\ddot{O}:^{\ominus}$ + $NH_4^{\oplus}$ ⇌ $H-O-S(=O)_2-\ddot{O}-H$ + :NH$_3$

   base   acid   acid   base

5. $CH_3-S(=O)_2-\ddot{O}-H$ + $H-C(=O)-CH_2:^{\ominus}$ ⇌ $CH_3-S(=O)_2-\ddot{O}:^{\ominus}$ + $H-C(=O)-CH_3$

   acid   base   base   acid

ACID AND BASE STRENGTHS

8. The experimental orders of acid and base strengths of the common functional groups are listed in Table 11.2. Note that the orders of acid strength and base strength run in opposite directions. *The stronger an acid, the weaker its conjugate base*, and vice versa.

Table 11.2 includes some $\rangle$C–H compounds that are not acids by our former definition. I prefer to leave $\rangle$C–H out of the definition of an acid because these compounds are *not appreciably acidic unless* some adjacent functional group makes them so. Two examples of such functional groups, $-C(=O)-CH\langle$ and $-C\equiv C-H$, are included. The first activates by virtue of resonance stabilization of the conjugate base (the negative charge is borne mainly on oxygen; see frame 5). The second activates as a result of the enhanced electronegativity of *sp*-hybridized carbon (see your textbook). The unactivated alkanes, for instance CH_4, are so weak as acids that their quantitative acidities are not really known.

The quantitative measure of acid strength is the equilibrium constant for the acid-base reaction with water acting as a base:

$$\begin{matrix} BH^{\oplus} \\ \text{or} \\ BH \end{matrix} + H_2O \xrightleftharpoons{K_a} \begin{matrix} BH \\ \text{or} \\ B^{\ominus} \end{matrix} + H_3O^{\oplus}$$

The value of K_a is known for a great number of acids, and it is handled most conveniently for many purposes as its negative logarithm, $pK_a = -\log K_a$. In view of this definition, the *stronger* the acid, the *smaller* its pK_a. The approximate pK_a values are given in Table 11.2.

192 HOW TO SUCCEED IN ORGANIC CHEMISTRY

Table 11.2 *Strengths of Organic Acids and Bases*

Acids	pK_a	Bases
$R_2C=\overset{\oplus}{O}H$	−7	$R_2C=O$
$HCl, H_2SO_4, HClO_4$	−4	$Cl^\ominus, HSO_4^\ominus, ClO_4^\ominus$
HNO_3, RSO_3H	ca. −3	$NO_3^\ominus, RSO_3^\ominus$
$H_3O^\oplus, R-\overset{\oplus}{O}H_2, R_2\overset{\oplus}{O}H$	−2	H_2O, ROH, R_2O
RCO_2H	+5	$RCO_2^\ominus$
H_2CO_3	6.4	$HCO_3^\ominus$
$NH_4^\oplus, RNH_3^\oplus, R_2NH_2^\oplus, R_3NH^\oplus$	ca. 10	NH_3, RNH_2, R_2NH, R_3N
ArOH	10	$ArO^\ominus$
$HCO_3^\ominus$	10.3	$CO_3^{2\ominus}$
H_2O, ROH	ca. 15	$OH^\ominus, OR^\ominus$
$R_2CH-\underset{O}{\overset{\|}{C}}-R'$	ca. 20	$R_2\overset{\ominus}{C}-\underset{O}{\overset{\|}{C}}-R'$
$R-C\equiv C-H$	ca. 25	$R-C\equiv C:^\ominus$
NH_3, RNH_2, R_2NH	ca. 36	$NH_2^\ominus, RNH^\ominus, R_2N^\ominus$
$R-H$ (e.g., CH_4)	>40	$R:^\ominus$ (e.g., $:CH_3^\ominus$)

increasing acid strength ↑   *increasing base strength* ↓

Exercises

Which is the stronger base?

1. $CH_3-\overset{\ominus}{N}-CH_3$ or $CH_3-O^\ominus$?
2. $CH_3-NH-CH_3$ or $CH_3-\overset{\ominus}{N}-CH_3$?
3. $CH_3CO_2^\ominus$ or $CH_3-O^\ominus$?

---

1. $CH_3-\overset{\ominus}{N}-CH_3$
2. $CH_3-\overset{\ominus}{N}-CH_3$
3. $CH_3-O^\ominus$

PREDICTING TO WHICH SIDE AN ACID-BASE EQUILIBRIUM WILL LIE

9. This problem is sometimes encountered by itself, but it is also implicit in the solution to a number of other problems. There are several ways to state the basic rule:

   1. The stronger base wins the proton.

2. The equilibrium will lie toward the side of the weaker acid.
3. If the four species are labeled weaker acid (wa), stronger acid (sa), weaker base (wb), and stronger base (sb), the equilibrium has the pattern

$$sa + sb \rightleftharpoons wb + wa$$

these unequal arrows are used to indicate that the right-hand side is favored at equilibrium

Example

To which side will this equilibrium lie?

$$H-Cl + H_2O \rightleftharpoons Cl^{\ominus} + H_3O^{\oplus}$$

Solution:

Via rule (1):

Label: $\quad H-Cl + H_2O \rightleftharpoons Cl^{\ominus} + H_3O^{\oplus}$
$\qquad\quad$ acid $\quad$ base $\quad$ base $\quad$ acid

Identify: H_2O is a stronger base than $Cl^{\ominus}$, according to Table 11.2.

Conclude: H_2O wins the proton: $H-Cl + H_2O \rightleftharpoons Cl^{\ominus} + H_3O^{\oplus}$

or via rule (2):

Label: as before

Identify: $H_3O^{\oplus}$ is a weaker acid than HCl (Table 11.2)

Conclude: Equilibrium lies toward $H_3O^{\oplus}$:

$$H-Cl + H_2O \rightleftharpoons Cl^{\ominus} + H_3O^{\oplus}$$

or via rule (3):

Label strong base, etc.: $H-Cl + H_2O \rightleftharpoons Cl^{\ominus} + H_3O^{\oplus}$
$\qquad\qquad\qquad\qquad\quad$ sa $\qquad$ sb $\qquad$ wb $\qquad$ wa

Check: if labeling has been done properly, sa and sb will always be on the same side of the equation.

Conclude from rule (3): $H-Cl + H_2O \rightleftharpoons Cl^{\ominus} + H_3O^{\oplus}$
$\qquad\qquad\qquad\qquad\qquad$ sa $\qquad$ sb $\qquad$ wb $\qquad$ wa

★ *Exercises*

To which side will the acid-base equilibria in the exercises in frame 7 lie?

- - - - - - - - -

1. left
2. right
3. left
4. left
5. right

USING ACID-BASE EQUILIBRIA
TO MAKE PRACTICAL CHEMICAL PREDICTIONS

10. The $H-Cl + H_2O \rightleftharpoons Cl^{\ominus} + H_3O^{\oplus}$ example demonstrates that the strong acids (HCl, H_2SO_4, $HClO_4$, etc.) always exist as $H_3O^{\oplus}$ in aqueous solution. This is an important fact about the structure of aqueous acids. Now consider some examples in which the position of acid-base equilibria governs the feasibility of certain organic reactions and others in which it is used as a diagnostic tool.

Example

A sample of $CH_3O^{\ominus} Na^{\oplus}$ is desired. Can it be prepared in aqueous solution?

Solution:

The answer is probably yes, unless it reacts with H_2O. This boils down to the question: To which side will $CH_3O^{\ominus} + H_2O \rightleftharpoons CH_3OH + OH^{\ominus}$ lie? Table 11.2 shows that CH_3OH and H_2O have approximately the same acid strength. Consequently, the equilibrium constant will be close to 1 for the equilibrium

$$CH_3O^{\ominus} + H_2O \underset{}{\overset{K \approx 1}{\rightleftharpoons}} CH_3OH + OH^{\ominus}$$

The concentration of H_2O in aqueous solutions is ~55 M versus 1 M or less for most solutes. Consequently, although the equilibrium constant for this reaction is near 1, the large excess of H_2O will push the equilibrium to the right. The solvent, water, thus destroys $CH_3O^{\ominus}$, and $CH_3O^{\ominus} Na^{\oplus}$ cannot be prepared satisfactorily in contact with water for this reason.

Example

Can the sodium salt of $CH_3-C{\equiv}C-H$ be prepared by treating $CH_3-C{\equiv}C-H$ with $Na^{\oplus}CH_3O^{\ominus}$?

Solution:

Translate the problem into an acid-base equilibrium:

$$CH_3-C{\equiv}C-H + CH_3O^{\ominus} \overset{?}{\rightleftharpoons} CH_3-C{\equiv}C{:}^{\ominus} + CH_3OH$$

Label: $\quad CH_3-C{\equiv}C-H + CH_3O^{\ominus} \rightleftharpoons CH_3-C{\equiv}C{:}^{\ominus} + CH_3OH$
$\qquad\qquad$ wa $\qquad\qquad$ wb $\qquad\qquad$ sb $\qquad\qquad$ sa

Identify and conclude: The weak acid, $CH_3-C{\equiv}C-H$, will be formed, that is, the equilibrium will lie to the left. This will not be a good way to prepare $CH_3-C{\equiv}C{:}^{\ominus}Na^{\oplus}$.

This last example also shows that the metal ions that balance off the charge on anionic bases in acid-base equilibria are just along for the ride and can be dispensed with. This is also often true of halide anions, as in $NH_4^{\oplus} Cl^{\ominus} + Na^{\oplus}OH^{\ominus} \rightleftharpoons NH_3 + H_2O + Na^{\oplus}Cl^{\ominus}$. The real acid-base part of this equilibrium is just $NH_4^{\oplus} + OH^{\ominus} \rightleftharpoons NH_3 + H_2O$. In
$\qquad\qquad\qquad\qquad\qquad\qquad\qquad\qquad\qquad\qquad$ a $\qquad$ b $\qquad$ b $\qquad$ a

other cases, for example the

$$HCl + H_2O \rightleftharpoons Cl^{\ominus} + H_3O^{\oplus}$$

equilibrium, halide ion is the actual base.

Example

The labels have fallen off samples of CH_3CO_2H and C_6H_5-OH. How can $NaHCO_3$ be used to distinguish which is which?

Solution:

Explore the reactions with $NaHCO_3$, that is, with $HCO_3^\ominus$:

$$HCO_3^\ominus + CH_3CO_2H \rightleftharpoons H_2CO_3 + CH_3CO_2^\ominus$$
  sb   sa   wa   wb

$$HCO_3^\ominus + C_6H_5-OH \rightleftharpoons H_2CO_3 + C_6H_5-O^\ominus$$
  wb   wa   sa   sb

Since one lies to the right, one to the left, they can be distinguished by sight since H_2CO_3 breaks down to CO_2, which bubbles off, and H_2O. So CH_3CO_2H (or any carboxylic acid) gives CO_2 on treatment with $NaHCO_3$, while C_6H_5OH (or most phenols) give none.

★ Exercises

1. Given that the equilibrium $CH_3OH + CH_3-C\equiv N-H^\oplus \rightleftharpoons CH_3-O\overset{\oplus}{\underset{H}{\diagup}}^H + CH_3-C\equiv N:$

   lies to the right, which is the stronger acid, $CH_3OH_2^\oplus$ or $CH_3-C\equiv N^\oplus H$? Which is the stronger base, $CH_3-C\equiv N:$ or CH_3OH?

2. Given that the following equilibria both lie to the *right*, what can you deduce about the base strength of $CH_3-\overset{O}{\overset{\parallel}{C}}-NH_2$? Can you find its place in the basicity-order chart?

   (a) $CH_3-\overset{O}{\overset{\parallel}{C}}-NH_2 + H_3O^\oplus \rightleftharpoons CH_3-C(NH_2)\overset{\oplus}{O}H + H_2O$

   (b) $CH_3-C(NH_2)OH^\oplus + CH_3COO^\ominus \rightleftharpoons CH_3-\overset{O}{\overset{\parallel}{C}}-NH_2 + CH_3CO_2H$

3. Will CH_3-O-H liberate CO_2 from a solution of $Na^\oplus HCO_3^\ominus$?

4. Will $CH_3-C\equiv C^\ominus Na^\oplus$ liberate NH_3 from a solution of $NH_4^\oplus Cl^\ominus$?

– – – – – – – – –

1. Stronger acid = $CH_3-C\equiv \overset{\oplus}{N}-H$
   Stronger base = CH_3OH

2. From (a), $CH_3-\overset{O}{\overset{\parallel}{C}}-NH_2$ is a stronger base than H_2O.

   From (b), $CH_3-\overset{O}{\overset{\parallel}{C}}-NH_2$ is a weaker base than $CH_3CO_2^\ominus$.

   Therefore, $CH_3-\overset{O}{\overset{\parallel}{C}}-NH_2$ falls between H_2O and $CH_3CO_2^\ominus$ in the basicity order.

3. The relevant equilibrium is $CH_3OH + HCO_3^\ominus \rightleftharpoons CH_3-O^\ominus + H_2CO_3$. Since this
   (wa)   (wb)   (sb)   (sa)

   equilibrium lies to the left, only low concentrations of H_2CO_3 will be present, and the equilibrium $H_2CO_3 \rightleftharpoons H_2O + CO_2$ will probably not produce enough CO_2 to exceed its solubility. So CO_2 will probably not bubble off.

4. The relevant equilibrium is $CH_3-C\equiv C^\ominus + NH_4^\oplus \rightleftharpoons CH_3-C\equiv C-H + NH_3$. Since
   (sb)   (sa)   (wa)   (wb)

   this equilibrium lies to the right, large amounts of NH_3 will be produced and should be smellable over the solution.

ROUGH-QUANTITATIVE PREDICTIONS

11. Knowledge of the approximate pK_a values from Table 11.2 allows you to solve quantitative versions of these problems, as in the next example.

Example

What is the approximate value of the equilibrium constant for the reaction $OH^\ominus$ + $-\overset{O}{\overset{\|}{C}}-CH\!\!\!<\; \rightleftharpoons\; OH_2\; +\; -\overset{O}{\overset{\|}{C}}-\overset{\ominus}{C}\!\!\!<\;$?

Solution:

First decide qualitatively in which direction the equilibrium will lie:

$OH^\ominus\; +\; -\overset{O}{\overset{\|}{C}}-CH\!\!\!<\;\; \overset{K}{\rightleftharpoons}\;\; HOH\; +\; -\overset{O}{\overset{\|}{C}}-\overset{\ominus}{C}\!\!\!<\;$
(wb)   (wa)   (sa)   (sb)

The absolute value of the logarithm of the equilibrium constant K is given by the difference in pK_a of the sa and wa species: $\pm \log K = 25 - 15 = 10$. Therefore, K is 10^{+10}, or 10^{-10}.

Decide: on the basis of the qualitative direction of the equilibrium, $\rightleftharpoons$, K must be less than 1; therefore, K is about 10^{-10}.

This example shows that one cannot prepare the ion $-\overset{O}{\overset{\|}{C}}-\overset{\ominus}{C}\!\!\!<\;$ in much of a yield by this reaction. Nevertheless, the $-\overset{O}{\overset{\|}{C}}-\overset{\ominus}{C}\!\!\!<\;$ formed in very low concentrations in this reaction can be an *intermediate* in the 100% conversion of $-\overset{O}{\overset{\|}{C}}-CH\!\!\!<\;$ to various reaction products. $OH^\ominus$ is an important reagent for this purpose despite the unfavorable equilibrium constant. As $-\overset{O}{\overset{\|}{C}}-\overset{\ominus}{C}\!\!\!<\;$ is consumed to give reaction products, more is formed from $-\overset{O}{\overset{\|}{C}}-CH\!\!\!<\;$ + $OH^\ominus$ to maintain the equilibrium concentration dictated by K. Eventually, all of the $-\overset{O}{\overset{\|}{C}}-CH\!\!\!<\;$ is consumed through the intermediacy of $-\overset{O}{\overset{\|}{C}}-\overset{\ominus}{C}\!\!\!<\;$.

Exercises

Predict the approximate equilibrium constants for each of the following equilibria.

1. $CH_3SO_3H + CH_3CO_2^{\ominus} \rightleftharpoons CH_3CO_2H + CH_3SO_3^{\ominus}$
2. $C_6H_5-OH + NH_3 \rightleftharpoons C_6H_5-O^{\ominus} + NH_4^{\oplus}$

— — — — — — — — —

1. Qualitatively, this equilibrium lies to the right (side of the weaker acid, CH_3CO_2H). Therefore, $K > 1$. pK_a values for the two acids are -3 and $+5$; difference $= \pm 8$. Therefore, $K \approx 10^8$.

2. C_6H_5-OH and $NH_4^{\oplus}$ are of approximately equal acidity. Therefore, $K \approx 1$.

UNIT TWELVE
Reactions of the Aliphatic Hydrocarbons

In this unit we begin to work on the 100 to 150 standard reactions that make up the core of the traditional organic chemistry course. This material breaks naturally into a number of packages, each of which contains the important reactions that one functional group, say $-C\overset{\displaystyle O}{\underset{\displaystyle H}{\diagup\!\!\!\diagdown}}$, undergoes in whatever compounds it occurs. The reactions of the common functional groups will be begun in this unit and completed in Units 14 to 19. Each of these units will have the same organization: (1) ways to make the list of reactions reasonable, coherent, and easy to learn; (2) methods of drilling on the reactions; and (3) solving problems based on the reactions.

The present unit follows this same pattern, but points 1 and 3 are expanded to include general guidance to functional group chemistry that will apply here and in Units 14 to 19.

OBJECTIVES

When you have completed this unit, you should be able to:

- Classify reactions as additions, eliminations, substitutions, oxidations, and/or reductions (frame 4).
- Write the equation and mechanism for each of the reactions of the aliphatic hydrocarbons, both in generic and specific form (frames 3 to 15, 17 to 19, 23 to 26).
- Supply, if given any two of the three elements: reactant $\xrightarrow{\text{reagent}}$ product, the structural formula of the third (frame 16).
- Decide which of two possible reactants for a given aliphatic hydrocarbon reaction would react fastest (frames 7, 8).
- Decide which of several possible products of a given reaction will predominate over the others (frames 20, 21).
- Use various clues and analogies to deduce products and mechanisms for less familiar reactions (frame 22).

If you think you may have already achieved these objectives and might skip all or part of this unit, take a self-test consisting of the exercises marked with a star in the margin. If you miss any problems, study the frames in question and rework the problems before proceeding. If you are not ready for a self-test, proceed to frame 1.

REACTIONS OF THE ALIPHATIC HYDROCARBONS

GENERAL PRINCIPLES FOR LEARNING REACTIONS

1. First we must make certain you can read and write the notation used in writing equations. The common formats and some examples are shown below. Since only rarely are there three or more reactants, the equations below show only two; more could be added. Items in parentheses may be either present or absent.

reactant 1 (+ reactant 2) $\longrightarrow$ product(s)

$$\text{Example: } CH_2=CH_2 + HBr \longrightarrow CH_2Br-CH_3 \tag{1}$$

reactant 1 (+ reactant 2) $\xrightarrow[\text{or conditions}]{\text{catalyst}}$ product(s)

$$\text{Examples: } CH_3-\underset{CH_3}{\overset{CH_3}{\underset{|}{C}}}-OH \xrightarrow{H_2SO_4} CH_3-\underset{CH_3}{\overset{}{\underset{|}{C}}}=CH_2 + H_2O \tag{2}$$

$$CH_4 + Br_2 \xrightarrow{\text{light}} CH_3Br + HBr \tag{3}$$

$$CH_3-CH=CH-CH_2-CH_3 + H_2O \xrightarrow{H_2SO_4}$$
$$CH_3-\underset{OH}{\underset{|}{CH}}-CH_2-CH_2-CH_3 + CH_3-CH_2-\underset{OH}{\underset{|}{CH}}-CH_2-CH_3 \tag{4}$$

reactant 1 $\xrightarrow[\text{(catalyst or conditions)}]{\text{reactant 2}}$ products

$$\text{Examples: } CH_2=CH_2 \xrightarrow{HBr} CH_3-CH_2Br \tag{5}$$

$$CH_2=CH_2 \xrightarrow[H_2SO_4]{H_2O} CH_3-CH_2-OH \tag{6}$$

reactant 1 $\xrightarrow[\text{(2) reactant 3}]{\text{(1) reactant 2}}$ product(s)

$$\text{Example: } CH_3-CH_2Br \xrightarrow[\text{(2) } H_2O]{\text{(1) Mg}} CH_3-CH_3 \tag{7}$$

The format in equation (7) means that reactant 2 and reactant 3 are added *sequentially*. Reactant 1 reacts with reactant 2 to give some product (called an *unisolated intermediate*) that is never taken out of the flask. Reactant 3 then is added to react with this intermediate, and the final product(s) results.

Equations (1) and (5) are equally legitimate ways of representing the same reaction. (The symbol Δ will often be used to indicate that heat is applied.)

The terms *reactant, reagent,* and *starting material* are all used for the compound(s) on the left side. "Starting material" often carries the connotation of being the reactant that is most expensive or has the largest carbon skeleton; "reagent" is often used for inorganic, cheap, or small molecules, but the terms are not really standardized.

The equation does not usually balance. One reason for writing unbalanced equations is that our reactions often give a mixture of products. This means that the "reaction" really consists of two or more reactions occurring simultaneously and giving different products. For example, equation (4) represents the actual result of two such *parallel* reactions:

$$CH_3-CH=CH-CH_2-CH_3 + H_2O \xrightarrow{H_2SO_4} CH_3-CHOH-CH_2-CH_2-CH_3 \quad (4a)$$

$$CH_3-CH=CH-CH_2-CH_3 + H_2O \xrightarrow{H_2SO_4} CH_3-CH_2-CHOH-CH_2-CH_3 \quad (4b)$$

Collapsing both together into equation (4) is a shorthand way of telling the whole story. Equation (4) doesn't balance, and it doesn't mean to say that 5 carbon atoms have turned into 10. It says that some of the molecules of $CH_3-CH=CH-CH_2-CH_3$ turn into $CH_3-CHOH-CH_2-CH_2-CH_3$ and others turn into $CH_3-CH_2-CHOH-CH_2-CH_3$, so that the *sum total* of moles or molecules of the *two* products equals the number of moles or molecules of $CH_3-CH=CH-CH_2-CH_3$ consumed.

An equation representing two or more parallel reactions carries with it an implied question: What are the relative amounts of the two or more products? This may be answered in the equation, for example:

$$\underset{\underset{CH_3}{|}}{\overset{\overset{CH_3}{|}}{CH_3-C-H}} \xrightarrow{\underset{light}{Cl_2}} \underset{\underset{CH_3}{|}}{\overset{\overset{CH_3}{|}}{CH_3-C-Cl}} (36\%) + \underset{\underset{CH_3}{|}}{\overset{\overset{CH_2-Cl}{|}}{CH_3-C-H}} (64\%) \quad (8)$$

This last example brings up two more points. First, inorganic products are often not written [HCl is omitted from equation (8)]. Second, the product percentages may or may not sum to 100%. When they don't, it means that the remainder of the starting material turned into some additional product(s). These percentage figures are called the percentage *yields* of the various products. When the percentages do sum to 100%, it usually means that they are not the actual yields of the products, but just represent the composition of the product mixture without stating how much of the mixture was obtained. Sometimes the products are labeled *major* and *minor* rather than giving actual percentage yields.

Exercises

Given the equation

$$CH_3-C\equiv C-H + Cl_2 \xrightarrow{60-70°C} \underset{\underset{20\%}{}}{\overset{}{\underset{Cl}{\overset{CH_3}{>}}C=C\underset{H}{\overset{Cl}{<}}}} + \underset{63\%}{CH_3-\overset{\overset{Cl}{|}}{\underset{\underset{Cl}{|}}{C}}-\overset{\overset{Cl}{|}}{\underset{\underset{Cl}{|}}{C}}-H} \quad (9)$$

1. Write the balanced equations for the two parallel reactions represented by equation (9). Call them (9a) and (9b).

2. What percentage of the reactant molecules turn into products by following path (9a)?

3. If you had to use these labels, what would you call the reagent and the starting material in equation (9)?

4. What is the minor product?

_ _ _ _ _ _ _ _ _

1. $CH_3-C\equiv C-H + Cl_2 \longrightarrow \underset{Cl}{\overset{CH_3}{>}}C=C\underset{H}{\overset{Cl}{<}} \quad (9a)$

$CH_3-C\equiv C-H + 2Cl_2 \longrightarrow CH_3-CCl_2-CCl_2H \quad (9b)$

2. 20% of the $CH_3-C\equiv C-H$ traverses path (9a), giving 20% of
$$\begin{array}{c}CH_3\\ \end{array}\!\!\!C\!\!=\!\!C\!\!\!\begin{array}{c}Cl\\ H\end{array}$$
with Cl below CH$_3$ and H below Cl.

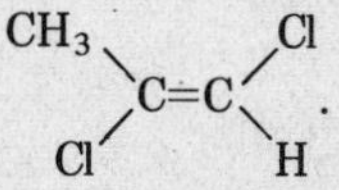

3. Starting material = $CH_3-C\equiv C-H$; reagent = Cl_2.

4. Minor product =
$$\begin{array}{c}CH_3\\ Cl\end{array}\!\!\!C\!\!=\!\!C\!\!\!\begin{array}{c}Cl\\ H\end{array}$$

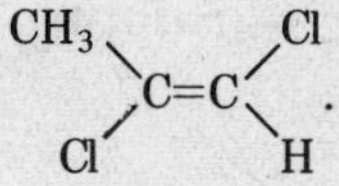

2. We can write equations for an unlimited number of reactions, but being able to write them on paper does not mean that they actually can be made to take place. By *allowed* reactions we mean the ones that are actually found to occur. The reaction CH_3-CH_3 + $H_2O \longrightarrow CH_3-CH_2-OH + H_2$ *has never been observed* under any experimental conditions tried, and consequently it is what I call a *forbidden* reaction—it is not a meaningful answer to any question. Forbidden reactions fail to occur either because the rate at which they occur is too small, or because the position of equilibrium favors the reactants rather than the products, or both.

Since there is no easy way to *figure out* whether the above reaction (or any reaction) would be allowed or forbidden, you must memorize the list of allowed reactions as you study the chemistry of each family of compounds. Most people can memorize these lists easily. The keys to doing this are (1) organizing the lists, and (2) drilling adequately. We talk about organization in the next few frames, and consider drill in frames 15 and 16.

3. The first step in organizing the unlimited number of reactions in organic chemistry is classification according to the functional group of the reactant. You practiced this classification in Unit 3, where we saw that there are only about 15 common functional groups. The result is that practically all organic compounds fall into about 15 major families. All the members of a family undergo the same set of reactions (to a first approximation). The neatest way of putting this is to write a *generic* structure (Unit 2, frame 10) representing all the compounds in a family, for instance R—Cl. Then the reactions of all the members of the family can be written as generic reactions that use generic structures for reactant and product. For example, the reaction R—Cl + Mg $\longrightarrow$ R—Mg—Cl sums up hundreds of reactions since it works no matter what kind of skeleton R represents. In fact, it takes only 10 to 20 of these generic reactions to summarize all the chemistry of a given family (function), and that means that only 100 to 200 generic reactions summarize all the common reactions of organic chemistry.

This is an economical way to store information compared to trying to learn the reactions of each specific compound. However, if you learn a reaction in generic form, you will have to practice "seeing the general in the specific" if you are going to apply that knowledge to specific reactants. This consists of recognizing the common functional groups in specific structural formulas, including cyclic and complex ones containing more than one functional group. Unit 5 gives you this kind of practice. Then, after locating the reactive functional group, you must be able to look at the generic reaction and visualize the same chemical change happening to the function in this specific molecule.

Example

All of the following are specific molecules implied by the general formula R—C≡C—R′ (R, R′ = alkyl, aryl, or H)*:

---

* This usage stretches the meaning of R considerably (Unit 2, frame 10), but this is often done.

202 HOW TO SUCCEED IN ORGANIC CHEMISTRY

H−C≡C−H

CH$_3$−CH$_2$−CH$_2$−CH$_2$−C≡C−H

[decalin]−C≡CH

[cyclooctyne structure]

[chromone with C≡CH substituent]

[Ph−C≡C−Ph]

CH$_2$=CH−CH$_2$
   |
H−C≡C−CH$_2$

[cyclopropyl]−C≡C−CH$_2$−CH$_3$

They are all expected to show the allowed reactions of the −C≡C− function, for instance adding two molecules of Cl$_2$, as in equation (9b), frame 1 (although other functions in the molecule may also react). If you know this reaction in generic form (R−C≡C−R′ + 2Cl$_2$ ⟶ R−CCl$_2$−CCl$_2$−R′) and are required to write the analogous process for one of the complex alkynes in the list above, merely locate the functional group (−C≡C−), operate on it, and drag the rest of the molecule along, for example

[chromone-C≡CH] = [chromone−C≡CH] $\xrightarrow{2Cl_2}$ [chromone−CCl$_2$−CHCl$_2$]

Exercise

Draw specific structures for the products of adding 2Cl$_2$ to the remaining alkynes in the structures displayed at the top of this page.

CHCl$_2$−CHCl$_2$

CH$_3$CH$_2$CH$_2$CH$_2$CCl$_2$CHCl$_2$

[cyclooctane with Cl, Cl, Cl, Cl]

[Ph−CCl$_2$−CCl$_2$−Ph]

[decalin]−CCl$_2$−CHCl$_2$

CH$_2$=CH−CH$_2$−CH$_2$−CCl$_2$−CHCl$_2$

[cyclopropyl]−CCl$_2$−CCl$_2$−CH$_2$−CH$_3$

(these functions also react)

4. The second step in organizing the reactions of the functions is classification by reaction type. Most organic reactions (and most individual steps in multistep reactions) can be classified as representing one or more of a few basic reaction types:

- Addition reactions (which turn double bonds into single bonds and triple bonds into single or double bonds—in other words, reactions that destroy multiple bonds).
- Elimination reactions (which create multiple bonds).
- Substitution reactions (in which a new group replaces some group in the starting material).
- Oxidation (in which one or more atoms of the starting material increase in oxidation state).

- Reduction (in which one or more atoms of the starting material decrease in oxidation state).

This classification is useful in many problems. At present we are interested in it as an aid in learning and recalling the basic set of reactions of a given functional group. One technique of recall is to ask yourself in turn "are there any additions," "are there any eliminations," and so on, among the set of reactions to be recalled. This technique works best if you have thought about the classification at the time of learning the set, and best of all if you can remember that there are, say, four additions among the set of reactions to be recalled.

Oxidation and reduction refer to the *organic* starting material. The reaction

$$CH_3-\underset{\underset{O}{\|}}{C}-CH_3 + 2Zn + 4H^{\oplus} \rightarrow CH_3-CH_2-CH_3 + 2Zn^{2+} + H_2O$$

involves reduction of CH_3COCH_3 and oxidation of Zn (every reduction must be accompanied by oxidation of something else), but since the organic reactant is reduced, we think of this as a *reduction*. You can identify most oxidations and reductions by looking at the molecular formulas of reactant and product and using these criteria:

- Oxidations usually increase the number of C—O or C—X bonds in the starting material and/or decrease the number of C—H or C—metal bonds.
- Reductions usually decrease the number of C—O or C—X bonds in the starting material and/or increase the number of C—H or C—metal bonds.

Examples

$$CH_3-CH_2OH + Cr_2O_7^{2-} \rightarrow CH_3-C\underset{H}{\overset{O}{\diagup}} + Cr^{3+} \qquad \text{oxidation}$$
$$-2H$$

$$CH_3-C\underset{OH}{\overset{O}{\diagup}} + LiAlH_4 \rightarrow CH_3-CH_2OH \qquad \text{reduction}$$
$$+2H - O$$

$$CH_2=CH_2 + Br_2 \rightarrow \underset{Br}{CH_2}-\underset{Br}{CH_2} \qquad \text{oxidation}$$
$$+ Br$$

$$2CH_3-CH_2-Br + 2Na \rightarrow CH_3-CH_2-CH_2-CH_3 + 2NaBr \qquad \text{reduction}$$
$$-Br$$

Many reactions are both reductions and additions, both oxidations and eliminations, and so on:

$$CH_3-CH_3 \underset{\text{addition (of } H_2 \text{) and reduction}}{\overset{\text{elimination (of } H_2 \text{) and oxidation}}{\rightleftarrows}} CH_2=CH_2$$

Many reactions are neither oxidations nor reductions:

$$CH_3-Br + OH^{\ominus} \rightarrow CH_3-OH + Br^{\ominus}$$

★ Exercises

Classify each reaction as an addition, elimination, substitution, oxidation, and/or reduction of the organic starting material.

1. $CH_4 + Cl_2 \xrightarrow{light} CH_3-Cl + HCl$

2. $\underset{H}{\overset{O}{\underset{\|}{H-C-H}}} + H_2 \longrightarrow \underset{H}{\overset{OH}{\underset{|}{H-CH-H}}}$

3. $\overset{O}{\underset{\|}{H-C-H}} \xrightarrow{Ag^{\oplus}} \overset{O}{\underset{\|}{H-C-O-H}}$

4. $CH_2Br-CH_2Br + Zn \longrightarrow CH_2=CH_2 + ZnBr_2$

5. $CH_3-CH_2-OH + HBr \longrightarrow CH_3-CH_2-Br + H_2O$

6. $CH_3-CH_2-Br + OH^{\ominus} \longrightarrow CH_2=CH_2 + H_2O + Br^{\ominus}$

7. $CH_3-Mg-Br + H_2O \longrightarrow CH_4 + HO-Mg-Br$

8. $CH_3-Br + Mg \longrightarrow CH_3-Mg-Br$

- - - - - - - - - -

1. Substitution and oxidation.

2. Addition and reduction.

3. Oxidation.

4. Elimination and reduction.

5. Substitution.

6. Elimination; the left carbon is oxidized (loses H), the right carbon reduced (loses Br).

7. Substitution.

8. Substitution and reduction.

5. The third classification is by mechanism. The common reactions take place in only a very limited number of ways, namely the nine mechanisms of Unit 10, frame 5. For a given type of reaction, there are at most three mechanisms that could possibly apply. Experiment always shows that only one applies in a given case.

Substitution can only be accomplished by:

- Radical substitution (mechanism I), or
- Electrophilic aromatic substitution (mechanism VII), or
- Nucleophilic substitution (mechanism IV, V, or IX)

Addition can only be accomplished by:

- Electrophilic addition (mechanism II), or
- Radical addition (mechanism III), or
- Nucleophilic addition (mechanism VIII)

Elimination can only be accomplished by the:

- $E1$ (mechanism IV) or
- $E2$ (mechanism VI) paths

Oxidation and reduction are not usefully subcategorized by mechanism.

Now, if we apply all three types of classification, we can put the unlimited number of individual organic reactions into the following form:

all organic reactions
⎧
⎨ reactions of the alkene family
⎩

⎧
│ R–CH=CH–R' + X$_2$ $\xrightarrow{\text{room temp.}}$ R–CH–CH–R' with X, X below
│ R–CH=CH–R' + HX ⟶ R–CH–CH–R' with H, X below
│ R–CH=CH–R' + H$_2$SO$_4$ ⟶ R–CH–CH–R' with H, O–SO$_3$H below
│ ⋮
⎫ electrophilic additions (mechanism II)

⎧
│ R–CH=CH–R' + H–CCl$_3$ $\xrightarrow{\text{peroxide}}$ R–CH–CH–R' with H, CCl$_3$ below
│ R–CH=CH–R' + R''–SH $\xrightarrow{\text{peroxide}}$ R–CH–CH–R' with H, S–R'' below
│ ⋮
⎫ additions — radical additions (mechanism III)

⎧
│ R–CH=CH–R' + KMnO$_4$ $\xrightarrow{\text{cold}}$ R–CH–CH–R' with OH, OH below
│ R–CH=CH–R' + H$_2$ $\xrightarrow{\text{Pt}}$ R–CH$_2$–CH$_2$–R'
│ ⋮
⎫ reactions of unknown or nongeneral mechanism

⎧
│ R–CH$_2$–CH=CH–R' + X$_2$ $\xrightarrow{\text{high temp.}}$ R–CH–CH=CH–R' with X below
│ ⋮
⎫ substitutions — radical substitutions (mechanism I)

⎧
│ R–CH$_2$–CH=CH–R' + O$_2$ ⟶ CO$_2$ + H$_2$O
│ ⋮
⎫ miscellaneous reactions — reactions of unknown or nongeneral mechanism

reactions of the alkyl halide family
⋮

This list is manageable because it is structured. Individual reactions can be remembered in terms of their position in the scheme. Reactions can be learned in small groups that have several things in common: the reactant's functional group, the reaction type, the reaction mechanism.

6. Some people grasp the scheme of allowed reactions more easily from a pictorial representation. These can take many forms; one of the most useful is an oxidation-level chart such as Chart 12.2 (p. 220). This type of chart serves you in three important ways. First, it shows (by the *absence* of arrows) the forbidden as well as the allowed reactions. So if you can remember the chart, you will not make the mistake of trying to solve problems using reactions that do not exist. Second, the chart emphasizes the allowed reactions as *one-step conversions* of one organic functional group into another. If you know these conversions, you can solve *synthetic* problems ("how do you make A from B," considered in Unit 21), whose solutions are strings of one or more one-step conversions. Third, the chart arranges the functional groups involved according to *oxidation state;* the top row contains compounds of the highest, the bottom row those of the lowest oxidation level. This simplifies recalling the reagent(s) required to bring about a given conversion. For example, the conversion $-CH_2-CH_2- \longrightarrow CO_2$ is an *upward* movement, consequently an *oxidation*, and requires an oxidizing reagent, namely O_2. Conversely, the conversion $-C\equiv C- \longrightarrow -CH=CH-$ is a *downward* movement (*reduction*) requiring a reducing agent (H_2).

Exercises

1. The one-step conversion $-CH_2-CH_2- \longrightarrow _H^{}\!\!>\!C\!=\!C\!<\!_H^{}$ in Chart 12.2 is a _____ conversion.

2. Does this mean that there is no way to convert $-CH_2-CH_2-$ into $_H^{}\!\!>\!C\!=\!C\!<\!_H^{}$?

― ― ― ― ― ― ― ― ―

1. Forbidden.

2. No, all it means is that there is no *one-step* conversion. The same result can be accomplished by the two-step conversion $-CH_2-CH_2- \longrightarrow -CH_2-CHX \longrightarrow _H^{}\!\!>\!C\!=\!C\!<\!_H^{}$.

7. Two different functional groups in the same molecule that happen to react with the same reagent will usually show some *selectivity* in their reactivity—that is, one group often reacts faster than the other. This can allow isolation of a product in which one has reacted but the other has not. For example, although both substituted ethenes and substituted benzenes react with H_2 in the presence of a platinum catalyst $\Big(R-CH=CH_2 + H_2 \xrightarrow{Pt} R-CH_2-CH_3$ and $R-\!\!\bigcirc \xrightarrow{3H_2}_{Pt} R-\!\!\bigcirc\Big)$, it is generally possible to reduce an alkene function while leaving a benzene function in the same molecule untouched:

$$CH_3-CH_2-CH=CH-CH_2-\!\!\bigcirc \xrightarrow{H_2}_{Pt} CH_3-CH_2-CH_2-CH_2-CH_2-\!\!\bigcirc$$

Many reactions of one functional group can be carried out selectively without altering other types of functional groups in the same molecule.

However, when the two functional groups in the same molecule are alike, less selectivity is displayed. For example, in the reaction

$$CH_3-CH=CH-CH_2-CH_2-CH=CH-C_2H_5 \xrightarrow[Pt]{H_2} CH_3-CH=CH-CH_2-CH_2-CH_2-CH_2-C_2H_5$$
$$+$$
$$CH_3-CH_2-CH_2-CH_2-CH_2-CH=CH-C_2H_5$$

the two isomeric products shown are formed in nearly equal amounts; the reaction is *unselective* with respect to these very similar double bonds. *Some* selectivity can be shown in the reaction of two functional groups of the same class, however, especially if the environment of one is very different from that of the other. In the hydrogenation reaction above, for example, monosubstituted double bonds (Unit 3, frame 2) react faster than disubstituted, disubstituted faster than trisubstituted, and so on. The *reactivity order* in this reaction is: tetra < tri < di < mono < unsubstituted. This reactivity order is an important additional piece of information beyond just knowing that >C=C< + $H_2 \xrightarrow{Pt} \text{>CH-CH<}$ is one of the allowed reactions of alkenes. To solve problems effectively, you should *learn the reactivity order* (if there is one) *along with the reaction*, just as in studying a language the best practice is to learn the gender of a noun at the same time you learn the noun.

The simplest functional group, the C—H bond, occupies an unusual position because there are usually many of these groups in every molecule. For example, the alkane

$$\overset{a}{CH_3}-\overset{b}{\underset{\underset{a}{CH_3}}{CH}}-\overset{c}{CH_2}-\overset{d}{CH_2}-\overset{e}{CH_3}$$

has two types of 1° (a, e), two types of 2° (c, d), and one type of 3° (b) C—H bonds (Unit 3, frame 5), each capable of reacting to give a different product and each reacting at somewhat different rates. As a result, the reactions of this alkane, and many others, are generally *unselective*, leading to a reaction of more than one site and giving mixtures of products in which no one compound may strongly predominate.

Exercise

There are three functional groups, A, B, and C, in a molecule. Represent it by A—B—C. Each functional group reacts with a certain reagent. A is transformed to A′, B to B′, and C to C′. Now when A—B—C is treated with the reagent for an appropriate time, it is transformed into a mixture of A—B′—C′, A—B′—C, and A′—B′—C′. Does this reaction show any selectivity toward A, B, and C? If so, what is the reactivity order of A, B, and C?

— — — — — — — — —

A, B, and C react at different rates; otherwise, the products would contain A′—B—C and A—B—C′ as well as A—B′—C. So there is some selectivity. Since B is transformed most frequently and A least frequently, the reactivity order is B > C > A.

8. Some of the standard reactions of a given functional group work well only on some, not all, members of the family. In other words, there are limitations on the type of environment the functional group can be in and still give good results in the reaction. A summary of any such limitations is often called the *scope* of the reaction. It is important that you learn the scope at the same time you learn the other aspects of each new reaction; otherwise, you will have difficulty applying your knowledge of the generic reaction to specific systems.

Example

The scope of the halogenation reaction of alkanes (R—H + X_2 $\xrightarrow{\text{light}}$ R—X + H—X) can be described in this way: all C—H bonds on saturated carbon atoms react, regardless of environment ($1°$, $2°$, $3°$, allylic, benzylic). Only X = Cl and X = Br are practical reactions (F is too reactive, I too inert).

The scope of a reaction tells you what species can play the various roles laid out in the generic equation. The equation in the example above defines two general roles: that of "halogen" (X) and that of "skeleton" (R). The statement of scope says that only Cl and Br can play the role of X, and any alkyl group can play the role of R. Or you can use the terms sometimes used in learning grammar: R and X are *slots* in the equation (sentence) that can be replaced by any of the permissible *slot fillers*, in this case any alkyl group and Cl or Br, respectively.

9. Frames 1 to 8 presented the strategy of learning the reactions of the functional groups. To summarize, the strategy is to:

- Study the reactions of one functional group at a time.
- Learn four basic facts about each reaction studied:
  (1) The generic equation (reactants, products, conditions).
  (2) The mechanism (which of the standard mechanisms applies; ability to write it out).
  (3) The scope (permissible slot fillers).
  (4) The reactivity order (how rate, equilibrium position, or yield varies with identity of the slot fillers).
- Use classification by reaction type to relate the reactions learned and the reagents they require to one another and to help you recall them or recognize specific examples of them.
- Use oxidation-level or other pictorial charts to keep the relationship of each functional group to the others straight.

The next few frames show how to apply this strategy to the reactions of the C—H, >C=C<, and —C≡C— functions.

Then frames 15 and 16 suggest a practical way of doing the memorization involved, practicing the method on the alkane, alkene, and alkyne reactions.

The last part of the unit is devoted to methods of solving problems that involve reaction products (frames 17 to 23) and mechanisms (frames 24 to 26), laying out the general principles that apply to all the functions and applying them to the C—H, >C=C<, and —C≡C— functions.

BASIC MODES OF REACTIVITY AND FUNCTIONAL GROUP INTERRELATIONSHIPS FOR THE ALIPHATIC HYDROCARBONS

10. Most reactions occur because the products are more stable than the reactants and because there exists some sequence of bond making and bond breaking that serves to bring about the reaction in steps that require only a modest temporary input of energy. But this "reason" doesn't help us predict what reactions will occur or even to classify or understand them better. A more practical way of "understanding" why reactions occur is to identify the reactive sites in a functional group and see what property the reagents that attack that site have in common. The most common understanding we arrive at is to look at many reactions as generalized acid-base reactions in which the

REACTIONS OF THE ALIPHATIC HYDROCARBONS

reactive site is a base, and the reagent plays the role of Lewis acid—or the reverse. We often call the base component a *nucleophile* and the Lewis acid component an *electrophile*.

For the aliphatic hydrocarbons, the various reactions flow from these basic modes of reactivity as shown below. Although this does not come close to *explaining* the occurrence of the reactions, it may make them seem less arbitrary. [Actually the corresponding diagrams for most of the other functions (Units 14 to 19) are considerably richer and more helpful than Chart 12.1.]

Chart 12.1

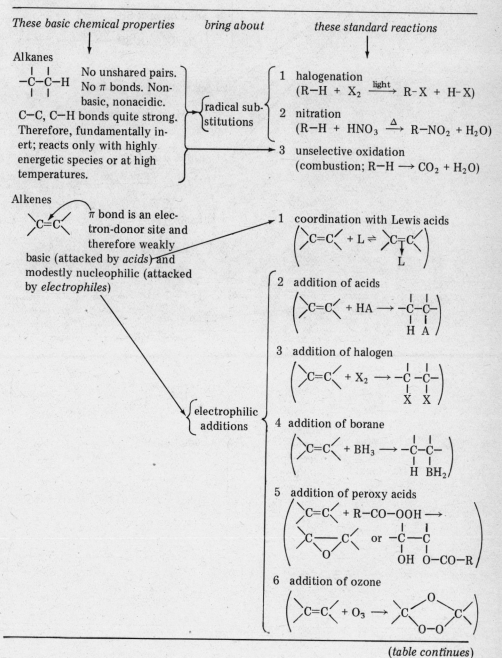

(table continues)

Chart 12.1 (*continued*)

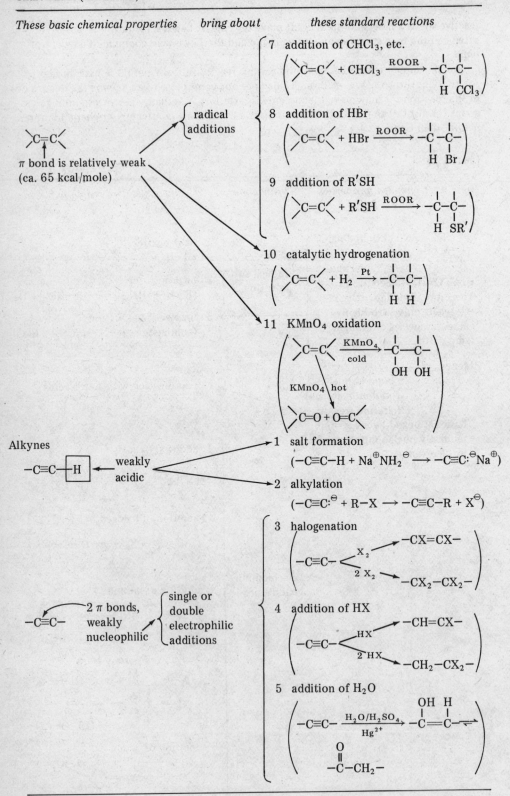

REACTIONS OF THE ALIPHATIC HYDROCARBONS 211

Chart 12.1 *(continued)*

These basic chemical properties   bring about   these standard reactions

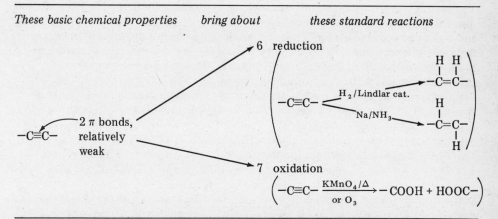

11. To see exactly how some of the reactions are generalized acid-base reactions requires looking at the details of the reaction mechanisms. We do this now for the three mechanisms involved in the reactions of the aliphatic hydrocarbons. The remaining standard mechanisms will be discussed as they come up in Units 14 to 19.

Mechanism I,* *radical substitution*, is the path followed in alkane halogenation and nitration, reactions 1 and 2 in Chart 12.1.

$$:\ddot{X}-\ddot{X}: \xrightarrow{light} :\ddot{X}\cdot + \cdot\ddot{X}: \qquad \text{initiation step}$$

$$\left. \begin{array}{l} :\ddot{X}\cdot \ \ H-CH_3 \longrightarrow :\ddot{X}-H + \cdot CH_3 \\ :\ddot{X}-\ddot{X}: \ \ \cdot CH_3 \longrightarrow :\ddot{X}\cdot + :\ddot{X}-CH_3 \end{array} \right\} \text{propagation steps}$$

$$\left. \begin{array}{l} CH_3\cdot \ \ \cdot CH_3 \longrightarrow CH_3-CH_3 \\ CH_3\cdot \ \ \cdot\ddot{X}: \longrightarrow CH_3-\ddot{X}: \end{array} \right\} \text{termination reactions}$$

A mechanism written in standard notation can stand alone; it tells the stepwise story of how the reaction is accomplished without the need for any written description. However, we now want to analyze mechanism I in words so that you can focus on its important features. These are the energy profile and the concepts of high-energy, short-lived intermediates and chain reactions.

The typical energy profile for mechanisms of more than one step is:

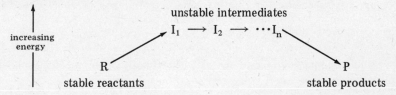

The first step involves absorption of energy from the surroundings, supplied either as heat or light. In the radical substitution mechanism the energy source is ultraviolet light.

*These are numbered arbitrarily (as in Table 10.1) in the order in which they usually appear in the organic course. If you are unfamiliar with the notation used in writing mechanisms in this frame, review Unit 10 before proceeding.

Each quantum of light supplies the energy to break the X—X bond in one X_2 molecule. This radiant energy is converted to chemical energy stored temporarily in $:\ddot{X}\cdot$, an unstable, high-energy species (Unit 8) with no future (lifetime perhaps 0.01 second). The breaking and making of bonds to carbon is done in reactions that do not require any additional large energy inputs; this is the high-energy plateau ($I_1 \longrightarrow I_2 \longrightarrow$ etc.) in the energy profile shown here. In radical substitution, each of these steps is energetically well balanced because it destroys one unstable intermediate and creates another. These are the propagation steps in the mechanism. Since the second propagation step creates the same radical intermediate ($X\cdot$) that the first propagation step destroys, the two are self-sustaining; this is called a *chain* reaction. One $:\ddot{X}\cdot$ starts a chain of propagations (1st, 2nd, 1st, 2nd, etc.), which in principle would suffice to convert all the CH_4 to CH_3X, were it not for the termination reactions that destroy the high-energy radical intermediates, $:\ddot{X}\cdot$ and $\cdot CH_3$ (the "chain carriers"). The radical intermediates are present only at minute concentrations. Nevertheless, every molecule of CH_4 in the entire sample of reactant passes through the $CH_3\cdot$ stage on its way to product. This means that the propagation steps must be very fast reactions if the product is to accumulate at a practical rate, as it does in this case. The initiation step is slow but each occurrence of it starts a chain that may produce thousands of product molecules.

In a chain reaction, the sum of the two propagation steps is the overall reaction. (Remember that chemical equations can be added like arithmetic ones; simplify by canceling out identical species on opposite sides.)

$$:\ddot{\cancel{X}}\cdot + H-CH_3 \longrightarrow :\ddot{X}-H + \cdot\cancel{CH_3}$$
$$+ :\ddot{X}-\ddot{X}: + \cdot\cancel{CH_3} \longrightarrow :\ddot{\cancel{X}}\cdot + :\ddot{X}-CH_3$$
$$= CH_4 + X_2 \longrightarrow X-H + X-CH_3$$

Mechanism II, *electrophilic addition*, applies to a large number of reactions. To emphasize its generality and bring out its general acid-base character, we express it in slot-and-filler form. The reagent, Nu-E, has two slots, or roles—a nucleophilic one, Nu, and an electrophilic one, E. The alkene has four slots (the open bonds on $\rangle C=C\langle$) that can be filled by any alkyl or aryl group or by hydrogen. The generic mechanism shows what each species does in the reaction:

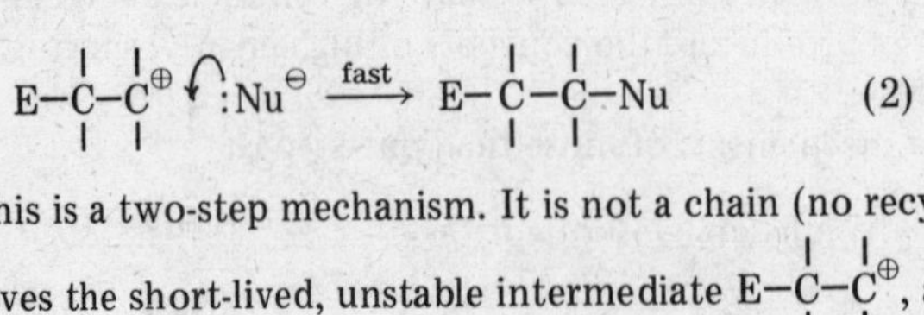

$$\overset{\frown}{Nu-E} \enspace \overset{\frown}{\rangle C=C\langle} \xrightarrow{slow} Nu:^{\ominus} + E-\overset{|}{\underset{|}{C}}-\overset{|}{\underset{|}{C}}{}^{\oplus} \quad (1)$$

$$E-\overset{|}{\underset{|}{C}}-\overset{|}{\underset{|}{C}}{}^{\oplus} \overset{\frown}{} :Nu^{\ominus} \xrightarrow{fast} E-\overset{|}{\underset{|}{C}}-\overset{|}{\underset{|}{C}}-Nu \quad (2)$$

This is a two-step mechanism. It is not a chain (no recycling of intermediates), but it involves the short-lived, unstable intermediate $E-\overset{|}{\underset{|}{C}}-\overset{|}{\underset{|}{C}}{}^{\oplus}$, a carbonium ion. This intermediate is an electrophile (sextet only on $-\overset{|}{C}{}^{\oplus}$) that is destroyed by the nucleophile $:Nu^{\ominus}$ in step (2). This is the typical electrophile + nucleophile or Lewis acid + base reaction discussed in Units 8 and 11; the base $:Nu^{\ominus}$ donates its unshared pair to form a new bond to the electrophile $E-\overset{|}{\underset{|}{C}}-\overset{|}{\underset{|}{C}}{}^{\oplus}$. If other nucleophiles (labeled $:Nu'^{\ominus}$) are also present, the product is a mixture of $E-\overset{|}{\underset{|}{C}}-\overset{|}{\underset{|}{C}}-Nu$ and $E-\overset{|}{\underset{|}{C}}-\overset{|}{\underset{|}{C}}-Nu'$. This happens, for example, whenever the reaction is run in aqueous solution—the solvent, H_2O, is the foreign nucleophile, Nu':.

To write the mechanism for any specific example of this reaction, we simply select specific slot fillers to fill the slots in Nu-E and >C=C<. The permissible slot fillers for Nu-E are shown in Table 12.1. Unless we are just writing arbitrary examples for practice, the choice of slot fillers is dictated by the problem at hand.

Table 12.1 *Reagents for Electrophilic Addition*

Nu — E
Cl — H
Br — H
I — H
$H_2\overset{\oplus}{O}$ — H ($H_3O^{\oplus}$)
$HO-SO_2-O$ — H (H_2SO_4)
Cl — Cl
Br — Br
HO — Cl
HO — Br
HO — I
$R-\overset{\overset{O}{\|}}{C}-O$ — OH
(none) | BH_3

Example

Write specific electrophilic addition mechanisms for the addition of (a) HI and (b) Br_2 to $CH_2=CH_2$.

Solution:

(a) First, identify the slot fillers:

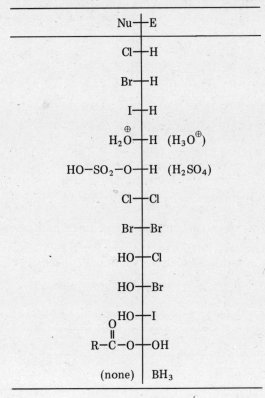

Second, substitute in the generic mechanism:

:Ï—H ↷CH$_2$=CH$_2$ ⟶ :Ï:$^{\ominus}$ + H—CH$_2$—CH$_2^{\oplus}$

CH$_3$—CH$_2^{\oplus}$ ↶:Ï:$^{\ominus}$ ⟶ CH$_3$—CH$_2$—Ï:

(b) Br_2. Identify: Br—Br
                       ↑   ↑
                       Nu   E

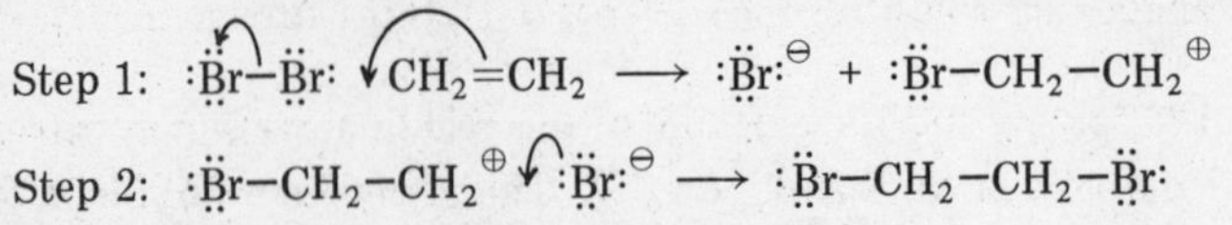

Step 1: $:\ddot{B}r-\ddot{B}r: \curvearrowright CH_2=CH_2 \longrightarrow :\ddot{B}r:^{\ominus} + :\ddot{B}r-CH_2-CH_2^{\oplus}$

Step 2: $:\ddot{B}r-CH_2-CH_2^{\oplus} \curvearrowleft :\ddot{B}r:^{\ominus} \longrightarrow :\ddot{B}r-CH_2-CH_2-\ddot{B}r:$

Mechanism III, *radical addition*, looks like this:

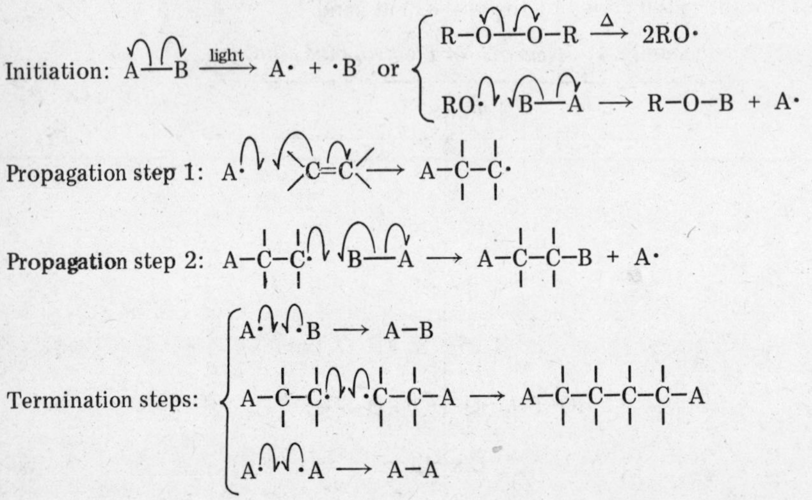

Initiation: $A\!\!-\!\!B \xrightarrow{\text{light}} A\cdot + \cdot B$ or $\begin{cases} R-O\!\!-\!\!O-R \xrightarrow{\Delta} 2RO\cdot \\ RO\cdot \curvearrowright B\!\!-\!\!A \longrightarrow R-O-B + A\cdot \end{cases}$

Propagation step 1: $A\cdot \curvearrowright \!\!>\!\!C\!\!=\!\!C\!\!<\, \longrightarrow A-\overset{|}{\underset{|}{C}}-\overset{|}{\underset{|}{C}}\cdot$

Propagation step 2: $A-\overset{|}{\underset{|}{C}}-\overset{|}{\underset{|}{C}}\cdot \curvearrowright B\!\!-\!\!A \longrightarrow A-\overset{|}{\underset{|}{C}}-\overset{|}{\underset{|}{C}}-B + A\cdot$

Termination steps: $\begin{cases} A\cdot \curvearrowright \cdot B \longrightarrow A-B \\ A-\overset{|}{\underset{|}{C}}-\overset{|}{\underset{|}{C}}\cdot \curvearrowright \cdot \overset{|}{\underset{|}{C}}-\overset{|}{\underset{|}{C}}-A \longrightarrow A-\overset{|}{\underset{|}{C}}-\overset{|}{\underset{|}{C}}-\overset{|}{\underset{|}{C}}-\overset{|}{\underset{|}{C}}-A \\ A\cdot \curvearrowright \cdot A \longrightarrow A-A \end{cases}$

Slots A and B in the reagent are difficult to describe. B must be easily abstracted by the initiating R–O· radicals (R–O· $\curvearrowright$ B—A $\longrightarrow$ R–O–B + A·), or A and B must be part of a compound in which the A—B bond is readily broken by the energy from a quantum of light (A—B $\xrightarrow{\text{light}}$ A· + ·B). The most common slot fillers are shown in Table 12.2.

Table 12.2 *Reagents for Radical Addition*

A	B
Cl_3C	Cl
Cl_3C	H
Br	H (*not* HCl or HI)
$R-S$	H

Mechanism III is another chain mechanism. The chain-carrying radical A· is consumed in the first propagation step and regenerated in the second propagation step to start a new cycle. Since only small amounts of A· are required to convert all the A—B + $>\!\!C\!\!=\!\!C\!\!<$ mixture to product, a great many molecules of product are formed for each quantum of light absorbed when light is the initiator. Similarly, a trace of peroxide suffices when it is used as a source of A·.

Exercises

Translate the indicated generic mechanism into a specific mechanism, using the indicated slot fillers and referring to Tables 12.1 and 12.2 where necessary.

1. Electrophilic addition of $CH_3-C\begin{smallmatrix}\diagup O \\ \diagdown O-O-H\end{smallmatrix}$ to $CH_2=CH_2$.

REACTIONS OF THE ALIPHATIC HYDROCARBONS   215

2. Radical substitution of CH_3-CH_3 by Br_2/light.

3. Radical addition of ⟨Ph⟩–SH to ⟨cyclohexene⟩ in the presence of $(CH_3)_3C-O-O-C(CH_3)_3$.

- - - - - - - -

1. Identify: $CH_3-\overset{O}{\underset{\underbrace{}_{Nu}}{C}}-\underbrace{O-H}_{E}$

Step 1: $CH_3-\overset{O}{\underset{\|}{C}}-O-O-H \quad CH_2{=}CH_2 \longrightarrow CH_3-\overset{O}{\underset{\|}{C}}-\ddot{\underset{..}{O}}:^{\ominus} + HO-CH_2-CH_2^{\oplus}$

Step 2: $HO-CH_2-CH_2^{\oplus} \quad :\ddot{\underset{..}{O}}-\overset{O}{\underset{\|}{C}}-CH_3 \longrightarrow HO-CH_2-CH_2-O-\overset{O}{\underset{\|}{C}}-CH_3$

2. $Br{-}Br \longrightarrow Br\cdot + \cdot Br$ ⎫ initiation

   $Br\cdot \quad H{-}CH_2-CH_3 \longrightarrow Br-H + \cdot CH_2-CH_3$ ⎫
   $Br{-}Br \quad \cdot CH_2-CH_3 \longrightarrow Br\cdot + Br-CH_2-CH_3$ ⎬ propagation

   $CH_3-CH_2\cdot \quad \cdot CH_2-CH_3 \longrightarrow CH_3-CH_2-CH_2-CH_3$ ⎫
   $CH_3-CH_2\cdot \quad \cdot Br \longrightarrow CH_3-CH_2-Br$ ⎬ terminations

3. $(CH_3)_3C-O{-}O-C(CH_3)_3 \longrightarrow 2\,(CH_3)_3C-O\cdot$ ⎫ initiation

   $(CH_3)_3C-O\cdot \quad H{-}S{-}\langle Ph\rangle \longrightarrow (CH_3)_3C-O-H + \cdot S{-}\langle Ph\rangle$ ⎭

   $\langle Ph\rangle{-}S\cdot \; \langle \text{cyclohexene}\rangle \longrightarrow \langle Ph\rangle{-}S{-}\langle \text{cyclohexyl}\rangle \cdot$ ⎫

   $\cdot\langle \text{cyclohexyl-S-Ph} \rangle \; H{-}S{-}\langle Ph\rangle \longrightarrow$
   $\langle Ph\rangle{-}S{-}\langle\text{cyclohex}\rangle + \cdot S{-}\langle Ph\rangle$ ⎬ propagation

   $\langle Ph\rangle{-}S\cdot \; \cdot S{-}\langle Ph\rangle \longrightarrow \langle Ph\rangle{-}S{-}S{-}\langle Ph\rangle$ ⎫

   $\langle Ph\rangle{-}S\cdot \; \cdot\langle\text{cyclohexyl-S-Ph}\rangle \longrightarrow \langle Ph\rangle{-}S{-}\langle\text{cyclohexyl-S-Ph}\rangle$ ⎬ termination

12. If the two ends of the alkene are not identical, the reagents Nu-E and A-B can add

in two different ways to give two different products.

Example

Write the possible isomeric products from addition of H_2O and $CHCl_3$ to propene.

Solution:

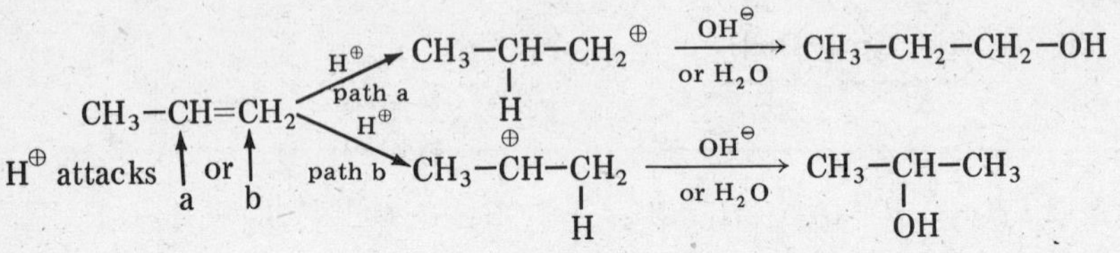

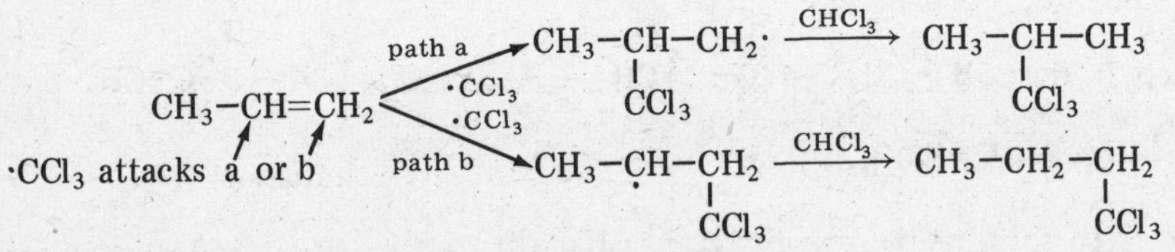

$$CH_3-CH=CH_2 \xrightarrow[\text{path b}]{\text{path a}} \begin{array}{c} H^\oplus \\ H^\oplus \end{array}$$

$H^\oplus$ attacks $\uparrow$ or $\uparrow$
$\quad\quad\quad\quad$ a $\quad$ b

path a: $CH_3-CH-CH_2^\oplus \xrightarrow[\text{or } H_2O]{OH^\ominus} CH_3-CH_2-CH_2-OH$
$\quad\quad\quad\quad\quad |$
$\quad\quad\quad\quad\quad H$

path b: $CH_3-CH-CH_2 \xrightarrow[\text{or } H_2O]{OH^\ominus} CH_3-CH-CH_3$
$\quad\quad\quad\quad |\quad\quad\quad\quad\quad\quad\quad\quad\quad\quad |$
$\quad\quad\quad\quad H\quad\quad\quad\quad\quad\quad\quad\quad\quad OH$

$$CH_3-CH=CH_2 \xrightarrow[\text{path b}]{\text{path a}} \begin{array}{c} \cdot CCl_3 \\ \cdot CCl_3 \end{array}$$

$\cdot CCl_3$ attacks $\overset{\uparrow}{a}$ or $\overset{\uparrow}{b}$

path a: $CH_3-CH-CH_2\cdot \xrightarrow{CHCl_3} CH_3-CH-CH_3$
$\quad\quad\quad\quad\quad |\quad\quad\quad\quad\quad\quad\quad\quad\quad |$
$\quad\quad\quad\quad\quad CCl_3\quad\quad\quad\quad\quad\quad\quad CCl_3$

path b: $CH_3-\overset{\cdot}{CH}-CH_2 \xrightarrow{CHCl_3} CH_3-CH_2-CH_2$
$\quad\quad\quad\quad\quad\quad |\quad\quad\quad\quad\quad\quad\quad\quad\quad\quad\quad |$
$\quad\quad\quad\quad\quad\quad CCl_3\quad\quad\quad\quad\quad\quad\quad\quad CCl_3$

The product from the faster path, a or b, accumulates faster and predominates in the final product mixture. To predict which path is faster, use these rules:

Rule 1: The most stable of two reactive, transient intermediates is formed fastest.
Rule 2: The cation- and radical-intermediate stability order is $3° > 2° > 1° > CH_3 \cdot$ or $\overset{\oplus}{CH_3}$.

Example

Which path predominates in the preceding reactions?

Solution:

The choice in each case is between a primary intermediate:

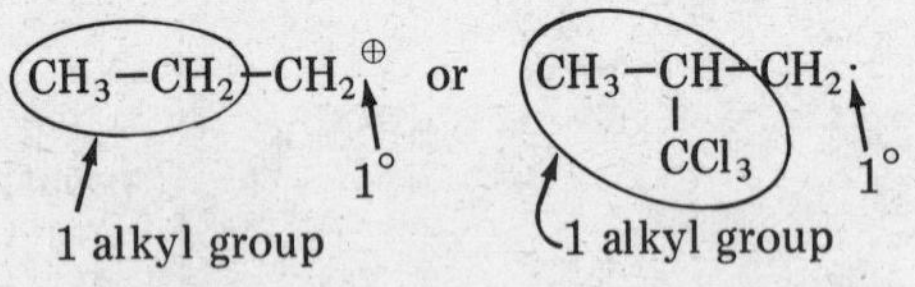

$(CH_3-CH_2)-CH_2^\oplus$ or $(CH_3-CH-CH_2 \cdot)$
$\quad\quad\quad\quad\quad\quad\quad\quad\quad\quad\quad\quad |$
$\quad\quad\quad\quad\quad\quad\quad\quad\quad\quad\quad\quad CCl_3$
$\quad\uparrow\quad\quad\quad\quad 1°\quad\quad\quad\quad\quad\quad\quad\quad\quad 1°$
1 alkyl group $\quad\quad\quad\quad$ 1 alkyl group

and a secondary one:

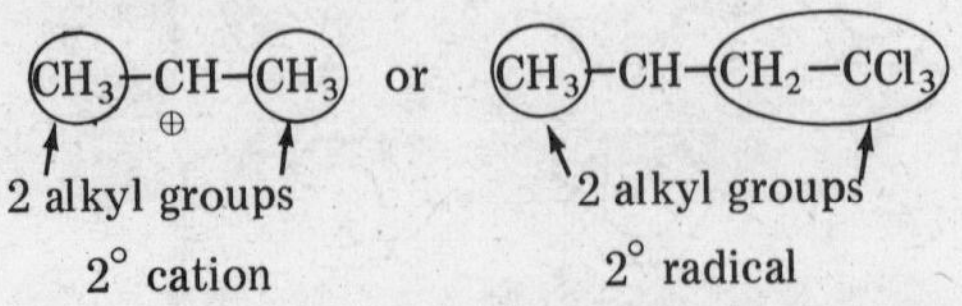

$(CH_3)-\overset{\oplus}{CH}-(CH_3)$ or $(CH_3)-CH-(CH_2-CCl_3)$
$\uparrow\quad\quad\quad\quad\uparrow\quad\quad\quad\quad\quad\uparrow\quad\quad\quad\quad\quad\uparrow$
2 alkyl groups $\quad\quad\quad\quad$ 2 alkyl groups
2° cation $\quad\quad\quad\quad\quad\quad\quad$ 2° radical

The 2° intermediates are more stable; therefore, they are formed faster, so the 2° products accumulate faster and predominate in the product mixture.

13. Before you start to learn the oxidation reactions that break $\diagup C=C \diagdown$ bonds (reactions 6 and 11 in Chart 12.1), let's simplify that task by organizing it as follows.

Three sets of reagents can be used: $\xrightarrow{\text{hot KMnO}_4}$, or $\xrightarrow[(2)\text{ Zn/H}_2\text{O}]{(1)\text{ O}_3}$, or $\xrightarrow[(2)\text{ H}_2\text{O}_2]{(1)\text{ O}_3}$. Each one cleaves the molecule into two fragments, $\mathrm{>C\!\!+\!\!C<}$, but the resulting functional groups in the product fragments depend on which reagent is used. The product functions also depend on the degree of alkyl substitution of the original $\mathrm{>C\!=\!C<}$ function. So there are many possible combinations. They can be put in the form of Table 12.3. Or they can be remembered in the form of rules:

1. The grouping $\mathrm{\begin{smallmatrix}R\\R'\end{smallmatrix}\!\!>\!C\!=}$ gives $\mathrm{\begin{smallmatrix}R\\R'\end{smallmatrix}\!\!>\!C\!=\!O}$ with all three reagents.

2. The grouping $\mathrm{\begin{smallmatrix}R\\H\end{smallmatrix}\!\!>\!C\!=}$ gives an aldehyde, $\mathrm{R\!-\!C\!\!\begin{smallmatrix}\diagup\!\!O\\\diagdown\!H\end{smallmatrix}}$, with O_3, Zn/H_2O, but an acid, $\mathrm{R\!-\!C\!\!\begin{smallmatrix}\diagup\!\!O\\\diagdown\!OH\end{smallmatrix}}$, with O_3, H_2O_2, or hot $KMnO_4$.

3. The grouping $\mathrm{\begin{smallmatrix}H\\H\end{smallmatrix}\!\!>\!C\!=}$ gives formaldehyde, $\mathrm{H\!-\!C\!\!\begin{smallmatrix}\diagup\!\!O\\\diagdown\!H\end{smallmatrix}}$, with O_3, Zn/H_2O; gives formic acid, $\mathrm{H\!-\!C\!\!\begin{smallmatrix}\diagup\!\!O\\\diagdown\!OH\end{smallmatrix}}$, with O_3, H_2O_2; and gives CO_2 with hot $KMnO_4$.

Or notice that if you can remember that $\xrightarrow{\text{hot KMnO}_4}$ and $\xrightarrow[(2)\ H_2O_2]{(1)\ O_3}$ give $-COOH$ functions but $\xrightarrow[(2)\ Zn/H_2O]{(1)\ O_3}$ gives $-C\!\!\begin{smallmatrix}\diagup\!\!O\\\diagdown\!H\end{smallmatrix}$ functions, you can work out the rest as a slot-and-filler problem:

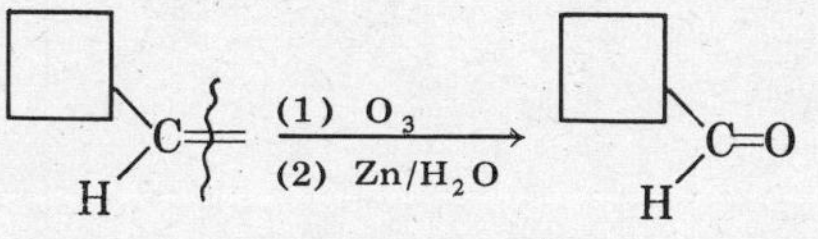

$\square\!\!>\!\!\begin{smallmatrix}\\H\end{smallmatrix}C\!\!\!+\!\!\!\begin{smallmatrix}\\ \end{smallmatrix}\xrightarrow[(2)\ Zn/H_2O]{(1)\ O_3}\square\!\!>\!\!\begin{smallmatrix}\\H\end{smallmatrix}C\!=\!O$ Whatever filled the $\square$ slot in the alkene still fills it in the product.

Notice that the product always consists of two molecules. The carbon chain is clipped at $\mathrm{>C\!\!+\!\!C<}$, forming two pieces, just as in cutting a string. Similarly, just as three fragments are the result of two cuts in the string, three molecules in the alkene cleavage products means that two $\mathrm{>C\!=\!C<}$ double bonds must be present in the alkene. ($\sim\!\!\sim\!\!C\!\!+\!\!C\!\!\sim\!\!\sim\!\!C\!\!+\!\!C\!\!\sim\!\!\sim$); the middle fragment will have a $\mathrm{>C\!=\!O}$ function on each end.

Table 12.3 Alkene Oxidation Products

Alkene	Products from: O_3, then Zn/H_2O	Products from: O_3, then H_2O_2	Products from: Hot $KMnO_4$
$H_2C=CH_2$	$H_2C=O + O=CH_2$	$H_2C=O + HOOC-H$	$CO_2 + CO_2$
$R-CH=CH_2$	$RCH=O + O=CH-H$	$RCOOH + HOOC-H$	$RCOOH + CO_2$
$R-CH=CH-R'$	$RCH=O + O=CH-R'$	$RCOOH + HOOC-R'$	$RCOOH + HOOC-R'$
$RR'C=CH_2$	$RR'C=O + O=CH-H$	$RR'C=O + HOOC-H$	$RR'C=O + CO_2$
$R_1R_2C=CHR_3$	$R_1R_2C=O + O=CHR_3$	$R_1R_2C=O + HOOC-R_3$	$R_1R_2C=O + HOOC-R_3$
$R_1R_2C=CR_3R_4$	$R_1R_2C=O + O=CR_3R_4$	$R_1R_2C=O + O=CR_3R_4$	$R_1R_2C=O + O=CR_3R_4$

REACTIONS OF THE ALIPHATIC HYDROCARBONS 219

Example

Write the product(s) of oxidation of

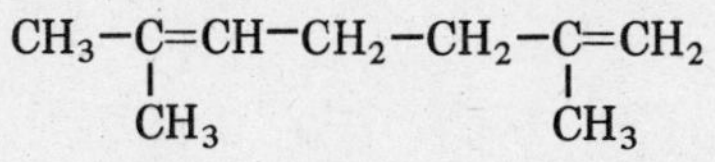

$$CH_3-\underset{\underset{CH_3}{|}}{C}=CH-CH_2-CH_2-\underset{\underset{CH_3}{|}}{C}=CH_2 \quad \text{with hot KMnO}_4$$

Solution:

You can adjust the oxygen functions first, then fill the alkyl and H slots:

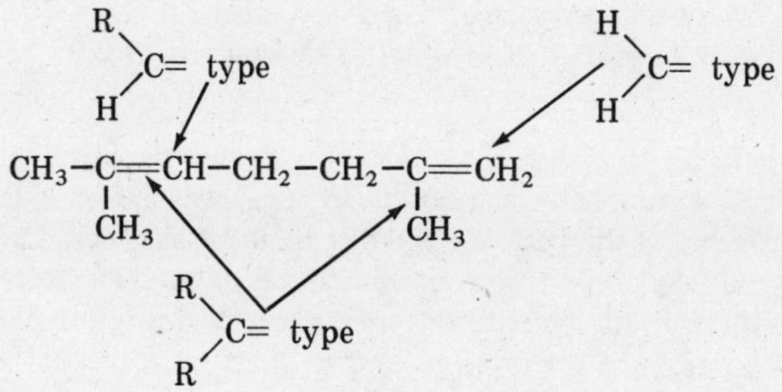

$$\underset{H}{\overset{R}{>}}C= \text{ type} \qquad \underset{H}{\overset{H}{>}}C= \text{ type}$$

$$CH_3-\underset{\underset{CH_3}{|}}{C}=CH-CH_2-CH_2-\underset{\underset{CH_3}{|}}{C}=CH_2$$

$$\underset{R}{\overset{R}{>}}C= \text{ type}$$

so that

[cyclobutyl]$_2$C=O + HO-C(=O)-[square]-[square]-C(=O) + O=C=O

$$\underset{CH_3}{\overset{CH_3}{>}}C=O \; + \; \underset{HO}{\overset{O}{>}}C-CH_2-CH_2-\underset{CH_3}{\overset{O}{<}}C \; + \; CO_2$$

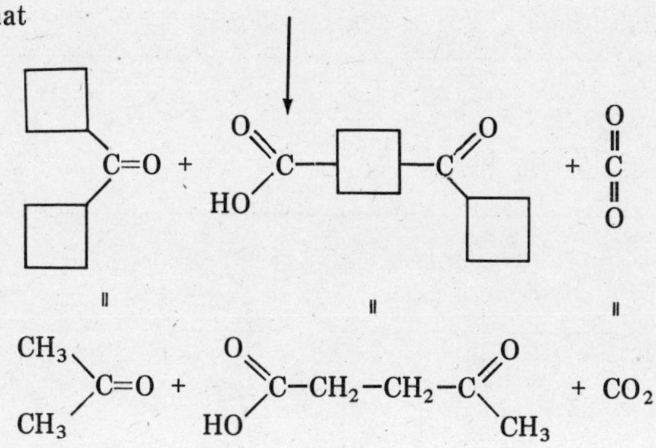

Or you can break the bonds, carefully maintaining the skeleton, then adjust the oxygen functions:

$$CH_3-\underset{\underset{CH_3}{|}}{C}=CH-CH_2-CH_2-\underset{\underset{CH_3}{|}}{C}=CH_2$$

cleave

$$CH_3-\underset{\underset{CH_3}{|}}{\overset{O}{C}} \quad \underset{\underset{CH_3}{|}}{\overset{O}{C}}H-CH_2-CH_2-\overset{O}{C} \quad \overset{O}{C}H_2$$

adjust

$$CH_3-\underset{CH_3}{\overset{O}{<}}C \; + \; \underset{HO}{\overset{O}{>}}C-CH_2-CH_2-\underset{CH_3}{\overset{O}{<}}C \; + \; CO_2$$

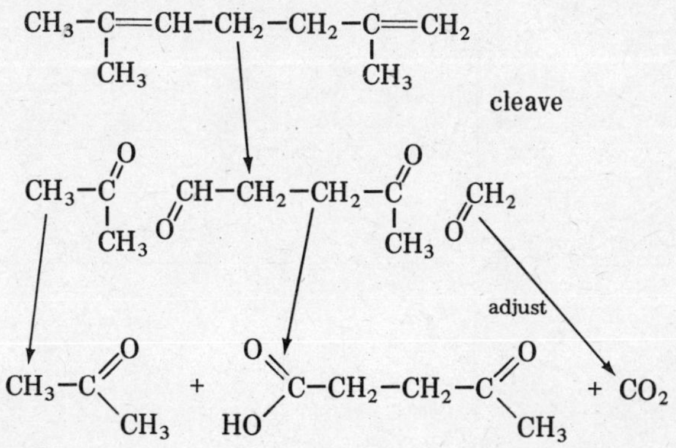

Notice that if only *one* product fragment is isolated, the alkene must have been cyclic, just as a single cut in a *loop* of string produces one piece.

Example

Write the product(s) of oxidation of cyclohexene with ozone followed by Zn/H_2O.

Solution:

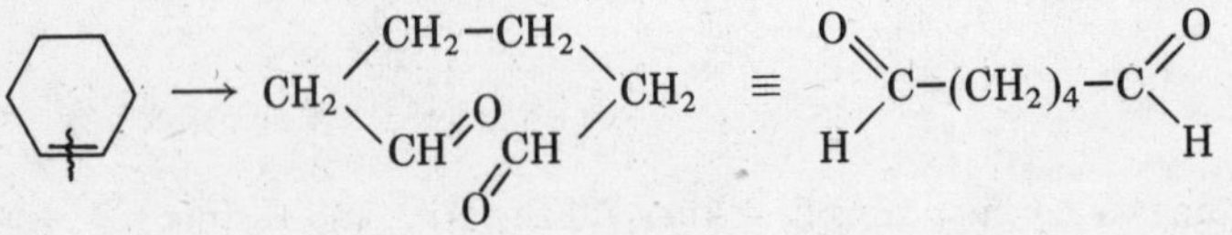

$$\text{cyclohexene} \rightarrow CH_2\begin{array}{c}CH_2-CH_2\\ \\CH\underset{O}{\overset{O}{\diagup\!\!\!\diagdown}}CH\end{array}CH_2 \equiv \underset{H}{\overset{O}{\diagdown}}C-(CH_2)_4-C\underset{H}{\overset{O}{\diagup}}$$

The reverse problem of reasoning back to the structure of the alkene if given the structures of the fragments is also an important one; we practice that in Unit 20.

14. Our last step in organizing the hydrocarbon reactions is to lay out graphically the way they connect various functional groups. We can do this in many ways; the oxidation-level chart shown in Chart 12.2 is one of the best. We will discuss it in some detail here and will use this format throughout the book. Not every reaction is entered in the chart. This reduces clutter. Alternative versions with different choices of reactions can be made. The charts in Unit 14 cover much the same ground as Chart 12.2.

Chart 12.2 *Relationship of* $\diagup\!\!\!\!C=C\!\!\!\diagdown$ *and* $-C\equiv C-$ *to Other Functions*

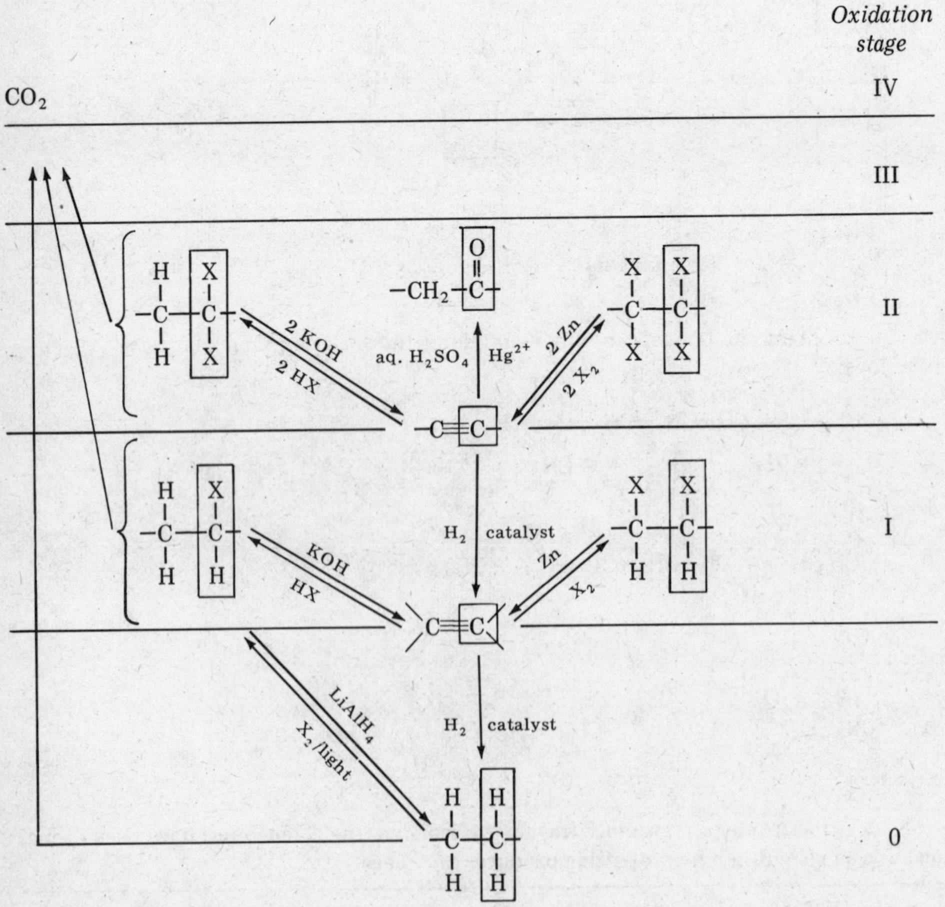

	Oxidation stage
CO_2	IV
	III
[chart showing interconversions between alkenes, alkynes, and substituted compounds with reagents: 2 KOH, 2 HX, aq. H_2SO_4, Hg^{2+}, 2 Zn, $2X_2$, KOH, HX, H_2 catalyst, Zn, X_2, $LiAlH_4$, X_2/light]	II
	I
	0

REACTIONS OF THE ALIPHATIC HYDROCARBONS

The number of oxidation stages has been minimized. The labeling is arbitrary, but it puts the carbon atom of CO_2 in stage IV corresponding to the +4 oxidation state it would have in the standard oxidation-state scheme of inorganic and general chemistry.

The alkanes, in stage 0, are the most reduced organic compounds. The alkenes and alkynes are in oxidation stages intermediate between 0 and I, I and II, respectively. Moving up is oxidation, down is reduction. Since hydrogenation of alkenes

$$\left(\diagup C=C \diagdown \xrightarrow{H_2}_{Pt} -\underset{|}{\overset{|}{C}}-\underset{H}{\overset{|}{C}}- \right)$$ makes new C—H bonds, the rules of frame 4 label this reaction

as a *reduction*; correspondingly, it is a downward movement in the chart.

The main difficulty in talking about oxidation states of carbon in organic molecules is that there are usually several (many) carbons, and each can be in a different oxidation state. Chart 12.2 simplifies this problem by assigning the whole compound to the oxidation stage of the carbon atom in (or bearing) the functional group. These carbons used for classification have been boxed in Chart 12.2.

If we halogenate an alkane $\left(-\overset{|}{\underset{|}{C}}-H \longrightarrow -\overset{|}{\underset{|}{C}}-X \right)$, we destroy one C—H bond and make one C—X bond. By the rules of frame 4, this is oxidation and so is shown as an upward movement in the chart. But notice that a neighboring $-CH_2-$ group that does not react

$\left(-CH_2-\overset{|}{\underset{|}{C}}-H \xrightarrow{X_2}_{light} -CH_2-\overset{|}{\underset{|}{C}}-X \right)$ is still in the alkane oxidation stage (0). When this

last halide is entered in the chart in stage I as $-\overset{H}{\underset{H}{C}}-\boxed{\overset{X}{\underset{H}{C}}}-$, the classification is being made

on the basis of the boxed $-\boxed{\overset{X}{\underset{H}{C}}}-$ carbon, which has indeed been oxidized to stage I;

the unoxidized $-\overset{H}{\underset{H}{C}}-$ group is just along for the ride as part of the skeleton.

If upward movements are oxidations, downward ones reductions, then all the upward and downward *reagents* should be oxidizing and reducing agents, respectively. In most cases the reagents in Chart 12.2 fit this pattern: oxidizing agents = X_2, O_2; reducing agents = H_2, $LiAlH_4$, Zn. The rest (H_2O, KOH, HX) are not oxidizing or reducing agents, but they appear to be playing that role in Chart 12.2. Actually, they are *not* participating in oxidation or reduction. Consider the upward reaction

$-\overset{H}{\underset{H}{C}}-\boxed{\overset{X}{\underset{H}{C}}}- \xleftarrow{HX} \diagup C=C\diagdown$

The boxed carbon has been oxidized. Something must have been reduced, but it is not the HX. It is the left-hand carbon atom, which actually went down in oxidation stage to stage 0. The alkene just rearranged the oxidation stages of its two carbons from both being intermediate between 0 and I to one in stage 0 and one in stage I. Since we classify the whole compound according to the oxidation stage of the functionalized carbon atom, the molecule goes in stage I. Whenever an upward or downward movement is accomplished by nonoxidizing, nonreducing reagents like H_2O, KOH, etc., it means some other carbon atom has also changed its oxidation stage. Reactions that stay within a single oxidation stage (horizontal movements) do not require an oxidizing or reducing agent. None of these are in Chart 12.2, but later charts contain many of them.

LEARNING THE REACTIONS

15. To get the reactions of the functions in your head in useful form you must (1) select the items to be memorized; (2) do a memory drill, and (3) do an application drill, usually in the form of problem solving with pencil and paper. Some approaches are more efficient than others. In this and the next frame we look at a detailed method of doing (1) and (2); frames 17 to 26 work on item (3). After some experience you may find drill techniques that suit you better; just be sure that they require active writing of structures, and practice each reaction in more than one way, as discussed below.

From textbook and lecture notes you will compile your list of common, widely applicable reactions to be memorized. Most textbooks set off these reactions in a summary outline or box somewhere in the chapter on each functional group. If you are studying on your own, you can use the textbook summaries or the lists in the "basic modes of reactivity" charts (all similar to Chart 12.1), located near the beginning of Units 14 to 19.

For two reasons you should physically compile your own list. First, it will be tailored to your own program and not contain extraneous material (no two books or instructors pick exactly the same core of basic reactions). Second, the compilation is a useful, *active* beginning (in contrast to looking over someone else's summary) that sets up the mood and materials for active drill.

Exercises

1. Compile the list of generic alkane reactions that you intend to study. Make four entries for each reaction:
   (a) Equation, including catalysts and necessary conditions.
   (b) Scope, if necessary.
   (c) The name of the mechanism that applies, and the full mechanism written out in generic form for the reaction in question.
   (d) Reactivity order, if any. This can be brief (e.g., rate: $1° > 2° > 3°$ or tetrasubstituted $>$ trisubstituted $>$, etc.).

2. Do the same for the alkenes.

3. Do the same for the alkynes.

16. For each family (function) set up a sheet of 8½- by 11-inch scratch paper with three columns:

Starting material	Reagent(s)	Product
$R-CH=CH_2$	H_2/Pt	$R-CH_2-CH_3$
$R-CH=CH_2$	X_2	$R-CHX-CH_2X$
.	.	.
.	.	.
.	.	.

Enter each of the reactions from your list in this simple three-item format.

REACTIONS OF THE ALIPHATIC HYDROCARBONS

Exercises

1. Fold a sheet of 8½- by 11-inch scratch paper lengthwise in thirds and use it to cover the *product* column of the table you just prepared. Now fill in the product of each reaction in turn using the starting material and reagent structures as cues. If you come to a product that you cannot recall, try to narrow down whether it is a product of substitution, addition, elimination, and so on, by looking at the reactant and reagent structures. Often this will immediately suggest the product structure. If this doesn't help, try to visualize Chart 12.2 so as to recall from the positions of reactant and reagent what the position and structure of the product must be. When you have filled in as many products as you can, slide your paper over and check your answers. Correct any wrong answers. Repeat the drill.

2. Next use a fresh side of your folded paper to cover the *reagent(s)* column. Using the starting material and product structures as cues, fill in the missing reagent (including any catalyst and necessary conditions, for instance Br_2/light). If you come to a reagent you cannot immediately recall, compare the starting material and product structures to see what atoms have been added or subtracted by the reaction; the reagent must be the source of atoms added and the means of removing any subtracted. If this does not cue the answer, try to decide if the reaction is an oxidation or reduction, in which case you know you are looking for an oxidizing or reducing agent. When you have supplied all the reagents you can, check and correct your answers. Repeat the drill.

3. Now cover the *starting material* column and fill in the missing items. This is the "from where can I get to here?" aspect of the material, and visualizing Chart 12.2 can be a good cueing device. Check and correct. Repeat the drill.

4. Note how many reactions are in your drill list. Try recalling them by name without looking ("hydrogenation, addition of halogen, ozonolysis," etc.). Do all the reactions of a type, then go to the next type (additions, oxidations, ...). Cue yourself from the list if you come up short. Repeat until you can recall all the items.

5. Repeat the preceding exercise, but now write down the four items (equation, mechanism, scope, reactivity order) associated with each reaction on your original list. Check and correct.

6. Look over Chart 12.2, then put it aside and see if you can reproduce it on scratch paper. Check and correct. Repeat the exercise.

If you are spending a week on a given family, set up your study lists during the first couple of days, then spend about half an hour on each of the succeeding five days running through the preceding drill exercises. If at any point you feel you are recalling missing items just by their order or position on the list, start drilling in reverse or random order.

SOLVING PROBLEMS BASED ON REACTIONS OF THE FUNCTIONS

17. In this and the remaining frames of this unit, we take up the standard problem types in which you apply the facility you have developed with the generic form of the reactions to making concrete chemical predictions about specific systems. These are the types of problem you find in textbooks and on exams. Experience shows that it is a waste of time to start working these problems before you have done an adequate amount of drill of the type in frames 15 and 16, or its equivalent. If you find yourself having to leaf back to look up something on almost every end-of-chapter problem in your text, you probably need more drill before tackling those problems.

The problem types and the solution methods developed here apply equally well to all the functional groups, and we will use them repeatedly in later units. In this unit we apply them to the alkane, alkene, and alkyne reactions.

First, let's examine the "Write all products of the following reactions" type of problem. The most common variation on this problem would call for organic products only. An example of the information given is

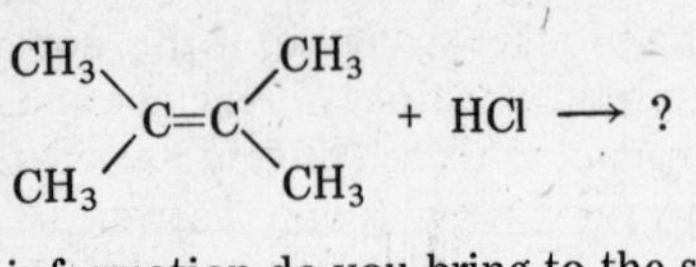

$$\begin{array}{c}CH_3\\ \\ CH_3\end{array}\!\!C\!=\!C\!\!\begin{array}{c}CH_3\\ \\ CH_3\end{array} + HCl \longrightarrow \ ?$$

What information do you bring to the solution of this problem? Nothing specific to this reaction, because it is unlikely that you have ever seen it before. What you have in your head is the complete set of *generic* reactions of the alkenes. To solve this specific problem with that general information you have to do three things:

- Recognize which generic reaction this is an example of.
- Recall the generic equation for the reaction.
- Insert the appropriate slot fillers to turn the generic equation into the equation for this specific reaction.

In a common variation, the problem is stated: "Write structural formulas for all the organic products of the following reactions, or write No reaction." This problem goes one step further and tests your ability to recognize whether the given reactants correspond to an allowed reaction or a fictitious one. To solve this problem reliably you must do one extra thing, namely scan (in your memory) *all* the generic reactions of the functional group in question. If the given reactants cannot be fitted to the generic equation for any allowed reaction, the reaction must be fictitious and you answer "No reaction."

A final step in either variation: writing the product structure you decide upon correctly and checking it for errors.

Examples

Write structural formulas for the important products of each reaction, or write "No reaction."

1. 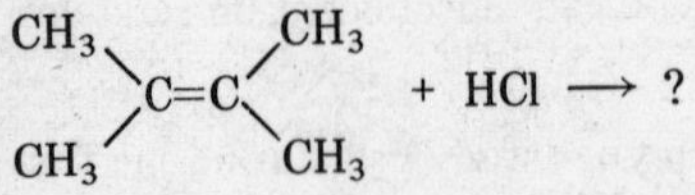 $\begin{array}{c}CH_3\\ \\ CH_3\end{array}\!\!C\!=\!C\!\!\begin{array}{c}CH_3\\ \\ CH_3\end{array} + HCl \longrightarrow \ ?$

2. 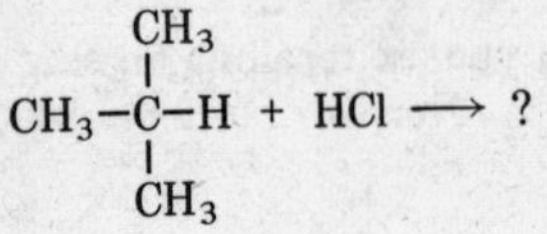 $CH_3\!-\!\underset{\underset{CH_3}{|}}{\overset{\overset{CH_3}{|}}{C}}\!-\!H + HCl \longrightarrow \ ?$

Solutions:

1. (a) Scan your memory to see if this reactant pair fits any of the allowed reactions in Chart 12.1 or 12.2. Find that with H—Cl = E—Nu, it fits the electrophilic-addition-to-alkene pattern: $>\!\!C\!=\!C\!\!< + \ E\!-\!Nu$.

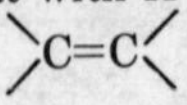

   (b) Write the product of the generic reaction: $-\underset{\underset{E}{|}}{C}\!-\!\underset{\underset{Nu}{|}}{C}\!-$.

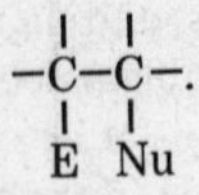

(c) Fit this to the specific problem at hand: if $\mathrm{\backslash C=C/}$ is $\mathrm{(CH_3)_2C=C(CH_3)_2}$ and

E—Nu is H—Cl, then the product is $\mathrm{CH_3-\underset{H}{\underset{|}{C}}(CH_3)-\underset{Cl}{\underset{|}{C}}(CH_3)-CH_3}$.

2. Scan your memory to see if this reactant pair fits any of the allowed reactions in Chart 12.1 or 12.2. You find that it does not. Therefore, write "No reaction."

★ *Exercises*

Draw the structures of all organic products of each reaction, or write "No reaction."

(alkane reactions)

1. $\mathrm{CH_3-\underset{\underset{CH_3}{|}}{\overset{\overset{CH_3}{|}}{C}}-CH_3 + Cl_2} \xrightarrow{\text{light}}$?

2. $\mathrm{CF_3-CCl_2H + Br_2} \xrightarrow{\text{light}}$?

3.  (cyclopentane) $+ \mathrm{Cl_2} \xrightarrow{\text{light}}$?

4. 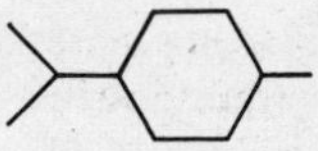 (1,4-di-tert-butyl cyclohexane) $+ \mathrm{O_2}$ (excess) $\xrightarrow{\Delta}$?

(alkene reactions)

5.  (cyclohexene) $+ \mathrm{H_2} \xrightarrow{\text{Pt}}$?

6. $\mathrm{CH_3-CH=CH-CH_3 + NaCl} \longrightarrow$?

7. $\mathrm{CH_3-CH=C\underset{CH_3}{\overset{CH-CH_3}{\diagup}}(CH_3)} \xrightarrow[\text{(2) Zn/H}_2\text{O}]{\text{(1) O}_3}$?

8. 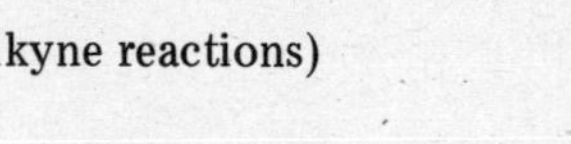 (Ph)—CH=CH—(Ph) $+ \mathrm{H_2O} \xrightarrow{\mathrm{H_2SO_4}}$?

(alkyne reactions)

9. $\mathrm{CH_3-CH_2-CH_2-C{\equiv}C-CH_3} \xrightarrow[\mathrm{Br_2, dark}]{\text{excess}}$?

10. 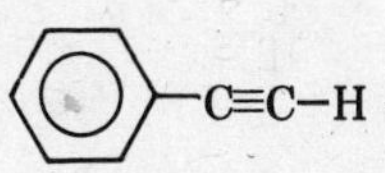 (Ph)—C≡C—H $\xrightarrow{\mathrm{Na^{\oplus} NH_2^{\ominus}}}$?

226 HOW TO SUCCEED IN ORGANIC CHEMISTRY

1. $CH_3-\underset{\underset{CH_3}{|}}{\overset{\overset{CH_3}{|}}{C}}-CH_2Cl$

2. CF_3-CCl_2Br

3. 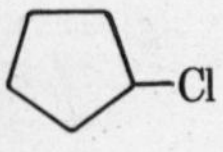 cyclopentyl-Cl

4. CO_2

5. 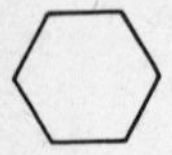 cyclohexane

6. No reaction

7. 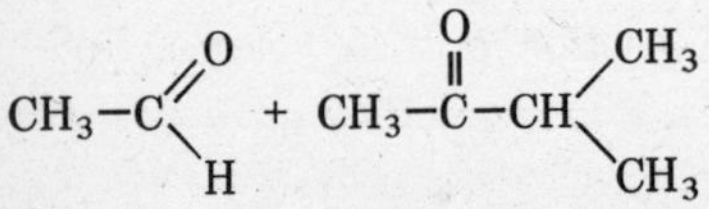 $CH_3-C\overset{\nearrow O}{\underset{\searrow H}{}} \ + \ CH_3-\overset{\overset{O}{\|}}{C}-CH\overset{\nearrow CH_3}{\underset{\searrow CH_3}{}}$

8. 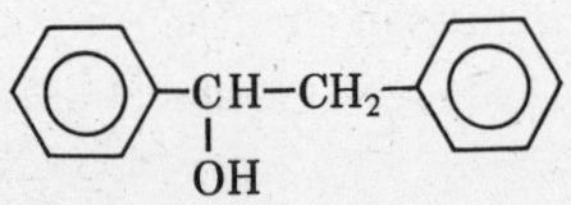 Ph$-\underset{\underset{OH}{|}}{CH}-CH_2-$Ph

9. $CH_3-(CH_2)_2-CBr_2-CBr_2-CH_3$

10. 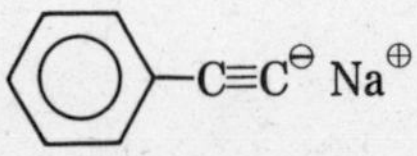 Ph$-C\equiv C^{\ominus} \ Na^{\oplus}$

COVERING ALL POSSIBILITIES

18. The answer you generate by following the steps above may or may not be complete. It is not complete if (a) there is another similar functional group in the starting material that could have reacted instead, or (b) if two orientations of the reagent were possible in its attack on the reactant. Your approach must be systematic if you are to find all products. You can apply methods similar to those used in Unit 6.

Example

Write structures for all the organic products of each reaction.

1. $CH_3-\underset{\underset{H}{|}}{\overset{\overset{CH_3}{|}}{C}}-CH_2-CH_3 \ + \ Cl_2 \xrightarrow{light} \ ?$ 2. $CH_3-CH=CH_2 \ + \ HCl \longrightarrow ?$
    (much)   (little)

Solutions:

1. (a) The reactants fit the general radical substitution pattern.

   (b) The generic reaction is $R-H \ + \ Cl_2 \xrightarrow{light} R-Cl \ + \ HCl$.

   (c) The specific reaction is $CH_3-\underset{\underset{H}{|}}{\overset{\overset{CH_3}{|}}{C}}-CH_2-CH_3 \ + \ Cl_2 \xrightarrow{light} CH_3-\underset{\underset{H}{|}}{\overset{\overset{CH_3}{|}}{C}}-CH_2-CH_2Cl$
                                                                                                                      + HCl.

But there are four types of $>$C-H function—each slightly different—in the reactant molecule:

REACTIONS OF THE ALIPHATIC HYDROCARBONS 227

$$\begin{array}{c} \text{d CH}_3 \\ | \\ \text{CH}_3-\underset{\underset{\text{c}}{|}}{\overset{|}{\text{C}}}-\text{CH}_2-\text{CH}_3 \\ \text{d} \quad \text{H} \quad \text{b} \quad \text{a} \end{array}$$

Therefore, you must expand your answer to give the complete list of organic products that the problem requires:

$$\begin{array}{cccc} \text{CH}_3 & \text{CH}_3 & \text{CH}_3 & \text{CH}_2\text{Cl} \\ | & | & | & | \\ \text{CH}_3-\text{C}-\text{CH}_2-\text{CH}_2\text{Cl} & \text{CH}_3-\text{C}-\text{CHCl}-\text{CH}_3 & \text{CH}_3-\text{C}-\text{CH}_2-\text{CH}_3 & \text{CH}_3-\text{C}-\text{CH}_2-\text{CH}_3 \\ | & | & | & | \\ \text{H} & \text{H} & \text{Cl} & \text{H} \end{array}$$

According to the statement of the problem, an excess of alkane is present, so there is little opportunity for di-, trichloro, etc., products to form.

2. (a) The reactants correspond to an electrophilic addition, >C=C< + E–Nu.

(b) The generic reaction is R–CH=CH$_2$ + Nu–E $\longrightarrow$
$$\begin{array}{c} \text{R–CH–CH}_2 \\ |\quad\;\; | \\ \text{Nu}\;\;\text{E} \end{array} \text{major product}$$
$$+$$
$$\begin{array}{c} \text{R–CH–CH}_2 \\ |\quad\;\; | \\ \text{E}\;\;\text{Nu} \end{array} \text{minor product}$$

(c) The specific reaction is CH$_3$–CH=CH$_2$ + H–Cl $\downarrow$

$$\begin{array}{cc} \text{CH}_3-\text{CH}-\text{CH}_3 & + \;\; \text{CH}_3-\text{CH}_2-\text{CH}_2-\text{Cl} \\ | & \\ \text{Cl} & \\ \text{major product} & \text{minor product} \end{array}$$

★ *Exercises*

Write the structural formulas of all the organic products of each reaction.

1. $\text{CH}_3-\underset{\underset{\text{CH}_3}{|}}{\text{CH}}-\text{CH}_2-\text{CH}_2-\underset{\underset{\text{CH}_2-\text{CH}_3}{|}}{\text{CH}}-\text{CH}_3$ + Br$_2$ (deficiency) $\xrightarrow{\text{light}}$? (consider monobromo products only)

2. CH$_3$–⟨◯⟩–CH$_2$–CH$_2$–⟨◯⟩ with CH$_3$ and CH$_3$ substituents + Cl (deficiency) $\xrightarrow{\text{light}}$? (consider monochloro products only)

3. ⟩◯⟨ + Br$_2$ (deficiency) $\xrightarrow{\text{light}}$? (monobromo products only)

4. ⟨◯⟩ + H$_2$SO$_4$ $\longrightarrow$?

- - - - - - - -

1. $\text{CH}_2\text{Br}-\underset{\underset{\text{CH}_3}{|}}{\text{CH}}-\text{CH}_2-\text{CH}_2-\underset{\underset{\text{CH}_2-\text{CH}_3}{|}}{\text{CH}}-\text{CH}_3$ $\quad$ $\text{CH}_3-\underset{\underset{\text{CH}_3}{|}}{\text{CBr}}-\text{CH}_2-\text{CH}_2-\underset{\underset{\text{CH}_2-\text{CH}_3}{|}}{\text{CH}}-\text{CH}_3$

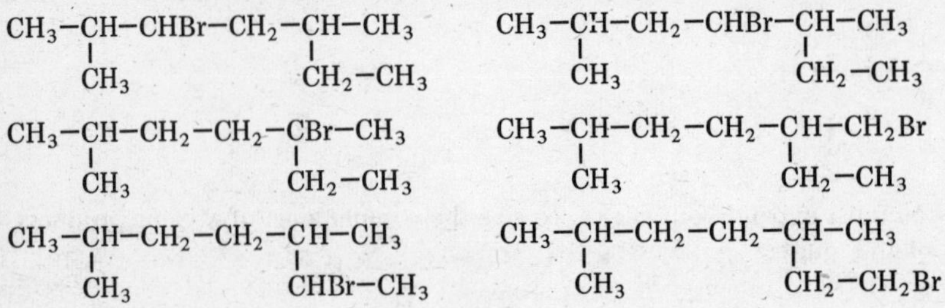

$CH_3-CH-CHBr-CH_2-CH-CH_3$
    $|$ $|$
    CH_3 CH_2-CH_3

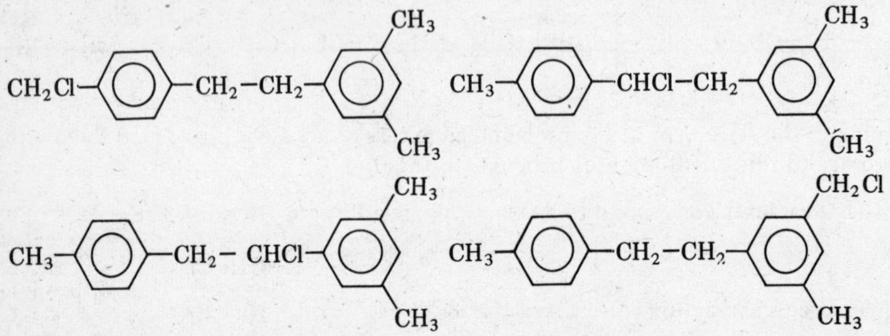

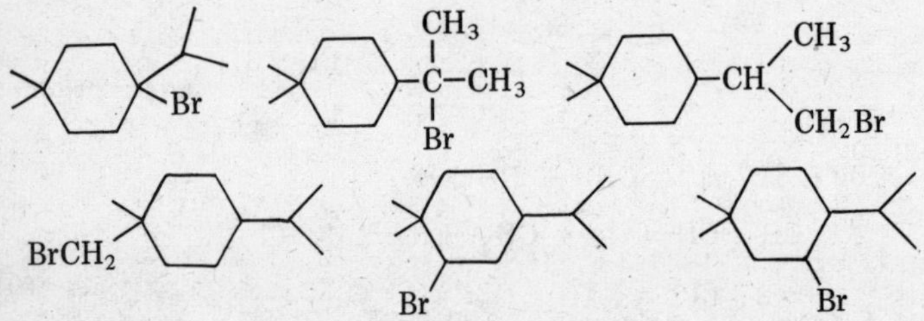

$CH_3-CH-CH_2-CHBr-CH-CH_3$
    $|$ $|$
    CH_3 CH_2-CH_3

$CH_3-CH-CH_2-CH_2-CBr-CH_3$
    $|$ $|$
    CH_3 CH_2-CH_3

$CH_3-CH-CH_2-CH_2-CH-CH_2Br$
    $|$ $|$
    CH_3 CH_2-CH_3

$CH_3-CH-CH_2-CH_2-CH-CH_3$
    $|$ $|$
    CH_3 $CHBr-CH_3$

$CH_3-CH-CH_2-CH_2-CH-CH_3$
    $|$ $|$
    CH_3 CH_2-CH_2Br

2. CH_2Cl—⟨ ⟩—CH_2-CH_2—⟨2,4-diCH_3⟩

   CH_3—⟨ ⟩—$CHCl-CH_2$—⟨3,5-diCH_3⟩

   CH_3—⟨ ⟩—CH_2-CHCl—⟨2,4-diCH_3⟩

   CH_3—⟨ ⟩—CH_2-CH_2—⟨3-CH_2Cl, 5-CH_3⟩

3. (cyclohexane with two methyls and Br)

   (cyclohexane with two methyls and $-C(CH_3)_2-Br$)

   (cyclohexane with two methyls and $-CH(CH_3)CH_2Br$)

   $BrCH_2$—(cyclohexane)—isopropyl

   (cyclohexane with Br and isopropyl)

   (cyclohexane with Br and isopropyl)

4. CH_3—(cyclohexane)—OSO_3H

   CH_3—(cyclohexane)—OSO_3H

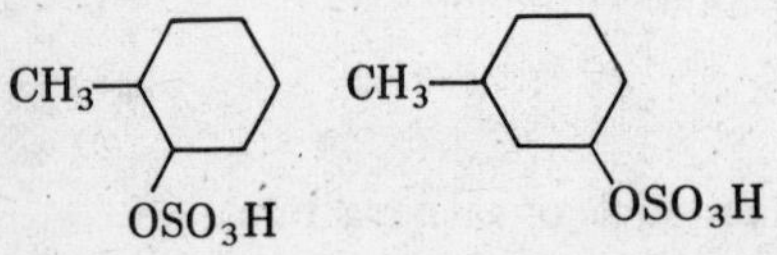

THE REACTANT RATIO QUESTION

19. When more than one functional group is present in one of the reactants, we have the additional problem of deciding how many of them will react. The answer to this depends partly on how much of the other reactant is present. This information is usually given in the problem.

Example

Write structural formulas for all the organic products of each reaction.

1. $CH_4 + Cl_2$ (deficiency) $\xrightarrow{light}$?
2. $CH_4 + Cl_2$ (excess) $\xrightarrow{light}$?
3. CH_4 (1 mole) $+ Cl_2$ (1 mole) $\xrightarrow{light}$?

Solutions:

1. (a) Identify the reactants as those for radical substitution.

(b) Generic reaction: $\ce{>C-H + Cl2 ->[light] >C-Cl + HCl}$.

(c) Specific reaction: $\ce{CH4 + Cl2 ->[light] CH3-Cl + HCl}$.

(d) Check for completeness: In principle each of the remaining C—H bonds of CH_3Cl could also react with Cl_2 to give CH_2Cl_2, $CHCl_3$, and so on. However, the problem specifies a deficiency of Cl_2, meaning fewer moles of Cl_2 than of CH_4. Consequently, there is not even enough Cl_2 to convert all the CH_4 to CH_3Cl, and the probability of converting CH_3Cl to higher chlorination products is slight. The product is thus a mixture of CH_3Cl and unreacted CH_4. The presence of unreacted CH_4 is trivial, and the best answer is just CH_3Cl.

2. Begin as in example 1. But with excess Cl_2 present, all CH_4 will be consumed. The CH_3Cl formed is now also exposed to excess Cl_2 and will react to form CH_2Cl_2:

$$\ce{CH3Cl + Cl2 ->[light] CH2Cl2 + HCl}$$

The replacement of further C—H's continues until all are consumed. The sole product is CCl_4.

3. Identify the reaction as radical halogenation: $\ce{CH4 + Cl2 ->[light] CH3-Cl + HCl}$. Check for completeness: Consider the situation after a few molecules of CH_4 have been converted to CH_3Cl. With a little less than 1 mole of Cl_2 remaining, there is no chance of all C—H bonds being replaced—Cl_2 is in deficiency versus all available C—H bonds, and the CH_4 and CH_3Cl compete for the available Cl_2. The rates are not very different and so, while more CH_3Cl is being formed from CH_4, smaller amounts of CH_2Cl_2 are being formed from the CH_3Cl. Now CH_4, CH_3Cl, and CH_2Cl_2 compete, and so on. The final result: the major product is CH_3Cl, accompanied by less CH_2Cl_2, still less $CHCl_3$, and still less CCl_4; some CH_4 remains unreacted. Since the problem requires all organic products, the answer is CH_3Cl, CH_2Cl_2, $CHCl_3$, and CCl_4. If, instead of equal quantities of CH_4 and Cl_2, you started with $CH_4:Cl_2 = 1:2$, you could make CH_2Cl_2 the major product, but CH_3Cl, $CHCl_3$, and CCl_4 would still be present. The only way to get a single product in a system like this is to use either a large excess of Cl_2 as in example 2, above, or a deficiency of Cl_2 (excess of CH_4, as in example 1).

Exercises

Write the structural formulas of all the organic products of each reaction.

1. $\ce{C6H5-CH3 + Cl2}$ (deficiency) $\xrightarrow{\text{light}}$?

2. $\ce{C6H5-CH3 + Cl2}$ (large excess) $\xrightarrow{\text{light}}$?

3. $\ce{C6H5-CH3}$ (1 mole) $+ \ce{Cl2}$ (1 mole) $\xrightarrow{\text{light}}$?

4. $\ce{CH3-CH=CH-C(CH3)(CH3)-CH2-CH=CH-CH3}$ (1 mole) + H_2O (1 mole) $\xrightarrow{H_3O^{\oplus}}$?

5. (1,4-di-isopropenyl... CH₃/CH₃ substituted 1,4-cyclohexadiene-like structure) (1 mole) + HI (1 mole) → ?

- - - - - - - -

1. C₆H₅−CH₂Cl

2. C₆H₅−CCl₃

3. C₆H₅−CH₂Cl + C₆H₅−CHCl₂ + C₆H₅−CCl₃

4.
$$CH_3-CH_2-\underset{OH}{\underset{|}{CH}}-\underset{CH_3}{\underset{|}{C}}-CH_2-CH=CH-CH_3 \quad\quad CH_3-\underset{OH}{\underset{|}{CH}}-CH_2-\underset{CH_3}{\overset{CH_3}{\underset{|}{C}}}-CH_2-CH=CH-CH_3$$

$$CH_3-CH=CH-\underset{CH_3}{\underset{|}{C}}\!\!\overset{CH_3}{}-CH_2-CH_2-\underset{OH}{\underset{|}{CH}}-CH_3 \quad\quad CH_3-CH=CH-\underset{CH_3}{\overset{CH_3}{\underset{|}{C}}}-CH_2-\underset{OH}{\underset{|}{CH}}-CH_2-CH_3$$

5.
(Three cyclohexane/cyclohexene structures bearing CH₃ groups and I substituents)

PREDOMINANT PRODUCTS

20. Now let's examine the "Write the structural formula of the major organic product" type of problem. To arrive at the correct solution to this problem, you must solve the "write structures of all the important organic products..." problem first (at least in your head). Then you select the one product that will predominate in the product mixture. This is a revealing exam problem, because to solve it you must know both the allowed reactions and the relevant reactivity orders.

When more than one product is possible, what determines which one will predominate? There are two possibilities:

(a) For reactions in which the two or more products, once formed, do not interconvert under the reaction conditions, that product predominates that is formed the fastest. This is called *rate-controlled* (or *kinetic control of*) *product identity.*

(b) For reactions in which the various products are in equilibrium with one another under the experimental conditions, the most stable product predominates. This is called *equilibrium control of product identity.*

Sometimes the product predicted by kinetic control is the same as that predicted by equilibrium control, sometimes not. That is, the product formed fastest may or may not be the most stable. (In any single mechanistic *step* that forms an *unstable, reactive intermediate*, the more stable intermediate is formed faster.)

REACTIONS OF THE ALIPHATIC HYDROCARBONS

Kinetic control is more common. Of the complete, *overall reactions* discussed so far, only proton-transfer (acid-base) reactions and the Na/NH_3 reduction of alkynes are equilibrium-controlled—all the rest are rate-controlled.

Clearly, for each new reaction, we must learn whether product identity is rate- or equilibrium-controlled, along with learning the limitations, mechanism, and reactivity order, as discussed previously. This can be simplified if you remember just the cases of equilibrium control, knowing that all the rest are rate-controlled.

First let's take examples of competitive reaction of two functional groups.

Example

Write the major organic product of the following reaction:

$$H_2\ddot{N}-CH_2-CH_2-CH_2-\ddot{O}-H \xrightarrow{H^\oplus} ?$$

Solution:

With two basic functions in the molecule, the question is which will compete most effectively for the proton. The possible products are $H_3\overset{\oplus}{N}-CH_2-CH_2-CH_2-\ddot{O}-H$ and $H_2\ddot{N}-CH_2-CH_2-CH_2-\overset{\oplus}{\underset{H}{O}}{\overset{H}{\diagdown}}$. This acid-base reaction is equilibrium-controlled; that is, the products are in rapid equilibrium ($H_3N^\oplus\text{-}\mathcal{W}\text{-}OH \rightleftharpoons H_2N\text{-}\mathcal{W}\text{-}OH_2^\oplus$), and the predominant product will be the most stable one. Put another way, the predominant product will be the same as that predicted by the rule: the strongest base wins the proton (Unit 11, frame 9). Reference to our order of base strengths (Unit 11, frame 8) shows that NH_2- is a stronger base than $-OH$. Therefore, the major product is $H_3\overset{\oplus}{H}-CH_2-CH_2-CH_2-OH$.

When the product is formed under kinetic control you can decide which product dominates by using the reactivity order you learned with the reaction. If you know the reason behind the reactivity order, you can decide which product dominates even if you forget the order. Let's look at the most common case in detail. This is the case of any kinetically controlled reaction whose mechanism has more than one step. The rate at which products accumulate is the rate of the slowest step, and this is the energetically uphill step that converts the starting material to an unstable intermediate, usually $R\cdot$ or $R^\oplus$. The least unstable intermediate is formed least slowly. In other words, the product formed fastest (predominant product) is the one derived from the more stable intermediate. For both $R\cdot$ and $R^\oplus$ intermediates the stability order is $3° > 2° > 1° > CH_3\cdot$ or $CH_3{}^\oplus$.

Example

Draw the structure of the predominant product of the reaction

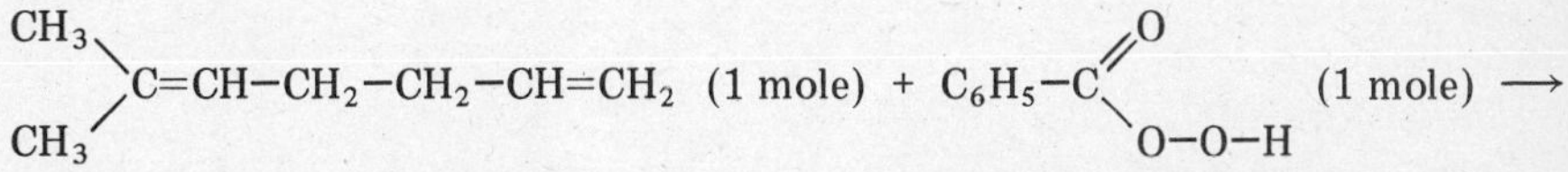

$$\begin{array}{c}CH_3\\ \\ CH_3\end{array}\!\!\!\!\diagdown\!\!\!\!\!\!\!\diagup C=CH-CH_2-CH_2-CH=CH_2 \text{ (1 mole)} + C_6H_5-C\!\!\!\diagup\!\!\!\!\!\!\!\!\diagdown\!\!\!\begin{array}{c}O\\ \\ O-O-H\end{array} \text{ (1 mole)} \longrightarrow$$

Solution:

The reaction is electrophilic addition, kinetically controlled. The competing paths are

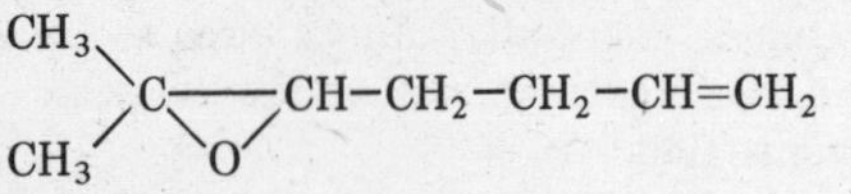

$$\begin{array}{c}CH_3\\ \diagdown\\ C\!\!\!-\!\!\!-\!\!\!-CH\!-\!CH_2\!-\!CH_2\!-\!CH\!=\!CH_2 \quad (1)\\ \diagup\diagdown\diagup\\ CH_3O\end{array}$$

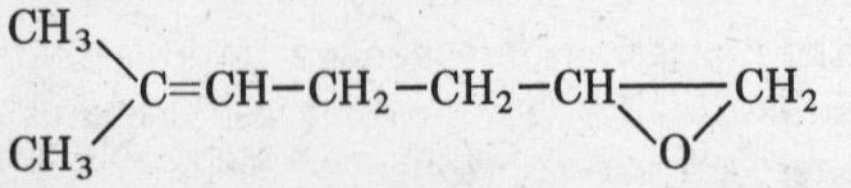

$$\begin{array}{c}CH_3\\ \diagdown\\ C\!=\!CH\!-\!CH_2\!-\!CH_2\!-\!CH\!\!\!-\!\!\!-\!\!\!-CH_2 \quad (2)\\ \diagup\diagdown\diagup\\ CH_3O\end{array}$$

The reactivity order for electrophilic addition: rate increases with increasing alkyl substitution of $\diagup\!\!\!C\!=\!C\!\!\!\diagdown$. The left-hand $\diagup\!\!\!C\!=\!C\!\!\!\diagdown$ is trisubstituted, the right-hand one monosubstituted. So path (1) is faster, and its product is the major one.

Or: Draw the mechanism of each path and see which reactive intermediate is more stable:

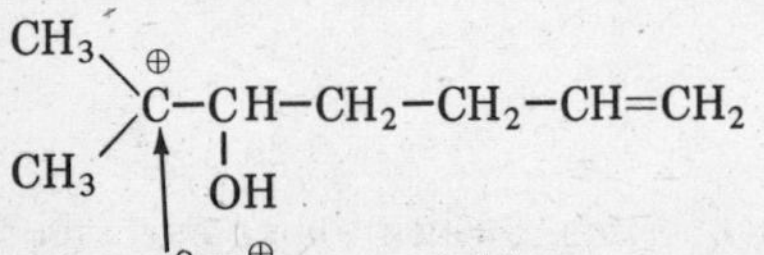

$$\begin{array}{c}CH_3\oplus\\ \diagdown\\ C\!-\!CH\!-\!CH_2\!-\!CH_2\!-\!CH\!=\!CH_2 \quad (1)\\ \diagup|\\ CH_3OH\end{array}$$

$3°$ $R^\oplus$, more stable

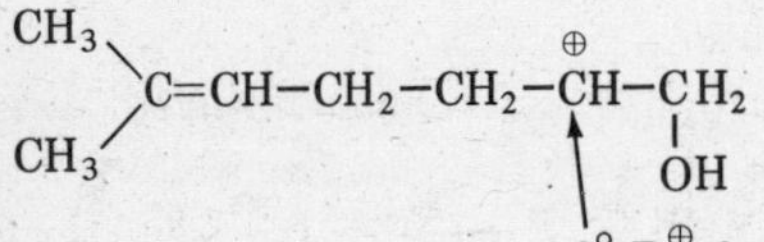

$$\begin{array}{c}CH_3\\ \diagdown\oplus\\ C\!=\!CH\!-\!CH_2\!-\!CH_2\!-\!CH\!-\!CH_2 \quad (2)\\ \diagup|\\ CH_3OH\end{array}$$

$2°$ $R^\oplus$, less stable

So the top $R^\oplus$ is formed faster and product (1) accumulates faster and becomes the major product.

The same reasoning lets you decide the major product when there are two possible directions of addition to a single $\diagup\!\!\!C\!=\!C\!\!\!\diagdown$, as discussed in frame 12.

Example

$$CH_3-CH=CH_2 + HBr \longrightarrow ?$$

Solution:

The mechanism is electrophilic addition, and the products are under kinetic control. The fastest-formed product is that which puts the H on the least substituted carbon because, of the two possible intermediates, $CH_3-\underset{H}{\underset{|}{CH}}-CH_2{}^\oplus$ and $CH_3-\overset{\oplus}{CH}-CH_2-H$, the latter is the more stable ($2°$ versus $1°$) and the fastest formed. Therefore, $CH_3-\underset{Br}{\underset{|}{CH}}-CH_3$ predominates.

Example

$$CH_3-CH=CH_2 + HBr \xrightarrow{ROOR} ?$$

Solution:

The presence of peroxide catalyst signals a free-radical reaction. This must be radical

REACTIONS OF THE ALIPHATIC HYDROCARBONS 233

addition, therefore kinetic control. As in the preceding examples, the fastest-formed product is derived from the most stable intermediate, namely $CH_3-\overset{+}{C}H-CH_2-Br$ rather than $CH_3-CH-CH_2\cdot$. Therefore, $CH_3-CH_2-CH_2Br$ predominates.
$\qquad\qquad\qquad\;\; |$
$\qquad\qquad\qquad\; Br$

★ *Exercises*

Write the structure of the predominant organic product of each reaction.

(acid-base reactions)

1.  $^{\ominus}O_3S$—⟨Ph⟩—$CO_2^{\ominus}$ (1 mole) + $HClO_4$ (1 mole) ⟶ ?

2. 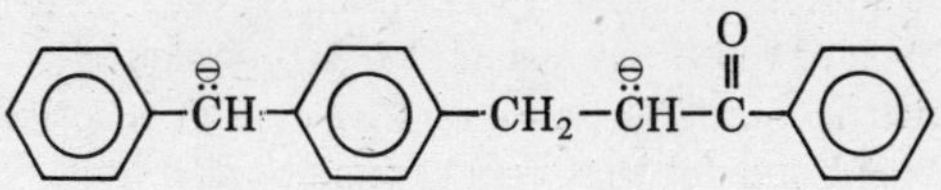 ⟨Ph⟩—$\overset{\ominus}{C}H$—⟨Ph⟩—$CH_2-\overset{\ominus}{C}H-\overset{O}{\overset{\|}{C}}$—⟨Ph⟩ (1 mole) $\xrightarrow[\text{(1 mole)}]{CH_3OH}$?

(alkene reactions)

3. 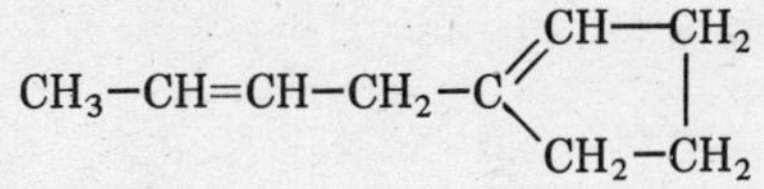 $CH_3-CH=CH-CH_2-C\begin{smallmatrix}\diagup CH-CH_2 \\ | \\ \diagdown CH_2-CH_2\end{smallmatrix}$ + Br_2 (deficiency) $\xrightarrow{dark}$?

4. 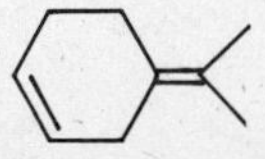 ⟨cyclohexene=⟩ (1 mole) + Br_2 (1 mole) $\xrightarrow{dark}$?

5. 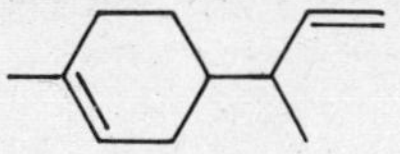 ⟨cyclohexene-⟩ (1 mole) + HCl (1 mole) ⟶ ?

- - - - - - - - -

1. 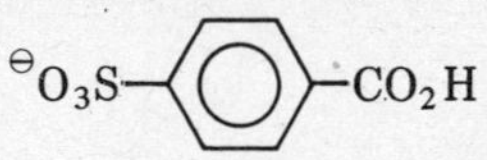 $^{\ominus}O_3S$—⟨Ph⟩—CO_2H (equilibrium control; the proton on the stronger base site is the most stable product)

2. 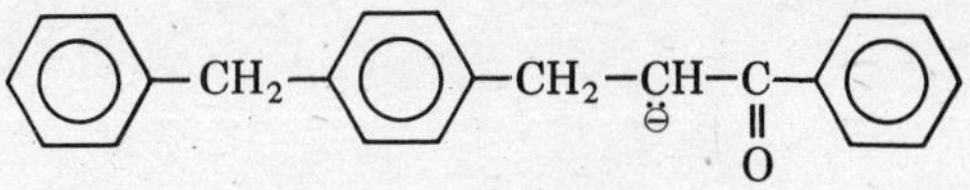 ⟨Ph⟩—CH_2—⟨Ph⟩—$CH_2-\overset{\ominus}{C}H-\overset{\|}{\underset{O}{C}}$—⟨Ph⟩ [same as (1)]

3. 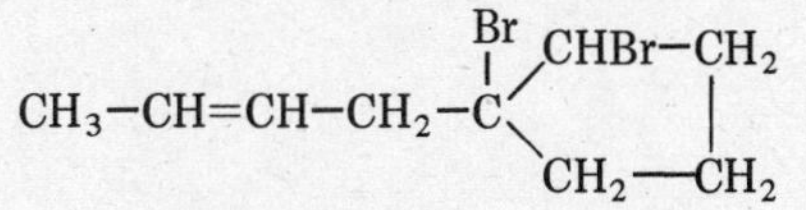 $CH_3-CH=CH-CH_2-C\begin{smallmatrix}Br & CHBr-CH_2 \\ | & | \\ & CH_2-CH_2\end{smallmatrix}$ (kinetic control; 3° $R^{\oplus}$ formation faster than 2°)

4. 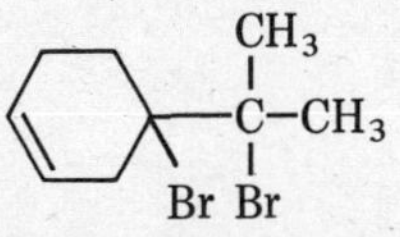 ⟨cyclohexene⟩—$\underset{Br}{\overset{CH_3}{\underset{|}{\overset{|}{C}}}}-CH_3$ (kinetic control; 3° $R^{\oplus}$ versus 2°)
$\qquad\quad Br$

5. 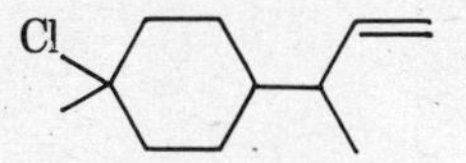 ⟨Cl-cyclohexane=⟩ (kinetic control; 3° $R^{\oplus}$ versus 1°, 2°, and 2° possibilities)

PREDOMINANT PRODUCT—QUANTITATIVE VERSION

21. Sometimes the method we have been using fails to make a definite prediction of the major product identity. Consider the reaction

$$\overset{a}{C}H_3-\overset{d}{C}H-\overset{c}{C}H_2-\overset{b}{C}H_3 \text{ (1 mole)} + Cl_2 \text{ (1 mole)} \xrightarrow{light} ?$$
$$\underset{a\ CH_3}{|}$$

Four similar functions (1° C—H bonds a and b, 2° C—H bonds c, and 3° C—H bonded) compete for the chlorine, giving a mixture of four products. This is a rate-controlled reaction, so the product will be that in which the rate of replacement of H by Cl is greatest. The relative rates of formation of 3°, 2°, and 1° R· in chlorination are 3°:2°:1° = 5.0:3.8:1.0, which would tend to make C—H bond d react 5.0 times faster than the 1° C—H bond of type a. However, solely because there are six times as many type a C—H bonds, their reaction is 6.0 times as probable and 6.0 times as fast. Both the reactivity and the probability factor must be taken into account. The true relative rate of reaction of any type of C—H bonds = (number of C—H bonds of that type) (relative rate of formation for that type). So

relative rate of reaction of type a C—H = number of type a C—H bonds × 1.0 = 6 × 1.0 = 6.0
relative rate of reaction of type b C—H = 3 × 1.0 = 3.0
relative rate of reaction of type c C—H = 2 × 3.8 = 7.6 ←—largest relative rate
relative rate of reaction of type d C—H = 1 × 5.0 = 5.0

Therefore, $CH_3-\underset{\underset{CH_3}{|}}{CH}-CH_2-CH_3$ reacts fastest at $\overset{a}{C}H_3-\overset{d}{C}H-\overset{c}{C}H_2-\overset{b}{C}H_3$, and the major
$$\underset{a\ CH_3}{|}$$

product is $CH_3-\underset{\underset{CH_3}{|}}{CH}-\underset{\underset{Cl}{|}}{CH}-CH_3$.

★ *Exercise*

If the rates of formation of R· in bromination fall in the ratio 3°:2°:1° = 1600:82:1, predict the major product of

$$CH_3-\underset{\underset{CH_3}{|}}{CH}-CH_2-CH_3 \text{ (1 mole)} + Br_2 \text{ (1 mole)} \xrightarrow{light} ?$$

- - - - - - - - - -

Relative rate of reaction of type a C—H = 6 × 1.0 = 6.0
Relative rate of reaction of type b C—H = 3 × 1.0 = 3.0
Relative rate of reaction of type c C—H = 2 × 82 = 164
Relative rate of reaction of type d C—H = 1 × 1600 = 1600

Therefore, the major product is $CH_3-\underset{\underset{CH_3}{|}}{\overset{\overset{Br}{|}}{C}}-CH_2-CH_3$.

UNFAMILIAR REACTIONS

22. Let's change one word in the statement of the problem to get "Predict the major organic product of the following reactions." Here *predict* means that the reaction you are being given is one that you have not previously studied, probably one you have never seen before. A problem of this type requires a "jump beyond" the specific material that you have learned in order to make a prediction about a system that is new to you.

That jump beyond requires reasoning by analogy. Analogies are made by taking the bones of an old problem/solution and adding the flesh of the new problem to get a new means of solution. The reaction mechanism is usually a constant element, applying to both the old and the new problems. The fundamental chemical analogy is: similar compounds have similar reactivity. The specific types of analogy are obtained by developing what we mean by similar compounds. Since changing the mass of the atomic nucleus has very little effect on the chemical properties (which are determined by the electronic structure), the very closest parallels in reactivity are between molecules that differ only in the isotopic identity of one or more atoms (e.g., between CH_4 and CD_4, or between $^{14}CH_4$ and $^{12}CH_4$).

The next closest parallels of reactivity we have already been using, in our rule that a compound's reactivity is determined mainly by the identity of the functional group. Thus all members of a family of compounds, RX, which differ only in the skeleton, R, have similar reactivity. This analogy actually breaks into two parts. The most reliable parallel is between homologs, that is between $CH_3(CH_2)_n X$ with different values of n. The somewhat less reliable parallel is between positional and structural isomers. Thus $CH_3-CH-CH_3$ and $CH_3-CH_2-CH_2-X$ have almost the same set of allowed reactions,
$\quad\quad\quad\;\;|$
$\quad\quad\quad\;\;X$

but as we have seen, they may differ drastically in relative rates or equilibrium constants for these reactions.

Next most reliable are the analogies based on the periodic table. These allow us to predict, for instance, that F, Cl, Br, and I will have similar reactivity in organic reactions, as would P and N, S and O, or Li, Na, K, and so on. This is a useful guide for prediction if more exact information is not available; however, we have already seen one limitation of this analogy, namely that F_2 and I_2 fail in the radical halogenation of alkanes—F_2 because it is too reactive so that other reactions intervene, I_2 because the equilibrium

$R-H + I_2 \underset{}{\overset{\text{light}}{\rightleftharpoons}} R-I + HI$ lies to the left.

All other chemical analogies are less exact—and more risky to use—than the ones above. Examples are (1) the reactivity of the C—H bond in the vicinity of various functional groups, for instance $-\underset{\underset{H}{|}}{\overset{\overset{O}{||}}{C}}-\underset{\underset{O}{|}}{\overset{}{C}}-$, $-\underset{\underset{H}{|}}{\overset{}{C}}-O-CH_3$, or $-\underset{\underset{H}{|}}{\overset{}{C}}-\overset{|}{C}=C\!\!\diagdown$ in comparison with that in CH_4; (2) $\diagup\!\!\ddot N-$ as an analog of $-\ddot O-$ because both have an unshared pair; (3) the $\diagup C=N-$ function as an analog of $\diagup C=O$ on the basis that both are double bonds between C and an electronegative atom; (4) $-O-O-$ bonds versus $\diagup N-N\!\!\diagdown$ bonds on the basis that both are weak. Table 12.4 summarizes these analogies.

Now, how do we use these chemical analogies?

Example

Predict the principal organic product of each reaction.

1. $\begin{array}{c}CH_3\diagdown\\ \quad\quad C=CH_2\\ CH_3\diagup\end{array} + D_2O \xrightarrow{D^\oplus}$?

2. $CH_2=CH_2 + CF_3I \xrightarrow{ROOR}$?

3. $\begin{array}{c} CH_3 \\ \diagdown \\ C=O \\ \diagup \\ CH_3 \end{array} + H_2 \xrightarrow{Pt}$?

Table 12.4 *Chemical Analogies (in order of decreasing reliability)*

Chemical behavior of:	Similar to that of:
^{18}O	^{16}O
$^{13}C, ^{14}C$	^{12}C
D	H
$CH_3-(CH_2)_{\overline{n}}X$	$CH_3-(CH_2)_{\overline{n+1}}X$
R–X	R'–X
that is:	
$CH_3-\underset{\underset{CH_3}{\vert}}{\overset{\overset{CH_3}{\vert}}{C}}-CH_2-CH_2-X$ vs.	$CH_3-\underset{\underset{CH_3}{\vert}}{CH}-CH_2-CH_2-CH_2-X$
$CH_3-CH_2-CH_2-CHX-CH_3$ vs.	$CH_3-CH_2-CHX-CH_2-CH_3$
$CH_3-CH_2-CH_2-CH_2X$ vs.	$CH_3-CH_2-CHX-CH_3$
$CH_3-\underset{\underset{CH_3}{\vert}}{\overset{\overset{CH_3}{\vert}}{C}}-X$ vs.	$CH_3-\underset{\underset{CH_3}{\vert}}{CH}-CH_2X$
R–I	R–Br
R–I	R–Cl
R–I	R–F, etc.
R–S̈–R	R–Ö–R
R–P̈H$_2$	R–N̈H$_2$, etc.
R$_3$N:	R$_2$Ö:
$\begin{array}{c}R_1\diagdown\\ C=N-R_3\\ R_2\diagup\end{array}$	$\begin{array}{c}R_1\diagdown\\ C=O\\ R_2\diagup\end{array}$
–O–O–	$\diagdown$N–N$\diagup$
$\diagup$C=O	$\diagup$C=C$\diagdown$
–C≡N	–C≡C–

Solutions:

1. The reactivity of $D_2O/D^\oplus$ is the same as that of $H_2O/H^\oplus$. Identify the reaction as electrophilic addition of water to the $\diagup\!\!\!\!\!C\!\!=\!\!C\diagdown$. With H_2O the product would be

$$\begin{array}{c}CH_3\\ \diagdown\\ \diagup\\ CH_3\end{array}\!\!\!\!\!C\!\!-\!\!\underset{H}{\underset{|}{\underset{OH}{C}}H_2}.$$ With D_2O this becomes $\begin{array}{c}CH_3\\ \diagdown\\ \diagup\\ CH_3\end{array}\!\!\!\!\!C\!\!-\!\!\underset{D}{\underset{|}{\underset{OD}{C}}H_2}.$

2. The peroxide catalyst signals a radical reaction. With no saturated $\diagup\!\!\!\!C\!-\!H$ bonds present, but with a $\diagup\!\!\!\!\!C\!\!=\!\!C\diagdown$ available, this must be not a radical substitution, but a radical addition: $\diagup\!\!\!\!\!C\!\!=\!\!C\diagdown$ + A–B $\longrightarrow$ $\underset{A\ B}{\underset{|\ \ |}{\diagup\!\!\!\!\!C\!-\!C\diagdown}}$. Is CF_3I analogous to any of the A–B reagents that we know to give this reaction? Yes, it is analogous to CCl_4, in which CCl_3 plays the role of A and Cl the role of B: CCl_3–Cl. But what will the role assignment in CF_3I be? Probably the bond broken will be the weakest one. Find C–I ($\sim$56 kcal/mole) weaker than C–F ($\sim$108 kcal/mole). So predict that the A–B bond will be CF_3–I and CF_3I will add as CF_3 and I:

$$CH_2\!\!=\!\!CH_2 + CF_3I \xrightarrow{ROOR} \underset{CF_3}{\underset{|}{CH_2\!-\!CH_2\!-\!I}}$$

3. In the absence of any specific knowledge of the reactions of the $\diagup\!\!\!\!\!C\!\!=\!\!O$ function, which we have not studied yet, we proceed by analogy. Using the only double bond we have studied, $\diagup\!\!\!\!\!C\!\!=\!\!C\diagdown$, as a model, we find catalytic hydrogenation to the alkane as an analogous process: $\diagup\!\!\!\!\!C\!\!=\!\!C\diagdown + H_2 \xrightarrow{Pt} \underset{H\ H}{\underset{|\ \ |}{\diagup\!\!\!\!\!C\!-\!C\diagdown}}$.

Therefore, predict: $\begin{array}{c}CH_3\\ \diagdown\\ \diagup\\ CH_3\end{array}\!\!\!\!\!C\!\!=\!\!O + H_2 \xrightarrow{Pt} \begin{array}{c}CH_3\\ \diagdown\\ \diagup\\ CH_3\end{array}\!\!\!\!\!\underset{H}{\underset{|}{C}}\!-\!OH.$

★ *Exercises*

Predict the principal organic product of each of the following reactions. Base your answers on the soundest possible analogies with the allowed reactions of $\diagup\!\!\!\!\!C\!\!=\!\!C\diagdown$.

1. $CH_2\!\!=\!\!CH_2 + Br\!-\!Cl \xrightarrow[=CCl_4]{solvent}$?

2. $CH_3\!-\!CH\!\!=\!\!CH_2 + CH_3OD \xrightarrow{D^\oplus}$?

3. $CH_3\!-\!CH\!\!=\!\!CH_2 + CH_3\!-\!\overset{\overset{O}{\|}}{C}\!-\!O\!-\!Br \longrightarrow$?

– – – – – – – – – –

1. $\underset{Br\ \ Cl}{\underset{|\ \ \ |}{CH_2\!-\!CH_2}}$

2. $\underset{OCH_3}{\underset{|}{CH_3\!-\!CH\!-\!CH_2D}}$

3. $CH_3-C(=O)-O-\overset{..}{\underset{..}{Br}}:$ $CH_2=CH-CH_3 \longrightarrow$ $\begin{matrix} CH_3-C(=O)-O^{\ominus} \\ \updownarrow \\ CH_3-C(-O^{\ominus})=O \end{matrix}$ $+ :\overset{..}{\underset{..}{Br}}-CH_2-\overset{\oplus}{CH}-CH_3$

$Br-CH_2-\overset{\oplus}{CH}-CH_3 \quad \longrightarrow \quad Br-CH_2-CH-CH_3$
$\overset{\ominus}{:}\overset{..}{\underset{..}{O}}-\overset{\parallel}{\underset{O}{C}}-CH_3 \qquad\qquad\qquad\quad \overset{|}{O}-\overset{\parallel}{\underset{O}{C}}-CH_3$

Although it is not asked for in these problems, writing the reaction mechanism is the best way to discover the role played by each part of the analog molecule.

ERROR CHECKS ON THE "WRITE ALL PRODUCTS" PROBLEM AND ITS VARIATIONS

23. Use of the following checks will eliminate the errors commonly made in response to these problems.

1. Is the question you answered the question that was asked?
   (a) Did you read correctly the structures or names of the reactants?
   (b) Did you note whether "all products," "all organic products," or "the major organic product" was required?
2. Do any elements appear in your products that are not available in the reactant/reagents?
3. Have you written an unstable compound as a product? Carbonium ions ($R^{\oplus}$), carbanions ($R:^{\ominus}$), free radicals ($R\cdot$), and carbenes ($R-\overset{..}{C}-R$) are almost never reasonable products of any reaction.
4. If one reactant is reduced in your equation, is another oxidized? Oxidation must always be accompanied by reduction.
5. Is the number of carbon atoms in your product the same as in the reactant? If not, the reagents must have brought in C, or unwritten or inorganic by-products must have carried some off—or else there is an error somewhere.

★ *Exercises*

Locate the error in the answer to each "Write all organic products of these reactions" problem.

1. $CH_3-CH_2-CH_3$ (1 mole) $+ Cl_2$ (1 mole) $\longrightarrow CH_3-CHCl-CH_3$

2. $CH_3-CH_3 + Br_2$ (large excess) $\longrightarrow CCl_3-CCl_3$

3. $C_6H_{10}=CH-C_6H_5 \xrightarrow[(2)\ H_2O_2]{(1)\ O_3} C_6H_{10}-C(=O)-OH + HO-C(=O)-C_6H_5$

4. $C_6H_5-Mg-Cl \xrightarrow{KOH} C_6H_5:^{\ominus} + K^{\oplus} + Mg^{2\oplus} + Cl^{\ominus} + H_2O$

5. $O=C=O + H_2O + KOH \longrightarrow \underset{H}{\overset{H}{>}}C=O + KHCO_3$

REACTIONS OF THE ALIPHATIC HYDROCARBONS   239

1. Minor product omitted: $CH_3-CH_2-CH_2Cl$.

2. CBr_3-CBr_3.

3. Reactants = C_{13}, products = C_{14}.

4. Carbanion is never a product.

5. CO_2 is reduced but nothing is oxidized. You can recognize fallacious equations of this sort by trying to balance them; it isn't possible.

PROBLEMS INVOLVING MECHANISMS

24. In this and the next few frames, we work on the "write all steps in the mechanism of the following reactions" problem, as applied to reactions you have already studied. This problem can also be posed for reactions you have not seen before; that is a more difficult problem that comes up after you have had more experience.

The steps in the solution of the present version are:

1. Identify which of the basic mechanisms, I to IX, the reaction goes by.
2. Identify the specific role played in the mechanism by each of the specific reactants in the problem.
3. Write the specific mechanism by inserting these slot fillers in the generic mechanism.
4. Check for errors.

We practice step 1 here, steps 2 and 3 in frame 25, and step 4 in frame 26.

If you have memorized which mechanism applies as one of the four basic points learned at the time of studying each reaction, then as soon as you recognize the reaction in this problem, you know which mechanism you are dealing with. However, if you can't recall which mechanism applies, or even if you can't recognize the reaction precisely, you can still solve this problem. You do it by using the clues provided in the (given) equation for the reaction.

The important clues are (a) the kind of reagent (nucleophile, electrophile, strong base, strong acid, etc.) that is present to attack the starting material, (b) the catalysts used, and (c) creation or destruction of multiple bonds in the reaction (which is conveniently expressed as change in the unsaturation number, Δ). The conclusions that you can draw are summarized in Table 12.5. (For convenience all nine mechanisms are included, although we haven't discussed them all yet.)

Only mechanisms I and III require the radical-generating conditions *light* or *ROOR + heat*. These two can be distinguished from one another easily because the substitution neither creates nor destroys double bonds (Δ unchanged), while the addition destroys a double bond, decreasing Δ. A reaction that destroys a double bond but requires no light or peroxide catalyst must be an electrophilic or nucleophilic addition. If the double bond is an ordinary $\ce{>C=C<}$ function, this must be electrophilic addition, since nucleophilic additions to $\ce{>C=C<}$ are rare. If it is $\ce{>C=O}$, the mechanism is nucleophilic addition, never electrophilic addition. And so on with other possibilities, which we will analyze in detail as we take them up.

240 HOW TO SUCCEED IN ORGANIC CHEMISTRY

Table 12.5 *Recognition Clues for the Nine Basic Mechanisms*

| | | Reaction characteristics | | | | |
| | | | | Reagent required | | |
	Mechanism	Effect on Δ	Catalyst required	Electrophile	Nucleophile	Base
I	Radical substitution	None	Light or ROOR			
II	Electrophilic addition	Decreases	Often Lewis acid	X	X	
III	Radical addition	Decreases	Light or ROOR			
IV	$S_N1/E1$	None/increases			X	
V	S_N2	None			X	
VI	E2	Increases				X
VII	Electrophilic aromatic substitution	None	Often Lewis acid	X		
VIII	Nucleophilic addition	Decreases			X	
IX	Acyl/aryl S_N2	None			X	

Exercises

Identify the mechanism type (I to IX) by which each reaction proceeds.

1. 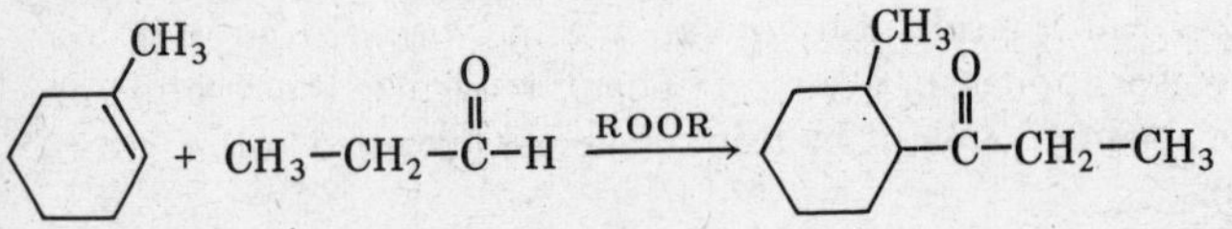 [methylcyclohexene] + $CH_3-CH_2-\overset{O}{\underset{\|}{C}}-H$ $\xrightarrow{ROOR}$ [methylcyclohexane with $-\overset{O}{\underset{\|}{C}}-CH_2-CH_3$ substituent]

2. $CH_3-CH_2-\overset{\oplus}{N}(CH_3)_3 Cl^{\ominus}$ $\xrightarrow{KOH}$ $CH_2=CH_2 + N(CH_3)_3 + K^{\oplus}Cl^{\ominus} + H_2O$

3. 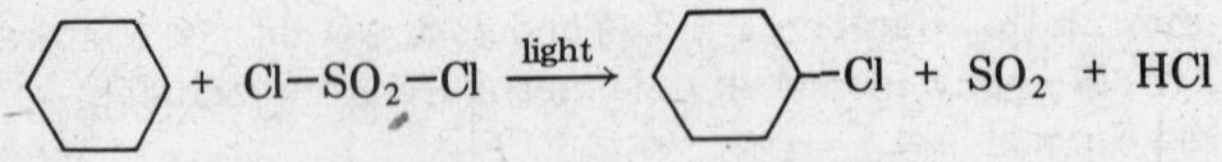 [benzene] + $Cl-SO_2-Cl$ $\xrightarrow{light}$ [benzene]$-Cl + SO_2 + HCl$

4. $CH_2=CCl_2 + CH_3-\overset{O}{\underset{\|}{C}}-Cl$ $\xrightarrow{AlCl_3}$ $Cl-CH_2-CCl_2-\overset{O}{\underset{\|}{C}}-CH_3$

5. $CH_3-\overset{O}{\underset{\|}{C}}-CH_3 + CH_3-OH$ $\xrightarrow{H^{\oplus}}$ $CH_3-\underset{OCH_3}{\overset{OH}{\underset{|}{\overset{|}{C}}}}-CH_3$

6. 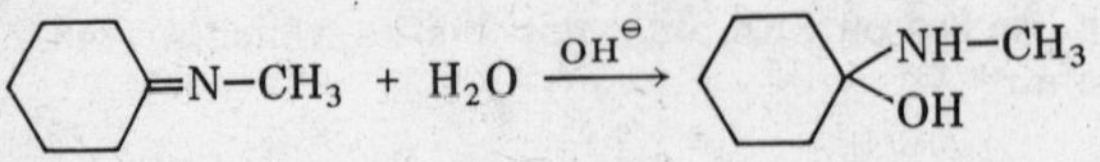 [cyclohexane]$=N-CH_3 + H_2O$ $\xrightarrow{OH^{\ominus}}$ [cyclohexane with $NH-CH_3$ and OH substituents]

- - - - - - - - - -

1. III (Δ decreases, peroxide required)   2. VI (Δ increases, base required)

3. I (Δ unchanged, light required)

4. II (Δ decreases, Lewis acid catalyst used, $\cee{>C=C<}$ reacts)

5. VIII (Δ decreases, reagent is nucleophile, $\cee{>C=O}$ reacting)

6. VIII (like exercise 5 but involves a nitrogen analog of $\cee{>C=O}$)

25. After you have decided which mechanism you are dealing with, recall the generic form of the mechanism. You do not necessarily need to write it out, but it is wise for beginners to do so. Then assign the roles to be played by the reactants, and plug the reactants into the generic mechanism.

Example

Write all steps in the mechanism of exercise 1 in frame 24, already identified as a radical addition, mechanism III.

Solution:

Write:
$$\left.\begin{array}{l} R-O-O-R \longrightarrow R-O\cdot + \cdot O-R \\ R-O\cdot + A-B \longrightarrow R-O-A + B\cdot \end{array}\right\} \text{initiation}$$

$$\left.\begin{array}{l} B\cdot + \mathrm{>C=C<} \longrightarrow B-C-C\cdot \\ B-C-C\cdot + A-B \longrightarrow B-C-C-A + B\cdot \end{array}\right\} \text{propagation}$$

$$\left.\begin{array}{l} B-C-C\cdot + \cdot B \longrightarrow B-C-C-B \\ B-C-C\cdot + \cdot O-R \longrightarrow B-C-C-O-R \end{array}\right\} \text{termination}$$

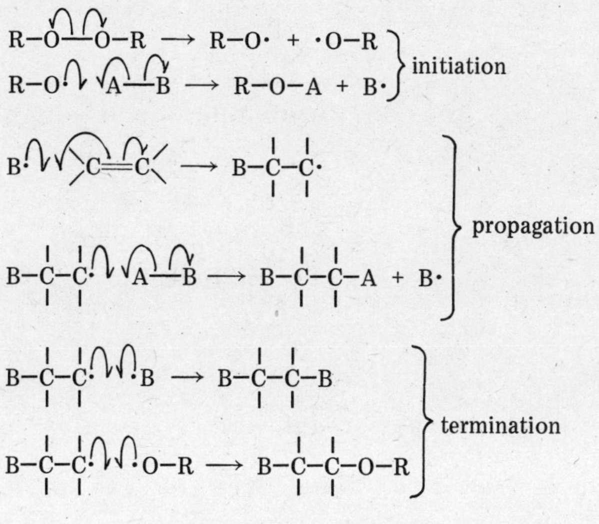

The role of ROOR is spelled out in the generic mechanism. $CH_3-CH_2-\overset{\overset{O}{\|}}{C}-H$ plays the role of A—B, but what is A and what is B? The groups added across $\cee{>C=C<}$ can be read from the product structure:

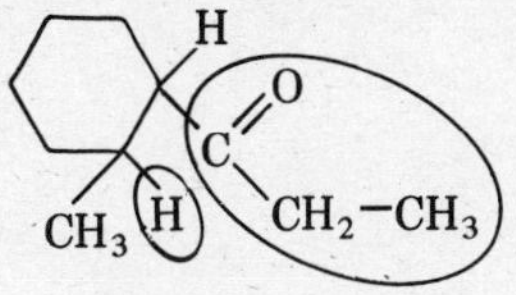

So $CH_3-CH_2-\overset{\overset{O}{\|}}{C}-H$ breaks here: $CH_3-CH_2-\overset{\overset{O}{\|}}{C} \dashv H$.

(We say that $CH_3-CH_2-\overset{\overset{O}{\|}}{C}-H$ "adds as $CH_3-CH_2-\overset{\overset{O}{\|}}{C}$ and H.") But do we identify CH_3-CH_2-CO = A and H = B, or the reverse? The direction of addition to

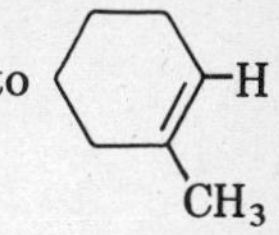

$\cee{<cyclohexene>-H}$ with CH_3

tells us. We know that the intermediate radical will be [cyclohexane with H, •B, CH₃], not [cyclohexane with •—H, —B, CH₃], because 3° radicals are formed in preference to 2° radicals. So we can identify B by seeing what is attached at the [cyclohexane with CH₃] site in the product. It turns out to be

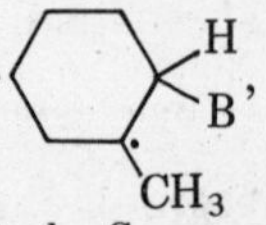

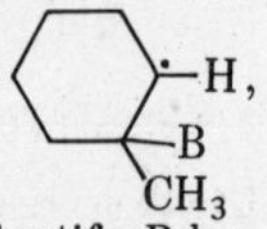

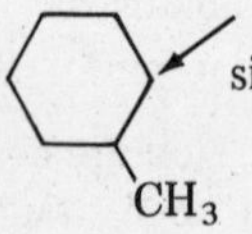

$$\overset{O}{\underset{\|}{C}}-CH_2-CH_3,$$ which is then B, and A must be H. Since the direction of attack has already been settled, we can translate the generic mechanism into

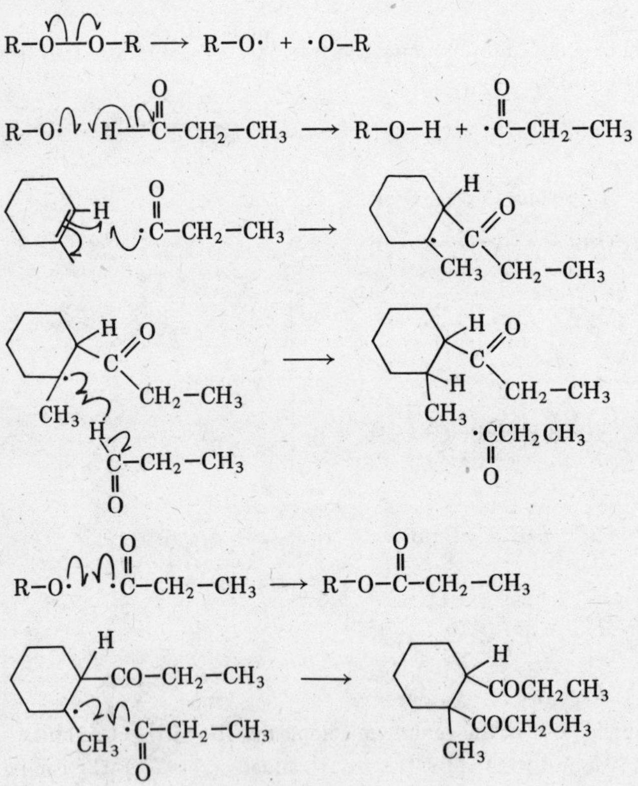

$$R-O\frown\frown O-R \longrightarrow R-O\cdot + \cdot O-R$$

$$R-O\cdot \frown H\frown \overset{O}{\underset{\|}{C}}-CH_2-CH_3 \longrightarrow R-O-H + \cdot\overset{O}{\underset{\|}{C}}-CH_2-CH_3$$

[cyclohexane-H + •C(=O)CH₂–CH₃ → cyclohexane with H, C(=O), CH₃, CH₂–CH₃]

[cyclohexane with H, C(=O)CH₂–CH₃, CH₃, H + •C–CH₂–CH₃ (=O) → cyclohexane with H, C(=O)CH₂–CH₃, CH₃, •CCH₂CH₃ (=O)]

$$R-O\cdot \frown \cdot \overset{O}{\underset{\|}{C}}-CH_2-CH_3 \longrightarrow R-O-\overset{O}{\underset{\|}{C}}-CH_2-CH_3$$

[cyclohexane with H, –CO–CH₂–CH₃, CH₃, •C–CH₂–CH₃ (=O) → cyclohexane with H, COCH₂CH₃, COCH₂CH₃, CH₃]

★ *Exercises*

Write all steps in the mechanism of each reaction.

1. $CF_3-CCl_2H + Br_2 \xrightarrow{light} CF_3-CCl_2Br + HBr$

2. $CH_3-CH=CH_2 + ICl \longrightarrow CH_3-CHCl-CH_2I$

3. $CCl_3Br + CH_2=CH_2 \xrightarrow{light} CCl_3-CH_2-CH_2Br$

— — — — — — — —

1. $Br\frown\frown Br \xrightarrow{light} Br\cdot + \cdot Br$

$$Br\cdot \curvearrowright H\!\!-\!\!CCl_2\!-\!CF_3 \longrightarrow Br\!-\!H + \cdot CCl_2\!-\!CF_3$$

$$Br\!-\!Br \curvearrowright \cdot CCl_2\!-\!CF_3 \longrightarrow Br\cdot + Br\!-\!CCl_2\!-\!CF_3$$

$$CF_3\!-\!CCl_2\cdot \curvearrowright \cdot CCl_2\!-\!CF_3 \longrightarrow CF_3\!-\!CCl_2\!-\!CCl_2\!-\!CF_3$$

$$CF_3\!-\!CCl_2\cdot \curvearrowright \cdot Br \longrightarrow CF_3\!-\!CCl_2Br$$

2. $CH_3\!-\!CH\!=\!CH_2 \curvearrowright I\!-\!Cl \longrightarrow CH_3\!-\!\overset{\oplus}{CH}\!-\!CH_2\!-\!I + Cl^\ominus$

$$:\!\ddot{C}l\!:^\ominus \curvearrowright \overset{\oplus}{C}H\!-\!CH_2\!-\!I \longrightarrow :\!\ddot{C}l\!-\!CH\!-\!CH_2\!-\!I$$
$$\quad\quad\quad\;\; |\quad\quad\quad\quad\quad\quad\quad\quad |$$
$$\quad\quad\quad\;\; CH_3 \quad\quad\quad\quad\quad\quad\; CH_3$$

3. $Cl_3C\!\!-\!\!Br \xrightarrow{\text{light}} Cl_3C\cdot + \cdot Br$

$$Cl_3C\cdot \curvearrowright CH_2\!=\!CH_2 \longrightarrow Cl_3C\!-\!CH_2\!-\!CH_2\cdot$$

$$Cl_3C\!-\!CH_2\!-\!CH_2\cdot \curvearrowright Br\!-\!CCl_3 \longrightarrow Cl_3C\!-\!CH_2\!-\!CH_2\!-\!Br + \cdot CCl_3$$

$$Cl_3C\cdot \curvearrowright CH_2\!-\!CH_2\!-\!CCl_3 \longrightarrow Cl_3C\!-\!CH_2\!-\!CH_2\!-\!CCl_3$$

$$Cl_3C\!-\!CH_2\!-\!CH_2\cdot \curvearrowright CH_2\!-\!CH_2\!-\!CCl_3 \longrightarrow Cl_3C\!-\!(CH_2)_4\!-\!CCl_3$$

26. The errors most commonly made by students in handling this problem are

- Departing from the scenario dictated by the relevant generic mechanism.
- Writing steps that do not balance.
- Writing mixtures of two mechanisms.
- Using incorrect or inadequate notation.
- Telescoping sequential steps.

You will avoid the first one of these errors by using the approach of frames 24 and 25, that is, by deciding what packaged mechanism is going to be used before starting to write. Beginning students who just start writing individual mechanistic steps heading in the general direction of products usually never get there, or else they get there via a mechanism that is not the one that experiment shows actually occurs in nature.

You should check the mechanisms you write for the presence of the other four errors. Let's look at these one by one.

Each individual step in a mechanism must balance—in atoms and electrons. Electron balance requires that the sum of charges on right and left sides be equal.

Example

Locate the error(s) in each mechanistic step.

(a) $R\cdot + CH_3OH \longrightarrow RH + CH_3O^\ominus$

(b) $R^\oplus + HI \longrightarrow RI + H\cdot$

(c) $CH_3\!-\!CH_2\!:^\ominus + :\!CH_2\!-\!CH_3 \longrightarrow CH_3\!-\!CH_2\!-\!CH_2\!-\!CH_3$

Solution:

In (a), (b), and (c) the atoms balance, but the electrons don't; (a) and (b) have created an electron, and (c) has destroyed two electrons. (a) transforms a radical (one unpaired electron) into an anion (paired electrons only); thus the number of electrons does not balance. (b) converts a carbonium ion to a radical. The error in all three can be spotted by the absence of charge balance: (a) neutral = −1; (b) +1 = neutral; (c) −2 = neutral.

Another common error that occurs is writing mixtures of two mechanisms. A given mechanism never involves more than one type of reactive intermediate—$R^\oplus$, $R^\ominus$, or $R\cdot$. A radical reaction creates and destroys radicals, $R\cdot$, but never also employs $R^\ominus$ or $R^\oplus$. Carbonium ion mechanisms have no place for $R\cdot$ or $R^\ominus$, and so on. Example b above should *look wrong* because it turns $R^\oplus$ into $H\cdot$, which makes it a forbidden, mixed mechanism.

Then there are the errors that originate in poor notation. Your practice in Unit 10 should give you an advantage here. If you confuse ⤻ and ⤺, you should review frame 2 of Unit 10. If you find yourself writing such steps as

$$\cdot CH_3 \;+\; I\!-\!H \longrightarrow CH_4 \;+\; I\cdot$$

you should stop and take the time to rewrite this equation (and all others) so that (1) all reactants are positioned so as to minimize the movements of atoms necessary to make reactants into products, and (2) all bonds made or broken in the step are explicitly shown. (If the + sign connecting the reactant molecules gets in the way, eliminate it.) The result is readable:

$$I\!-\!H \quad \cdot CH_3 \longrightarrow I\cdot \;+\; H\!-\!CH_3$$

Example

Improve the notation in this step:

$$R\!-\!O\cdot \;+\; CCl_4 \longrightarrow R\!-\!OCl \;+\; \cdot CCl_3$$

Solution:

Faults: $RO\cdot$ and Cl on opposite sides of C, reactants too far apart, C—Cl bond being broken not shown.

Rewrite: $R\!-\!O\cdot \quad Cl\!-\!CCl_3 \longrightarrow R\!-\!O\!-\!Cl \;+\; \cdot CCl_3$

Don't write lasso mechanisms. A *lasso mechanism* is a pseudomechanism whose arrows move atoms instead of (or in addition to) electrons; it really only shows what becomes bonded to what in the products—something the equation itself shows. Consider the example

$$H_2C\!=\!CH_2 \longrightarrow \begin{array}{c} CH_2\!-\!CH_2 \\ |\quad\;\; | \\ H \quad I \end{array}$$
$$(H)(I)$$

This "mechanism" fails to show that the reaction occurs in steps, what the structure of the intermediate is, and in which direction the electron flow occurred.

Don't telescope sequential steps. By drawing enough arrows one can often crowd all the things happening in a two- or three-step mechanism into one step. Why is this incorrect? Because it forces unrealistically large numbers (>2) of molecules to collide simultaneously, makes/breaks too many bonds at once, and ignores the way nature achieves energetically costly results by using less costly, smaller steps. Writing

$$Nu{:}^{\ominus} \quad R\!-\!OH \quad H\!-\!A \longrightarrow Nu\!-\!R \;+\; HOH \;+\; A{:}^{\ominus}$$

in place of

$$R-\overset{..}{\underset{..}{O}}H \quad H-A \longrightarrow R-\overset{\oplus}{O}H_2 \quad A:^{\ominus}$$

$$Nu:^{\ominus} \quad R-\overset{\oplus}{O}H_2 \longrightarrow Nu-R + OH_2$$

not only requires a three-way collision, but also makes the C—O bond breaking occur without much help from the acid—a process that probably does not happen in nature.

Examples

Analyze the errors in each mechanism.

1. $CH_2{=}CH_2 \longrightarrow \underset{I\ H}{\overset{\ominus}{|}}\underset{\ }{CH_2}-\underset{I}{\overset{}{|}}CH_2 \quad H^{\oplus}$

   I–H   I–H

2. $CH_2{=}CH_2 \longrightarrow \underset{H}{\overset{}{|}}CH_2-\underset{I}{\overset{}{|}}CH_2$

   H–I

Solutions:

1. This representation of the mechanism has some correct aspects; for example, it gives $H^{\oplus}$ and $I^{\ominus}$ the correct roles as electrophile and nucleophile in the addition. But this answer can be spotted as wrong by checking the rules of Unit 10: it involves three reactant molecules and makes or breaks five bonds.

2. This answer no longer can be spotted as wrong in the way just mentioned, since it involves two molecules and four bonds made or broken. However, it is still wrong because it telescopes a two-step mechanism,

$$CH_2{=}CH_2 \longrightarrow \underset{H}{\overset{\oplus}{|}}CH_2-CH_2 \quad I^{\ominus}$$
$$H-I$$

$$I^{\ominus} \quad \overset{\oplus}{C}H_2-\underset{H}{\overset{}{|}}CH_2 \longrightarrow I-CH_2-\underset{H}{\overset{}{|}}CH_2$$

into a single step.

One more caution: in nonradical mechanisms the sum of all the steps must be the given overall reaction. In radical chain mechanisms (I, III), the sum of the two *propagation* steps is the overall equation. Verify this for the exercises in frame 25.

★ *Exercises*

1. Analyze the errors in each answer to the "Write all steps. . ." problem.

   (a) $Cl-Cl \longrightarrow :\overset{..}{\underset{..}{Cl}}{}^{\oplus} + :\overset{..}{\underset{..}{Cl}}{}^{\ominus}$

   $CH_2{=}CH_2 \quad \overset{..}{\underset{..}{Cl}}{:}^{\oplus} \longrightarrow \overset{\oplus}{C}H_2-CH_2-Cl$

   $Cl-CH_2-CH_2 \overset{\oplus}{} + :\overset{..}{\underset{..}{Cl}}{:}^{\ominus} \longrightarrow Cl-CH_2-CH_2-Cl$

(b) $\overset{\curvearrowleft}{CH_3-H} \longrightarrow :CH_3^{\ominus} + H^{\oplus}$ initiation

$H_3C:^{\ominus} + Br\overset{\curvearrowright}{-}Br \longrightarrow CH_3-Br + Br\cdot$
$Br\cdot + H\overset{\curvearrowleft}{-}CH_3 \longrightarrow Br-H + CH_3\cdot$ propagation

$2 \cdot CH_3 \longrightarrow CH_3-CH_3$ termination

(c) $:NH_2-OH + H_3O^{\oplus} \rightleftharpoons NH_2\overset{\oplus}{-}\overset{\curvearrowleft}{OH_2} \longrightarrow :NH_2^{\oplus} + H_2O$

$\underset{\oplus:NH_2}{\overset{:\ddot{O}}{\underset{\|}{CH_3-C}}\!\!\curvearrowright\!\! CH_3} \longrightarrow \underset{:NH_2}{\overset{\ddot{O}:^{\oplus}}{\underset{|}{CH_3-C-CH_3}}} = \underset{\oplus:\ddot{O} \curvearrowright NH_2}{CH_3-C-CH_3} \longrightarrow \underset{\underset{\oplus}{O-NH_2}}{CH_3-C-CH_3}$

$\underset{\underset{H}{O\overset{\oplus}{-}N-H}}{\overset{:\overset{\ominus}{\ddot{O}H}}{\underset{|}{CH_3-C-CH_3}}} \longrightarrow \underset{\underset{OH}{NH}}{\overset{OH}{\underset{|}{CH_3-C-CH_3}}}$

2. Write all steps in the mechanism of each reaction.

(a) $CH_3-CH=CH_2 + H_2O \xrightarrow{H_2SO_4} CH_3-\underset{OH}{\underset{|}{CH}}-CH_3$

(b) Exercise 3, frame 24.

— — — — — — — —

1. (a) The writer of this mechanism has apparently confused and mixed elements of the radical addition and electrophilic addition mechanisms. In the first step the half-head arrows indicate homolytic cleavage to Cl· + ·Cl, but the fragments written are ions. In the absence of light, Cl–Cl breaks by itself only at high temperatures. If the mechanism is ionic (electrophilic), Cl–Cl should break heterolytically only when attacked by the alkene. One final error: in the second step the flow of electrons is in the wrong direction; the indicated transfer from $\ddot{C}l:^{\oplus}$ to alkene would result in the nonsense products $\overset{\ominus}{C}H_2-CH_2-Cl:^{\oplus\ominus}$.

(b) Another mixed mechanism. There are no useful conditions under which the CH_3-H bond can break, solo, as written. The first "propagation" step involves destruction of one electron. Arrows in the second "propagation" step are not correct for radical abstraction. Chain ionic mechanisms are very rare.

(c) $\dot{N}H_2^{\oplus}$ is extremely unstable, even for an intermediate. The $NH_2^{\oplus}-CH_3COCH_3$ reaction is electrophilic addition to $>C=O$—forbidden because it generates the very unlikely intermediate $R-\ddot{O}:^{\oplus}$ (see Unit 8). The last step uses $OH^{\ominus}$, unavailable in acidic solution, and is unlikely because it breaks/makes five bonds.

UNIT THIRTEEN
Stereochemistry

In Unit 6 we defined isomers as compounds having the same molecular formula but different structural formulas. We applied the criterion of superimposability to decide if structures that have the same molecular formula are indeed different (represent isomers), or if they represent the same compound. But we did not ask many questions about all the possible manners in which a pair of isomeric structures differ. In this unit we look at the three fundamental ways in which a pair of isomers can differ. We have already encountered structural and conformational isomers; we will now consider stereoisomers, which differ only by arrangement of groups in space. In the rest of this unit we take up their recognition, classification, and properties.

In frame 2 we find that of all the many common structural features in organic molecules, only a few ("stereochemical centers") are capable of causing stereoisomers to exist. Frame 3 shows that stereoisomers tend to occur in pairs because there are just two possible arrangements of groups in space (configurations around any stereochemical center).

Before we can take the next step of learning how the chemical and physical properties of stereoisomers compare (frame 7), we must develop the concepts of chirality (frame 4) and enantiomerism (frame 6). These concepts not only provide the key to understanding the properties of stereoisomers, they allow us to take the classification of isomers one step further at this point.

Frame 8 shows you how to name stereoisomers, and frame 9 outlines the experimental methods for determining their three-dimensional structures. The increased possibilities for stereoisomerism caused by the presence of more than one stereochemical center in a molecule are considered in frames 10 and 11.

The remainder of the unit is devoted to the effects of chemical reactions on the configuration of stereochemical centers.

OBJECTIVES

When you have completed this unit, you should be able to:

- Given any pair of structural formulas, decide whether they represent the same compound, unrelated compounds, conformers, structural isomers, enantiomers, or diastereomers (frames 1, 6, 11).
- Draw the two possible configurations about a suitably substituted $\diagup C = C \diagdown$ center and give them their correct E and Z labels (frames 2, 8).
- Draw the two possible configurations about a suitably substituted $\diagup \overset{*}{C} \diagdown$ center, and give each its proper R or S label (frames 2, 8).
- Explain the difference between conformation and configuration (frame 1).
- Read the configuration of an asymmetric carbon atom drawn in either the

notation, and redraw the same or the opposite configuration using each of the other notations (frame 3).
- Classify objects, molecules, and environments as chiral or achiral (frame 4).
- Compare the properties and the conditions required for separation of pairs of (1) enantiomers and (2) diastereomers (frames 7, 11).
- Identify *meso* compounds and describe their properties (frame 11).
- Predict the configuration of the products of any given example of a radical substitution, electrophilic addition, or radical addition reaction, and show how the reaction mechanism determines this stereochemical outcome (frames 12 to 14).

If you think you may already have accomplished some of these objectives and might skip all or part of this unit, take a self-test consisting of the exercises marked with a star in the margin. If you miss any of these, reread the frames in question and rework the exercise before proceeding. If you are not ready for a self-test, proceed to frame 1.

CONFORMATIONAL ISOMERS, STRUCTURAL ISOMERS, AND STEREOISOMERS

1. Let's review what you have learned about the possible relationship of one structure, A, to another, B. The following flowchart is a good way to summarize the decisions that go into classifying these relationships.

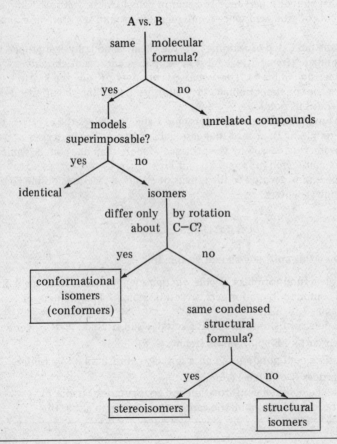

Suppose that you have used the first two questions to decide that some pair of structures is a pair of isomers. Two more questions will allow you to decide which of the three types of isomerism is involved. If rotation around carbon-carbon single bonds suffices to turn A into B, then the two are conformational isomers. Since these rotations occur at very rapid rates (except at very low temperatures), conformational isomers are continuously interconverting and cannot be separated and studied individually. Consequently structures that represent different conformational isomers we call "equivalent," just as structures that represent different views of the same conformer are equivalent—for practical purposes they all describe a single compound. That compound has a well-defined, dependable set of properties, even though these properties are weighted-average properties of the individual, rapidly converting conformational isomers.

The other two types of isomerism are very different from conformational isomerism in being effectively permanent—interconversion of isomers of these types does not occur at a measurable rate, and the individual isomers are stable and can be isolated.

Now consider the last question in the flowchart, which distinguishes the last two classes of isomers. What structural difference does it detect? In other words, what does it mean that two molecules are not superimposable when their three-dimensional models are compared, but appear identical when their condensed structural formulas are compared? It means that the two differ not in what is bonded to what (which is what the condensed structural formula shows), but in some aspect of the arrangement of the atoms in space.

The three classes of isomers have these characteristics:

Class and subclass	Defining properties
Isomers	Molecular formulas same, three-dimensional models different
Conformational isomers	Differ only by rotation about single bonds
Structural isomers	Differ in what is bonded to what (condensed structural formulas differ)
Stereoisomers	Differ only in arrangement of groups in space (identical condensed structures)

Examples

Determine the relationship between the members of each pair.

1. CH$_3$—CH—CH$_3$ and CH$_3$—CH—CH$_3$
        | |
       CH$_3$ CH$_3$
        a   b

2. CH$_3$—CH—CH$_3$ and CH$_3$—CH$_2$—CH$_2$—CH$_3$
        |
       CH$_3$ b
        a

250 HOW TO SUCCEED IN ORGANIC CHEMISTRY

Solutions:

1. Apply the classification flowchart:
   - Both are C_4H_{10}.
   - Models of the two superimpose.
   - ∴ a and b represent the same compound.

2. Apply the classification flowchart:
   - Both a and b are C_4H_{10}.
   - Models of a and b do not superimpose.
   - a cannot ⟶ b by rotation about single bonds.
   - a and b do not have the same condensed structural formula:

   $CH_3CH(CH_3)CH_3$ $CH_3(CH_2)_2CH_3$
           a   b

   ∴ a and b are structural isomers.

★ Exercises

Determine the relationship between the members of each pair.

1.
$$\underset{a}{\overset{H}{\underset{H}{>}}C=C\overset{Cl}{\underset{Cl}{<}}} \quad \text{and} \quad \underset{b}{\overset{H}{\underset{Cl}{>}}C=C\overset{Cl}{\underset{H}{<}}}$$

2.
$$\underset{a}{\overset{H}{\underset{Cl}{>}}C=C\overset{Cl}{\underset{H}{<}}} \quad \text{and} \quad \underset{b}{\overset{H}{\underset{Cl}{>}}C=C\overset{H}{\underset{Cl}{<}}}$$

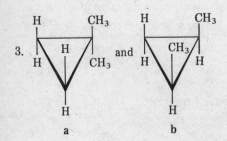

Note: Condensed structures for cyclic compounds are drawn using a long bond to close the ring:

3 a = $\underline{CH_2CH_2C(CH_3)_2}$

or $\underline{CH_2C(CH_3)_2CH_2}$ or $\underline{C(CH_3)_2CH_2CH_2}$.

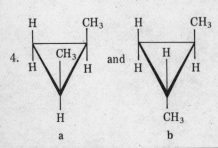

5. 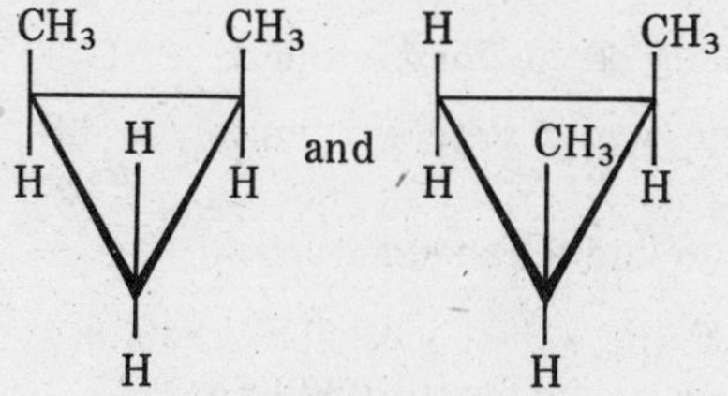 [structures a and b: triangular/tetrahedral representations of C₅H₁₀ isomers]

6. 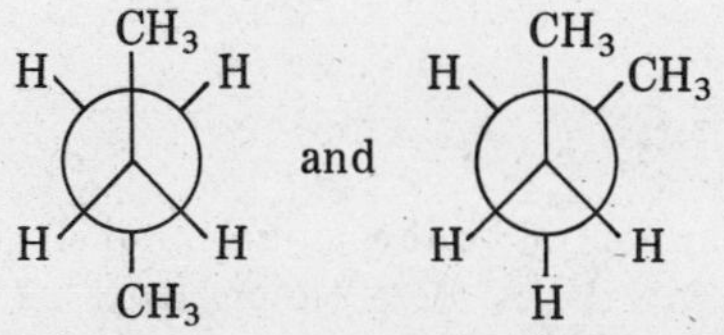 [structures a and b: Newman projection representations]

---

1. Both are $C_2H_2Cl_2$; models nonsuperimposable; don't differ by C—C rotation; condensed structures differ: $CH_2{:}CCl_2$ versus $CHCl{:}CHCl$; ∴ structural isomers.

2. Both are $C_2H_2Cl_2$; models nonsuperimposable; don't differ by C—C rotation; condensed structures identical: $CHCl{:}CHCl$; ∴ stereoisomers.

3. Both are C_5H_{10}; models nonsuperimposable; don't differ by C—C rotation; condensed structures differ:

$$\underbrace{CH_2{-}CH_2{-}C(CH_3)_2}_{a} \qquad \underbrace{CH_2CH(CH_3)CH(CH_3)}_{b}$$

∴ structural isomers.

4. Both are C_5H_{10}; models nonsuperimposable; don't differ by C—C rotation; condensed structures identical: $CH_2CH(CH_3)CH(CH_3)$ represents a and b; ∴ stereoisomers.

5. Both are C_5H_{10}; models superimposable (rotate the triangle in a 120° clockwise); ∴ represent the same compound.

6. Both are C_4H_{10}; models not superimposable; differ only by rotation about $CH_3{-}CH_3$; ∴ conformational isomers.

SOURCES OF STEREOISOMERISM

2. To understand the occurrence and consequences of stereoisomerism, we must first analyze the ways in which an organic molecule can achieve "different arrangements of groups in space" and consequently exist as two or more stereoisomers. There are only two common ways,* neither of which involves distorting the bond angles or distances but rather are just alternative ways of arranging the available groups on normal tetrahedral and trigonal-planar carbon atoms. They are

---

* Several more exotic structural situations produce stereoisomerism, but these are rarely taken up in the beginning organic course.

- Alternative arrangements on the fixed directions of the bonds around $\ce{>C=C<}$, $\ce{>C=N-}$, and $\ce{-N=N-}$ ($\ce{-N=N-}$). This is the type of stereoisomerism you discovered in exercise 2 of frame 1.
- Alternative arrangements on the four bonds around tetrahedral carbon.

In the first case the two arrangements differ in having a distinguishable group on each carbon arranged either close together (called *cis*) or far apart (called *trans*):

$$\underset{\text{cis}}{{}^a\!\!\diagdown_{\diagup}\!\!C=C\!\!{}^{\diagup c}_{\diagdown}} \qquad \underset{\text{trans}}{{}^a\!\!\diagdown_{\diagup}\!\!C=C\!\!{}^{\diagup}_{\diagdown c}}$$

This source of stereoisomerism applies only when the central bond is double, not single, because the double bond provides *rigidity* that prevents the cis stereoisomer from being turned into the trans stereoisomer by rotation:

$${}^a\!\!\diagdown_{\diagup}\!\!C\!\!\not\!\!\circlearrowright\!\!C\!\!{}^{\diagup c}_{\diagdown} \;\;\not\to\;\; {}^a\!\!\diagdown_{\diagup}\!\!C=C\!\!{}^{\diagup}_{\diagdown c}$$

Permanant cis and trans isomers can't exist around an ordinary single bond because in this case rotation is rapid, as we have seen, and the two arrangements turn into one another:

$$-\overset{a}{\underset{|}{C}}\!\!\circlearrowright\!\!\overset{c}{\underset{|}{C}}- \;\longrightarrow\; -\overset{a}{\underset{|}{C}}-\overset{|}{\underset{c}{C}}-$$

We have already split this situation off for separate treatment under the name "conformational isomerism" (frame 1). (In fact, we know that rotation about C—C produces more than two conformational isomers or conformers; see frame 5 of Unit 4.)

Not every alkene exists as a pair of stereoisomers. The requirement for stereoisomerism in ${}^a\!\!\diagdown_{b\diagup}\!\!C=C\!\!{}^{\diagup c}_{\diagdown d}$ is that a must not = b and c must not = d. It is not necessary that a and/or b be different from c and/or d.

Example

CHCl=CHCl exists as cis and trans isomers: $\underset{\text{cis}}{{}^{Cl}\!\!\diagdown_{H\diagup}\!\!C=C\!\!{}^{\diagup Cl}_{\diagdown H}}$ and $\underset{\text{trans}}{{}^{Cl}\!\!\diagdown_{H\diagup}\!\!C=C\!\!{}^{\diagup H}_{\diagdown Cl}}$.

CH$_2$=CHCl does not, because the two atoms on the left carbon are identical.

The second source of stereoisomerism differs from the first because on a single tetrahedral carbon atom, $\diagdown\!\!\underset{\diagup}{C}\!\!{}^{\cdot\cdot}$, no two bonds are any farther apart than any other two.

The two possible arrangements here differ in *order* or *handedness*. It is easy to prove that two distinct arrangements, corresponding to two nonsuperimposable models, exist only when no two of the four groups on $\diagdown\!\!\underset{\diagup}{C}\!\!{}^{\cdot\cdot}$ are the same.

STEREOCHEMISTRY 253

Example

Determine the relationship between these molecules:

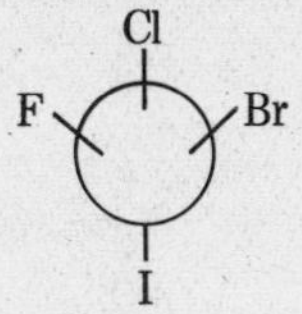

```
    Cl              Cl
 F ⊕ Br        Br ⊕ F
    |               |
    I               I
    a               b
```

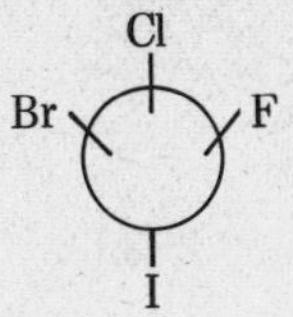

Solution:

Both have molecular formula CFClBrI. They are not superimposable—pushing one onto the other makes Cl, C (hidden in the center of the sphere), and I coincide, but Br matches F and vice versa:

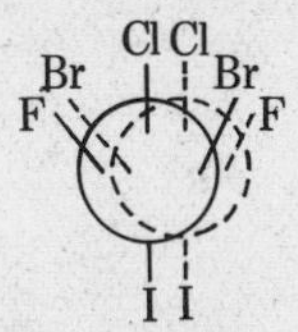

```
      Cl Cl
   Br ╎╎ Br
  F ⊕⊕ F
      ╎╎
      I I
```

Therefore, a and b are isomers. The condensed structure CFClBrI represents both. Therefore, they are stereoisomers.

Now in what sense do these structures differ in order of groups or handedness? When a and b are stood on the I atom and the top face of the tetrahedron is "read," a has the order F ⟶ Cl ⟶ Br *clockwise* and b has it *counterclockwise*. However, when two or more of the four groups on the central C are the same, the two possible "orders" prove to be identical, the molecules superimpose, and no stereoisomerism exists, as in the following examples.

CF_2ClBr:

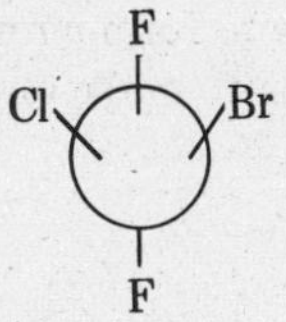

```
    F              F
 Cl ⊕ Br      Br ⊕ Cl
    |    and       |
    F              F
    c              d
```

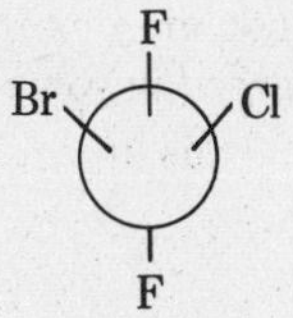

don't seem to superimpose:

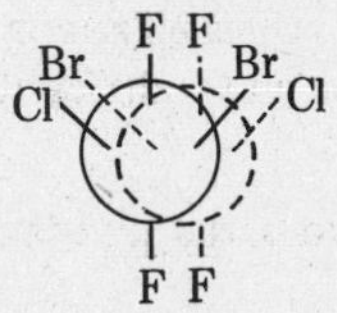

```
       F F
   Br ╎╎ Br
   Cl ⊕⊕ Cl
       ╎╎
       F F
```

But if d is flipped upside down first  (↻), we get d'

which is just a slightly more topside view of c. Thus c and d are identical.

A tetrahedral carbon bearing four different groups is called an *asymmetric carbon atom*; it is sometimes identified by an asterisk, for instance C*FClBrI.

How different must the groups on >C< be to make this an asymmetric carbon? Any difference suffices. For instance, this carbon is asymmetric, although the two groups at the right differ only in having *m* versus *p* substitution on the phenyl groups:

Such cyclic compounds as and (a, c) are also stereoisomers, and they are given the labels cis and trans in analogy with cis and trans alkenes. However, these stereoisomers are actually of the >C*< type and are not really analogous to cis and trans alkenes. We will defer the proof of this statement until frame 10, when we consider compounds containing two >C*< centers.

Identifying asymmetric carbon atoms, >C*<, in cyclic compounds requires special care. If two of the groups on a potential asymmetric carbon are tied together to form a ring, they count as different groups only if the ring is *marked* in some fashion to make one path around the ring differ from the opposite path. Thus in

the ring is marked by the C=O group, so that the path C_4, C_5, C_6, C_1, C_2 is clearly different from the path C_2, C_1, C_6, C_5, C_4; this makes the four groups on C_3 different, and C_3 is an asymmetric carbon. In the unmarked ring the two paths are identical, these two groups on C_3 are identical, and C_3 is not asymmetric.

A similar criterion holds for cyclic alkenes of type  . Such alkenes can exist in distinguishable cis and trans arrangements only if the ring is marked by some feature to make its two halves differ.

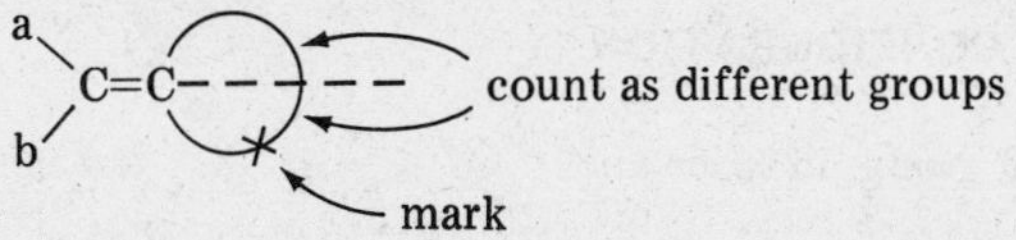

Example

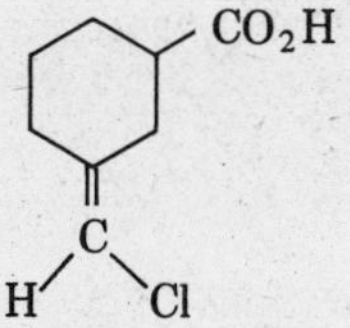 and 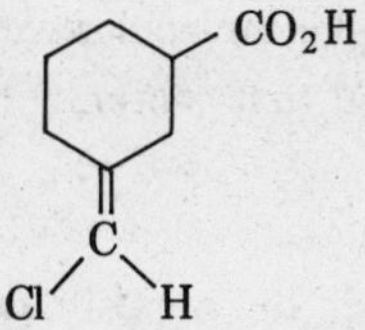 are stereoisomers, but 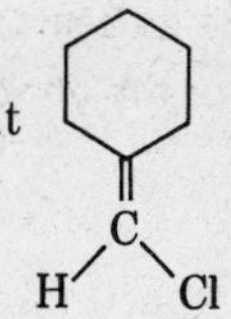 is a single compound.

★ *Exercises*

Label each $>C<$ (with *) and each $>C{=}C<$ group (with a circle) that is capable of existing in two stereoisomeric forms.

1. 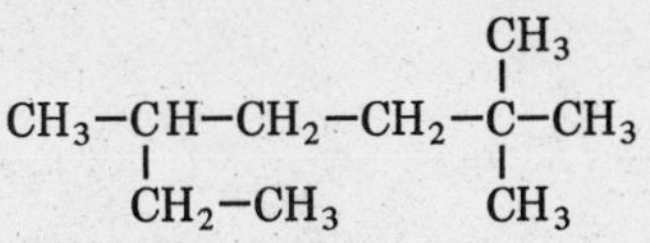

2. $CFCl{=}CH{-}CH{=}CF_2$

3. 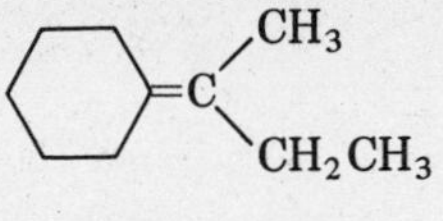

4. 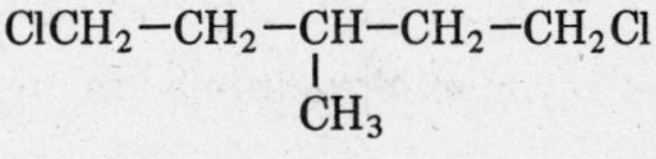

5. $CH_3CF{=}CFCH_3$

6. 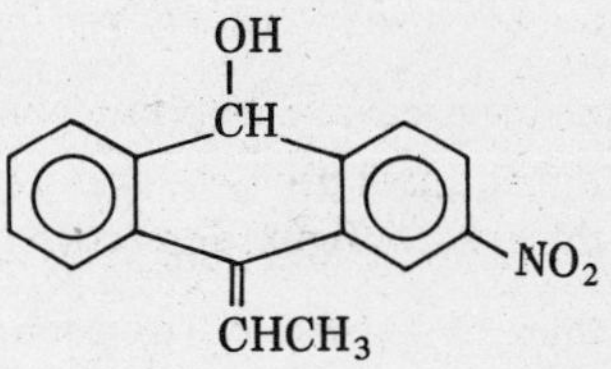

- - - - - - - - - - -

1. 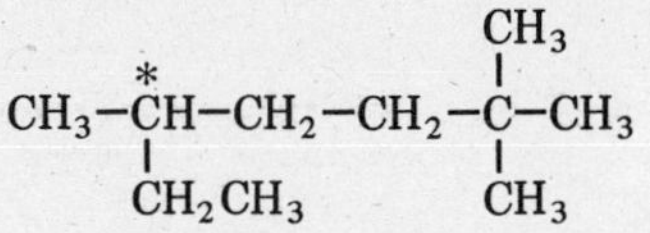

2. 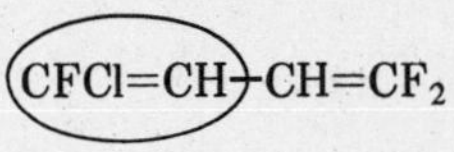

3. 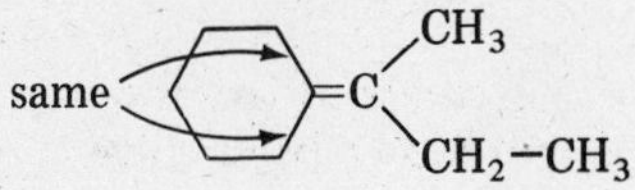

4. 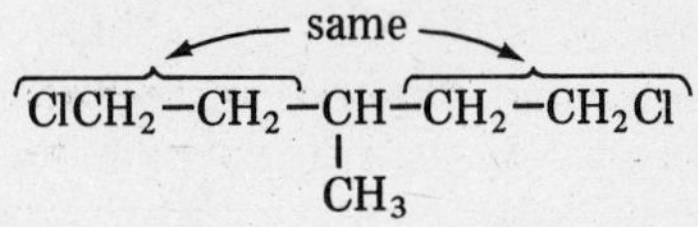

5. CH₃(CF=CF)CH₃

6. 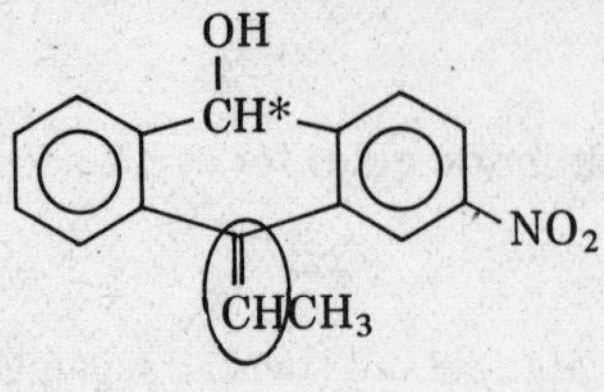 Structure: biphenyl system with CH(OH)* group at top connecting two phenyl rings (one circled, the other bearing NO₂), and a (CHCH₃) group circled at bottom connecting the rings.

CONFIGURATION

3. The alternative arrangements of groups in space about $\mathrm{\overset{}{\underset{}{C}}=\overset{}{\underset{}{C}}}$ or $\mathrm{\overset{*}{\underset{}{C}}}$ are called *configurations*. The number of possible configurations at any $\mathrm{\overset{}{\underset{}{C}}=\overset{}{\underset{}{C}}}$ or $\mathrm{\overset{*}{\underset{}{C}}}$ center is always *two*.

It is not difficult to tell whether two similar molecules with a $\mathrm{\overset{}{\underset{}{C}}=\overset{}{\underset{}{C}}}$ center have the same or the opposite configuration about that center.

Example

Which structures have the same, and which the opposite configurations?

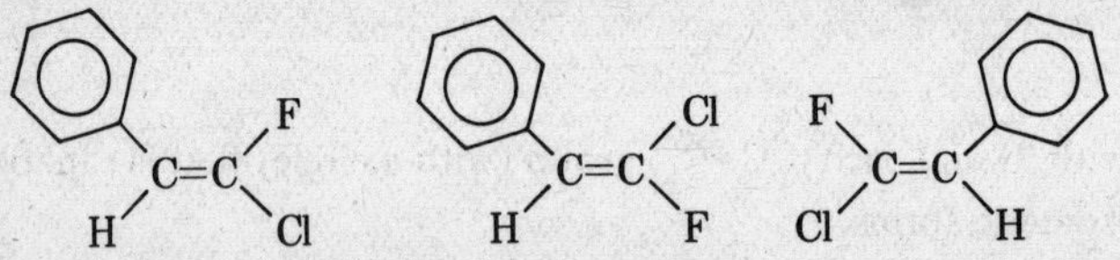

a: Ph and F on one carbon (top), H and Cl on other (structure with C=C)
b: Ph and Cl on top, H and F on bottom
c: F and Ph, Cl and H

Solution:

Structures a and c have the same configuration (⟨Ph⟩— and F cis) and represent the same compound. Structure b has the opposite configuration (⟨Ph⟩— and F trans); it represents a stereoisomer of a or c.

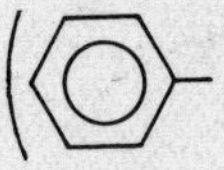

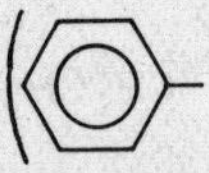

Identifying identical and opposite configurations about $\mathrm{\overset{*}{\underset{}{C}}}$ is not so easy because it is a three-dimensional problem that we usually handle in two dimensions. As we saw in frame 2, the two configurations correspond to the different orders of groups about $\overset{*}{C}$. Why are there two and only two such orders? Taking the $\mathrm{C F C l B r I}$ molecule again, let's assign F to the bottom vertex of a tetrahedron: ◇ with F at bottom. Now in how many ways can the remaining groups occupy their three positions? If they were on a line, the answer would be $3! = 3 \cdot 2 \cdot 1 = 6$:

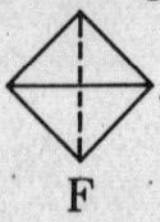

Cl Br I   Cl I Br   Br Cl I   Br I Cl   I Cl Br   I Br Cl

Such permutation problems are the basis of many puzzles and parlor tricks. However, our tetrahedral problem is different—the three groups are arranged in an endless circle rather than in a line: . A little thought shows that this arrangement cuts the number of possibilities to *two* orders of encounter when passing clockwise around the circle:

(a) F Cl Br F Cl Br F Cl Br...

(b) F Br Cl F Br Cl F Br Cl...

Alternatively, we could describe these two configurations as

(a) F ⟶ Cl ⟶ Br clockwise (= F Cl Br | F Cl Br | F Cl Br...).

(b) F ⟶ Cl ⟶ Br counterclockwise (= F Br Cl F | Br Cl F | Br Cl...).

So there are exactly *two* configurations because there are exactly two ways to travel a circle: clockwise and counterclockwise.

This is the point where we begin to really depend upon the four types of notation for tetrahedral carbon that we developed in Unit 4. For some problems one notation is better than another, but in most cases you can choose the one that you find easiest to use. The exercises at the end of this frame will give you some practice; you may also want to look at frame 9 of Unit 4 again.

You won't find it difficult to compare the order of groups (configuration) of two structures that are written in the same notation and oriented the same way on the page. For example,

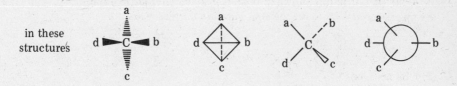

are perfectly set up for comparison. Structure a has F ⟶ Cl ⟶ Br clockwise; b has F ⟶ Cl ⟶ Br, also clockwise. Therefore, they represent the *same* configuration and the same compound. The difficulty comes when either the notation or the orientation (or both) are different in two structures being compared. The problem is very much simplified if you always look directly on the *outside surface* of a face of the tetrahedron. Never read the order of groups by looking through the tetrahedron at one of its inside surfaces—this always gives the opposite configuration.

In each of the standard notations, two faces of the tetrahedron face forward and are easy to read. The faces of the tetrahedron that can be read directly because they are facing you are labeled below:

you see these faces directly

The remaining two faces are turned to the rear, and to read them correctly you must imagine yourself behind the paper. The rear faces are the ones that look like this:

Example

Read the clockwise order of groups on each face of this tetrahedral molecule:

Solution:

Top face, read directly: Br–Cl–F or Cl–F–Br or F–Br–Cl
Bottom face, read directly: F–Cl–I or Cl–I–F or I–F–Cl
Left face, read from behind paper: Br–F–I or F–I–Br or I–Br–F
Right face, read from behind paper: Br–I–Cl or Cl–Br–I or I–Cl–Br

Finally, we need to note that *configuration* and *conformation* are parallel terms:

- *Configuration* describes a *permanent* arrangement of groups in space.
- *Conformation* describes a *momentary* arrangement of groups in space.

The only way one can change the configuration (order of groups) about ▶C◀ is by breaking one of the bonds to carbon. The only way one can change the configuration about C=C is to break a bond to carbon or to rotate internally about the C=C bond. Both of these are energetically costly processes that happen only under unusual circumstances, for example during chemical reactions or at very high temperatures. They do not occur during storage. On the other hand, a given molecule is continually changing its conformation because the energy barrier to rotation about C–C is so small that the thermal energy possessed by the molecules at room temperature is sufficient to overcome it.

★ *Exercises*

1. Translate each structure shown at the left into the alternative notation by supplying the missing groups. *Preserve* the configuration in translation.

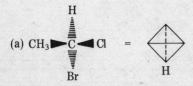

STEREOCHEMISTRY 259

(b) 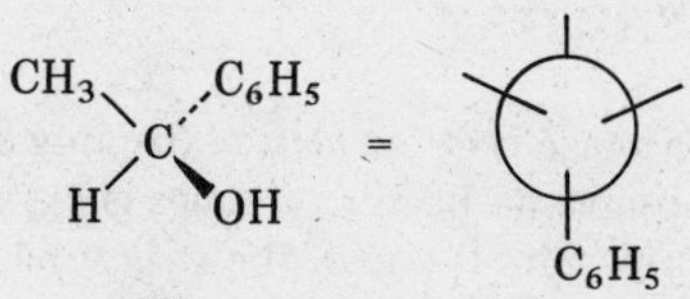
2. Translate each structure, and in the process change the molecule into the opposite configuration.

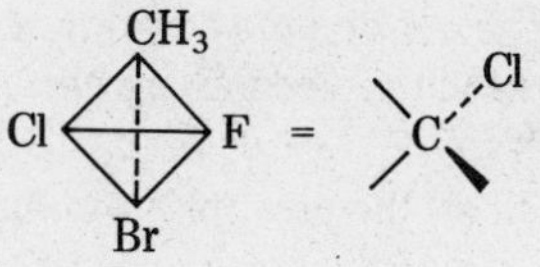

(a)

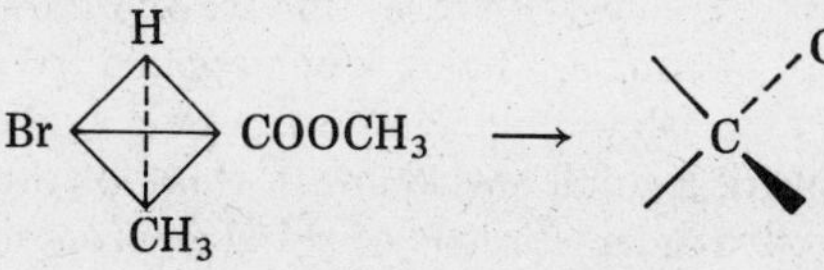

(b)

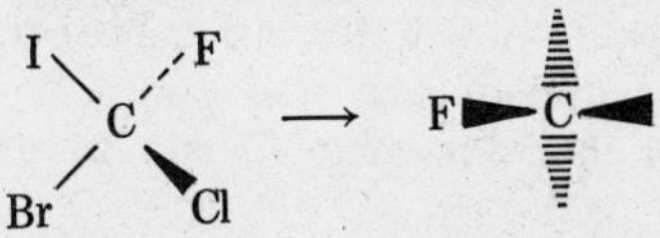

- - - - - - - -

1. (a) 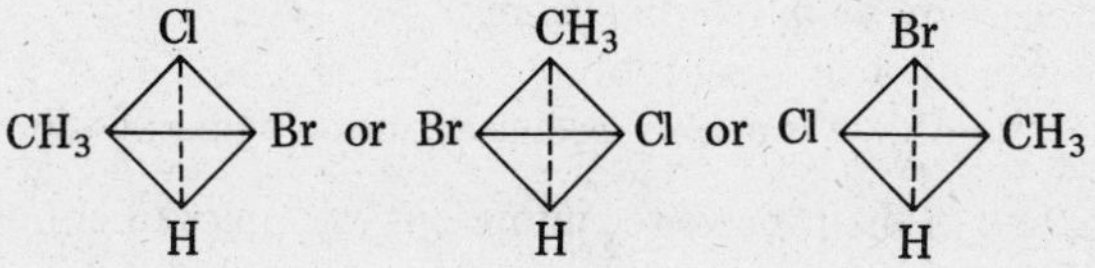 CH_3 ⇔ Br or Br ⇔ Cl or Cl ⇔ CH_3 (with Cl/H, CH_3/H, Br/H respectively)

(b) 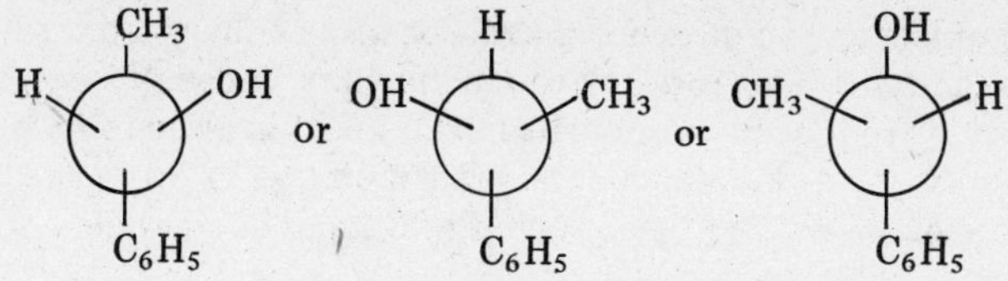 three Newman projection alternatives with CH_3, OH, H, C_6H_5

(c) three structures with F/Cl/Br/CH_3, Br/Cl/CH_3/F, CH_3/Cl/F/Br

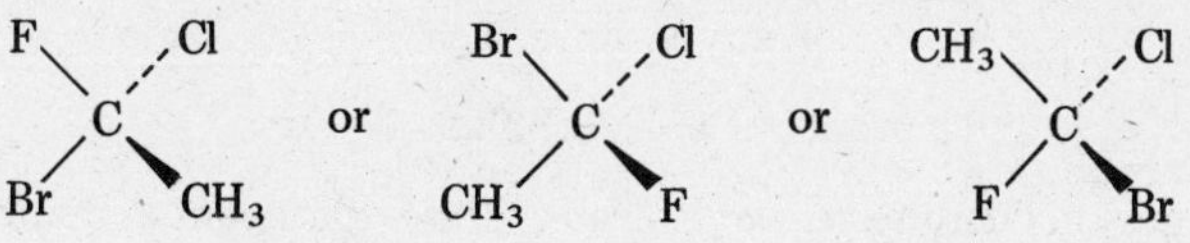

2. (a) 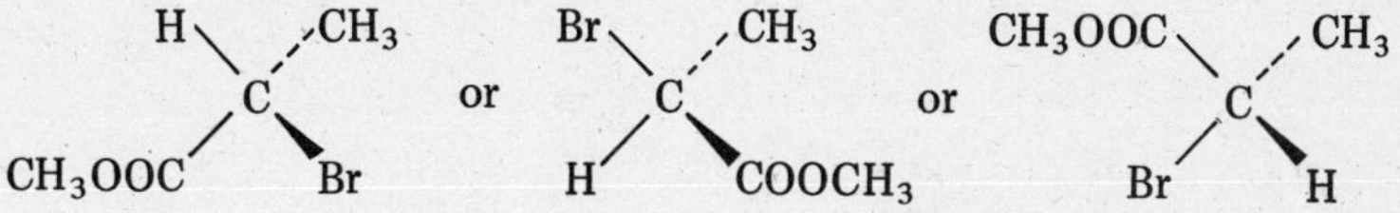 H/CH_3/CH_3OOC/Br or Br/CH_3/H/COOCH$_3$ or CH_3OOC/CH_3/Br/H

(b) 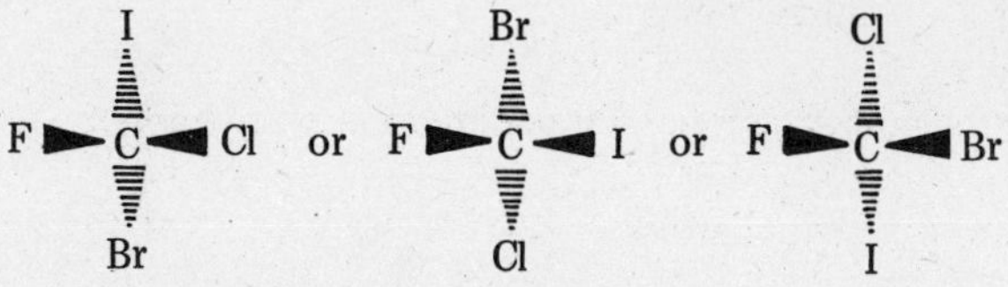 F▶C◀Cl with I up, Br down or F▶C◀I with Br up, Cl down or F▶C◀Br with Cl up, I down

CHIRALITY AND SYMMETRY

4. An object that is superimposable on its mirror image is called *achiral* (lacking chirality) or *symmetric*. An object that is not superimposable on its mirror image is *chiral* or *dissymmetric*. Chirality (or dissymmetry) of the molecules is one of the structural features that determine the chemical and physical properties of organic compounds. Your main purpose in this frame is to learn to recognize which molecules are chiral, which achiral, so that later you can predict their properties.

Chirality is the property of possessing "handedness." In all the examples of chiral objects that we will encounter, whether on a human-sized (macroscopic) scale or on the molecular scale, an element of handedness will be present. If we do not specifically comment on it, you should look for it. Chiral objects must be right- or left-handed; they can possess both right- and left-handed parts in certain cases that we will look into later. The prime example of a chiral object is, of course, a hand. When applied to objects, such as hands, chiral and dissymmetric mean the same thing. However, chiral and achiral apply to environments as well. As an example of a chiral environment, think of the unoccupied space in an empty glove or shoe. It is characteristic of chiral environments that they interact differently with right- and left-handed objects (as you can readily prove by putting your right foot in your left shoe) but identically with achiral objects (such as socks). Later we will see close chemical analogs of such interactions. Chirality is thus the concept better suited to chemical work than dissymmetry (and it is easier to spell).

To determine if an object is chiral, you can use any of three methods:
- See if the object superimposes on its mirror image; or
- Determine if handedness is present or absent; or
- (in some cases) Prove that the object is achiral because it has a plane of symmetry.

Now we practice these in turn, and in the process see numerous examples of chiral and achiral objects.

The mirror-image superimposability test can be carried out with a real object and a real mirror, with a real object and an imagined mirror, or with an imagined (or drawn-on-paper) object and an imaginary mirror. You should now practice with some real objects. Place the object in front of the mirror. Its mirror image appears to be behind the mirror. In your mind's eye, pick up the image, bring it around in front, and see if it superimposes on the original object. By doing so you will find that a needle, a sock, a pair of sunglasses, a fork, the domino ⟦∷|∷⟧, and a model of CH_2ClBr are all achiral. But you will find a woodscrew, a penny, a glove, a shoe, a grand piano, the ace of spades, the domino ⟦•|∵⟧, and a model of $CHClBrI$ all to be chiral.

Now let's switch to the pencil-and-paper equivalent of this experiment and apply it to drawings of the dominos and molecules noted above. Draw a domino before a mirror. Visualize what the image behind the mirror would look like and draw it in:

    domino   mirror   mirror image

When one tries to pick up the mirror image and superimpose it on the domino, this proves to be impossible:

mirror image

domino

Therefore, this domino is chiral.

Repeat the experiment with CH_2ClBr and $CHClBrI$:

CH_2ClBr — [Cl/H–C–H/Br] | [Cl/H–C–H/Br] superimposable and achiral

$CHClBrI$ — [Cl/I–C–H/Br] | [Cl/H–C–I/Br] nonsuperimposable, therefore chiral

Next, we can use the second method of testing these same objects for chirality. The search for handedness obviously turns out positive for the glove and shoe. The woodscrew, like any thread or wormgear, is also either right- or left-handed. The penny has handedness because the writing on it has direction—left to right. The ace of spades and the

domino have handedness in the sense that both have pips on a diagonal that descends from left to right, not right to left.

We have already looked at the handedness of CFClBrI in frame 3; it must have either a clockwise or a counterclockwise order of groups. All the objects above are thus chiral, in agreement with the mirror-image tests. Similarly in agreement are the sock, the needle, the sunglasses, the fork, and , all of which lack any element of handedness and are achiral. To see the absence of handedness in CH_2ClBr, set it up like this:

[Br/H–C–H/Cl]

The order of groups Br—H—H around the top face of the tetrahedron can be achieved by setting out from Br either in a clockwise *or* a counterclockwise direction.

Now turn to the last test for chirality. An object with one or more planes of symmetry is always superimposable on its mirror image and thus must be achiral. A plane of symmetry is a plane that cuts through the object in such a way that for every characteristic feature on one side of the plane there exists an exactly equivalent feature at an equal distance away from the plane on the exact opposite side. The domino

has two planes of symmetry:

Both are perpendicular to the plane of the domino; one cuts it parallel to the long side, the other parallel to the short side. Now test each characteristic point, such as the corners of the domino and each spot, to see if an equivalent corner or spot is found precisely opposite on the other side of the plane. The result is shown below, where the characteristic pairs of points are labeled.

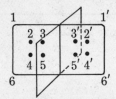 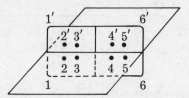

In both cases, each point has its opposite number, and consequently the two planes shown are planes of symmetry. Therefore, this domino is achiral. Do the same for the domino:

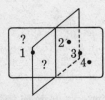 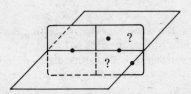

With the plane arranged as in the left drawing, no equivalent or mirror-image point exists opposite spots 2 or 4; the same holds in the right drawing. All other planes through this domino give a similar result. Since we can find no plane of symmetry, the third method does not apply to this domino.

By searching the other objects, you will find planes of symmetry in the needle, sock, sunglasses, and fork, which are consequently achiral, in agreement with what we found using the other methods.

Since any linear or planar object has one or more planes of symmetry, *linear and all planar molecules can immediately be identified as achiral.* Examples are H—Br and H—C≡C—H (linear),

$$\underset{H}{\overset{H}{\diagdown}}C=C\underset{Cl}{\overset{Cl}{\diagup}}, \quad \underset{Cl}{\overset{H}{\diagdown}}C=C\underset{Cl}{\overset{H}{\diagup}}, \quad \underset{Cl}{\overset{H}{\diagdown}}C=C\underset{H}{\overset{Cl}{\diagup}},$$

and ⌬ (planar). Tetrahedral molecules must be investigated individually. CH_2ClBr has a plane of symmetry (the plane of the paper in this representation:

$$\underset{Br}{\overset{Cl}{\diagdown}}C\underset{H}{\overset{H}{\diagup}}$$

) and can be declared achiral on that basis. CHClBrI lacks a plane of symmetry; the third rule doesn't help us, but we have already classified this molecule chiral using the other rules.

To sum up, the three methods agree in their classifications of all the test objects. Since we could have worked with any other $\overset{*}{>}\!\!C\!\!<$ molecule in place of $\overset{*}{C}HClBrI$, our results prove that the *presence of an asymmetric carbon atom suffices to make a molecule chiral.*

(Actually, this rule needs a hedge attached to it: molecules with *paired* $\overset{*}{>}\!\!C\!\!<$ can have a plane of symmetry and be achiral; this is considered in frame 11.)

★ *Exercises*

1. We just pronounced $\underset{H}{\overset{H}{\diagdown}}C=C\underset{Cl}{\overset{Cl}{\diagup}}$, $\underset{Cl}{\overset{H}{\diagdown}}C=C\underset{Cl}{\overset{H}{\diagup}}$, and $\underset{Cl}{\overset{H}{\diagdown}}C=C\underset{H}{\overset{Cl}{\diagup}}$ achiral because they have a plane of symmetry. Verify their achirality using the mirror-image test.

2. Some of the following molecules are achiral, some chiral. Identify the molecules that are necessarily achiral *because they are planar*.

   (a) CH_3-CH_2-Br (b) $H-\overset{\overset{O}{\|}}{C}-H$

   (c) C₆H₅—Cl   (d) $CH_3-CHBr-CH=CH_2$

   (e) $CH_2=CH-CH=CH_2$ (f) $Br-C\equiv C-H$

3. Identify the molecules in each set that have one or more planes of symmetry and are consequently achiral.

   (a) CH_4 (b) CH_3-CH_3

   (c) CH_3-CH_2Cl (d) $CH_3-CHClBr$

   (e) $CH_3-C\underset{H}{\overset{\diagup\!\!\!\!O}{\diagdown}}$ (f) $CHD_2-C\underset{H}{\overset{\diagup\!\!\!\!O}{\diagdown}}$

   (g) cyclohexyl with $-CH_3$ and $-Cl$

- - - - - - - -

1. $\underset{Cl}{\overset{H}{\diagdown}}C=C\underset{H}{\overset{Cl}{\diagup}}$ $\underset{H}{\overset{Cl}{\diagdown}}C=C\underset{Cl}{\overset{H}{\diagup}}$ is the tricky case. The mirror image cannot be made to superimpose on the original by shuffling it around in the plane of the paper. However, flipping it on its back, $\underset{H}{\overset{Cl}{\diagdown}}C=C\underset{Cl}{\overset{H}{\diagup}}$ $\underset{Cl}{\overset{H}{\diagdown}}C=C\underset{H}{\overset{Cl}{\diagup}}$, makes it superimpose on the original.

2. All except (a) and (d), which have a saturated, tetrahedral carbon, are planar and necessarily achiral. [(a) is achiral, but not because it is planar.]

3. All except (d) have at least one plane of symmetry and are achiral. It is sufficient that any single conformation have a plane of symmetry, as in $D\cdots\overset{H}{\underset{D}{C}}-C\underset{H}{\overset{\diagup\!\!\!\!O}{\diagdown}}$, in which

the plane of the paper is the plane of symmetry, or in [Newman projection with H, H, H, H and Cl]. In (g) the

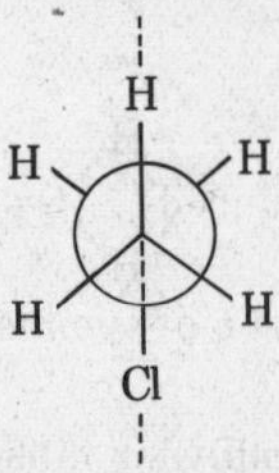

plane of symmetry cuts the ring like this:

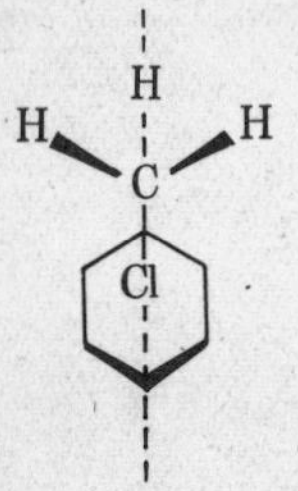

[structure: cyclohexane with CHCl group, H's shown]

5. The property that sets all chiral molecules apart experimentally is their ability to alter the polarization of light passing through them. If the light is plane-polarized, passage through a sample of some compound whose molecules are chiral causes a rotation in the light's plane of polarization. This rotation has a handedness (clockwise or counterclockwise) and a magnitude. The *handedness* of the rotation is determined by the *configuration* of the chiral center; if one configuration rotates clockwise, the opposite configuration will always rotate counterclockwise. The magnitude of rotation is determined by the identity of the groups attached to the chiral center and the size and concentration of the sample through which the light passes. You should refer to your textbook for a description of the experimental method for measuring the rotation and the calculation by which the effects of sample size and concentration are eliminated. The resulting quantity, which depends only on the structure of the chiral compound, is given the name *specific rotation* and the symbol $[\alpha]$. Clockwise direction is indicated by giving the value of $[\alpha]$ a plus sign, counterclockwise rotation by a minus sign. Substances that have nonzero values of $[\alpha]$ are called *optically active*.

★ *Exercises*

1. If $[\alpha]$ measured for an appropriate solution of pure compound X is found to be $-45.5°$, which of the following structures is possible for X?

   (a) 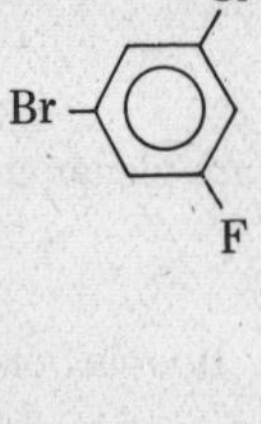 [benzene ring with Br, Cl, F substituents]

   (b) 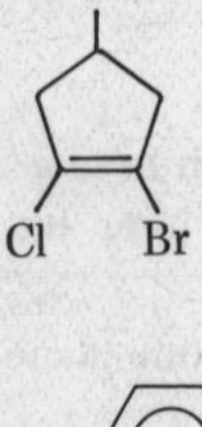 [cyclopentene with F, Cl, Br substituents]

   (c) 4-chloroheptane

   (d)  [Cl–C₆H₄–C(CH=CH₂)(C≡CH)–C₆H₄–Br]

   (e) CF_3-CF_2Cl

2: If X in question 1 actually turns out to be 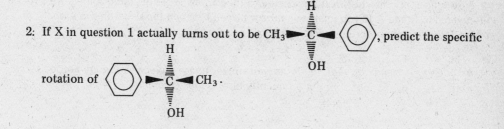, predict the specific rotation of

---

1. Only (b) (the carbon bearing F is asymmetric) and (d) (central carbon asymmetric) are chiral, so only these could have [α] different from zero.

2. +45.5° (the second structure has all groups identical to those of X but arranged in the opposite configuration).

ENANTIOMERS AND DIASTEREOMERS

6. A chiral object and its nonsuperimposable mirror image constitute two versions of the object—the right-handed and the left-handed versions. In many cases both versions actually exist; your two hands are an example. In the case of chiral molecules, too, we often can isolate and study both the right- and left-handed versions. Such a pair of mirror-image molecules is called a pair of *enantiomers*. Now a chiral molecule and its enantiomer are nonsuperimposable, but they have the same molecular formula and condensed structural formula. That means they fit the definition of stereoisomers (frame 1). So there is a relationship between chirality and stereoisomerism. In this and the next few frames we will work out that relationship as completely as possible.

Stereoisomers that are mirror images of one another are enantiomers. Stereoisomers that are not enantiomers are called *diastereomers*. Including this subclassification of stereoisomers in our scheme from frame 1, we get

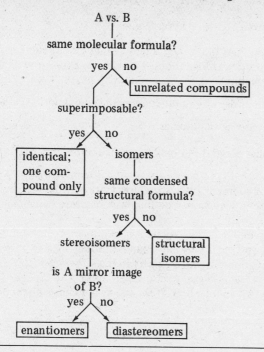

Now let's apply the classification scheme to the various types of stereoisomers we discovered in frame 2. If you imagine a mirror placed between the structures in each of the following nonsuperimposable pairs, you will find that

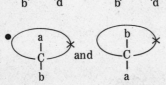

- $\begin{array}{c}a\\ \diagdown\\ b\end{array}C=C\begin{array}{c}a\\ \diagup\\ b\end{array}$ and $\begin{array}{c}a\\ \diagdown\\ b\end{array}C=C\begin{array}{c}b\\ \diagup\\ a\end{array}$ are not mirror-image molecules and are consequently always *diastereomers*.
   cis   trans

- are mirror images of one another and are consequently *enantiomers*.

- are mirror images and therefore *enantiomers*. To see this, flip the right structure like a pancake to get and compare it with the left structure.

The distinction between enantiomers and diastereomers is important. It allows one to make explicit predictions about the similarities and differences in the chemical and physical properties of a pair of enantiomeric or diastereomeric molecules. We consider these properties in the next frame. Here we concentrate on learning to recognize closely similar structures as diastereomers, enantiomers, or identical compounds.

Distinguishing enantiomers and diastereomers is simplified by noting that in compounds containing a single $\overset{*}{C}$ or $C=C$ stereochemical center:

- Enantiomerism results only from $\overset{*}{C}$, never from $C=C$.
- Diastereomerism results only from $C=C$, never from $\overset{*}{C}$.

Examination of the pairs of enantiomers drawn above tells us that *a compound and its enantiomer differ only in having opposite configurations*. This gives us a tool for recognizing or generating enantiomeric structures:

1. Any operation that reverses the order of groups (changes the configuration) at a $\overset{*}{C}$ center turns a structure into its enantiomer.

2. If a pair of structures containing one $\overset{*}{C}$ differ only in configuration, they are enantiomers.

The following three techniques of comparing two structures work equally well, but one may be easier for you than another. Which one you find best may vary depending on the type of notation used.

Method 1: You mentally pick up one structure, rotate it into proper orientation, set it down on the second structure, and see if they superimpose. Rotate in such a way as to get C* and two of the attached groups oriented exactly in the same way in the two structures to be compared. If the last two groups also superimpose when the test molecules are brought together, the two are identical. If the last two do not superimpose, the molecules are enantiomers.

STEREOCHEMISTRY

Examples

Are these structures identical or enantiomeric?

1. 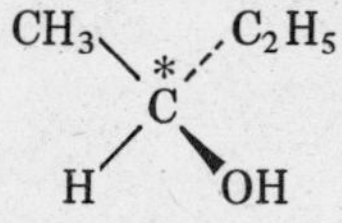 $\underset{H}{\overset{CH_3}{\diagdown}}\underset{OH}{\overset{*}{C}\diagup}\overset{C_2H_5}{}$ and $\underset{HO}{\overset{H_5C_2}{\diagdown}}\underset{CH_3}{\overset{*}{C}\diagup}\overset{H}{}$

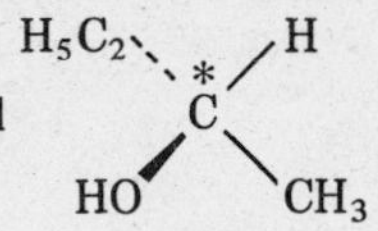

    a   b

2. 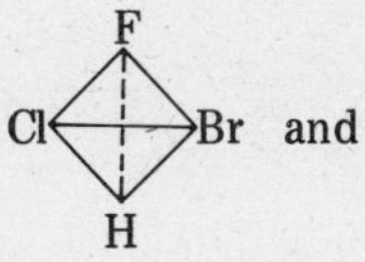 Cl–[F/H]–Br and Br–[Cl/H]–F

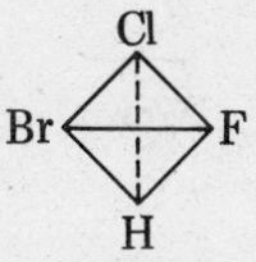

    a   b

3. 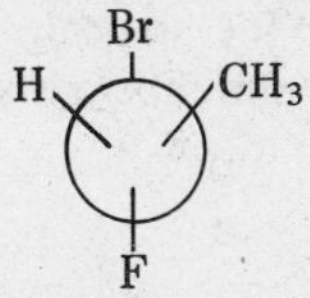 H–(Br/CH_3/F) and F–(Br/CH_3/H)

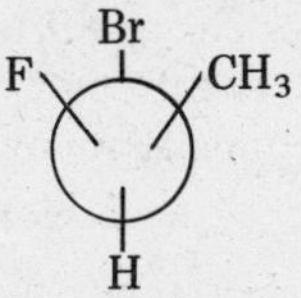

    a   b

Solutions:

1. Rotate b 180° clockwise in the plane of the paper to get its in-plane bonds on the left as they are in a:

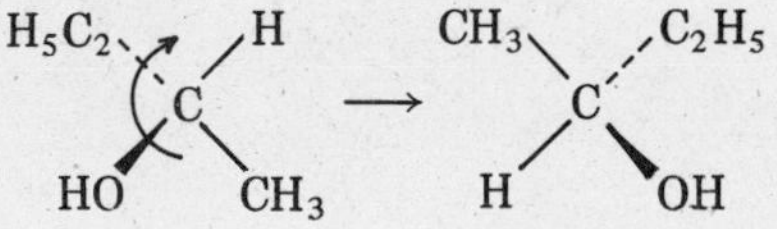

   $\underset{HO}{\overset{H_5C_2}{\diagdown}}\underset{CH_3}{\overset{}{C}\diagup}\overset{H}{}$ → $\underset{H}{\overset{CH_3}{\diagdown}}\underset{OH}{\overset{}{C}\diagup}\overset{C_2H_5}{}$

   b   new b

   Comparison of new b with a shows that they are identical (superimposable).

2. C* (center of tetrahedra) and H are oriented similarly; we need one more group oriented the same in the two structures to make comparison easy. Try to get F on the right in both. To do this, rotate a about the $\overset{C}{\underset{H}{|}}$ axis, 120° in this direction: ↻, that is,

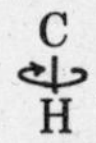

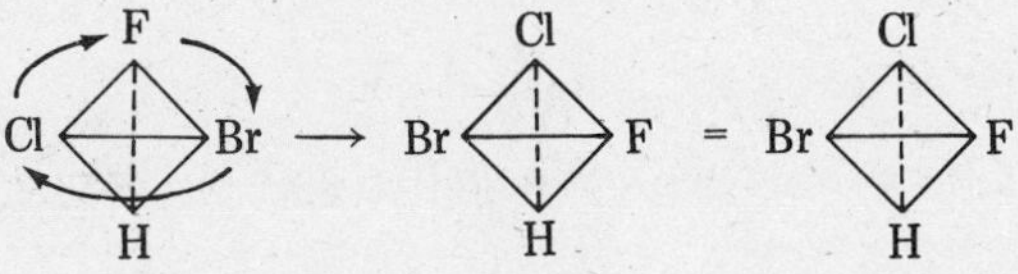

   a → new a = b

   We now have C, H, and F lined up in new a, and when new a is moved over to b to put F on F, C on C, and H on H, we see that Cl goes on Cl, and Br on Br. b and new a are identical, therefore b and a are identical, since the order of groups in a was not altered in making new a.

268 HOW TO SUCCEED IN ORGANIC CHEMISTRY

3. C (center of spheres), CH$_3$, and Br are already aligned. Try to superimpose:

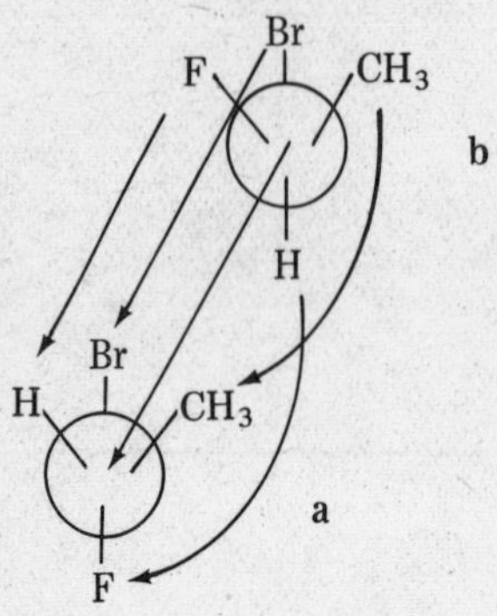

CH$_3$, Br, C superimpose, but F goes on H and H on F. Therefore, a and b are enantiomers.

Method 2: Interchanging any two groups on an asymmetric carbon atom turns a molecule into its enantiomer (true only for compounds containing one C*). Doing it a second time with the same or a different pair of groups changes it back to the original. More generally, an odd number of exchanges gives the enantiomer, an even number gives the original. So to see if a and b are identical or enantiomers, count the number of interchanges necessary to convert a into b.

Examples

Are these structures identical or enantiomeric?

1.

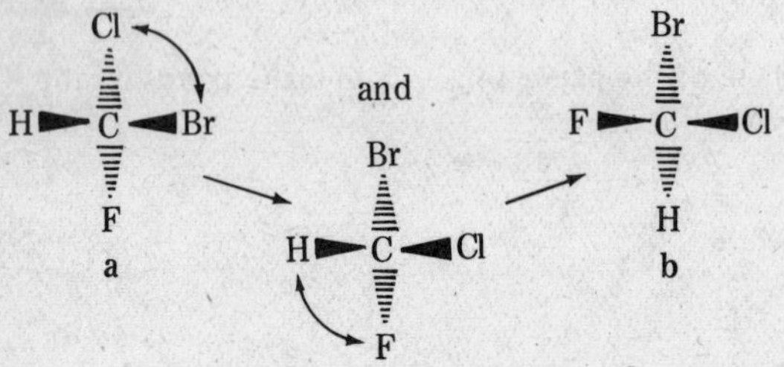

2.

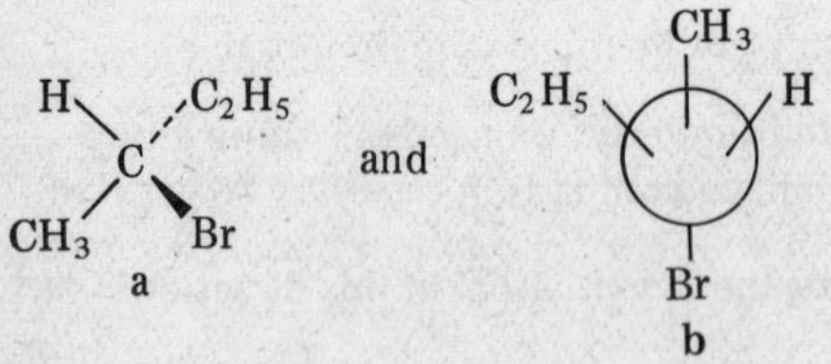

Solutions:

1. Two interchanges make a into b; therefore, a and b are identical.

2. First, translate one to get both in the same notation. To do this, grab the Br of a and pull down:

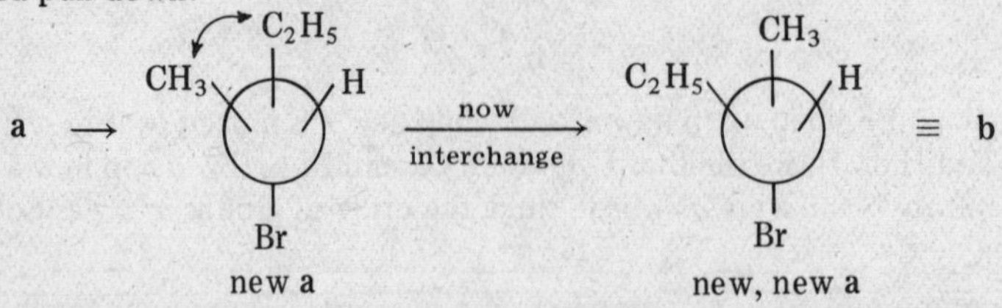

After translation, one interchange is required to make a into b; therefore, a and b are enantiomers.

Method 3: Pick one group attached to the C* as a reference. Look down on the face of the tetrahedron that is opposite the reference group and read the clockwise order of groups in structure a. Do the same in structure b, using the same reference group. If the orders are the same, a and b are identical; if the orders are opposite, a and b are enantiomers.

Examples

Are these structures identical or enantiomeric?

1. 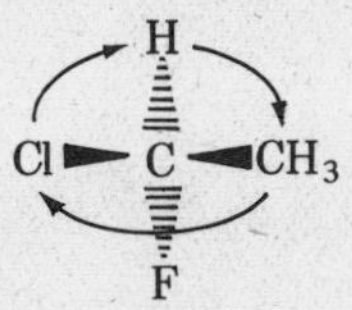 Cl—C—CH_3 (with H up, F down)   and CH_3—C—H (with Cl up, F down)
       a   b

2. 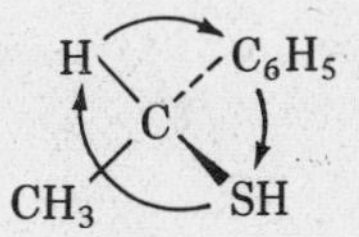 (H, C_6H_5, CH_3, SH around C)   and   (SH, CH_3, H, C_6H_5 around C)
       a   b

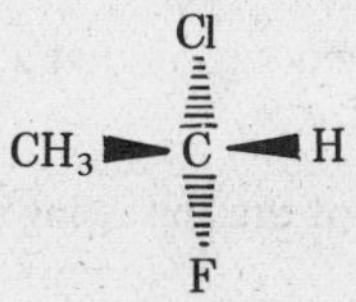

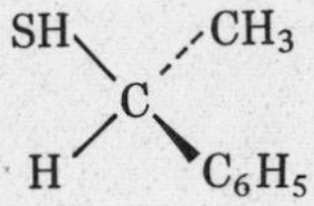

Solutions:

1. In this notation, you can see and read directly only the faces opposite the top and bottom groups—for example, in a, the faces opposite H (bottom face) and opposite F (top face). By chance the face opposite F is visible in both a and b, so use that face:

   clockwise order of groups in a = Cl, H, CH_3
   clockwise order of groups in b = Cl, H, CH_3

   Identical orders, therefore identical molecules.

2. Read the faces opposite CH_3 (clockwise):

   (a) H, C_6H_5, SH

   (b) H, SH, C_6H_5

   Opposite orders, therefore enantiomers. Note that you get the same result by reading any other face, but some faces are hard to read because they require you to imagine yourself behind the paper. This is good practice. Try the face opposite SH.

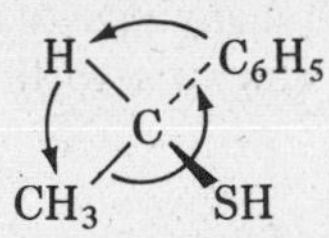

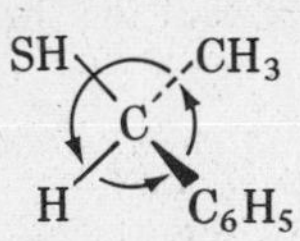

   (H, C_6H_5, CH_3, SH around C)   (SH, CH_3, H, C_6H_5 around C)
          a   b

   Clockwise orders: a: H, CH_3, C_6H_5 (read from behind the paper); b: H, C_6H_5, CH_3 (read from the bottom right-hand corner of the paper).

★ *Exercises*

Given the four molecules

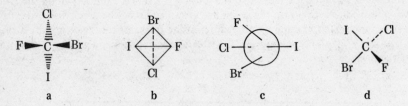

1. Identify the relationship (identical or enantiomers) of each to the other three.
2. Draw the enantiomer of a, using the same notation that a is written in.
3. State the relationship of [structure] to each of the following structures.

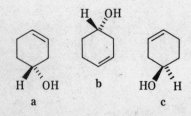

----

1. [diagram showing a-b-c-d with e and i labels]

   e = enantiomers
   i = identical

2. Easiest way: interchange any two groups, which gives

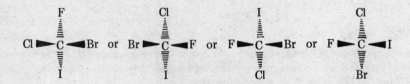

You get the same result by reversing the order of the three groups on any face of the tetrahedron. For instance, the top face of a has F ⟶ Cl ⟶ Br clockwise, ∴ the enantiomer has them counterclockwise: F►C◄Cl (with Br up, I down), which is the same as the first structure drawn above, just rotated about the C–I bond.

3. a = enantiomer; ring position unchanged, OH and H interchanged; b = identical; order of groups unchanged, ring flipped on its back; c = enantiomer; ring flipped on its back *and* OH and H interchanged.

PROPERTIES OF ENANTIOMERS AND DIASTEREOMERS

7. Diastereomers have different properties, like any pair of structural isomers. As a result, mixtures of diastereomers can be separated by any of the standard methods—distillation, chromatography, and so on.

Enantiomers have the unique characteristic of possessing *identical* physical properties, with one exception. We saw that the magnitude of $[\alpha]$ is determined by the identity of the four groups on the chiral center, so this property, too, must be the same for both enantiomers. Only the sign of $[\alpha]$, which is determined by the configuration of the chiral center, differs for a pair of enantiomers; $[\alpha]$ is positive for one enantiomer, negative for the other. Enantiomers are necessarily chiral. The opposite handedness that is built into their structures by means of opposite orders of groups expresses itself physically in opposite handedness of interaction with polarized light, rotating it clockwise or counterclockwise.

All our separation techniques are based on differences in some property displayed by the substances separated; for example, a difference in boiling point permits separation by distillation. It seems at first that separation of enantiomers must be impossible if their properties are identical (the sign of $[\alpha]$ is not a property on which you can base a separation). However, there is a loophole. Although enantiomers interact with *achiral* objects or environments (e.g., solvents or surfaces) in *identical* fashion and with an identical energy change, enantiomers interact *differently* with a *chiral* object or environment. This means that if a pair of enantiomers is chromatographed on a finely divided solid whose surface is chiral, one enantiomer will be more strongly attracted to the surface than the other and the two will travel down a chromatographic column of this chiral solid at different rates, leading to their separation. A human-scale analogy helps make the difference in interactions clear. Right- and left-handed batters (both chiral) can use the same bat (they "interact equally" with it) because the bat is *achiral*. However, a right- and a left-handed fielder cannot use the same glove—they interact differently with it because the glove is *chiral*—it is either right- or left-handed. The unequal interactions of enantiomers with a chiral surface are analogous.

Since the process of separating enantiomers is not ordinary separation, it gets a special name: *resolution*. Another trick is available for resolving a pair of enantiomers. It involves conversion of the pair of enantiomers to a pair of diastereomers, separation of these, and reconversion of the separated diastereomers to (separate) enantiomers. Since this scheme uses compounds containing two chiral centers, we are not ready to take it up in detail. Study your textbook's discussion of resolution after you have finished frame 11.

Enantiomers have identical chemical properties in all interactions with achiral molecules or ions. But they have different energies of interaction with any chiral species, which means their reactions with a chiral molecule or ion have different rate constants and different equilibrium constants. When these differences are large enough, they result in preferential reaction with one enantiomer over the other. This behavior is precisely the same as the difference in interaction of enantiomers with a chiral environment or surface—the approach of the chiral reagent molecule makes the local environment chiral, and the two enantiomers respond differently.

If the specific rotation, $[\alpha]$, of a chiral compound is $+x$, the specific rotation of its enantiomer is precisely $-x$. This means that a mixture consisting of 50 percent of a compound and 50 percent of its enantiomer must always have a specific rotation of exactly zero: $[\alpha] = 0.5(+x) + 0.5(-x) = 0$. The clockwise rotation due to each molecule of the compound is canceled out by the counterclockwise rotation produced by a molecule of its enantiomer. The mixture thus does not rotate polarized light. This is the third peculiar property of a 50-50 mixture of a pair of enantiomers—by measurement of $[\alpha]$ we cannot tell the difference between such a mixture of chiral molecules and a sample of a pure achiral compound. Such mixtures get a special name: *racemic mixtures*. The first two peculiarities of racemic mixtures we have already noted: they come through ordinary separation procedures unseparated and their component enantiomers are transformed to precisely equal extents and in analogous fashion by any achiral reagent.

Exercises

Compound A has $[\alpha] = +100°$. What will $[\alpha]$ be for:

1. A racemic mixture of A and its enantiomer?
2. A mixture of 60 percent A and 40 percent of its enantiomer?
3. A mixture of 60 percent A and 40 percent of racemic mixture?

— — — — — — — —

1. $[\alpha] = 0.5(+100) + 0.5(-100) = 0$
2. $[\alpha] = 0.6(+100) + 0.4(-100) = 60 - 40 = +20$
3. $[\alpha] = 0.6(+100) + 0.4(0) = +60$

SPECIFICATION OF CONFIGURATION

8. In frame 7 we used the clumsy term "mixture of compound A and its enantiomer." Since a compound and its enantiomer are identical except for the subtle feature of configuration, they should have very similar names. The natural extension of our nomenclature system would be to use the systematic name for each enantiomer and add some prefix to designate either the right- or left-handed version of the compound.

This system has been adopted, and the prefixes used to denote configuration at a $\overset{*}{C}$ center are R (Latin *rectus* = right) and S (*sinister* = left). Now suppose that you are a chemist writing to a colleague about the configuration of a certain compound. If you are planning to use names instead of structures, the system of naming must include a scheme whereby both you and the reader will visualize the same enantiomer when the prefix is F. This act of figuring out whether a given structure has the R or the S configuration is called *specifying* the configuration.

Separate notations are used for configurations at the $C=C$ and C centers. Both are based on the idea of order of groups (frame 2) and require a system by which any four groups can be assigned a priority order. Priority is based on highest atomic number. One first considers the "first atoms"—those directly bonded to $C=C$ or to C. These are arranged in decreasing order of atomic number.

Example

What is the priority order of the following groups? Cl, F, Br, and I.

Solution:

It is the order of decreasing atomic number, namely $I > Br > Cl > F$.

If ties are encountered, you look at the second atoms out from $\ce{>C=C<}$ or $\ce{>C<}$ to break the tie.

Example

What is the priority order of CH_3, F, H, and C_2H_5?

Solution:

The priority order based on first atoms is:

$$F > \left(-\overset{H}{\underset{H}{C}}-H \quad \text{and} \quad -\overset{H}{\underset{H}{C}}-\overset{H}{\underset{H}{C}}-H \right) > H$$

It contains a tie for second place between CH_3 and C_2H_5, both of whose first atoms are C, designated by an arrow. To break the tie, look at the second atoms, designated here with an arrow:

$$-\overset{H\leftarrow}{\underset{H\leftarrow}{C}}-H\leftarrow \quad \text{vs.} \quad -\overset{\rightarrow H}{\underset{\rightarrow H}{C}}-\overset{H}{\underset{H}{C}}-H \quad \text{or} \quad H, H, H \quad \text{vs.} \quad H, H, C$$

or

$$H, H, H \quad \text{vs.} \quad H, H, C$$
$$\uparrow \quad \uparrow \quad \underline{}$$
$$\text{tie} \quad \text{tie} \quad C$$

C wins over H; therefore, $-CH_2-CH_3$ has a higher priority than $-CH_3$. Therefore, the whole priority order is $F > C_2H_5 > CH_3 > H$.

We need one more elaboration of the rule. If a "first atom" is doubly bonded to one of its "second atoms," we replace the double bond with two single bonds to the same second atom.

Example

What is the priority order of $-CH_3$, $-C_2H_5$, $-H$, and $-CH=CH_2$?

Solution:

The priority order based on first atoms contains a three-way tie:

$$(-CH_3 \text{ and } -C_2H_5 \text{ and } -CH=CH_2) > H$$

We replace $-CH=CH_2$ by $-\overset{H}{\underset{C}{C}}-C$. So the choice is now between

$$-\overset{H\leftarrow}{\underset{H\leftarrow}{C}}-H\leftarrow, \quad -\overset{H\leftarrow}{\underset{CH_3}{C}}-H\leftarrow, \quad \text{and} \quad -\overset{C\leftarrow}{\underset{H\leftarrow}{C}}-C$$

to be decided on the basis of atomic-number priority of the second atoms (arrowed). C, C, and H wins over C, H, and H, which wins over H, H, and H. Therefore, the total order is

$$-\overset{C}{\underset{H}{C}}-C > -\overset{C}{\underset{H}{C}}-H > -\overset{H}{\underset{H}{C}}-H > -H \quad \text{or} \quad -CH=CH_2 > -CH_2-CH_3 > -CH_3 > -H$$

What do we do with the priority order once it is established? We use it to rank the groups attached to $\diagdown C=C \diagup$ and $\diagdown \overset{*}{C} \diagup$ centers.

In the $\underset{b}{\overset{a}{\diagdown}} C=C \underset{d}{\overset{c}{\diagup}}$ case, we rank the two groups at the left, then the two at the right end. If the top-priority group at the left is on the same side of the double bond as the top-priority group at right, the molecule is called Z. If the top-priority groups, left and right, are on opposite sides, the molecule is called E.

Examples

Specify the configuration of each compound.

1. $\underset{H}{\overset{Cl}{\diagdown}} C=C \underset{H}{\overset{Cl}{\diagup}}$

2. $\underset{H}{\overset{CH_3}{\diagdown}} C=C \underset{C_2H_5}{\overset{H}{\diagup}}$

Solutions:

1. Priority order of left-side groups: Cl > H

   Priority order of right-side groups: Cl > H

   Top-priority group left = Cl

   Top-priority group right = Cl

   Top-priority groups are on the same side: $\underset{H}{\overset{\boxed{Cl}}{\diagdown}} C=C \underset{H}{\overset{\boxed{Cl}}{\diagup}}$

   Therefore, this is a Z molecule; this information is included in the name: (Z)-1,2-dichloroethene.

2. Priority order of left-side groups: CH₃ > H

   Priority order of right-side groups: C₂H₅ > H

   Top-priority group left = CH₃

   Top-priority group right = C₂H₅

   Top-priority groups are on opposite sides: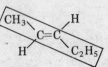

   Therefore, this is an E molecule: (E)-2-pentene.

   To specify the configuration at a $\underset{b}{\overset{a}{\diagdown}} \overset{*}{C} \underset{d}{\overset{c}{\diagup}}$ center, we rank the four groups attached to C*, then we read the order of groups on the face opposite the lowest-priority group. If the 1, 2, 3 (top, second, third priority) groups are arranged clockwise, the configuration is specified as R; if counterclockwise, it is S.

Examples

Specify the configuration of each compound.

1.
$$\begin{array}{c} Cl \quad\quad H \\ \diagdown \;\; \nearrow \\ C \\ \diagup \;\; \searrow \\ F \quad\quad I \end{array}$$

2. H—[◇ CH₃ / CH₂–CH₃]—⟨⚬⟩

Solutions:

1. The priority order is: I > Cl > F > H

   Lowest priority group = H.

   Can we read the face opposite the H group in the drawing as is? Yes, that face faces front (H to the rear):

   Recite the priority order for this face: I > Cl > F.

   Is this a clockwise or a counterclockwise rotation?

   It is counterclockwise. So this is the S configuration—the compound is (S)-chlorofluoroiodomethane.

2. The priority order about C* is:

   $(-C_6H_5 \text{ and } -CH_3 \text{ and } -CH_2-CH_3) > H$

   $$\left(-C\overset{\|}{=}C\diagup^C \text{ and } -\overset{H}{\underset{H}{C}}-H \text{ and } -\overset{H}{\underset{H}{C}}\diagup^H_C\right) > H$$

   $$\downarrow$$

   $$\left(-C\overset{}{=}C\diagup^C_C \text{ and } -\overset{H}{\underset{H}{C}}-H \text{ and } -\overset{H}{\underset{H}{C}}\diagup^H_C\right) > H$$

Therefore,

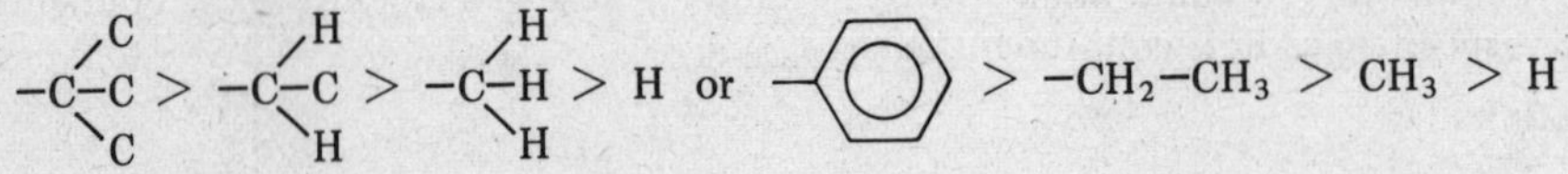

$$-\overset{C}{\underset{C}{C}}-C > -\overset{H}{\underset{H}{C}}-C > -\overset{H}{\underset{H}{C}}-H > H \quad \text{or} \quad -\bigcirc > -CH_2-CH_3 > CH_3 > H$$

Low-priority group = H.

Is the face of the tetrahedron opposite H visible? No. Therefore, transform the drawing to make it visible.

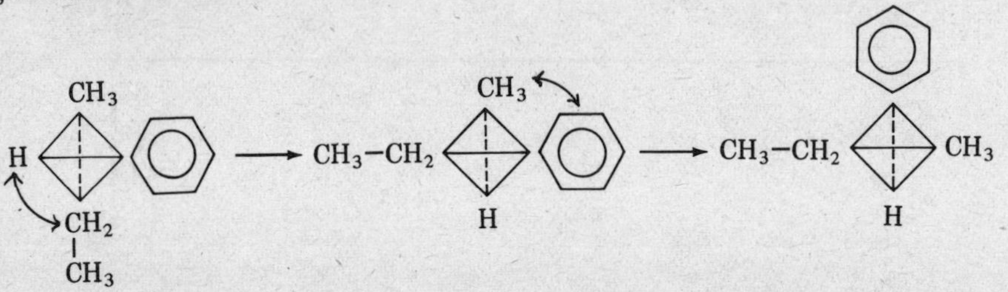

[Diagram: three tetrahedron structures showing interchanges of groups, starting with H, CH₃, phenyl, CH₂CH₃ and rearranging to CH₃CH₂, phenyl(top), CH₃, H(bottom)]

Two interchanges of groups leaves the configuration unchanged (frame 6).

Now we can read the order of groups about the top face of the tetrahedron, opposite H.

[Diagram: tetrahedron with phenyl on top, CH₃-CH₂ on left, CH₃ on right, H on bottom]

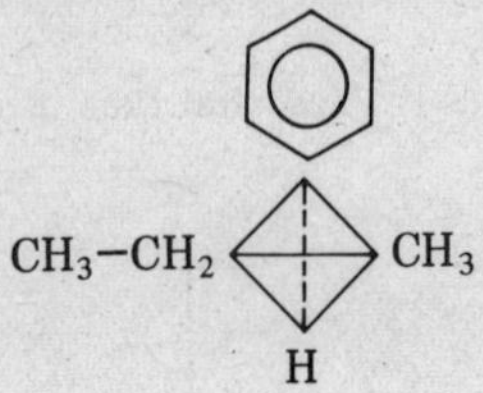

Recite 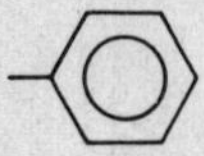 $-\bigcirc > -CH_2CH_3 > -CH_3$ and find this order of groups to be counter-clockwise:

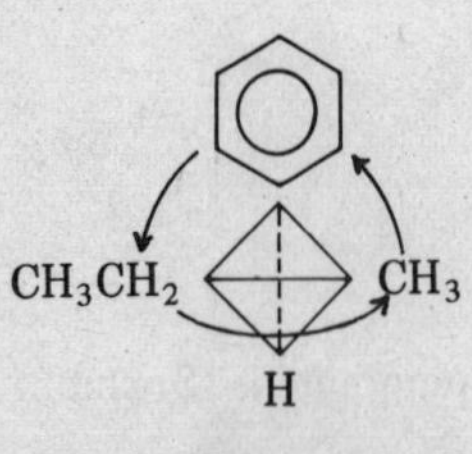

[Diagram showing counterclockwise arrows around the tetrahedron with CH₃CH₂, phenyl, CH₃, H]

Therefore, this is an S molecule: (S)-2-phenylbutane.

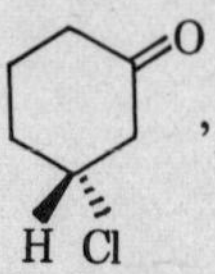

The configurations of cyclic compounds can be shown in $\overset{\diagdown}{\diagup}C\overset{\diagup}{\diagdown}$ notation as we have been doing, for instance in [cyclohexane with =O and H, Cl substituents], where the ring lies in the plane of the paper. Often, though, cyclic compounds are drawn with the ring perpendicular to the paper. If one edge of the ring is drawn with heavy lines to identify it as the front edge, rising out of the paper, the configuration can be drawn unambiguously without using heavy and dashed bonds for substituents.

Example

The C* in [cyclic structure with OH, CH₃, CH₃, H] translates as [3D structure with ③CH₂, ②C(CH₃)₂, HO①, H④]

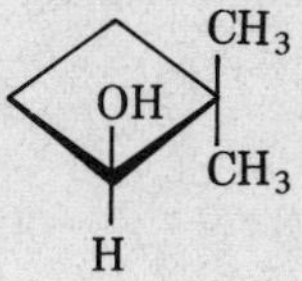

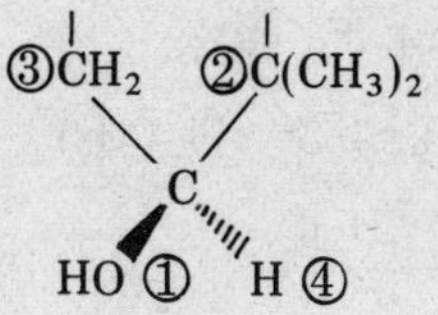

(priorities indicated), and therefore is the S enantiomer. Note that to turn a cyclic structure into its enantiomer, you need only exchange the "up" (OH) and "down" (H) groups:

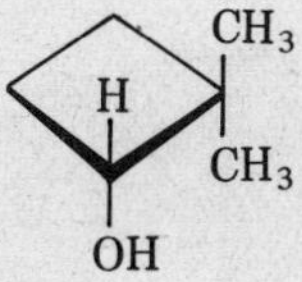

[cyclic structure with H, CH₃, CH₃, OH] = R

★ *Exercises*

1. Specify the configuration of each compound.

    (a) 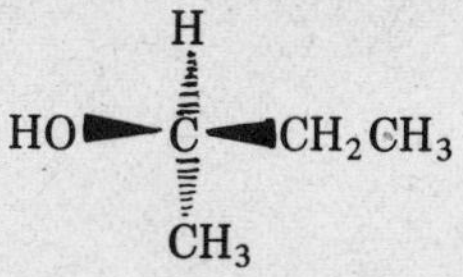 HO—C—CH₂CH₃ (with H up, CH₃ down)

    (b) 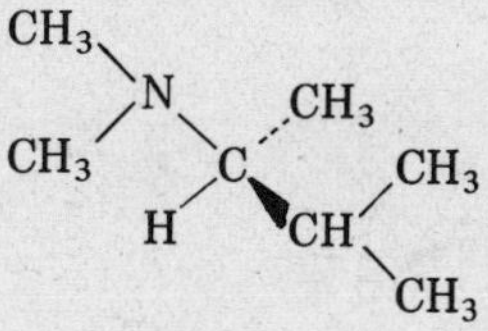 CH₃, CH₃ on N; C bonded to H, CH, CH₃, CH₃ (with CH branching to CH₃, CH₃)

    (c) CH₃—CH₂ and H on one carbon; Cl and CCl₂—CH₃ on other carbon of C=C

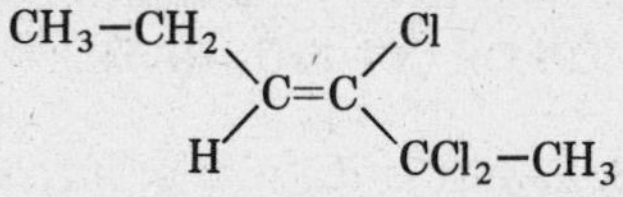

    (d) 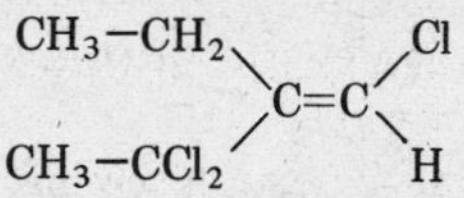 CH₃—CH₂ and CH₃—CCl₂ on one carbon; Cl and H on other carbon of C=C

    (e) 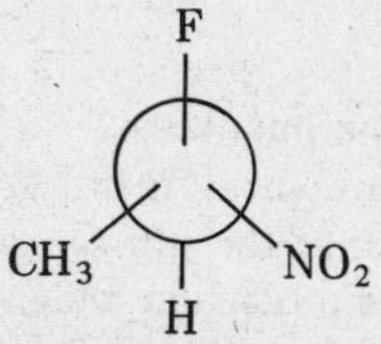 benzene ring with F, NO₂, CH₃, H substituents

    (f) cyclobutane with Cl, Cl, H, Cl, =C, Cl, H (both centers)

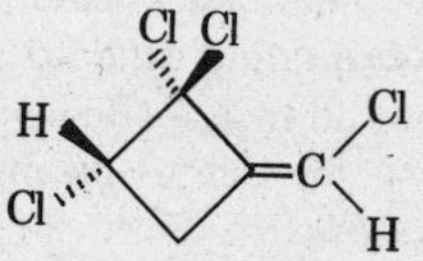

2. Draw three-dimensional structural formulas for each molecule.

    (a) (S)-CFHClBr
    (b) (R)-(CH₃CHCl—CH₂—CH₃)
    (c) (E)-(CFCl=CH—CH₃)
    (d) (Z)-1-phenyl-2-chloroethene
    (e) (R)- [indane structure with H, Cl at * carbon]

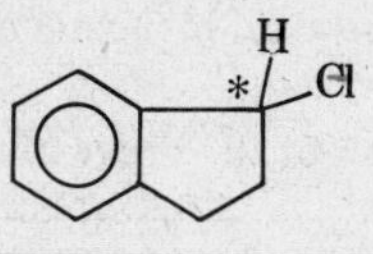

278 HOW TO SUCCEED IN ORGANIC CHEMISTRY

3. Redraw the compound in exercise 1(f) using the alternative, ring-perpendicular-to-paper notation.

— — — — — — — —

1. (a) R ($OH > CH_2CH_3 > CH_3$). (b) S ($N(CH_3)_2 > CH(CH_3)_2 > CH_3$).
   (c) Z ($CH_3CH_2 > H$, $Cl > CCl_2CH_3$). (d) E ($CH_3CCl_2 > CH_3CH_2$, $Cl > H$).
   (e) R ($F > NO_2 > CH_3$). (f) Z ($Cl > H$, $-CCl_2- > -CH_2-$) and S ($Cl > -CCl_2- > -CH_2-$).

2. (a) [structure: square with F, Cl, Br, H]
   (b) [structure: square with Cl, C_2H_5, CH_3, H]
   (c) [structure: $C=C$ with Cl, H on one side; F, CH_3 on the other]
   (d) [structure: $C=C$ with φ, Cl; H, H]
   (e) [bicyclic structure with Cl, H]

3. [structure with Cl, H, Cl, =C, Cl, H]

DETERMINATION VERSUS SPECIFICATION OF CONFIGURATION

9. The process of specification—connecting a specific configuration with a designating symbol (R or S, Z or E)—is just a naming process. Like all naming operations, it is arbitrary, based on a man-made convention.

The process of determining what structural formula is correct for a given solid or liquid organic compound in a bottle is quite different. It is an experimental scientific operation whose outcome cannot be anticipated before the experiment; it is not arbitrary. It generally follows this path: (1) isolation and purification of the substance; (2) analysis to get the empirical formula; (3) determination of the molecular weight to refine the empirical into the molecular formula; (4) spectroscopic and chemical study to determine the skeleton, identify the functional groups, and deduce how they are bonded together. The result is a structural formula. If the molecule contains no or ⟩C⟨ centers, the process is complete at this stage. However, if such centers are present, a fifth experimental stage is required: determination of the order of groups about this center. One has samples of two compounds, which are distinguished experimentally by positive and negative values of [α]. One also has the two structural formulas with their opposite configurations. The fifth stage determines which structure, R or S, goes with which compound, the + or the −. This process is called the *determination of configuration*.

The situation is thus described as follows:

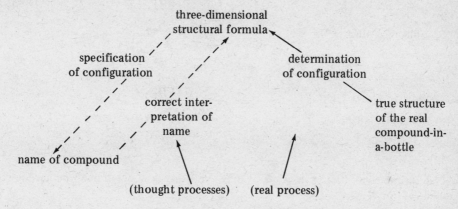

Exercise

Here are the three structures, the three correct names, and the three measured values of [α] for three real compounds. The entries are scrambled, however. Draw as many lines connecting entries in the three columns as is possible without additional information:

Name	Structure	[α]
2-Phenyl-2-propanol	Ph–C(H)(CH₃)–OH	+45.5°
(R)-1-phenyl-1-ethanol	Ph–C(CH₃)(CH₃)–OH	−45.5°
(S)-1-phenyl-1-ethanol	Ph–C(CH₃)(H)–OH	0.0°

2-Phenyl-2-propanol — [structure: Ph–C(H)(CH₃)–OH shown with wedge/dash] — +45.5°

(R)-1-phenyl-1-ethanol — [structure: Ph–C(CH₃)(CH₃)–OH] — −45.5°

—— 0.0°

(S)-1-phenyl-1-ethanol — [structure: Ph–C(CH₃)(H)–OH]

No further connections can be made; only experiment can determine whether the + isomer is R or S.

MOLECULES CONTAINING TWO C*

10. Putting two chiral centers in the same molecule has some consequences that are just extended applications of what we have already done (covered in this frame), and it has some that are new (frame 11). In this frame we will take up in turn:

- Drawing structures for compounds of type $\overset{a}{\underset{c}{b}}\text{C*–C*}\overset{d}{\underset{f}{e}}$.

- Naming (specifying) these compounds.

- The number of stereoisomers possible in >C*–C*< systems.

In principle, any of the types of perspective drawings that we used for >C*< molecules can also be used when there is more than one $\overset{*}{C}$ per molecule. They do all work well when the $\overset{*}{C}$ are separated by intervening atoms, as in

[structures: H₂N–C(CH₂OH)(H)–C(=O)–NH–C(CH₂Ph)(H)–CO₂H = H₂N–[C with CH₂OH up, H down]–C(=O)–NH–[C with HOCH₂ up, H down]–CO₂H (shown with cyclic/perspective representations)]

But when the C* are directly connected, some of the drawings produce odd-looking bent bonds; for instance,

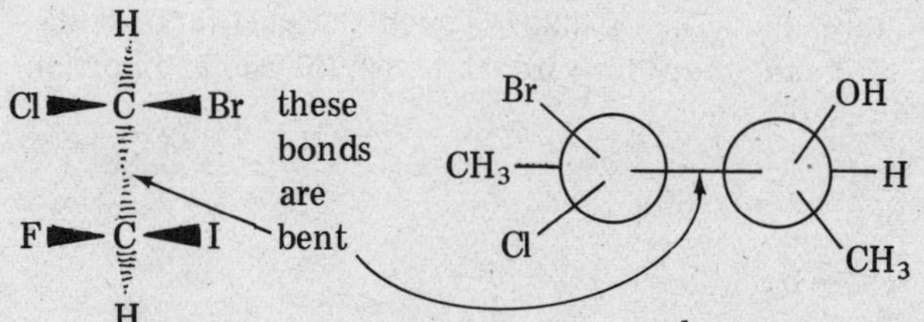

Although you will sometimes see structures of type b written, most chemists find some other solution. In earlier times stacked tetrahedra were popular:

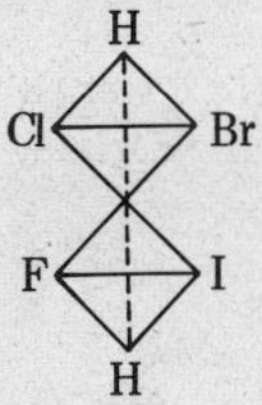

$a' = a$

This structure is identical with a above; it distorts the C—C bond just as badly, but the bond doesn't show. Nowadays the tendency is to use a nondistorting combination of $\diagup\!\!\!\!C\!\diagdown$ centers. Or one can abandon perspective drawings altogether and use planar projection formulas that lie flat on the page and represent configuration symbolically. Structure a or a' translated into $\diagup\!\!\!\!C\!\diagdown$ notation looks like this:

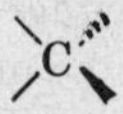

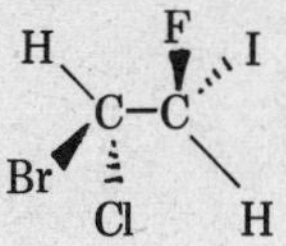

$a'' = a$

This notation can handle any number of C*, and it is condensible to line-segment notation:

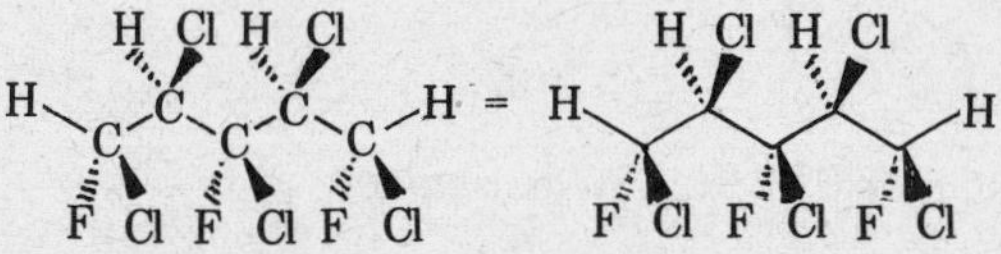

The projection formulas are called *Fischer projections*; they take

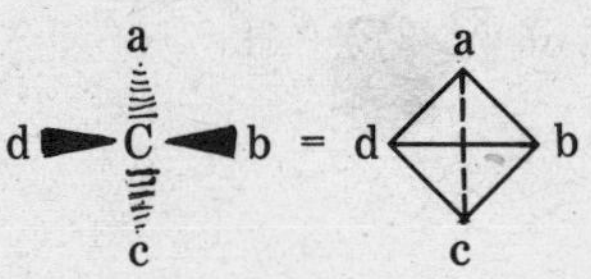

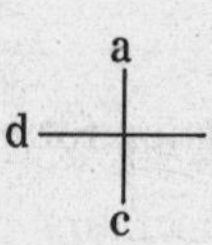

and squash it into the plane of the paper, getting d——b (the central carbon is omitted). To reconstruct the three-dimensional shape of such a projection formula, mentally pull the right- and left-hand groups forward, and push the top and bottom groups back:

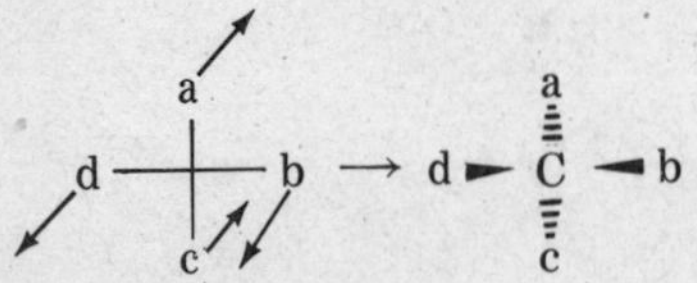

$$d\!\!-\!\!\!\!\underset{c}{\overset{a}{+}}\!\!\!\!-\!\!b \longrightarrow d\blacktriangleright\underset{c}{\overset{a}{\underset{\vdots}{C}}}\blacktriangleleft b$$

Fischer projections look like a simplification, but the price paid for the simplicity is loss of flexibility. Fischer projections cannot be moved about in any fashion except 180° rotation in the plane of the paper. A 90° rotation (CW or CCW) or flipping over on its back converts the projection formula into its enantiomer. The even/odd exchange rule does still work.

Specifying configuration for these more complex compounds requires a letter (R or S) for each C* present. Each letter is preceded by the number of the C* in the carbon chain to identify the center being specified.

Example

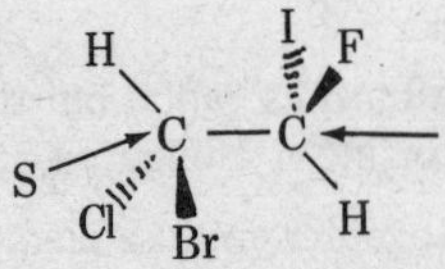

R is (1S,2R)-1-bromo-1-chloro-2-fluoro-2-iodoethane.

It should be apparent that since each C* can be either R or S, the number of possible combinations of configurations increases rapidly as the number of R* per molecule increases. The total number of possible stereoisomers (possible combinations) can be determined by use of a table. For compounds of type $C_1^*\!-\!C_2^*$, we have:

possible con- figurations of C_1^*	possible con- figurations of C_2^*	
	R	S
R	R-R	R-S
S	S-R	S-S

} possible configurations of $C_1^*\!-\!C_2^*$

It turns out that a compound containing x C* has 2^x stereoisomers.

★*Exercises*

1. Draw perspective structures of all four stereoisomers of ⬡—CHBr—CHCl—⬡ using the 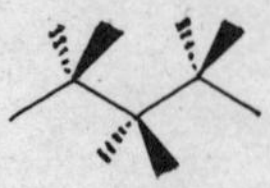 ⤳⤴ notation.

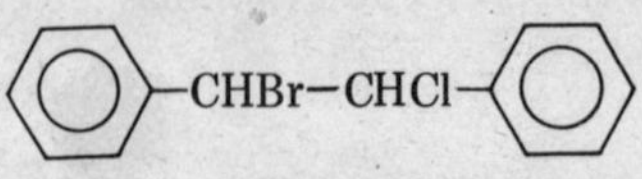

2. Repeat using Fischer projection formulas.

3. Name and specify the configuration of the stereoisomers in exercise 1.

4. Which 2-methylcyclohexanol is this?

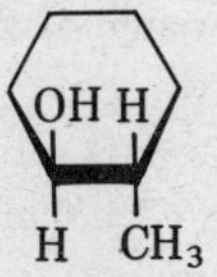

---

1. 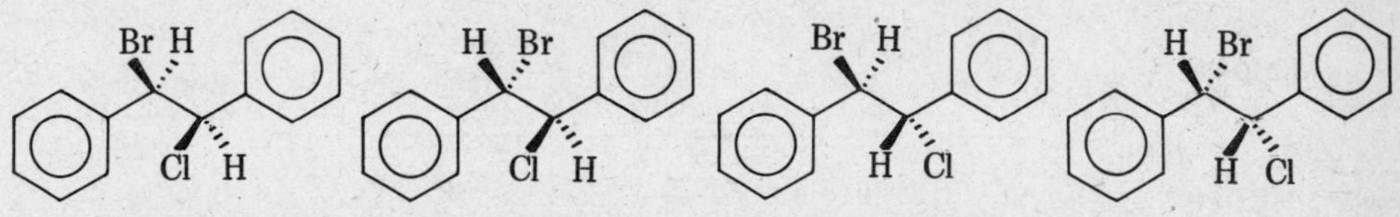 [four structures: Ph-CHBr-CHCl-Ph with various wedge/dash configurations: (Br H / Cl H), (H Br / Cl H), (Br H / H Cl), (H Br / H Cl)]

2. 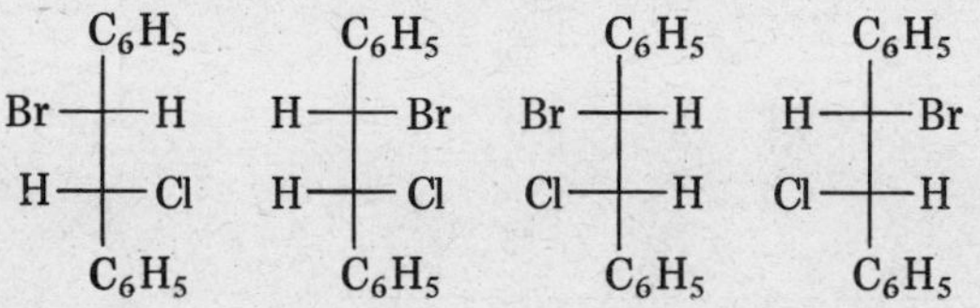 Fischer projections:

```
   C6H5        C6H5        C6H5        C6H5
Br—┼—H     H—┼—Br     Br—┼—H     H—┼—Br
 H—┼—Cl     H—┼—Cl    Cl—┼—H    Cl—┼—H
   C6H5        C6H5        C6H5        C6H5
```

3. Left to right: (1R,2R)-, (1S,2R)-, (1R,2S)-, and (1S,2S)-1-bromo-2-chloro-1,2-diphenylethane

4. The four groups on the $\overset{OH}{\underset{H}{|}}$ center have the following priorities:

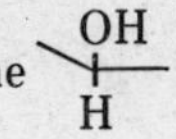

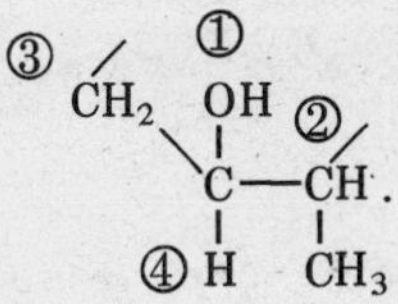

③ CH₂   ① OH
       \\ /
        C—CH ②. Read the configuration from above the ring (face of tetrahedron
       / \\
   ④ H   CH₃   opposite group ④):   ③ CH₂   ②
                                    ① HO ▶—CH(CH₃)
                                         ④ H

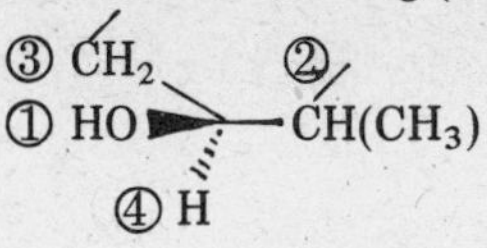

The ① → ② → ③ order is counterclockwise, so this center is S. Similarly, the second C* looks like this (from below the ring):

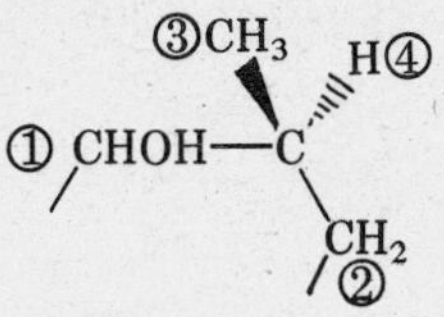

③ CH₃   H ④
      \\ /
① CHOH—C   , ① → ② → ③ = counterclockwise, ∴ S. So (1S,2S)-
       / \\   2-methylcyclohexanol
        CH₂
        ②

11. The new features that need discussion for molecules containing two C* are:
- Stereochemical relationships between R-R, R-S, S-R, and S-S isomers.
- Properties of these four stereoisomers.

To change a molecule with a single chiral center into its enantiomer, you must reverse its order of groups (*invert* its configuration). Working with the perspective or projection formula, you have learned to produce the enantiomer most easily by interchanging any two groups on C*. Working with the name of the compound, you form the enantiomer by changing R to S, or vice versa.

To do the same conversion for a C*—C* molecule, you operate in the same way, but *to get the enantiomer, you must invert the configuration of every C* present.* If you invert only part of the chiral centers present, you get a diastereomer of the original, not the enantiomer.

Examples

The enantiomer of	is
(R)-2-chlorobutane [CH₃, H, HO, C₆H₅ around C]	(S)-2-chlorobutane [HO, H, CH₃, C₆H₅ around C]
(2R,3S)-2-bromo-3-chlorobutane	(2S,3R)-2-bromo-3-chlorobutane

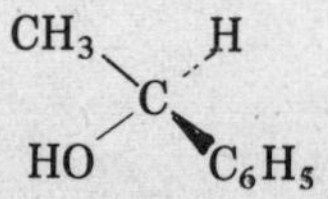

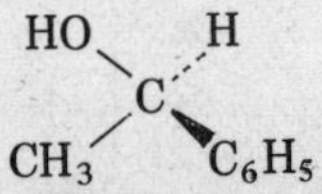

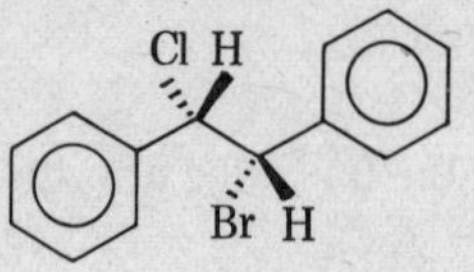

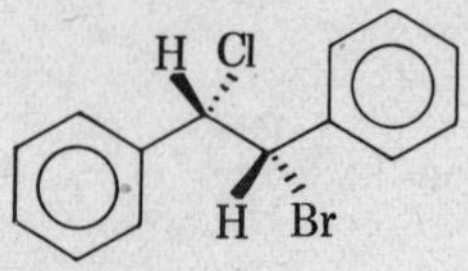

Keeping these results in mind makes it easy to classify any pair of C*—C* stereoisomers:

- If both C* are inverted in structure a relative to structure b, the structures are enantiomers.
- If only one C* is inverted, a and b are diastereomers.
- If neither C* is inverted, a and b are identical.

Example

How are [ring with OH OH up, H H down] (a) and [ring with OH H up, H OH down] (b) related?

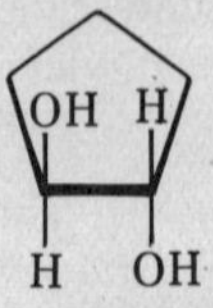

Solution:

Easiest way: Left-hand C* has same configuration in both ("OH up"); right C* has opposite configuration in a and b ("OH up" and "OH down"). So a ⟶ b is inversion of one C* out of two C* present, ∴ a and b are diastereomers.

Alternative: Are they stereoisomers? Yes.
Is b the mirror image of a? No.
∴ not enantiomers, must be diastereomers.

Alternative: a is the S-R isomer, b is the S-S isomer, ∴ diastereomers (enantiomer of a would be R-S; enantiomer of b would be R-R).

So with two different C* present, the relationships can be summarized this way:

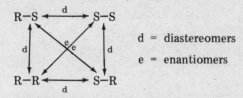

d = diastereomers
e = enantiomers

This means that the R-S isomer has different properties than the R-R and S-S and can be separated from them by normal means. R-S has properties identical with S-R except for the sign of [α], and the two can be separated only by use of the special methods mentioned in frame 7.

R-S, S-R, R-R, and S-S above are all *chiral*. How can I be certain? By testing superimposability on the mirror image, for instance

nonsuperimposable, ∴ R-R and S-S are chiral.

The same holds for R-S and S-R.

New relationships and properties appear when we consider a C*—C* compound in which *exactly the same set of four groups is attached to each C**, as in

R-S: [structural formulas showing R-S configuration in three representations]

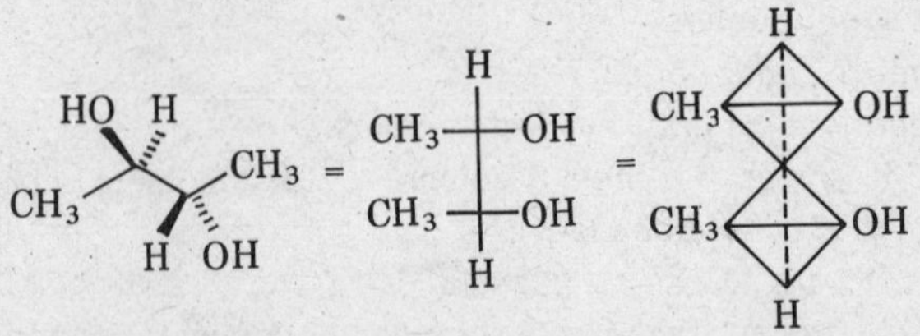

S-R: [structural formulas showing S-R configuration in three representations]

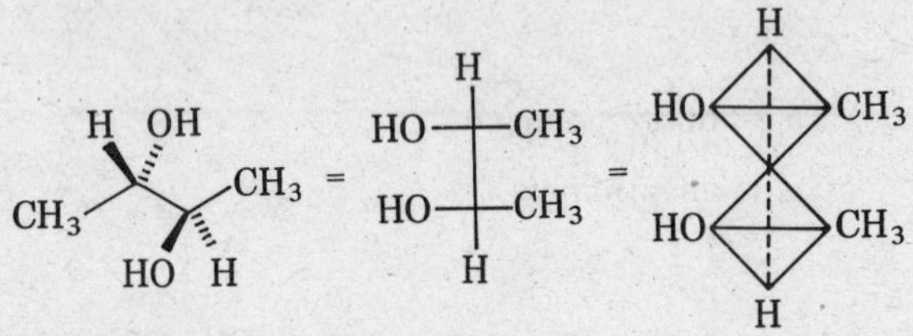

The R-R and S-S isomers are again enantiomers. The difference lies in R-S and S-R, which are now identical. This is readily seen by rotating the S-R drawing 180° in the plane of the paper:

S-R [structure] → R-S [structure]

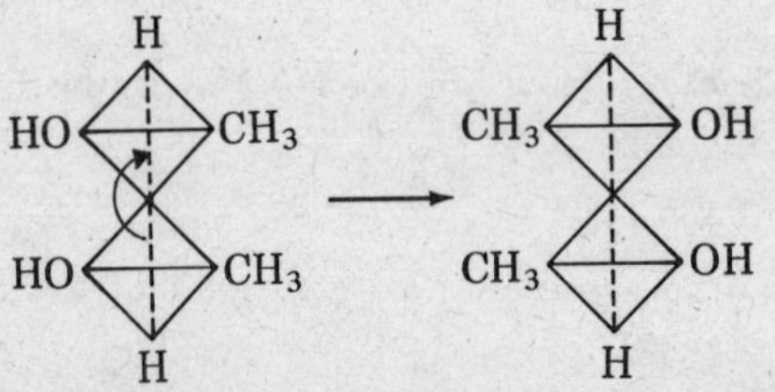

The new structure superimposes on the R-S structure above. So there are just three stereoisomers in this two-identical-C* case. The S-R (or R-S) isomer is again diastereomeric with S-S and R-R. The relationships can be summarized as

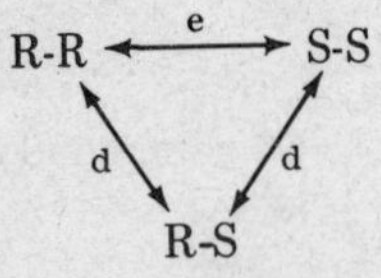

$$R\text{-}R \xleftrightarrow{e} S\text{-}S$$
$$\searrow d \quad d \swarrow$$
$$R\text{-}S$$
$$S\text{-}R$$

The R-R and S-S isomers are chiral and optically active, but the R-S (= S-R) isomer has the peculiarity of being *optically inactive* ($[\alpha] = 0$) because it is *achiral*. You can determine the absence of chirality in two ways:

- Compare the compound with its mirror image. The mirror image of R-S turns out to be S-R:

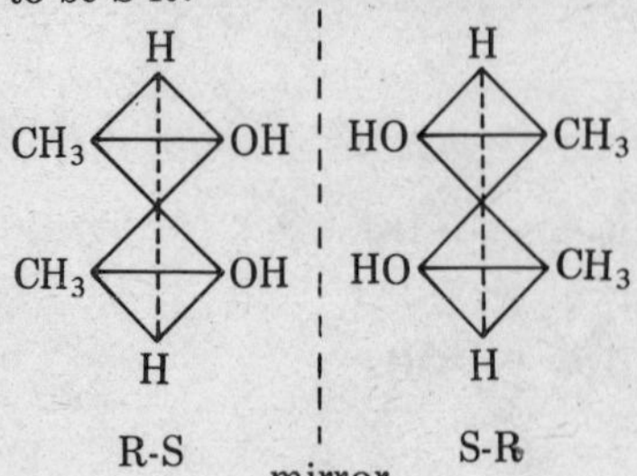

[structures showing R-S and S-R on either side of mirror plane]

R-S | mirror plane | S-R

Since we have already determined that S-R superimposes on R-S, we have here a compound superimposable on its mirror image, proving it to be symmetric or achiral.

- You can show that the R-S isomer has a plane of symmetry, which guarantees that it is achiral. Using the Fischer projection or stacked-tetrahedra representation, a horizontal plane perpendicular to the paper is found to be a plane of symmetry:

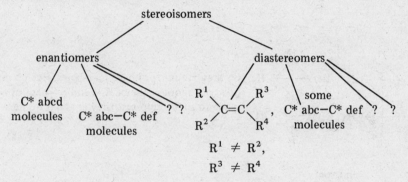

Compounds that, like this R-S isomer, possess chiral centers but are overall achiral because of internal symmetry are called *meso* compounds.

To conclude this frame, we can now extend our list of the types of structures that can display the different types of stereoisomerism:

```
                        stereoisomers
                       /             \
              enantiomers            diastereomers
             /         \             /            \
      C* abcd      C* abc–C* def   R¹    R³    some
      molecules    molecules        \C=C/      C* abc–C* def
                          ? ?       /    \  ,  molecules    ? ?
                                   R²    R⁴
                                  R¹ ≠ R²,
                                  R³ ≠ R⁴
```

★ *Exercises*

1. In the discussion above we did not actually prove that the R-R and S-S isomers of CH_3–CHOH–CHOH–CH_3 are enantiomers. Do so now by finding an arrangement that shows them to be mirror images of one another, and then show that they cannot be superimposed. In manipulating the structures for these or any other purposes, you can transport, twist, or flip the perspective structures at will—their configurations are unchanged by such treatment just as if they were models. However, remember that the projection formula is turned into its enantiomer by flipping it over on its back or rotating it 90°. Use

   d—│—b (a on top, c on bottom) to prove that this is so.

2. Each of the following molecules has two asymmetric carbon atoms. Identify those that are nonetheless achiral because they possess a plane of symmetry.

(a) [structure with C₂H₅, H, CH₃, H, CH₃, C₂H₅]

(b) [cyclohexane with Cl, Cl, H, H]

(c) [cyclohexane with CH₃, Cl, Cl, CH₃, H, H]

(d) [structure with CH₃, H, H, CH₃]

(e) [structure: CH₃, H, Br, C—C, Br, H, CH₃]

(f) CH_3—C—C—CH_3 with F, F, D, H substituents

1.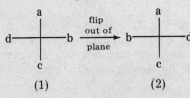

   R-R | mirror plane | S-S

   So they are mirror-image molecules. But no matter how you reorient them, R-R and S-S cannot be made to superimpose; ∴ they are enantiomers.

   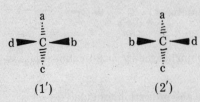

   1′ and 2′ are the three-dimensional reconstructions of 1 and 2 (frame 10) Since one interchange of two groups (b and d) converts 1′ to 2′, 1′ and 2′ are enantiomers. Therefore, 1 and 2 are enantiomers.

2. a, b, and e have a plane of symmetry and are achiral:

   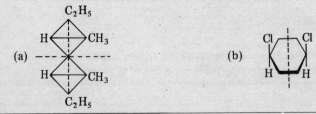

(e) [Newman projection: front carbon with BrBr and CH₃, back carbon with H, H, CH₃]

STEREOCHEMISTRY OF THE STANDARD MECHANISMS: RADICAL SUBSTITUTION

12. When either the reactant or the product of a reaction (or both) contain a stereochemical center, a new question arises: What happens to the configuration as a result of the reaction? The answer (the "stereochemical outcome" of the reaction) can only be obtained experimentally by running the reaction in the laboratory, using a reactant of known configuration, then determining the configuration of the product and comparing the two. Since this relationship between reactant and product configurations obviously depends on the manner in which the old bonds are broken and the new bonds made in the reaction, *the stereochemical outcome of the reaction depends on the reaction mechanism.* This is a potent relationship. It provides a tool for experimentally testing proposed reaction mechanisms and for predicting product configuration once the mechanism is known.

So we must now look at the stereochemical results of those basic mechanisms that we have covered so far. From now on you should learn the stereochemical outcome of a reaction when you learn the reaction mechanism.

The four general stereochemical outcomes of organic reactions are listed in Table 13.1.

Table 13.1 *Stereochemical Outcomes of Reactions*

Case	Configurational possibilities: (C*abcd or Cab=Cab) present?		Stereochemical result
	Reactant	*Product*	
1	No	No	No stereochemistry involved
2	Yes	No	Stereochemical possibilities destroyed
3	No	Yes	Stereochemical possibilities created
4	Yes	Yes	Problem in stereochemical *changes*

To begin with, we consider reactions involving only chiral centers, :

Case 1 involves no chiral centers. The reaction mechanism and the structures of reactant and product can be adequately drawn in two dimensions, without regard for stereochemistry.

Case 2 destroys chiral center(s). The outcome is the same regardless of the original configuration of the center destroyed, so again there is no need to pay attention to three-dimensional geometry in writing the mechanism.

Case 3 creates a new C* and can do so with either the R or S order of groups (configurations) or a mixture of the two. We need to write the reaction mechanism in such a way that it shows how the observed product configurations came about, and we need to develop rules capable of predicting these product configurations.

Case 4 includes four possibilities (Table 13.2).

Table 13.2 *Detailed Stereochemical Outcomes for Case 4*

Stereochemical change	Characteristics	Name
(a) R $\longrightarrow$ R (or S $\longrightarrow$ S)	Product retains the original order of groups around C*	Retention
(b) R $\longrightarrow$ S (or S $\longrightarrow$ R)	Product has opposite order of groups from reactant	Inversion
(c) R (or S) $\longrightarrow$ random mixture of R and S	Reactant configuration is lost	Racemization
(d) Random mixture of R and S $\longrightarrow$ R or S	Configurational purity is created	Induction

Retention of configuration (case 4a) occurs whenever none of the bond breaking and bond making occurs at the C*, for instance in

$$H \blacktriangleright \underset{CH_3}{\overset{CH_2CH_3}{C}} \blacktriangleleft CH_2-CH=CH_2 \xrightarrow[H_2]{Pt} H \blacktriangleright \underset{CH_3}{\overset{CH_2CH_3}{C}} \blacktriangleleft CH_2-CH_2-CH_3$$

In cases like this, you do not need to know the detailed mechanism of the reaction to predict that the product will retain the original configuration—you only need to know that the four bonds to C* are not disturbed by that mechanism. A few reactions break a bond to C* and yet preserve the original configuration, but they are not common.

Inversion of configuration (case 4b) and racemization (randomization of configuration, case 4c) are the ordinary outcomes of reactions making/breaking bonds at C*. Inversion occurs via a mechanism yet to be discussed ($S_N 2$: mechanism V, Unit 14, frame 4). Racemization can occur via several mechanisms. We will illustrate it here with radical substitution, mechanism I.

No known reaction or mechanism is capable of converting racemic compound A into pure enantiomeric compound B, as in

$$d \blacktriangleright \underset{c}{\overset{a}{C}} \blacktriangleleft b + b \blacktriangleright \underset{c}{\overset{a}{C}} \blacktriangleleft d \longrightarrow d \blacktriangleright \underset{c}{\overset{x}{C}} \blacktriangleleft b$$

However, some known processes are related to this. First, a racemic mixture of C*abcd enantiomers can be *separated* into the pure enantiomers (by carrying out the separation process in a chiral environment, which gives them different properties). This is called *resolution*. Resolution does not create chirality, which must be present already, but it creates optical activity (see frame 7). Second, biological systems sometimes can be found that selectively destroy one enantiomer in a racemic mixture, leaving the other behind, pure and optically active. Third, if a reaction takes place not at, but near, an asymmetric carbon atom, C*, and if it makes a new C*, then one or the other configuration at the new C* generally predominates—an excess of one kind of handedness is created in the chiral environment provided by the original C*.

Now before we examine the stereochemistry of mechanisms I, II, and III in detail, consider the following rule, which summarizes a lot of experimental facts: *optically inactive reactants give optically inactive products.* The converse is not always true; optical activity (nonzero $[\alpha]$) can be conserved *or lost* in chemical reactions. Is there a corresponding rule for changes in *configuration? Achiral reactants either give achiral products or racemic mixtures of chiral products.* Chiral reactants can give either chiral or achiral products. Chiral centers can be destroyed or created in chemical reactions, but they are always created with equal numbers of molecules in each of the two possible configurations, except when they are created in a chiral environment. This is closely analogous to the principle that enantiomers can only be separated in chiral environments. In each case the chiral environment is provided by the presence of chiral centers nearby, in the same molecule or in neighboring molecules.

We now investigate the stereochemistry of the radical substitution (this frame), electrophilic addition (frame 13), and radical addition (frame 14) mechanisms.

The typical problem that frames 12 to 14 are preparing you to handle may read "Predict [or "Account for"] the stereochemistry of the products of this reaction." In the one case you need only write perspective structures for the products; in the other case you need to make perspective drawings for each step in the mechanism to show *how* the final product configurations came about. Whether you are required to draw out the mechanism in three dimensions or not, you will ordinarily have to be able to think it through in order to predict the product stereochemistry. The steps in the process are:

- Determine if the product can indeed exist in more than one stereoisomeric form; if not, the question is moot.
- Identify the reaction type.
- Identify the basic mechanism involved.
- Identify the reaction stereochemistry expected (inversion, racemization, etc.).
- Plug the specific structures of the problem at hand into this stereochemical scenario.

Some of the sample solutions in the remainder of this unit are cast in this form. First, we will work through each mechanism in detail to discover what its stereochemistry must be. Once you have learned what the basic stereochemical result of each standard mechanism is, these problems require just recall and application. Each time you learn a new mechanism you should learn its stereochemistry at the same time.

Mechanism I: Radical Substitution

The radical substitution reactions occur at a saturated (sp^3) carbon atom. Thus the stereochemistry of these reactions is

$$\begin{matrix} a & & c \\ & \overset{*}{C} & \\ b & & d \end{matrix}$$

stereochemistry, and the various possibilities are those shown in Table 13.1.

Case 1 is perhaps the most common. Substitution reactions occurring at $-CH_3$, $-CH\begin{matrix}CH_3\\CH_3\end{matrix}$, $-CCl_2H$, $-CHCl_2$, and similar centers involve no chiral compounds, since both these reactant groups and the products of substitution of H with

X ($-CH_2X$, $-CX\begin{smallmatrix}CH_3\\CH_3\end{smallmatrix}$, $-CCl_2X$, $-CXCl_2$, etc.) have at least two identical groups on the central carbon atom. Thus neither reactant nor product can be chiral, and neither can exist in more than one arrangement in space, that is, as more than one isomer.

In case 2, regardless of whether one begins the reaction with R- or S- reactant (e.g., Br–C(Cl)(CH₃)(H) or CH₃–C(Cl)(Br)(H)), the product has lost its possibilities for stereoisomerism due to the newly acquired identity of two of the groups attached to the originally asymmetric carbon. For example:

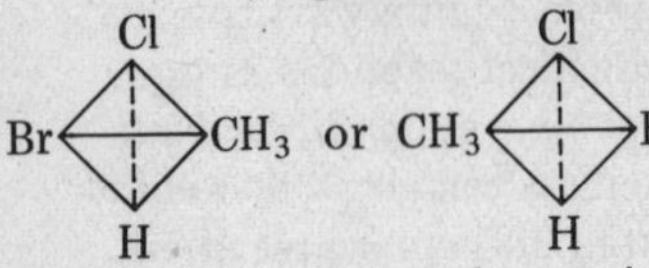

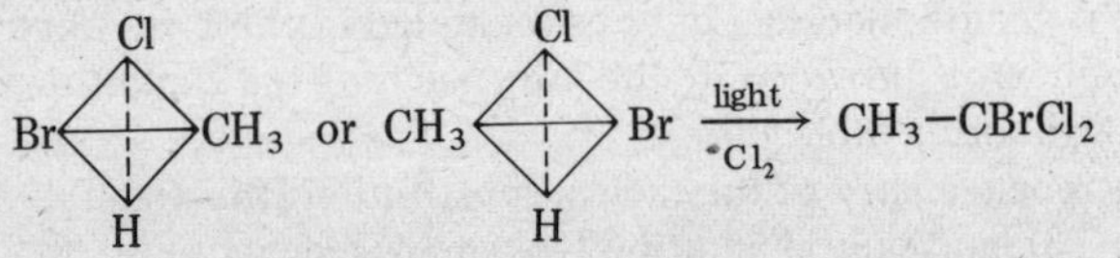

Br–C(Cl)(CH₃)(H) or CH₃–C(Cl)(Br)(H) $\xrightarrow[\cdot Cl_2]{light}$ $CH_3-CBrCl_2$

Both enantiomers of the reactant give the same product. This process fits the rules stated above; optical activity is destroyed and chirality is destroyed.

To examine case 3, we choose a reactant with an achiral center that is turned into a chiral center by the reaction, for instance

$$CH_3-CH_2-Br + Cl_2 \xrightarrow{light} CH_3-\overset{*}{C}HCl-Br + HCl$$

The experimental observation in this and analogous reactions is formation of racemic product. The accepted mechanism is:

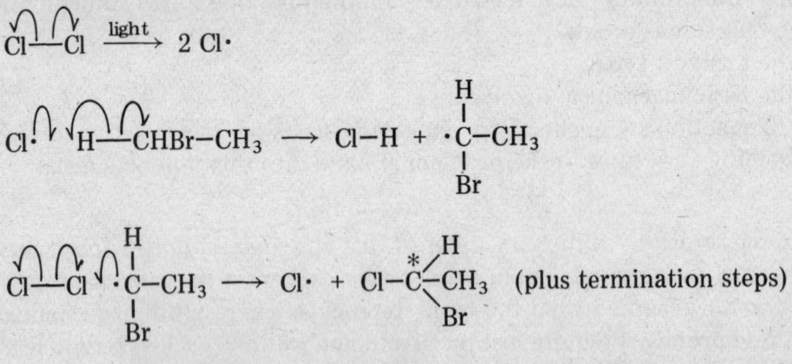

$Cl-Cl \xrightarrow{light} 2\ Cl\cdot$

$Cl\cdot + H-CHBr-CH_3 \longrightarrow Cl-H + \cdot\underset{Br}{\overset{H}{C}}-CH_3$

$Cl-Cl + \cdot\underset{Br}{\overset{H}{C}}-CH_3 \longrightarrow Cl\cdot + Cl-\overset{*}{\underset{Br}{C}}\overset{H}{-}CH_3$ (plus termination steps)

To be correct, this mechanism must predict the stereochemical outcome that is actually observed.

To discover what product configuration these steps predict, we must answer two questions. (1) When exactly is the configuration (order of groups about C*) of the product fixed? *Answer:* In the last step above, where C* first appears. (2) What is the three-dimensional geometry at that instant? *Answer:* Since the radical intermediate is planar, the approach of Cl—Cl must be either from above or below; *above* leads to one configuration in the product, *below* leads to the opposite configuration:

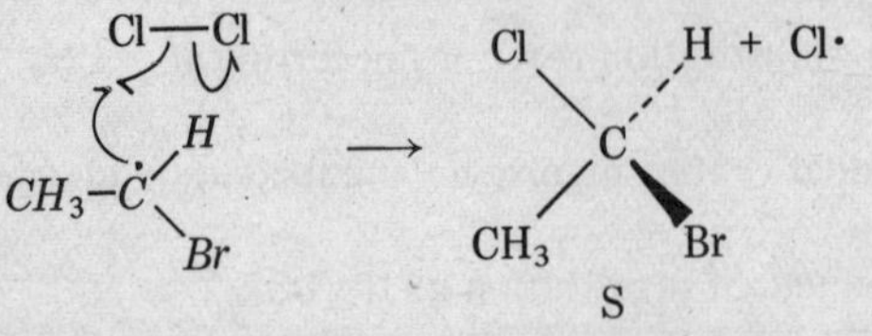

$CH_3-\dot{C}\begin{smallmatrix}Cl-Cl\\H\\Br\end{smallmatrix} \longrightarrow CH_3-\underset{Br}{\overset{Cl}{C}}-H + Cl\cdot$

S

$$\text{CH}_3-\overset{H}{\underset{\underset{\text{Cl}-\text{Cl}}{\text{Br}}}{C}} \longrightarrow \overset{\text{CH}_3\quad H}{\underset{\underset{R}{\text{Cl}}}{C}}\text{Br} + \text{Cl}\cdot$$

Furthermore, since the top of the radical is in all ways (except handedness) equivalent to the bottom, the probabilities of the reactions proceeding at top and at bottom are precisely equal. Thus the rates of accumulation of S and R enantiomers are precisely equal, and the product is an exactly random (50-50) mixture of R and S, a *racemic mixture*. Practically all physical properties of R and S enantiomers are identical; therefore, the mixture has these same properties (boiling point, solubility, etc.). The R and S enantiomers differ in one property—the sign of the specific rotation. But a precisely 50-50 mixture of a pair of enantiomers rotating polarized light by $+x$ and $-x$ degrees, respectively, has a net rotation of exactly zero. This process thus obeys the general rules: optically inactive reactants give optically inactive products; achiral reactants give a racemic mixture of chiral products. The mechanism predicts the observed result, and it explains precisely how this result comes about.

Suppose now that you start with a preformed chiral center and change one of the groups on C* by radical substitution, for instance by chlorination of a racemic, 50-50 mixture of $\text{CH}_3\langle\overset{H}{\underset{Br}{|}}\rangle F$ and $F\langle\overset{H}{\underset{Br}{|}}\rangle\text{CH}_3$. The radical substitution mechanism predicts
$\quad\quad\quad\quad\quad\quad\quad\quad\quad\quad\quad\;\; A \quad\quad\quad\quad\quad A'$

the same result as above—a racemic product. To see this, imagine a racemic reactant sample consisting of 0.5 mole of R + 0.5 mole of S. The S reactant goes to 0.25 mole of R product and 0.25 mole of S product (mechanism above). The R reactant goes to 0.25 mole of R product and 0.25 mole of S product. Thus the total product is 0.5 mole of R plus 0.5 mole of S = racemic mixture. This is also the observed result.

A simpler way to see all these results is this: no matter whether the starting material is pure R, pure S, any mixture (including the racemic mixture) of the two, or lacks an asymmetric center altogether, the intermediate radical has the same structure—planar and lacking any trace of its configurational history. It can give only racemic products. Thus we have also the result of applying mechanism I to pure enantiomers, such as A or A' above, as reactants; this mechanism always gives result 4c—racemization—never 4a or 4b. Other mechanisms are capable of retention or inversion of configuration, but radical substitution can only racemize.

These conclusions apply to the radical substitution of all systems in which reactant and/or product contain a single asymmetric carbon atom, C*. If there is already a chiral center in the molecule, the creation of a new chiral center (case 3) need no longer give equal amounts of R and S configurations at that center. This is also true if the reactant is a racemic mixture. The formation of a second asymmetric center can result in unequal quantities of R and S products—the restriction to equal formation of both configurations is removed for a second C*. The existing asymmetric carbon atom creates a chiral environment that can induce an excess of one configuration over the other at a new chiral center created in its neighborhood. The last task of this section is to confirm that the radical substitution mechanism in fact predicts this result and to understand why it does so.

Consider the system

$$\text{C}_6\text{H}_5-\text{CH}_2-\overset{*}{\text{CHCl}}-\text{C}_6\text{H}_5 + \text{Cl}_2 \xrightarrow{\text{light}} \text{C}_6\text{H}_5-\overset{*}{\text{CHCl}}-\overset{*}{\text{CHCl}}-\text{C}_6\text{H}_5 + \text{HCl}$$

294 HOW TO SUCCEED IN ORGANIC CHEMISTRY

The mechanistic step determining the configuration at the new C* is

$$C_6H_5\text{-}\overset{\cdot}{C}H\text{-}\overset{*}{C}HCl\text{-}C_6H_5 + Cl\text{-}Cl \longrightarrow C_6H_5\text{-}\overset{*}{C}HCl\text{-}\overset{*}{C}HCl\text{-}C_6H_5 + Cl\cdot$$

The situation of this radical is now different from that considered above in the absence of the neighboring asymmetric carbon. The radical is planar, but now the space above that plane is no longer identical to that below the plane. This is best seen from the Newman projections, suitably modified to show the radical center as planar. To give a detailed picture of the radical, 12 conformations are shown; the sequence $1 \to 10 \to 6 \to 8 \to 2 \to 11 \to 4 \to 9 \to 3 \to 12 \to 5 \to 7 \to 1$ carries the C—C through one complete rotation.

In no conformation is the top of this radical equivalent to the bottom. At first sight conformations 1 and 4 seem to average out to identity of the top and bottom, as do 7 and 11, and some other pairs. This is not so, because 1 and 4 have different nonbonded interactions (e.g., Cl and C_6H_5 farther apart in 1 than in 4), so they have different energies and are populated to different extents. Therefore, they do not contribute equally to the average molecular structure and do not balance out from the point of view of radical top versus radical bottom. Thus *in the real radical (= weighted average of all conformers), the top and bottom are structurally different.* This is true of all radicals possessing an asymmetric carbon, and an approaching Cl—Cl molecule will find top and bottom structurally distinguishable. The transition states for top versus bottom attack by Cl—Cl will have different energies, and therefore the rates of top and bottom attack will differ. This means that the product derived from top attack will accumulate faster or slower than that from bottom attack, and at the end of the reaction these products will be present in different amounts. But what are the products of top versus bottom attack? Find out by bringing Cl—Cl up to conformer 1:

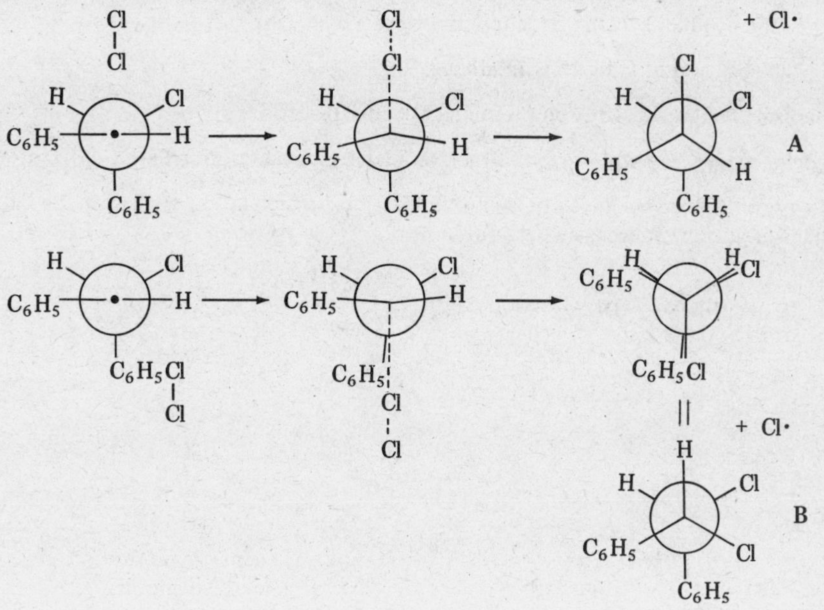

The two products, A and B, have the same order of groups about the far carbon atom. But they have opposite orders about the near carbon atom, the one that is made newly asymmetric in this reaction. Since the two products are formed in different amounts, the two orders of groups or configurations at the new C* are formed in unequal amounts. This is always the case when a chiral center is already present in the reactant.

Case 3 now shapes up like this: If the reactant is achiral (no C* present anywhere in the molecule), creation of a new C* in a chemical reaction gives exactly equal quantities of the two possible configurations (orders of groups about C*). If the reactant is already chiral (already has a C* someplace in the molecule), creation of another C* in a chemical reaction gives unequal quantities of the two configurations at the new C*.

What is the relationship of the two products formed in the latter case (A and B in the example above)? They have the same molecular formula, are not superimposable, are not structural isomers, and are not mirror images of one another. They are therefore diastereomers. The easiest way to identify them as diastereomers is to compare configurations at each C* in turn (frame 11). The far C* has the same order of groups (same configuration) in A and B; the near C* has Cl, H, C_6H_5 clockwise in A, counterclockwise in B (opposite configuration); therefore, A and B are diastereomers. They have different properties, can be separated by physical means (e.g., gas chromatography), and have different magnitudes of specific rotation.

All of what we have learned about the creation of a new asymmetric center has general application. Much of it can be boiled down into a rule: *enantiomers are formed in equal quantities, diastereomers are formed in unequal quantities.*

Example

Predict the stereochemistry of the products of the reaction

$$CH_3-CH_2-\phi \xrightarrow[\text{gas phase}]{HNO_3/\text{heat}} CH_3-CH(NO_2)-\phi + H_2O$$

Solution:

Since the product has one C*, there is a stereochemical result to discuss.

Type: substitution of H by NO_2 in alkane.

Mechanism: radical substitution (as are all alkane substitutions).

Stereochemistry: top and bottom attack on planar radical intermediate.

Application: even if we have not seen this reaction before and do not know the detailed mechanism, it must follow the pattern

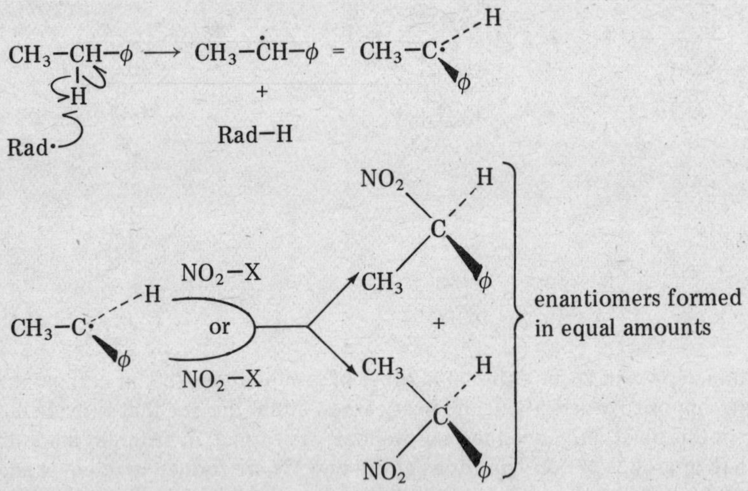

$$CH_3-CH(H)-\phi \xrightarrow{Rad\cdot} CH_3-\dot{C}H-\phi = CH_3-\dot{C}(\cdots H)(\phi) + Rad-H$$

$CH_3-\dot{C}(\cdots H)(\phi)$ $\xrightarrow[NO_2-X]{NO_2-X \text{ or}}$ $\begin{Bmatrix} C(NO_2)(CH_3)(H)(\phi) \\ + \\ C(CH_3)(NO_2)(H)(\phi) \end{Bmatrix}$ enantiomers formed in equal amounts

The products are formed in equal amounts because the top and bottom sides of the planar radical are indistinguishable (except for handedness) and suffer attack at identical rates.

★ *Exercises*

Predict the stereochemistry of the organic product of each reaction.

1. $R-O-O-R \longrightarrow 2\ R-O\cdot$

   initiation:

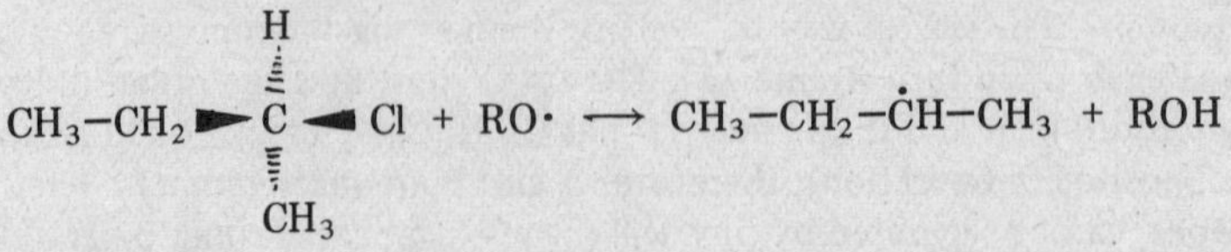

   $CH_3-CH_2-\overset{H}{\underset{CH_3}{C}}-Cl + RO\cdot \longrightarrow CH_3-CH_2-\dot{C}H-CH_3 + ROH$

$(n\text{-}C_4H_9)_3Sn\text{-}H + CH_3\text{-}CH_2\text{-}\overset{\cdot}{C}H\text{-}CH_3 \longrightarrow CH_3\text{-}CH_2\text{-}CH_2\text{-}CH_3 + (n\text{-}C_4H_9)_3Sn\cdot$

propagation:

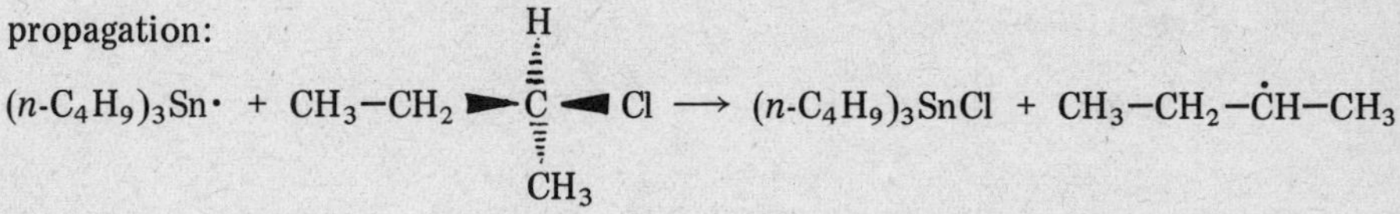

$(n\text{-}C_4H_9)_3Sn\cdot + CH_3\text{-}CH_2\blacktriangleright\underset{CH_3}{\overset{H}{C}}\blacktriangleleft Cl \longrightarrow (n\text{-}C_4H_9)_3SnCl + CH_3\text{-}CH_2\text{-}\overset{\cdot}{C}H\text{-}CH_3$

(plus termination steps)

overall:

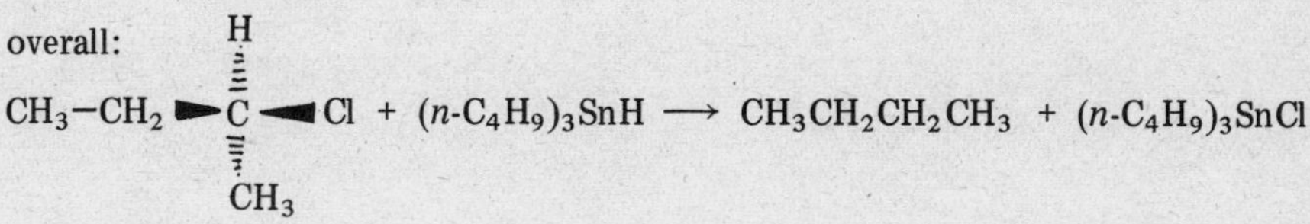

$CH_3\text{-}CH_2\blacktriangleright\underset{CH_3}{\overset{H}{C}}\blacktriangleleft Cl + (n\text{-}C_4H_9)_3SnH \longrightarrow CH_3CH_2CH_2CH_3 + (n\text{-}C_4H_9)_3SnCl$

2.

$CH_3CH_2CH_2\blacktriangleright\underset{Cl}{\overset{CH_3}{C}}\blacktriangleleft CH_2CH_3 \xrightarrow[(n\text{-}C_4H_9)_3SnH]{ROOR} CH_3CH_2CH_2\underset{CH_3}{CH}CH_2CH_3$

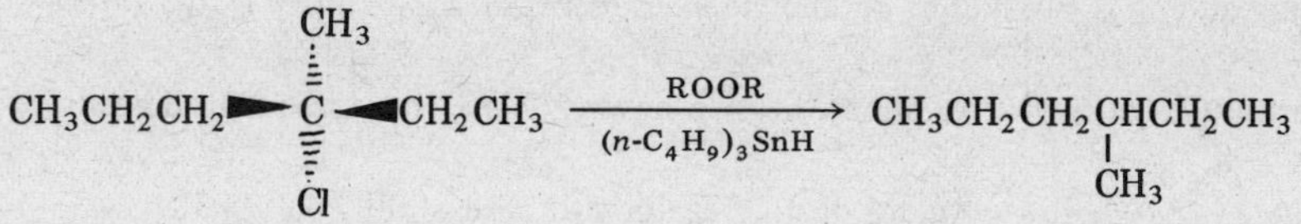

(mechanism same as in exercise 1)

3. $\langle\bigcirc\rangle\blacktriangleright\underset{H}{\overset{CH_3}{C}}\blacktriangleleft\underset{CH_3}{\overset{CH_3}{C}}\text{-}CCl_2\text{-}CH_3 \xrightarrow[(n\text{-}C_4H_9)_3SnH]{ROOR} \langle\bigcirc\rangle\text{-}\underset{CH_3}{\overset{CH_3}{CH}}\text{-}\underset{CH_3}{\overset{CH_3}{C}}\text{-}CHCl\text{-}CH_3$

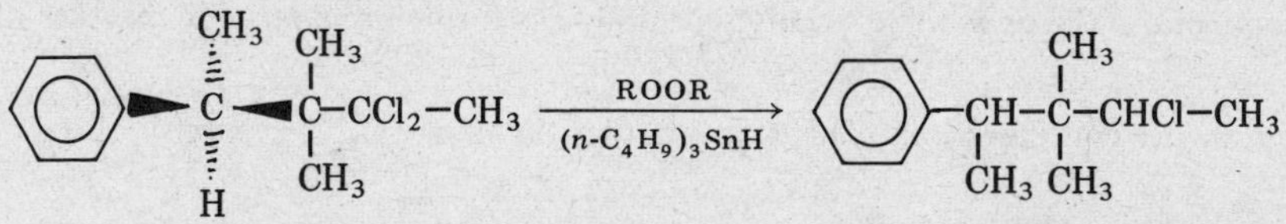

- - - - - - - - -

1. No product stereochemistry; reaction destroyed the chirality originally present (case 2).

2. $CH_3\text{-}CH_2\text{-}CH_2\blacktriangleright\underset{H}{\overset{CH_3}{C}}\blacktriangleleft CH_2\text{-}CH_3 + CH_3\text{-}CH_2\text{-}CH_2\blacktriangleright\underset{CH_3}{\overset{H}{C}}\blacktriangleleft CH_2\text{-}CH_3 =$

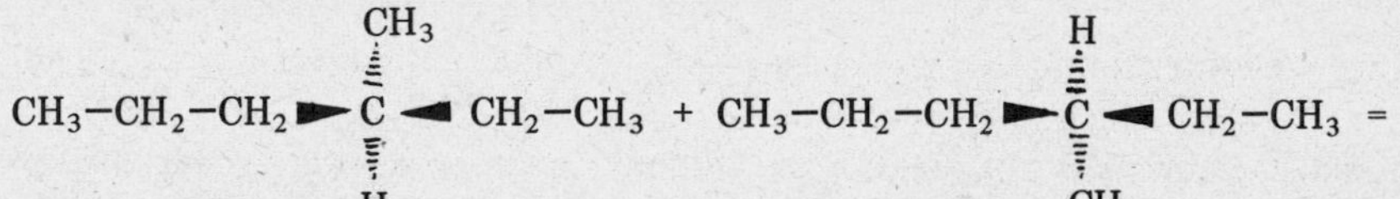

       50%   50%

                       racemization

Attack on the planar, achiral radical $C_3H_7\text{-}\overset{\cdot}{C}\underset{C_2H_5}{\overset{CH_3}{\diagup}}$ occurs top and bottom.

3. $\langle\bigcirc\rangle\blacktriangleright\underset{H}{\overset{CH_3}{C}}\blacktriangleleft\underset{CH_3}{\overset{CH_3}{C}}\blacktriangleright\underset{Cl}{\overset{H}{C}}\blacktriangleleft CH_3 + \langle\bigcirc\rangle\blacktriangleright\underset{H}{\overset{CH_3}{C}}\blacktriangleleft\underset{CH_3}{\overset{CH_3}{C}}\blacktriangleright\underset{H}{\overset{Cl}{C}}\blacktriangleleft CH_3$

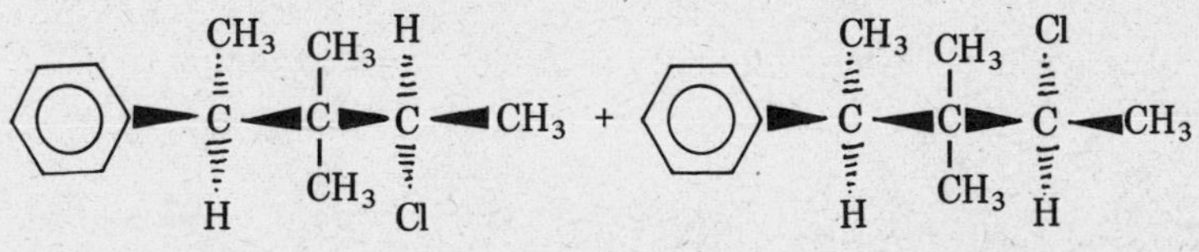

       $x\%$ $y\%$

Diastereomers formed in unequal amounts from the chiral radical

ELECTROPHILIC ADDITION

13. The two carbon atoms involved in addition reactions go from sp^2 to sp^3 hybridization.

$$\text{>C=C<} + \text{E-Nu} \longrightarrow -\underset{\underset{E}{|}}{\overset{|}{C}}-\underset{\underset{Nu}{|}}{\overset{|}{C}}-$$

If the starting alkene is capable of geometric isomerism, that capability is lost in the saturated product. On the other hand, the two new sp^3 >C< centers created could be chiral centers, $\overset{a}{\underset{b}{>}}\overset{*}{C}\overset{c}{\underset{d}{<}}$. If so, we are dealing again with case 3 of Table 13.1.

Since the alkene is planar, it is equally probable that attack will be from the top or the bottom; and, if the new >C<_E center is chiral, it is formed as a racemic mixture of two enantiomeric configurations. This is so whether the new >C<_{Nu} center is chiral or not, since it is formed later:

$$\underset{Nu-C}{\text{>C=C<}} \longrightarrow \underset{E}{\overset{\oplus}{\text{>C-C<}}}_{:Nu^{\ominus}} \longrightarrow \underset{E \ Nu}{\text{>C-C<}}$$

If the second center formed is asymmetric $\left(\overset{*}{\underset{Nu}{>C<}}\right)$, but the first one $\left(\underset{E}{>C<}\right)$ is not, then again equal quantities of enantiomers are formed; the product is racemic and optically inactive. This is because the intermediate carbonium ion center is planar, and its top side is identical with the bottom side, just as in the case of the radical intermediate we discussed in frame 12. If both new centers are chiral, the carbonium ion is chiral in exactly the same way that the $\phi-\overset{*}{\text{CHCl}}-\overset{\cdot}{\text{CH}}-\phi$ radical was (p. 294). Top and bottom are different, and the two products are formed in unequal amounts.

Example

Give an explicit stereochemical description of the products of HI addition to (Z)-2,3-dichloro-2-butene.

$$\underset{\underset{CH_3}{}}{\overset{\overset{Cl}{}}{>}}C=C\underset{\underset{CH_3}{}}{\overset{\overset{Cl}{}}{<}} + \text{HI} \longrightarrow CH_3-\overset{*}{C}HCl-\overset{*}{C}lCl-CH_3$$

Solution:

Two intermediate carbonium ions are formed in equal amounts:

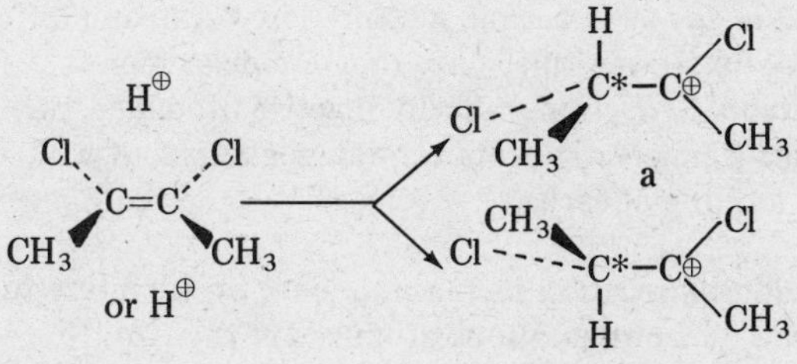

These ions are enantiomers, as seen by turning b upside down to get a different view of the same ion, b':

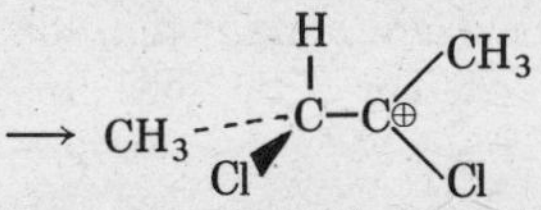

The view down the C—C bond best shows the mirror-image relationship:

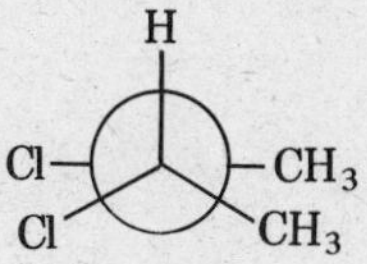

Top and bottom attack on the $-\overset{\oplus}{\text{C}}\!\!<$ center by $I^{\ominus}$ proceed at different rates, giving

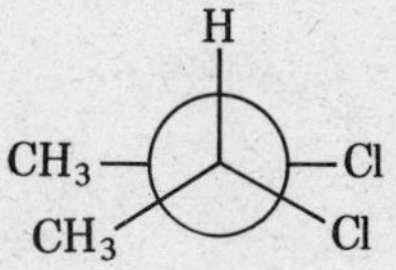

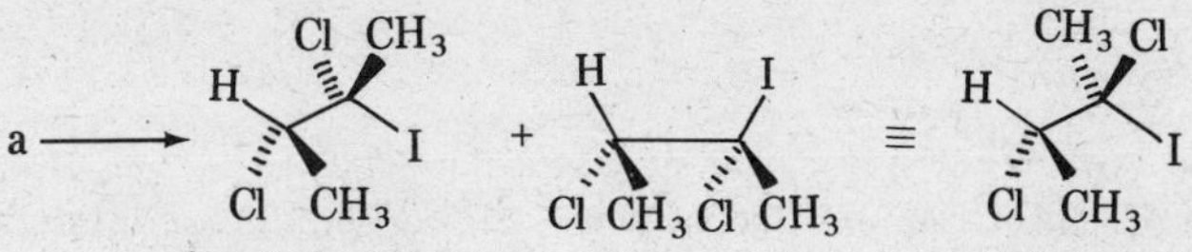

a ⟶   x% (via bottom)   +   ≡   y% (via top)

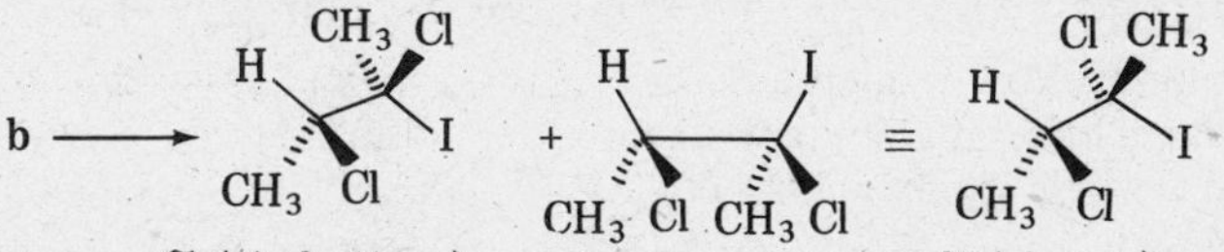

b ⟶   x% (via bottom)   +   ≡   y% (via top)

Compare the x% product derived from a and from b':

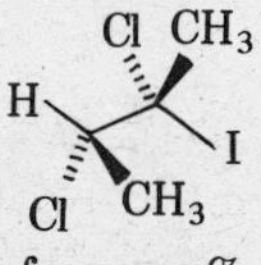

from a, x% 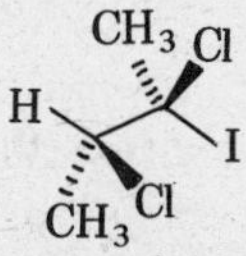   from b', x%

They are enantiomers, formed in identical yields, as they must always be if formed from symmetric reagents. The y% product is also a racemic mixture:

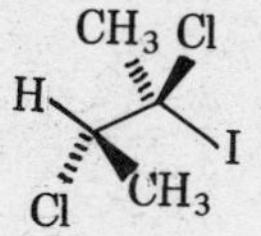

from a, y% 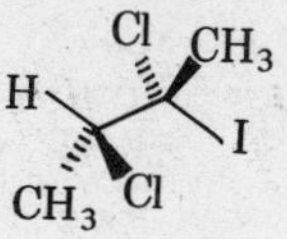   from b', y%

Except by coincidence, $x \neq y$; that is, the two racemic mixtures are formed in different amounts. This is as it should be according to our rule, since they are diastereomers of one another. Optically inactive reactants have given optically inactive products (two racemic mixtures). Two chiral centers were created from achiral reactants, but with equal numbers of molecules in each configuration at each C*.

When two C* are formed in an addition reaction such as this one, we are interested to know if there is a correlation between the configurations produced at the two. Notice in the example above that if $E^{\oplus}$ and $Nu^{\ominus}$ both come on from the same side (top attack of $H^{\oplus}$ and $I^{\ominus}$ on a or b'), the $y\%$ diastereomer is formed; this is called *syn addition*. If they approach from opposite sides, the $x\%$ diastereomer results; this is *anti addition*. The addition of acids to alkenes in general gives a mixture of syn ($y\%$) and anti ($x\%$) products in which either can predominate, depending on the alkene and the reaction conditions. Syn addition predominates when the dielectric constant of the solvent is low and substituents stabilize the carbonium ion intermediate. This is because under these conditions the cation ($-\overset{\oplus}{C}\diagdown$) and anion ($Nu^{\ominus}$) fragments of the intermediate are held close by electrostatic attraction, and the second step of the addition tends to occur before they rotate or diffuse apart:

$$\underset{E-Nu}{\diagup\!\!C\!\!=\!\!C\!\diagdown} \;\longrightarrow\; \underset{E \;\; :Nu^{\ominus}}{\diagup\!\!C\!-\!\overset{\oplus}{C}\!\diagdown} \;\longrightarrow\; \underset{E \;\; Nu}{\diagup\!\!C\!-\!C\!\diagdown} \qquad \text{syn addition}$$

Anti addition predominates when the electrophile temporarily adopts a central position between the alkene carbons, bonding partly to each:

$$\underset{E-Nu}{\diagup\!\!C\!\!=\!\!C\!\diagdown} \;\longrightarrow\; \underset{\underset{\oplus}{E} \; + \; Nu^{\ominus}}{\diagup\!\!C\!-\!C\!\diagdown}$$

This happens in addition of acids to simple alkenes where the carbonium ion intermediate is not especially stable, probably via a π complex: $\diagup\!\!C\!\overset{\overset{H^{\oplus}}{\vdots}}{=\!=\!=}C\!\diagdown$. And it nearly always happens in halogen addition due to internal stabilization of the carbonium ion by covalent bonding with the neighboring halogen atom (a weak nucleophile) to form a cyclic halonium ion:

$$\underset{Br-Br}{\diagup\!\!C\!\!=\!\!C\!\diagdown} \;\longrightarrow\; \underset{:\ddot{B}r:}{\diagup\!\!C\!-\!\overset{\oplus}{C}\!\diagdown} \;\rightleftharpoons\; \underset{\underset{\oplus}{:\ddot{B}r:}}{\diagup\!\!C\!-\!C\!\diagdown}$$

These *bridged intermediates* can be attacked by $Nu^{\ominus}$ only from the back side:

$$\underset{\underset{\oplus}{Br}}{\overset{^{\ominus}\!:\ddot{B}r:}{\diagup\!\!C\!-\!C\!\diagdown}} \;\longrightarrow\; \underset{Br}{\diagup\!\!C\!-\!C\!\diagdown} \qquad \underset{\underset{\oplus}{H}}{\overset{:\ddot{X}:^{\ominus}}{\diagup\!\!C\!=\!\!=\!\!C\!\diagdown}} \;\longrightarrow\; \underset{H}{\diagup\!\!C\!-\!C\!\diagdown}^{X}$$

Thus the result is anti addition. In the bromination it is 100 percent anti. Such a reaction is called *stereoselective*.

STEREOCHEMISTRY

Examples

Predict the stereochemistry of the products of each reaction.

1. $CH_2=CH_2 + Cl_2 \longrightarrow CH_2Cl-CH_2Cl$

2. 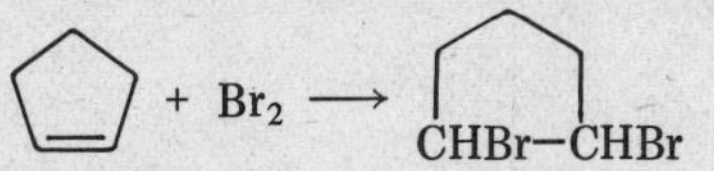 cyclopentene $+ Br_2 \longrightarrow$ cyclopentane-CHBr—CHBr

Solutions:

1. Neither reactant nor product has stereochemical possibilities:

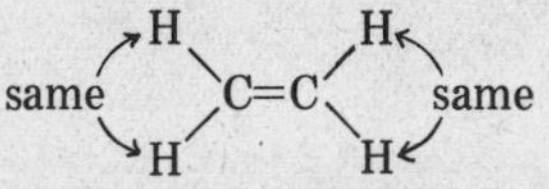

   same $\begin{array}{c}H\\H\end{array}$ C=C $\begin{array}{c}H\\H\end{array}$ same $CH_2Cl-CH_2Cl \longleftarrow$ no C*

2. The product has stereochemical possibilities (two C*).

   Reaction type: electrophilic addition.

   Mechanism: $-CH=CH- \longrightarrow$ bridged $\overset{+}{C}-C$ with Br $+ Br^{\ominus} \longrightarrow -CHBr-CHBr-$.

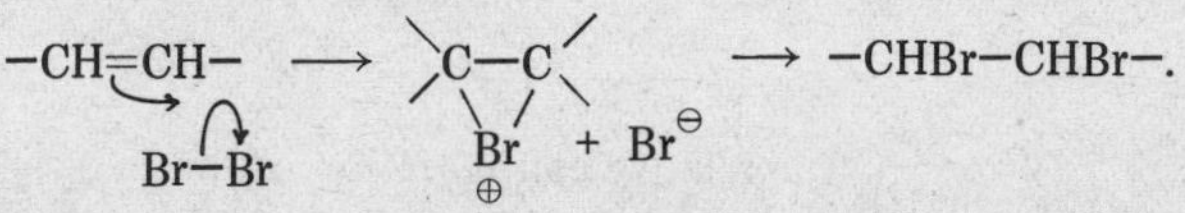

   Stereochemistry: rear attack on bridged intermediate always gives overall result of anti addition.

   Application to present case: cyclopentene + Br—Br $\longrightarrow$ bromonium ion $+ Br^{\ominus}$

   Br$^{\ominus}$ attacks bromonium ion $\longrightarrow$ Br/Br (50%) + Br/Br (50%)   racemic mixture

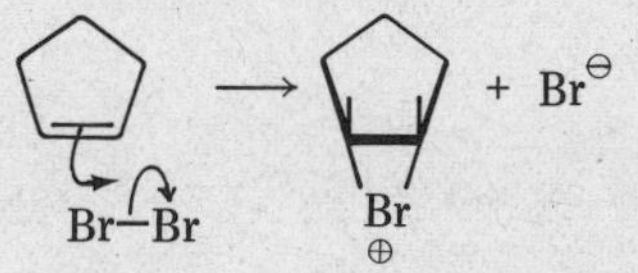

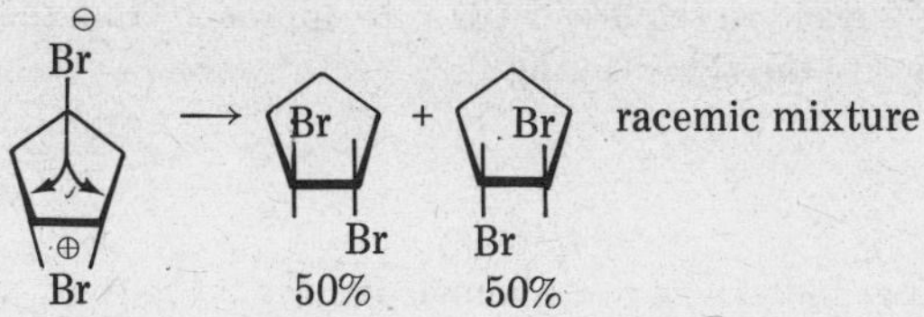

   As usual, in the cyclic case it is easy to see that addition has been anti—the two Br atoms end up on opposite sides of the ring; the product is *trans*-1,2-dibromocyclohexane. Attack of Br$^{\ominus}$ occurs in two equally probable ways, and the product is a racemic mixture.

★ *Exercises*

Predict the stereochemistry of the organic product of each reaction.

1. 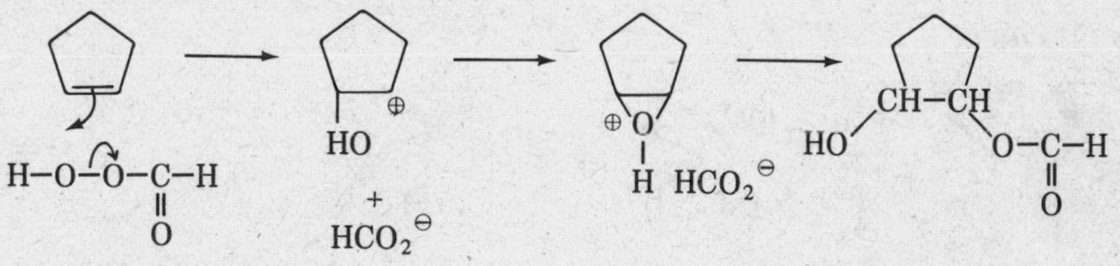 cyclopentene with H—O—O—C—H (O) $\longrightarrow$ protonated epoxide, HO $+$ HCO$_2^{\ominus}$ $\longrightarrow$ epoxide with $\overset{+}{O}$, H, HCO$_2^{\ominus}$ $\longrightarrow$ HO-CH—CH-O-C-H (O)

302 HOW TO SUCCEED IN ORGANIC CHEMISTRY

2. 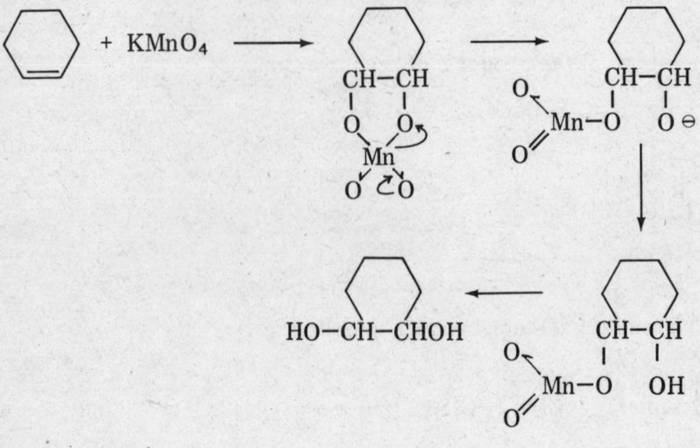 Cyclohexene + KMnO₄ → [cyclic Mn intermediate with CH-CH, O, O, Mn, O, O] → [intermediate with Mn-O, O⁻] → [CH-CH, Mn-O, OH] → HO-CH-CHOH (cyclohexane diol)

3. 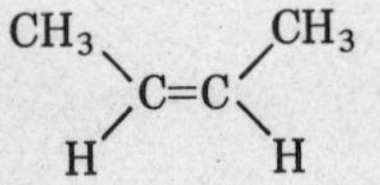 $\begin{array}{c}CH_3\\H\end{array}C=C\begin{array}{c}CH_3\\H\end{array}$ + Cl_2 + H_2O → C_4H_9OCl + HCl

- - - - - - - - -

1. 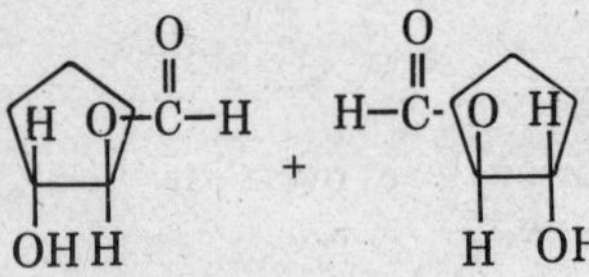 $H\ O-\overset{O}{\overset{\|}{C}}-H$ + $H-\overset{O}{\overset{\|}{C}}-O\ H$ The relative configuration of the two centers is fixed in
   $\ \ \ OH\ H$ $\ \ \ H\ OH$ the last step, which is S_N2 displacement by $HCO_2^\ominus$ on 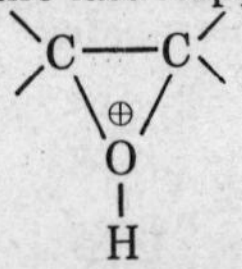
   50%   50% $\overset{}{\underset{\underset{H}{\overset{\oplus}{\underset{O}{|}}}}{C---C}}$. A racemic mixture results from the equally

probable attack on either carbon of the $\underset{\underset{O}{\diagup}}{C---C}$ ring. Only trans products result,

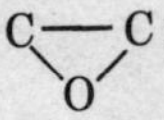

because rearside attack in the S_N2 reaction requires $HCO_2^\ominus$ to approach from the
side of the cyclohexane ring opposite the $\underset{\underset{O}{\diagup}}{C---C}$ ring.

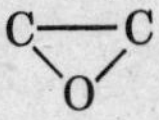

2. 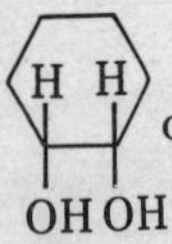 [cyclohexane with H H up, OH OH down] only. The configurations at both the new asymmetric carbons are fixed in the
first step. Although the detailed mechanism of this step is not available, it can only
lead to cis product, a, since there is not enough length to the O-Mn-O chain
to reach from one side of the cyclohexane ring to the other (b). Since the
two C-O bonds formed in the first step are never broken subsequently,
the configurations cannot subsequently change.

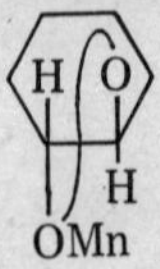

   b [structure with H/O up, H/OMn down]   a [structure with H H up, O O Mn down]

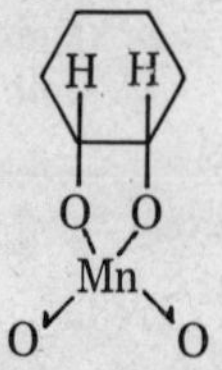

3. 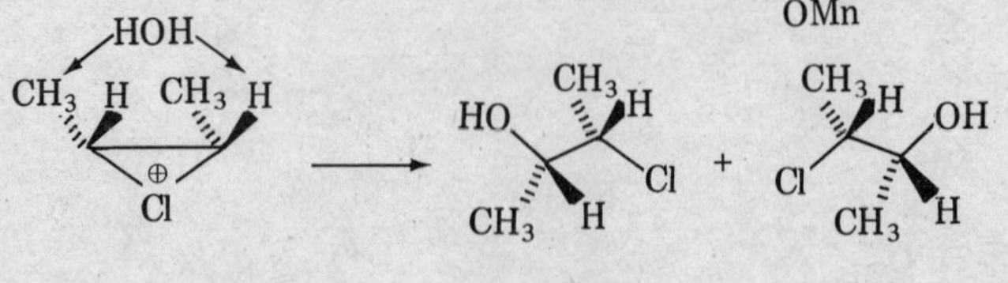 $CH_3\ H\ CH_3\ H$ with HOH arrow, Cl, ⊕ → $\begin{array}{c}CH_3\ H\\HO\diagdown\diagup\\\diagup\diagdown Cl\\CH_3\ H\end{array}$ + $\begin{array}{c}CH_3\ H\\Cl\diagdown\diagup OH\\\diagup\diagdown\\CH_3\ H\end{array}$

   50%   50%

RADICAL ADDITION

14. The stereochemical situation for radical addition is just the same as that with simple electrophilic addition, treated in frame 13. The alkene is attacked equally from above and below to give enantiomeric radicals that are attacked in the second propagation step at unequal rates from above and below. For example:

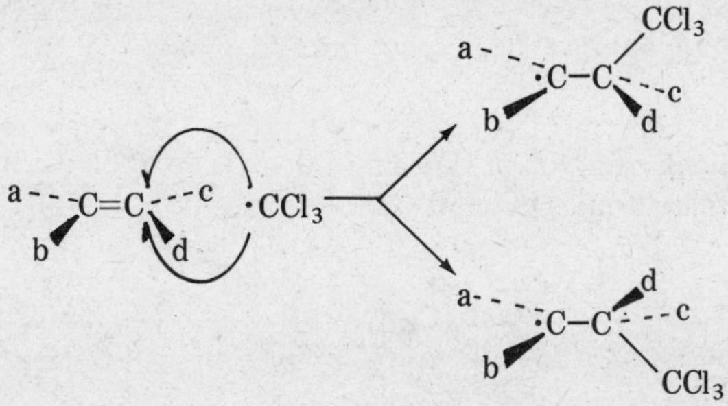

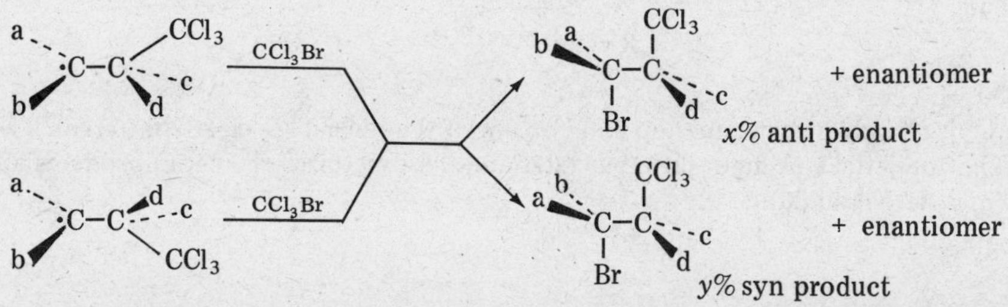

Again the diastereomeric syn and anti products are formed in different amounts. Bridged intermediates are not present to control the addition stereochemistry, and usually neither the syn nor the anti product greatly predominates; in other words, the reactions are not very stereoselective.

Exercise

Predict the stereochemistry of the product of the reaction

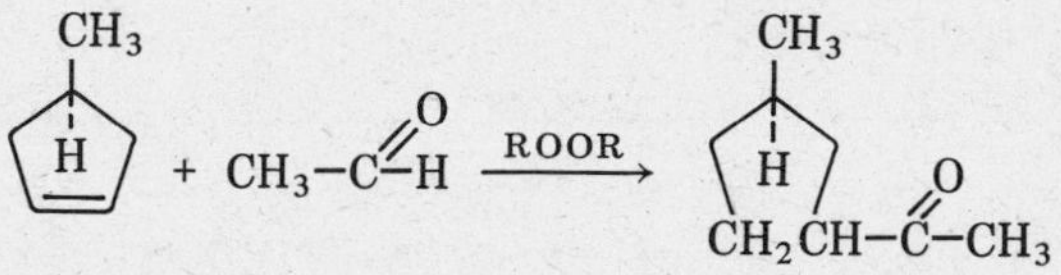

$$\text{cyclopentene-CH}_3\text{-H} + CH_3-C(=O)H \xrightarrow{ROOR} \text{cyclopentyl-CH}_3\text{-H with } CH_2CH-\overset{O}{\underset{\|}{C}}-CH_3$$

- - - - - - - -

Stereochemical possibilities: The reactant has one C*, which is not disturbed in the reaction. One new C* is generated by the reaction. Reaction type: radical addition. [Addition because a double bond becomes saturated, radical because the peroxide catalyst (radical source) gives the radical mechanism away.]

Mechanism: $RO\frown\frown OR \longrightarrow 2\ RO\cdot$

$$RO\cdot\ +\ H-\overset{O}{\underset{\|}{C}}-R\ \longrightarrow\ R-O-H\ +\ R-\overset{O}{\underset{\|}{C}}\cdot$$

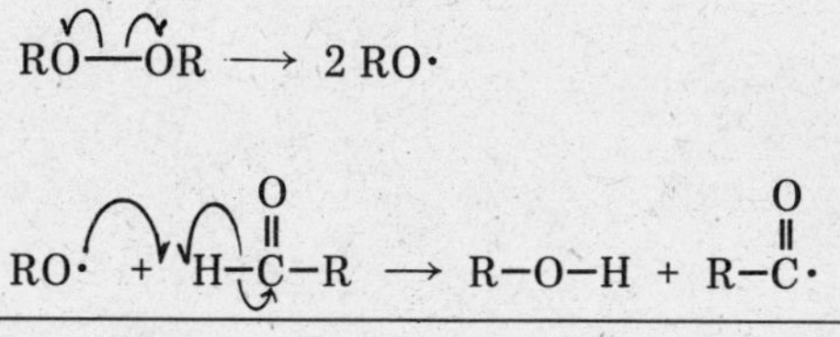

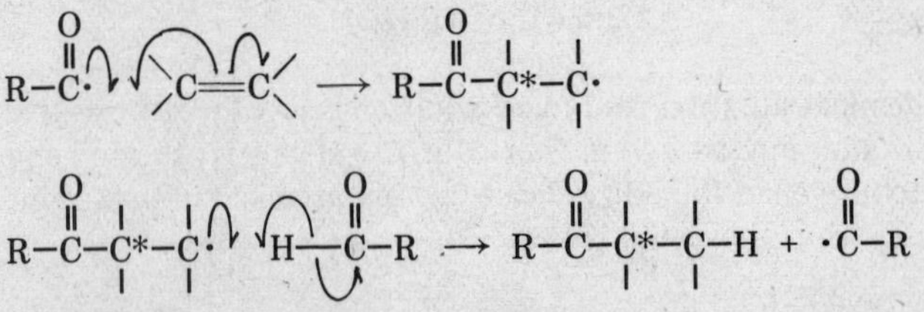

$$R-\overset{O}{\underset{\|}{C}}\cdot \quad \underset{\diagdown}{\overset{\diagup}{C}}=\underset{\diagdown}{\overset{\diagup}{C}} \quad \longrightarrow \quad R-\overset{O}{\underset{\|}{C}}-\overset{|}{\underset{|}{C^*}}-\overset{|}{C}\cdot$$

$$R-\overset{O}{\underset{\|}{C}}-\overset{|}{\underset{|}{C^*}}-\overset{|}{C}\cdot \quad H-\overset{O}{\underset{\|}{C}}-R \quad \longrightarrow \quad R-\overset{O}{\underset{\|}{C}}-\overset{|}{\underset{|}{C^*}}-\overset{|}{\underset{|}{C}}-H \;+\; \cdot\overset{O}{\underset{\|}{C}}-R$$

(plus termination steps)

Stereochemistry: attack from top or bottom of the alkene by $R-\overset{O}{\underset{\|}{C}}\cdot$ in the third step determines the configuration of the new C*:

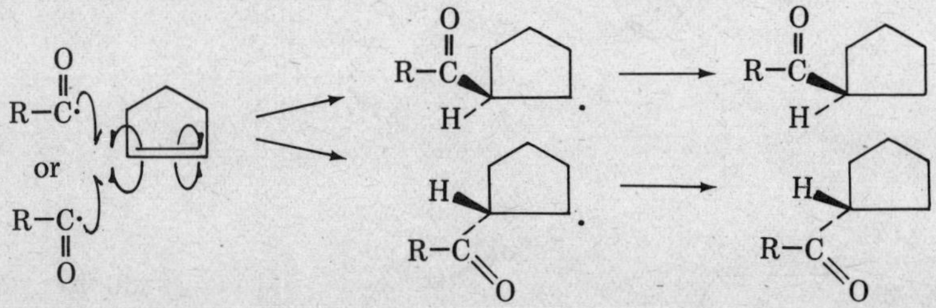

Since a C* is already present, the top and bottom of the radical are clearly different; top and bottom attack occur at different rates and the two (diastereomeric) products are formed in different amounts.

UNIT FOURTEEN
Halides

The generic reactions of the alkyl halides are few in number but exceedingly important. They represent a very large number of specific reactions that connect all the important functional groups. These halide reactions involve three of the nine standard mechanisms, which must be learned at this point. We take up in turn: understanding and organizing the halide reactions, learning the reactions, and solving problems based on these reactions and mechanisms.

OBJECTIVES

When you have completed this unit, you should be able to:

- Predict the occurrence of a large number of viable nucleophilic substitution reactions and write their equations (frames 1, 2).
- Predict whether elimination or substitution products (or both) will result from reaction of a given halide under given conditions, and predict the structure of the alkene formed (frames 6, 9, 10).
- Explain why and how conversion of a halide to an organometallic compound reverses the reactivity of the compound (frame 1).
- Write the mechanism for the $S_N 2$, $S_N 1$, $E2$, and $E1$ reactions of the halides and organometallic compounds and predict the way reactivity varies with the structure of the halide (frames 4, 5, 7, 8).

If you think you may have already achieved these objectives and might skip all or part of this unit, take a self-test consisting of the exercises marked with a star in the margin. If you miss any problems, study the frames in question and rework the problems before proceeding. If you are not ready for a self-test, proceed to frame 1.

UNDERSTANDING THE REACTIVITY OF R—X

1. Chart 14.1 organizes the generic reactions of the alkyl halides around the three basic modes of reactivity. It also emphasizes the ability of the R—X (unique among all the functional groups) to make the alkyl group R either a nucleophile or an electrophile. All the reactions are easy to identify as Lewis acid + base reactions. The ability of the R—X to react with *both* nucleophiles and electrophiles accounts for their great versatility and utility— —X can be transformed in one or a few steps into any other function.

In reactions of RX, the halogen atom always ends up as $:\ddot{X}:^{\ominus}$, which means that it gains an electron and is reduced. In the reaction with metals, the metal is the reducing agent. In the absence of metals (or other added reducing agents) X snatches the needed electron from the R—X bond, departing with both of what was formerly a shared electron pair. This act defines the role of a *leaving group*, played in this case by X:

R−X. The immediate fragments resulting are $:\ddot{X}:^{\ominus}$ and the carbonium ion $R^{\oplus}$, a strong electrophile.

In some cases $R^{\oplus}$ is actually formed, then rapidly destroyed by reactions with nucleophiles or bases to give the products of reactions ① and ②:

$$>\!\!CH-\overset{|}{\underset{|}{C}}{}^{\oplus} + :Nu^{\ominus} \longrightarrow >\!\!CH-\overset{|}{\underset{|}{C}}-Nu \quad \text{(substitution product)}$$

$$-\overset{|}{\underset{H}{C}}-\overset{|}{\underset{|}{C}}{}^{\oplus} \longrightarrow >\!\!C\!=\!C\!< + B-H^{\oplus} \quad \text{(elimination product)}$$
:B

In other cases the leaving group never departs spontaneously. The substitution and elimination reactions can nevertheless occur, but Nu: and B: must be considerably

---

Chart 14.1

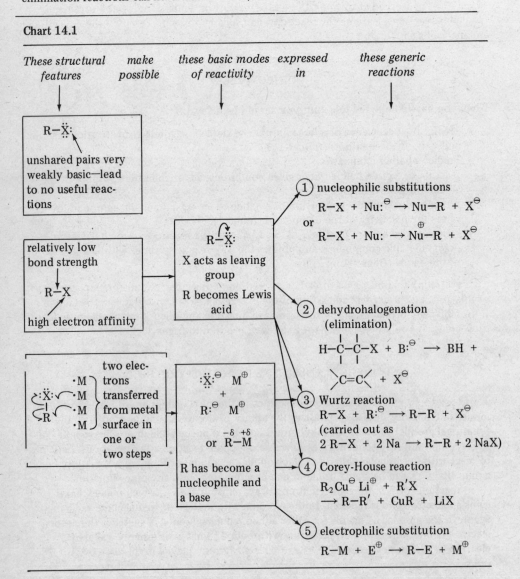

stronger to make it happen. The leaving group departs at the same time Nu: or B: attacks:

$$\text{>CH-C-X} \longrightarrow \text{>CH-C-Nu} + X^{\ominus}$$
$$\text{Nu:}$$

$$^{\ominus}\text{B:} \quad -\underset{H}{\overset{|}{C}}-\underset{|}{\overset{|}{C}}-X \longrightarrow \text{>C=C<} + X^{\ominus}$$
$$\text{B-H}$$

Spontaneous departure of the leaving group ($R-X \longrightarrow R^{\oplus} + X^{\ominus}$) is energetically uphill. It happens only when $R^{\oplus}$ and $X^{\ominus}$ are at least modestly stabilized. $X^{\ominus}$ is not an inherently unstable species; in solution the most effective way of making it as stable as possible is to provide weak acid molecules to H bond to it. These are usually the solvent molecules (e.g., $X^{\ominus}\cdots$H-O-H in water, $X^{\ominus}\cdots$H-O-R in an alcohol as solvent, etc.). $R^{\oplus}$ is inherently unstable and, as a Lewis acid, is stabilized best by basic solvents, usually with a $-\ddot{\text{O}}-$ or a $=\ddot{\text{O}}:$ bonding unit.

There is also an internal influence on the stability of $R^{\oplus}$. The positive charge is most stable on a 3° carbon atom $\left(\begin{array}{c}C\\|\\C-\overset{|}{C}\oplus\\|\\C\end{array}\right)$ and least stable on a methyl carbon $\left(\begin{array}{c}H\\|\\H-\overset{|}{C}\oplus\\|\\H\end{array}\right)$. This is because alkyl groups can donate electron density better than H can. Thus alkyl groups stabilize the cation by reducing the electron deficit on the central C:

$$\underset{C}{\overset{C}{\overset{\downarrow}{C\rightarrow \overset{|}{C}\delta+}}} \quad \underset{C}{\overset{C}{C\rightarrow \overset{|}{C}\delta+}} \quad \underset{H}{\overset{H}{C\rightarrow \overset{|}{C}\delta+}} \quad \underset{H}{\overset{H}{H-\overset{|}{C}\delta+}}$$

←—— smaller charge ——— increasing stability ——— larger charge ——→

A more complete way to look at the stabilization by alkyl groups is this: if the alkyl groups donate electron density to the central carbon so that the positive charge on central carbon decreases, the rest of the positive charge must appear in the alkyl groups. That is, the ⊕ charge cannot be destroyed but only shifted:

$$\text{delocalized charge} \longrightarrow \left\{\begin{array}{c}C\delta+\\\delta+\;|\\C-\overset{|}{C}\delta+\\|\\C\delta+\end{array}\right. \text{vs.} \quad H-\overset{H}{\underset{H}{\overset{|}{C}}}\oplus \longleftarrow \text{localized charge}$$

The word describing this shift is *delocalization* of the plus charge. *Delocalization of charge stabilizes any ion*, just as physics tells us that the energy of a charged sphere decreases (stability increases) as the size of the sphere increases (delocalization of charge over a larger surface area).

The actual mechanism by which charge is delocalized into saturated alkyl groups is still not completely understood. The delocalization of a minus charge on an anion by the electron-withdrawing effect of electronegative atoms is easier to visualize:

$\text{F}-\underset{F}{\overset{F}{\overset{|}{\underset{|}{C}}}}:\ominus$ is really $\text{F}\overset{\delta-}{-}\underset{F\,\delta-}{\overset{F\,\delta-}{\overset{|}{\underset{|}{C}}}}:\delta-$. For most purposes it is sufficient to look at alkyl groups as behaving just the opposite of F—like electropositive atoms.

Another mechanism for delocalization of charge in ions is provided by the resonance phenomenon (Unit 9). The allyl cation, $\overset{\oplus}{C}H_2-CH=CH_2 \leftrightarrow CH_2=CH-\overset{\oplus}{C}H_2$, achieves perfect delocalization of the charge over two carbons: $\overset{+/2}{CH_2}\text{-----}CH\text{-----}\overset{+/2}{CH_2}$. In general, the allylic ($R-CH=CH-\overset{\oplus}{C}H_2 \leftrightarrow$ etc.) and benzylic $\left(\langle\bigcirc\rangle-\overset{\oplus}{C}H_2 \leftrightarrow \text{etc.}\right)$ cations are as stable as 2° or 3° alkyl cations.

Reactions ① and ⑤ in Chart 14.1 are written in slot-and-filler generic form. They actually represent a large number of reactions, corresponding to all the possible combinations of the permissible slot fillers. The nucleophilic substitution reaction has three slots:

$$\underset{\text{slot}}{\text{nucleophile}}\quad \underset{\text{slot}}{\text{alkyl}}\quad \underset{\text{slot}}{\text{leaving-group}}$$

$$Nu:^{\ominus} + R-X \rightarrow Nu-R + X^{\ominus} \tag{1}$$

There are, in fact, other groups that can fill the leaving group slot besides halogen; they are usually studied along with the alkyl halides. Let's use —L to represent any leaving group. The permissible slot fillers are listed in Table 14.1. In one equation and one

Table 14.1 *Permissible Slot Fillers for the $Nu:^{\ominus}$ (or $Nu:$) $+ R-L \rightarrow Nu-R + L^{\ominus}$ Reaction*

Permissible nucleophiles, Nu:	Permissible leaving groups, —L	Permissible alkyl groups, R
$:\ddot{F}:^{\ominus}, :\ddot{C}l:^{\ominus}, :\ddot{B}r:^{\ominus}, :\ddot{I}:^{\ominus}$	$-\ddot{F}: \rightarrow :\ddot{F}:^{\ominus}$	CH_3-
$:\ddot{O}-H^{\ominus}, :\ddot{O}-R^{\ominus}$	$-\ddot{C}l: \rightarrow :\ddot{C}l:^{\ominus}$	$R-CH_2-$ (1°)
$:C\equiv N:^{\ominus}$	$-\ddot{B}r: \rightarrow :\ddot{B}r:^{\ominus}$	R_2CH- (2°)
$:N{\overset{R}{\underset{R}{\diagdown}}}R, :P{\overset{R}{\underset{R}{\diagdown}}}R$	$-\ddot{I}: \rightarrow :\ddot{I}:^{\ominus}$	R_3C- (3°)
$:\ddot{S}-H^{\ominus}, :\ddot{S}-R^{\ominus}$	$-\overset{O}{\underset{O}{\overset{\|}{\underset{\|}{S}}}}-R \rightarrow :\overset{O}{\underset{O}{\overset{\|}{\underset{\|}{S}}}}-R$	$R-CH=CH-CH_2-$ (allylic)
$^{\ominus}:\ddot{O}-C-R$ with O		$Ar-CH_2-$ (benzylic)
$:R^{\ominus}, R-C\equiv C:^{\ominus}$	$-\ddot{O}-N\overset{O}{\diagdown}_O \rightarrow :\ddot{O}-N\overset{O}{\diagdown}_O$	
$:H^{\ominus}$ (latent in $\sqrt{H}-AlH_3^{\ominus}$)	$-\overset{\oplus}{N}{\overset{R}{\underset{R}{\diagdown}}}R \rightarrow :N{\overset{R}{\underset{R}{\diagdown}}}R$	
$H_2\ddot{O}:, R-\ddot{O}-H$	$-\overset{\oplus}{O}H_2 \rightarrow OH_2$	

table you actually have 17 × 8 × 6 = 816 generic reactions, corresponding to the 816 possible three-way combinations of slot fillers. An example of one such combination is the generic reaction

$$Br^\ominus + R-CH_2-OSO_2-\underset{}{\bigcirc} \rightarrow Br-CH_2-R + \underset{}{\bigcirc}-SO_3^\ominus$$

The number of specific reactions represented by this equation is unlimited, since we can conceive as many alkyl groups to fill the role of R— in the generic equation as we please, for instance $(CH_3)_2CH-(CH_2)_{15}-CH_2-$.

Table 14.2 shows the permissible slot fillers for the elimination reaction

$$:B^\ominus + H-\overset{|}{\underset{|}{C}}-\overset{|}{\underset{|}{C}}-L \rightarrow B-H + \hspace{-2pt}>\hspace{-4pt}C=C\hspace{-4pt}<\hspace{-2pt} + L^\ominus \tag{2}$$

Table 14.2 *Permissible Slot Fillers for the Base-Promoted Elimination Reaction*

$$B:^\ominus + H-\overset{|}{\underset{|}{C}}-\overset{|}{\underset{|}{C}}-L \rightarrow B-H + \hspace{-2pt}>\hspace{-4pt}C=C\hspace{-4pt}<\hspace{-2pt} + L^\ominus$$

Permissible $B:^\ominus$	Permissible $-L$
$OH^\ominus$	$-X$
$OR^\ominus$	$-O-\overset{O}{\underset{O}{\overset{\uparrow}{\underset{\downarrow}{S}}}}-R$
	$-\overset{\oplus}{N}R_3$

The permissible slot fillers for the electrophilic substitution reaction, ⑤, of the organometallic compounds,

$$\begin{array}{ccc} \text{alkyl} & \text{metal} & \text{electrophile} \\ \text{slot} & \text{slot} & \text{slot} \\ \searrow & \downarrow & \swarrow \\ R^\ominus & M^\oplus & + \quad E^\oplus \rightarrow R-E + M^\oplus \end{array} \tag{3}$$

are listed in Table 14.3. In most cases the electrophile that actually becomes bonded to R does not have an independent existence. It is transferred from the reagent to $R^\ominus$, as for instance $Br^\oplus$ is in the possible combination of slot fillers

$$Li^\oplus \underset{}{\bigcirc}:^\ominus \quad Br-Br \rightarrow \underset{}{\bigcirc}-Br + Br^\ominus Li^\oplus$$

Many more organometallic compounds (compounds with carbon-metal bonds) are known than are listed in Table 14.3. The three metals listed are the most useful. Mg is divalent and does not fit into the equations in the same way Li and Na do. The common organomagnesium compounds are R—Mg—X, the Grignard reagents. You can look at the magnesium slot filler as —Mg—X or $\overset{\oplus}{Mg}-X$, which acts like a monovalent metal and fills

Table 14.3 Permissible Slot Fillers for the $RM + E^{\oplus} \rightarrow R-E + M^{\oplus}$

Permissible M	Permissible R	Permissible $E^{\oplus}$	Equation	Reaction name
Li	Alkyl	$H^{\oplus}$ (latent in acids, HA)[a]	$R:\overset{\frown}{} H\overset{\frown}{}A \rightarrow R-H + A^{\ominus}$	Protonation
Na	Allylic	$X^{\oplus}$ (latent in $X-X$)	$R:\overset{\frown}{} X\overset{\frown}{}X \rightarrow R-X + X^{\ominus}$	Halogenation
Mg	Benzylic			
	Vinylic	$M'^{\oplus}$ (other metal ions)[b]	$R-M + M'^{\oplus} \rightarrow R-M' + M^{\oplus}$	Metal interchange
	Aryl	$\overset{+\delta}{C}=\overset{-\delta}{O} \leftrightarrow \overset{}{C}-\overset{\oplus}{O}$ (aldehydes, ketones)	$R:\overset{\frown}{}\overset{}{C}=O \rightarrow R-\overset{}{C}-O\,M^{\oplus}$	Nucleophilic addition
		$R'^{\oplus}$ latent in $R'-X$	$R:\overset{\frown}{} R'\overset{\frown}{}X \rightarrow R-R' + X^{\ominus}$	Alkylation

[a] These acids can be as weak as H_2O, ROH, etc.
[b] M' must lie to the left of M in the electropositivity order

$Hg < Sn < Cd < Zn < Al < Mg < Li$

slots similarly to —Li or Li$^\oplus$. The most commonly encountered metals, after Li, Na, and Mg, are those listed in the second footnote to Table 14.3.

The carbon-metal bond in these organometallic compounds is a strongly polar covalent bond (covalent bond with considerable ionic character):

$$R-M \leftrightarrow R{:}^{\ominus} M^{\oplus} \quad \text{or} \quad \overset{\delta-}{R}\!-\!\overset{\delta+}{M} \quad \text{or} \quad \overleftrightarrow{R-M}$$

The ionic character of the common C—M bonds increases in the order

$$Hg < Al < Mg < Li < Na$$

(nearly nonpolar)   (nearly pure ionic)

Thus these compounds possess variable amounts of *carbanion* ($R{:}^{\ominus}$) reactivity.

The great importance of the alkyl halides, their highly versatile reactivity, and their position as key intermediates in the synthesis of a great variety of other compounds all stem from (a) the natural polarity of the C—X bond: $\overset{\delta+}{\diagdown\!C}\!-\!\overset{\delta-}{X}$ and (b) its reversal when R—X is converted to R—M: $\overset{\delta-}{\diagdown\!C}\!-\!\overset{\delta+}{M}$. Consequently, the R—X are unique in combining high reactivity toward nucleophiles [equation (1)] with high reactivity toward electrophiles [equation (3)].

The electrophiles vary in their reactivity. The stronger ones (H—A, X$_2$) react with R$_3$Al, RMgX, RLi, and RNa. The weakest (such as $\overset{\delta+}{R}\!-\!\overset{\delta-}{X}$) react only with RNa. This last reaction looks complicated because R is playing two roles. R—Na is so extensively ionized that it is a good source of the potent nucleophile R$:^{\ominus}$, which fills the Nu: slot in the nucleophilic substitution reaction of a second molecule of R—X:

$$R{:}^{\ominus} Na^{\oplus} + R'-X \longrightarrow R-R' + Na^{\oplus} X^{\ominus}$$

This reaction occurs automatically if you attempt to make R—Na from R—X + Na. It is called the *Wurtz reaction*, reaction ③ in Chart 14.1.

$$R-X + 2\,Na \longrightarrow R-Na \leftrightarrow R{:}^{\ominus} Na^{\oplus} + 2\,Na^{\oplus} X^{\ominus}$$

$$Na^{\oplus} R{:}^{\ominus} + R-X \longrightarrow R-R + Na^{\oplus} X^{\ominus}$$

sum: $2\,R-X + 2\,Na \longrightarrow R-R + 2\,Na^{\oplus} X^{\ominus}$

Reaction ④ of Chart 14.1 is a more useful way to bring about the alkylation of an organometallic compound with an alkyl halide, because the alkyl groups in the two reagents can be different:

$$Li^{\oplus} R-\overset{\ominus}{Cu}-R + R'-X \longrightarrow R-R' + Li^{\oplus} X^{\ominus} + Cu-R$$

The mechanism is not known, but it cannot be the same as with R$^{\ominus}$ Na$^{\oplus}$ because the halide can be vinyl or aryl as well as alkyl.

★ *Exercises*

1. Write the nucleophilic substitution reaction, equation (1), for the following choices of slot fillers:

   (a) Nu$:^{\ominus}$ = CH$_3$—CH$_2$—O$^{\ominus}$,   L = $-\text{O}-\overset{\overset{\text{O}}{\uparrow}}{\underset{\underset{\text{O}}{\downarrow}}{\text{S}}}-\text{CH}_3$,   R = CH$_2$=CH—CH$_2$—

(b) Nu: = $(CH_3)_2N-CH_3$, L = $-I$, R = $C_6H_5-CH_2-$

2. Write the base-promoted elimination reaction, equation (2), for the following choices of slot fillers:

    (a) B = $C_6H_5-CH_2-O^\ominus$, R = $(CH_3)_3C-$, L = CH_3-SO_2-O-

    (b) B = $OH^\ominus$, R = cyclohexyl, L = $(CH_3)_3\overset{\oplus}{N}-$

3. Write the electrophilic substitution reaction, equation (3), for the following choices of slot fillers:

    (a) M = Li, R' = phenyl, $E^\oplus$ = $H^\oplus$, obtained from $H-O-\underset{\underset{O}{\|}}{C}-CH_3$

    (b) M = MgBr, R' = cyclopentyl, $E^\oplus$ = $\overset{\oplus}{C}(CH_3)_2-O^\ominus \leftrightarrow (CH_3)_2C=O$

4. Use equations (1) and (3) with Tables 14.1 and 14.3 to supply the missing species in these reactions:

    (a) $CH_3-CH_2-CH_2-CH_2-Li + H_2O \rightarrow$?

    (b) ? + $O=C=O \rightarrow CH_3-CH_2-C(=O)(O^\ominus)\, MgCl^\oplus$

    (c) $C_6H_5:^\ominus Na^\oplus$ + ? $\rightarrow C_6H_5-CH_2-CH_2-CH_3$

    (d) ? + $I_2 \rightarrow CH_2=CH-CH_2-I + Mg^{2\oplus} + 2I^\ominus$

    (e) $CH_3-CHBr-CH_3 \xrightarrow{\text{excess Na}}$?

    (f) $C_6H_5-CH_2-Cl + Na^\oplus CN^\ominus \rightarrow$?

    (g) $Na^\oplus S^\ominus-CH_3 + CH_3-O-\underset{\underset{O}{\|}}{\overset{\overset{O}{\|}}{S}}-C_6H_5 \rightarrow$?

    (h) ? + $Li^\oplus AlH_4^\ominus \rightarrow$ cyclopentane

    (i) $CH_3-CH(CH_3)-CH_2-Cl + ? \rightarrow CH_3-CH(CH_3)-CH_2-I + K^\oplus Cl^\ominus$

(j) $C_6H_5-\underset{\underset{C_6H_5}{|}}{\overset{\overset{C_6H_5}{|}}{P}}$ + ? → $C_6H_5-\underset{\underset{C_6H_5}{|}}{\overset{\overset{C_6H_5}{|}}{\overset{\oplus}{P}}}-CH_2-CH_2-CH_3$ $Br^{\ominus}$

(k) $CH_3-CH=CH-CH_2-Br + CH_3OH \longrightarrow ?$

(l) 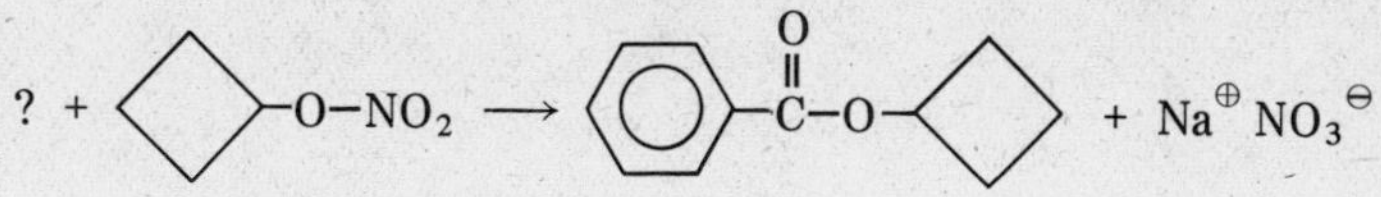 ? + ◇–O–NO$_2$ → ⌬–C(=O)–O–◇ + $Na^{\oplus} NO_3^{\ominus}$

- - - - - - - - -

1. (a) $CH_3-CH_2-O^{\ominus} + CH_2=CH-CH_2-O-SO_2-CH_3 \longrightarrow CH_3-CH_2-O-CH_2-CH=CH_2$
    + $^{\ominus}O-SO_2-CH_3$

(b) 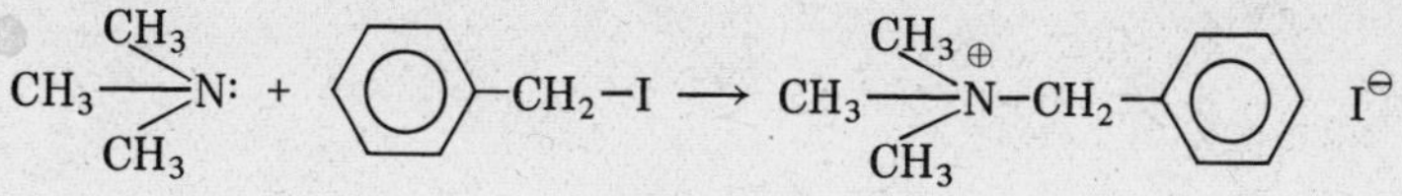 $CH_3-\underset{CH_3}{\overset{CH_3}{N:}}$ + ⌬–CH_2-I → $CH_3-\underset{CH_3}{\overset{CH_3}{\overset{\oplus}{N}}}-CH_2$–⌬ $I^{\ominus}$

2. (a) 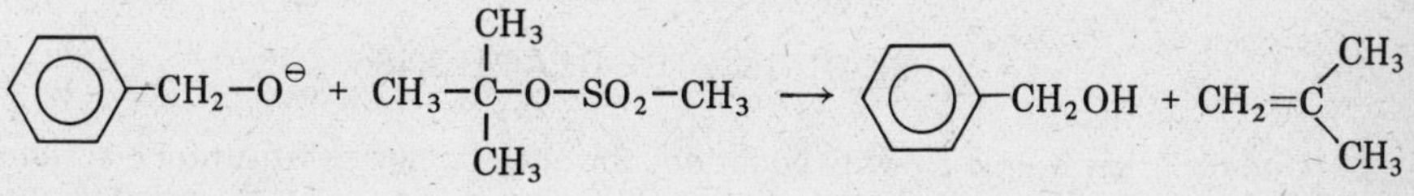 ⌬–$CH_2-O^{\ominus} + CH_3-\underset{CH_3}{\overset{CH_3}{\overset{|}{C}}}-O-SO_2-CH_3 \longrightarrow$ ⌬–$CH_2OH + CH_2=\underset{CH_3}{\overset{CH_3}{C}}$
    + $CH_3-SO_2-O^{\ominus}$

(b) 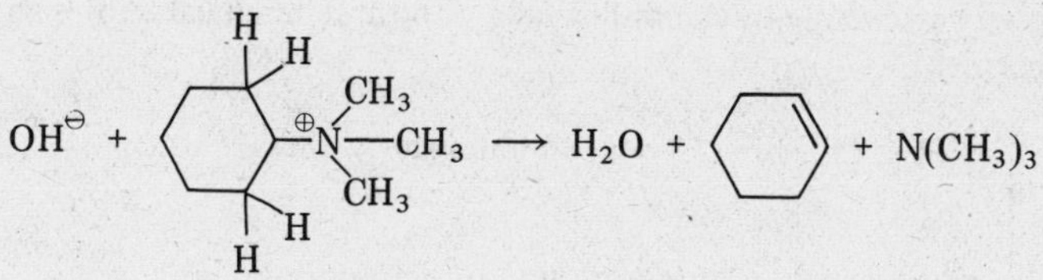 $OH^{\ominus}$ + [cyclohexyl-N(CH$_3$)$_3^{\oplus}$ with H's] → H_2O + ⬡ + $N(CH_3)_3$

3. (a) 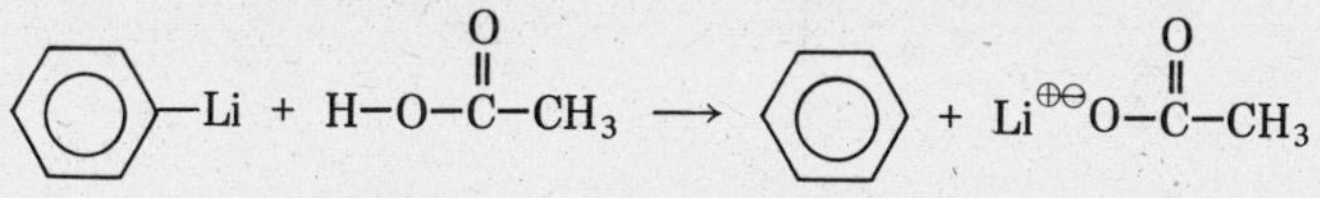 ⌬–$Li + H-O-\overset{O}{\overset{||}{C}}-CH_3 \longrightarrow$ ⌬ + $Li^{\oplus} O-\overset{O}{\overset{||}{C}}-CH_3$

(b) 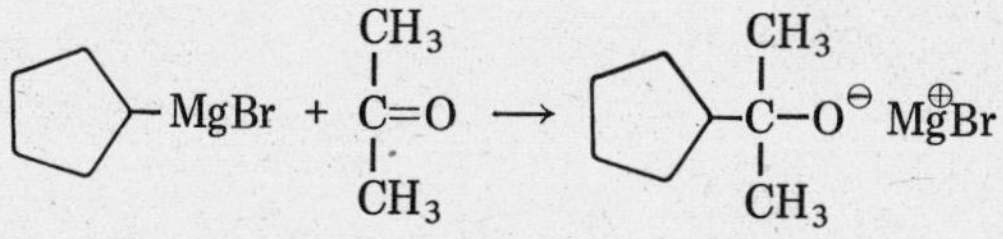 ⬠–$MgBr + \underset{CH_3}{\overset{CH_3}{\overset{|}{C}}}=O \longrightarrow$ ⬠–$\underset{CH_3}{\overset{CH_3}{\overset{|}{C}}}-O^{\ominus} M\overset{\oplus}{g}Br$

4. (a) $CH_3-CH_2-CH_2-CH_2-H + Li^{\oplus} OH^{\ominus}$

(b) $CH_3-CH_2-Mg-Cl$

(c) $CH_3-CH_2-CH_2-Cl, CH_3-CH_2-CH_2-Br$, etc.

(d) $CH_2=CH-CH_2-Mg-I$

(e) $\underset{CH_3}{\overset{CH_3}{>}}CH-CH\underset{CH_3}{\overset{CH_3}{<}}$

(f) 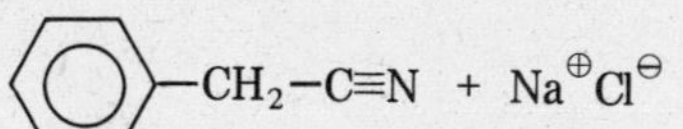 ⌬–$CH_2-C\equiv N + Na^{\oplus} Cl^{\ominus}$

314 HOW TO SUCCEED IN ORGANIC CHEMISTRY

(g) CH_3-S-CH_3 + Ph−S(=O)(=O)−O$^{\ominus}$ Na$^{\oplus}$

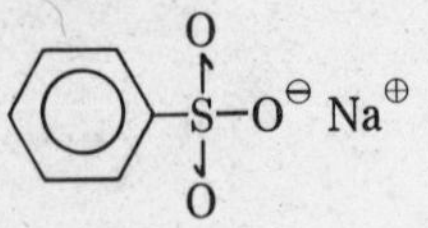

(h) 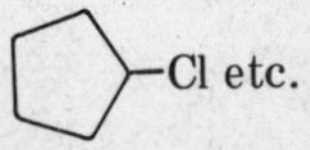 cyclopentyl−Cl etc.

(i) $K^{\oplus} I^{\ominus}$

(j) $CH_3-CH_2-CH_2-Br$

(k) $CH_3-CH=CH-CH_2-\overset{\oplus}{O}(H)(CH_3)$ Br$^{\ominus}$

(l) 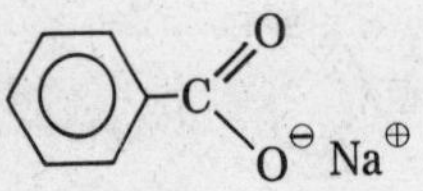 Ph−C(=O)−O$^{\ominus}$ Na$^{\oplus}$

LEARNING THE REACTIONS

2. Once you know how to write equations for nucleophilic substitution reactions and electrophilic substitution reactions using Tables 14.1 and 14.3, all you need to do is remember the contents of the tables. To make this easier, classify where possible. For example, the list of permissible nucleophiles falls into three main categories that break into smaller ones:

anionic nucleophiles
- $F^{\ominus}, Cl^{\ominus}, Br^{\ominus}, I^{\ominus}$ } $^{\ominus}$ on halogen
- $OH^{\ominus}, OR^{\ominus}$
- $SH^{\ominus}, SR^{\ominus}$ } $^{\ominus}$ on O or S
- $R-CO_2^{\ominus}$
- $:C\equiv N^{\ominus}$
- $R:^{\ominus}, R-C\equiv C:^{\ominus}$ } $^{\ominus}$ on C

electrically neutral nucleophiles
- $H_2\ddot{O}:, R\ddot{O}H$
- $R_3N:, R_3P:$

latent nucleophiles
- $H-\overset{\ominus}{Al}(H)(H)H$
- $H-\overset{\ominus}{B}(H)(H)H$ } source of H:$^{\ominus}$
- $F-\overset{\ominus}{B}(F)(F)F$
- $Cl-\overset{\ominus}{Al}(Cl)(Cl)Cl$ } source of $:\ddot{X}:^{\ominus}$

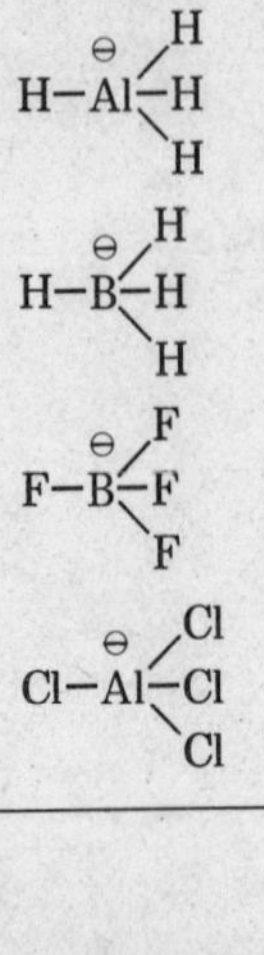

Lists are more easily memorized when grouped so that the smallest group contains no more than half a dozen items. Furthermore, the categorization usually has chemical significance. In this case the reactivity order of the nucleophiles is expressed (*roughly*) in the categorization: anionic > neutral > latent nucleophiles—provided that one does not use the anionic nucleophiles in aqueous or alcoholic solvents, which degrade their reactivity by H bonding: $Nu:^\ominus \cdots H-O-R$.

In this discussion the latent nucleophile class has been expanded to contain some species not included in Table 14.1. We will meet these species in many later contexts. The donation of nucleophile $H:^\ominus$ or $:\ddot{X}:^\ominus$ by the latent nucleophile occurs simultaneously with the reaction of the nucleophile:

$$H_3Al-H \; R-Cl \longrightarrow H_3Al + H-R + Cl^\ominus$$

$$F_3B-F \; R-NR_3 \longrightarrow F_3B + F-R + :NR_3$$

The list of leaving groups can be also be grouped for memorization.

Note that the list of permissible R is more restricted in nucleophilic than in electrophilic substitution (Table 14.1 versus 14.3). Nucleophilic substitution is not ordinarily shown by the vinyl and aryl halides, which have electron-poor halogen due to resonance of the types

You can remember this inertness in either of two ways: The halogen atom is electron-poor because it is acting as an electron donor within the R—X molecule; or the normal polarity of the C—X bond $\left(\overset{\delta+}{C}-\overset{\delta-}{X}\right)$ has been erased by donation of electrons from X toward R.

Exercises

1. (a) (If you are in a classroom situation, first adjust the lists in Table 14.1 to match your lecture notes or textbook coverage. Omit species that your course does not cover and add species if necessary.) In memorizing these (adjusted) lists, and future lists, first memorize the category titles (e.g., anionic nucleophiles, neutral nucleophiles, latent nucleophiles). Then work on one category at a time. If there are subcategories, memorize their titles next (e.g., ⊖ on halogen; ⊖ on O, S; ⊖ on C), then memorize the individual members of each subcategory (e.g., $F^\ominus$, $Cl^\ominus$, $Br^\ominus$, $I^\ominus$).

    When you want to recall the whole list, recall the main categories first, then the subcategories. You will find that most of the individual items suggest themselves immediately; the missing ones can usually be figured out by asking yourself what is needed to fill out the category or subcategory logically.

   (b) Practice reproducing the lists of Nu:, L, and R slot fillers in your (adjusted)

316 HOW TO SUCCEED IN ORGANIC CHEMISTRY

Table 14.1 from memory with paper and pencil. Check, correct, and/or complete if necessary.

2. Repeat exercise 1 for Tables 14.2 and 14.3.

3. Make sure you can write the completely generalized form of the nucleophilic substitution reaction, equation (1), and the elimination and electrophilic substitution equations, (2) and (3), from memory without cues. Now, without referring to Table 14.1, write balanced equations for specific examples of the nucleophilic substitution reaction that conform to these patterns:

(a) anionic Nu:, neutral —L

(b) anionic Nu:, cationic —L

(c) neutral Nu:, neutral —L

Verify the charge balance (sum of charges on left = sum of charges on right) of each equation written.

— — — — — — — —

3. Illustrative answers:

(a) $HO^\ominus + CH_3-Br \rightarrow CH_3-OH + Br^\ominus$

(b) $I^\ominus + \bigcirc\!\!\!\!\triangle\!-OH_2^\oplus \rightarrow \bigcirc\!\!\!\!\triangle\!-I + H_2O$

(c) $\langle\!\bigcirc\!\rangle\!-N\!\begin{smallmatrix}CH_3\\CH_3\end{smallmatrix} + CH_3-CH_2-CH_2-I \rightarrow \langle\!\bigcirc\!\rangle\!-\overset{\oplus}{N}\!\begin{smallmatrix}CH_3\\|\\|\\CH_3\end{smallmatrix}\!-CH_2-CH_2-CH_3\ I^\ominus$

3. Chart 14.2 is an alternative way of organizing the reactions that connect the alkyl halides with the other common functional groups. Like Chart 12.2, this one organizes according to oxidation stage. As before, the absence of an arrow means that there is no known reaction (or at least no generally useful reaction) connecting the two compounds. You can use this chart in several ways:

- As an alternative, pictorial way to recall the alkyl halide reactions.
- To understand and recall the reagents required.
- To single out the reactions that can be used to synthesize alkyl halides.

Note that the nucleophilic substitution reactions take place largely within oxidation stage I. This is because most of the nucleophiles react to replace C—X bonds with C—O, C—S, or C—N bonds, which are all bonds to electronegative elements and consequently leave the central carbon atom in the same oxidation state. The major exceptions are the nucleophiles $H:^\ominus$, $R:^\ominus$, and $R-C\equiv C:^\ominus$, which turn C—X bonds into C—H or C—C bonds and are reductions. The $H:^\ominus$ and $R:^\ominus$ reactions are written separately in the chart. The reactions of RX with metals are also reductions: the —CH$_2$—X carbon is reduced and M is oxidized from the 0 to the +1 oxidation state.

One other nucleophilic substitution is written out specifically in Chart 14.2—the $R-X + H_2O \rightarrow ROH + HX$ reaction. This is called *hydrolysis* of RX; like all hydrolysis reactions it is not an oxidation or reduction. Water is not an oxidizing or a reducing agent under ordinary conditions. Neither are proton-transfer reactions oxidations or reductions. The $RM + HA \rightarrow RH + MA$ reaction is a proton transfer; it takes place

Chart 14.2

Oxidation stage

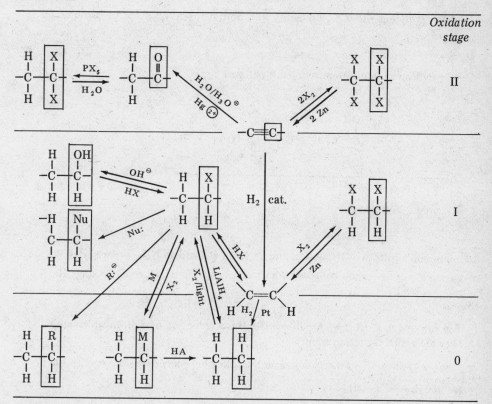

within oxidation stage 0. This gives us another viewpoint on the organometallic compound, R—M. It is effectively the *salt of an alkane*. Since alkanes are extraordinarily weak acids, $R:^\ominus$ is an extraordinarily strong base (Unit 11, frame 8). The protonation reaction

$$R^\ominus M^\oplus + H_2O \rightleftharpoons R-H + M^\oplus OH^\ominus$$
$$\text{sb} \quad\quad \text{sa} \quad\quad \text{wa} \quad\quad \text{wb}$$

is an acid-base equilibrium that lies far to the right. Thus water destroys organometallic compounds. This reaction is sometimes called a *hydrolysis reaction*. It should not be confused with the hydrolysis of RX, which gives the alcohol ROH. These are both non-redox reactions with water, but they take place in different oxidation stages:

$$R-X \xrightarrow{H_2O} R-OH \quad\quad\quad I$$

$$R-M \xrightarrow{H_2O} R-H \quad\quad\quad 0$$

Since R—X is a source of carbonium ions, $R^\oplus$, and R—M is a source of carbanions, $R:^\ominus$, this is a good way to see that $R^\oplus$ falls in oxidation stage I, $R:^\ominus$, in stage 0:

$$R^\oplus \curvearrowleft :OH_2 \longrightarrow R-OH_2^\oplus \xrightarrow{-H^\oplus} R-OH \quad\quad I$$

$$R:^\ominus \curvearrowleft H-\overset{\frown}{OH} \longrightarrow R-H + OH^\ominus \quad\quad 0$$

318 HOW TO SUCCEED IN ORGANIC CHEMISTRY

Finally, note that $-\underset{\underset{H}{|}}{\overset{\overset{H}{|}}{C}}-\underset{\underset{X}{|}}{\overset{\overset{X}{|}}{C}}-$, a dihalide with both halogens on the same carbon, and

$-\underset{\underset{H}{|}}{\overset{\overset{H}{|}}{C}}-\underset{\underset{H}{|}}{\overset{\overset{X}{|}}{C}}-$ hydrolyze to different oxygen functions:

$$-CX_2- \xrightarrow{H_2O} -\underset{\underset{OH}{|}}{\overset{\overset{OH}{|}}{C}}- \rightleftharpoons -\overset{\overset{O}{\|}}{C}- + H_2O \qquad\qquad II$$

$$-CHX- \xrightarrow{H_2O} -\underset{\underset{H}{|}}{\overset{\overset{OH}{|}}{C}}- \qquad\qquad I$$

All compounds with two OH, an OH and an X, an OH and an NH_2, or two NH_2 groups on the same carbon collapse spontaneously to $\rangle C=O$; all reside in stage II.

★ *Exercises*

1. The conversion R—X ⟶ R—M ⟶ R—H is a reduction overall. In which step does the actual reduction occur?

2. Using only allowed reactions, how can the following conversions be accomplished?

   (a) R—H ⟶ R—OH

   (b) $\rangle C=C\langle$ ⟶ —C≡C—

   (c) $-CH_2-\underset{\underset{OH}{|}}{CH}-$ ⟶ $-CH_2-CH_2-$ (three ways)

— — — — — — — — —

1. In the first step.

2. (a) R—H ⟶ R—X ⟶ R—OH

   (b) $\rangle C=C\langle$ ⟶ $-\underset{\underset{|}{}}{\overset{\overset{X}{|}}{C}}-\underset{\underset{|}{}}{\overset{\overset{X}{|}}{C}}-$ ⟶ —C≡C—

   (c) $-CH_2-CHOH-$ ⟶ $\rangle C=C\langle$ ⟶ $-CH_2-CH_2-$; or $-CH_2-CHOH-$ ⟶

   $-CH_2-CHX- \xrightarrow{LiAlH_4} -CH_2-CH_2-$; or $-CH_2-CHOH-$ ⟶ $-CH_2-CHX-$

   ⟶ $-CH_2-CHM-$ ⟶ $-CH_2-CH_2-$

SOLVING PROBLEMS BASED ON MECHANISM

4. First, you need to know what the standard mechanisms consist of. Second, you must be able to decide, in a given specific case, which standard mechanism applies. We treat

these questions in this frame. Third, you must identify which species will play the various roles in the generic mechanism, and write the resulting specific mechanism correctly (frame 5). Finally, you must be able to reason from the mechanism you write or to use rules to predict the reactivity order(s) for the reaction.

The halide reactions involve three mechanisms:

1. The S_N2 ←(requiring collision of two species in the slowest step) mechanism
   ←(substitution)
   ←(nucleophilic, that is by a nucleophile)

(mechanism V in Table 10.1). Nucleophile attacks rearside, on a line 180° from the C—X bond direction. The new Nu—C bond is forming while the C—L bond is breaking (*concerted* process).

This mechanism has two consequences that are not readily apparent from the mechanism as written in ordinary notation: $\text{Nu:}^{\ominus} \downarrow \quad \text{CH}_2\text{—X} \longrightarrow \text{Nu—CH}_2\text{—R} + \text{X}^{\ominus}$.
$\qquad\qquad\qquad\qquad\qquad\qquad\qquad\qquad | \atop \text{R}$

(a) The reaction *inverts* the configuration of the central carbon atom, much as a high wind inverts an umbrella. The *transition state* (highest point on the energy profile:

[energy diagram showing RL and RNu with t.s. between, vs. progress of reaction]

has Nu and L arranged on a line through C; the three remaining groups lie in a plane perpendicular to this line:

$$\overset{\ominus}{\text{Nu:}} \quad \overset{a}{\underset{c}{\text{b---C—L}}} \longrightarrow \left[\overset{a}{\underset{\underset{c}{b'|}}{\text{Nu}\cdots\overset{\delta-}{C}\cdots\overset{\delta-}{L}}} \right] \longrightarrow \text{Nu—}\overset{a}{\underset{c}{C\text{---}b}} + \text{L}^{\ominus}$$
$$\text{transition state}$$

The brackets are used to remind you that no transition state can ever be isolated. The dotted lines denote partially made/broken bonds.
In this example, the clockwise order of groups a, b, and c, viewed from their face of the tetrahedron, in the reactant is a, c, b (or c, b, a or b, a, c). In the product, the order is a, b, c (or b, c, a or c, a, b), that is, the opposite of the reactant, so that inversion has occurred. Expressed in alternative notation:

[stereochemistry diagrams showing reactant and product with inversion]

reactant → product

(b) The rearside attack of Nu: on the central carbon must be made through a corridor obstructed by the three other groups attached tetrahedrally to C:

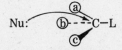

The more hydrogen atoms (small) among a, b, and c, the faster this process proceeds.

320 HOW TO SUCCEED IN ORGANIC CHEMISTRY

The more alkyl groups (large), the slower. Therefore, the reactivity order is

$$CH_3-X > 1° \; RX > 2° \; RX > 3° \; RX$$

2. The $S_N1/E1$ (elimination) mechanism (mechanism IV in Table 10.1) (requiring one molecule only in the slowest step; no collisions)

The substitution part of this mechanism accomplishes the same thing as the S_N2 mechanism, but in two steps: bond breaking, then bond making.

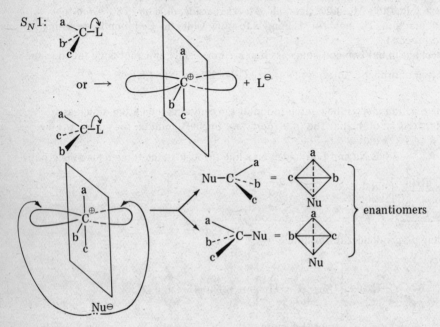

The first step, $R-L \longrightarrow R^\oplus + L^\ominus$, forms a high-energy intermediate, $R^\oplus$, with a short lifetime. The carbonium ion, $R^\oplus$, has three important properties:

(a) It is a planar species whose top and bottom present identical targets to the nucleophile.

(b) It is a sufficiently reactive Lewis acid to react with almost any nucleophile (base) it collides with.

(c) In competition with its reactions with nucleophiles, the carbonium ion stabilizes itself by giving up a proton to form an alkene.

Owing to the planar structure of $R^\oplus$, regardless of which enantiomer of R—X one begins with, the product R—Nu should be a racemic mixture of both enantiomers. The third property results in substantial elimination as a byproduct; this is the $E1$ branch of the reaction.

$E1$:

$$-\overset{|}{\underset{|}{C}}-\overset{|}{\underset{H}{C}}-L \longrightarrow -\overset{|}{\underset{|}{C}}-\overset{|}{\underset{H}{C}}\oplus + L^\ominus$$

$$-\overset{|}{\underset{H}{C}}-\overset{|}{\underset{|}{C}}\oplus \longrightarrow \;\; >C=C< \; + \; H-B^\oplus$$
:B

3. The $E2$ ← (two species collide in the slowest step) mechanism (mechanism VI (elimination)
in Table 10.1). This is another concerted mechanism. It is one of the "busiest" mechanisms, making or breaking a total of four bonds simultaneously. The reaction is rapid only when the halide can attain an anti-peri-planar conformation (L and H departing in directions 180° apart):

$E2$:

[Newman projection and sawhorse diagrams showing E2 elimination mechanism with base B: removing H while L leaves, forming alkene with H–B]

or

[Second Newman projection showing the same E2 process from a different perspective]

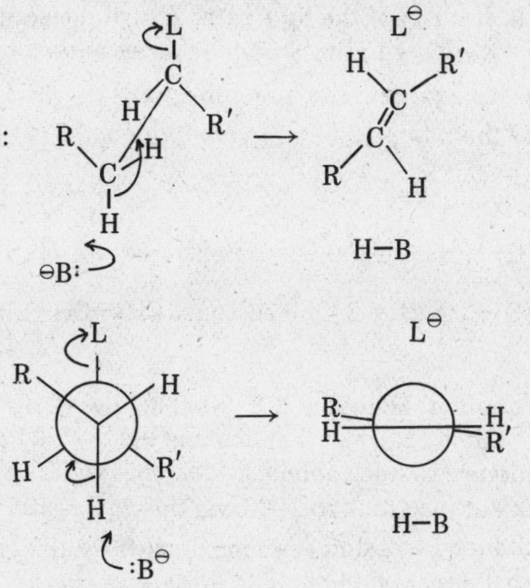

The function of the base is to pluck $H^{\oplus}$ from a β C–H bond of the halide. This requires a much stronger base than the deprotonation of the carbonium ion, $R^{\oplus}$ (last step of the $E1$ mechanism, p. 320).

When the leaving group is $-X$ (halides), $-OSO_2Ar$, $-OSO_2R$, or $-ONO_2$, the double bond is well developed in the transition state:

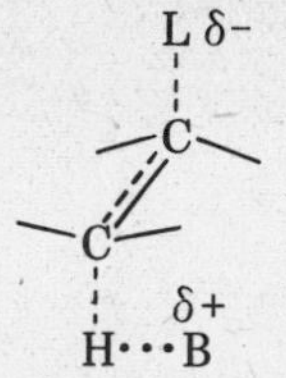

[Transition state diagram showing L with partial negative charge $\delta-$ leaving, C=C double bond forming, and H···B with partial positive charge $\delta+$]

This causes the transition-state energy to parallel the energy of the alkene product. The more stable the alkene, the lower in energy the transition state leading to it, the lower the activation energy for its formation, and the more rapidly it will be formed. Since the most substituted alkene is most stable, this means that the β H removed is the one on the most substituted carbon atom (3° > 2° > 1° C–H), if there is a choice.

When the leaving group is $-NR_3^{\oplus} \rightarrow :NR_3$, this is not true. β–H is removed preferentially from the least substituted carbon, leading to the least substituted alkene.

Since you know from the equation whether a given reaction is substitution or elimination, the decision about which mechanism applies is $E1$ versus $E2$ or S_N1 versus S_N2. The first case is easy: if the *base* is a *strong* one ($OH^{\ominus}$, $RO^{\ominus}$, $NH_2^{\ominus}$, or $R^{\ominus}$) the elimination goes by the $E2$ mechanism.

The decision in nucleophilic substitution involves more factors:
1. How stable the carbonium ion, $R^{\oplus}$, is. This determines how fast the $R-L \rightarrow R^{\oplus} + L^{\ominus}$ step takes place.

2. How hindered the rear side of RL is. This partly determines how fast the S_N2 reaction goes.

3. How reactive the nucleophile is. This is the other determinant of the rate of the S_N2 process. The $S_N1/E1$ path does not depend on the nucleophile's reactivity. $OH^\ominus$ is a strong nucleophile that promotes S_N2 reactions. H_2O is a weak nucleophile that promotes $S_N1/E1$ reaction by slowing the S_N2 path.

4. How much the reaction medium (solvent + solutes) assists the charge separation involved in the ionization step (R–L $\longrightarrow$ $R^\oplus$ + $L^\ominus$) of the S_N1 path. A high dielectric constant (symbol: D) lowers the energetic cost of separating $\oplus$ and $\ominus$ charges and increases the S_N1 rate. H-bonding of solvent to the leaving group, which requires $-OH$ or $>NH$ groups, stabilizes departing $L^\ominus$ and increases the rate:

$$R\text{–}L + \text{H-Solv} \longrightarrow R^\oplus + L^\ominus \cdots \text{H-Solv}$$

Water (H–O–H, $D = 78$) and formic acid (H–C(=O)–O–H, $D = 58$) are examples of solvents that promote the $S_N1/E1$ paths. Dioxane (O⌬O, $D = 2$) and acetone (CH_3–CO–CH_3, $D = 56$) are solvents that inhibit them.

The first two items above reinforce one another. Methyl and 1° alkyl halides (CH_3–X and R–CH_2–X) provide the least stable $R^\oplus$ (Unit 12, frame 12), making the $S_N1/E1$ reaction slow, and have minimal rearside hindrance to nucleophilic attack, making the S_N2 reaction fast. It is likely that their displacement reactions go 100% via the S_N2 route. 3° alkyl halides have maximal rearside hindrance (S_N2 slow) and maximal $R^\oplus$ stability (S_N1 fast). They go 100% via $S_N1/E1$. 2° halides are intermediate in all respects and probably undergo parallel reaction by both paths. The allyl and benzyl halides react rapidly with nucleophiles by both the S_N2 and S_N1 mechanisms because they are relatively unhindered and ionize to produce resonance-stabilized carbonium ions.

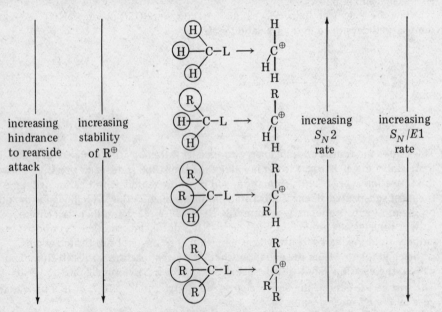

In some circumstances two mechanisms compete—some of the reactant molecules follow one mechanism, some follow the other. Obviously, for this to happen the two paths must proceed at comparable rates. The most common cases are

- $E1$ always accompanies S_N1 (in variable amounts), as already discussed.
- Secondary halides may substitute by simultaneous operation of the S_N1 and S_N2 mechanisms.
- When strong bases are used to bring about the $E2$ reaction of a tertiary halide, the $S_N1/E1$ reactions usually proceed simultaneously with the $E2$.

Examples

Predict for each reaction whether it goes by the S_N2 or the S_N1 mechanism.

1. $\triangleleft\!\!\!\!\diagup\!\!\!\!\triangleright\!-CH_2-I \;+\; Na^\oplus\, CN^\ominus \xrightarrow[\text{solvent}]{CH_3C\equiv N} \triangleleft\!\!\!\!\diagup\!\!\!\!\triangleright\!-CH_2-CN$

2. $CH_3-CH_2-\underset{\underset{CH_3}{|}}{\overset{\overset{Br}{|}}{C}}-CH\!\!\begin{array}{c}\diagup CH_3 \\ \diagdown CH_3\end{array} \xrightarrow{H_2O} CH_3-CH_2-\underset{\underset{CH_3}{|}}{\overset{\overset{OH}{|}}{C}}-CH\!\!\begin{array}{c}\diagup CH_3 \\ \diagdown CH_3\end{array}$

Solutions:

1. R–I is 1°; solvent is nonacidic and of intermediate dielectric constant (D = 40). A potent nucleophile is present. Therefore, S_N2.

2. R–Br is 3°; solvent is a weak acid (H-bonds to departing $Br^\ominus$) and has a high dielectric constant. Therefore, S_N1.

Exercises

Predict the mechanism by which each indicated product is formed.

1. $\bigcirc\!\!-\underset{\underset{Cl}{|}}{CH}\!-\bigcirc \;+\; CH_3OH \;\rightarrow\; \bigcirc\!\!-\underset{\underset{OCH_3}{|}}{CH}\!-\bigcirc \;+\; HCl$

2. $\bigcirc\!\!-\underset{\underset{CH_3}{|}}{\overset{\overset{CH_3}{|}}{C}}\!-Br \xrightarrow{CH_3-CH_2-O^\ominus K^\oplus} \bigcirc\!\!=\!C\!\!\begin{array}{c}\diagup CH_3 \\ \diagdown CH_3\end{array}$

3. $CH_3-(CH_2)_8-CH_2-O-\underset{\underset{O}{\|}}{\overset{\overset{O}{\|}}{S}}-\bigcirc \;+\; CH_3-C\equiv C\!:^\ominus K^\oplus \xrightarrow{NH_3(\text{liquid})}$

$CH_3-(CH_2)_8-CH_2-C\equiv C-CH_3 \;+\; \bigcirc\!\!-SO_3^\ominus K^\oplus$

4. $CH_3-\underset{\underset{CH_3}{|}}{\overset{\overset{CH_3}{|}}{C}}-Cl \xrightarrow[D=58]{H-\overset{\overset{O}{\|}}{C}-OH} CH_2=C\!\!\begin{array}{c}\diagup CH_3 \\ \diagdown CH_3\end{array} \;+\; HCl$

- - - - - - - -

1. S_N1 ($R^\oplus$ is strongly resonance stabilized)

2. $E2$ 3. S_N2 4. $E1$

5. Once you have decided which mechanism applies, writing out the mechanism is just a question of replacing the species Nu: (or Nu:$^\ominus$), B: (or B:$^\ominus$), R, R$^\oplus$, and/or L in the generic mechanisms (reactions 1 to 3, pp. 308, 309) with the **appropriate slot fillers**. This is easy because, except for R$^\oplus$, they are exactly the same slot fillers you use in writing product structures. R$^\oplus$ is what you get by removing —L from the reactant:

$$R\!-\!\!L \longrightarrow R^\oplus$$

Examples

Write all steps in the mechanisms of the examples from frame 4.

Solutions:

1. Mechanism = S_N2. Identify: Nu:$^\ominus$ = :C≡N:$^\ominus$ —L = —I → I$^\ominus$

$$R-L = \text{(cyclobutyl)}-CH_2-L$$

Substitution in Nu:$^\ominus$ ↷ R—L ⟶ Nu—R + L:$^\ominus$ gives

:N≡C:$^\ominus$ ↷ CH$_2$—I(cyclobutyl) ⟶ :N≡C—CH$_2$(cyclobutyl) + I$^\ominus$

2. Mechanism = S_N1. Identify: Nu: = H$_2$Ö: —L = —Br → Br$^\ominus$

$$R-L = CH_3-CH_2-\underset{\underset{CH_3}{|}}{\overset{\overset{Br}{|}}{C}}-CH\underset{CH_3}{\overset{CH_3}{\diagup}} \qquad R^\oplus = CH_3-CH_2-\underset{\underset{CH_3}{|}}{\overset{\oplus}{C}}-CH\underset{CH_3}{\overset{CH_3}{\diagup}}$$

Substitution in R—L ⟶ R$^\oplus$ + L$^\ominus$

Nu:$^\ominus$ ↷ R$^\oplus$ ⟶ Nu—R

gives

$$CH_3-CH_2-\underset{\underset{CH_3}{|}}{\overset{\overset{\curvearrowleft Br}{|}}{C}}-CH\underset{CH_3}{\overset{CH_3}{\diagup}} \longrightarrow CH_3-CH_2-\underset{\underset{CH_3}{|}}{\overset{\oplus}{C}}-CH\underset{CH_3}{\overset{CH_3}{\diagup}} + Br^\ominus$$

$$\overset{H-\ddot{O}-H}{\underset{\curvearrowright}{}}$$
$$CH_3-CH_2-\underset{\underset{CH_3}{|}}{\overset{\oplus}{C}}-CH\underset{CH_3}{\overset{CH_3}{\diagup}} \longrightarrow CH_3-CH_2-\underset{\underset{CH_3}{|}}{\overset{\overset{H\diagdown\!\!\overset{\oplus}{O}\!\!\diagup H}{|}}{C}}-CH\underset{CH_3}{\overset{CH_3}{\diagup}}$$

$$:\!\ddot{B}r:^\ominus \curvearrowright \overset{H\diagdown\!\!\overset{\oplus}{O}\!\!\diagup H}{\underset{\curvearrowleft}{}}$$
$$CH_3-CH_2-\underset{\underset{CH_3}{|}}{\overset{|}{C}}-CH\underset{CH_3}{\overset{CH_3}{\diagup}} \longrightarrow :\!\ddot{B}r\!-\!H \quad \overset{:\!\ddot{O}\!-\!H}{\underset{|}{}} CH_3-CH_2-\underset{\underset{CH_3}{|}}{\overset{|}{C}}-CH\underset{CH_3}{\overset{CH_3}{\diagup}}$$

Note that this mechanism uses three steps instead of two because, to get the real product and not its conjugate acid, one must remove the extra $H^\oplus$ brought in with the nucleophile. (In the S_N1 and S_N2 reactions involving $-L = -OH_2^\oplus$, an extra step is also necessary, but it comes first, to make $-OH$ into $-OH_2^\oplus$ so that it can play the role of leaving group; this variation on the substitution mechanisms is discussed in detail in Unit 15.)

★ *Exercises*

Write all steps in the mechanism of each of the reactions in the exercises of frame 4.

- - - - - - - - - -

1. S_N1.

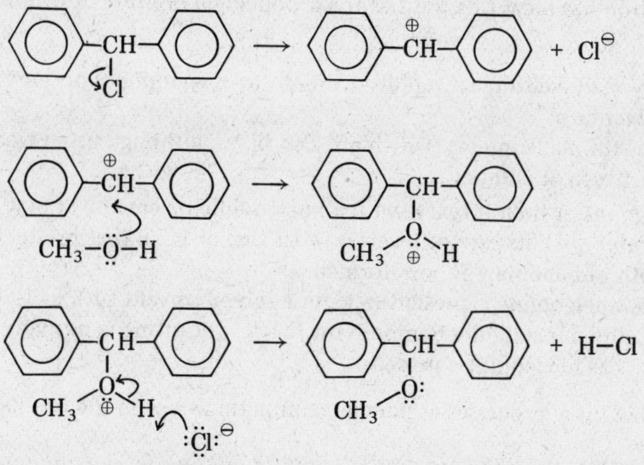

Ph–CH(Cl)–Ph → Ph–CH$^\oplus$–Ph + Cl$^\ominus$

Ph–CH$^\oplus$–Ph + CH$_3$–Ö–H → Ph–CH(O(CH$_3$)H$^\oplus$)–Ph

Ph–CH(O(CH$_3$)H$^\oplus$)–Ph + :Cl:$^\ominus$ → Ph–CH(O–CH$_3$)–Ph + H–Cl

Or the third step could be written

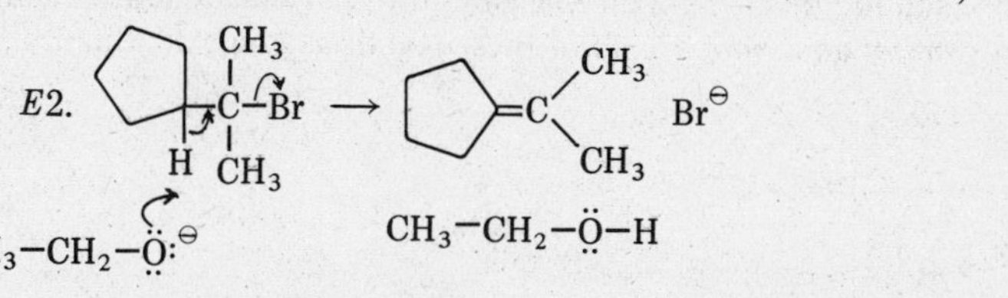

Ph–CH(O(CH$_3$)H$^\oplus$)–Ph + :Ö(CH$_3$)H → Ph–CH(O–CH$_3$)–Ph + H–Ö(CH$_3$)–H$^\oplus$

Which is closer to the physical reality? (See Unit 11, frame 10.)

2. $E2$.

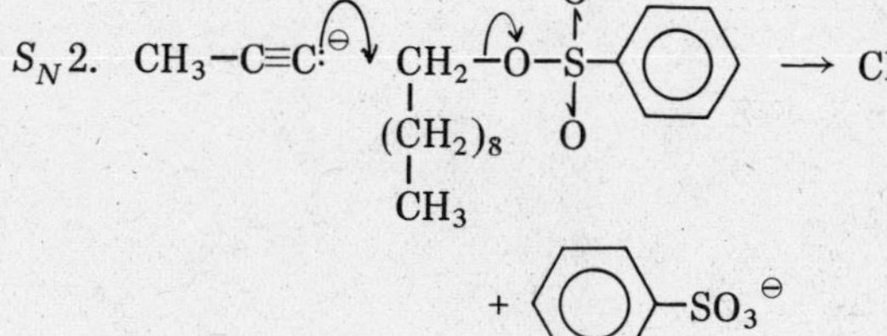

cyclopentyl-C(CH$_3$)(CH$_3$)–Br with H, attacked by CH$_3$–CH$_2$–Ö:$^\ominus$ → cyclopentylidene=C(CH$_3$)(CH$_3$) + Br$^\ominus$ + CH$_3$–CH$_2$–Ö–H

3. S_N2. CH$_3$–C≡C:$^\ominus$ + CH$_2$(–(CH$_2$)$_8$–CH$_3$)–O–S(=O)(=O)–Ph → CH$_3$–C≡C–CH$_2$–(CH$_2$)$_8$–CH$_3$ + Ph–SO$_3^\ominus$

4. E1.

$$CH_3-\underset{\underset{CH_3}{|}}{\overset{\overset{CH_3}{|}}{C}}-Cl \longrightarrow CH_3-\underset{\underset{CH_3}{|}}{\overset{\overset{CH_3}{|}}{C}}\oplus + Cl^{\ominus}$$

redrawn to show the β C—H bond

$$CH_2-\underset{\underset{CH_3}{|}}{\overset{\overset{CH_3}{|}}{C}}\oplus \longrightarrow CH_2=C\underset{CH_3}{\overset{CH_3}{\diagup}}$$

$:\ddot{Cl}:^{\ominus} \qquad H-\ddot{Cl}:$

6. The $S_N1/E1$ reaction has great possibilities for producing a mixture of products because

- $R^{\oplus}$ is sufficiently reactive to react rapidly with all the nucleophiles present, usually including the solvent.
- When $R^{\oplus}$ reacts with nucleophiles (which are also bases, although often weak ones), it can either substitute or eliminate.
- $R^{\oplus}$ may rearrange faster than it reacts with a nucleophile or base, or it may rearrange at a rate comparable with its rate of reaction with Nu: or B, so that products derived from both old and new $R^{\oplus}$ are formed.
- If more than one nucleophile is present (e.g., in a mixed solvent such as H_2O–CH_3CH_2OH), all the nucleophiles compete for the $R^{\oplus}$, resulting in as many substitution products as nucleophiles present.

You should visualize these processes as parallel, competing steps in the mechanism:

$R-L \longrightarrow R^{\oplus} + L^{\ominus}$ generation of $R^{\oplus}$

$R^{\oplus} \longrightarrow R'^{\oplus}$ rearrangement of $R^{\oplus}$

$$\begin{array}{c} R^{\oplus} \\ \text{B:}\diagup \quad |\text{Nu:} \quad \diagdown \text{Nu':} \\ \text{alkene} \quad R-Nu \quad R-Nu' \end{array} \qquad \begin{array}{c} R'^{\oplus} \\ \text{B:}\diagup \quad |\text{Nu:} \quad \diagdown \text{Nu':} \\ \text{alkene'} \quad R'-Nu \quad R'-Nu' \end{array} \Bigg\} \begin{array}{l} \text{destruction} \\ \text{of } R^{\oplus} \text{ and } R'^{\oplus} \end{array}$$

7. We did not list any separate mechanisms in frame 4 for the organometallic reactions. Most of them are just examples of the S_N2 mechanism, in which the atom attacked is usually some element other than carbon. In all of these reactions you can visualize $R:^{\ominus}$ as the nucleophile:

Protonation: $R-M \longrightarrow R:^{\ominus} + M^{\oplus}$

$R:^{\ominus} \; H-A \longrightarrow R-H + :A^{\ominus}$

Halogenation: $R-M \longrightarrow R:^{\ominus} + M^{\oplus}$

$R:^{\ominus} \; X-X \longrightarrow R-X + X^{\ominus}$

Alkylation: $R-M \longrightarrow R:^{\ominus} + M^{\oplus}$

$R:^{\ominus} \; R-X \longrightarrow R-R + X^{\ominus}$

The metal interchange can be written simply as

$$\overset{\frown}{R-M} \longrightarrow R{:}^{\ominus} + M^{\oplus}$$

$$R{:}^{\ominus}\overset{\frown}{}M'^{\oplus} \longrightarrow R-M'$$

We discuss the mechanism of the nucleophilic addition reactions of R—M in Unit 17.

★ *Exercises*

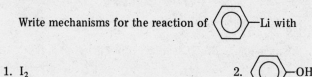

Write mechanisms for the reaction of ⟨Ph⟩—Li with

1. I_2 2. ⟨Ph⟩—OH

- - - - - - - - -

1. ⟨Ph⟩—Li ⇌ ⟨Ph⟩:⁻ + Li⁺

   ⟨Ph⟩:⁻ ⤹ I—I ⟶ ⟨Ph⟩—I + I⁻

2. ⟨Ph⟩—Li ⇌ ⟨Ph⟩:⁻ + Li⁺

   ⟨Ph⟩:⁻ ⤹ H—O—⟨Ph⟩ ⟶ ⟨Ph⟩—H + ⟨Ph⟩—O⁻

8. Problems involving the *relative reactivity* of two reactants must be answered by way of the reaction mechanism. Since the rate of production of products is the rate of the slowest step (the "rate-limiting step"), questions of reactivity are solved by looking at the reactant and transition state of the rate-limiting step. The stability of these two species determines the height of the energy barrier to the reaction, E_a:

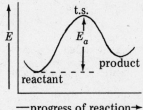

⟶ progress of reaction ⟶

The larger E_a, the slower the reaction. Students sometimes try to predict reactivity with the rule: the more stable the reactant, the slower it reacts. This does not work. As the diagram shows, the rate and E_a depend on both t.s. and reactant energies (or stabilities), and in fact the t.s. stability is the more important factor because it is the more variable. The way in which the rate of reaction changes when the reactant structure changes depends strongly upon the mechanism because the structure of t.s. is determined by the mechanism. For the $S_N1/E1$, S_N2, and $E2$ mechanisms the reaction rate depends on the structure, as follows:

$S_N1/E1$: slow step = $R\overset{\curvearrowright}{-}L \longrightarrow R^{\oplus} + L^{\ominus}$.

Experiments show that the t.s. is similar to $R^{\oplus} + L^{\ominus}$ structurally and energetically, so its stability follows the order of $R^{\oplus}$ stability and also the order of $L^{\ominus}$ stability:

rate dependence on R: $3° > 2° > 1° > CH_3$ (extent of alkyl substitution at

this carbon $\longrightarrow \ -\underset{|}{\overset{|}{C}}-L$)

rate dependence on $L^{\ominus}$: $\overset{\ominus}{O_3S-R} > I^{\ominus} > Br^{\ominus} \approx H_2O > Cl^{\ominus} > R_3N$

There is no rate dependence on the nucleophile, Nu:, because it is not involved in the rate-limiting step:

S_N2: $Nu:\overset{\curvearrowright}{} R\overset{\curvearrowright}{-}L \longrightarrow Nu-R + L^{\ominus}$

The t.s. has been found to look like

$\overset{-\delta}{Nu}---\overset{+\delta}{R}---\overset{-\delta}{L}$ (detailed drawing on p. 319).

Rate dependence on R: $1° > 2° > 3°$ (due to hindered rearside approach of Nu:; see frame 4)

Rate dependence on L: as in $S_N1/E1$

Rate dependence on Nu:

$I^{\ominus} > SCN^{\ominus} > C_6H_5NH_2 > OH^{\ominus} > Br^{\ominus} > Cl^{\ominus} > RCO_2^{\ominus} > H_2O$ (in H-bonding solvents: H_2O, ROH, RCO_2H, $RCONH_2$, etc.)

$CN^{\ominus} > RCO_2^{\ominus} > N_3^{\ominus} > Cl^{\ominus} > Br^{\ominus} > SCN^{\ominus}$ (in solvents containing *no* $-OH$ or $>N-H$ functions)

$E2$: t.s. looks more like the products than anything else.

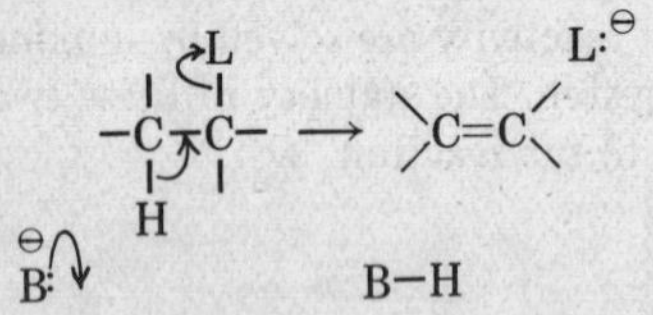

$\underset{B:^{\ominus}\curvearrowleft}{-\underset{|}{\overset{\overset{\curvearrowleft L}{|}}{C}}\underset{H}{\overset{|}{\underset{|}{\overset{\curvearrowright}{C}}}}-} \longrightarrow \underset{B-H}{>C=C<} \quad L:^{\ominus}$

Rate dependence on R: rate increases with increasing alkene stability

(tetra- > tri- > di- > monosubstituted and conjugated > nonconjugated)

Rate dependence on $B:^{\ominus}$: rate increases with increasing base strength

Rate dependence on $L:^{\ominus}$: $I^{\ominus} > Br^{\ominus} > R-SO_3^{\ominus} > Cl^{\ominus} > F^{\ominus}$

Example

Which site reacts faster in the S_N2 reaction of $Cl-\underset{\underset{CH_3}{|}}{\overset{\overset{CH_3}{|}}{C}}-CH_2-CH_2-CH_2-Cl$ with $CN^{\ominus}$?

Solution:

The primary ($-CH_2-Cl$) chloro group, not the tertiary one $\left(Cl-\overset{\overset{CH_3}{|}}{\underset{\underset{CH_3}{|}}{C}}- \right)$.

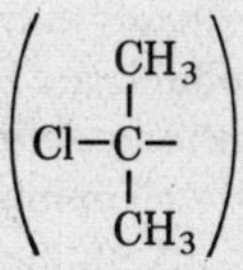

★ *Exercises*

Which site reacts faster in these reactions?

1. S_N2: $CH_3O^{\ominus}$ + Cl–CH$_2$–⟨C$_6$H$_4$⟩–CH$_2$–I

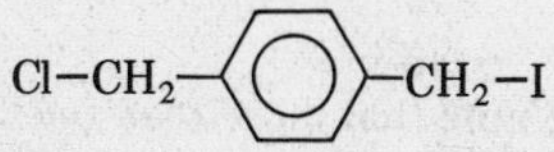

2. $E2$: $OH^{\ominus}$ + (cyclohexane with CH$_3$, Br on one carbon and CBr(CH$_3$)–CH$_3$ on another)

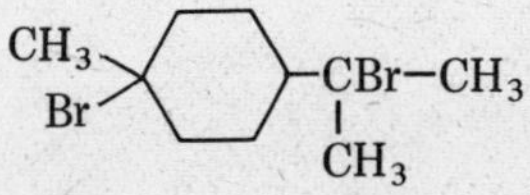

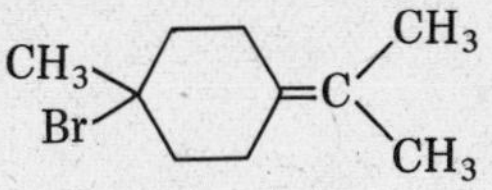

3. S_N1: H_2O + (cyclohexane with CH$_3$, Br on one carbon and –CH$_2$Br on another)

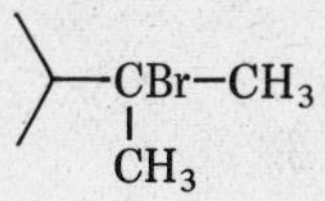

— — — — — — — — — —

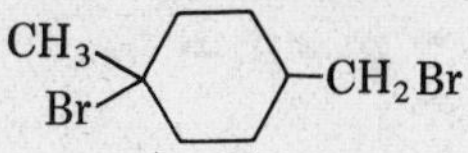

1. $-CH_2-I$

2. 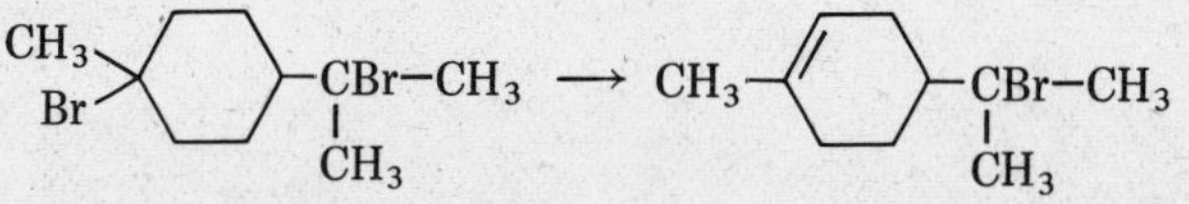 (CH$_3$/Br cyclohexane)–CBr(CH$_3$)–CH$_3$ → CH$_3$–(cyclohexene)–CBr(CH$_3$)–CH$_3$ (trisubstituted)

   or CH$_3$/Br(cyclohexane)=C(CH$_3$)(CH$_3$) (tetrasubstituted); ∴ ⟩–CBr(CH$_3$)–CH$_3$ reacts faster.

3. 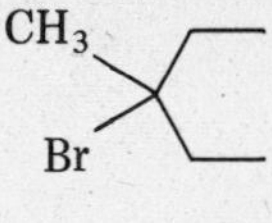 (CH$_3$)$_3$C–Br (3°) faster than $-CH_2Br$ (1°).

SOLVING PROBLEMS BASED ON REACTION PRODUCTS

9. Your practice with the standard RX and RM reactions in frame 2 should allow you to immediately recognize what *kind* of product applies in the "Supply the missing product(s)" type of problem. However, you still need to practice the correct writing of the specific product structure for that particular choice of reactants. The exercises in this frame provide this practice. Keep in mind the following two general points when working with halide reactions:

1. Reactions of organometallic compounds, R–M, are often written with R–X as the starting material because the R–M are usually made from R–X just prior to use rather than being stored. The reactions are written in tandem:

$$R-X \xrightarrow[(2)\ \text{reagent}]{(1)\ M}, \text{e.g.,} \quad C_6H_5-Br \xrightarrow[(2)\ CO_2]{(1)\ Mg} C_6H_5-\overset{O}{\underset{\|}{C}}-O-Mg-Br$$

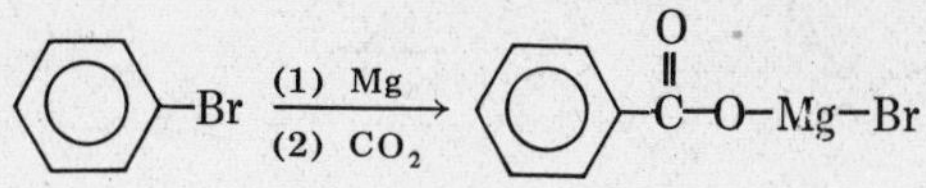

2. Substitution and elimination usually accompany one another, because nucleophiles are always bases (although they may be weak ones), and vice versa. This means that side-reaction products are usually present, and it may be difficult to predict in many cases whether substitution or elimination products will dominate. Substitution usually dominates with nucleophiles that are reactive but not very basic ($X^{\ominus}$, $SCN^{\ominus}$, $CN^{\ominus}$, R_3N, $AlH_4^{\ominus}$). The relative amount of elimination is increased by increasing the strength and concentration of base present ($OR^{\ominus}$ or $NH_2^{\ominus}$ are best), increasing the temperature and increasing the extent of alkyl substitution (3° > 2° > 1°). The result (with $RO^{\ominus}$ in ROH solvent) is high yields of alkene from 3° RX and reasonable yields from 2° RX, while 1° RX give predominant substitution. To get high yields of alkene from 1° systems, the leaving group must be changed to $-NR_3^{\oplus} \rightarrow NR_3$.

If "all organic products" are to be given, note that more than one alkene can usually result since there is usually more than one β hydrogen available:

$$R-CH_2-CHX-CH_2-R'$$

$$(CH_3)_3CO^{\ominus} | K^{\oplus}$$

$$R-CH=CH-CH_2-R' \qquad\qquad R-CH_2-CH=CH-R'$$

★ *Exercises*

Draw the structural formulas of all the organic products of each reaction.

1. 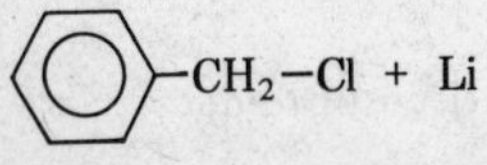 Ph–CH_2–Cl + Li

2. Ph–Cl + Mg

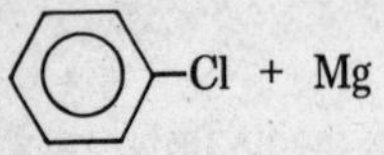

3. CH_3–CH_2–Br + Na

4. (cyclopentyl)–I + $Li^{\oplus} AlH_4^{\ominus}$

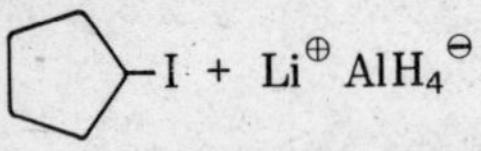

5. CH_3Cl $\xrightarrow[(2)\ CH_3CO_2H]{(1)\ Mg}$

6. CH_3Li + Br_2

7. 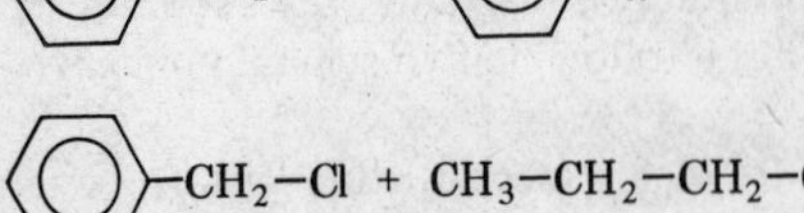 Ph–CH_2–Cl + Ph–$\ddot{O}$:$^{\ominus}$ $Na^{\oplus}$

8. 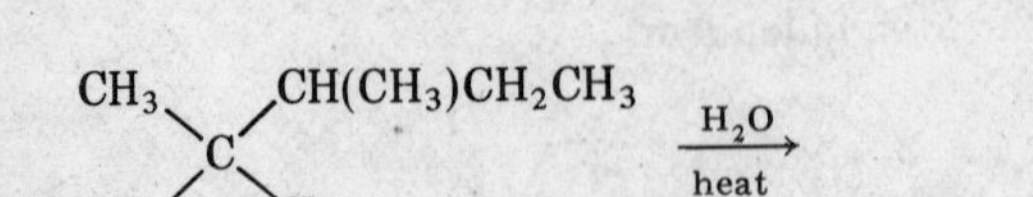 Ph–CH_2–Cl + CH_3–CH_2–CH_2–CH_2:$^{\ominus}$ $Na^{\oplus}$

9. 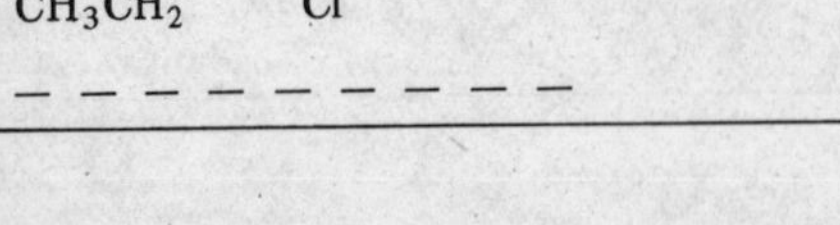 $\begin{array}{c} CH_3 \\ CH_3CH_2 \end{array} \!\!\! C \!\!\! \begin{array}{c} CH(CH_3)CH_2CH_3 \\ Cl \end{array}$ $\xrightarrow[heat]{H_2O}$

1. 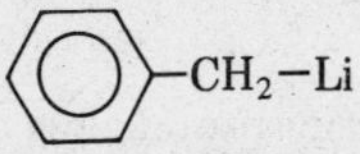 ⬡–CH₂–Li

2. 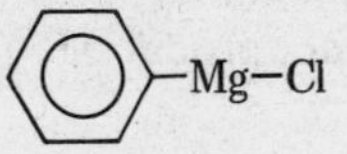 ⬡–Mg–Cl

3. CH₃–CH₂–CH₂–CH₃

4. 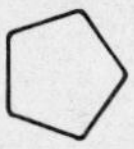 ⬠

5. CH₄ + CH₃–C(=O)–O⁻ Mg⁺–Cl

6. CH₃–Br

7. 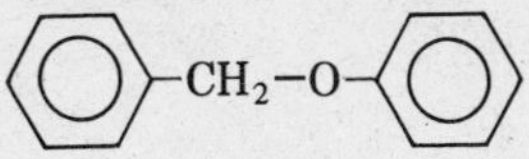 ⬡–CH₂–O–⬡

8. 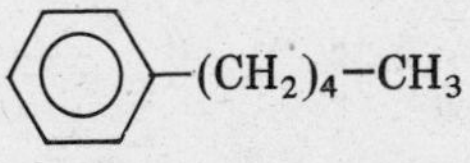 ⬡–(CH₂)₄–CH₃

9.

$$\underset{CH_3}{\overset{CH_3}{H}}C=C\underset{CH_2CH_3}{\overset{CH_3}{CH}} \quad + \quad \underset{H}{\overset{CH_3}{CH_3}}C=C\underset{CH_2CH_3}{\overset{CH_3}{CH}} \quad + \quad \underset{CH_3CH_2}{\overset{CH_2}{}}C=CH\underset{CH_2CH_3}{\overset{CH_3}{}}$$

$$+ \quad \underset{CH_3CH_2}{\overset{CH_3}{}}C=C\underset{CH_2CH_3}{\overset{CH_3}{}} \quad + \quad \underset{CH_3CH_2}{\overset{CH_3}{}}C=C\underset{CH_3}{\overset{CH_2CH_3}{}}$$

$$+ \quad CH_3CH_2-\underset{OH}{\overset{CH_3}{\underset{|}{C}}}-CH\underset{CH_2CH_3}{\overset{CH_3}{}}$$

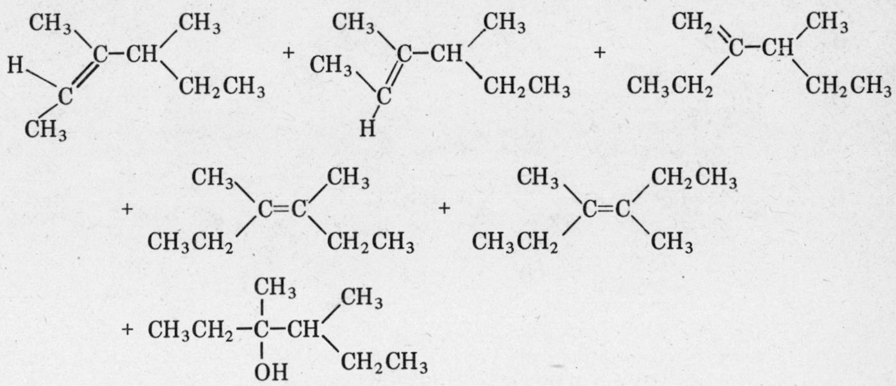

10. Both the *E*2 and *E*1 elimination reactions of RX produce a predominance of the more stable alkene:

    tetrasubstituted > trisubstituted > disubstituted > monosubstituted

and

    conjugated > nonconjugated

Examples

Write the structural formula of *the predominant organic product* of each reaction, or write "No reaction."

1. CH₃–CH₂–CBr(CH₃)(CH₃) + KOH

2. 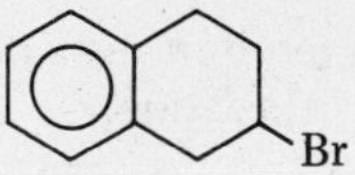 [tetralin with Br] + KOH

Solutions:

1. 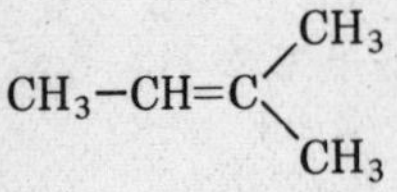 $CH_3-CH=C{\overset{CH_3}{\underset{CH_3}{\Big<}}}$ trisubstituted, more stable $\qquad$ ($CH_3-CH_2-C{\overset{CH_3}{\underset{CH_2}{\Big<}}}$ disubstituted, less stable, would be a minor product)

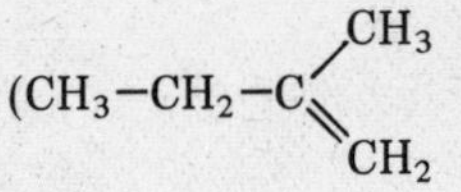

2. 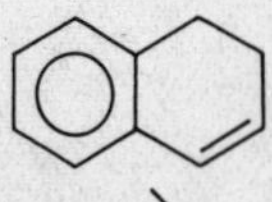 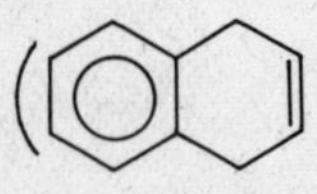 [bicyclic alkene with conjugated double bond in ring] conjugated $>C=C<$, more stable $\qquad$ [bicyclic alkene with isolated double bond] unconjugated, less stable, would be a minor product)

★ Exercises

Write the structural formula of the predominant organic product of each reaction, or write "No reaction."

1. 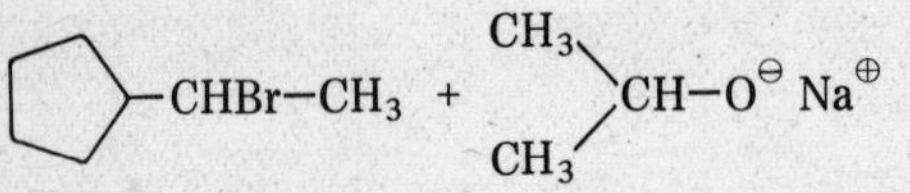 [cyclopentyl]–CHBr–CH$_3$ + ${\overset{CH_3}{\underset{CH_3}{\Big>}}}CH-O^{\ominus}\ Na^{\oplus}$

2. [cyclohexyl]–Br + [cyclohexyl]–O$^{\ominus}$ K$^{\oplus}$

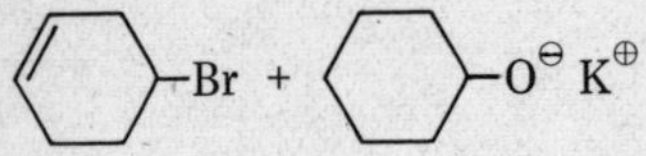

3. ${\overset{CH_3}{\underset{CH_3}{\Big>}}}C{\overset{CH_2Br}{\underset{CH_3}{\Big<}}}$ + KOH

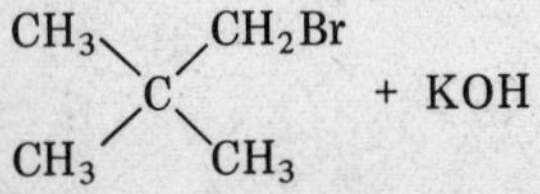

4. 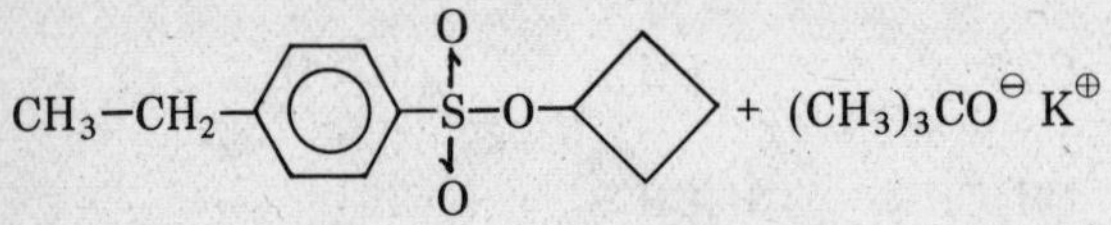 CH_3-CH_2-[phenyl]$-\overset{\overset{O}{\uparrow}}{\underset{\underset{O}{\downarrow}}{S}}-O-$[cyclobutyl] + $(CH_3)_3CO^{\ominus}\ K^{\oplus}$

– – – – – – – –

1. 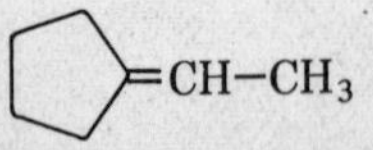 [cyclopentyl]=CH–CH$_3$ (tri- vs. monosubstituted alkene)

2.  [cyclohexadiene] (conjugated vs. nonconjugated alkene)

3. 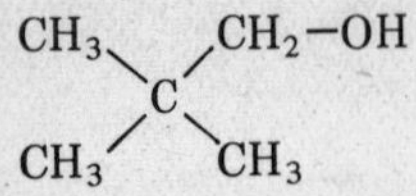 ${\overset{CH_3}{\underset{CH_3}{\Big>}}}C{\overset{CH_2-OH}{\underset{CH_3}{\Big<}}}$ (no β hydrogen available for elimination)

4. 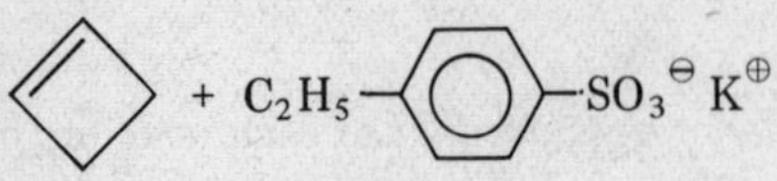 [cyclobutene] + C_2H_5-[phenyl]$-SO_3^{\ominus}\ K^{\oplus}$

11. The stereochemical result of the S_N2 reaction is inversion. The S_N1 reaction usually results in predominant racemization.

Examples

Draw structures that show the configuration of the products of each reaction.

1. 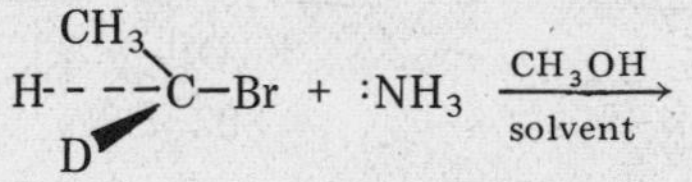 $\text{H}\text{-}\text{-}\text{-}\overset{\text{CH}_3}{\underset{\text{D}}{\text{C}}}\text{-Br} + :\text{NH}_3 \xrightarrow[\text{solvent}]{\text{CH}_3\text{OH}}$

2. 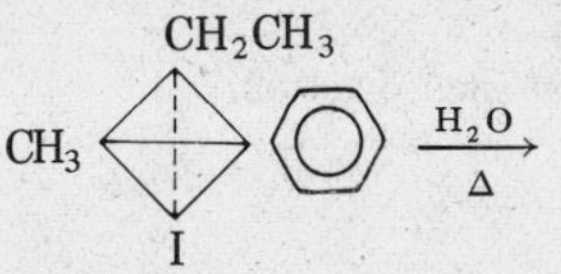 $\text{CH}_3\overset{\text{CH}_2\text{CH}_3}{\underset{\text{I}}{\diamondsuit}}\bigcirc \xrightarrow[\Delta]{\text{H}_2\text{O}}$

Solutions:

1. First identify mechanism as in frame 4 = $S_N 2$ (1° RX, good nucleophile).

   Second, identify stereochemical outcome = inversion.

   Third, write the $S_N 2$ product = $\text{CH}_3\text{-CHD-}\overset{\oplus}{\text{NH}_3}\text{Br}^{\ominus}$.

   Fourth, put it in the inverted configuration. This can be done in a number of ways:

   (a) Actually, invert the reactant:

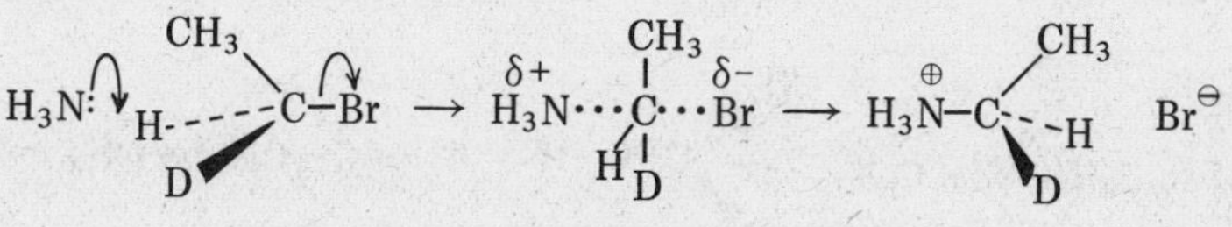

   $\text{H}_3\text{N}: \curvearrowright \text{H-}\text{-}\text{-}\overset{\text{CH}_3}{\underset{\text{D}}{\text{C}}}\overset{\curvearrowright}{\text{-Br}} \rightarrow \text{H}_3\text{N}\overset{\delta+}{\cdots}\overset{\text{CH}_3}{\underset{\text{H}\,\text{D}}{\text{C}}}\overset{\delta-}{\cdots}\text{Br} \rightarrow \overset{\oplus}{\text{H}_3\text{N}}\text{-}\overset{\text{CH}_3}{\underset{\text{D}}{\text{C}}}\text{-}\text{-}\text{-H}\quad \text{Br}^{\ominus}$

   (b) Draw the product in the same configuration as the reactant and then draw its mirror image:

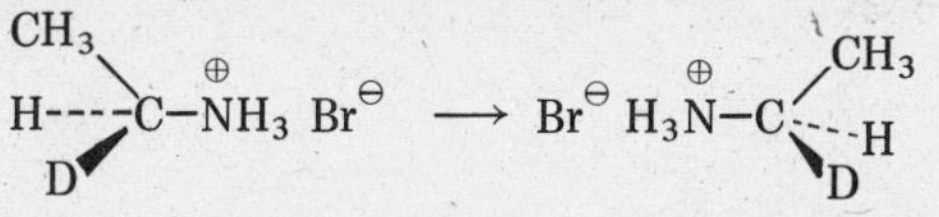

   $\text{H-}\text{-}\text{-}\overset{\text{CH}_3}{\underset{\text{D}}{\text{C}}}\text{-}\overset{\oplus}{\text{NH}_3}\ \text{Br}^{\ominus} \rightarrow \text{Br}^{\ominus}\ \overset{\oplus}{\text{H}_3\text{N}}\text{-}\overset{\text{CH}_3}{\underset{\text{D}}{\text{C}}}\text{-}\text{-}\text{-H}$

   (c) Draw the product in the starting configuration and exchange any two groups on the asymmetric carbon:

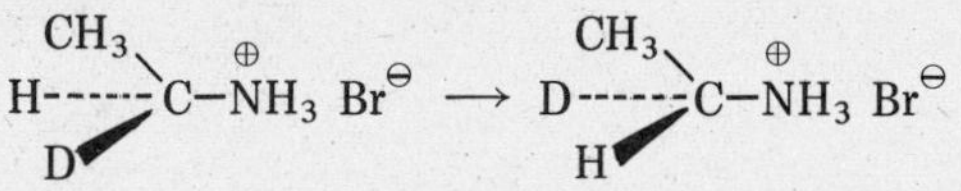

   $\text{H-}\text{-}\text{-}\text{-}\overset{\text{CH}_3}{\underset{\text{D}}{\text{C}}}\text{-}\overset{\oplus}{\text{NH}_3}\ \text{Br}^{\ominus} \rightarrow \text{D-}\text{-}\text{-}\text{-}\overset{\text{CH}_3}{\underset{\text{H}}{\text{C}}}\text{-}\overset{\oplus}{\text{NH}_3}\ \text{Br}^{\ominus}$

2. Mechanism = $S_N 1/E1$ (benzylic and 3° RX → stable $\text{R}^{\oplus}$; poor nucleophile; decent $S_N 1$ solvent).

   Stereochemical result: racemization.

   $S_N 1$ product: $\bigcirc\!\!-\!\!\underset{\text{CH}_3}{\overset{\text{CH}_2\text{-CH}_3}{\underset{|}{\overset{|}{\text{C}}}}}\!\!-\!\text{OH}$

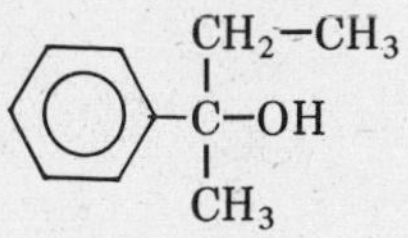

334 HOW TO SUCCEED IN ORGANIC CHEMISTRY

Configuration of the S_N1 product:

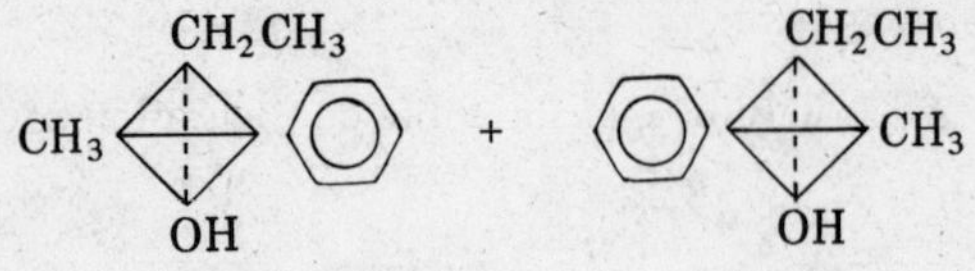

CH_3—[C with CH_2CH_3 up, C_6H_5 right, OH down] + [C with CH_2CH_3 up, C_6H_5 left, CH_3 right, OH down]

50%   50%

★ *Exercises*

Draw the structural formulas of all the organic products of each reaction.

1. 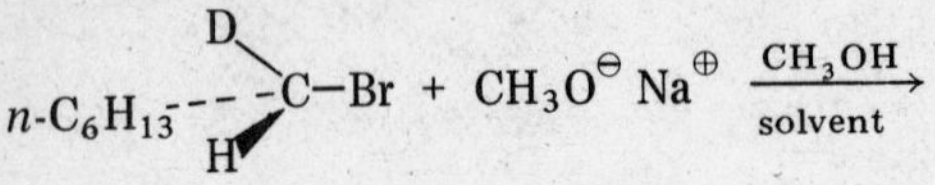 $n\text{-}C_6H_{13}\!-\!\overset{D}{\underset{H}{C}}\!-\!Br + CH_3O^{\ominus} Na^{\oplus} \xrightarrow[\text{solvent}]{CH_3OH}$

2. 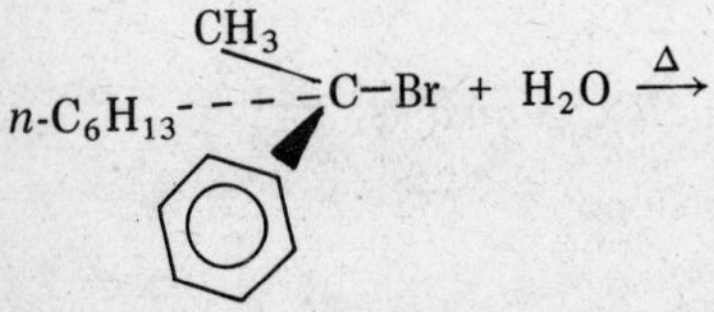 $n\text{-}C_6H_{13}\!-\!\overset{CH_3}{\underset{C_6H_5}{C}}\!-\!Br + H_2O \xrightarrow{\Delta}$

- - - - - - - - - -

1. 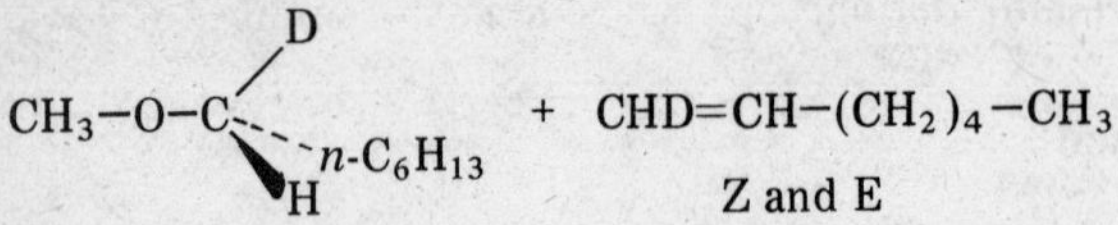 $CH_3\!-\!O\!-\!\overset{D}{\underset{H}{C}}\!-\!n\text{-}C_6H_{13}$ + $CHD=CH-(CH_2)_4-CH_3$

    Z and E

2. $H\!-\!O\!-\!\overset{CH_3}{\underset{C_6H_5}{C}}\!-\!n\text{-}C_6H_{13}$ + $n\text{-}C_6H_{13}\!-\!\overset{CH_3}{\underset{C_6H_5}{C}}\!-\!O\!-\!H$ + $CH_2=C-(CH_2)_4-CH_3$ +

    $\quad\quad\quad\quad\quad\quad\quad\quad\quad\quad\quad\quad\quad\quad\quad\quad\quad\quad\quad\quad\quad\quad\quad\quad\quad\quad\quad\quad\quad\quad\quad\quad\quad\quad\quad\quad\quad\quad\quad\quad\quad\quad\quad\quad\quad\quad\quad\quad\quad\quad C_6H_5$

$CH_3-C=CH-(CH_2)_4-CH_3$
$\quad\quad\;\;|$
$\quad\quad\;\;C_6H_5$

E and Z

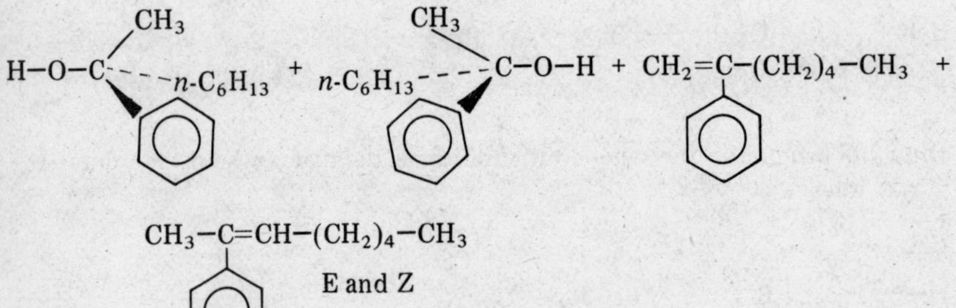

UNIT FIFTEEN
Alcohols

The alcohol function, $-\overset{|}{\underset{|}{C}}-O-H$, is in the same oxidation stage as the halide function, $-\overset{|}{\underset{|}{C}}-X$, and it is nearly as versatile. The $-OH$ and $-X$ functions complement one another; alcohols are oxidizable but not easily reduced, halides are reducible but not so readily oxidized. The $-X$ and $-OH$ functions are interconvertible. Their reactions are mainly nucleophilic substitutions and eliminations; the mechanisms learned for $R-X$ in Unit 14 apply with few modifications to the $R-OH$. The principal new features in this unit result from the weak-acid and weak-base properties of $-OH$.

OBJECTIVES

When you have completed this unit, you should be able to:

- Explain how the acidic and basic properties of $-\ddot{O}H$ lead to primary reaction products (salts, H-bonded complexes) and also influence much of the rest of the chemistry of the alcohols (frame 1).
- Predict the product(s) of reaction of a given alcohol with a variety of specific reagents drawn from these general classes: strong acids, strong bases, nucleophiles, electrophiles, oxidizing agents (frames 2 to 4).
- Predict the structure of the predominant alkene formed by alcohol dehydration when various directions of elimination and carbonium ion rearrangement are possible (frame 5).
- Write all steps in the mechanism of most of the common alcohol reactions (frames 3 to 5).

If you believe that you may be able to skip part or all of this unit, take a self-test consisting of the exercises marked with a star in the margin. Check your answers and study the frames covering any problems missed, then rework those problems before proceeding.

If you are not ready for a self-test, proceed to frame 1.

UNDERSTANDING THE CHEMISTRY OF THE $-OH$ GROUP

1. Chart 15.1 shows how the standard reactions of the $-OH$ function relate back to its fundamental properties.

$R-O-H$ is weakly nucleophilic but becomes strongly nucleophilic when converted to its conjugate base, $R-\ddot{O}:^{\ominus}$, in which the electron density on O is greatly increased.

The $-OH$ group is never a leaving group in S_N1, S_N2, $E1$, or $E2$ reactions: $\overset{\frown}{-\ddot{O}}-H$

335

336 HOW TO SUCCEED IN ORGANIC CHEMISTRY

✳ :ÖH$^\ominus$. This contrasts with the behavior of $-\ddot{\text{X}}:$. If we break the leaving process into two stages, we can see the reason for the difference. The numbers in Table 15.1 are the energies required for the two stages shown. A negative number means that the step is

Chart 15.1

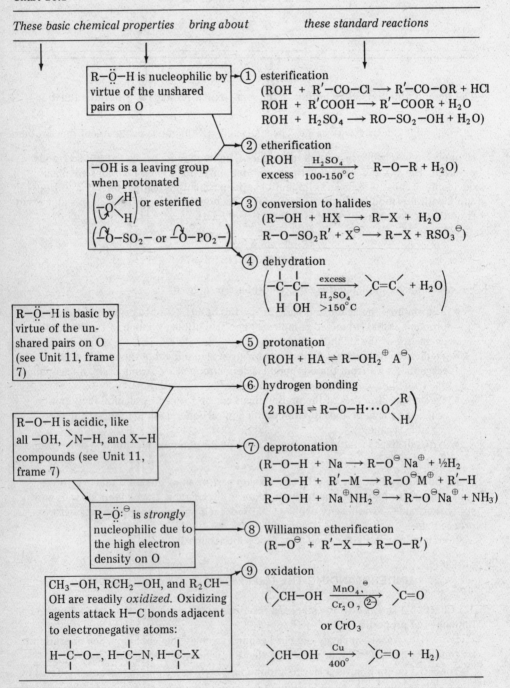

Table 15.1 *Leaving-Group Departure Energies (kcal/mole)*

$-L$	Energy for stage 1, $>\!\!C-L \longrightarrow -C\cdot + L\cdot$	Energy for stage 2, $L\cdot \longrightarrow L:$	Sum
$-OH$	$+91$	-42	$+91$
$-F$	$+116$	-82	$+34$
$-Cl$	$+81$	-87	-6
$-Br$	$+68$	-81	-13
$-I$	$+51$	-73	-22
$-\overset{\oplus}{O}H_2$	$+119$	-291	-172

energetically downhill. Positive values indicate energetically unfavorable processes. We see that $-OH$ is a poor leaving group partly because it forms one of the stronger bonds to carbon (stage 1 energy input is large), but mainly because OH accepts a negative charge less readily than the halogens do (stage 2 is less favorable).

There are two ways to make $-OH$ into a good leaving group. The first is simply to protonate it, making $R-\overset{\oplus}{O}H_2$. This leaving group is compared with $-OH$ and $-X$ in Table 15.1. It accepts the electron pair in the σ bond very much more readily than does $-OH$, as shown by the more favorable energy of stage 2. The numbers suggest it should be the best of all these leaving groups, but this is because the numbers are for the gas phase. In solution, where our reactions are carried out, much of the advantage of $-\overset{\oplus}{O}H_2$ versus $-X$ is lost, and $-\overset{\oplus}{O}H_2$ proves to be about as reactive a leaving group as $-Br$.

The second way to increase the reactivity of $-OH$ as a leaving group is to esterify it with an acid of part structure
$$H-O-\overset{\overset{O}{\uparrow}}{\underset{\downarrow O}{S}}- \quad \text{or} \quad H-O-\overset{\overset{O}{\uparrow}}{\underset{O^\ominus}{P}}-.$$
In the laboratory we often use $H-O-SO_2-OH$, $HO-SO_2-R$, or $HO-SO_2-Ar$ as the acid. Living organisms often use
$$HO-\overset{\overset{O}{\uparrow}}{\underset{O^\ominus}{P}}-O-\overset{\overset{O}{\uparrow}}{\underset{O^\ominus}{P}}-O^\ominus.$$
The esterified $-OH$ group is a better leaving group for the same reason $-\overset{\oplus}{O}H_2$ is; the $-SO_2-$ or $-PO_2-$ group helps the O atom of the alcohol bear the extra negative charge that it receives in the process $-\ddot{O}-SO_2- \longrightarrow :\overset{\ominus}{\ddot{O}}-SO_2-$. These esters remove electron density from O by virtue of the electron-attracting power of the positively charged $-\overset{|}{\underset{|}{S}}-{\textcircled{2+}}$ or $-\overset{|}{\underset{|}{P}}-\oplus$ centers.

This electron-attracting power is also what makes the $HO-SO_2-$ and $HO-PO_2-$ structures strong acids—it stabilizes the leaving group when a proton is given up to some base:

$$\text{B:} \curvearrowright \text{H} \overset{\curvearrowleft}{-} \overset{\oplus}{\text{O}} - \text{SO}_2- \longrightarrow \text{B-H} + \overset{\ominus}{\text{O}} - \text{SO}_2-$$

As a result, a rule is often stated as follows: a group $-L$ will be a good leaving group if $:L^{\ominus}$ is the conjugate base of a strong acid. The rule makes $-O-SO_2-$ and $-O-PO_2-$ good leaving groups for the reasons just discussed; it also summarizes the behavior of $-X$ and $-OH_2^{\oplus}$ (good leaving groups, since HX and $H_3O^{\oplus}$ are strong acids) and $-OH$ and $-NH_2$ (poor leaving groups, since H_2O and NH_3 are weak acids).

Reactions ②, ③, and ⑧ in Chart 15.1 are nucleophilic substitution (S_N1 or S_N2) reactions. They can be learned as examples of the single generic reaction

$$\text{Nu:} + \text{R-L} \longrightarrow \text{Nu-R} + \text{L:}$$

via the slot-and-filler technique used in Unit 14. The permissible slot fillers are summarized in Table 15.2. The table is divided into two parts. With one exception, all the possible combinations of Nu: + L in the upper part of the table work well. R−O−H is such a weak nucleophile that it gives few practical reactions and has been segregated in the bottom part of the table. R−OH reacts with those $R'X$, $R'-O-SO_2-R''$, and $R'-O-SO_2-Ar$ that use the S_N1 mechanism (mainly 3° allylic or benzylic R) to give substantial amounts of $R-O-R'$, always accompanied by considerable alkene. Its most important reaction is the one in the table:

$$\text{R-OH} + \text{R-OH}_2^{\oplus} \longrightarrow \text{R-O-R} + \text{H}_2\text{O}$$

Table 15.2 *Permissible Slot Fillers in Nucleophilic Substitutions Involving Alcohols, Their Salts, and Their Esters*
$$\text{Nu:} + \text{R-L} \longrightarrow \text{Nu-R} + \text{L:}$$

Permissible Nu:	Permissible −L	Permissible R−
$R-\ddot{O}:^{\ominus\ a}$	$-\overset{\oplus}{O}H_2{}^a$	CH_3-
$:\ddot{X}:^{\ominus}$	$-X$	$R-CH_2-$
	$-O-SO_2-$	R_2CH-
	$-O-PO_2-$	$R-CH=CH-CH_2-$
		$Ar-CH_2-$
$R-\ddot{O}-H$	$-\overset{\oplus}{O}H_2$	Same as above

[a] The combination $R-O^{\ominus} + R-OH_2^{\oplus}$ would be highly reactive, but it cannot be realized because one reactant is stable only in strongly acidic, the other only in strongly basic solutions.

Reaction ② in Chart 15.1 belongs to the potentially confusing class of reactions in which the same compound plays two separate roles. Actually, this reaction fits the Nu: + R-L ⟶ Nu-R + L: pattern perfectly, with Nu: = R−O−H and R−L = $R-\overset{\oplus}{O}H_2$. The function of the H_2SO_4 catalyst is to prepare the leaving group; the mechanism is

$$\text{R-OH} + \text{H}_2\text{SO}_4 \rightleftharpoons \text{R-}\overset{\oplus}{\text{O}}\text{H}_2\text{HSO}_4^{\ominus}$$

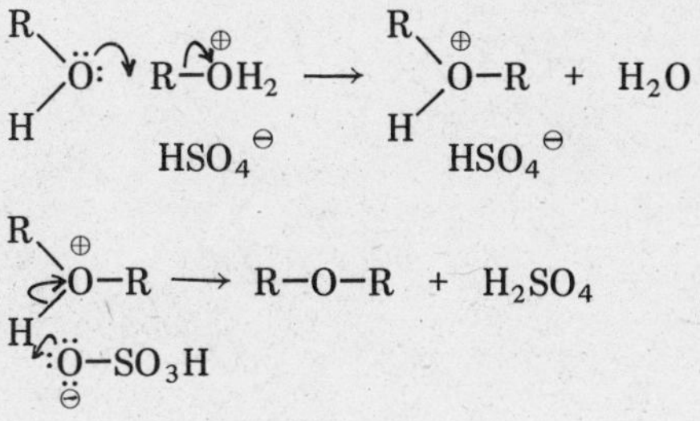

$$\begin{array}{c} R \\ \diagdown \\ \ddot{O}: \curvearrowright R\overset{\oplus}{-}OH_2 \\ / \\ H \qquad HSO_4{}^{\ominus} \end{array} \longrightarrow \begin{array}{c} R \\ \diagdown \overset{\oplus}{O}-R \\ / \\ H \qquad HSO_4{}^{\ominus} \end{array} + H_2O$$

$$\begin{array}{c} R \\ \diagdown \overset{\oplus}{O}-R \\ \curvearrowleft \\ H \\ \overset{\ominus}{:\ddot{O}}-SO_3H \end{array} \longrightarrow R-O-R + H_2SO_4$$

Obviously, the acidity of the reaction medium must be such that some of the ROH is free and some protonated.

Exercises

1. The ethers, R−Ö−R, are closely related chemically to the alcohols. In which of the reactions, ① through ⑨, in Chart 15.1, could an ether take the place of the alcohol?

2. The $E2$ and S_N2 mechanisms are closely related. Suppose that we make a first draft of a table similar to Table 15.2 but applying to $E2$ reactions involving alcohols:

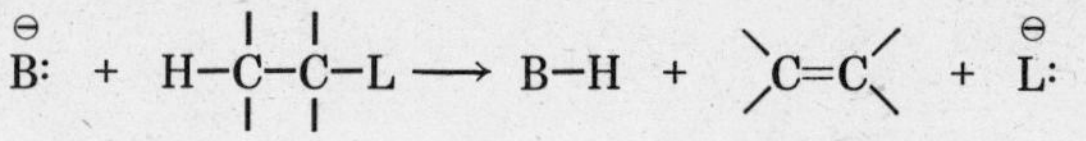

$$\overset{\ominus}{B:} + H-\overset{|}{\underset{|}{C}}-\overset{|}{\underset{|}{C}}-L \longrightarrow B-H + \diagdown C=C \diagup + \overset{\ominus}{L:}$$

Permissible $\overset{\ominus}{B:}$	Permissible $\diagdown CH-\overset{|}{\underset{|}{C}}-L$
$:\overset{..}{\underset{..}{O}}-H^{\ominus}$	$\diagdown CH-\overset{|}{\underset{|}{C}}-X$
$:\overset{..}{\underset{..}{O}}-R^{\ominus}$	$\diagdown CH-\overset{|}{\underset{|}{C}}-O-H$
	$\diagdown CH-\overset{|}{\underset{|}{C}}-O-R$
	$\diagdown CH-\overset{|}{\underset{|}{C}}-\overset{\oplus}{O}H_2$
	$\diagdown CH-\overset{|}{\underset{|}{C}}-O-SO_2-Ar$

Edit the "Permissible −L" column to weed out the impractical combinations.

3. All the following reactions have been observed experimentally. Which of the alcohol reaction types ① to ⑨ in Chart 15.1 does each one most resemble?

(a) $CH_3-CH_2-O^{\ominus} + CH_3-O-SO_2-O-CH_3 \longrightarrow CH_3-CH_2-O-CH_3 +$

$$\overset{\ominus}{O}-SO_2-O-CH_3$$

(b) 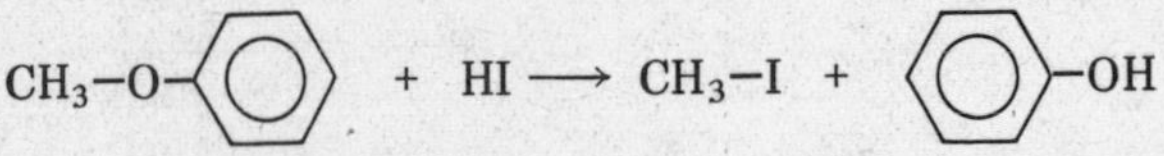 $CH_3-O-C_6H_5$ + HI ⟶ CH_3-I + C_6H_5-OH

(c) $CH_3-O-\overset{O}{\underset{\|}{C}}-C_6H_3(O_2N)(NO_2)(O_2N)$ + KI ⟶ CH_3I + $K^⊕\,{}^⊖O_2C-C_6H_3(O_2N)(NO_2)(O_2N)$

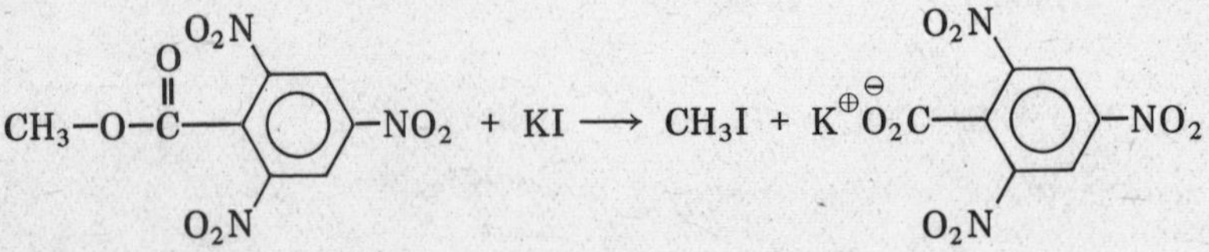

(d) $CH_3\underset{\underset{CH_3}{|}}{\overset{⊕}{O}}CH_3\ BF_4^⊖\ \xrightarrow{\text{heat}}\ CH_3F + CH_3-O-CH_3$

- - - - - - - - -

1. Those in which $-\overset{⊕}{O}H_2$ plays the role $-L$ can be mimicked by $-\underset{\underset{H}{|}}{\overset{⊕}{O}}-R$, so reactions ③ and ④ should have ether analogs. Reaction ⑤ requires only $-\ddot{O}-$; it works on ethers: $R-\ddot{O}-R + HA \rightleftharpoons R-\underset{\underset{H}{|}}{\overset{⊕}{O}}-R\ A^⊖$. All the rest require the presence of $-O-H$.

2. Remove $\diagdown CH-\underset{|}{\overset{|}{C}}-OH$ and $\diagdown CH-\underset{|}{\overset{|}{C}}-OR$ because $\frown OH \longrightarrow OH^⊖$ and $\frown OR \longrightarrow OR^⊖$ ought to be just as impractical leaving groups here as they are in nucleophilic substitution. Remove $\diagdown CH-\underset{|}{\overset{|}{C}}-\overset{⊕}{O}H_2$ because it can't exist in the same solution as $OH^⊖$ or $OR^⊖$; a rapid proton transfer would destroy both reactants:

$R-O^⊖ + \diagdown CH-\underset{|}{\overset{|}{C}}-\overset{⊕}{O}H_2 \rightleftharpoons R-OH + \diagdown CH-\underset{|}{\overset{|}{C}}-OH$

3. (a) is reaction ⑧ with $L = -O-SO_2-O-CH_3$ in place of $-X$.

(b) is ③ with $L = -\underset{\underset{H}{|}}{\overset{⊕}{O}}-C_6H_5$ in place of $-\overset{⊕}{O}-H$ with H.

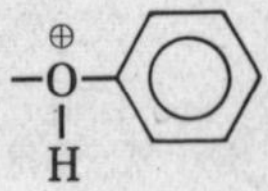

(c) is ③ with Nu: $= I^⊖$ and $-L = -O-CO-C_6H_3(O_2N)(NO_2)(O_2N)$ in place of $-O-SO_2-Ar$.

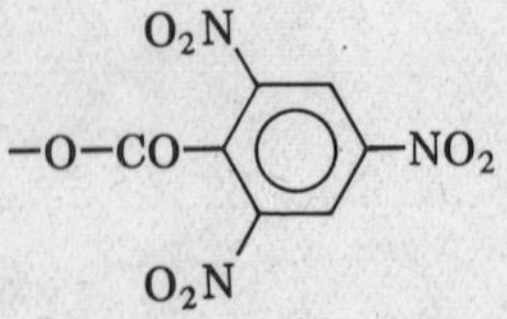

(d) is ③ with $-L = -\overset{⊕}{O}\underset{\diagdown CH_3}{\diagup CH_3}$ in place of $-\overset{⊕}{O}H_2$ and Nu: $= F^⊖$, latent in $BF_4^⊖$ and released at the moment of attack via $\overset{\frown}{F}-\overset{⊖}{B}\underset{\diagdown F}{\diagup F}-F$.

LEARNING THE ALCOHOL REACTIONS

2. The slot-and-filler summary in frame 1 gives you a means of organizing and remembering all the *substitution* reactions of the alcohols (②, ③, and ⑧ in Chart 15.1). Complications in the R—OH ⟶ R—X reaction get special consideration in frame 4.

The esterification reaction (① in Chart 15.1) is best studied in detail with the carboxylic acid derivatives (Unit 18); it usually suffices at this point if you know just the equations for the reaction.

Unit 11 showed you how the acid and base strengths of ROH compare with those of other functions. The alcohol function is a sufficiently weak base that conversion to its conjugate acid (reaction ⑤ in Chart 15.1) is never stoichiometrically (i.e., 100%) achieved in any of the reaction mixtures ordinarily encountered. However, solutions containing ROH and any strong acid contain a certain equilibrium concentration of the conjugate acid, $R-\overset{\oplus}{O}H_2$. ROH is also quite weak as an acid, but conversion to the conjugate base, $R-O^{\ominus}$, can be accomplished stoichiometrically via the reactions listed as ⑦ in Chart 15.1.

The dehydration reaction (④ in Chart 15.1) requires some special handling; this is done in frame 5.

This leaves the oxidation reactions (⑨ in Chart 15.1). In this frame we organize them in a way that puts the ROH in context with the other oxygen-containing functional groups. Then we develop drills for learning all the alcohol reactions.

First we construct an oxidation-level chart of the type used for the alkyl halide reactions (Chart 14.2). In the case of the alcohols, the higher oxidation states are more important and the lower ones less so. Since the nature of the higher oxidation states depends strongly on the number of hydrogen atoms α to the —OH group $\left(\underset{\gamma \;\; \beta \;\; \alpha}{-C-C-C-OH} \right)$, it is useful to spread the various cases out horizontally according to what I call the "alkylation state" of the α carbon. The various oxidation states are stacked vertically as before. The result is Chart 15.2.

The chart contains arrows for all the reactions that convert the alcohols to the other oxygenated functional groups. These are all oxidation reactions. The chart also includes the oxidation reactions of the other oxygen functions since they are involved in the present story. In succeeding units we will gradually log in the various reduction reactions that complete the chart. If your class or text treats synthesis of alcohols together with alcohol reactions, you may wish at this time to add connecting arrows and reagents for these alcohol-forming reactions to Chart 15.2.

You should focus on these points:

- The less alkylated the alcohol, the greater the number of oxidation states to which it can be oxidized.
- We are considering primarily *selective* oxidations that leave the carbon skeleton intact. Of course, all these compounds can be unselectively oxidized to CO_2 + H_2O under vigorous conditions (e.g., combustion). Except for CH_3OH, these uncontrolled oxidations are not shown in the chart, since they are not chemically useful.
- There is no selective oxidation of a tertiary alcohol.
- The same reagents attack all the oxidizable alcohols: $MnO_4^{\ominus}$, $Cr_2O_7^{2-}$, CrO_3; however, the next item introduces one restriction on their use.
- There are two oxidation stages above the 1° alcohols, $R-CH_2-OH$. Under most oxidation conditions the aldehyde, $R-\overset{\overset{O}{\parallel}}{C}-H$, is further oxidized to the carboxylic

Chart 15.2[a]

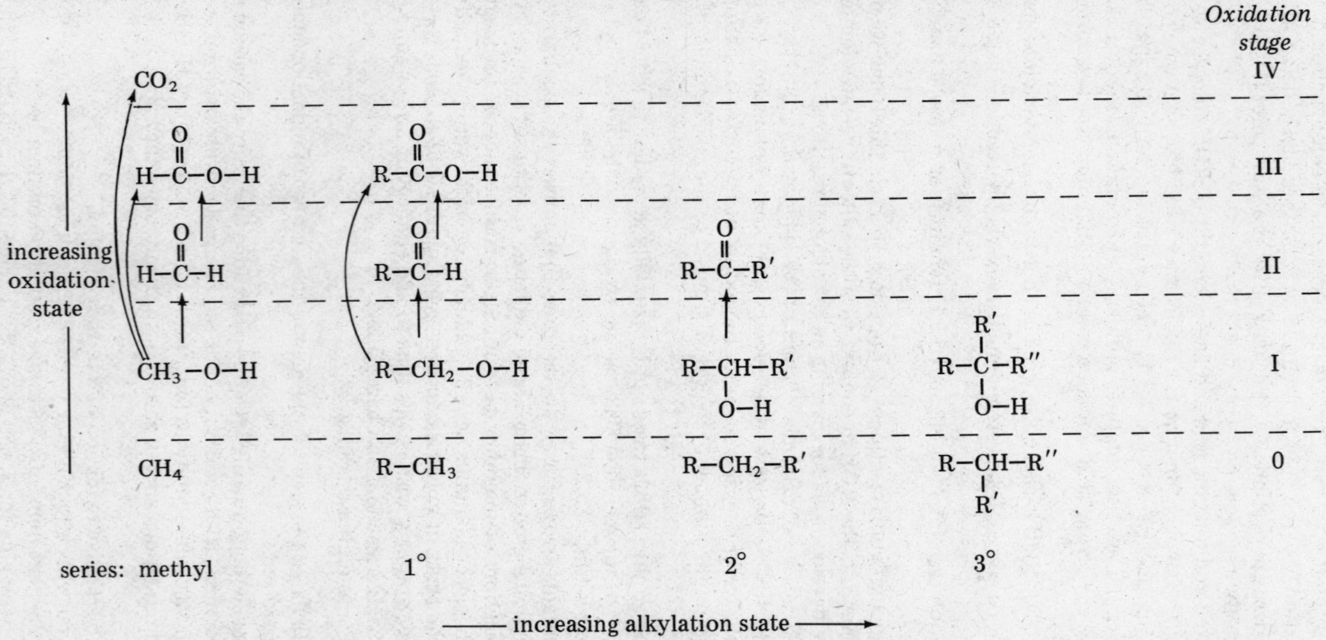

Oxidation stage	methyl	1°	2°	3°
IV	CO_2			
III	$\underset{H}{\overset{O=}{C}}-O-H$	$\underset{R}{\overset{O=}{C}}-O-H$		
II	$\underset{H}{\overset{O=}{C}}-H$	$\underset{R}{\overset{O=}{C}}-H$	$\underset{R}{\overset{O=}{C}}-R'$	
I	CH_3-O-H	$R-CH_2-O-H$	$R-\underset{O-H}{CH}-R'$	$R-\underset{\underset{R'}{\mid}}{\overset{\overset{R'}{\mid}}{C}}-R''$ with $O-H$
0	CH_4	$R-CH_3$	$R-CH_2-R'$	$R-\underset{R'}{CH}-R''$

← increasing oxidation state

increasing alkylation state →

[a] Extended versions appear in Unit 17, frame 8, and Unit 18, frame 11.

acid as rapidly as it is formed. The only generally applicable means of stopping at the aldehyde stage is to use CrO_3 under anhydrous conditions, usually in a mixture of chloroform and pyridine as solvent:

$$3R-CH_2-OH + 2CrO_3 \xrightarrow[\text{pyridine}]{CHCl_3} 3R-C{\overset{O}{\underset{H}{\diagup\!\!\!\diagdown}}} + 2Cr(OH)_3$$

- Oxidation (upward movement in the chart) corresponds to addition of oxygen to the molecular formula, or subtraction of hydrogen, or both. Dehydrogenation is thus equivalent to oxidation, and catalytic dehydrogenation of alcohols can indeed be carried out:

$$1°: \quad R-CH_2-OH \xrightarrow[>250°C]{Cu\,catalyst} R-C{\overset{O}{\underset{H}{\diagup\!\!\!\diagdown}}} + H_2$$

$$2°: \quad \underset{R}{\overset{R}{\diagdown}}CH-OH \xrightarrow[>250°C]{Cu\,catalyst} R-\overset{O}{\underset{\|}{C}}-R + H_2$$

Complete reduction of the alcohols to the corresponding alkanes cannot be done directly. It requires stepping horizontally out of Chart 15.2 into Chart 14.2 via the reaction $R-OH \xrightarrow{HX} R-X$ or equivalent. The halide can be reduced to $R-H$ as in Chart 14.2. Alternatively, the sequence

$$-\underset{H}{\overset{|}{C}}-\underset{OH}{\overset{|}{C}}- \xrightarrow{H_2SO_4} \diagup C=C \diagdown \xrightarrow{H_2}_{Pt} \diagup CH-CH \diagdown$$

can be used if at least one β C—H bond is present.

Exercises

1. If you are in a classroom situation, add or delete reactions to adjust the right-hand column of Chart 15.1 to include just those reactions covered in your course. Connect any newly added reactions to the appropriate basic chemical properties.

2. Following the plan of Unit 12, frames 9 and 15, make an expanded list of the reactions in your edited Chart 15.1. Use your textbook and lecture notes to include four characteristics for each reaction:

   (a) Generic equation (e.g., ROH + HX ⟶ RX + H_2O).

   (b) Scope (e.g., R = 1°, 2°, or 3° alkyl, allylic, or benzylic, but not vinylic or aryl; 2° and hindered 1° ROH may give RX with rearranged skeleton. X = Cl, Br, I).

   (c) Mechanism followed; give its name and write it out for the reaction in generic form. For the nucleophilic substitution reactions, the mechanism will be S_N2 for unhindered 1° systems, S_N1 for 3° systems, for instance

$$R-CH_2-\overset{\oplus}{\underset{\curvearrowleft H-X}{\ddot{O}-H}} \longrightarrow R-CH_2-\overset{\oplus}{O}{\overset{H}{\underset{H}{\diagup\!\!\!\diagdown}}} \quad X^{\ominus}$$

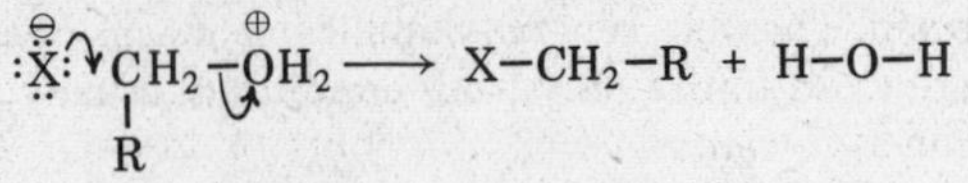

$$\overset{\ominus}{\underset{R}{:\ddot{X}:}} \curvearrowright CH_2 \overset{\oplus}{-}\underset{\downarrow}{O}H_2 \longrightarrow X-CH_2-R + H-O-H$$

and

$$\underset{R''}{\overset{R}{\underset{|}{R'-C}}}\overset{|}{-}\ddot{O}-H \quad \curvearrowright \underset{H-X}{} \longrightarrow \underset{R''}{\overset{R}{\underset{|}{R'-C}}}\overset{|}{-}\overset{\oplus}{\underset{H}{O}} \quad X^{\ominus}$$

$$\underset{R''}{\overset{R}{\underset{|}{R'-C}}}\overset{|}{-}\overset{\oplus}{\underset{\downarrow}{O}}H_2 \longrightarrow \underset{R''}{\overset{R}{\underset{|}{R'-C\oplus}}} + H_2O$$

$$\underset{R''}{\overset{R}{\underset{|}{R'-C\oplus}}} \curvearrowright :\ddot{X}:^{\ominus} \longrightarrow \underset{R''}{\overset{R}{\underset{|}{R'-C-X}}}$$

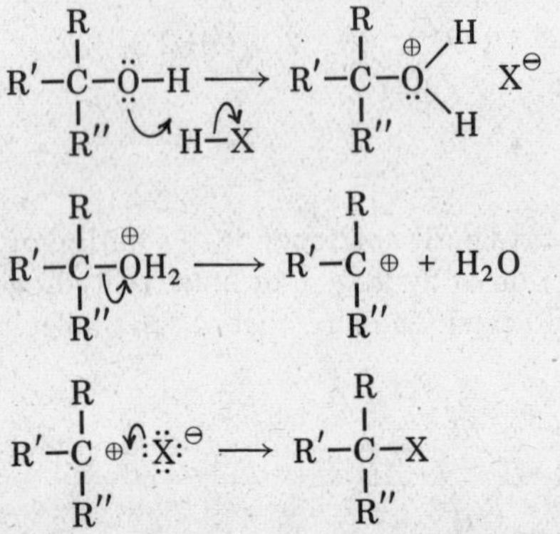

(d) Reactivity order (e.g., HX: HI > HBr > HCl
R: 3° > 2° > 1°, and unhindered 1° > hindered 1°).

3. Using the alcohol lists you prepared above, carry out the drills described in exercises 1 to 5 of frame 16, Unit 12.

4. Look over Chart 15.2, then put it aside and see if you can reproduce it on scratch paper. Check and correct. Repeat the exercise.

SOLVING PROBLEMS BASED ON REACTIONS OF THE ALCOHOLS

3. The basic problem types "write the products of these reactions" and "write all steps in the mechanisms of these reactions" are handled in three steps:

(a) Identify the generic equation or mechanism involved. For example, the problem CH₃—⟨ ⟩—OH $\xrightarrow{CrO_3}$? is recognized as oxidation of a secondary alcohol to a ketone. Or the generic mechanism for ⟨ ⟩⟨CH₃/OH⟩ $\xrightarrow{HI}$ ⟨ ⟩⟨CH₃/I⟩ is recognized as S_N1. If you cannot make the identification of applicable generic reaction type or mechanism type when working these problems, your drill in frame 2 was inadequate.

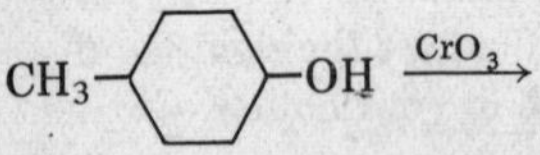

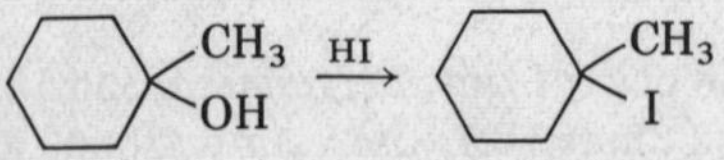

(b) By comparing the given structures with the generic equation or mechanism, identify the specific slot filler that is to play each role in the reaction.

(c) Insert the chosen slot fillers in the generic equation or mechanism to get the required specific product structures or specific mechanism.

Examples

(continued from above):

The generic equation R—CH(OH)—R' + CrO₃ ⟶ R—C(=O)—R', applied to the given

reactant $CH_3-\langle\bigcirc\rangle-OH$, makes $R-\overset{|}{C}H-R' = \langle\bigcirc\rangle$, so the desired specific reaction
                                                                          $|$
                                                                         CH_3
is

$$CH_3-\langle\bigcirc\rangle-OH \xrightarrow{CrO_3} CH_3-\langle\bigcirc\rangle=O$$

Or, to apply the generic S_N1 mechanism

$$R\overset{\curvearrowright}{-}L \longrightarrow R^{\oplus} + L: \text{ (from Table 10.1 or Unit 14, frame 4)}$$

$$R^{\oplus} \overset{\curvearrowleft}{:}Nu^{\ominus} \longrightarrow R-Nu$$

to the reaction

$$\langle\bigcirc\rangle\!\!\begin{array}{c}CH_3\\OH\end{array} \xrightarrow{HI} \langle\bigcirc\rangle\!\!\begin{array}{c}CH_3\\I\end{array}$$

set R = $\langle\bigcirc\rangle\!\!\begin{array}{c}CH_3\\ \end{array}$. L cannot be $-OH$, but must $= -\overset{\oplus}{O}H_2$. Nu: $= I^{\ominus}$. To make these slot fillers fit the generic mechanism requires an additional, initial step to put $-OH$ into the form of the leaving group, $-\overset{\oplus}{O}H_2$:

$$\langle\bigcirc\rangle\!\!\begin{array}{c}CH_3\\ \ddot{O}-H\\ \curvearrowleft H\!-\!I\end{array} \longrightarrow \langle\bigcirc\rangle\!\!\begin{array}{c}CH_3\\ \overset{\oplus}{O}H_2 \; I^{\ominus}\end{array}$$

Then

$$\langle\bigcirc\rangle\!\!\begin{array}{c}CH_3\\ \overset{\curvearrowright}{\overset{\oplus}{O}H_2}\end{array} \longrightarrow \langle\overset{\oplus}{\bigcirc}\rangle\!\!\begin{array}{c}CH_3\\ \end{array} + H_2O$$

$$\langle\overset{\oplus}{\bigcirc}\rangle\!\!\begin{array}{c}CH_3\\ \overset{\curvearrowleft}{:\ddot{I}:^{\ominus}}\end{array} \longrightarrow \langle\bigcirc\rangle\!\!\begin{array}{c}CH_3\\ I\end{array}$$

★ *Exercises*

1. Write the structural formula of the predominant organic product of each reaction, or write "No reaction."

   (a) $\langle\bigcirc\rangle-CH_2-Br + CH_3-O^{\ominus} Na^{\oplus} \longrightarrow$

   (b) $\langle\bigcirc\rangle-CH_2-OH + CH_3Li \longrightarrow$

   (c) $\langle\bigcirc\rangle-CH_2-CH_2OH \xrightarrow[H_2SO_4]{K_2Cr_2O_7}$

(d) [cyclohexane]⟨H, OH⟩ $\xrightarrow{H_2SO_4}$

(e) [phenyl]–CH$_2$–CH$_2$–OH + CrO$_3$ $\xrightarrow{\text{pyridine} / CHCl_3}$

(f) CH$_3$–C(CH$_3$)(CH$_3$)–OH $\xrightarrow{HCl}$

(g) [cyclohexane]⟨H, OH⟩ $\xrightarrow{K}$

(h) [phenyl]–C(=O)–Cl + CH$_3$OH ⟶

(i) (CH$_3$)$_2$CH–CH$_2$–OH (excess) $\xrightarrow[100°C]{H_2SO_4}$

(j) CH$_3$–C(CH$_3$)(CH$_3$)–Br + CH$_2$=CH–CH$_2$–O$^\ominus$Na$^\oplus$ ⟶

2. Write stepwise mechanisms for each reaction.

(a) [phenyl]–CH$_2$OH + HI ⟶ [phenyl]–CH$_2$–I + H$_2$O

(b) [phenyl]–CH(OH)–CH$_3$ $\xrightarrow{H_2SO_4}$ [phenyl]–CH=CH$_2$ + H$_2$O

(c) [cyclobutane with H]–O$^\ominus$ Na$^\oplus$ + [cyclobutyl]–CH$_2$–Br ⟶ [cyclobutyl]–O–CH$_2$–[cyclobutyl] + Na$^\oplus$ Br$^\ominus$

3. Supply the missing reagents and intermediates.

(a) [cyclopentane]–OH $\xrightarrow{?}$ [cyclopentene]

(b) CH$_3$–CH$_2$–CH(CH$_3$)–CH$_2$–CH$_2$OH $\xrightarrow{?}$ CH$_3$–CH$_2$–CH(CH$_3$)–CH$_2$–CH$_2$Cl

(c) 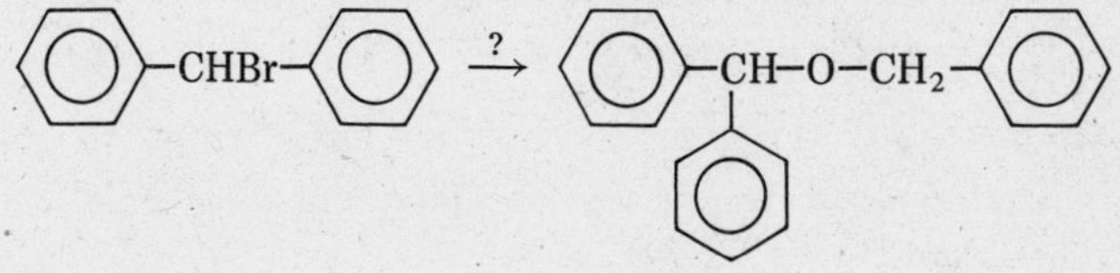 Ph–CHBr–Ph $\xrightarrow{?}$ Ph–CH(–O–CH$_2$–Ph)–Ph

(d) $CH_3-\underset{\underset{CH_3}{|}}{\overset{\overset{CH_3}{|}}{C}}-CH_2OH \xrightarrow{?} CH_3-\underset{\underset{CH_3}{|}}{\overset{\overset{CH_3}{|}}{C}}-CO_2H$

(e) $CH_3-\underset{\underset{CH_3}{|}}{\overset{\overset{CH_3}{|}}{C}}-CH_2OH \xrightarrow{?} CH_3-\underset{\underset{CH_3}{|}}{\overset{\overset{CH_3}{|}}{C}}-\overset{O}{\overset{\|}{C}}-H$

(f) 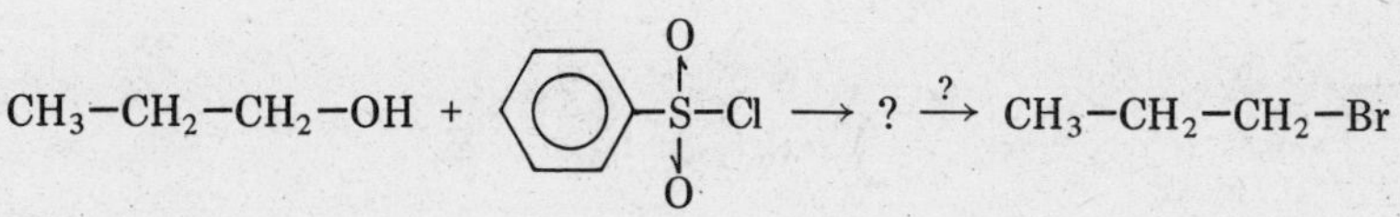 $CH_3-CH_2-CH_2-OH$ + Ph–S(=O)$_2$–Cl $\longrightarrow$? $\xrightarrow{?}$ $CH_3-CH_2-CH_2-Br$

4. By analogy with the alcohol reaction mechanisms, write a mechanism for this ether cleavage reaction:

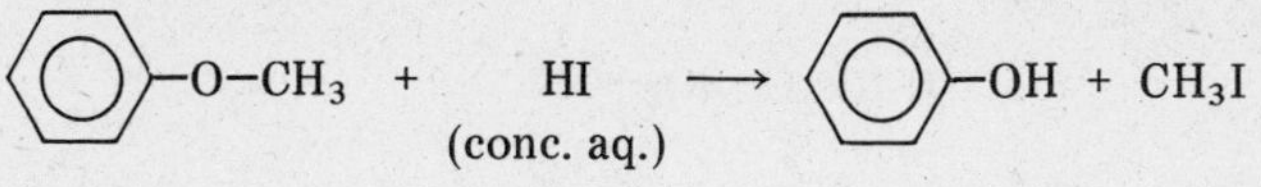

Ph–O–CH$_3$ + HI (conc. aq.) $\longrightarrow$ Ph–OH + CH$_3$I

– – – – – – – – –

1. (a) 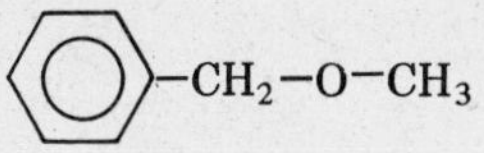 Ph–CH$_2$–O–CH$_3$

(b) 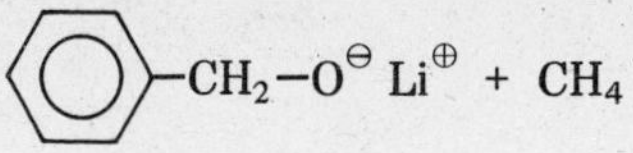 Ph–CH$_2$–O$^\ominus$ Li$^\oplus$ + CH$_4$

(c) 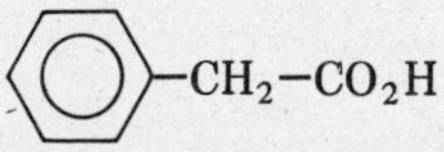 Ph–CH$_2$–CO$_2$H

(d)  (cyclohexane)

(e) 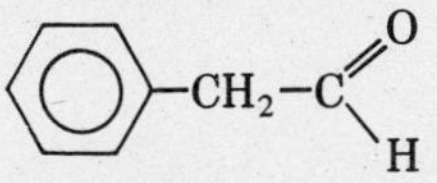 Ph–CH$_2$–C(=O)H

(f) $CH_3-\underset{\underset{CH_3}{|}}{\overset{\overset{CH_3}{|}}{C}}-Cl$

(g) 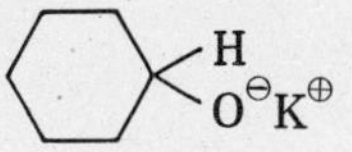 cyclohexyl–O$^\ominus$K$^\oplus$ (with H)

(h) Ph−CO−O−CH₃

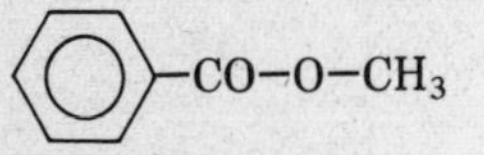

(i) (CH₃)₂CH−CH₂−O−CH₂−CH(CH₃)₂

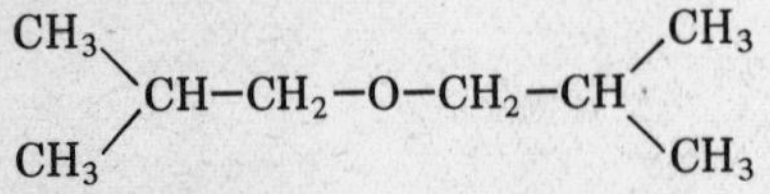

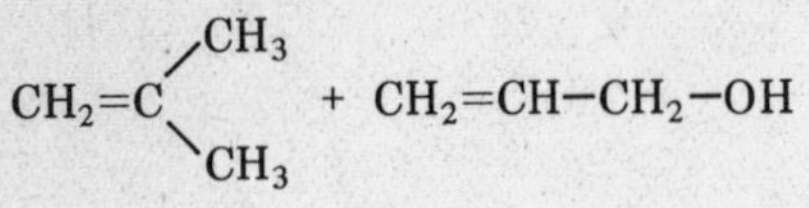

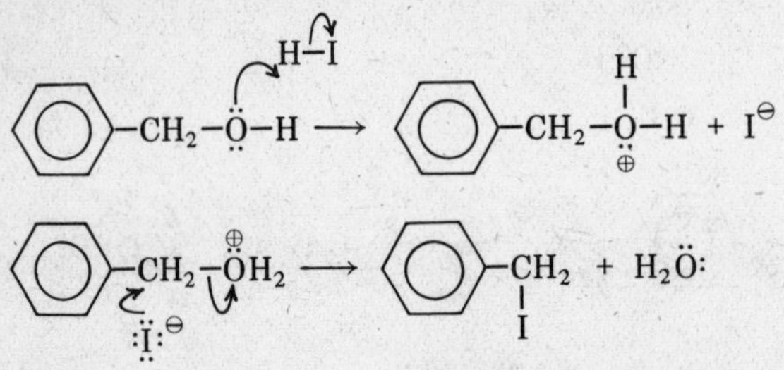

(j) CH₂=C(CH₃)₂ + CH₂=CH−CH₂−OH

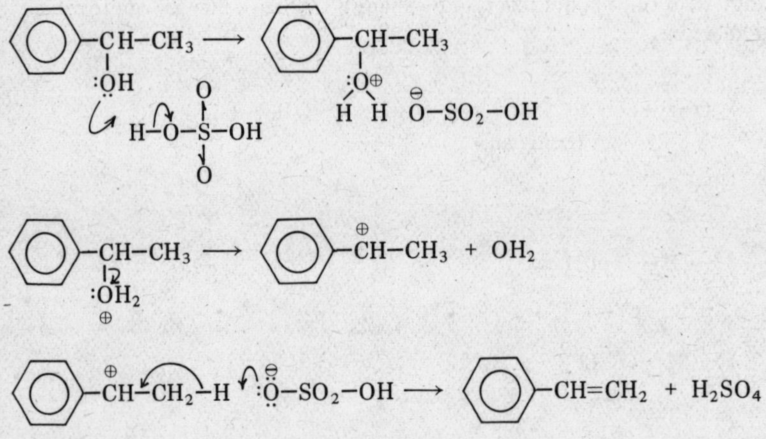

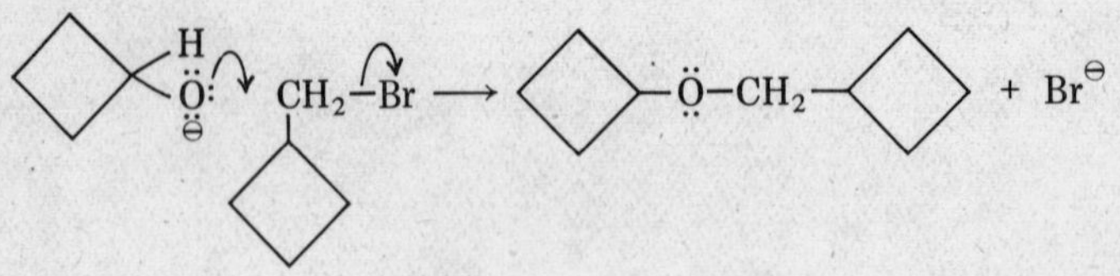

2. (a) Ph−CH₂−Ö−H + H−I → Ph−CH₂−Ö(H)−H + I⁻

Ph−CH₂−ÖH₂⁺ (with :I⁻) → Ph−CH₂−I + H₂Ö:

(b) Ph−CH(OH)−CH₃ + H−O−S(=O)₂−OH → Ph−CH(O⁺H H)−CH₃ + ⁻O−SO₂−OH

Ph−CH(ÖH₂⁺)−CH₃ → Ph−CH⁺−CH₃ + OH₂

Ph−CH⁺−CH₂−H ⁻Ö−SO₂−OH → Ph−CH=CH₂ + H₂SO₄

(c) cyclobutanol Ö: + CH₂−Br (cyclobutyl) → cyclobutyl−Ö−CH₂−cyclobutyl + Br⁻

3. (a) H₂SO₄, 150°

(b) HCl

(c) Ph−CH₂−O⁻ Na⁺

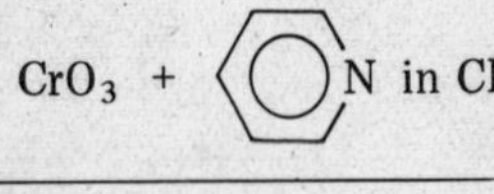

(d) K₂Cr₂O₇ or CrO₃ + H₂SO₄

(e) CrO₃ + pyridine (N) in CHCl₃

(f) $\text{C}_6\text{H}_5-\text{SO}_2-\text{O}-\text{CH}_2-\text{CH}_2-\text{CH}_3 \xrightarrow{\text{Na}^\oplus \text{ Br}^\ominus}$

4. $\text{C}_6\text{H}_5-\overset{..}{\text{O}}-\text{CH}_3 \longrightarrow \text{C}_6\text{H}_5-\overset{\oplus}{\text{O}}-\text{CH}_3 \ \ :\overset{..}{\text{I}}{:}^\ominus \longrightarrow \text{C}_6\text{H}_5-\overset{..}{\text{O}}-\text{H} + \text{CH}_3-\text{I}$

$\quad\ \ \ \ \underset{\text{acid}}{\text{H}-\text{I}} \qquad\qquad \underset{\text{L}-}{\ \ \ \ \ \ \ \ } \underset{\text{R}}{\ \ \ \ } \underset{:\text{Nu}^\ominus}{\ \ \ \ }$

4. The results obtained from reaction ③ of Chart 15.1 (R—OH → R—X) depend on the structure of the alcohol and the choice of reagent.

Unhindered 1° alcohols give good results (good yield of a single product) with all three of the reagents listed in Chart 15.1. These are $S_N 2$ reactions:

$$\text{R}-\text{CH}_2-\text{OH} \xrightarrow{\text{HX}} \text{R}-\text{CH}_2-\overset{\oplus}{\text{OH}}_2 \xrightarrow{S_N 2} \text{R}-\text{CH}_2-\text{X} + \text{H}_2\text{O}$$
$$\qquad\qquad\qquad\qquad :\overset{..}{\text{X}}{:}^\ominus$$

$$\text{R}-\text{CH}_2-\text{OH} \xrightarrow[:\text{B}]{\text{SOCl}_2} \text{R}-\text{CH}_2-\text{O}-\text{SO}-\text{Cl} \longrightarrow \text{R}-\text{CH}_2-\text{Cl} + \text{BH}^\oplus \text{ Cl}^\ominus + \text{SO}_2$$
$$\qquad\qquad\qquad\qquad :\overset{..}{\text{Cl}}{:}^\ominus \text{ BH}^\oplus$$

$$\text{R}-\text{CH}_2-\text{OH} \xrightarrow[\text{B}:]{\text{PX}_3} \text{R}-\text{CH}_2-\text{O}-\text{PX}_2 \longrightarrow \text{R}-\text{CH}_2-\text{X} + \text{PX}_2\text{O}^\ominus \text{ BH}^\oplus$$
$$\qquad\qquad\qquad\qquad :\overset{..}{\text{X}}{:}^\ominus \text{ BH}^\oplus$$

In the last two equations B: plays the role of ionizing the HX formed in the first step, furnishing $\text{X}^\ominus$ as a nucleophile for the second step. B: can be a trace of water, the alcohol itself, or an added amine (e.g., pyridine, $\text{C}_5\text{H}_5\text{N}$).

Secondary alcohols may react by these $S_N 2$ mechanisms, or by the corresponding $S_N 1$ paths. SOCl_2 + pyridine promotes $S_N 2$, and HX promotes the $S_N 1$ reaction:

$$\text{R}-\text{OH} + \text{HX} \rightleftharpoons \text{R}-\overset{\oplus}{\text{OH}}_2 \text{X}^\ominus$$

$$\text{R}-\overset{\oplus}{\text{OH}}_2 \longrightarrow \text{R}^\oplus + \text{H}_2\text{O}$$

$$\text{R}^\oplus \ :\overset{..}{\text{X}}{:}^\ominus \longrightarrow \text{R}-\text{X}$$

Tertiary alcohols and some highly hindered primary alcohols such as $(\text{CH}_3)_3\text{CCH}_2\text{OH}$ can use only the $S_N 1$ mechanism because the rate of rearside attack by nucleophile is negligible.

The mechanism followed influences the product structure because the $S_N 1$ path permits *carbonium ion rearrangements*. The rearrangement can be a shift of R, Ar, or H, from an adjacent carbon, with its pair of bonding electrons:

$$-\overset{\oplus}{\text{C}}-\overset{}{\text{C}}\underset{\text{H}}{\nwarrow} \longrightarrow -\overset{}{\text{C}}-\overset{\oplus}{\text{C}}-\ \ \text{or}\ \ -\overset{\oplus}{\text{C}}-\overset{}{\text{C}}\underset{\text{R}}{\nwarrow} \longrightarrow -\overset{}{\text{C}}-\overset{\oplus}{\text{C}}-\ \ \text{or}\ \ -\overset{\oplus}{\text{C}}-\overset{}{\text{C}}\underset{\text{C}_6\text{H}_5}{\nwarrow} \longrightarrow -\overset{}{\text{C}}-\overset{\oplus}{\text{C}}-$$

Significant amounts of rearranged carbon skeleton appear in the $S_N 1$ products whenever rearrangement of $R^\oplus$ could produce an equally stable or a more stable carbonium ion.

The following rearrangements are energetically downhill and occur rapidly, leading to predominantly rearranged products:

$1° R^\oplus \xrightarrow{\Omega} 2° R^\oplus$

$1° R^\oplus \xrightarrow{\Omega} 3° R^\oplus$ (the symbol $\xrightarrow{\Omega}$ means rearrangement)

$2° R^\oplus \xrightarrow{\Omega} 3° R^\oplus$

The rearrangements

$2° R^\oplus \xrightarrow{\Omega} 2° R^\oplus$

$3° R^\oplus \xrightarrow{\Omega} 3° R^\oplus$

are slower. The rate of rearrangement approximates the rate of reaction with nucleophile, and comparable amounts of rearranged and unrearranged halide are often observed.

Examples

1. 2-Bromopentane prepared from 2-pentanol is always contaminated with 3-bromopentane:

$$CH_3-CH_2-CH_2-\underset{\underset{OH}{|}}{CH}-CH_3 + HBr \longrightarrow CH_3-CH_2-CH_2-\underset{\underset{\overset{\oplus}{OH_2}}{|}}{CH}-CH_3 \; Br^\ominus$$

$$CH_3-CH_2-CH_2-\underset{\underset{\overset{\oplus}{OH_2}}{\curvearrowleft|}}{CH}-CH_3 \longrightarrow CH_3-CH_2-CH_2-\overset{\oplus}{CH}-CH_3 + H_2O$$

$$CH_3-CH_2-\underset{\underset{H}{|\nearrow}}{CH}-\overset{\oplus}{CH}-CH_3 \longrightarrow CH_3-CH_2-\overset{\oplus}{CH}-\underset{\underset{H}{|}}{CH}-CH_3$$

$$CH_3-CH_2-\overset{\oplus}{CH}-CH_2-CH_3 \longrightarrow CH_3-CH_2-\underset{\underset{Br}{|}}{CH}-CH_2-CH_3$$
$$\underset{:\ddot{Br}:^\ominus}{\curvearrowleft}$$

$$CH_3-CH_2-CH_2-\overset{\oplus}{CH}-CH_3 \longrightarrow CH_3-CH_2-CH_2-\underset{\underset{Br}{|}}{CH}-CH_3$$
$$\underset{:\ddot{Br}:^\ominus}{\curvearrowleft}$$

2. Treatment of neopentyl alcohol, $CH_3-\underset{\underset{CH_3}{|}}{\overset{\overset{CH_3}{|}}{C}}-CH_2-OH$, with HBr gives only

$CH_3-\underset{\underset{CH_3}{|}}{\overset{\overset{Br}{|}}{C}}-CH_2-CH_3$ via the rearrangement

$$CH_3-\underset{\underset{CH_3}{|\nearrow}}{\overset{\overset{CH_3}{|}}{C}}-CH_2^\oplus \longrightarrow CH_3-\underset{\underset{\overset{\oplus}{}CH_3}{|}}{\overset{\overset{CH_3}{|}}{C}}-CH_2$$

To decide if rearrangement is involved in one of these S_N1 R—OH $\longrightarrow$ R—X reactions, proceed as follows:

- Write the structure of the $R^{\oplus}$ intermediate.
- Explore the structures of all the isomeric $R^{\oplus}$ that can be obtained from the original $R^{\oplus}$ by one shift of adjacent R, Ar, or H.
- If one of these new $R^{\oplus}$ is more stable than the original, derive the predominant product from the more stable $R^{\oplus}$.

With some experience, you will not need to write out all possible rearranged $R^{\oplus}$ because you can anticipate which rearrangements will produce the most stable new $R^{\oplus}$. For example, you will never shift H out of a neighboring —CH_3 group because this gives a 1° $R^{\oplus}$:

$$-\underset{\underset{H}{\overset{\oplus}{|}}}{\overset{H}{\underset{|}{C}}}-\underset{\underset{H}{|}}{\overset{H}{\underset{|}{C}}}-H \quad \not\longrightarrow \quad -\underset{\underset{H}{|}}{\overset{|}{C}}-\overset{\oplus}{\underset{H}{\overset{H}{C}}}\diagdown H$$

You are really looking for an adjacent carbon bearing more alkyl groups than the carbon presently bearing the positive charge:

$$CH_3-\underset{\underset{3°}{\oplus}}{\overset{CH_3}{\underset{|}{C}}}-CH-CH_3 \longleftarrow CH_3-\underset{\underset{CH_3}{|}}{\overset{CH_3}{\underset{|}{C}}}-\overset{\oplus}{\underset{|}{CH}}-\underset{\underset{CH_3}{|}}{\overset{H}{\underset{|}{C}}}-H \quad \underset{2°}{\not\longrightarrow} \begin{array}{c} CH_3 \; H \\ CH_3-\underset{\underset{CH_3}{|}}{\overset{|}{C}}\!-\!-\!-CH-\overset{\oplus}{\underset{\underset{CH_3}{|}}{C}}-H \; 2° \\ \\ CH_3 \quad H \\ CH_3-\underset{\underset{CH_3}{|}}{\overset{|}{C}}\!-\!-\!-CH-\underset{\underset{\oplus}{|}}{\overset{|}{C}}-H \; 1° \\ CH_3 \; CH_3 \end{array}$$

The only way to avoid carbonium ion rearrangements in those secondary and hindered primary cases that are rearrangement-prone is to change the leaving group. The process

$$:\overset{\ominus}{X} \curvearrowright R\!-\!\overset{\frown}{O}\!-\!SO_2\!-\!Ar \longrightarrow X\!-\!R + \overset{\ominus}{O}\!-\!SO_2\!-\!Ar$$

is pure S_N2 in these cases and yields very pure halides.

★ Exercises

1. Which of these $R^{\oplus}$ can rearrange to a more stable carbonium ion? Write the more stable ion.

   (a) $CH_3-CH_2-CH_2^{\oplus}$ (b) $CH_3-CH_2^{\oplus}$

   (c) [bicyclic structure with ⊕] (d) $\langle\bigcirc\rangle-\overset{\oplus}{CH}-\langle\bigcirc\rangle$

2. Draw the structure of the predominant halide formed in these reactions:

   (a) $\langle\bigcirc\rangle\!\!<\!\!\overset{CH_3}{\underset{OH}{}}$ + HBr $\longrightarrow$?

(b) 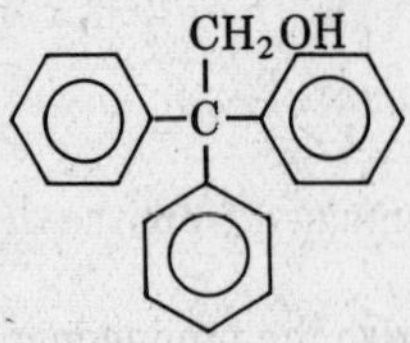 Ph-C(CH₂OH)(Ph)(Ph) + HCl ⟶ ?

(c) 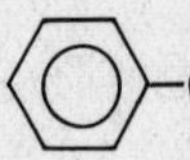 Ph-CH₂-CH₂-CH₂-OH + HI ⟶ ?

- - - - - - - -

1. (a) $CH_3-\underset{H}{CH}-CH_2^\oplus \longrightarrow CH_3-\overset{\oplus}{CH}-\underset{H}{CH_2}$

(b) Can't; only possible shift is 1° $R^\oplus \longrightarrow$ identical 1° $R^\oplus$.

(c) 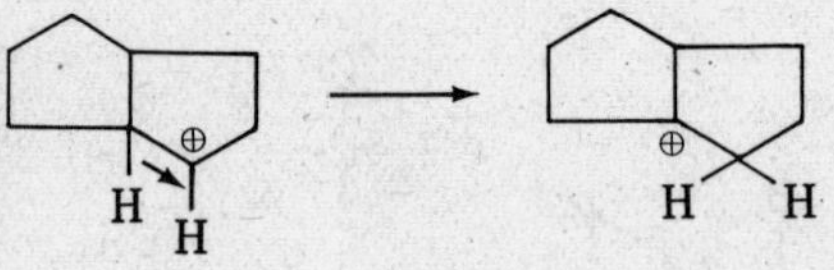 [bicyclic carbocation rearrangement figure]

(d) Can't; there are no adjacent sp^3 carbons.

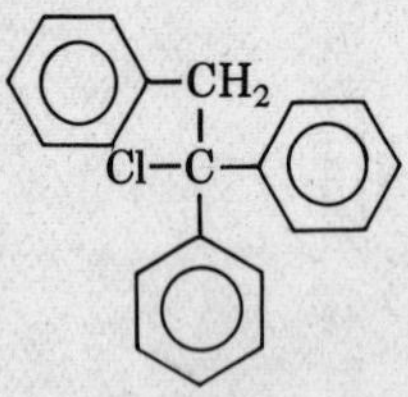

2. (a) 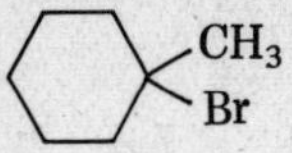 cyclohexyl-C(CH₃)(Br)

(b) Ph-C(Cl)(CH₂Ph)(Ph)

(c) Ph-CH₂-CH₂-CH₂-I; the carbonium ion Ph-CH₂-CH₂-CH₂$^\oplus$ is not formed in this S_N2 reaction.

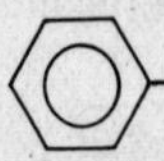

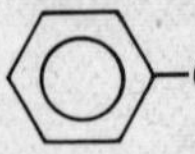

5. The only alcohol reaction that involves regioselectivity is dehydration to an alkene. Often more than one type of β H is present, and competitive removal of each leads to two or more alkenes.

Example

3-Hexanol, on heating with H_2SO_4, gives a mixture of 2-hexene and 3-hexene:

ALCOHOLS 353

$$CH_3-CH_2-CH_2-CH(OH)-CH_2-CH_3 + H_2SO_4 \rightleftharpoons CH_3-CH_2-CH_2-CH(\overset{\oplus}{O}H_2)-CH_2-CH_3 + HSO_4^{\ominus}$$

$$\downarrow$$

$$CH_3-CH_2-\overset{H}{C}H-\overset{\oplus}{C}H-\overset{H}{C}H-CH_3 + H_2O$$

$$CH_3-CH_2-CH=CH-CH_2-CH_3$$
$$+$$
$$CH_3-CH_2-CH_2-CH=CH-CH_3$$

choice of two kinds of β H for removal by $HSO_4^{\ominus}$

This reaction is further complicated by carbonium ion rearrangements. Not only can the resulting alkene have a skeleton different from the starting alcohol, but some alcohols can yield dehydration products even though they have no β C—H bonds available. *To predict the product of dehydration of any alcohol, follow these three steps:*

1. Protonate —OH and let OH_2 leave, generating a carbonium ion.
2. If a more stable carbonium ion can be formed by shifting H or R from an adjacent carbon, do it. (This converts 1° to 2°, 1° to 3°, or 2° to 3° $R^{\oplus}$.)
3. Remove $H^{\oplus}$ from that β carbon atom that leads to the most stable (most highly alkylated) alkene.

Example

Predict the principal dehydration product of $(CH_3)_3C-CH_2-OH$.

Step 1: $CH_3-\underset{CH_3}{\overset{CH_3}{\underset{|}{\overset{|}{C}}}}-CH_2-OH \xrightarrow{H_2SO_4} CH_3-\underset{CH_3}{\overset{CH_3}{\underset{|}{\overset{|}{C}}}}-CH_2-\overset{\oplus}{O}H_2 \longrightarrow CH_3-\underset{CH_3}{\overset{CH_3}{\underset{|}{\overset{|}{C}}}}-CH_2^{\oplus}$

$CH_3-\underset{CH_3}{\overset{CH_3}{\underset{|}{\overset{|}{C}}}}-\overset{\oplus}{C}H_2 \xrightarrow{\text{step 2}} CH_3-\underset{CH_3}{\overset{CH_3}{\underset{|}{\overset{|}{\overset{\oplus}{C}}}}}-CH_2 \xrightarrow[\text{step 3}]{HSO_4^{\ominus}}$

$CH_3-\underset{}{\overset{CH_3}{\underset{|}{\overset{|}{C}}}}=CH-CH_3$ trisubstituted

$CH_2=\underset{}{\overset{CH_3}{\underset{|}{\overset{|}{C}}}}-CH_2-CH_3$ disubstituted

★ *Exercises*

1. Predict the predominant alkene formed by dehydration of each alcohol.

(a) $\underset{CH_3}{\overset{CH_3}{\diagdown}}CH-\underset{}{\overset{CH_3}{\underset{|}{C}}}OH-CH_2-CH_3 \xrightarrow[175°C]{H_2SO_4}$

(b) [cyclohexane ring]$\underset{CH_3}{\overset{OH}{\diagup}} \xrightarrow[175°C]{H_2SO_4}$

(c) $CH_3-\underset{\underset{CH_3}{|}}{\overset{\overset{CH_3}{|}}{C}}-\underset{\underset{OH}{|}}{CH}-CH_3 \xrightarrow{H_2SO_4}{175°C}$

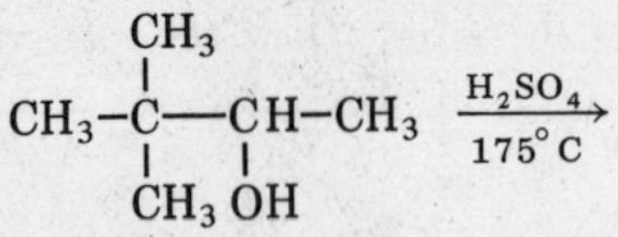

(d) 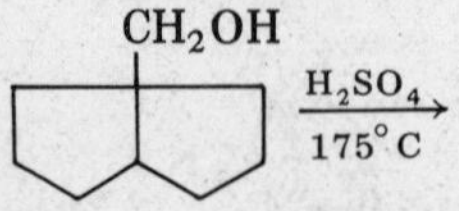 [bicyclic structure with CH$_2$OH] $\xrightarrow{H_2SO_4}{175°C}$

2. Write out the mechanism for reaction 1d.

- - - - - - - - -

1. (a) 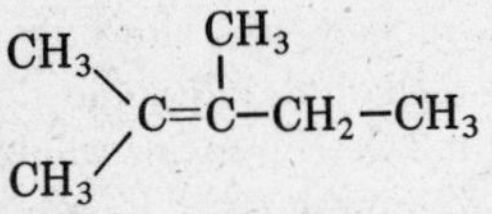 $\underset{CH_3}{\overset{CH_3}{\diagdown}}C=\underset{\underset{}{}}{\overset{\overset{CH_3}{|}}{C}}-CH_2-CH_3$ (b) [cyclohexene]–CH$_3$

(c) $\underset{CH_3}{\overset{CH_3}{\diagdown}}C=C\underset{CH_3}{\overset{CH_3}{\diagup}}$ (d) [bicyclic alkene structure]

2. $\underset{}{CH_2-\overset{..}{\underset{..}{O}}H} \curvearrowright H-\overset{\curvearrowright}{O}-SO_2OH$ [bicyclic] $\longrightarrow$ $CH_2-\overset{\oplus}{O}H_2\ HSO_4^{\ominus}$ [bicyclic] $\longrightarrow$ $CH_2^{\oplus}\ HSO_4^{\ominus}$ [bicyclic]

$\downarrow$

[bicyclic alkene] $\longleftarrow$ [bicyclic cation with H's labeled, $\oplus$, and $\overset{..}{\underset{..}{O}}^{\ominus}-SO_3H$]

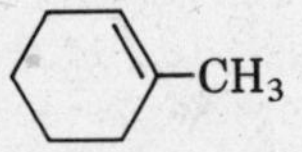

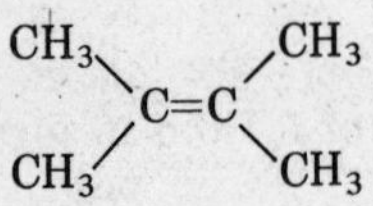

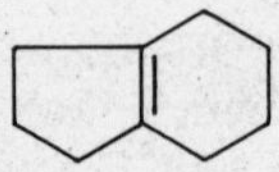

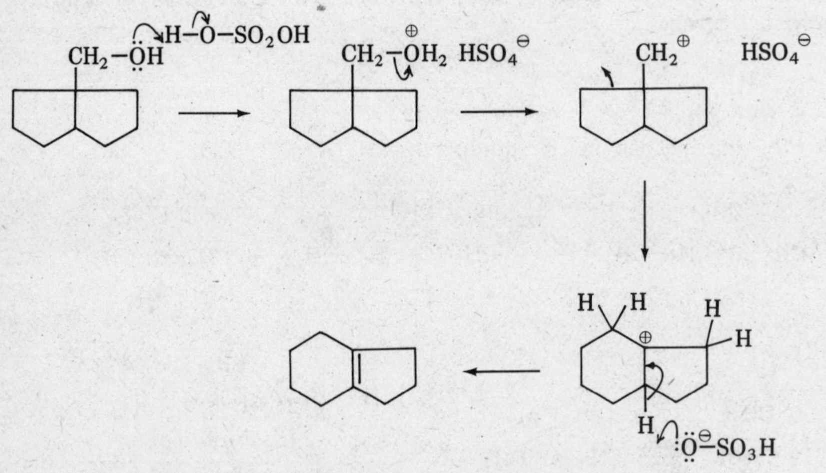

UNIT SIXTEEN
Aromatic Hydrocarbons and Their Derivatives

Aromatic hydrocarbons, for instance benzene (⌬), naphthalene (⌬⌬), biphenyl (⌬—⌬), and pyrene (⌬⌬⌬⌬), have one or more rings made up entirely of sp^2-hybridized carbon atoms. Such rings have a completely conjugated system of alternating double and single bonds. These aromatic compounds are characterized by:

- Bond lengths and strengths that differ from expectations based on aliphatic compounds.
- An unexpectedly great stability (low energy content).
- A pronounced tendency to undergo substitution reactions at $-\overset{H}{\underset{|}{C}}=\overset{H}{\underset{|}{C}}-$ rather than the additions characteristic of alkenes.

These characteristics make up what is often called *aromatic character* (aromaticity). Although the aromatic hydrocarbons look like polyalkenes, their aromatic character dictates an entirely different reactivity pattern. The properties of functional groups attached to these aromatic rings are also not what they would be if attached to an aliphatic skeleton. This unit covers aromatic reactivity in three sections: understanding its basis (frames 1 to 3), learning the reactions and mechanisms involved (frames 4 to 6), and practicing applications to specific problem types (frame 7).

OBJECTIVES

When you have completed this unit, you should be able to:

- Give an explanation of aromatic character based on the theory of resonance (frame 1).
- Predict the effect of resonance on the properties of functional groups attached to an aromatic ring (frame 1).
- Predict the products of the common electrophilic substitution reactions applied to benzene and substituted benzenes (frames 4, 5, 7).

- Predict qualitatively the effect of a ring substituent on the rate of electrophilic substitution (frame 5).
- Describe the conditions (substituents present, type of reagent, temperature) under which a leaving group on an aromatic ring can be displaced by nucleophiles (frame 6).
- Write the appropriate stepwise mechanisms for the common cases of electrophilic and nucleophilic aromatic substitution (frames 3, 5).

If you believe you have already achieved some of these objectives and may be able to skip part or all of this unit, take a self-test consisting of the exercises marked with a star in the margin. If you miss any of these exercises, go over the frames in question and rework the missed exercises before proceeding.

If you are not yet ready for a self-test, proceed to frame 1.

UNDERSTANDING THE REACTIVITY OF AROMATIC COMPOUNDS

1. In this frame we examine the effects of resonance on the properties of aromatic compounds.

The conjugation in an aromatic ring is endless and unique. Correspondingly, resonance of the type a ⟷ b reflects a unique degree of π-electron delocalization, and it results in a large resonance stabilization. This resonance accounts for the three facets of aromatic character, as follows:

- Since structures a and b are exactly equal contributors to the hybrid structure, all the C—C bonds in benzene are identical and intermediate in length between single and double.
- The large resonance energy (some 36 kcal/mole) stabilizes the molecule nearly as much as an extra covalent bond would. The resonance energy (resonance stabilization) is energy the real compound doesn't, in fact, have, but would have if it had the simple structure ⬡ or ⬡. This is the origin of the extra-low energy content that one detects by measuring the heat of hydrogenation or combustion. The resonance stabilization is also the reason why the aromatic ring remains intact when the aliphatic portion of the molecule is destroyed in such reactions as

$$\text{Ph--CH}_2\text{--CH}_2\text{--CH}_3 \xrightarrow[\text{heat}]{\text{KMnO}_4} \text{Ph--CO}_2\text{H} + CO_2 + H_2O$$

- The large resonance stabilization of benzene also accounts for the preference for substitution over addition. If benzene were simply ⬡, the reaction

$$\text{Ph} + Br_2 \longrightarrow \text{cyclohexadiene-Br,H,H,Br}$$

would be energetically downhill by about 25 kcal/mole:

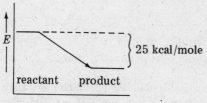

But *real* benzene is about 36 kcal/mole more stable than or , owing to resonance stabilization. The product, , like any conjugated diene, is only slightly (about 5 kcal/mole) resonance-stabilized. If we now incorporate these resonance stabilizations of reactant and product in the energy profile (as done by the heavy vertical arrows in the next figure), the picture changes considerably:

The addition reaction for benzene is really energetically *uphill* by approximately 6 kcal/mole. The resonance stabilization of aromatic rings acts as a large energy barrier to any reaction that destroys the aromaticity.

Resonance in aromatic compounds also affects the properties of groups attached to the ring, provided that the atom directly attached to the ring has unshared pair(s) or π bond(s).

An aryl group (e.g., the phenyl group,) can act both as an electron donor and as an electron acceptor (see Table 9.1). As a result, both electron-acceptor and electron-donor functions have a resonance interaction with the ring.

1. The common electron-donor functions $-\ddot{\underset{..}{X}}:$, $-\ddot{\underset{..}{O}}-R$, and $-\underset{R}{\overset{R}{N}}$ interact as follows:

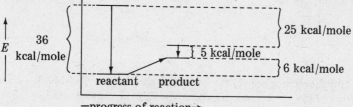

[Resonance structures of PhNR₂ showing delocalization into ring]

The consequences of this resonance interaction can be read directly from the contributing structures:

(a) The C–X, C–O, and C–N bonds attain some double-bond character and are thus shortened and strengthened compared to pure single bonds in, say, CH_3-X, CH_3-O-CH_3, and $(CH_3)_3N$. This shows up in the measured C–Cl bond lengths: CH_3Cl, 1.77 Å; Ph–Cl, 1.69 Å.

(b) The donor atom (X, O, N) suffers loss of electron density. Consequently, it is less basic than in aliphatic compounds. For instance, Ph–NH_2 is weaker as a base than CH_3NH_2 by a factor of about 10^4. [We develop a more accurate way of applying resonance theory to this case in exercise 2(b) below.]

(c) The diminished electron density on $-\ddot{X}$: and the increased strength of the C–X bond in the aryl halides are part of the reason why halogen on an aromatic ring does not ordinarily act as a leaving group:

Ph–$\ddot{X}$: $\not\leftrightarrow$:$\ddot{X}$:$^\ominus$

The other causes of this behavior are (1) the geometric situation: a nucleophile cannot approach the rearside of the C–X bond to initiate the S_N2 mechanism because the ring blocks its path; (2) aryl cations such as [phenyl cation with H's] $\oplus$ are much less stable than the aliphatic cations ($R^\oplus$), which rules out the S_N1 mechanism:

Ph–$\ddot{X}$: $\not\leftrightarrow$ Ph$^\oplus$ + $X^\ominus$

Unlike the [cyclohexadienyl cation with X, Y substituents] ions discussed in frame 3, such aryl cations are not resonance-stabilized, except for the normal benzene resonance, which the reactant Ph–X also has:

Ph$^\oplus$ ↔ Ph$^\oplus$ and Ph–X ↔ Ph–X

Only in special circumstances does X on an aromatic ring act as a leaving group, and it does so via two special mechanisms. The special circumstances are described in frame 2 and the mechanisms in frame 3.

AROMATIC HYDROCARBONS AND THEIR DERIVATIVES 359

(d) The aromatic ring is more electron-rich than in benzene itself. This increases its reactivity toward electrophiles (see frame 5).

2. The common electron-acceptor functions, $-\overset{\overset{O}{\|}}{C}-R$, $-C\equiv N$, $-N\overset{\nearrow O}{\searrow O}$, etc., interact like this:

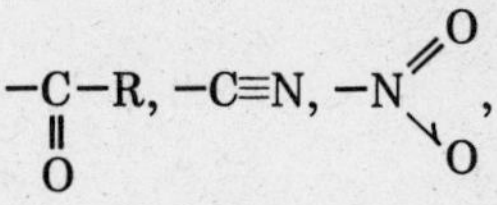

[Resonance structures showing benzene ring with –C(=O)–R group, displaying the four resonance forms with negative charge on oxygen and positive charge distributed on ortho and para ring positions]

[Resonance structures showing benzene ring with –C≡N group, displaying the four resonance forms with negative charge on nitrogen and positive charge distributed on ortho and para ring positions]

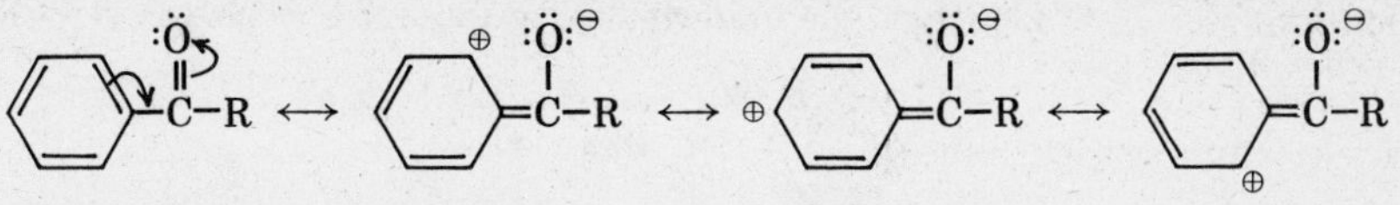

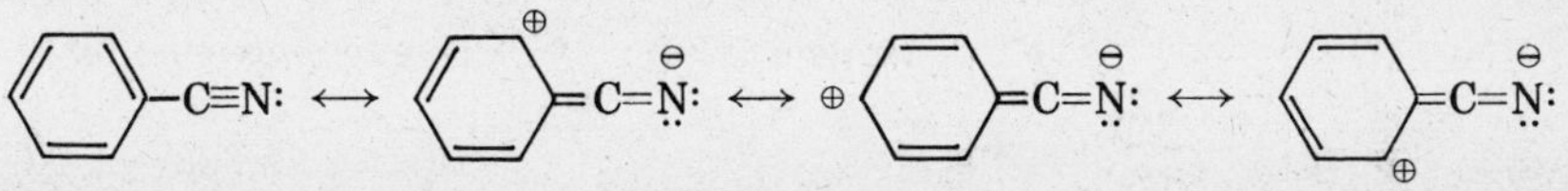

This resonance interaction decreases electron density in the ring and gives the C=O group some single-bond character.

Finally, the aromatic ring acts as a transmitter of resonance interactions. An electron-acceptor function and an electron-donor function attached *ortho* or *para* to one another can interact by resonance just as if they were directly connected by a single bond. For example:

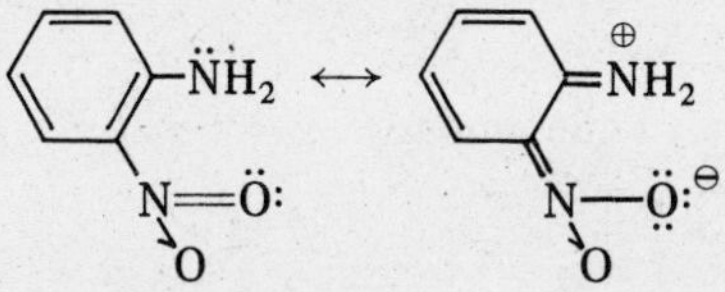

[Resonance structures showing o-nitroaniline and p-acyl phenoxide interactions]

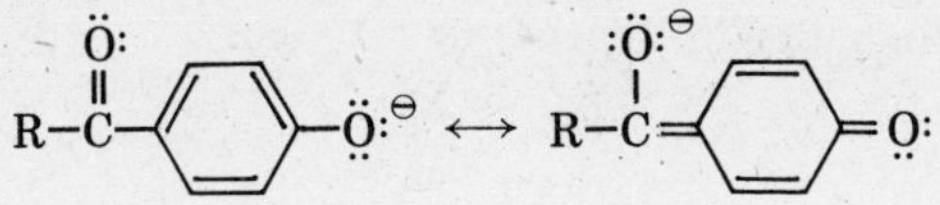

★ *Exercises*

1. Write resonance-hybrid structures for the following molecules. Review Unit 9 if you have difficulty.

(a) 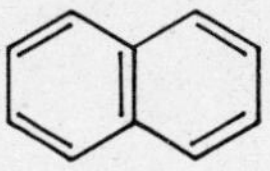 [naphthalene]

(b) 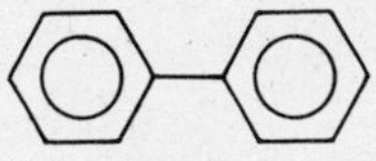 [biphenyl]

(c) 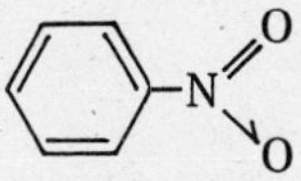 [C$_6$H$_5$–NO$_2$, nitrobenzene]

(d) [C$_6$H$_5$–NH$_2$, aniline]

(e) [C$_6$H$_5$–$\overset{\oplus}{N}$H$_3$]

(f) [C$_6$H$_5$–ÖH]

(g) [C$_6$H$_5$–Ö:$^\ominus$]

(h) [N≡C–C$_6$H$_4$–Ö:$^\ominus$]

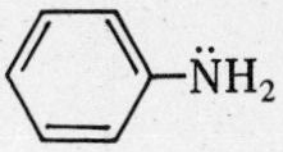

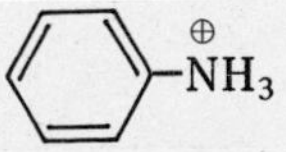

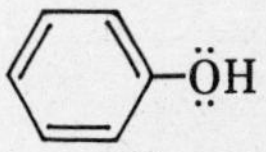

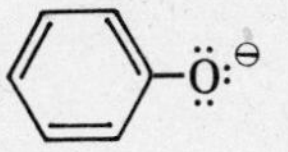

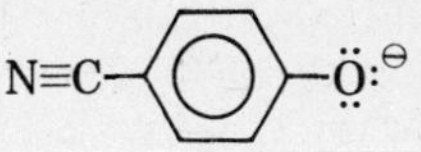

(i) 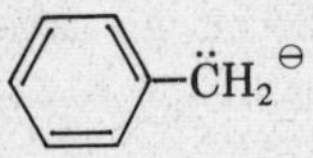 Ph–CH₂⁻   (j) 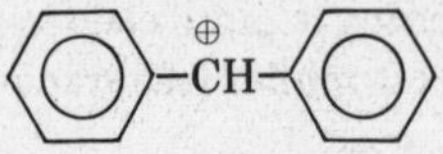 Ph–CH⁺–Ph

(k) 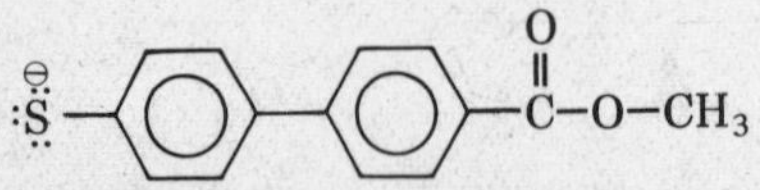 :S⁻–C₆H₄–C₆H₄–C(=O)–O–CH₃   (l) 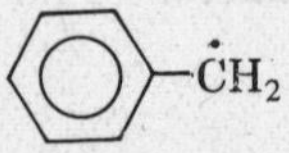 Ph–ĊH₂

2. Predict the *effect of resonance* on the following (refer to your answers to exercise 1):

   (a) The electron density in the Ph–NO₂ ring.

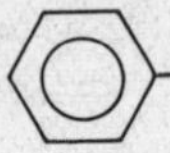

   (b) The stability of Ph–NH₂ vs. that of Ph–NH₃⁺, and consequently the basicity of Ph–NH₂ vs. that of CH₃NH₂.

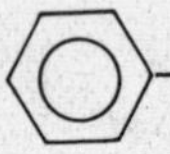

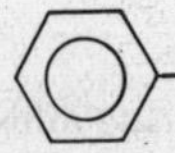

   (c) The stability of the benzyl anion, Ph–CH₂⁻, vs. that of :CH₃⁻ and consequently the acidity of Ph–CH₃ vs. that of CH₄.

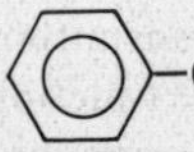

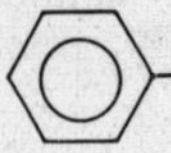

   (d) The S_N1 reactivity of Ph–CHBr–Ph in comparison with that of CH₃–CHBr–CH₃.

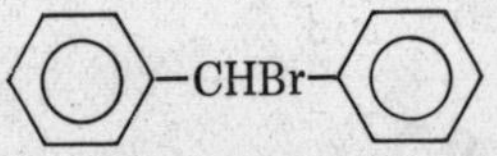

   (e) The rate of gas-phase radical chlorination of toluene, Ph–CH₃.

---

1. (a) 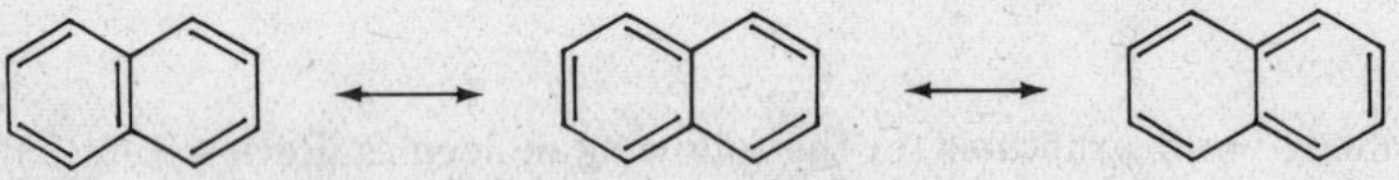 [naphthalene resonance structures]

   (b) 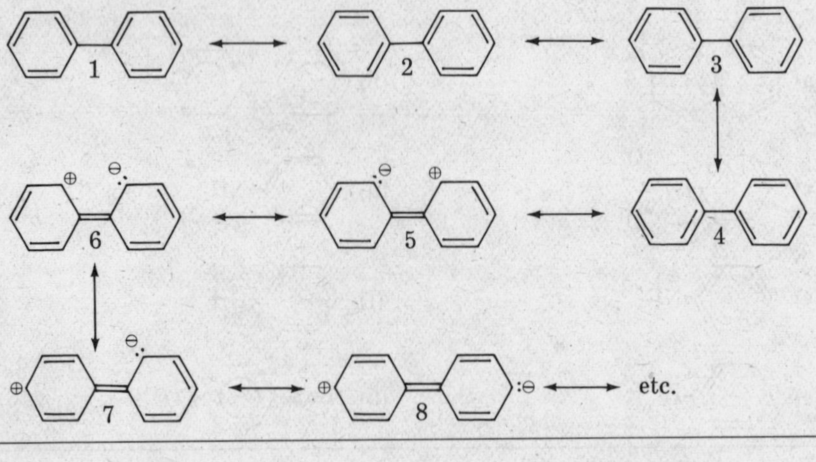 [biphenyl resonance structures 1–8]

Structures 1 to 4 are the major contributors; structures 5 to 8 are each much less stable and consequently contribute much less weight to the hybrid structure.

(c) [resonance structures of nitrobenzene showing six contributors with various charge distributions on the ring and nitro group]

(d) [resonance structures of aniline: Ph–NH$_2$ ↔ Ph–NH$_2$ ↔ (ring with –) =NH$_2^+$ ↔ ⊖(ring)=NH$_2^+$ ↔ (ring)=NH$_2^+$ with ⊖ on ring]

(e) Ph–$\overset{\oplus}{N}$H$_3$ ↔ Ph–$\overset{\oplus}{N}$H$_3$ (no unshared pair, not a donor atom)

(f) Ph–ÖH ↔ Ph–ÖH ↔ (ring)=Ö–H ↔ ⊖(ring)=O–H ↔ (ring)=O–H with ⊖
        1   2   3   4
                        5

Structures 3 to 5 again are small contributors compared with 1 and 2.

(g) Ph–Ö:$^\ominus$ ↔ Ph–Ö:$^\ominus$ ↔ (ring)=Ö: ↔ ⊖(ring)=O: ↔ (ring)=O:$^\ominus$

In this case, the last three structures do not involve charge separation, and are comparable in stability to the first two structures.

(h) :N≡C–(ring)–Ö:$^\ominus$ ↔ :N≡C–(ring)–Ö:$^\ominus$ ↔ :N≡C–(ring)=O:

$^\ominus$:N=C=(ring)=O ↔ :N≡C–:(ring)=O ↔ :N≡C–(ring)=O:$^\ominus$

This ion has all the contributing structures of (ring)–O$^\ominus$ plus an additional one that puts the charge on N, an electronegative atom better able to bear it than is C.

(i) [Ph-ĊH₂ ↔ Ph-CH₂⁻ ↔ (cyclohexadienylidene)=CH₂ with ⁻ on ring ↔ ⁻: (cyclohexadienylidene)=CH₂ ↔ (cyclohexadienylidene)=CH₂ with ⁻ on ring]

(j) [Ph-CH⁺-Ph ↔ Ph-CH⁺-Ph ↔ Ph-CH⁺-Ph]
    ↕
    [Ph-CH=(ring⁺) ↔ (ring⁺)=CH-Ph ↔ Ph-CH⁺-Ph] ↔ etc.

(16 limiting structures in all)

(k) :S⁻–⟨ring⟩–⟨ring⟩–C(=O)–O–CH₃   :S=⟨ring⟩=⟨ring⟩=C(O⁻)–O–CH₃

structures anal-   a new structure, more stable
ogous to all those   and contributing more than
in (b)   the dipolar structures in (b)
                      because this one involves
                      no charge separation

(l) [Ph-ĊH₂ ↔ (ring•)−CH₂ ↔ (ring)=CH₂ with • ↔ •(ring)=CH₂ ↔ (ring•)=CH₂]

2. (a) Smaller than in ⟨benzene⟩, according to exercise 1(c).

(b) According to exercises 1(d) and (e), ⟨Ph⟩–NH₂ has additional resonance stabili-

zation due to interaction of the —NH₂ group with the ring, but ⟨Ph⟩–NH₃⁺ does

not. So resonance stabilizes the left side of

⟨Ph⟩–NH₂ + H₂O ⇌ ⟨Ph⟩–NH₃⁺ + OH⁻

preferentially, which means the equilibrium constant is smaller and the base

(⟨Ph⟩–NH₂) weaker than in the absence of resonance (e.g., with CH₃–NH₂).

(c) $Ph\text{-}\overset{\ominus}{C}H_2$ has resonance stabilization that $:\overset{\ominus}{C}H_3$ does not have, so the equilibrium

$$Ph\text{-}CH_3 + :B^{\ominus} \rightleftharpoons Ph\text{-}\overset{\ominus}{C}H_2 + BH$$

goes farther to the right than does

$$CH_4 + :B^{\ominus} \rightleftharpoons :\overset{\ominus}{C}H_3 + BH$$

That is, $Ph\text{-}CH_3$ is the stronger acid.

(d) The S_N1 reaction rate is the rate of $R\text{-}Br \longrightarrow R^{\oplus} + Br^{\ominus}$. It increases with increasing stability of $R^{\oplus}$. $Ph\text{-}\overset{\oplus}{C}H\text{-}Ph$ is resonance-stabilized (more stable than $CH_3\text{-}\overset{\oplus}{C}H\text{-}CH_3$) and $Ph\text{-}CHBr\text{-}Ph$ ionizes faster than $CH_3CHBrCH_3$.

(e) $Ph\text{-}CH_3$ should chlorinate faster than alkanes of type $R\text{-}CH_3$ do because $Ph\text{-}\overset{\cdot}{C}H_2$ is more stable than $R\text{-}\overset{\cdot}{C}H_2$ and is consequently formed faster in the first propagation step of mechanism I.

2. The common reactions of the aromatic ring and of the ordinary functional groups attached to aromatic rings are shown in Chart 16.1, on the following two pages.
Reactions ①, ④, ⑤, ⑦, and ⑨ are general slot-and-filler reactions. The permissible slot fillers for the electrophilic addition ⑦ are the same as those given in Table 12.1 (p. 209). Those for the other cases are listed in Tables 16.1 to 16.4, on pages 366-368.

Exercises (use Chart 16.1 and Tables 16.1 to 16.4 freely):

1. Write the specific equations for reactions ①, ③, ④, ⑤, ⑦, and ⑨ using the following choices of slot fillers:

   (a) reaction ① with G = H and E-Nu = $Ph\text{-}\overset{O}{\underset{\|}{C}}\text{-}Cl$

   (b) reaction ③ wtih R = $-CH-CH_2-OH$
                                  $|$
                                  CH_3

   (c) reaction ④ with W = $-C{\equiv}N$, L = $-O-SO_2-CH_3$, and Nu = $Ph\text{-}CH_2\text{-}S^{\ominus}$

   (continued on page 366)

364 HOW TO SUCCEED IN ORGANIC CHEMISTRY

Chart 16.1

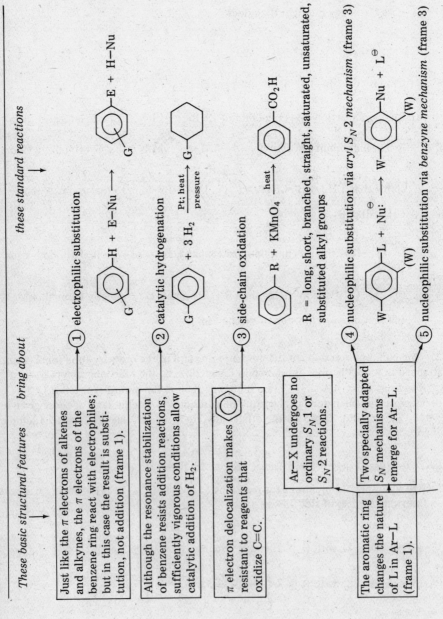

AROMATIC HYDROCARBONS AND THEIR DERIVATIVES 365

⑥ metallation

[Ar-L with G] + B: → [Ar-B with G] and [Ar-H with G]

But reactions of Ar—X with metals still work.

[Ar-X with G] → [Ar-MgX with G]

[Ar-X with G] → [Ar-Li with G] + LiX

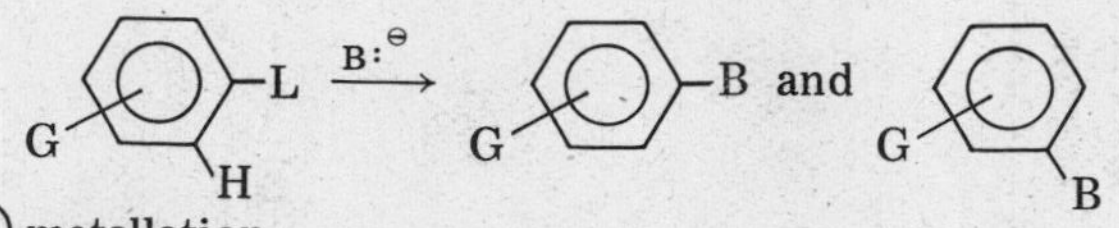

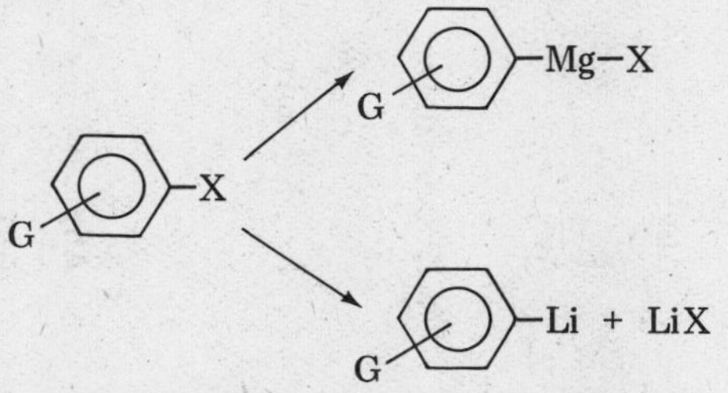

G not = $-NO_2$, $-CO-$, $-COOH$, $-COOR$, $-OH$, $-SO_3H$

⑦ electrophilic addition

$Ar-C=C$ + E-Nu ⟶ $Ar-\overset{|}{C}-\overset{|}{C}-$
 NuE

Electrophilic additions to $Ar-C=C$ always go via the resonance-stabilized benzylic cation,

$Ar-\overset{\oplus}{\underset{|}{C}}-\overset{|}{\underset{|}{C}}-E$

⑧ reduction of $Ar-NO_2$

$Ar-NO_2 \xrightarrow[\text{or } H_2/Pt]{\text{Fe/aq. HCl}} Ar-NH_2$

⑨ nucleophilic substitutions on $Ar-N_2^\oplus$

$Ar-N_2^\oplus$ + Nu:⁻ ⟶ $Ar-Nu + N_2$

The $-N\equiv N$: function is much more stable when attached to Ar than to R, so it lives long enough to react with a chosen, added nucleophile.

Table 16.1 Permissible Slot Fillers for the Electrophilic Aromatic Substitution Reaction

$$\text{G-C}_6\text{H}_4\text{-H} + \text{E-Nu} \xrightarrow{\text{catalyst}} \text{G-C}_6\text{H}_4\text{-E} + \text{H-Nu}$$

Permissible E+Nu	Required catalyst	Permissible substituents, G — May be present	Permissible substituents, G — Must be present
O₂N+O-H	Conc. H$_2$SO$_4$	Any	
Cl+Cl, Br+Br	Fe (FeCl$_3$, FeBr$_3$)	Any	
R+X	AlX$_3$	R−, X− only	
R-CO+X	AlX$_3$	R−, X− only	
Ar−N≡N:⁺ | HSO$_4$⁻			−OH or −NR$_2$
Tl(O-CO-CF$_3$)$_2$⁺ | CF$_3$COO⁻		Any	
HO-SO$_2$+OH		Any	
D+O-D	D$_2$SO$_4$	Any	

(d) reaction ⑤ with G = CH$_3$− (para), X = F, and B:⁻ = C$_6$H$_5$:⁻

(e) reaction ⑥ with X = Br, G = −C$_2$H$_5$ (meta), M = Mg

(f) reaction ⑨ with G = NO$_2$ (para) and Nu:⁻ = CN⁻

★ 2. Supply the missing reactant, reagent, or product.

(a) [naphthalene]−N≡N:⁺ HSO$_4$⁻ + H$_3$PO$_2$ ⟶ ?

(b) CH$_3$−C$_6$H$_4$−CH$_3$ + ? ⟶ CH$_3$−C$_6$H$_3$(SO$_3$H)−CH$_3$

(continued on page 368)

Table 16.2 *Permissible Slot Fillers for the Nucleophilic Aromatic Substitution Reaction Occurring via the Aryl S_N2 Mechanism*

$$W\text{-}\underset{}{\bigcirc}\text{-}L + Nu\text{:} \longrightarrow W\text{-}\underset{}{\bigcirc}\text{-}Nu + L\text{:}$$

Permissible leaving groups, $\curvearrowleft L \longrightarrow L$:	Activating (electron-withdrawing) substituents, W; at least one must be present ortho or para to L	Permissible nucleophiles, Nu:
$\curvearrowleft F \longrightarrow F^\ominus$	$-NO_2$	$F^\ominus, Cl^\ominus, Br^\ominus, I^\ominus$
$\curvearrowleft Cl \longrightarrow Cl^\ominus$	$-C\equiv N$	$OH^\ominus, OR^\ominus$
$\curvearrowleft Br \longrightarrow Br^\ominus$	$-SO_2-OH$	$SH^\ominus, SR^\ominus$
$\curvearrowleft I \longrightarrow I^\ominus$	$-SO_2-R$	$RCO_2^\ominus$
$\curvearrowleft O-SO_2-R \longrightarrow RSO_3^\ominus$	$\overset{O}{\underset{\parallel}{-C-R}}$	$CN^\ominus$
$\curvearrowleft O-SO_2-Ar \longrightarrow Ar-SO_3^\ominus$		$R\text{:}^\ominus, R-C\equiv C\text{:}^\ominus$
		$R_2\overset{..}{C}\text{-}\overset{O}{\underset{\parallel}{C}}\text{-}R$
		$R_3N\text{:}$

Table 16.3 *Permissible Slot Fillers for the Nucleophilic Aromatic Substitution Reaction Occurring via the Elimination/Addition (Benzyne) Mechanism*

$$\underset{G}{\bigcirc}\text{-}X + B\text{:}^\ominus \longrightarrow \underset{G}{\underset{H}{\bigcirc}}\text{-}B \text{ and } \underset{G}{\underset{B}{\bigcirc}}\text{-}H + \text{:}\ddot{X}\text{:}^\ominus$$

Permissible leaving groups, $-X$	Permissible superbases, $B\text{:}^\ominus$	Permissible substituents, $-G$
$-F$	$NH_2^\ominus$	$-R$
$-Cl$	$R\text{:}^\ominus$	$-OR$
$-Br$	$Ar\text{:}^\ominus$	$-Ar$
$-I$	$RO^\ominus$ in CH_3SOCH_3 solvent	
	$OH^\ominus$ in H_2O under pressure above $300°C$	

368 HOW TO SUCCEED IN ORGANIC CHEMISTRY

Table 16.4 *Permissible Slot Fillers for Nucleophilic Substitution Reactions ($S_N 1$) of $Ar-N_2^{\oplus}$*

$$G-C_6H_4-N{\equiv}N: + Nu:^{\ominus} \xrightarrow{catalyst} G-C_6H_4-Nu + N{\equiv}N$$

Permissible nucleophiles, Nu:	Catalyst required	Permissible substituents, $-G$
$I^{\ominus}$	None	
$Cl^{\ominus}, Br^{\ominus}$	$Cu^{\oplus}$	
$F^{\ominus}$ (latent in $F-BF_3^{\ominus}$)	Heat	almost any, except $-OH$, $-NH_2$, or $-NR_2$
$CN^{\ominus}$	$Cu^{\oplus}$	
$NO_2^{\ominus}$	Cu	
H_2O	Heat	
$H{:}^{\ominus}$ (latent in $H_2P(O)(OH)$ or $H-BH_3^{\ominus}$)	None	

(c) $O_2N-C_6H_3(NO_2)-F$ + cyclopentanone enolate $\longrightarrow$?

(d) $O_3S^{\ominus}-C_6H_4-N_2^{\oplus}$ + ? $\longrightarrow$ $O_3S^{\ominus}-C_6H_4-N{=}N-C_6H_4-OH$

(e) ? $\xrightarrow{heat}$ $C_6H_5-C_6H_4-C_6H_4-F$

- -

1. (a) C_6H_6 + $C_6H_5-C(=O)-Cl$ $\xrightarrow{AlCl_3}$ $C_6H_5-C(=O)-C_6H_5$ + HCl

(b) $C_6H_5-CH(CH_3)-CH_2OH$ $\xrightarrow[heat]{KMnO_4}$ $C_6H_5-CO_2H$

(c) $N{\equiv}C-C_6H_3(C{\equiv}N)-O-SO_2-CH_3$ + $C_6H_5-CH_2-S^{\ominus}$ $\longrightarrow$

$N{\equiv}C-C_6H_3(C{\equiv}N)-S-CH_2-C_6H_5$

AROMATIC HYDROCARBONS AND THEIR DERIVATIVES 369

(d) CH$_3$—⟨C$_6$H$_4$⟩—F + ⟨C$_6$H$_5$⟩:$^\ominus$ Li$^\oplus$ ⟶ CH$_3$—⟨C$_6$H$_4$⟩—⟨C$_6$H$_5$⟩ + CH$_3$—⟨C$_6$H$_4$⟩—⟨C$_6$H$_4$⟩—⟨C$_6$H$_5$⟩

(e) CH$_3$CH$_2$—⟨C$_6$H$_4$⟩—Br + Mg ⟶ CH$_3$CH$_2$—⟨C$_6$H$_4$⟩—Mg—Br

(f) O$_2$N—⟨C$_6$H$_4$⟩—N$\overset{\oplus}{\equiv}$N HSO$_4$$^\ominus$ + Cu$^\oplus$ CN$^\ominus$ ⟶ O$_2$N—⟨C$_6$H$_4$⟩—C≡N + N$_2$
   + CuHSO$_4$

2. (a) [naphthalene]   (b) conc. H$_2$SO$_4$

(c) O$_2$N—⟨C$_6$H$_3$(NO$_2$)⟩—C(=O)—cyclopentyl   (d) ⟨C$_6$H$_5$⟩—OH

(e) ⟨C$_6$H$_5$⟩—⟨C$_6$H$_4$⟩—⟨C$_6$H$_4$⟩—N$\overset{\oplus}{\equiv}$N BF$_4$$^\ominus$

3. The three new mechanisms introduced in Chart 16.1 are similar to one another—they all accomplish substitution in two steps, either via addition followed by elimination or via elimination followed by addition. They can be handled in slot-and-filler format:

   1. *Electrophilic aromatic substitution, mechanism VII (addition-elimination):*

   E—Nu ⟶ E$^\oplus$ + Nu:$^\ominus$ or   E—Nu A ⟶ E$^\oplus$ + Nu-A (where A is some Lewis acid catalyst)

   ⟨C$_6$H$_5$⟩—H   E$^\oplus$ ⟨arenium⟩—H ⟶ ⟨arenium⟩^{E,H} ⟷ ⟨arenium⟩^{E,H} ⟷ ⟨arenium⟩^{E,H} addition

   etc. ⟷ ⟨arenium⟩$^{E,H}_{:Nu^\ominus}$ ⟶ ⟨C$_6$H$_5$⟩—E + H-Nu   elimination

Note that all the catalysts involved in these reactions (Table 16.1) are Lewis acids. Their function is to aid in generation of E$^\oplus$ by binding Nu: E—Nu + A ⇌ E$^\oplus$ + Nu-A$^\ominus$. The specific instances of this step are

$$\text{HNO}_3 + \text{H}_2\text{SO}_4 \rightleftharpoons \underset{O=}{\overset{E-Nu}{\underset{O}{N}}}\overset{\oplus}{-}\text{OH}_2 \text{HSO}_4^\ominus \longrightarrow O=\overset{\oplus}{N}=O \text{ HSO}_4^\ominus + H_2O \xrightarrow{H_2SO_4} H_3\overset{\oplus}{O} \text{HSO}_4^\ominus$$

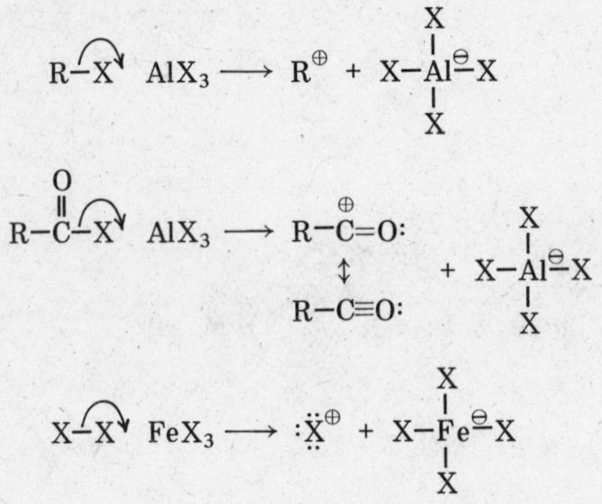

$$R-X \quad AlX_3 \longrightarrow R^{\oplus} + X-\overset{X}{\underset{X}{Al^{\ominus}}}-X$$

$$R-\overset{O}{\overset{\|}{C}}-X \quad AlX_3 \longrightarrow R-\overset{\oplus}{C}=O: \updownarrow R-C\equiv O: \quad + X-\overset{X}{\underset{X}{Al^{\ominus}}}-X$$

$$X-X \quad FeX_3 \longrightarrow :\overset{\oplus}{X} + X-\overset{X}{\underset{X}{Fe^{\ominus}}}-X$$

The sulfonation reaction (E–Nu = $HOSO_2$–OH) deviates slightly from this pattern. The actual electrophile is SO_3, which is present at low concentrations in concentrated sulfuric acid:

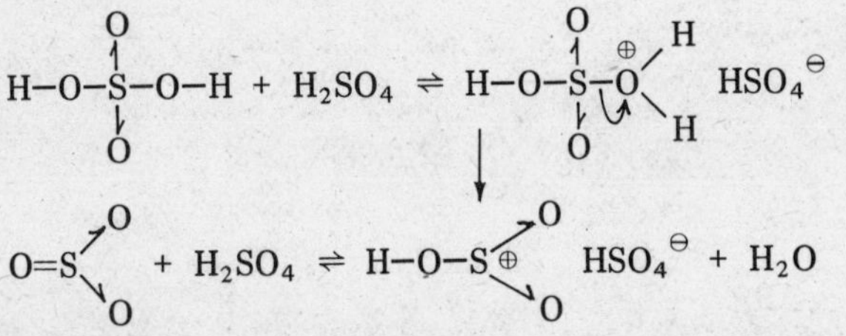

$$H-O-\overset{O}{\underset{O}{\overset{\|}{S}}}-O-H + H_2SO_4 \rightleftharpoons H-O-\overset{O}{\underset{O}{\overset{\|}{S}}}-\overset{\oplus}{O}\overset{H}{\underset{H}{<}} \quad HSO_4^{\ominus}$$

$$\downarrow$$

$$O=S\overset{O}{\underset{O}{<}} + H_2SO_4 \rightleftharpoons H-O-\overset{\oplus}{S}\overset{O}{\underset{O}{<}} \quad HSO_4^{\ominus} + H_2O$$

Extra SO_3 can be dissolved in H_2SO_4 to make a more reactive reagent (fuming sulfuric acid). The addition and elimination steps look like this:

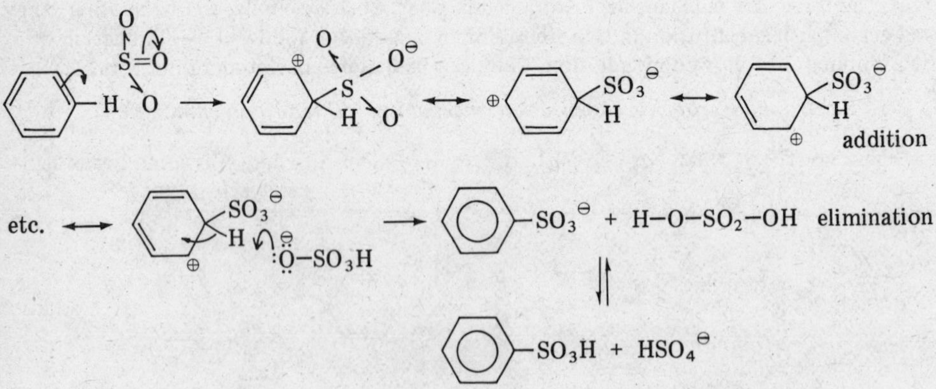

[Mechanism diagram showing benzene + SO₃ addition, formation of sigma complex with resonance structures, then elimination to give PhSO₃⁻ + H-O-SO₂-OH, equilibrating with PhSO₃H + HSO₄⁻]

The first step (addition) in the electrophilic aromatic substitution mechanism raises two questions: (1) Since this step destroys the aromatic character of the benzene ring, why is it not prohibited by the 36-kcal/mole barrier described above? (2) Since this step is identical with the first step in the electrophilic addition mechanism (mechanism II), where and how does the differentiation occur that leads to substitution product instead of addition? The answers are as follows.

1. Addition of electrophile is not energetically prohibitive because, although the resonance stabilization of benzene is lost, much of it is recovered as a resonance stabilization of the resulting carbonium ion:

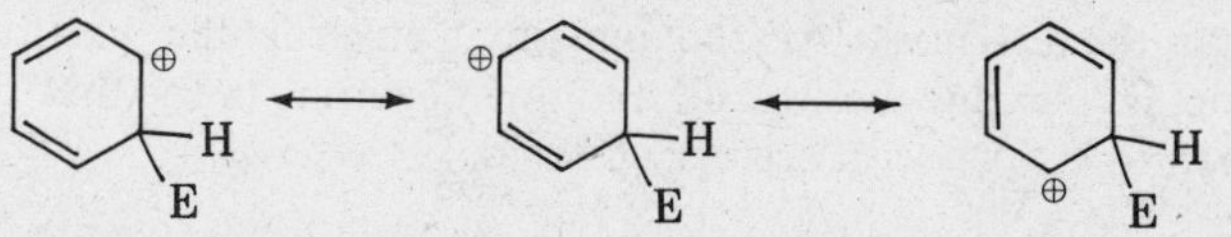

AROMATIC HYDROCARBONS AND THEIR DERIVATIVES

As a result, the energy barrier to this step is only about the same magnitude as in the first step in electrophilic addition to a simple alkene, $\ce{>C=C<} \; E^\oplus \longrightarrow \; \ce{>C-C-E}$.

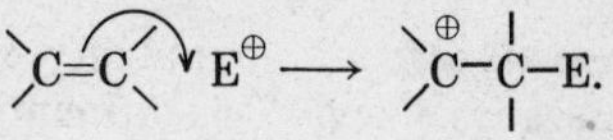

2. Discrimination between overall addition and substitution comes in the second step. For instance in the reaction with bromine,

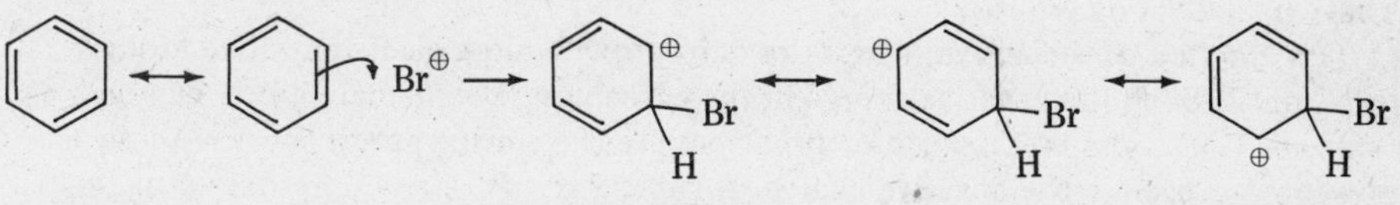

etc. 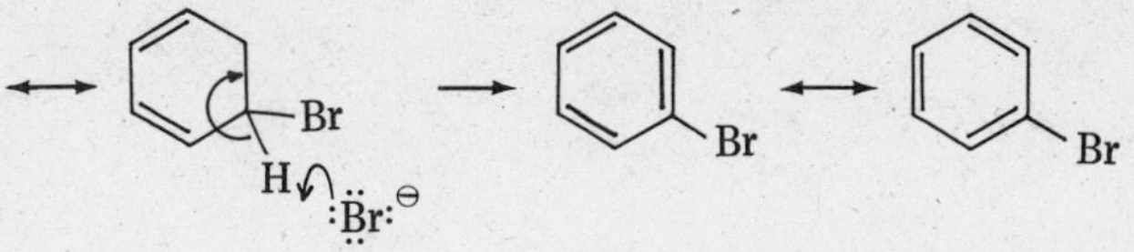 ⟷ [structure] substitution

versus

etc. 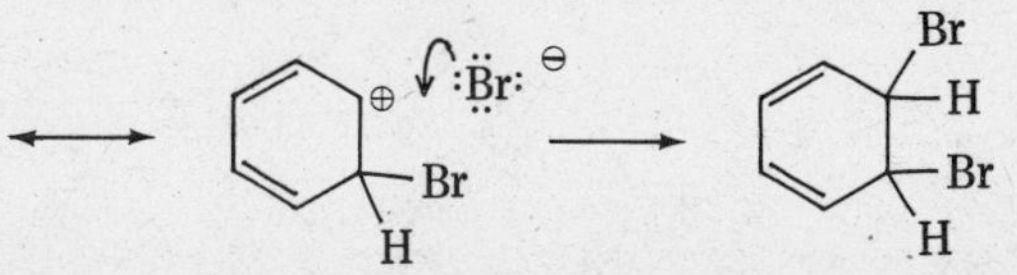 ⟷ [structure] addition

The first path (leading to substitution product) restores the aromatic ring, recollects the benzene resonance stabilization, and is consequently exothermic and rapid. The second path (leading to addition product) proceeds with loss of resonance stabilization, is less exothermic, and is slower. Thus the decision in favor of substitution comes in the second step of the reaction, and the electrophilic addition step is common to both processes.

2. *The aryl $S_N 2$ mechanism, mechanism IXb (addition-elimination):*

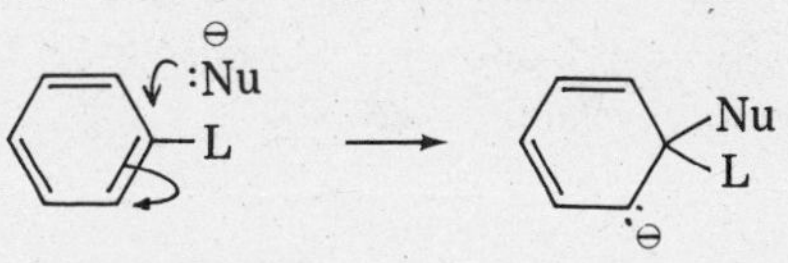

[structure] addition

[structure] elimination

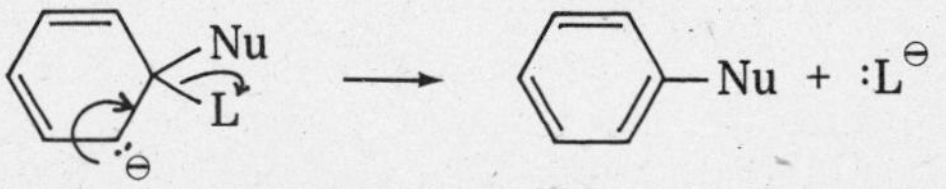

The first step disrupts electron delocalization by resonance in [Ph]–L, but this should not prohibit the reaction, since we saw that the analogous addition of electrophile is possible in the preceding mechanism. The intermediate,

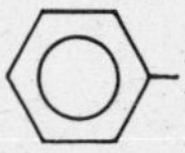

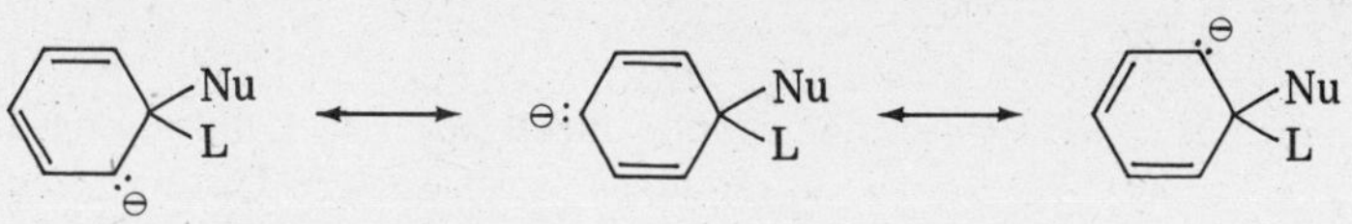

372 HOW TO SUCCEED IN ORGANIC CHEMISTRY

is resonance-stabilized just as the intermediate carbonium ion in electrophilic substitution is. A more suspicious feature is this: the nucleophilic addition occurring in the first step has never been observed to happen with ordinary alkenes:

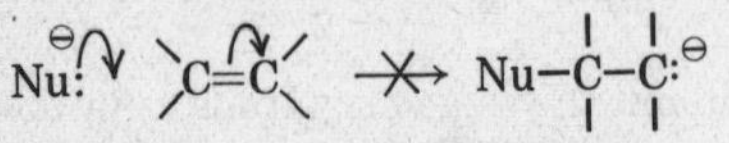

$$Nu:^{\ominus} \quad C=C \quad \nrightarrow \quad Nu-C-C:^{\ominus}$$

The reason is carbon's poor ability, as an element of intermediate electronegativity, to bear a negative charge. This difficulty does indeed rule out the aryl S_N2, addition-elimination mechanism when activating groups are absent; aromatic nucleophilic substitution (ordinary S_N2 or aryl S_N2) does *not* occur on simple aryl halides, like chlorobenzene, that have no activating groups.

The function of the activating groups is to provide more electronegative atoms (O or N) onto which the negative charge in the carbanion intermediate above can be delocalized. They must be attached ortho or para to the leaving group because these are the carbons bearing the negative charge. A typical aryl S_N2 reaction that actually works then looks like this:

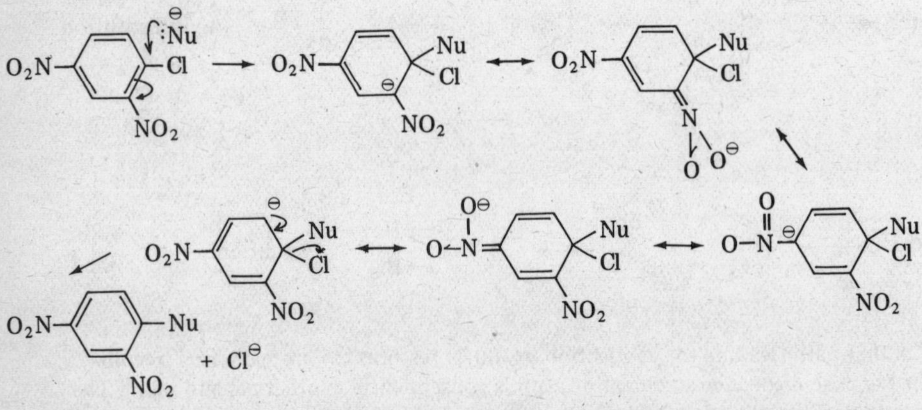

[Reaction scheme showing nucleophilic aromatic substitution on p-nitro, o-nitro chlorobenzene with resonance structures delocalizing the negative charge onto the nitro groups, ultimately yielding the substituted product plus $Cl^{\ominus}$]

The aryl S_N2 mechanism above is also important because it is very closely related to the acyl S_N2 mechanism (mechanism IXa, Unit 18) for nucleophilic displacement of leaving groups attached to $C=O$, $C=S$, or $C=N-$. These reactions are important in many organic and biochemical systems. The first step of the aryl S_N2 mechanism above is also the first step of the nucleophilic addition mechanism (mechanism VIII, Unit 17) for addition reactions to $C=O$, $C=S$, $C=N$, and $-C\equiv N$ bonds.

3. *The elimination-addition (benzyne) mechanism:*

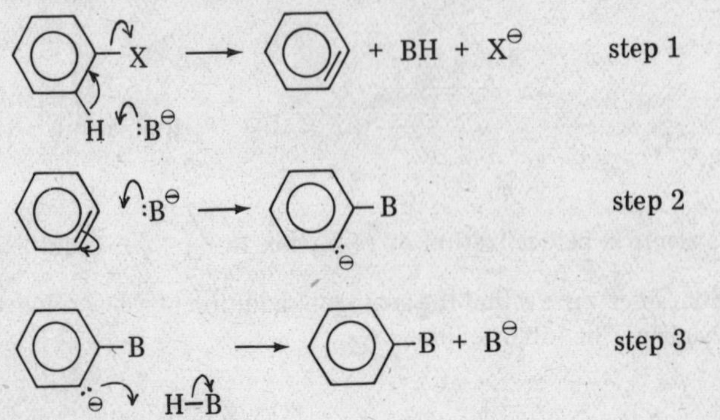

[Step 1: aryl halide with ortho H removed by base $B^{\ominus}$ to give benzyne + BH + $X^{\ominus}$]

[Step 2: benzyne attacked by $:B^{\ominus}$ to give aryl carbanion]

[Step 3: aryl carbanion abstracts H from H–B to give aryl–B + $B^{\ominus}$]

This is the only mechanism that succeeds on simple, unactivated aryl halides. The first step is just like the $E2$ process, except that it requires a stronger base. The intermediate

⬡ (benzyne) has a unique electronic structure that is actually not much like an alkyne because linear geometry is impossible in the ring:

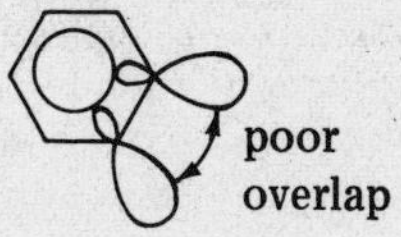

poor overlap

This weak new π bond *is* susceptible to attack by nucleophiles (step 2).

★ *Exercises*

1. Referring when necessary to the preceding tables and generic mechanisms, write all steps in the mechanisms of these reactions:

   (a) 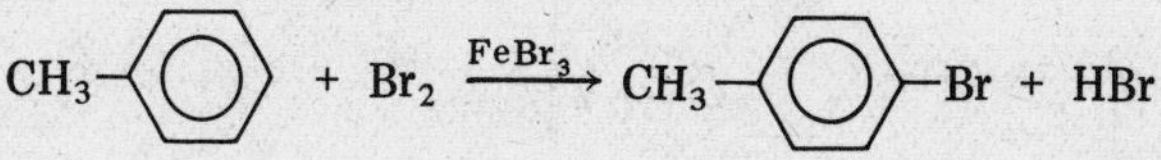 $CH_3-\bigcirc + Br_2 \xrightarrow{FeBr_3} CH_3-\bigcirc-Br + HBr$

   (b) 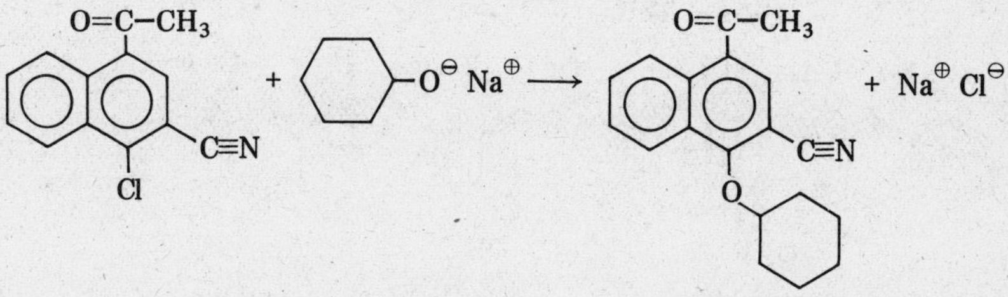 [naphthalene with O=C-CH₃, C≡N, Cl substituents] + cyclohexyl-O⁻ Na⁺ ⟶ [naphthalene with O=C-CH₃, C≡N, O-cyclohexyl substituents] + Na⁺ Cl⁻

   (c) $I-\bigcirc + CH_3-O^{\ominus} Na^{\oplus} \xrightarrow[CH_3SOCH_3]{heat} CH_3-O-\bigcirc + Na^{\oplus} I^{\ominus}$

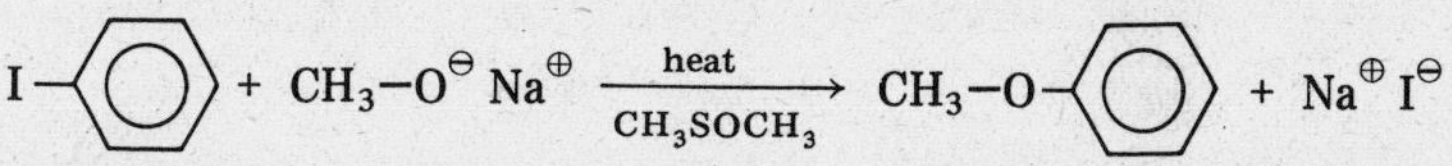

2. Reaction ⑨, Nu: + Ar—$\overset{\oplus}{N_2}$ ⟶ Ar—Nu + N_2, is aromatic nucleophilic substitution, but it differs from the two types discussed in this frame (reactions ④ and ⑤) in two ways: (a) neither activating groups nor superbases are required; (b) the mechanism is the standard S_N1 mechanism studied in Unit 14, frame 4. The S_N1 mechanism works in this case in spite of the instability of the aryl carbonium ion $\bigcirc^{\oplus}$ because the most reactive of all leaving groups, $\overset{\oplus}{N}\equiv N: \longrightarrow :N\equiv N:$, is involved. The $-\overset{\oplus}{N_2}$ group is so reactive because on leaving it produces the extremely stable nitrogen molecule, N_2.

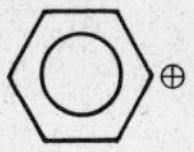

Write out the mechanism for these reactions of Ar—$\overset{\oplus}{N_2}$. Refer back to Unit 14 if necessary.

   (a) 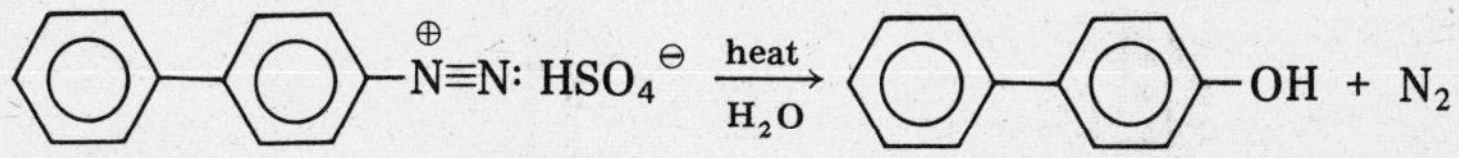 $\bigcirc-\bigcirc-\overset{\oplus}{N}\equiv N: HSO_4^{\ominus} \xrightarrow[H_2O]{heat} \bigcirc-\bigcirc-OH + N_2$

   (b) 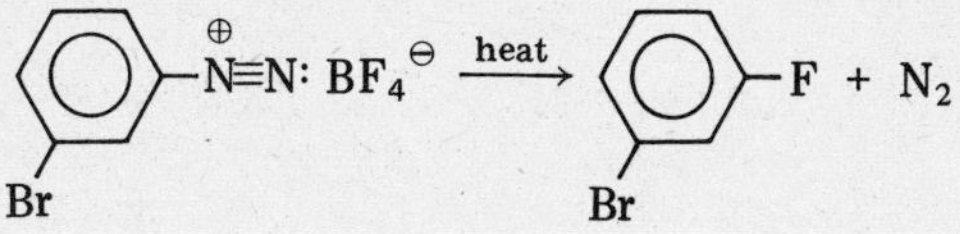 [Br-substituted phenyl]$-\overset{\oplus}{N}\equiv N: BF_4^{\ominus} \xrightarrow{heat}$ [Br-substituted phenyl]$-F + N_2$

3. Make use of the elimination-addition (benzyne) mechanism to explain why two products are obtained in most of these reactions. Work with the specific example

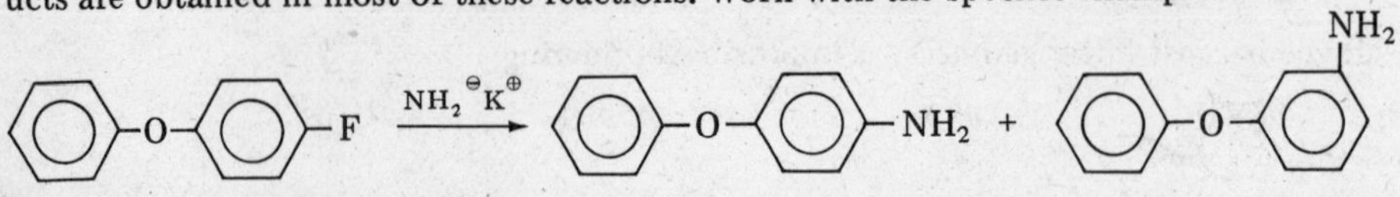

Ph–O–C₆H₄–F $\xrightarrow{NH_2^- K^+}$ Ph–O–C₆H₄–NH₂ + Ph–O–C₆H₄(NH₂)

---

1. (a) $:\ddot{Br}-\ddot{Br}:$ FeBr₃ ⟶ $:\ddot{Br}:^{\oplus}$ + FeBr₄⁻

   CH₃–C₆H₅ + Br⁺ ⟶ CH₃–[C₆H₅(Br)(H)]⁺ ⟷ CH₃–[C₆H₅(Br)(H)]⁺ ⟷

   CH₃–[C₆H₅(Br)(H)]⁺ etc. ⟷ CH₃–C₆H₄(Br)(H) $\xrightarrow{Br-FeBr_3}$ CH₃–C₆H₄–Br

   + H–Br + FeBr₃

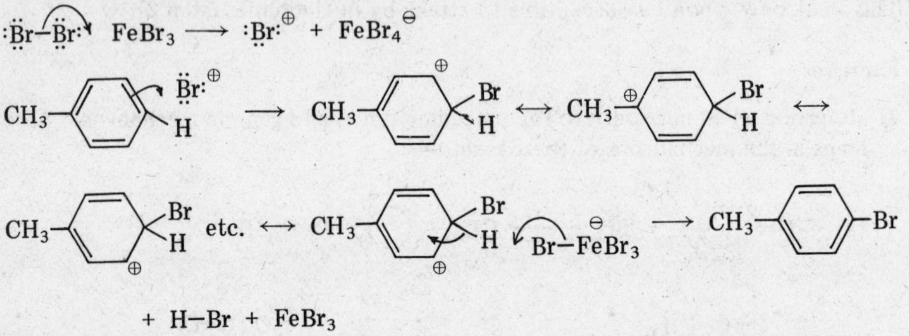

   (b) [naphthalene with O=C-CH₃, C≡N, Cl, :Ö-R substituents] ⟶ [intermediate with O=C-CH₃, CN, Cl, OR] ⟷ [with C=N:⁻] ⟷

   [enolate ⁻O-C=CH₃ form with CN, Cl, OR]

   etc. ⟷ [intermediate with O=C-CH₃, CN, Cl, OR] ⟶ [naphthalene with O=C-CH₃, C≡N, OR] + Cl⁻

   (c) I–C₆H₅ ⟶ I⁻ + C₆H₅·

   CH₃–Ö:⁻ ↷ H CH₃OH

   CH₃–Ö:⁻ ↷ [benzyne] ⟶ :⁻[C₆H₅ with CH₃–O]

   CH₃–Ö–H ↷ :⁻[C₆H₄ with CH₃–O] ⟶ CH₃–O⁻ + [C₆H₅–OCH₃]

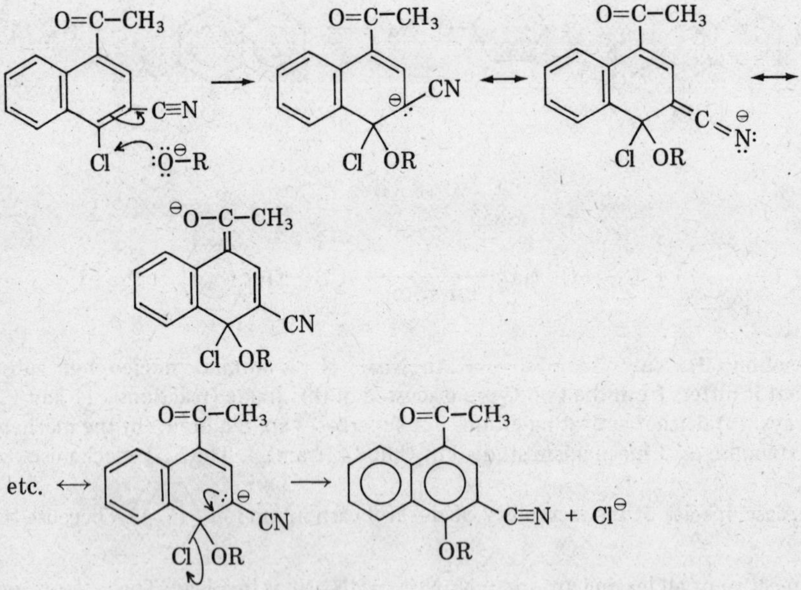

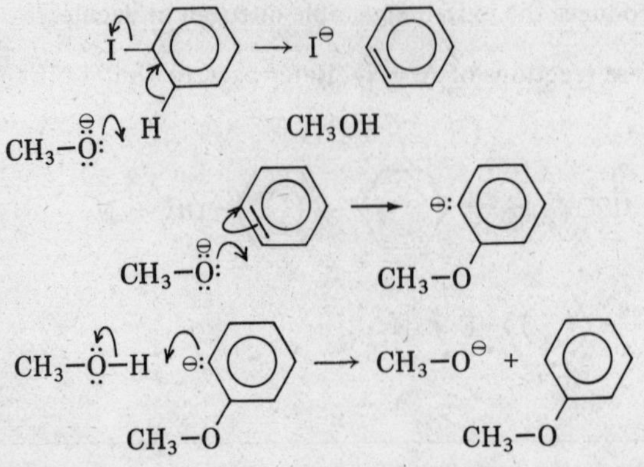

AROMATIC HYDROCARBONS AND THEIR DERIVATIVES

2. (a) Ph–Ph–N≡N:⁺ → Ph–Ph⁺ + :N≡N:

Ph–Ph⁺ + :ÖH₂ → Ph–Ph–Ö⁺(H)(H)

Ph–Ph–Ö⁺(H)(H) + :ÖH₂ → Ph–Ph–Ö–H + H₃O⁺

(b) (Br)Ph–N≡N:⁺ → (Br)Ph⁺ + :N≡N:

(Br)Ph⁺ + F–BF₃⁻ → (Br)Ph–F + BF₃

3. Ph–O–Ph–F + :N̈H₂⁻ → Ph–O–Ph(⁻) (two nonequivalent sites) + NH₃

Ph–O–Ph + :N̈H₂⁻ → Ph–O–Ph(⁻) → Ph–O–Ph–NH₂⁻ → Ph–O–Ph–NH₂

Ph–O–Ph + :N̈H₂⁻ → Ph–O–Ph(NH₂)(⁻) → Ph–O–Ph–NH₂

LEARNING THE REACTIONS AND MECHANISMS AND REACTIVITY ORDERS

4. In this frame we concentrate on learning the main reactions of aromatic rings as expressed in the generic reactions of Chart 16.1. And we conduct a refresher drill on electrophilic addition and apply it to $\diagdown C=C \diagup$ conjugated with aromatic rings (reaction ⑦). Frames 5 and 6 then develop specific drills for learning the various forms taken by reactions ① and ④, the electrophilic substitution and aryl $S_N 2$ reactions.

Exercises

1. If you are in a classroom situation, use your lecture notes and textbook to edit Chart 16.1 so that it contains just the reactions actually covered in your course. Since many books take up aromatic hydrocarbons, aryl halides, $ArN_2^{\oplus}$, and so on, in successive chapters, you may well want to use only a segment of the chart at a time. If you do this it will be worthwhile to put it back together after all the segments have been covered and do some drilling on the whole thing.

2. Reduce Chart 16.1 to a simple list containing reaction name, reactant, reagent, and product in generic form (using R, G, Nu, etc.). Arrange these entries in four columns on a single sheet of paper:

Name	Reactant	Reagent	Products
Electrophilic substitution	G–C₆H₄–H	E–Nu	G–C₆H₄–E + NuH
Catalytic hydrogenation	C₆H₅–G	H_2/Pt	C₆H₁₁–G
⋮	⋮	⋮	⋮

3. Cover the product column with scratch paper and fill in the missing product of each reaction in turn. Remove the answer sheet, check, and correct. Repeat the drill. Repeat, covering successively the reagent, reactant, and name columns.

4. Practice recalling from memory all the reaction names in your edited list. Do it aloud or in writing until you no longer omit any.

5. Refer to Unit 12 to refresh your memory on the electrophilic addition reaction of alkenes:

   (a) List the electrophiles that attack the C=C function.

   (b) Compare this list with the list of electrophiles that attack (benzene) (Table 16.1).

   Note that many of the electrophiles that attack C=C don't react with benzene. Bromine and chlorine attack benzene only in the presence of an iron catalyst, which is not needed for reaction with C=C. Aqueous acids exchange $H^{\oplus}$ with ring hydrogens on (benzene), but this leads to no new product. Only H_2SO_4 reacts with both C=C and (benzene), and conditions can normally be found that make the addition rapid and the ring substitution slow. Consequently, if (benzene) and C=C are present in the same molecule, one can carry out electrophilic additions on C=C selectively, without doing electrophilic substitution on the ring.

   (c) Write the overall equation and the mechanism for addition of HI to (indene).

   (d) Use the theory of resonance to explain why the addition in part (c) has the regioselectivity it does.

---

5. (c) indene + H–I → protonated intermediate + $I^{\ominus}$ → product with H, I, H, H

(d) The ion [resonance structure of protonated indane cation with H H and H] is resonance-stabilized,

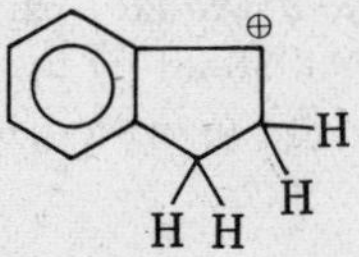

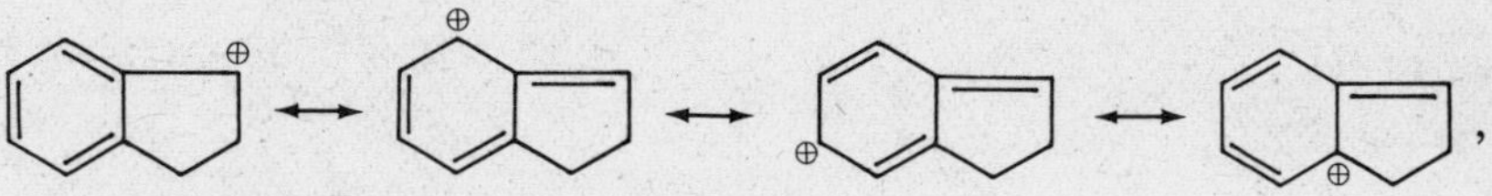

[four resonance structures of the bicyclic cation]

so it is formed faster than the less stable ion,

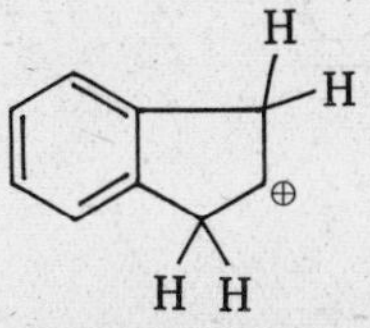

[structure with H H and localized positive charge]

which has a localized positive charge.

5. In this frame we drill on the eight common varieties of electrophilic substitution (Table 16.1), first the overall reaction (reactant, reagent, product), then the mechanism.

When the reactant is an aromatic ring that already carries some substituent group, G, the rate and usefulness of these reactions depends upon the structure of G. Substitution could occur at three different positions:

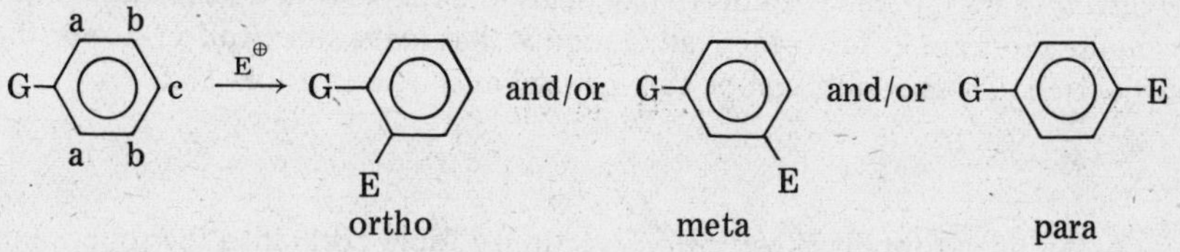

G—[benzene with positions a,b,c labeled] $\xrightarrow{E^\oplus}$ G—[benzene]—E (ortho, positions a,b) and/or G—[benzene]—E (meta) and/or G—[benzene]—E (para)

You need to learn to predict the approximate composition of such product mixtures. Fortunately, the facts can be boiled down to a few simple rules, which we will now develop.

The position at which a new group enters is mainly determined by the identity and location of the substituent groups already present, not by the identity of the electrophile. This means that a given monosubstituted benzene gives approximately the same mixture of isomers in reaction with each of the standard electrophiles. For instance,

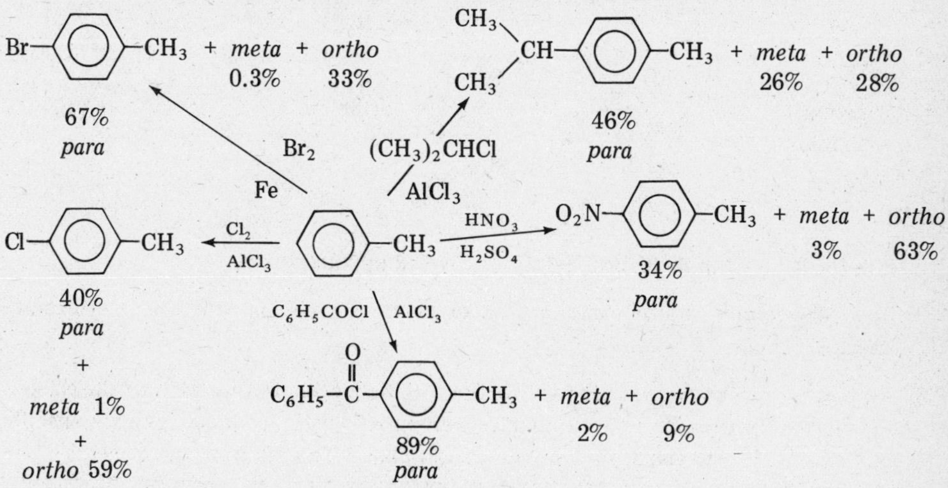

Br—[C$_6$H$_4$]—CH$_3$ + meta + ortho
                0.3%   33%
67% para

(CH$_3$)$_2$CH—[C$_6$H$_4$]—CH$_3$ + meta + ortho
                26%   28%
46% para

Br$_2$/Fe   (CH$_3$)$_2$CHCl/AlCl$_3$

Cl—[C$_6$H$_4$]—CH$_3$ $\xleftarrow{\text{Cl}_2/\text{AlCl}_3}$ [C$_6$H$_5$]—CH$_3$ $\xrightarrow{\text{HNO}_3/\text{H}_2\text{SO}_4}$ O$_2$N—[C$_6$H$_4$]—CH$_3$ + meta + ortho
                                                                          3%   63%
40% para   C$_6$H$_5$COCl / AlCl$_3$ 34% para
+
meta 1%   C$_6$H$_5$—C(=O)—[C$_6$H$_4$]—CH$_3$ + meta + ortho
+   2%   9%
ortho 59%   89% para

The product isomer distribution is not random (which would give 20% para, 40% meta, 40% ortho), but selective. In this case the selectivity brought about by the $-CH_3$ group can be described as "low meta." We usually say that $-CH_3$ is an ortho-para directing group. Similarly:

1.5% Cl—⟨◯⟩—NO_2 O_2N—⟨◯⟩—NO_2 0.3%

+

80.9% ⟨◯⟩—NO_2 ⟵$\frac{Cl_2}{FeCl_3}$ ⟨◯⟩—NO_2 $\xrightarrow{HNO_3, H_2SO_4}$ ⟨◯⟩—NO_2 93.2%
       | |
       Cl O_2N

+   +

17.6% ⟨◯⟩—NO_2 (Cl ortho)   ⟨◯⟩—NO_2 (NO_2 ortho) 6.4%

Consequently, $-NO_2$ is a "high meta" or *meta-directing group*. Study of a large number of reactions shows that the percentages of ortho and para taken together run opposite the percentage of meta—all groups fall into one of two classes: predominantly *o/p* directors or predominantly *m* directors. It turns out that with one exception (halogen) the *o/p* directors are the same groups that *activate* the ring (speed up substitution), and the *m* directors are in fact deactivating groups that make substitution slower than in benzene itself. These results can be summarized as shown in Table 16.5.

Exercises

1. Put the eight electrophilic substitution reactions (Table 16.1) into the four-column format used in frame 4, exercise 2:

Name	Reactant	Reagent	Product
Nitration	G—⟨◯⟩	conc. H_2SO_4 + conc. HNO_3	G—⟨◯⟩—NO_2
Chlorination, bromination	G—⟨◯⟩	Cl_2 or Br_2 + Fe	G—⟨◯⟩—X
⋮	⋮	⋮	⋮

   Edit the list to include just the reactions covered in your course.

2. Cover one column at a time and practice supplying the missing structures. Check and repeat.

3. All you need to know in order to write the specific mechanism for each of the electrophilic substitutions in your list is the generic mechanism, the identity of the specific electrophile, $E^\oplus$, involved, and how the electrophile is formed from the reagents

AROMATIC HYDROCARBONS AND THEIR DERIVATIVES 379

Table 16.5

Group		Common structural feature

A. *Ortho/para*-directing and Activating Groups

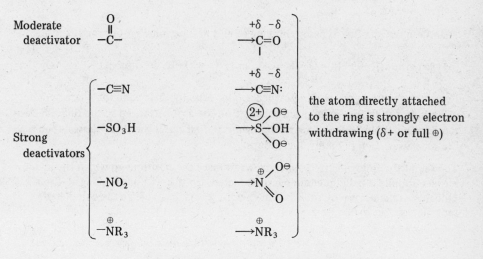

Super activator: $-\ddot{\underset{..}{O}}{:}^{\ominus}$

Strong activators: $-\ddot{\underset{..}{O}}-H$; $-\ddot{N}H_2, -\ddot{N}R_2$ — unshared pair(s) and no $\oplus$ charge on O or N (strong electron donor by resonance)

Moderate activators: $-\ddot{\underset{..}{O}}-R$; $-\ddot{N}H-CO-R$

Weak activator: $-R$ — electron donor by polar effect, has no resonance interaction

B. *Meta*-directing and Deactivating Groups

Moderate deactivator: $-\underset{\underset{\|}{O}}{C}-$; $\overset{+\delta\ -\delta}{\longrightarrow}\underset{\|}{C}=O$

Strong deactivators:
$-C\equiv N$; $\overset{+\delta\ -\delta}{\longrightarrow}C\equiv N{:}$

$-SO_3H$; $\longrightarrow \overset{(2+)}{S}\underset{O^\ominus}{\overset{O^\ominus}{\diagup}}-OH$

$-NO_2$; $\longrightarrow N\underset{O}{\overset{O^\ominus}{\diagup}}^{\oplus}$

$-\overset{\oplus}{N}R_3$; $\longrightarrow \overset{\oplus}{N}R_3$

— the atom directly attached to the ring is strongly electron withdrawing ($\delta+$ or full $\oplus$)

C. *Ortho/para*-directing and Deactivating Groups

Moderate deactivators: $-\ddot{\underset{..}{F}}{:}$; $-\ddot{\underset{..}{Cl}}{:}$; $-\ddot{\underset{..}{Br}}{:}$; $-\ddot{\underset{..}{I}}{:}$ — weak electron *donors* by resonance, strong electron *withdrawers* by polar effect

and catalysts used. Make a list containing the reaction E—Nu $\longrightarrow$ E$^\oplus$ for each of the specific electrophilic substitution reactions. Note that in the reagent
$Ar-\overset{\oplus}{N}\equiv N{:}\ HSO_4^{\ominus} \longleftrightarrow Ar-\ddot{N}=\overset{\oplus}{N}{:}\ HSO_4^{\ominus}$ the cation is the actual electrophile. The reagent $Tl(O-CO-CF_3)_3$ only needs to ionize to $\overset{\oplus}{Tl}(O-CO-CF_3)_2 + \overset{\ominus}{O}-CO-CF_3$ to form

the actual electrophile, $\overset{\oplus}{Tl}(O-CO-CF_3)_2$. The equations for the remaining E—Nu $\longrightarrow$ $E^{\oplus}$ processes are given in frame 3. Practice reproducing the list of E—Nu $\longrightarrow$ $E^{\oplus}$ processes on scratch paper until you can do it without cues.

★ 4. Write the complete mechanism for each of the specific electrophilic substitution reactions on your list. Use benzene as the reactant, and use only your memory of the generic mechanism and the various E—Nu and $E^{\oplus}$ slot fillers.

5. To learn Table 16.5, first memorize the names of the three major classes:

   o,p-directors/activators
   o,p-directors/deactivators
   m-directors/deactivators

   The groups belonging to the middle class (*o,p*-directors/deactivators) are easy to memorize—they are the four halogens: —F, —Cl, —Br, —I. To recall the groups in the *o,p*-directors/activators class, memorize the two types:

   Oxygen or nitrogen with unshared pair(s) and no $\oplus$ charge.
   —R or —Ar.

   Then flesh out the first type by thinking of all the functional groups that fit, e.g.,

   (oxygen groups), $-\ddot{O}-R$, $-\ddot{O}-H$, $-\ddot{O}{:}^{\ominus}$, and $-\ddot{O}-CO-R$, but not $-\overset{\oplus}{O}\!\!<\!\!\begin{smallmatrix}R\\R\end{smallmatrix}$ or $-\overset{\oplus}{O}\!\!<\!\!\begin{smallmatrix}H\\H\end{smallmatrix}$.

   Recall the groups in the *m*-director/deactivator class by their common characteristic of having a full or fractional positive charge on the atom attached to the ring. Practice reproducing Table 16.5 on paper from memory; first recall the three classes, then the subclasses, then the individual groups.

★ 6. To each of the following rings, add arrows to mark the positions to which the incoming electrophile is being predominantly directed. Do this without looking at Table 16.5; instead make use of your drill in exercise 5—try to visualize the completed table just as you wrote it before:

   Example: ⟨O⟩—Br. Br is in the *o,p*-director/deactivator class, so mark all positions *o* or *p* to Br: ⟶ ⟨O⟩—Br

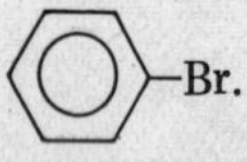

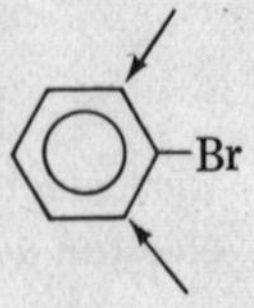

   (a) 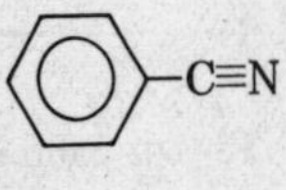 ⟨O⟩—C≡N

   (b) 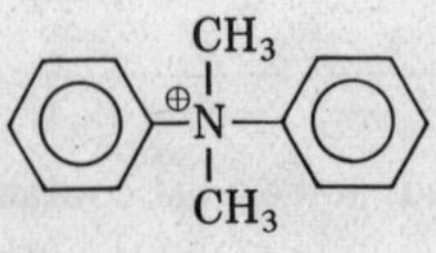 ⟨O⟩—$\overset{\oplus}{N}$(CH₃)₂—⟨O⟩

   (c) ⟨O⟩—O—CH₂—⟨O⟩

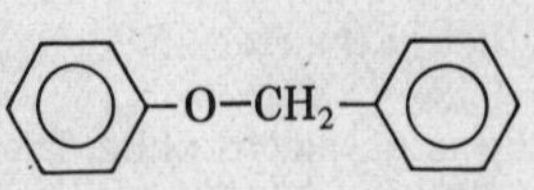

   (d) 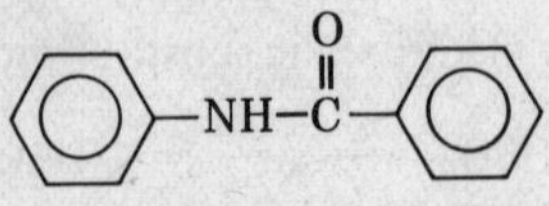 ⟨O⟩—NH—$\overset{O}{\overset{\|}{C}}$—⟨O⟩

3.

E–Nu	E$^\oplus$
$O_2N-OH \longrightarrow$	$O_2N^\oplus$
$X-X \longrightarrow$	$X^\oplus$
$R-X \longrightarrow$	$R^\oplus$
$R-\overset{O}{\underset{\|}{C}}-X \longrightarrow$	$R-\overset{O}{\underset{\|}{C}} \oplus$
$Ar-N^\oplus \atop HSO_4^\ominus \longrightarrow$	$Ar-N_2^\oplus$
$Tl(O-CO-CF_3)_3 \longrightarrow$	$\overset{\oplus}{Tl}(O-CO-CF_3)_2$
$H-O-SO_2-OH \longrightarrow$	SO_3
$D-O-D \longrightarrow$	$D^\oplus$ latent in $D-\overset{\oplus}{O}\overset{D}{\underset{D}{\diagdown}}$

4. (a) $H-\ddot{O}-N\overset{\nearrow O}{\searrow_O}$ + H_2SO_4 ⇌ $\overset{H}{\underset{H}{\diagdown}}\overset{\oplus}{O}-N\overset{\nearrow O}{\searrow_O}$ + $HSO_4^\ominus$

   base   acid   acid   base

   $\overset{H}{\underset{H}{\diagdown}}\overset{\oplus}{O}-N\overset{\nearrow \ddot{\ddot{O}}:^\ominus}{\searrow_O} \longrightarrow H_2\ddot{O}: + \overset{\oplus}{O=N=O}$

   [aromatic ring mechanism diagrams showing nitration]

   etc. ⟷ [intermediate] ⟶ [Ph–NO$_2$ + H_2SO_4]

   :$\overset{\ominus}{O}SO_3H$

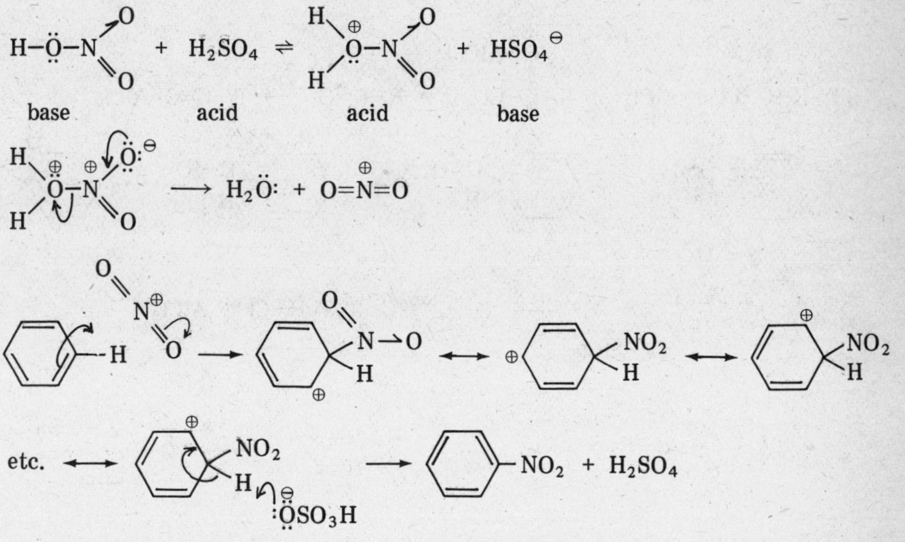

(b) $R\overset{\frown}{-Cl}$ $AlCl_3$ ⇌ $R^\oplus$ $AlCl_4^\ominus$

   [aromatic ring mechanism diagrams for Friedel–Crafts alkylation]

   etc. ⟷ [intermediate] ⟶ [Ph–R + H–Cl + $AlCl_3$]

   $Cl^\ominus-AlCl_3$

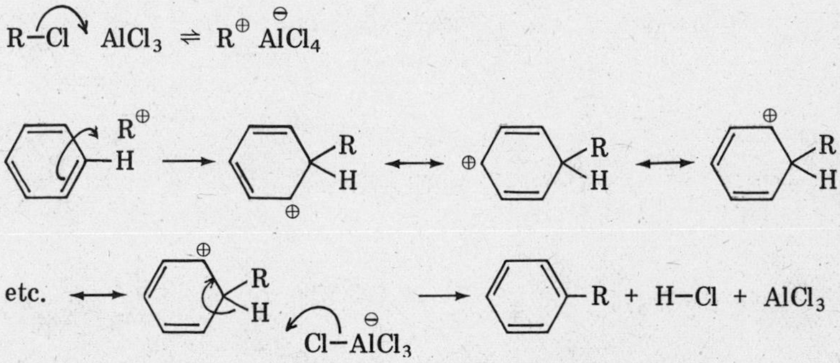

   or, in the case of less stable (e.g., 1°) carbonium ions:

(c) $3\ Cl_2 + 2\ Fe \longrightarrow 2\ FeCl_3$

(d) $R-\overset{O}{\overset{\|}{C}}-Cl + AlCl_3 \rightleftharpoons R-\overset{\overset{\ominus}{O}-AlCl_3}{\overset{|}{\underset{\|}{C}}}-Cl \rightleftharpoons R-\overset{\oplus}{C}=\overset{..}{O}: \longleftrightarrow R-C\equiv\overset{\ominus}{O}:\ AlCl_4^{\ominus}$

(f) $Tl(O-\overset{O}{\overset{\|}{C}}-CF_3)_3 \rightleftharpoons \overset{\oplus}{Tl}(O-\overset{O}{\overset{\|}{C}}-CF_3)_2 + CF_3CO_2^{\ominus}$

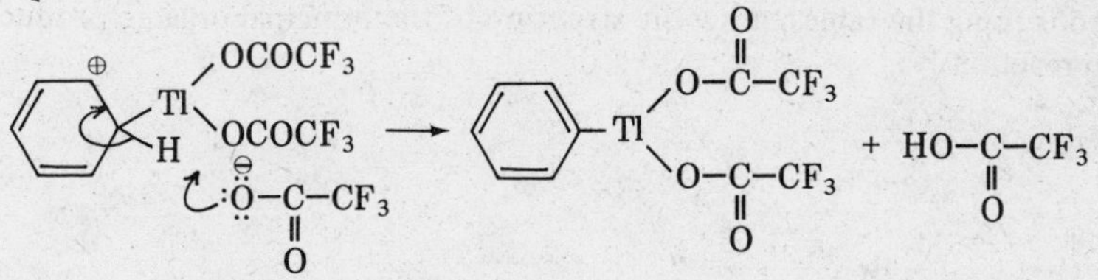

AROMATIC HYDROCARBONS AND THEIR DERIVATIVES 383

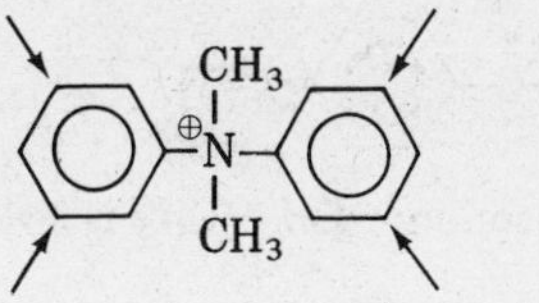

(g) $2 H_2SO_4 \rightleftharpoons H_3O^\oplus HSO_4^\ominus + SO_3$

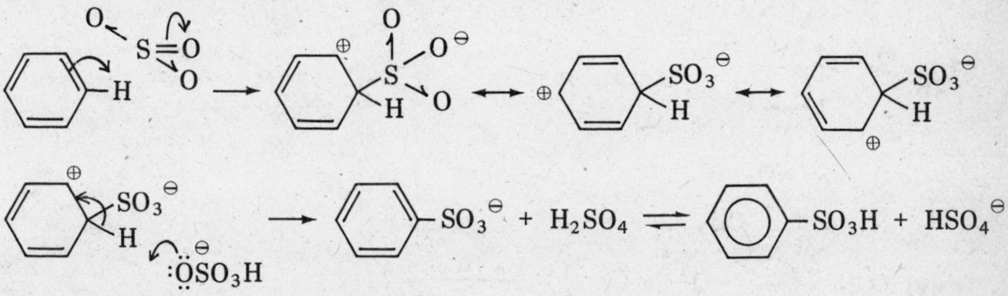

(h) $D_2O + D_2SO_4 \rightleftharpoons D_3O^\oplus + DSO_4^\ominus$

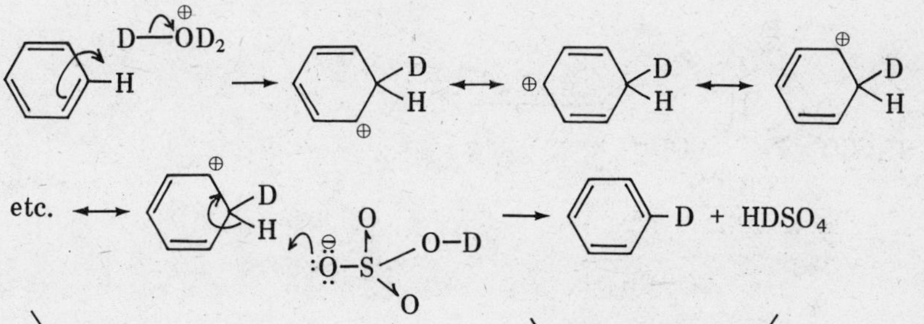

6. (a) 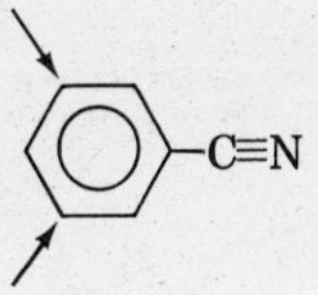 Ph—C≡N

(b) Ph—N$^\oplus$(CH$_3$)$_2$—Ph

(c) → Ph—O—CH$_2$—Ph ←

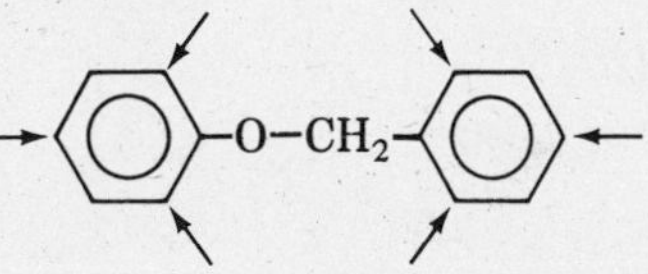

(d) 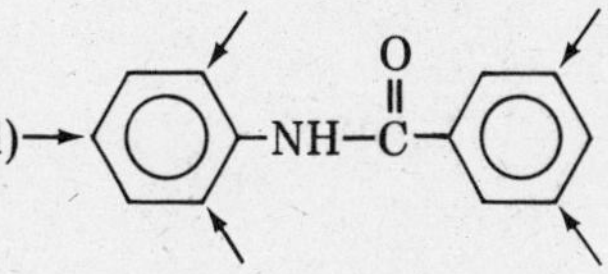 → Ph—NH—C(=O)—Ph

6. The aryl S_N2 reactions are easier to internalize than the electrophilic substitutions. First, you have already learned the permissible nucleophiles and leaving groups when you studied the S_N2 reactions of alkyl halides (Unit 14). Second, there is no ambiguity about which position on the ring is attacked by the incoming nucleophile—the nucleophile can only take up the position originally occupied by the leaving group. Third, *the activating groups for aryl S_N2 reactions are just the deactivators for electrophilic substitution* (minus $-NR_3^\oplus$).

Exercises

1. Practice reproducing the slot-filler lists in Table 16.2 using the techniques developed in frames 4 and 5.

★ 2. Without consulting the tables, draw the structure of the principal organic product or write "No reaction."

(a) 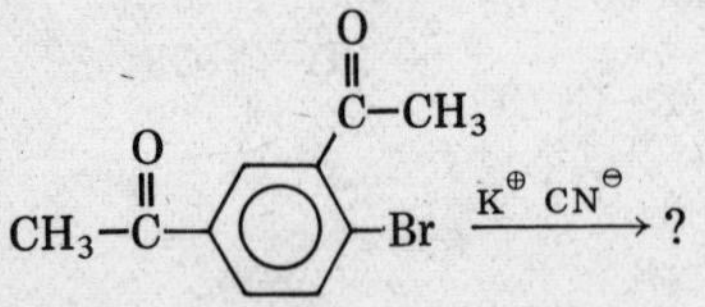 CH₃–CO–[C₆H₃(COCH₃)]–Br $\xrightarrow{K^\oplus CN^\ominus}$?

(b) 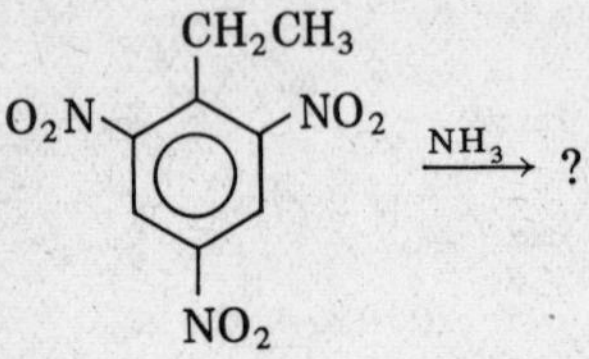 [2,4-dinitro-(ethyl)benzene: O₂N, NO₂, CH₂CH₃ substituents] $\xrightarrow{NH_3}$?

(c) 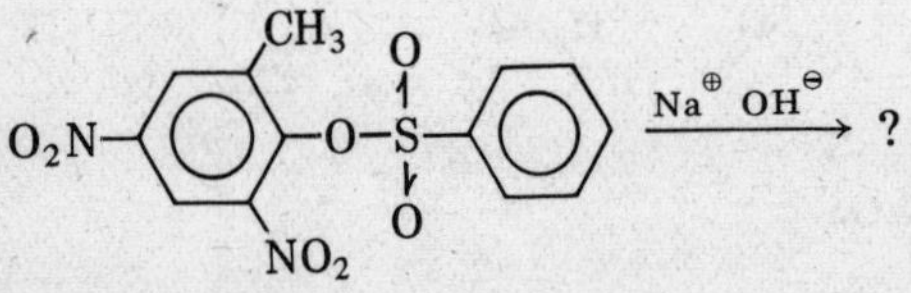 O₂N–[C₆H₃(CH₃)(NO₂)]–O–S(=O)₂–C₆H₅ $\xrightarrow{Na^\oplus OH^\ominus}$?

(d) 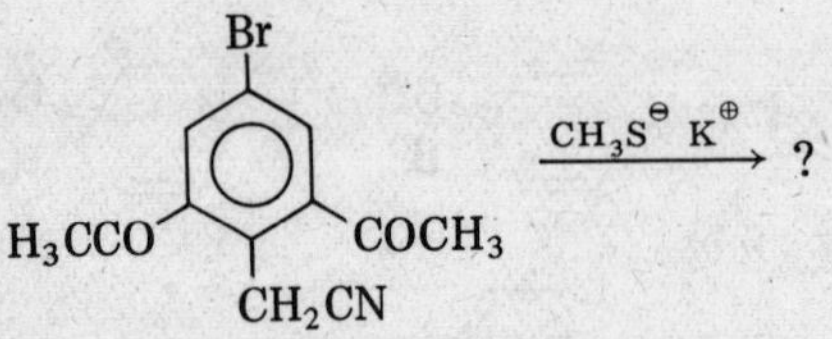 [benzene ring with Br, H₃CCO, COCH₃, CH₂CN substituents] $\xrightarrow{CH_3S^\ominus K^\oplus}$?

– – – – – – – –

2. (a) 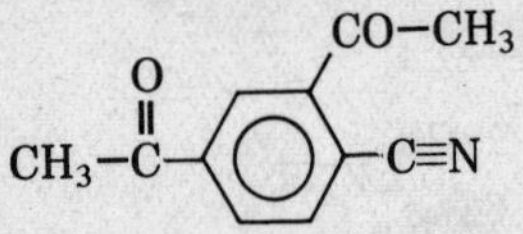 CH₃–CO–[C₆H₄(CO–CH₃)]–C≡N

(b) No reaction; –CH₂–CH₃ is not a leaving group.

(c) 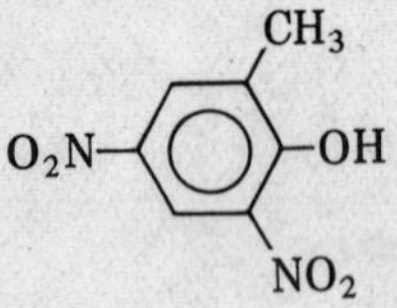 O₂N–[C₆H₃(CH₃)(NO₂)]–OH

(d) No reaction; the $-\overset{\overset{O}{\|}}{C}-O-CH_3$ groups activate the aryl S_N2 process only when they are ortho or para to the leaving group. The –CH₂–C≡N group is not an activator; the saturated –CH₂– group insulates –C≡N from any resonance interaction with the ring.

SOLVING SPECIFIC PROBLEM TYPES

7. The problem types most commonly encountered in the study of aromatic compounds are:
   (a) Predict reaction products for . . . .
   (b) Write reaction mechanisms for . . . .

(c) Devise a sequence of reactions that will convert compound A into compound B (synthesize B from A).

(d) Give a fundamental explanation of the behavior of a given group in (1) activating or deactivating electrophilic or nucleophilic aromatic substitution, or (2) directing incoming electrophiles to the m or o/p positions.

Problem type (c) is treated in detail in Unit 21, frames 3 to 10. For problems of type (d), you should rely on the theoretical development in your textbook. Type (b) you have practiced sufficiently in earlier frames. We concentrate here on the regioselectivity problems associated with predicting the position of electrophilic substitution of a benzene ring that already bears some substituent group(s).

One Group Already on the Ring

The identity of the group tells you if it orients to the o, p, or m positions. You practiced marking these positions of predominant attack in frame 5. Here are concrete directions for writing the structure of *the predominant reaction product:*

1. Mentally locate the substituent already present on the ring in Table 16.5 and note its directive influence: *meta* or *ortho, para*.
2. If the substituent is not listed, classify it in one of the three classes of Table 16.5 on the basis of its structural features. For example, the group $-\overset{\oplus}{S}(CH_3)_2$ is not listed, but clearly it should go in the m-director, deactivator class because it has a full positive charge on the atom directly attached to the ring.
3. After deciding if the group is a m or an o,p director, tag these positions on the reactant's ring.
4. If the result is o,p, the predominant product will be the para isomer in practically all cases.
5. Identify the incoming electrophile.
6. Write the product structure by plugging the electrophile into a tagged position in place of H.

Examples

Predict the predominant product of each reaction.

1. $O_2N-\bigcirc \xrightarrow{H_2SO_4}$?

2. $CH_3-CH_2-O-\bigcirc \xrightarrow[H_2SO_4]{HNO_3}$?

3. $CH_3-N=CH-\bigcirc \xrightarrow[Fe]{Cl_2}$?

Solutions:

1. Substituent = NO_2, a meta director. Therefore, $O_2N-\bigcirc$.

Incoming electrophile = SO₃, leading to —SO₃H. Therefore, O₂N—⟨ring⟩ —H₂SO₄→

O₂N—⟨ring⟩—SO₃H

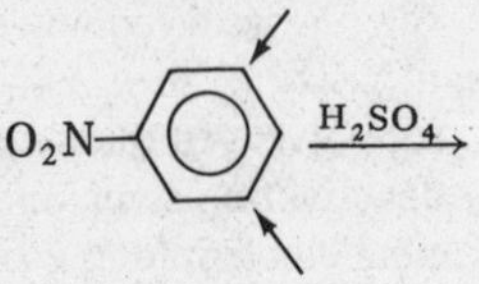

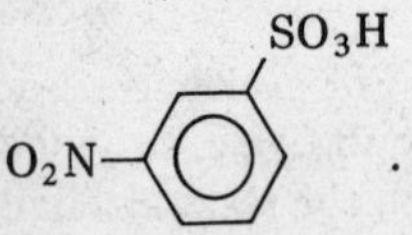

2. Substituent = R—O—, an *o/p* director.

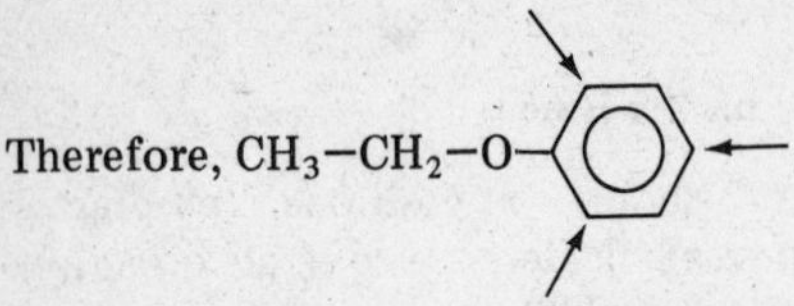

Therefore, CH₃—CH₂—O—⟨ring⟩ ←. Incoming electrophile = NO₂⁺, leading to —NO₂.

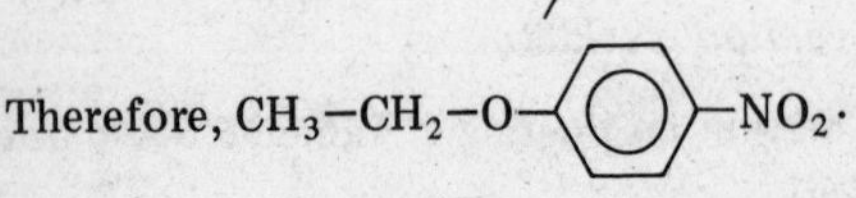

Therefore, CH₃—CH₂—O—⟨ring⟩—NO₂.

3. Substituent = R—N=C(H)—, not in the tables. Analyze it: atom attached to ring = C; this C has no unshared pairs but is doubly bonded to an electronegative atom and is therefore R—N=CH— (−δ on N, +δ on C). It belongs in the same class with —C(=O)— and is therefore a meta director. Hence

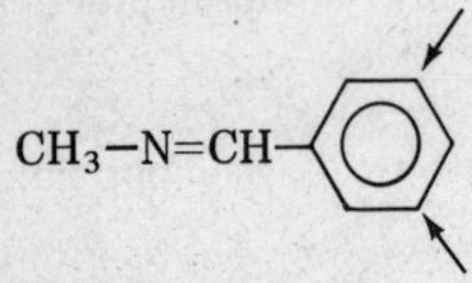

CH₃—N=CH—⟨ring⟩

Incoming electrophile = Cl⁺, leading to —Cl.

Therefore,

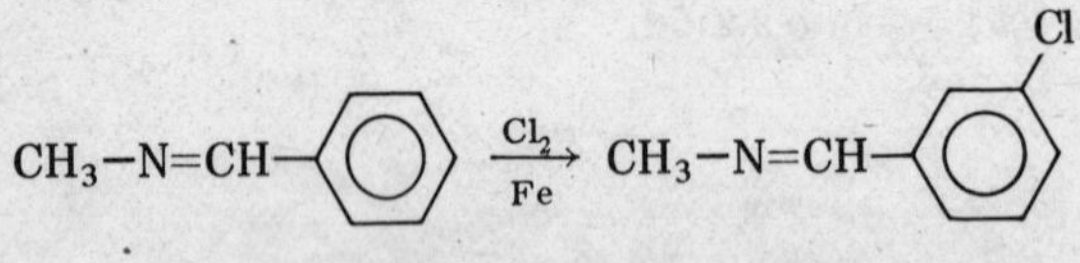

CH₃—N=CH—⟨ring⟩ —Cl₂/Fe→ CH₃—N=CH—⟨ring⟩—Cl

In these examples, although two equivalent meta (or ortho) positions are present and available for reaction, we put an incoming group into either one, *but not both:*

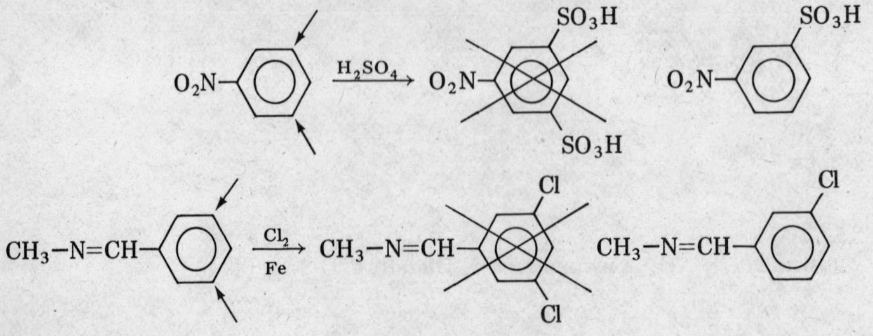

O₂N—⟨ring⟩ —H₂SO₄→ O₂N—⟨ring with SO₃H and SO₃H crossed out⟩   O₂N—⟨ring⟩—SO₃H

CH₃—N=CH—⟨ring⟩ —Cl₂/Fe→ CH₃—N=CH—⟨ring with Cl and Cl crossed out⟩   CH₃—N=CH—⟨ring⟩—Cl

First, as soon as one group goes in, the ring is deactivated, and introduction of a second group is slow compared to the first. Second, if another group were to go in, it might very well not go to the remaining originally marked position—the situation is now altered by the first group going in, and the two substituents now present would both have to be considered in a prediction of the preferential point of attack in the final substitution, as illustrated in the next section.

More Than One Group Already on the Ring

1. Identify the directive properties of each group present, and tag the position(s) to which each directs, just as in steps 1 to 3 of the preceding case. Don't tag any position that bears some group other than hydrogen—only H is ordinarily replaceable.

2. Discard any tag on a position that has *two ortho* groups: [structure with G, G on ring]. These positions are so hindered that they are only very slowly attacked.

3. If the groups already present agree on at least one of the position(s) to which they will send the incoming group, write the product corresponding to the most preferred of these positions.

4. If the groups present do not agree on any position(s), the directive influence of one wins out over another according to the priority order: strong activator > moderate activator > weak activator > deactivators.

5. If ties still exist, the product of substitution at a position that has no ortho substituents will predominate over ortho substitution products.

Examples

Write the structural formula of the predominant product of each reaction.

1. [o-nitrotoluene] $\xrightarrow[HNO_3]{H_2SO_4}$?

2. $HO-\bigcirc-CH_3$ $\xrightarrow{H_2SO_4}$?

3. $Cl-\bigcirc-CH_3$ $\xrightarrow[Fe]{Br_2}$?

Solutions:

1. Identify CH_3: *o,p*; NO_2: *m*. Therefore, [structure of CH₃, NO₂ on ring with arrows]. (The second position ortho to CH_3 is blocked.) The predictions agree. The most probable of the two positions

is the one not ortho to any group: [benzene with CH₃ at top, NO₂ at position 2, arrow pointing up from bottom]. Therefore, [benzene with CH₃ at top, NO₂ at position 2, SO₃H at bottom].

2. Identify: OH: *o,p*; CH₃: *o,p*. Therefore, HO—[benzene]—CH₃. The predictions do not agree. The strong activator (OH, —→) takes precedence over the weak activator (CH₃, --→). Therefore,

HO—[benzene]—CH₃ $\xrightarrow{H_2SO_4}$ HO—[benzene with SO₃H ortho to OH]—CH₃

3. Identify Cl = *o,p*; CH₃ = *o,p*. Therefore, Cl—[benzene]—CH₃. The predictions do not agree. Weak activator CH₃ takes precedence over deactivator Cl. Therefore,

Cl—[benzene]—CH₃ $\xrightarrow[Fe]{Br_2}$ Cl—[benzene with Br ortho to CH₃]—CH₃

This scheme predicts the predominant product. Remember that the other positions tagged in step 1 will also be substituted, to a smaller degree, and the product will be a mixture of isomers.

If the two directing groups are in different classes (e.g., moderate activator versus deactivator or strong activator versus moderate activator), the predominance of the favored product is substantial, except for the case of a weak activator versus a deactivator. The case of weak activator versus deactivator gives nearly equal amounts of the two products. Approximately equal amounts, of course, result when both groups belong to the same class.

★ *Exercises*

For each reaction predict (a) the predominant product, and (b) whether the substitution occurs slower or faster than the corresponding reaction of benzene.

1. Cl—[benzene] + [benzene]—C(=O)—Cl $\xrightarrow{AlCl_3}$

2. (CH₃)₂CH—[benzene]—CH(CH₃)₂ $\xrightarrow{HNO_3, H_2SO_4}$

3.  $CH_3-\langle\bigcirc\rangle-NO_2 \xrightarrow{H_2SO_4}$

4. $HO_2C-\langle\bigcirc\rangle(CO_2H)-CO_2H \xrightarrow{Br_2, Fe}$ (with CO₂H at ortho position)

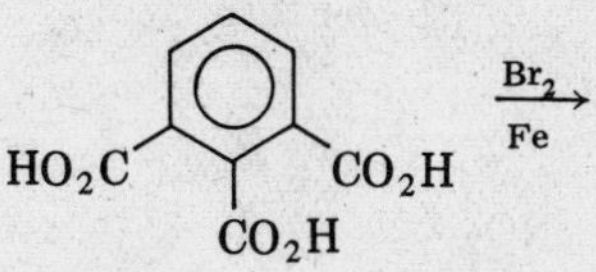

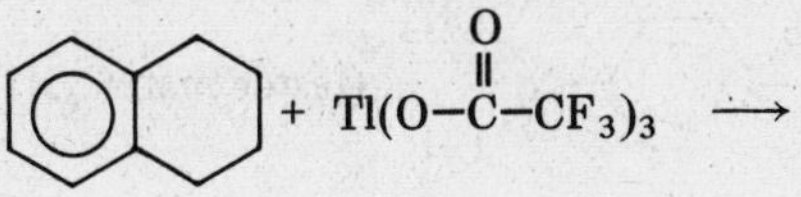

5. 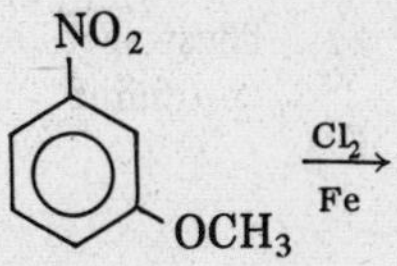 $\langle\bigcirc\rangle(NO_2)(OCH_3) \xrightarrow{Cl_2, Fe}$

6. 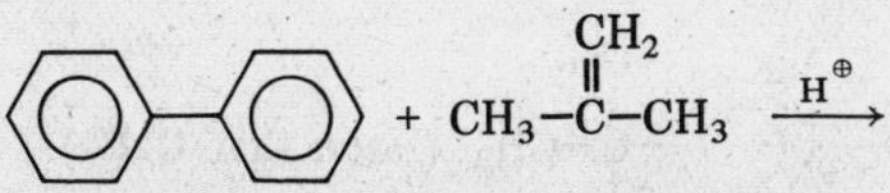 $\langle\bigcirc\rangle-\langle\bigcirc\rangle + CH_3-\underset{\overset{\|}{CH_2}}{C}-CH_3 \xrightarrow{H^\oplus}$

7. 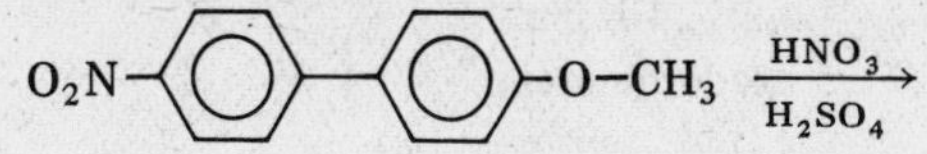 $O_2N-\langle\bigcirc\rangle-\langle\bigcirc\rangle-O-CH_3 \xrightarrow{HNO_3, H_2SO_4}$

8. 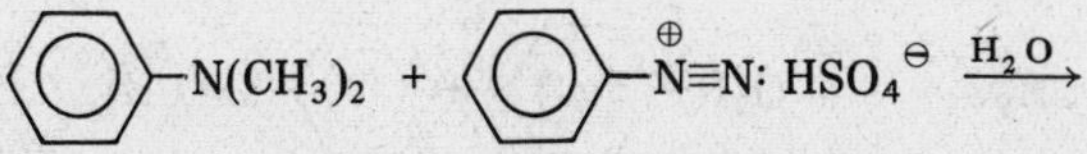 $\langle\bigcirc\rangle-N(CH_3)_2 + \langle\bigcirc\rangle-\overset{\oplus}{N}\equiv N: HSO_4^{\ominus} \xrightarrow{H_2O}$

9. 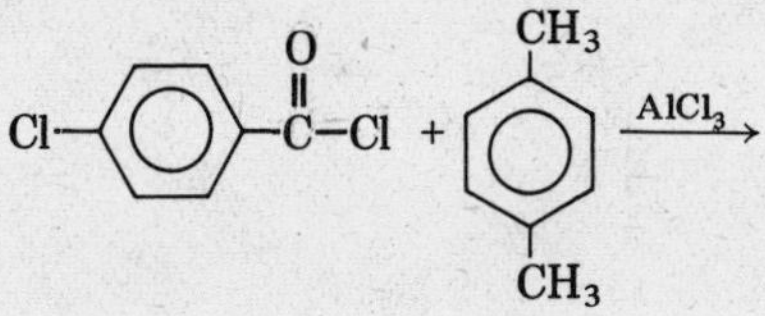 $Cl-\langle\bigcirc\rangle-\overset{\overset{O}{\|}}{C}-Cl + \langle\bigcirc\rangle(CH_3)(CH_3) \xrightarrow{AlCl_3}$ (CH₃ groups at 1,4 positions)

10. $\langle\bigcirc\bigcirc\rangle + Tl(O-\overset{\overset{O}{\|}}{C}-CF_3)_3 \longrightarrow$ (tetralin/naphthalene)

- - - - - - - - - - -

1. $Cl-\langle\bigcirc\rangle$ (arrows at ortho/para) ← para predominates over ortho → $Cl-\langle\bigcirc\rangle-\overset{\overset{O}{\|}}{C}-\langle\bigcirc\rangle$

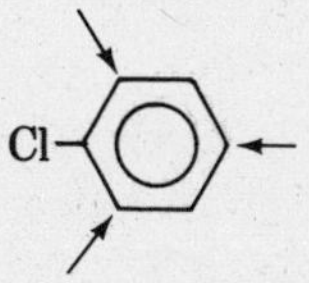

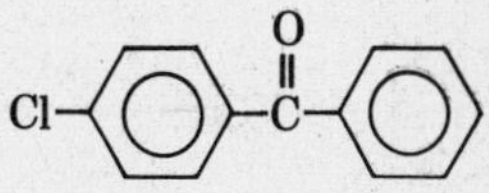

   (slower than C_6H_6)

2. $\rangle-\langle\bigcirc\rangle-\langle \longrightarrow \rangle-\langle\bigcirc\rangle(NO_2)-\langle$

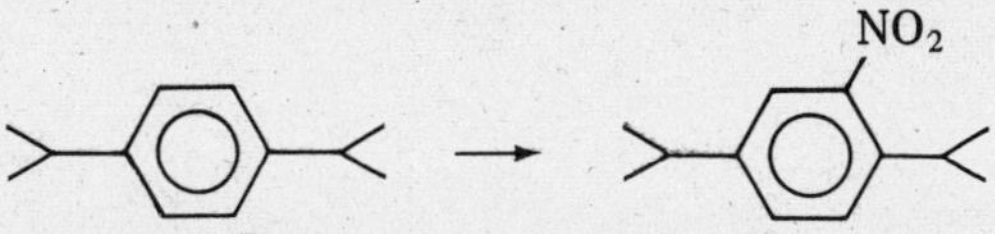

   the four open positions are identical   faster than C_6H_6

3. 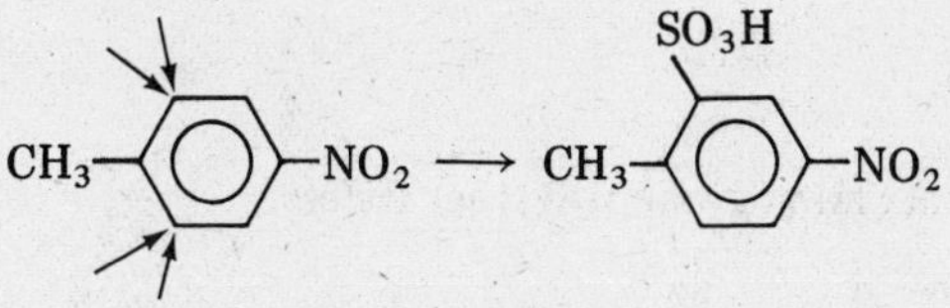 $CH_3-\langle\bigcirc\rangle-NO_2 \longrightarrow CH_3-\langle\bigcirc\rangle(SO_3H)-NO_2$ (slower than C_6H_6)

(arrows on positions ortho/meta to CH₃ on left structure)

4. [structure: benzene with HO₂C, CO₂H, CO₂H substituents with arrows] → [brominated product: HO₂C, CO₂H, CO₂H with Br] (slower than C₆H₆)

5. [structure with NO₂ and OCH₃, position marked with ✗] → [product with NO₂, OCH₃, Cl] or [product with Cl, NO₂, OCH₃]   These seem equally probable.

6. [biphenyl with arrows] → [biphenyl-C(CH₃)₃] (faster than C₆H₆)
   p predominates

7. O₂N—[ring]—[ring]—O—CH₃ → O₂N—[ring]—[ring](NO₂)—O—CH₃ (faster than C₆H₆)
   this ring deactivated   this ring activated— the action is here

8. [ring]—N(CH₃)₂ → [ring]—N=N—[ring]—N(CH₃)₂ (faster than C₆H₆)

9. CH₃—[ring]—CH₃ → Cl—[ring]—C(=O)—[ring](CH₃)(CH₃) (faster than C₆H₆)
   CH₃
   all four open positions identical

10. [tetralin-like bicyclic with arrows] →[CF₃COO—Tl—OOCCF₃ on ring]
    just a dialkyl benzene; the identical positions not ortho to either alkyl group will react fastest

UNIT SEVENTEEN
Aldehydes and Ketones (Carbonyl Compounds)

At this point in the organic course you are faced with a sizable increase in the amount of factual material (about 20 new reactions) that you must bring under control. This trend continues in the study of carboxylic acids and their derivatives (Unit 18). The first step in organizing this material is to identify the reactive sites in the structural formula and trace the general mechanisms and particular reactions that each participates in. These connections, outlined in Chart 17.1, provide a key to recalling the reactions and associating the applicable mechanisms with them. In frames 1 to 6 we take up in turn the six basic modes of reactivity shown in the chart. Frames 7 to 10 then provide drills for internalizing this material. Finally, solution techniques for particular problem types are taken up in frames 11 to 14.

The functional group involved here is the carbonyl group, $-\overset{\overset{\ddot{O}:}{\|}}{C}-$. It is present in aldehydes, $R-C\overset{O}{\underset{H}{\diagdown}}$ or $Ar-C\overset{O}{\underset{H}{\diagdown}}$, and ketones, $R-C\overset{O}{\underset{R}{\diagdown}}$, $Ar-C\overset{O}{\underset{Ar'}{\diagdown}}$, and $Ar-C\overset{O}{\underset{R}{\diagdown}}$. All these compounds together are given the generic name carbonyl compounds.

OBJECTIVES

When you have completed this unit, you should be able to:

- Explain, in simple terms, why the C=O group shows different reactions from C=C (frame 2).
- Describe the nucleophilic addition mechanism and explain why and how it often involves acid catalysis (frame 2).
- Write the mechanism for a variety of specific examples of the nucleophilic addition mechanism, in many cases both in the uncatalyzed and in the acid-catalyzed versions (frame 10).
- Draw the structures of all the enols to which a given carbonyl compound can tautomerize, and vice versa; decide which of all the tautomers is most stable; write a plausible mechanism for any of these tautomerization reactions (frame 4).
- Draw the structures of the conjugate acid and the conjugate base, if any, for any given carbonyl compound, represent them as resonance hybrids, and indicate the conditions necessary for their formation; discuss in general terms the reactivity of these species in comparison with the original carbonyl compound (frames 1, 3).
- Write structures for the various compounds to which a given carbonyl compound

392 HOW TO SUCCEED IN ORGANIC CHEMISTRY

can be oxidized or reduced, and indicate the reagents necessary to carry out these reactions (frame 8).
- Given any two of the three species (reactant, reagent, product) involved in a wide variety of carbonyl compound reactions, write the structure of the missing third species (frames 8, 9).
- Draw the structure of a carbonyl compound and an alkyl halide from which any of a wide variety of alkenes can be produced via the Wittig reaction (frame 12).
- Draw the structure of a carbonyl compound and a Grignard reagent from which a given alcohol can be made (frame 11).
- Predict whether a given carbonyl compound will undergo the aldol condensation or the Cannizzaro reaction when treated with alkali (frame 6).
- Identify the carbonyl compound from which a given aldol condensation product (or related compound) could be derived (frame 13).

If you believe you may have already achieved some or all of these objectives, take a self-test consisting of the exercises marked with a star in the margin. If you miss any problems, study the frames in question and rework the problem before proceeding. If you are not ready for a self-test, proceed to frame 1.

Chart 17.1

Structural features	Basic modes of reactivity	Specific reactions

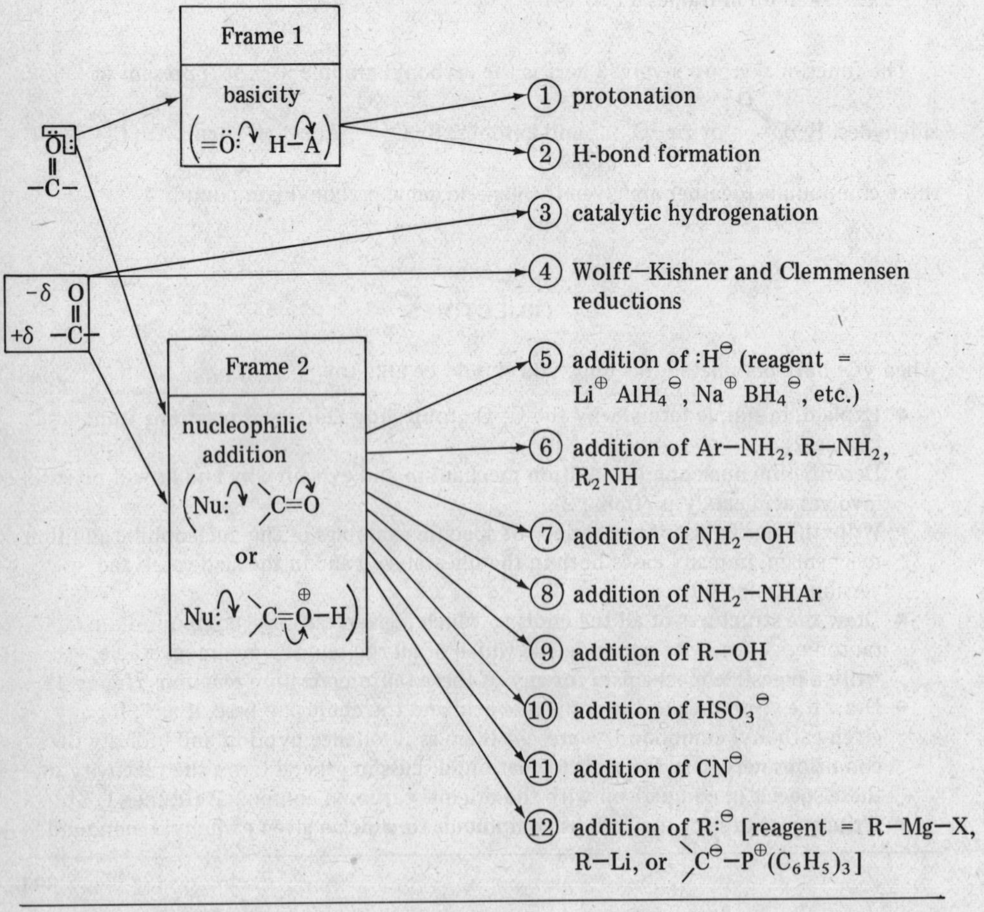

Frame 1: basicity $(=\ddot{O}: \curvearrowright H-A)$

1. protonation
2. H-bond formation
3. catalytic hydrogenation
4. Wolff–Kishner and Clemmensen reductions

Frame 2: nucleophilic addition $(Nu: \curvearrowright \overset{}{C}=\overset{\curvearrowright}{O})$ or $Nu: \curvearrowright \overset{}{C}=\overset{\oplus}{O}-H$

5. addition of $:H^{\ominus}$ (reagent = $Li^{\oplus} AlH_4^{\ominus}$, $Na^{\oplus} BH_4^{\ominus}$, etc.)
6. addition of $Ar-NH_2$, $R-NH_2$, R_2NH
7. addition of NH_2-OH
8. addition of NH_2-NHAr
9. addition of $R-OH$
10. addition of $HSO_3^{\ominus}$
11. addition of $CN^{\ominus}$
12. addition of $R:^{\ominus}$ [reagent = $R-Mg-X$, $R-Li$, or $\overset{}{C}^{\ominus}-P^{\oplus}(C_6H_5)_3$]

ALDEHYDES AND KETONES (CARBONYL COMPOUNDS) 393

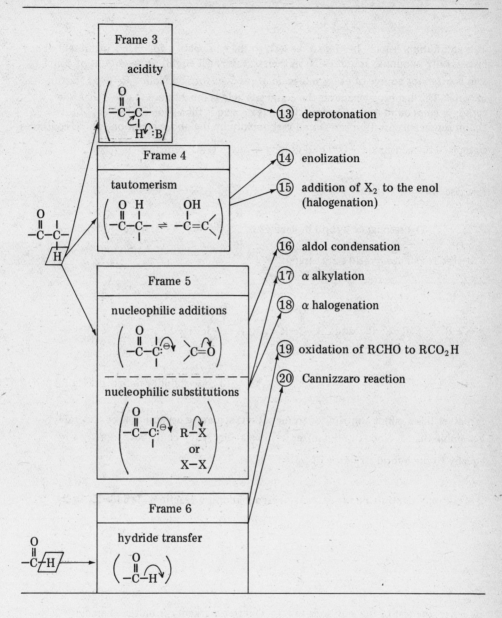

UNDERSTANDING THE BASIC MODES OF REACTIVITY

The first few frames are keyed to the basic modes of reactivity identified in the center column of Chart 17.1.

1. The $\diagup\!\!\!\!C=\ddot{O}\!:$ function is a *base* by virtue of its unshared electron pairs, but it is more weakly basic than the —OH or —OR functions (see Unit 11, frame 8). As in the case of ROH, the salts formed by $\diagup\!\!\!\!C=\ddot{O}\!:$ compounds with strong acids are almost never isolated. But they are important reaction intermediates that can be formed rapidly and reversibly in solutions containing strong acids:

$$\underset{\substack{\|\\ -\text{C}-}}{\ddot{\text{O}}\!:} + \text{H}-\text{A} \rightleftharpoons \underset{\substack{\|\\ -\text{C}-}}{\overset{\oplus}{\ddot{\text{O}}}\!-\text{H}} + :\text{A}^{\ominus}$$

This equilibrium usually lies far to the left, so the conjugate acid, $-\overset{\overset{\oplus}{\ddot{O}}-H}{\underset{\|}{C}}-$, is present at any instant only in minute amounts. Nevertheless, this small equilibrium amount of conjugate acid is often the source of every molecule of product formed from $\diagup\!\!\!\!\!\diagdown\!C{=}\ddot{O}{:}$ as starting material. This happens whenever the conjugate acid is much more reactive than the carbonyl compound itself. The initial conjugate acid is then soon destroyed, but its equilibrium concentration is restored very rapidly by the above equation, which continues to replace the consumed $\diagup\!\!\!\!\!\diagdown\!C{=}\overset{\oplus}{O}{-}H$ as long as any $\diagup\!\!\!\!\!\diagdown\!C{=}\ddot{O}{:}$ remains.

Exercise

Is $\overset{R}{\underset{R'}{\diagdown}}C{=}O$ a resonance-hybrid molecule? Is $\overset{R}{\underset{R'}{\diagdown}}C{=}\overset{\oplus}{O}{-}H$ a resonance-hybrid ion? If so, write the resonance-hybrid structure(s).

— — — — — — — —

$$\underset{a}{\underset{0}{R-\overset{\overset{\ddot{O}:}{\|}}{C}-R'}} \leftrightarrow \underset{b}{\underset{4}{R-\overset{\overset{:\ddot{O}:^{\ominus}}{|}}{\underset{\oplus}{C}}-R'}} \quad \text{and} \quad \underset{c}{\underset{1}{R-\overset{\overset{\oplus\ddot{O}-H}{\|}}{C}-R'}} \leftrightarrow \underset{d}{\underset{3}{R-\overset{\overset{:\ddot{O}-H}{|}}{\underset{\oplus}{C}}-R'}}$$

$\qquad\qquad\qquad\qquad\qquad\qquad\qquad\qquad\qquad$ 3 ← stability deficit (Table 8.1)

Structure b is a minor contributor to the $\diagdown\!C{=}O\!\diagup$ structure because it is considerably less stable than a. Since c and d differ less in stability, they contribute more nearly equally to the hybrid structure of $\diagdown\!C{=}\overset{\oplus}{O}{-}H$.

2. Like any multiple bond, $\diagdown\!C{=}O\!\diagup$ undergoes addition reactions. It adds H_2 in the presence of catalysts just as $\diagdown\!C{=}C\!\diagup$ does:

$$\diagdown\!\!C{=}O\!\!\diagup + H_2 \xrightarrow{\text{Pt or Ni}} -\underset{\underset{H}{|}}{\overset{\overset{O}{|}}{C}}-\underset{H}{}$$

However, the rest of the additions to $\diagdown\!C{=}O\!\diagup$ differ markedly from the characteristic electrophilic (and radical) additions to $\diagdown\!C{=}C\!\diagup$:

- The kinds of reagents that add are different.
- The first step in the mechanism is attack by *nucleophile* rather than electrophile.
- The additions are often acid-catalyzed.

Reagents such as $H_2O/H_3O^{\oplus}$, X–X, H–X, and $H-O-SO_2-OH$ are ineffective because they lead to unstable products:

$$\begin{matrix} \text{O} \\ \| \\ -\text{C}- \end{matrix} \begin{cases} + \text{ H-X} \rightleftarrows \begin{matrix} \text{OH} \\ | \\ -\text{C}- \\ | \\ \text{X} \end{matrix} \\ + \text{ X-X} \rightleftarrows \begin{matrix} \text{OX} \\ | \\ -\text{C}- \\ | \\ \text{X} \end{matrix} \\ + \text{ H-A} \rightleftarrows \begin{matrix} \text{OH} \\ | \\ -\text{C}- \\ | \\ \text{A} \end{matrix} \quad (\text{A = OH, OR, NH}_2, \text{NR}_2, \text{OSO}_3\text{H}) \end{cases}$$

All these compounds contain an —OH or —NH— group on a carbon singly bonded to a second electronegative atom (—O—, —N⟨, —X). This is not a stable structural feature, and all such compounds

$$\begin{matrix} \text{OH} \\ | \\ -\text{C}- \\ | \\ \text{X} \end{matrix} \quad \begin{matrix} \text{OH} \\ | \\ -\text{C}- \\ | \\ \text{OH} \end{matrix} \quad \begin{matrix} \text{OH} \\ | \\ -\text{C}- \\ | \\ \text{OR} \end{matrix} \quad \begin{matrix} \text{OH} \\ | \\ -\text{C}- \\ | \\ \text{N} \end{matrix} \quad \begin{matrix} \text{NH}- \\ | \\ -\text{C}- \\ | \\ \text{N} \end{matrix} \quad \begin{matrix} \text{NH}- \\ | \\ -\text{C}- \\ | \\ \text{X} \end{matrix} , \text{etc.}$$

decompose to ⟩C=O (or ⟩C=N—), so that the equilibria above lie back to the left. Note that when —O—R and —N⟨^R_R replace —OH and —NH— in these systems, stable compounds result, e.g., $\begin{matrix}\text{OR}\\|\\-\text{C}-\\|\\\text{OR}\end{matrix}$ and $\begin{matrix}\text{OR}\\|\\-\text{C}-\\|\\\text{X}\end{matrix}$, which, although highly reactive, are easily isolated.

Of all these compounds that contain two electronegative groups on the same carbon, the $\text{C}\begin{smallmatrix}\text{OH}\\\text{OR}\end{smallmatrix}$ function (hemiacetal, $\text{C}\begin{smallmatrix}\text{R}\quad\text{OH}\\\text{H}\quad\text{OR}'\end{smallmatrix}$, or hemiketal, $\text{C}\begin{smallmatrix}\text{R}\quad\text{OH}\\\text{R}'\quad\text{OR}''\end{smallmatrix}$) and the $\text{C}\begin{smallmatrix}\text{OR}\\\text{OR}\end{smallmatrix}$ function (full acetal or ketal) are the most important. They are present in many compounds of biological importance, especially in carbohydrates. The hemiacetal or hemiketal is present in equilibrium with any mixture of a carbonyl compound with an alcohol, ⟩C=O + ROH ⇌ ⟩C$\begin{smallmatrix}\text{OH}\\\text{OR}\end{smallmatrix}$, but usually the equilibrium lies to the left. The major exception occurs when the —OH and ⟩C=O functions are part of the same molecule and so spaced that the hemiacetal (hemiketal) contains a five- or six-membered ring: [tetrahydrofuran-OH structure] or [tetrahydropyran-OH structure]. These HO∼∼∼C=O ⇌ [cyclic hemiacetal] equilibria favor the hemiacetal or hemiketal.

The $\mathrm{>C=O}$ group is a target for nucleophiles because it is strongly *polar* $\left(\overset{+\delta}{\mathrm{>C=O}}{}^{-\delta}\right)$, both because oxygen is more electronegative than carbon and due to the $\mathrm{>C=O} \longleftrightarrow \mathrm{>\overset{\oplus}{C}-\overset{\ominus}{O}}$ resonance (exercise, frame 1). The carbon atom has some carbonium ion character, and it reacts with nucleophiles:

$$\mathrm{\overset{\ominus}{Nu:} \curvearrowright \underset{}{>C=\ddot{O}:} \longrightarrow Nu-\underset{|}{\overset{|}{C}}-\ddot{\ddot{O}}{:}^{\ominus}}$$

$$\mathrm{Nu-\underset{|}{\overset{|}{C}}-\ddot{\ddot{O}}{:}^{\ominus} \ H-A \longrightarrow Nu-\underset{|}{\overset{|}{C}}-\ddot{O}H + :A^{\ominus}}$$

This is the basic form of the *nucleophilic addition mechanism* (mechanism VIII, Table 10.1). This is a standard mechanism followed by many addition reactions to $\mathrm{>C=O}$, $\mathrm{>C=N-}$, and $\mathrm{-C\equiv N}$ groups. It has more little variations than the standard mechanisms you have learned before. We take up techniques for mastering them in frame 10.

Acid catalysis occurs when the conjugate acid of a reactant reacts faster than the reactant itself, as in this case:

$$\text{reacts more slowly with Nu:} \left\{ \begin{array}{c} \mathrm{>C=O} \\ \updownarrow a \\ \mathrm{>\overset{\oplus}{C}-O^{\ominus}} \\ b \end{array} + HA \rightleftarrows \begin{array}{c} \mathrm{>C=\overset{\oplus}{O}-H} \\ \updownarrow c \\ \mathrm{>\overset{\oplus}{C}-O-H} \\ d \end{array} \right\} \text{reacts faster with Nu:}$$

The reason for this behavior is simple: we decided in the exercise in frame 2 that structure d contributes more to the hybrid structure of the conjugate acid than b contributes to the hybrid structure of the unprotonated carbonyl compound. The conjugate acid thus has more carbonium ion character; its C atom makes a better target for nucleophiles. So we can add a second mechanism for attack by nucleophile:

$$\mathrm{>C=\ddot{O}: + HA \rightleftarrows >C=\overset{\oplus}{O}-H \longleftrightarrow >\overset{\oplus}{C}-\ddot{O}-H + A:^{\ominus}}$$

$$\mathrm{\overset{\ominus}{Nu:} \curvearrowright >\overset{\oplus}{C}-O-H \longleftrightarrow >C=\overset{\oplus}{O}-H \longrightarrow Nu-\underset{|}{\overset{|}{C}}-O-H}$$

This is the *acid-catalyzed* version of mechanism VIII.

★ *Exercises*

1. Which of the following would you expect to undergo nucleophilic addition? Why?

   (a) $\mathrm{\underset{R}{\overset{R}{>}}C=\ddot{N}-R'}$ (b) $\mathrm{:\ddot{O}=\ddot{O}:}$

   (c) $\mathrm{R-C\equiv N:}$

2. Which compound in each pair would you expect to react faster with nucleophile $\mathrm{Nu:^{\ominus}}$?

   (a) $\mathrm{C_6H_5-\overset{\overset{O}{\|}}{C}-H}$ or $\mathrm{CH_3-\overset{\overset{O}{\|}}{C}-H}$ (b) $\mathrm{CH_3-\overset{\overset{O}{\|}}{C}-H}$ or $\mathrm{CH_3-\overset{\overset{\oplus}{O}-CH_3}{\|}}{C}-H}$?

3. Is HO−C(NH$_2$)(NH$_2$)−OH likely to be a stable compound? If not, write the structures of the possible decomposition products.

– – – – – – – – –

1. (a) R$_2$C=N̈−R' ⟷ R$_2$C−N̈−R'⁻ (with ⊕ on C)   polar, adds

   (b) :Ö=Ö: nonpolar, probably doesn't add

   (c) R−C≡N: ⟷ R−C=N̈:⁻ (with ⊕ on C)   polar, adds

2. (a) Ph−C(=O)−H ⟷ [resonance structures into ring with O⁻ and ⊕ on ring carbons] ⟷ ...

   In this compound the positive charge on the carbonyl carbon is diminished by delocalization into the ring; therefore, it should react more slowly with nucleophiles than
   $$\begin{matrix}CH_3\\H\end{matrix}\!\!>\!\!C=O \longleftrightarrow \begin{matrix}CH_3\\H\end{matrix}\!\!>\!\!\overset{\oplus}{C}-\overset{\ominus}{O}.$$

   (b) $CH_3-\overset{\overset{\displaystyle\oplus}{O-CH_3}}{\underset{\|}{C}}-H$ should react faster for the same reason that the conjugate acid reacts faster.

3. HO−C(NH$_2$)(NH$_2$)−OH, whose carbon bears four −NH− or −OH groups, should be unstable:

   H$_2$O + NH$_2$−C(=O)−NH$_2$ ⟵ HO−C(NH$_2$)(NH$_2$)−OH ⟶ NH$_2$−C(=O)−OH + NH$_3$

   H$_2$O + HO−C(=NH)−NH$_2$ ⟵   ⟶ HO−C(=NH)−OH + NH$_3$

398 HOW TO SUCCEED IN ORGANIC CHEMISTRY

3. Hydrogens α to the $\diagdown$C=O group are weakly acidic:

$$-CH_2-CH_2-CH_2-\overset{O}{\underset{\|}{C}}-H$$

$\underbrace{\cdots \gamma \quad \beta}_{\text{not acidic}} \quad \underset{\text{acidic}}{\alpha\uparrow} \quad \underset{\text{aldehydic H is not acidic}}{\uparrow}$

This behavior contrasts with that of most C—H bonds, which are *not* usually sufficiently acidic to form salts with bases. The only exception encountered in previous units is R—C≡C—H, which owes its weak-acid property to the enhanced electronegativity of sp-hybridized carbon. Another example is $Cl-\underset{Cl}{\overset{Cl}{C}}-H$, where the electron-withdrawing property of the chlorines delocalizes the negative charge in the conjugate base and stabilizes it:

$$\underset{Cl}{\overset{Cl}{Cl\leftarrow}} C{:}\,^{\ominus}$$

The position of a chemical equilibrium always favors the more stable side:

$$W \rightleftarrows X \qquad Y \rightleftarrows Z$$
less more   more less
stable stable   stable stable

So anything that stabilizes the conjugate base ($A^{\ominus}$) of an acid (AH) shifts this equilibrium,

$$A-H + :B^{\ominus} \rightleftharpoons A^{\ominus} + H-B$$
acid   conjugate
              base

to the right, making AH a stronger acid. The α hydrogen in $-\overset{O}{\underset{\|}{C}}-CH\diagdown$ is made acidic by *resonance stabilization of the conjugate base:*

$$\overset{\ominus|\overset{\ddot{O}:}{\|}}{-C{-}C-} \longleftrightarrow \overset{:\ddot{O}{:}^{\ominus}}{-C{=}C-}$$

electron donor   electron acceptor

This conjugate base is called an *enolate ion*. The second limiting structure puts the negative charge on electronegative O, a stable situation. This structure is the main contributor, and a large percentage of the negative charge is on O. Only the α hydrogens are acidic because only they form a conjugate base with its electron pair separated from the electron acceptor function, $\diagup$C=O, by a single bond, making resonance stabilization possible.

How far to the right a given $\diagup$CH—CO— + $B^{\ominus} \rightleftharpoons \overset{\ominus}{\diagup}$C-C=O $\longleftrightarrow \diagup$C=C-O$^{\ominus}$ + BH

equilibrium lies depends on the strength of the base $:B^{\ominus}$. From the order of base strengths in Unit 11, frame 8, we see that the only common bases stronger than the enolate ion are $:\ddot{N}H_2^{\ominus}$, $R{:}^{\ominus}$, and $:H^{\ominus}$. Only these are capable of shifting the equilibrium above well to the right so as to convert practically all the carbonyl compound to enolate ion, as in

ALDEHYDES AND KETONES (CARBONYL COMPOUNDS) 399

$$R-CH_2-\overset{O}{\underset{\|}{C}}-R' + Na^{\oplus} H{:}^{\ominus} \rightleftharpoons R-\overset{\ominus}{C}H-\overset{\overset{\ddot{O}:}{\|}}{C}-R' + H-H$$

stronger acid   stronger base $\updownarrow$ weaker acid

$$R-CH=\overset{:\overset{\ominus}{\ddot{O}:}}{\underset{|}{C}}-R'$$

weaker base

However, only rarely do we need to prepare the enolate ion in quantity. In many applications a small equilibrium concentration of enolate ion is sufficient. It can be achieved using the weaker bases $OH^{\ominus}$ or $OR^{\ominus}$:

$$R-CH_2-\overset{\ddot{O}:}{\underset{\|}{C}}-R' + {:}\overset{\ominus}{\underset{..}{O}}-R'' \; Na^{\oplus} \rightleftharpoons R-\overset{\ominus}{C}H-\overset{\overset{\ddot{O}:}{\|}}{C}-R' + R''\ddot{O}H$$

wa   wb $\updownarrow$ sa

$$R-CH=\overset{:\overset{\ominus}{\ddot{O}:}}{\underset{|}{C}}-R'$$

sb

When enolate ion is used up in some other reaction, this equilibrium produces more of it from the waiting pool of $R-CH_2-CO-R'$.

If we set $R'' = H$ in the last equation written above, the equilibrium will still lie far to the left, since H_2O is just about as acidic as an alcohol. This tells us one more important thing, namely that if we have the enolate ion, from whatever source, contact with water will destroy it, except for a small equilibrium concentration.

The enolate ion is a strong base and a potent nucleophile. We discuss these properties in the next two frames.

★ *Exercises*

1. Draw resonance-hybrid structures for the enolate ions produced by using an appropriate base to remove $H^{\oplus}$ from the indicated carbon atom of each carbonyl compound.

   (a) $CH_3-\underline{CH_2}-\overset{O}{\underset{\|}{C}}-H$

   (b) cyclohexanone $=O$

   (c) $C_6H_5-\underline{CH_2}-\overset{O}{\underset{\|}{C}}-H$

   (d) $CH_3-\overset{O}{\underset{\|}{C}}-\underline{CH_2}-\overset{O}{\underset{\|}{\underset{}{C}}}-\underset{\underset{CH_3}{|}}{\overset{CH_3}{\underset{|}{C}}}-CH_3$ and

   (e) $CH_2=CH-\underline{CH_2}-\overset{O}{\underset{\|}{C}}-H$

   (f) $CH_3-\overset{O}{\underset{\|}{C}}-\underline{CH_2}-N\overset{\diagup O}{\diagdown O}$

2. Apply the theory of resonance (Unit 9) to predict which of the two enolate ions in exercise 1(d) is the more stable.

3. Arrange the three types of H in $CH_3-\overset{O}{\underset{\|}{C}}-CH_2-\overset{O}{\underset{\|}{C}}-\underset{\underset{CH_3}{|}}{\overset{CH_3}{\underset{|}{C}}}-CH_3$ in order of decreasing acidity.

- - - - - - - - - -

400 HOW TO SUCCEED IN ORGANIC CHEMISTRY

1. (a) $CH_3-\overset{\ominus}{\underset{..}{C}}H-\overset{\overset{..}{\overset{..}{O}}}{\underset{||}{C}}-H \longleftrightarrow CH_3-CH=\overset{:\overset{..}{O}:^\ominus}{\underset{|}{C}}-H$

(b) ⬡=O ⟷ ⬡−O^⊖ (not ⟷ ⬡−O^⊖ or ⬡=O, which moves H atoms and is a chemical equilibrium, not resonance).

(c) Ph−$\overset{\ominus}{C}$H−C(=O:)−H ⟷ Ph−CH=C(−O:^⊖)−H ⟷ (cyclohexadienylidene)=CH−C(=O:)−H

↕

Ph−$\overset{\ominus}{C}$H−C(=O:)−H ⟷ (cyclohexadienylidene)=CH−C(=O:)−H ⟷ ⊖:(cyclohexadienylidene)=CH−C(=O:)−H

(d) $\overset{\ominus}{C}H_2-\overset{\overset{..}{O}:}{\underset{||}{C}}-CH_2-\overset{\overset{..}{O}:}{\underset{||}{C}}-C(CH_3)_3 \longleftrightarrow CH_2=\overset{:\overset{..}{O}:^\ominus}{\underset{|}{C}}-CH_2-\overset{\overset{..}{O}:}{\underset{||}{C}}-C(CH_3)_3$
     a

and

$CH_3-\overset{\overset{..}{:O}}{\underset{||}{C}}-\overset{\ominus}{\underset{..}{C}}H-\overset{\overset{..}{O}:}{\underset{||}{C}}-C(CH_3)_3 \longleftrightarrow CH_3-\overset{:\overset{..}{O}:^\ominus}{\underset{|}{C}}=CH-\overset{\overset{..}{O}:}{\underset{||}{C}}-C(CH_3)_3 \longleftrightarrow$
                                                    b

$CH_3-\overset{\overset{..}{O}:}{\underset{||}{C}}-CH=\overset{:\overset{..}{O}:^\ominus}{\underset{|}{C}}-C(CH_3)_3$

(e) $CH_2=CH-\overset{\ominus}{\underset{..}{C}}H-\overset{\overset{..}{O}:}{\underset{||}{C}}-H \longleftrightarrow CH_2=CH-CH=\overset{:\overset{..}{O}:^\ominus}{\underset{|}{C}}-H \longleftrightarrow :\overset{\ominus}{C}H_2-CH=CH-\overset{\overset{..}{O}:}{\underset{||}{C}}-H$

(f) $CH_3-\overset{\overset{..}{:O}}{\underset{||}{C}}-\overset{\ominus}{\underset{..}{C}}H-N(=\overset{..}{O}:)(−O^{-}) \longleftrightarrow CH_3-\overset{:\overset{..}{O}:^\ominus}{\underset{|}{C}}=CH-N(=\overset{..}{O}:)(−O^{-}) \longleftrightarrow CH_3-\overset{\overset{..}{O}:}{\underset{||}{C}}-CH=N(−\overset{..}{O}:^\ominus)(−O^{-})$

2. $CH_3-\overset{O}{\underset{||}{C}}-CH_2-\overset{O}{\underset{||}{C}}-C(CH_3)_3$ The enolate ion a [see the answer to exercise 1(d)]
     a   b

obtained by removing H_a has the ⊖ charge delocalized over one C and one O atom. The isomeric enolate ion (b) from H_b has ⊖ delocalized over one C and *two* O atoms. Resonance theory says the more extensively delocalized system (more limiting structures of roughly equal energy) has the greatest resonance stabilization. Thus ion b is more stable.

3. $CH_3-\overset{O}{\underset{||}{C}}-CH_2-\overset{O}{\underset{||}{C}}-C(CH_3)_3$

singly α H's— removal leads to ordinary ion

doubly α H's—removal leads to superstable enolate ion

β H's—not more acidic than ordinary alkane type

Therefore, the acidity order is $H_b > H_a > H_c$.

4. Since the unshared pair of electrons and the negative charge in the enolate ion are spread over the α-C and the carbonyl O, both these sites are nucleophilic and basic, and reaction can occur at either site:

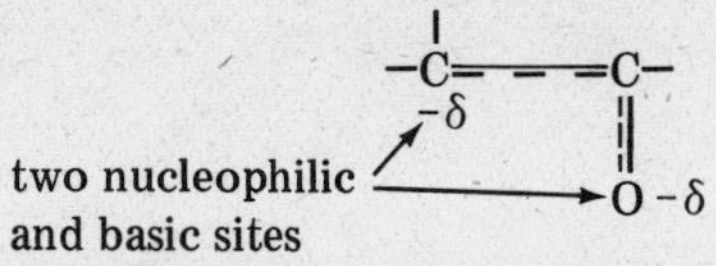

two nucleophilic and basic sites

$$-\overset{|}{C}=\cdots=\overset{|}{C}-$$
$$\overset{\|}{O}-\delta$$

If this enolate ion is treated with a source of $H^\oplus$, it is converted back to its conjugate acid. The question is, which of the two basic sites gets the $H^\oplus$? The answer is that both do. When the protonation occurs at C, the original carbonyl compound is formed. This can be visualized using either of the limiting structures:

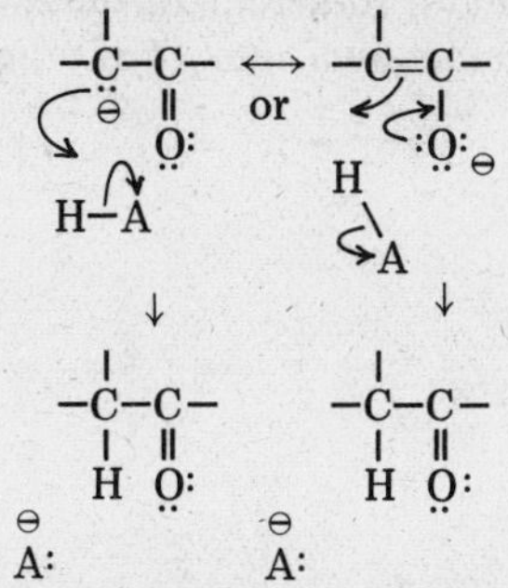

When the protonation occurs at O, a new compound is formed:

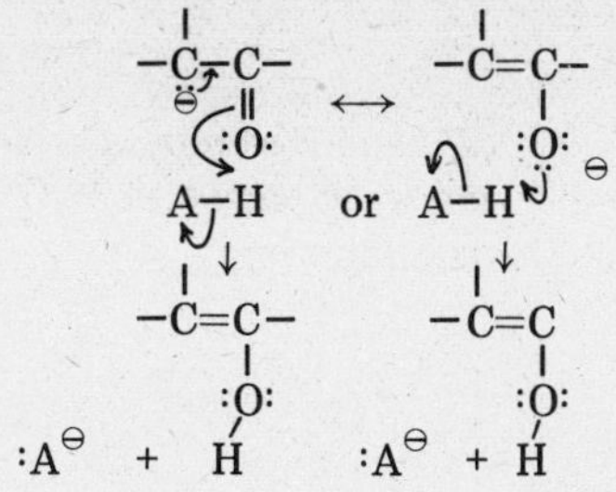

This composite functional group, $\underset{\underset{OH}{|}}{C}=C-$, is called an *enol* function (combining *-ene* and *-ol*). The enol molecule is a structural isomer of the original carbonyl compound.

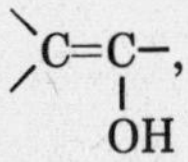

Which is formed fastest? The enol is, but this means very little because the two products are in rapid equilibrium with one another:

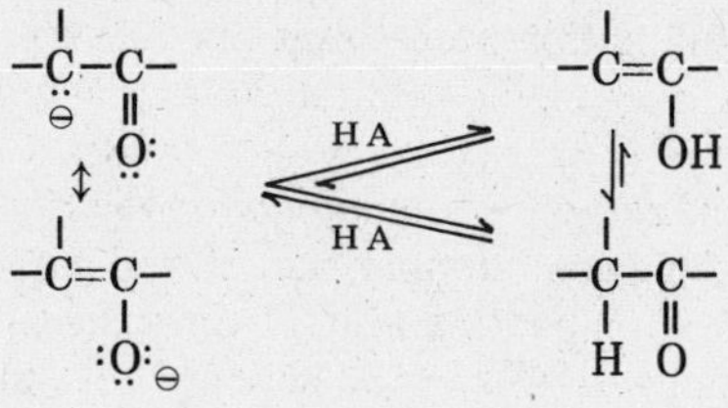

The product one actually gets is the equilibrium mixture of the two, but this proves to be >99% carbonyl compound (called the *keto* form) and <1% *enol* for most simple compounds. This is an example of a pair of isomers that rapidly interconvert. Since most pairs of isomers do not interconvert, even slowly, the ones that do are given a special name: *tautomers*. The phenomenon is called tautomerism. One form tautomerizes to the other.

In the present case we say that the keto form enolizes and the enol form ketonizes, setting up the equilibrium

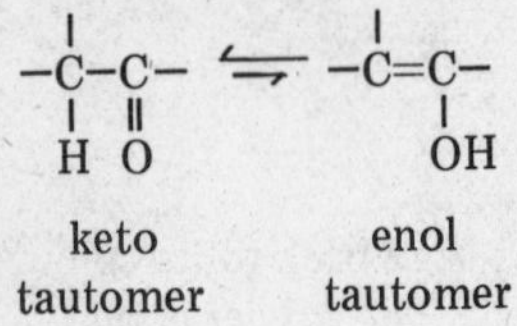

$$-\overset{|}{\underset{H}{C}}-\underset{O}{\overset{||}{C}}- \;\rightleftharpoons\; -\overset{|}{C}=\underset{OH}{\overset{|}{C}}-$$

keto tautomer   enol tautomer

The fact that the equilibrium position favors the keto form tells us that it is more stable than its enol.

Protonation of the enolate ion is not the only mechanism for formation of the enol. Keto and enol can interconvert via the conjugate acid, or in fact by a number of other mechanisms, some of which are illustrated below:

(a)

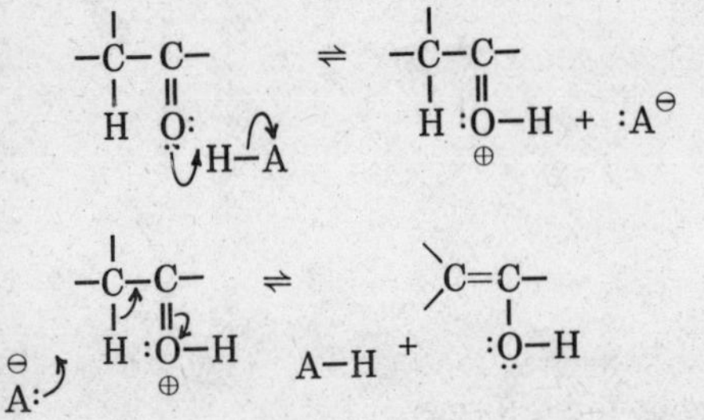

$$-\overset{|}{\underset{H}{C}}-\underset{O:}{\overset{||}{C}}- \;\; \overset{H-A}{\curvearrowleft} \;\rightleftharpoons\; -\overset{|}{\underset{H}{C}}-\underset{\overset{\oplus}{O}-H}{\overset{||}{C}}- \;+\; :A^{\ominus}$$

$$\underset{A:^{\ominus}}{} -\overset{|}{\underset{H}{C}}\overset{\curvearrowleft}{-}\underset{\overset{\oplus}{O}-H}{\overset{||}{C}}- \;\rightleftharpoons\; \underset{}{\overset{|}{>}}C=\underset{:O-H}{\overset{|}{C}}- \;\;A-H \;+\; :\ddot{O}-H$$

(b)

$$-\overset{|}{\underset{H}{C}}-\underset{O}{\overset{||}{C}}- \;+\; H-A \;\rightleftharpoons\; -\overset{|}{\underset{H}{C}}-\underset{O\cdots H-A}{\overset{||}{C}}- \;\;\text{H bond}$$

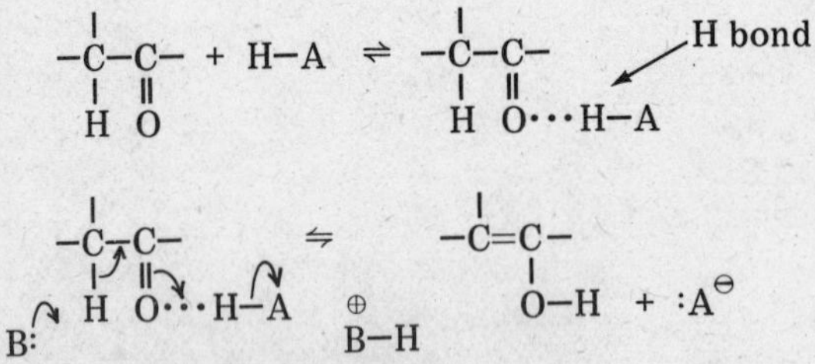

(AH and B can be weak acid and weak base, e.g., H_2O molecules.)

$$\underset{B:}{} -\overset{|}{\underset{H}{C}}\overset{\curvearrowleft}{-}\underset{O\cdots H-A}{\overset{||}{C}}- \;\rightleftharpoons\; -\overset{|}{C}=\underset{O-H}{\overset{|}{C}}- \;\;\overset{\oplus}{B}-H \;+\; :A^{\ominus}$$

(c)

$$-\overset{|}{\underset{H}{C}}-\underset{O}{\overset{||}{C}}- \;+\; 2H_2O \;\rightleftharpoons\; -\overset{|}{C}-\overset{|}{\underset{O}{C}}- \;\rightleftharpoons\; -\overset{|}{C}=\underset{O}{\overset{|}{C}}-$$

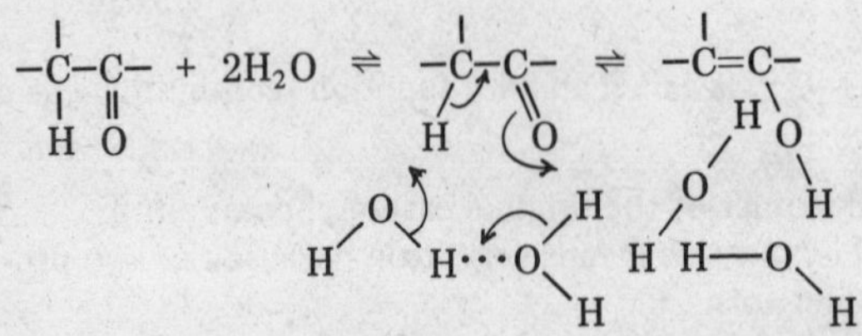

One consequence of the rapid keto ⇌ enol equilibration processes is this: in reactions that are expected to yield the enol tautomer, the more stable keto tautomer is actually isolated because enol ⟶ keto during the workup.

A second consequence is that some reactions of carbonyl compounds are really reactions of the enol tautomer. Such reactions would soon destroy the original equilibrium concentration of enol, except that the tautomerization continuously restores the enol and in fact keeps it constant at the equilibrium value until keto and enol are both ultimately consumed. An example is reaction ⑮—electrophilic addition to the enol. This occurs, for example, in the bromination of ketones in acidic solution:

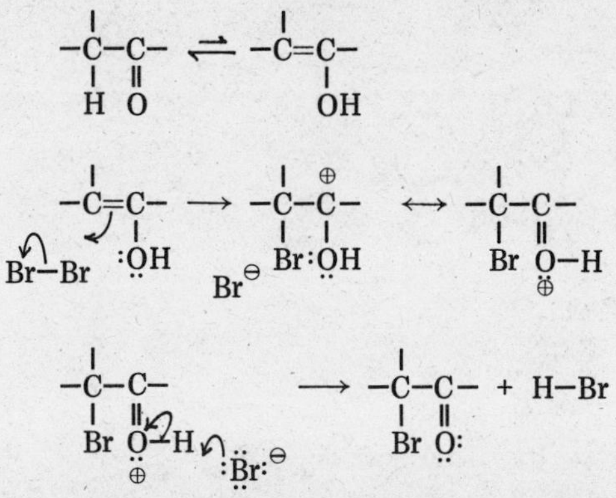

$$-\underset{H}{\overset{|}{C}}-\underset{O}{\overset{||}{C}}- \rightleftharpoons -\underset{}{\overset{|}{C}}=\underset{OH}{\overset{|}{C}}-$$

$$\underset{Br-Br}{-\overset{|}{C}=\overset{|}{C}-} \underset{:\ddot{O}H}{} \longrightarrow \underset{Br^{\ominus}}{-\overset{|}{\underset{Br}{C}}-\overset{\oplus}{\underset{:\ddot{O}H}{C}}-} \longleftrightarrow -\overset{|}{\underset{Br}{C}}-\overset{|}{\underset{\overset{\oplus}{O-H}}{C}}-$$

$$-\overset{|}{\underset{Br}{C}}-\overset{|}{\underset{\overset{\oplus}{O-H}}{C}}- \longrightarrow -\overset{|}{\underset{Br}{C}}-\overset{||}{\underset{O:}{C}}- + H-Br$$

To draw the structure of the enol of some carbonyl compound:

- Locate all the α hydrogens; if there is more than one type, there will be more than one enol.
- Remove an α-hydrogen atom and single-bond it to oxygen.
- Move the double bond to the α-carbon atom.

To generate a keto from an enol structure, reverse the second and third steps.

Example

Draw the enol form of ⬡=O. Since all four α hydrogens are equivalent, there is one enol:

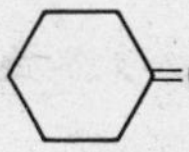

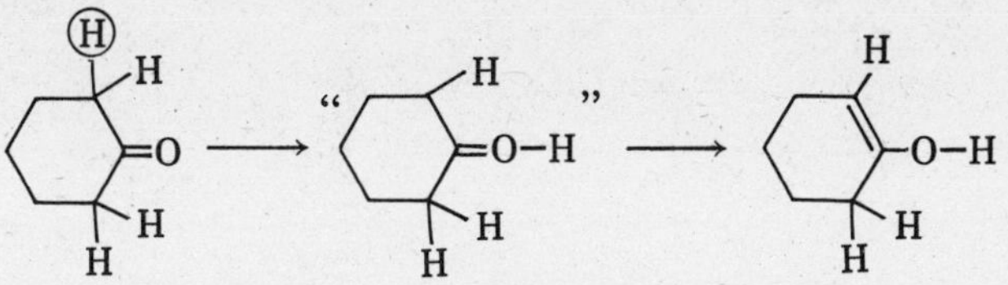

It is the ability to interconvert rapidly that attaches the special title tautomers to a pair of structural isomers. The only common isomerization processes that are, in fact, so rapid are the *making and breaking of bonds to acidic hydrogen atoms*. This class includes only —O—H, ⟩N—H, —S—H, and α C—H bonds; however, there are several functional groups that acidify α hydrogens besides just ⟩C=O:

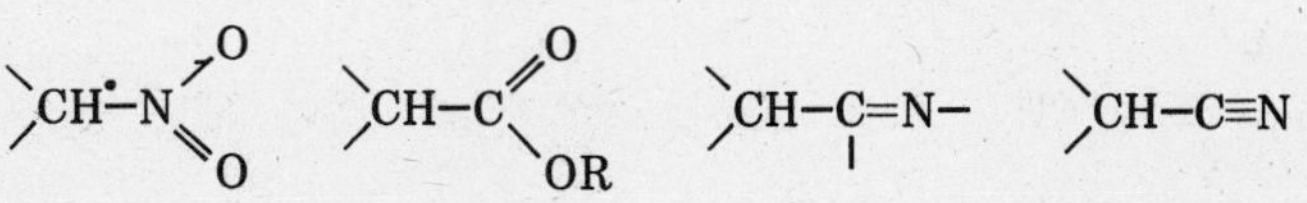

$$\underset{}{\overset{}{\rangle}CH-N\underset{O}{\overset{O}{\lessgtr}}} \quad \underset{}{\overset{}{\rangle}CH-C\underset{OR}{\overset{O}{\lessgtr}}} \quad \underset{}{\overset{}{\rangle}CH-C=N-} \quad \underset{}{\overset{}{\rangle}CH-C\equiv N}$$

Examples

These pairs are isomeric and tautomeric:

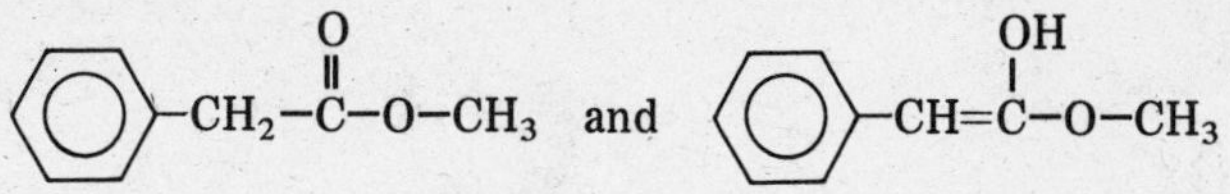

⬡—CH$_2$—$\overset{O}{\overset{||}{C}}$—O—CH$_3$ and ⬡—CH=$\overset{OH}{\overset{|}{C}}$—O—CH$_3$

Ph-C(=O)-NH_2 and Ph-C(OH)=NH

These pairs are isomeric but *not* tautomeric:

$$\text{CH}_3\text{-CH(H)-C(CH}_3\text{)(CH}_3\text{)-Ph} \quad \text{and} \quad \text{CH}_3\text{-C(CH}_3\text{)(H)-C(CH}_3\text{)(CH}_3\text{)-Ph}$$

(with H and CH₃ substituents as drawn)

$\text{CH}_3\text{-CH=C=CH}_2$ and $\text{CH}_3\text{-CH}_2\text{-C}\equiv\text{C-H}$

★ *Exercises*

1. Which of the following pairs are tautomers?

   (a) $\text{CH}_3\text{-C(=O)-H}$ and $\text{CH}_2\text{=C(OH)-H}$

   (b) $\text{CH}_3\text{-CH}_2\text{-C}\equiv\text{N}$ and $\text{CH}_3\text{-CH=C=N-H}$

   (c) $\text{CH}_3\text{-CH}_2\text{-CH=CH}_2$ and $\text{CH}_3\text{-CH=CH-CH}_3$

   (d) $\text{CH}_3\text{-C(OH)=CH-CH}_3$ and $\text{CH}_2\text{=C(OH)-CH}_2\text{-CH}_3$

   (e) $:\text{CH}_2^{\ominus}\text{-C(=O)-CH}_3$ and $\text{CH}_2\text{=C(-}\ddot{\text{O}}\text{:}^{\ominus}\text{)-CH}_3$

   (f) $\text{Ph-}\ddot{\text{O}}\text{:}^{\ominus}$ and cyclohexadienone anion

   (g) Ph-O-H and 2,4-cyclohexadien-1-one (with H,H at sp³ carbon)

2. Draw the structural formulas for all the tautomers of each of the following.

   (a) $\text{CH}_3\text{-CH}_2\text{-C(=O)-CH}_3$

   (b) 2-methylcyclohexanone (with =O and CH₃)

   (c) $\text{CH}_3\text{-C(=O)-CH}_2\text{-C(=O)-CH}_2\text{-CH}_3$

   (d) $\text{CH}_3\text{-N(=O)(=O)}$ (i.e., $\text{CH}_3\text{-NO}_2$)

   (e) $\text{CH}_3\text{-CH=N-CH}_3$

   (f) $\text{CH}_3\text{-C(=O)-O-CH}_3$

   (g) $\text{CH}_2\text{=C(OH)-CH(CH}_3\text{)(CH}_3\text{)}$

   (h) $\text{CH}_3\text{-C(OH)=C(OH)-H}$

(i) [phenanthrene with OH at 4-position, aromatic enol form]

---

1. The pairs of molecules in (a), (b), (d), and (g) are tautomers. If you thought those in (d) were not, it's probably because the keto form, $CH_3-\overset{O}{\overset{\|}{C}}-CH_2-CH_3$, is missing; but if each of the possible enol forms is in equilibrium with the keto form, they are also in equilibrium with one another. In (c) the two alkenes are isomers that do *not* spontaneously interconvert and are thus not tautomers (they *can* be interconverted by certain chemical *treatments*). In (e) and (f) the two structures differ only in the position of electrons and are thus limiting structures of a resonance-hybrid molecule.

2. (a) $CH_3-CH_2-\overset{O}{\overset{\|}{C}}-CH_3 \rightleftharpoons CH_3-CH=\overset{OH}{\overset{|}{C}}-CH_3 \rightleftharpoons CH_3-CH_2-\overset{OH}{\overset{|}{C}}=CH_2$

(b) [cyclohexane with CH₃ and =O] ⇌ [cyclohexene with CH₃ and —OH] ⇌ [cyclohexene with CH₃ and —OH, different double bond position]

(c) $CH_3-\overset{O}{\overset{\|}{C}}-CH_2-\overset{O}{\overset{\|}{C}}-CH_2-CH_3 \rightleftharpoons CH_2=\overset{OH}{\overset{|}{C}}-CH_2-\overset{O}{\overset{\|}{C}}-CH_2-CH_3 \rightleftharpoons$

$CH_3-\overset{O}{\overset{\|}{C}}-CH_2-\overset{OH}{\overset{|}{C}}=CH-CH_3 \rightleftharpoons CH_3-\overset{OH}{\overset{|}{C}}=CH-\overset{O}{\overset{\|}{C}}-CH_2-CH_3 \rightleftharpoons$

$CH_3-\overset{O}{\overset{\|}{C}}-CH=\overset{OH}{\overset{|}{C}}-CH_2-CH_3$

(d) $CH_3-N\overset{\nearrow O}{\searrow_O} \rightleftharpoons CH_2=N\overset{\nearrow OH}{\searrow_O}$

(e) $CH_3-CH=N-CH_3 \rightleftharpoons CH_2=CH-NH-CH_3$

(f) $CH_3-\overset{O}{\overset{\|}{C}}-O-CH_3 \rightleftharpoons CH_2=\overset{OH}{\overset{|}{C}}-O-CH_3$

(g) $CH_2=\overset{OH}{\overset{|}{C}}-CH\overset{\nearrow CH_3}{\searrow CH_3} \rightleftharpoons CH_3-\overset{O}{\overset{\|}{C}}-CH\overset{\nearrow CH_3}{\searrow CH_3} \rightleftharpoons CH_3-\overset{OH}{\overset{|}{C}}=C\overset{\nearrow CH_3}{\searrow CH_3}$

(h) $CH_3-\underset{OH\ OH}{\overset{}{C=C}}-H \rightleftharpoons CH_3-\underset{O\ OH}{\overset{}{\underset{\|}{C}-CH_2}} \rightleftharpoons CH_2=\underset{OH\ OH}{\overset{}{C-CH_2}} \rightleftharpoons CH_3-\underset{OH\ O}{\overset{}{CH-\overset{\|}{C}-H}}$

(i) [phenanthrone keto form] ⇌ [phenanthrol enol form]

5. The next basic mode of reactivity is the reaction type in which the enolate ion acts as nucleophile. It proves to be a strong nucleophile, and it is able to participate in nucleophilic substitutions (S_N2 reactions, mechanism V, Unit 14, frame 4) and nucleophilic additions (mechanism VIII, this unit, frames 2 and 10). In the substitutions, we may observe products derived from reaction at each of the two nucleophilic sites in the enolate ion:

$$X^\ominus + \underset{R\ \ O}{-\overset{|}{\underset{|}{C}}-\overset{\|}{C}-} \longleftarrow X-R\ \ \underset{:O:}{-\overset{|}{\underset{\ominus}{C}}-\overset{\|}{C}-} \longleftrightarrow \underset{:O:^\ominus}{-\overset{|}{C}=\overset{|}{C}-}\ \ R-X \longrightarrow \underset{O-R}{\overset{}{\diagup}C=C\overset{}{\diagdown}} + X^\ominus$$

Nucleophiles with more than one nucleophilic site are called *ambident nucleophiles*. Unlike the protonation of the enolate ion at these two different sites, the two products of nucleophilic displacement usually do not rapidly interconvert and may both be isolated as stable compounds. Under the usual conditions reaction at the α-carbon atom predominates.

When the enolate ion brings about nucleophilic *addition*, it is usually attacking the fractionally positive carbon atom in a $\diagup C=O$, $\diagup C=N-$, or $-C\equiv N$ function. This is really no different from the addition reactions produced by a Grignard reagent or an alkyl lithium; for example, compare these reactions:

[Reaction scheme a: PhCH₂–CH₂–Br →(Li) PhCH₂–CH₂–Li → PhCH₂–CH₂:⁻ (a) Li⁺ + CH₃–C(=O)–H → PhCH₂–CH₂–C(O⁻Li⁺)H–CH₃]

[Reaction scheme b: PhC(=O)–CH₃ →(NaOH) PhC(O⁻Na⁺)=CH₂ ⇌ PhC(=O)–CH₂:⁻ (b) Na⁺ + CH₃–C(=O)–H → PhC(=O)–CH₂–C(O⁻Na⁺)H–CH₃]

Reagents a and b are both carbanions, $R:^\ominus$, and react the same way. Ion b is resonance-stabilized, and it can be made by the action of base on its conjugate acid,

$R-H$ $\left(\text{Ph}-CO-CH_2-H\right)$. Ion a is an unstabilized carbanion that must be

made by action of a metal on the alkyl halide, R–X.

One of the nucleophilic addition reactions that often confuses students unnecessarily is the case where an enolate ion adds to the carbonyl group of its own conjugate acid (aldol condensation). With a little practice, however, it is not difficult to keep the two roles being played by the carbonyl compound in this reaction straight. In frame 13 we take up methods of handling the standard problems involving this reaction.

Exercises

1. Using the enolate ion-alkyl halide reaction written out above as a model, predict

ALDEHYDES AND KETONES (CARBONYL COMPOUNDS) 407

all the possible products of the reaction of $CH_3-CO-\overset{\ominus}{CH}-CO-C_6H_5$ $Na^{\oplus}$ with CH_3I.

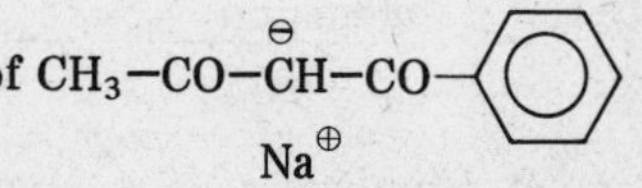

2. When the enolate ion above is treated instead with $CH_3-\underset{CH_3}{\overset{CH_3}{\underset{|}{C}}}-Br$, the products are

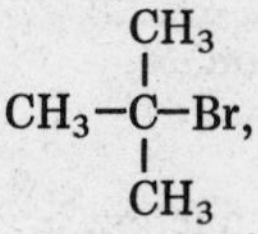

$CH_2=C\overset{CH_3}{\underset{CH_3}{\diagdown}}$, $CH_3-CO-CH_2-CO-C_6H_5$, and NaBr.

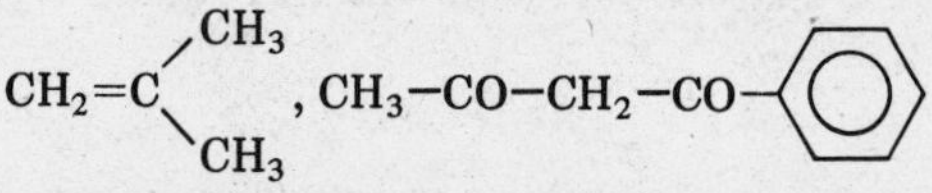

What causes this system to behave differently? What role is the enolate ion playing in this reaction? Write the mechanism.

3. Given that the reaction $C_6H_{10}=O + I_2 + NaOH \longrightarrow C_6H_9(I)=O + Na^{\oplus}I^{\ominus} +$

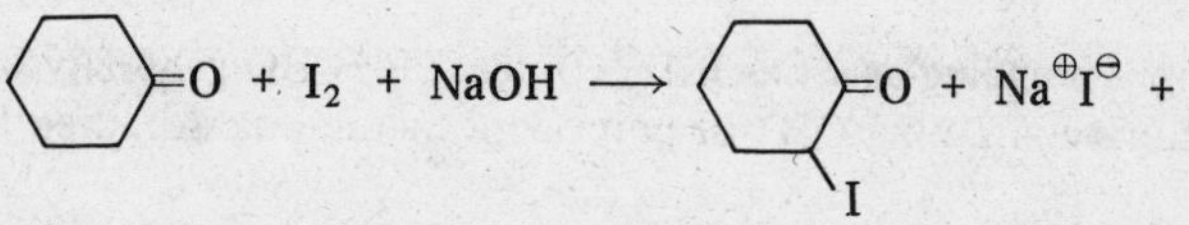

H_2O is an enolate ion reaction and resembles the S_N2 reaction mechanistically, write a mechanism for this reaction.

— — — — — — — — —

1. 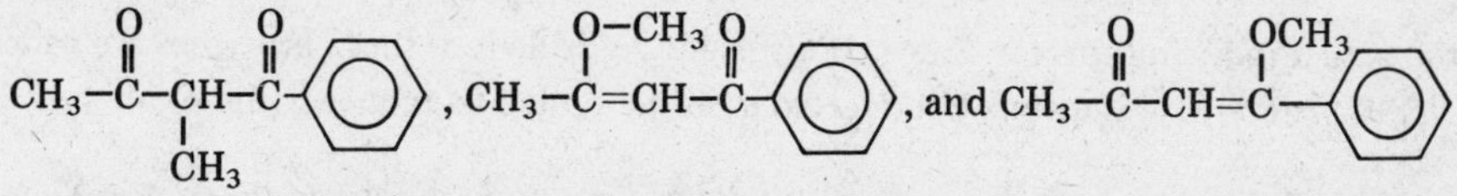 $CH_3-\overset{O}{\overset{\|}{C}}-\underset{CH_3}{\overset{|}{CH}}-\overset{O}{\overset{\|}{C}}-C_6H_5$, $CH_3-\overset{O-CH_3}{\overset{|}{C}}=CH-\overset{O}{\overset{\|}{C}}-C_6H_5$, and $CH_3-\overset{O}{\overset{\|}{C}}-CH=\overset{OCH_3}{\overset{|}{C}}-C_6H_5$

2. Exercise 1 is an S_N2 reaction of a methyl (superprimary) halide with enolate ion acting as nucleophile. The halide $(CH_3)_3CBr$ is tertiary, and tertiary halides give more elimination than substitution when treated with strongly basic nucleophiles. Since $CH_2=C\overset{CH_3}{\underset{CH_3}{\diagdown}}$, the product of elimination from $(CH_3)_3Br$, is indeed observed, this must be an $E2$ elimination brought about by enolate ion acting as base:

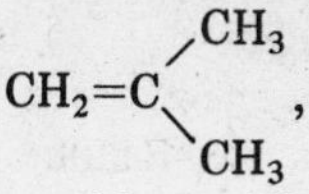

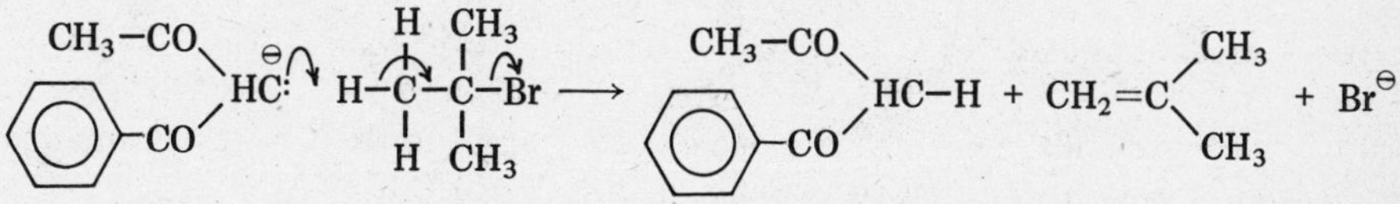

$\underset{C_6H_5-CO}{\overset{CH_3-CO}{\diagdown}}\overset{\ominus}{HC}: \curvearrowright H-\underset{H}{\overset{H}{\underset{|}{\overset{|}{C}}}}-\underset{CH_3}{\overset{CH_3}{\underset{|}{\overset{|}{C}}}}-Br \longrightarrow \underset{C_6H_5-CO}{\overset{CH_3-CO}{\diagdown}}HC-H + CH_2=C\overset{CH_3}{\underset{CH_3}{\diagdown}} + Br^{\ominus}$

3. [cyclohexanone with α-H] $\longrightarrow$ [enolate] $\longleftrightarrow$ [enolate resonance] $+ H_2O$

   $:\overset{\ominus}{O}H\ Na^{\oplus}$

   etc. $\longleftrightarrow$ $Na^{\oplus}$ [enolate + I-I] $\longrightarrow$ [α-iodocyclohexanone] $+ I^{\ominus} Na^{\oplus}$

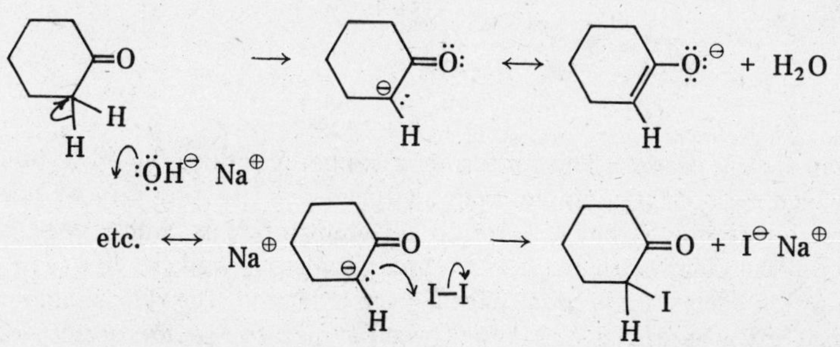

6. In the last basic reactivity mode of Chart 17.1 the aldehyde function plays the role of reducing agent. This means the aldehyde is itself oxidized:

$$\underbrace{R-\overset{\overset{O}{\|}}{C}-H + Ox \longrightarrow R-\overset{\overset{O}{\|}}{C}-OH + Ox-H}_{\text{oxidation}} \quad Ox = \text{oxidizing agent}$$

with "reduction" labeled over the arrow from R-CHO to the right-hand side.

Chart 17.1 represents reduction by $R-C\overset{O}{\underset{H}{\diagdown}}$ as transfer of H with both its bonding

electrons: $R-\overset{\overset{O}{\|}}{C}-H \curvearrowright Ox^{\oplus} \longrightarrow H-Ox$, where Ox is the oxidizing agent. This makes

the reaction analogous to reduction by $H_3B-H^{\ominus}$; however, $R-C\overset{O}{\underset{H}{\diagdown}}$ is not nearly so

potent a reducing agent as $BH_4^{\ominus}$. Indeed, R-CHO generally behaves in this way only in solutions containing $OH^{\ominus}$ or polyvalent metal ions. Apparently, species such as

$$R-\underset{OH}{\overset{\overset{\ddot{O}:^{\ominus}}{|}}{C}}-H \qquad R-\underset{OR'}{\overset{\overset{\ddot{O}:^{\ominus}}{|}}{C}}-H \qquad R-\underset{OR'}{\overset{\overset{\ddot{O}-Al^{\ominus}}{|}}{C}}-H$$

are the actual reducing species. Since $:H^{\ominus}$ is called hydride ion, these processes are called *hydride transfers*. $H:^{\ominus}$ transfers two electrons to the molecule reduced (the oxidizing agent). The fragment resulting from removal of $H:^{\ominus}$ from $R-\overset{\overset{O}{\|}}{C}-H$ ($R-\overset{\oplus}{C}=O$) is not stable and not observed. The isolated product is generally the product of reaction of $R-\overset{\oplus}{C}=O$ with water, which is $R-C\overset{O}{\underset{OH}{\diagdown}}$. This should be an oxidation product of $R-\overset{\overset{O}{\|}}{C}-H$; Charts 15.2 and 17.2 show that this is so.

Actually, many reductions by aldehydes (oxidations of aldehydes) are *one*-electron transfers instead of hydride transfers. This is the case with $Ag^{\oplus}$ and $Cu^{2\oplus}$, which do not accept more than one electron per ion.

The oxidations of $R-C\overset{O}{\underset{H}{\diagdown}}$ to $R-C\overset{O}{\underset{OH}{\diagdown}}$ by $Ag^{\oplus}$, $Cu^{2\oplus}$, $MnO_4^{\ominus}$, CrO_3, $Cr_2O_7^{2\ominus}$,

etc., are easy to understand. There is one oxidizing agent for R-CHO that sometimes makes the equation confusing to students, namely R-CHO itself:

$$2\ R-C\overset{O}{\underset{H}{\diagdown}} \xrightarrow{\text{NaOH}} R-C\overset{O}{\underset{OH}{\diagdown}} + R-CH_2-OH$$

This is called the *Cannizzaro reaction*. The confusion arises because this is an alkali-promoted reaction in which a carbonyl compound plays two roles—a situation very similar to the aldol condensation (frame 5), but with totally different products. Since both the aldol condensation and the Cannizzaro reaction are brought about by alkali, you need a rule to determine which occurs preferentially for a given aldehyde. The rule is: aldol condensation occurs if any α hydrogens are present; the Cannizzaro reaction occurs only if α hydrogens are absent. To get the products of a Cannizzaro reaction, just write one molecule of the aldehyde in its oxidized (RCO_2H) state and one molecule in its reduced (RCH_2OH) state.

★ *Exercises*

Which compounds would give the Cannizzaro reaction on treatment with concentrated NaOH?

1. C$_6$H$_5$—CHO

2. C$_6$H$_5$—C(=O)—CH$_3$

3. C$_6$H$_5$—CH$_2$—C(=O)—H

4. H—C(=O)—H

5. CH$_3$—C(=O)—H

6. CH$_3$—C(CH$_3$)(CH$_2$CH$_3$)—CHO

- - - - - - - - -

1, 4, and 6 only; the rest have α hydrogens.

LEARNING THE REACTIONS AND MECHANISMS

7. In frames 7 to 10 we set up and apply some learning strategies for the carbonyl compound reactions and the nucleophilic addition mechanism.

Exercises

1. Compile the list of reactions that you will use in the drills by editing the right-hand column of Chart 17.1 to correspond to the coverage in your course. Write the generic equation for each reaction tabulated; include each major variation [e.g., RMgX, RLi, and $-\overset{\ominus}{C}-\overset{\oplus}{P}(C_6H_5)_3$ under addition of R:$^\ominus$].

2. Drill on the list compiled in exercise 1, using the methods laid out in Unit 16, frame 4, exercises 3 and 4.

8. Chart 17.2 is an oxidation state-alkylation state chart that extends the previous ones (Charts 12.2 and 15.2). Chart 17.2 omits the aldehyde and ketone reactions that take place within oxidation stage II. All the >C=O group reactions that cross to other oxidation stages are represented there by arrows.

Chart 17.2 brings out these basic points:
1. Like the alcohols, the carbonyl compounds occupy a strategic, central position. They can be converted to many different compounds.
2. The ketones, like the 3° alcohols, have no higher oxidation state above them; they are oxidized only nonselectively—under conditions that destroy the whole molecule. The aldehydes, on the other hand, can go up or down in oxidation state via highly selective reactions.
3. The diagonal arrows leading from the carbonyl compounds to the various alcohols are the only reactions you have encountered up to this point in the organic course that produce a horizontal movement in the chart; they increase the alkylation state of the

410 HOW TO SUCCEED IN ORGANIC CHEMISTRY

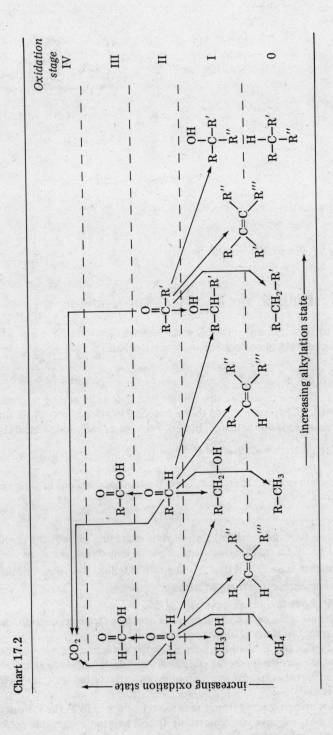

Chart 17.2

ALDEHYDES AND KETONES (CARBONYL COMPOUNDS) 411

compounds by forming new C—C bonds. Consequently, these reactions are important in synthetic problems (discussed in detail in Unit 21).

4. The Clemmensen and Wolff-Kishner reductions reduce $-\overset{\overset{O}{\|}}{C}-$ directly to $-CH_2-$ (oxidation stage II to 0). But no reaction (at least no *selective* reaction) exists for moving upward *from* the alkane level (oxidation stage 0). This illustrates a general principle: alkanes make awful starting materials or intermediates in the synthesis of other compounds. This is partly due to the inertness of the alkanes—they undergo very few reactions. Partly the difference is that alkanes usually have several different types of C—H bonds that cannot be selectively attacked. Once you get to the 0 oxidation stage, you are stuck.

Having the layout of Chart 17.2 in mind helps you in several ways:

- It makes the reagents required for carrying out the reactions seem natural and easy to recall. Upward movements require oxidizing agents; the common ones are $MnO_4^{\ominus}$, CrO_3, $Cr_2O_7^{2\ominus}$, O_2, O_3, H_2O_2, X_2, HOX, $IO_4^{\ominus}$, HNO_3, $Ag^{\oplus}$, $Fe^{3\oplus}$. Downward movement requires reducing agents, such as H_2, $BH_4^{\ominus}$, $AlH_4^{\ominus}$, metals, and N_2H_4.
- The reactions of the $\diagdown C=O$ compounds that do not show up in Chart 17.2 are neither oxidations nor reductions. Most of the nucleophilic additions fall in this category; the two exceptions are reductions: the addition of $H{:}^{\ominus}$ and $R{:}^{\ominus}$.
- The chart helps you recognize the functional group conversions that *can't* be done (indicated by absence of arrows) as well as those that are feasible. If you can recall the pattern of connections in the chart you will be unlikely to try to use nonexistent reactions in solving synthetic problems, a common student error.

Exercises

1. Without using your list of reactions, textbook, and so on, fill in the reagent required for each arrow in Chart 17.2. Check your answers and repeat.

2. Practice constructing Chart 17.2 from memory on blank paper, including arrows and reagents. Check and repeat.

3. If your course treats synthesis of carbonyl compounds along with their reactions, compile another version of Chart 17.2 in which only the reactions *forming* R—CHO and R—CO—R' are entered. Drill on this chart using exercises analogous to 1 and 2.

The following exercises test the effectiveness of the drills conducted in exercises 1 and 2 and those of frame 7.

★ 4. Supply the reagent(s) that will bring about each of the following transformations.

(a)

(continues on next page)

(a) continued

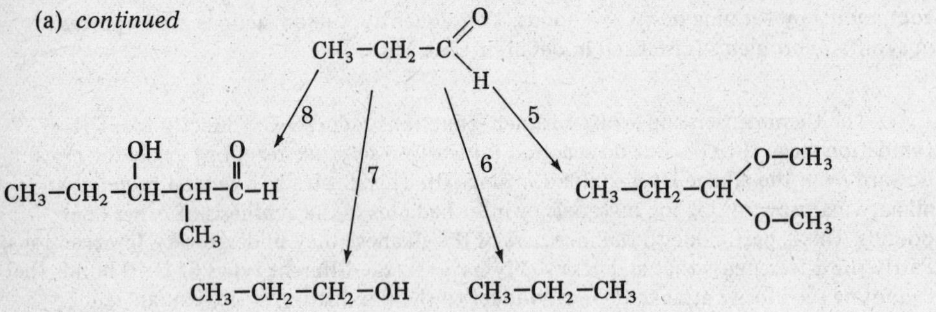

CH₃—CH₂—C(=O)H

8 → CH₃—CH₂—CH(OH)—CH(CH₃)—C(=O)—H

7 → CH₃—CH₂—CH₂—OH

6 → CH₃—CH₂—CH₃

5 → CH₃—CH₂—CH(O—CH₃)(O—CH₃)

(b)

(Ph)C(CH₃)(—S—CH₂—CH₂—S—) (dithiane)

Ph—C(=O)—OH

CH₃—C=CH—C(=O)—Ph

Ph—C(=O)—CH₂—CH₂—Ph

Ph—C(OH)(CH₃)(CH₃) [i.e., Ph—C(OH)(CH₃)₂]

Ph—C(OH)(CH₃)—SO₃⁻ Na⁺

Ph—C(=O)—CH₃ (center)

Ph(4-O₂N)—C(=O)—CH₃

Ph(4-O₂N)—CH(OH)—CH₃

Ph—C(OH)(CH₃)—C≡N

Arrows labeled 1, 2, 3, 4, 5, 6, 7, 8, 9 radiating from central acetophenone.

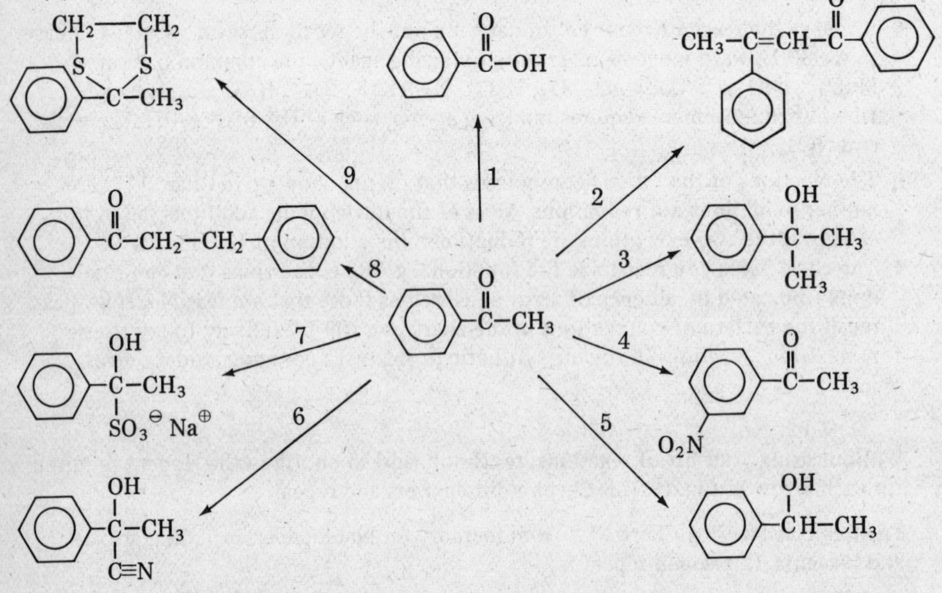

★ 5. Draw the structural formula of the product of each reaction, or write "No reaction."

(a) 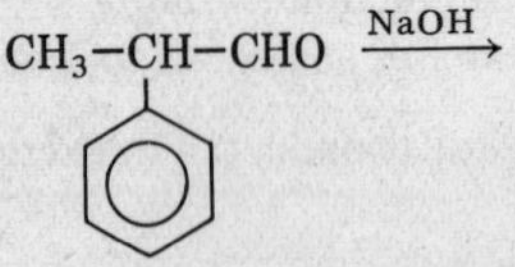 CH₃—CH(Ph)—CHO $\xrightarrow{\text{NaOH}}$

(b) 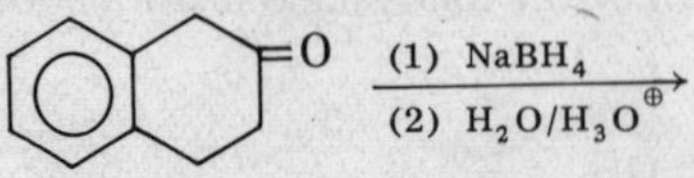 β-tetralone (bicyclic, =O) $\xrightarrow{\text{(1) NaBH}_4 \text{ (2) H}_2\text{O/H}_3\text{O}^+}$

(c) 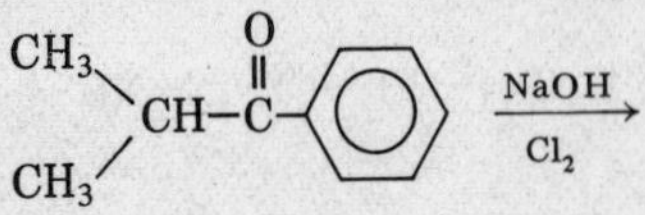 (CH₃)₂CH—C(=O)—Ph $\xrightarrow[\text{Cl}_2]{\text{NaOH}}$

(d) 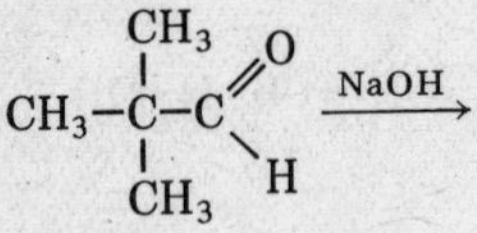 CH₃—C(CH₃)₂—C(=O)H $\xrightarrow{\text{NaOH}}$

(e) 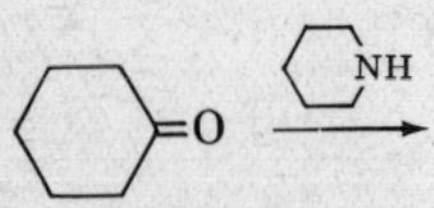 cyclohexanone + piperidine (NH) →

(f) $CH_3-\overset{O}{\underset{\|}{C}}-\!\!\bigcirc\!\!\!-\overset{O}{\underset{\|}{C}}-CH_3 \xrightarrow[\text{aq HCl}/\Delta]{\text{Zn·Hg}}$

(g) $CH_3-\overset{O}{\underset{\|}{C}}-CH\underset{CH_3}{\overset{CH_3}{\diagup\!\!\!\diagdown}} \xrightarrow[(2)\ H_2O/H_3O^+]{(1)\ CH_3Li}$

(h) $\bigcirc\!\!=\!O + HO-CH_2-CH_2-OH \xrightarrow{HA}$

(i) $\bigcirc\!\!=\!O \xrightarrow[Pt]{H_2\ (excess)}$

(j) $CH_3-\overset{N-NH_2}{\underset{\|}{C}}-CH_2-CH_3 \xrightarrow{NaOH/\Delta}$

(k) $\bigcirc\!\!-\overset{O}{\underset{\|}{C}}-CH_3 \xrightarrow{Ag^\oplus}$

(l) $\square\!\!-C\underset{H}{\overset{O}{\diagup\!\!\!\diagdown}} \xrightarrow{NH_2-OH}$

(m) $\bigcirc\!\!-\overset{O}{\underset{\|}{C}}-CH_3 \xrightarrow{NaHSO_3}$

(n) $CH_3-\overset{CHO}{\underset{|}{CH}}-CH_3 \xrightarrow{Cu^{2+}}$

(o) $\square\!\!=\!O \xrightarrow{NaOH}$

(p) $H-\overset{O}{\underset{\|}{C}}-\!\!\bigcirc\!\!\!-\overset{O}{\underset{\|}{C}}-H \xrightarrow[Na^\oplus\ CN^\ominus]{HCN\ (excess)}$

4. (a) 1, $KMnO_4$, $Cr_2O_7^{2\ominus}$, etc.; 2, $NH_2-NH-C_6H_5$; 3, NH_2OH; 4, C_6H_5MgBr, then $H_2O/H_3O^\oplus$; 5, $CH_3OH + HCl$ (dry); 6, $N_2H_4 + OH^\ominus/\Delta$; 7, $NaBH_4$, then $H_2O/H_3O^\oplus$ or H_2/Ni; 8, $OH^\ominus$; 9, $HN\frown O$; 10, $OH^\ominus + I_2$; 11, $C_6H_5NH_2$.

(b) 1, hot $KMnO_4$; 2, $OH^\ominus$; 3, CH_3MgBr, then $H_2O/H_3O^\oplus$; 4, $HNO_3 + H_2SO_4$; 5, H_2/Ni or $LiAlH_4$, then $H_2O/H_3O^\oplus$, etc.; 6, $HCN + NaCN$; 7, $NaHSO_3$; 8, $\bigcirc\!\!NH$, then $C_6H_5CH_2Br$, then $H_2O/H_3O^\oplus$; 9, $HS-CH_2-CH_2-SH$.

5. (a) $\bigcirc\!\!-\overset{CH_3}{\underset{\underset{CHO}{|}}{\overset{|}{C}}}-\overset{OH}{\underset{|}{CH}}-\overset{CH_3}{\underset{|}{CH}}-\!\!\bigcirc$

(b) [tetrahydronaphthalen-2-ol: naphthalene with one ring saturated, bearing an OH]

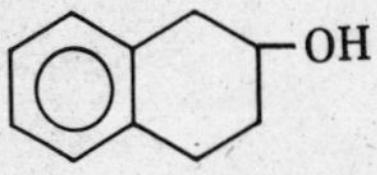

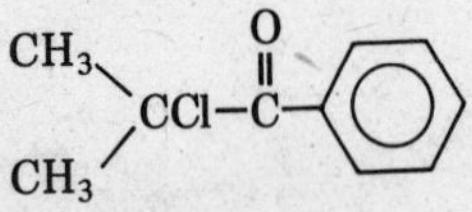

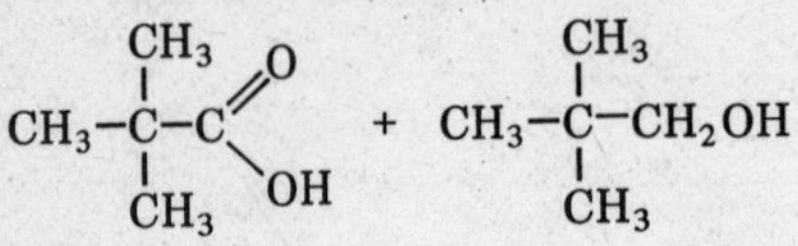

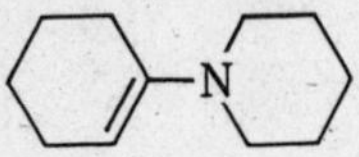

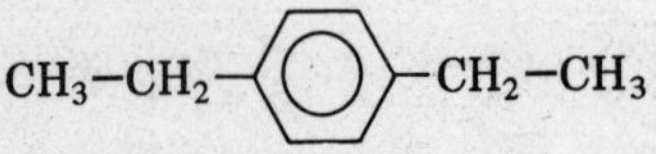

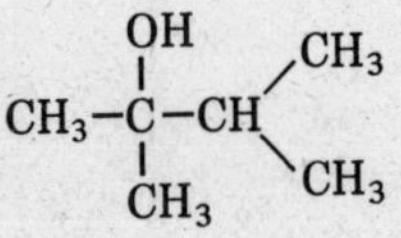

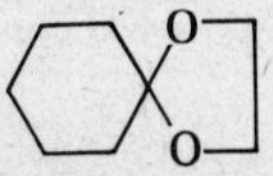

(c) (CH₃)₂CCl–C(=O)–C₆H₅

(d) CH₃–C(CH₃)₂–C(=O)OH + CH₃–C(CH₃)₂–CH₂OH

(e) [cyclohexenyl-piperidine enamine]

(f) CH₃–CH₂–C₆H₄–CH₂–CH₃ (para)

(g) CH₃–C(OH)(CH₃)–CH(CH₃)₂

(h) [cyclohexanone 1,2-ethanediol ketal (spiro dioxolane)]

(i) [cyclohexyl–OH]

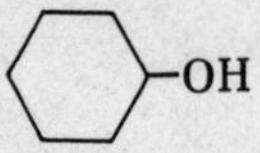

(j) CH₃–CH₂–CH₂–CH₃

(k) no reaction

(l) [cyclobutyl]–CH=N–OH

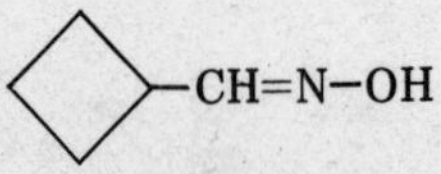

(m) [C₆H₅]–C(OH)(CH₃)–SO₃⁻ Na⁺

(n) (CH₃)₂CH–CO₂H

(o) [1-(2-oxocyclopentyl)cyclopentan-1-ol]

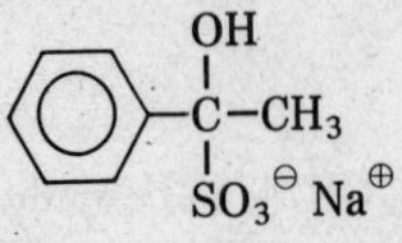

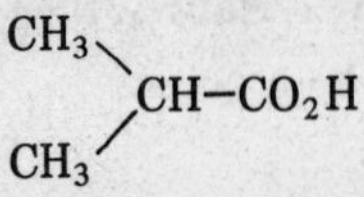

(p) N≡C–CH(OH)–[C₆H₄]–CH(OH)–C≡N

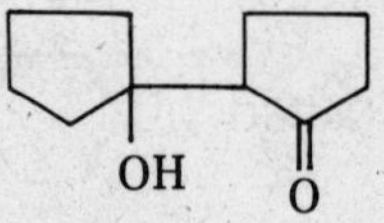

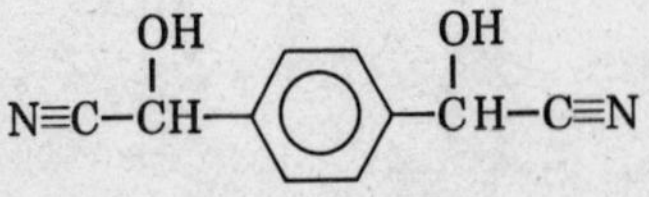

ALDEHYDES AND KETONES (CARBONYL COMPOUNDS) 415

9. The nucleophilic addition reactions of the carbonyl compounds can be learned by the slot-and-filler method. They follow one of two patterns, depending on whether elimination follows addition.

- The simple pattern looks like this:

$$\begin{array}{c} R \\ R' \end{array}\!\!C=O + Nu: \longrightarrow \begin{array}{c} R \\ R' \end{array}\!\!C\!\!\begin{array}{c} OH \\ Nu \end{array}$$

The H for the product's —OH group ordinarily comes from the solvent, the acid catalyst, or an aqueous acid workup, for example

reaction proper:

$$\underset{\underset{H-AlH_3\ Li^{\oplus}}{}}{\overset{:\ddot{O}:}{\underset{\|}{R-C-R'}}} \longrightarrow \underset{H}{\overset{Li^{\oplus}\ :\ddot{O}:^{\ominus}}{\underset{|}{R-C-R'}}}\ AlH_3 \longrightarrow \underset{H}{\overset{\overset{\ominus}{O}-AlH_3\ Li^{\oplus}}{\underset{|}{R-C-R'}}} \xrightarrow{\text{repeat}}$$

$$\left(\begin{array}{c} R \\ R' \end{array}\!\!CH-O-\right)_4 Al^{\ominus}\ Li^{\oplus}$$

workup:

$$\left(\begin{array}{c} R \\ R' \end{array}\!\!CH-O-\right)_4 Al^{\ominus}\ Li^{\oplus} + 4\ HA \xrightarrow{H_2O} \begin{array}{c} R \\ R' \end{array}\!\!CH-OH + Al^{3\oplus} + Li^{\oplus} + 4\ A^{\ominus}$$

or

$$\underset{\underset{}{\overset{\|}{R-C-R'}}}{\overset{:\ddot{O}:}{}} \xrightarrow{H-A} \underset{R-C-R'}{\overset{\overset{\oplus}{\ddot{O}}-H}{\underset{\|}{}}}\ A^{\ominus} \longleftrightarrow \underset{R-C-R'}{\overset{:\ddot{O}-H}{\underset{\overset{\oplus}{:\ddot{O}}\ \ \ \ CH_3}{|}}}\!\!H \longrightarrow \underset{R-C-R'}{\overset{:\ddot{O}-H}{\underset{\underset{CH_3}{\overset{\oplus}{\underset{|}{O}}}\ H}{|}}}\!\!A^{\ominus}$$

$$\longrightarrow \underset{\underset{:\ddot{O}-CH_3}{\overset{:\ddot{O}-H}{\underset{|}{R-C-R'}}}}{} + HA$$

- The addition/elimination most often takes the form

$$G-NH_2 + \begin{array}{c} R \\ R' \end{array}\!\!C=O \longrightarrow \begin{array}{c} R \\ R' \end{array}\!\!C\!\!\begin{array}{c} OH \\ NH-G \end{array} \longrightarrow \begin{array}{c} R \\ R' \end{array}\!\!C=N-G + H_2O$$

The permissible slot fillers for the two patterns are listed in Tables 17.1 and 17.2. The easiest way to keep the two sets straight is to note that elimination can follow addition only if the nucleophilic atom is trivalent, namely nitrogen $\left(\begin{array}{c} R \\ R' \end{array}\!\!C\!=\!N\!-\!G\right)$, and if the nitrogen is in the form —NH$_2$ in the reagent. The other nucleophile in Table 17.2 is unique, structurally and mechanistically:

$$\underset{\underset{}{\overset{\|}{R-C-R'}}}{\overset{:\ddot{O}:}{}} \longrightarrow \underset{\underset{R'\ \ R'''}{\overset{\overset{\oplus}{P(C_6H_5)_3}}{\underset{|}{R-C-C-R''}}}}{\overset{{}^{\ominus}:\ddot{O}:}{}} \longrightarrow \underset{\underset{R'\ \ R'''}{\overset{:\ddot{O}-P(C_6H_5)_3}{\underset{|}{R-C-C-R''}}}}{} \longrightarrow \underset{R}{\overset{O-P(C_6H_5)_3}{}}\ \begin{array}{c} R'' \\ C=C \\ R''' \end{array}$$

an ylide $\longrightarrow R''\!\!\overset{\ominus}{\underset{\underset{R'''}{|}}{C}}\!\!-\overset{\oplus}{P}(C_6H_5)_3$

416 HOW TO SUCCEED IN ORGANIC CHEMISTRY

Table 17.1 *Permissible Slot Fillers for the Simple Nucleophilic Addition Reaction*

$$\underset{R'}{\overset{R}{>}}C=O + Nu: \longrightarrow \underset{R'}{\overset{R}{>}}C\underset{Nu}{\overset{OH}{<}}$$

Permissible −Nu	Nu: delivered by:	Permissible R−, R′−
−H	$H{-}BH_3$, $H{-}AlH_3$	Any
−O−R	$:\ddot{O}\underset{H}{\overset{R}{<}}$	Product stable only if −OH, >C=O in same molecule
$-SO_3^{\ominus} Na^{\oplus}$	$:S(=O)(O^{\ominus}Na^{\oplus})OH$	R or R′ must = H or CH_3
−C≡N:	$:C{\equiv}N:^{\ominus}$	Any
−R	$R{-}Li$, $R{-}Mg{-}X$	Any

Table 17.2 *Permissible Slot Fillers for Nucleophilic Addition/Elimination*

$$\underset{R'}{\overset{R}{>}}C=O + NuH_2 \longrightarrow \underset{R'}{\overset{R}{>}}C=Nu + H_2O$$

Permissible =Nu	Reagent	Permissible R−, R′−
=N−Ar	$H_2N{-}Ar$	Any
=N−OH	$H_2N{-}OH$	Any
=N−NHAr	$H_2N{-}NH{-}Ar$	Any
$=C\underset{R'''}{\overset{R''}{<}}$	$\underset{R'''}{\overset{R''}{>}}C^{\ominus}{-}P^{\oplus}(C_6H_5)(C_6H_5)(C_6H_5)$	Best when R″ or R‴ = H

Once you have these nucleophiles from Table 17.2 in mind, you can assign all others to the $\underset{R'}{\overset{R}{>}}C=O \longrightarrow \underset{R'}{\overset{R}{>}}C\underset{Nu}{\overset{OH}{<}}$ class.

★ Exercises

1. Write the equations for the nucleophilic addition reaction of C₆H₅−C(=O)−CH₃ with each nucleophilic reagent.

(a) NH_2OH

(b) 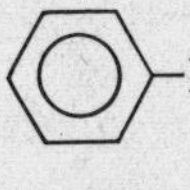 C_6H_5—Li, then H_2O + HCl

(c) 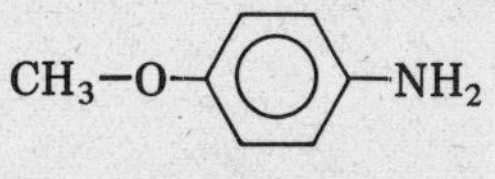 CH_3—O—C_6H_4—NH_2

(d) 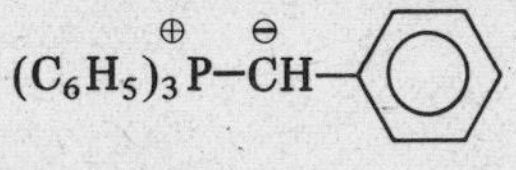 $(C_6H_5)_3\overset{\oplus}{P}$—$\overset{\ominus}{CH}$—$C_6H_5$

2. What nucleophile and carbonyl compound combination would produce each product?

(a) 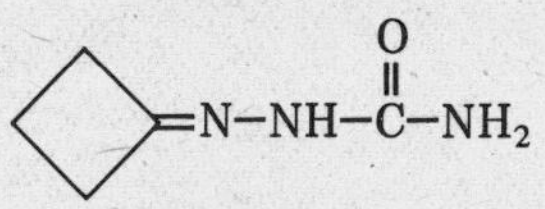 cyclobutylidene=N—NH—C(=O)—NH_2

(b) $(C_6H_5)_2C(OH)(CN)$

(c) CH_3—CH_2—CH(OH)—C_6H_5

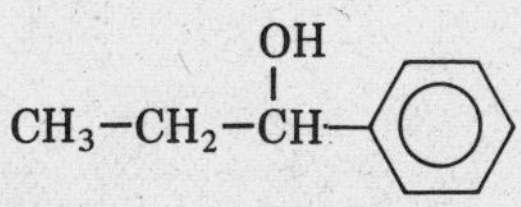

(d) 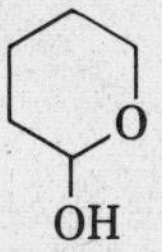 tetrahydropyran-2-ol

- - - - - - - - - -

1. (a) 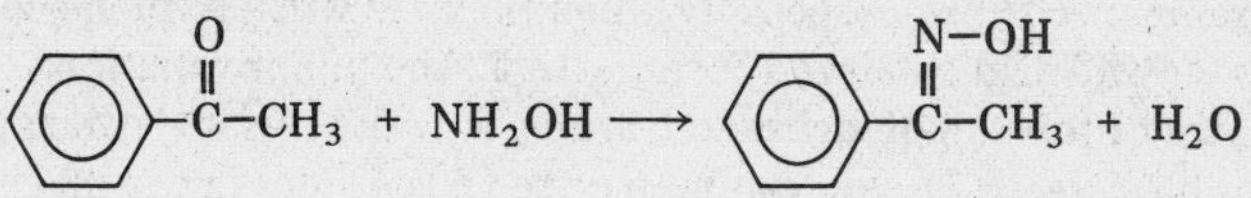 C_6H_5—C(=O)—CH_3 + NH_2OH ⟶ C_6H_5—C(=N—OH)—CH_3 + H_2O

(b) C_6H_5—C(=O)—CH_3 + C_6H_5—Li ⟶ C_6H_5—C($O^{\ominus} Li^{\oplus}$)(C_6H_5)—CH_3

C_6H_5—C($O^{\ominus} Li^{\oplus}$)(C_6H_5)—CH_3 + $H_3O^{\oplus} Cl^{-}$ ⟶ C_6H_5—C(OH)(C_6H_5)—CH_3 + $Li^{\oplus} Cl^{\ominus}$ + H_2O

(c) CH_3—O—C_6H_4—NH_2 + C_6H_5—C(=O)—CH_3 ⟶ CH_3—O—C_6H_4—N=C(C_6H_5)(CH_3) + H_2O

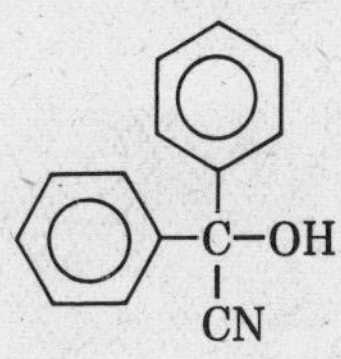

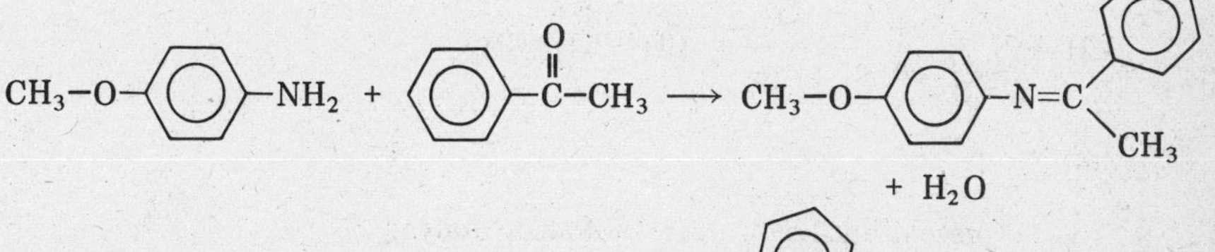

(d) 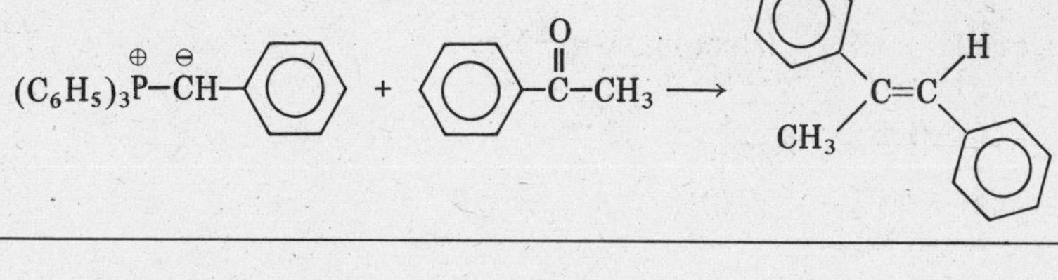 $(C_6H_5)_3\overset{\oplus}{P}$—$\overset{\ominus}{CH}$—$C_6H_5$ + C_6H_5—C(=O)—CH_3 ⟶ (CH_3)(C_6H_5)C=C(H)(C_6H_5) and/or

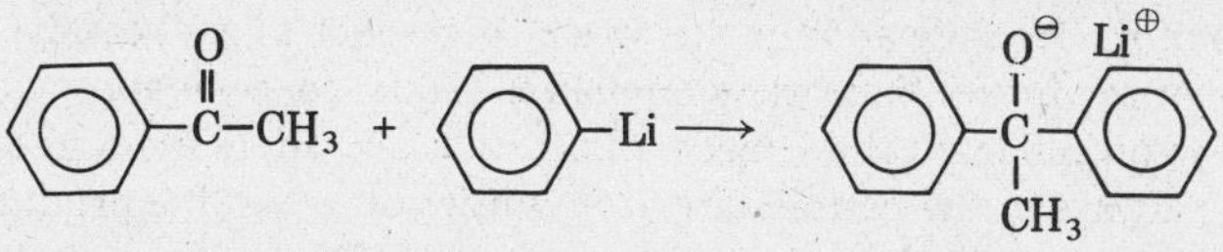

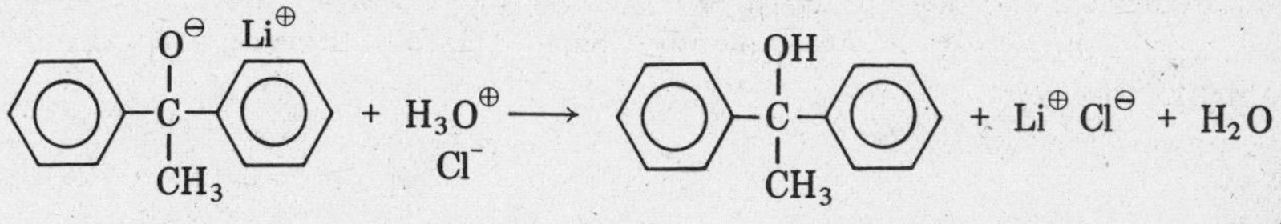

418 HOW TO SUCCEED IN ORGANIC CHEMISTRY

$$\text{Ph}\diagup \underset{\text{CH}_3}{\text{C}}=\underset{\text{H}}{\overset{\text{Ph}}{\text{C}}}$$

2. (a) cyclobutanone =O + NH$_2$–NH–C(=O)–NH$_2$

(b) Ph–C(=O)–Ph + H–C≡N

(c) CH$_3$CH$_2$–C(=O)–Ph + LiAlH$_4$, then H$_2$O; or CH$_3$CH$_2$–C(=O)–H + Ph–MgBr, then H$_2$O; or Ph–C(=O)H + CH$_3$CH$_2$–Mg–Br, then H$_2$O

(d) H–C(=O)–CH$_2$–CH$_2$–CH$_2$–CH$_2$–OH $\xrightarrow{\text{HCl, etc.}}$

10. The same basic mechanism—nucleophilic addition—applies to reactions ⑤ through ⑫ and ⑯ from Chart 17.1, and to many minor variations on them and to reaction of many less common nucleophiles. Nevertheless, the beginning student has a hard time seeing all these as the same mechanism, especially when they are written for specific cases, in both uncatalyzed and acid-catalyzed form, with all the minor variations this entails. You have a choice of three approaches to learning to write this mechanism for any of these many possible cases: (1) you can memorize all the steps in the mechanism for every single choice of nucleophile and for the presence/absence of acid catalyst; or (2) you can learn a series of rules that, when applied in sequence, generate the mechanism for any case; or (3) you can learn three mechanistic steps and apply them in open-ended fashion, heading always in the desired direction, until you get to the product. Only approaches 2 and 3 are practical; we practice them in this frame.

The steps in the "method" are shown in Chart 17.3; directions on each step are given at the right.

Example

$$\text{CH}_3-\text{C}(=O)\text{H} \xrightarrow[\text{(2) H}_2\text{O/H}_3\text{O}^\oplus]{\text{(1) CH}_3\text{MgCl}} \text{CH}_3-\underset{\text{OH}}{\text{CH}}-\text{CH}_3$$

Solution:

- No acid present until workup; ∴ not acid-catalyzed
- Nucleophile = CH$_3$:$^\ominus$, latent in CH$_3$–Mg–X
- So CH$_3$–C(=O)–H $\longrightarrow$ CH$_3$–C(–O$^\ominus$)(–CH$_3$)–H MgX$^\oplus$

  CH$_3$–MgX

ALDEHYDES AND KETONES (CARBONYL COMPOUNDS)

- No double bond in product, so dehydration not required
- Protonation of $-\ddot{O}:^{\ominus}$ must come in workup:

$$\begin{array}{c} {}^{\oplus}:\ddot{O}: \quad H-\overset{\oplus}{O}H_2 \\ | \\ CH_3-C-H \\ | \\ CH_3 \end{array} \longrightarrow \begin{array}{c} OH \\ | \\ CH_3-C-H + H_2O \\ | \\ H \end{array}$$

product

Example

$$\bigcirc\!\!=\!O + NH_2OH \longrightarrow \bigcirc\!\!=\!N-OH + H_2O$$

Solution:

- No acid present
- Nucleophile = $:NH_2-OH$

- So $\bigcirc\!\!=\!O \longrightarrow \bigcirc\!\!\begin{array}{c} O^{\ominus} \\ \overset{\oplus}{NH_2}-OH \end{array}$

  $\overset{\curvearrowright}{C}:NH_2-OH$

- Product is $\rangle C=Nu$, so dehydration is required

- Protonate $-O^{\ominus}$: $\bigcirc\!\!\begin{array}{c} \ddot{O}:^{\ominus} \, H-A \\ NH_2-OH \\ \oplus \end{array} \longrightarrow \bigcirc\!\!\begin{array}{c} OH \\ N \\ H \quad H \end{array} \overset{OH}{\underset{:A^{\ominus}}{}} \longrightarrow \bigcirc\!\!\begin{array}{c} OH \\ NH-OH \end{array}$

  $\bigcirc\!\!\begin{array}{c} :\ddot{O}H \, H-A \\ NHOH \end{array} \longrightarrow \bigcirc\!\!\begin{array}{c} \overset{\oplus}{O}H_2 A^{\ominus} \\ NHOH \end{array}$

- H_2O leaves: $\bigcirc\!\!\begin{array}{c} \overset{\oplus}{O}H_2 \\ NHOH \end{array} \longrightarrow \bigcirc\!\!\overset{\oplus}{-}NH-OH + H_2O$

- Deprotonate Nu: $\bigcirc\!\!\overset{\oplus}{=}N-OH \longrightarrow \bigcirc\!\!=\!N-OH + HA$
  $\quad\quad\quad\quad\quad\quad H \underset{:A^{\ominus}}{\curvearrowleft}$

Example

$$\begin{array}{c} O \\ \| \\ CH_3-C-CH_3 \end{array} + H_2N-NH_2 \xrightarrow{H_2SO_4} \begin{array}{c} CH_3 \\ \diagdown \\ \diagup \\ CH_3 \end{array} C=N-NH_2 + H_2O$$

Solution:

- Strong acid present, $\therefore$ acid-catalyzed

- So $\begin{array}{c} :\ddot{O}: \\ \| \\ CH_3-C-CH_3 \end{array} \overset{\curvearrowright}{H-O-SO_3H} \longrightarrow \begin{array}{c} \overset{\oplus}{:O}-H \\ \| \\ CH_3-C-CH_3 \end{array} HSO_4^{\ominus}$

- Nu: = $:NH_2-NH_2$

420 HOW TO SUCCEED IN ORGANIC CHEMISTRY

Chart 17.3

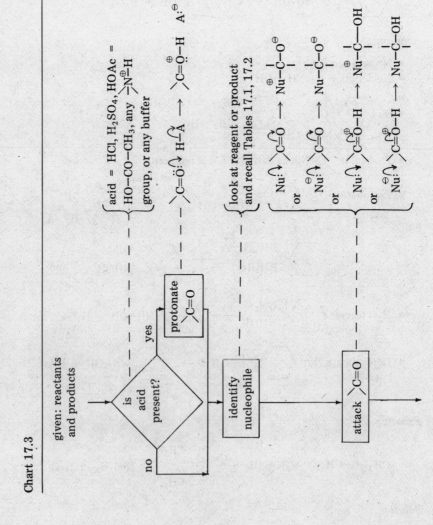

ALDEHYDES AND KETONES (CARBONYL COMPOUNDS)

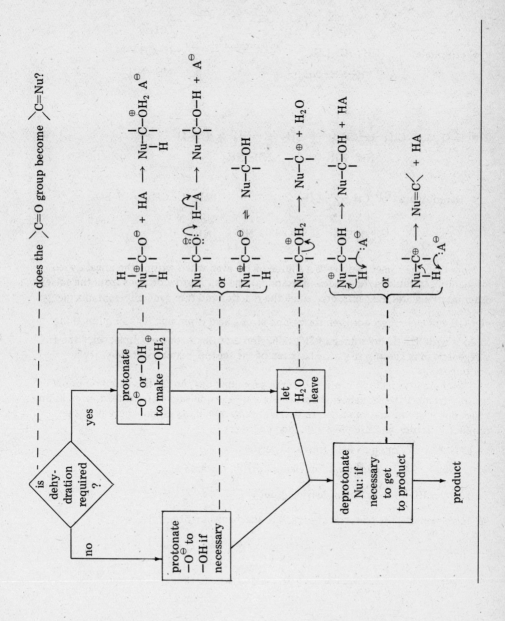

422   HOW TO SUCCEED IN ORGANIC CHEMISTRY

- Attack $>$C=O: $CH_3-\overset{\overset{\oplus}{O-H}}{\underset{\underset{:NH_2-NH_2}{\curvearrowleft}}{\overset{\|}{C}}}-CH_3 \longrightarrow CH_3-\underset{\overset{|}{\oplus NH_2-NH_2}}{\overset{\overset{O-H}{|}}{C}}-CH_3$

- Product has $>$C=Nu, so dehydration is required

- Protonate: $CH_3-\underset{\underset{\oplus}{H-\overset{\curvearrowright}{N}H-NH_2}}{\overset{\overset{:\ddot{O}H^{\curvearrowleft}\,H-A}{|}}{C}}-CH_3 \xrightarrow[\text{steps}]{\text{two successive}} CH_3-\underset{\underset{}{NH-NH_2}}{\overset{\overset{\oplus}{OH_2}}{\underset{|}{C}}}-CH_3$

  $^\ominus A:\curvearrowleft$

- H_2O goes: $CH_3-\underset{\underset{}{NH-NH_2}}{\overset{\overset{\oplus OH_2}{\curvearrowleft |}}{C}}-CH_3 \longrightarrow CH_3-\underset{\underset{}{NH-NH_2}}{\overset{\overset{\oplus}{|}}{C}}-CH_3 + H_2O$

- Deprotonate Nu: $CH_3-\underset{\underset{NH_2}{\overset{|}{N}}\curvearrowleft}{\overset{\overset{\oplus}{|}}{C}}-CH_3 \longrightarrow CH_3-\underset{\underset{NH_2}{\overset{}{N}}}{\overset{\overset{}{\|}}{C}}-CH_3 + H_2SO_4$

  $H \curvearrowright \overset{\ominus}{\underset{}{\ddot{O}}}-SO_3H$

Note that one sometimes needs a source of $H^\oplus$ even when writing the uncatalyzed mechanism. In this case (e.g., second example) one can borrow protons from the solvent, other intermediates, reagents, etc., since the reaction mixture generally contains plenty of —OH and $\overset{\oplus}{\underset{/}{\diagdown}}$N—H groups even though no strong acid is present. The simplest thing to do is write HA as the source of $H^\oplus$, as in step 2 of the second example; since the HA is regenerated in the end (e.g., the last step of the second example), its identity is not important.

The alternative method is much like the preceding one, but in this case you don't try to remember the sequence of operations after the addition step. Instead, keep asking what remains to be accomplished to shape the intermediate you now have into the product structure, and use only the individual steps

1. B:$\curvearrowright$ H$\overset{\curvearrowright}{-}$A   or equivalent (proton transfer)
2. Nu:$\curvearrowright$ $>$C=O   (addition of nucleophile)
3. $-\underset{|\curvearrowleft}{\overset{|}{C}}-\overset{\oplus}{O}H_2$ (departure of leaving group)

We might call this the freestyle approach to mechanism writing.

Example

$CH_3-C\overset{\overset{O}{\diagup\hspace{-0.3em}\diagdown}}{\underset{H}{}}\quad\xrightarrow[(2)\ H_2O/H_3O^\oplus]{(1)\ CH_3MgCl}\quad CH_3-\underset{\underset{OH}{|}}{\overset{}{CH}}-CH_3$

Solution:

- No acid present until workup, so add $CH_3:^\ominus$ in uncatalyzed manner:

ALDEHYDES AND KETONES (CARBONYL COMPOUNDS)

$$CH_3-\underset{\underset{CH_3-Mg-Cl}{}}{\overset{\overset{\ddot{O}:}{\|}}{C}}-H \longrightarrow CH_3-\underset{\underset{CH_3}{|}}{\overset{\overset{:\ddot{O}:^{\ominus}}{|}}{C}}-H \quad \overset{\oplus}{M}gCl$$

- Product of first step differs from final product only by $H^{\oplus}$; ∴ protonate it using the acid available in the workup stage:

$$CH_3-\underset{\underset{CH_3}{|}}{\overset{\overset{:\ddot{O}:^{\ominus}}{|}}{C}}H \quad H-\overset{\oplus}{O}H_2 \longrightarrow CH_3-\underset{\underset{CH_3}{|}}{\overset{\overset{O-H}{|}}{C}}H + H_2O$$

Example

$$\bigcirc{=}O + NH_2{-}OH \longrightarrow \bigcirc{=}N{-}OH + H_2O$$

Solution:

- No acid, add Nu: uncatalyzed:

$$\bigcirc{=}O \quad \overset{\curvearrowleft}{:}NH_2{-}OH \longrightarrow \bigcirc\overset{O^{\ominus}}{\underset{\overset{\oplus}{N}H_2{-}OH}{}}$$

- Compare:

$$\bigcirc\overset{O^{\ominus}}{\underset{\overset{\oplus}{N}(H_2){-}OH}{}} \quad \text{versus} \quad \bigcirc{=}N{-}OH. \text{ Need to remove } H_2O.$$

Check steps 1 to 3. Need step 3, but must convert $-O^{\ominus}$ to $-OH_2^{\oplus}$ first, and remove excess $H^{\oplus}$ from N. To do this use as many proton-transfer steps as necessary:

$$\bigcirc\overset{O^{\ominus}}{\underset{\underset{\underset{:\ddot{O}H_2}{H}}{\overset{\oplus}{N}H{-}OH}}{}} \longrightarrow \bigcirc\overset{O^{\ominus}}{\underset{NH{-}OH}{}} + H_3O^{\oplus}$$

$$\bigcirc\overset{\ddot{O}:^{\ominus}}{\underset{NH{-}OH}{}} \quad H{-}\overset{\oplus}{O}H_2 \longrightarrow \bigcirc\overset{OH}{\underset{NH{-}OH}{}} + OH_2$$

$$\bigcirc\overset{:\ddot{O}H}{\underset{NH{-}OH}{}} \quad H{-}\overset{\oplus}{O}H_2 \longrightarrow \bigcirc\overset{\overset{\oplus}{O}H_2}{\underset{NH{-}OH}{}} + OH_2$$

- Now use step 3.

$$\bigcirc\overset{\overset{\oplus}{O}H_2}{\underset{NH{-}OH}{}} \longrightarrow \bigcirc{-}\overset{\oplus}{N}H{-}OH + H_2O$$

- Compare: last species differs from final product by $H^{\oplus}$, so use step 1 to remove $H^{\oplus}$:

424 HOW TO SUCCEED IN ORGANIC CHEMISTRY

$$\text{Ph-}\overset{\oplus}{\underset{H}{N}}\text{-OH} \longrightarrow \text{Ph-}\overset{\oplus}{\underset{..}{N}}\text{-OH} \longleftrightarrow \text{Ph-N=OH}$$
$$\overset{}{\underset{:OH_2}{}} \qquad H_3O^{\oplus}$$

Example

$$CH_3-\overset{O}{\underset{\|}{C}}-CH_3 + NH_2-NH_2 \xrightarrow{H_2SO_4} CH_3-\overset{N-NH_2}{\underset{\|}{C}}-CH_3 + H_2O$$

Solution:

- Acid is present; ∴ use step 1 to protonate $>$C=O before using step 2:

$$CH_3-\overset{\ddot{O}:}{\underset{\|}{C}}-CH_3 \quad H-\overset{}{O}-SO_3H \longrightarrow CH_3-\overset{\overset{\oplus}{:O}-H}{\underset{\|}{C}}-CH_3 \quad HSO_4^{\ominus}$$

$$CH_3-\overset{\oplus O-H}{\underset{\|}{C}}-CH_3 \longrightarrow CH_3-\overset{O-H}{\underset{|}{C}}-CH_3 \quad HSO_4^{\ominus}$$
$$\overset{}{\underset{:NH_2-NH_2}{}} \qquad \overset{}{\underset{\oplus NH_2-NH_2}{}}$$

- Compare: last species differs from product by $H_3O^{\oplus}$, so need to remove one $H^{\oplus}$ + one H_2O. To get −OH in shape to apply step 3, protonate it first using step 1:

$$CH_3-\overset{H-O}{\underset{|}{C}}-CH_3 \longrightarrow CH_3-\overset{O-H}{\underset{|}{C}}-CH_3 + HO-SO_2-OH$$
$$\overset{}{\underset{H_2N-NH-H}{\overset{\oplus}{}}}\overset{\ominus}{:O}-SO_2-OH \qquad \overset{}{\underset{NH-NH_2}{}}$$

$$CH_3-\overset{H-\ddot{O}:}{\underset{|}{C}}-CH_3 \quad H-\overset{}{O}-SO_2-OH \longrightarrow CH_3-\overset{\oplus OH_2}{\underset{|}{C}}-CH_3 + HSO_4^{\ominus}$$
$$\overset{}{\underset{NH-NH_2}{}} \qquad \overset{}{\underset{NH-NH_2}{}}$$

- Now let H_2O go:

$$CH_3-\overset{\overset{\oplus}{\underset{|}{OH_2}} \; HSO_4^{\ominus}}{\underset{|}{C}}-CH_3 \longrightarrow CH_3-\overset{\oplus}{\underset{|}{C}}-CH_3 + H_2O$$
$$\overset{}{\underset{NH-NH_2}{}} \qquad \overset{}{\underset{NH-NH_2}{}}$$

- Remaining to be done: lose H from $-\overset{H}{\underset{|}{N}}-NH_2$; lose ⊕ from carbonyl carbon; make π bond. So remove $H^{\oplus}$ via step 1:

$$CH_3-\overset{\oplus}{\underset{|}{C}}-CH_3 \longrightarrow CH_3-\overset{\oplus}{\underset{|}{C}}-CH_3 \longleftrightarrow CH_3-\overset{N-NH_2}{\underset{\|}{C}}-CH_3 + H_2SO_4$$
$$\overset{}{\underset{:N-NH_2}{}} \qquad \overset{}{\underset{\ominus:N-NH_2}{}}$$
$$\overset{}{\underset{H}{}} $$
$$\ominus:\overset{}{O}-SO_2-OH$$

Several comments on these solutions are in order:
(a) The very last step written could have been given the form

ALDEHYDES AND KETONES (CARBONYL COMPOUNDS)

$$\underset{\underset{H}{\overset{\curvearrowright}{N}-NH_2}}{\overset{\oplus}{\underset{|}{CH_3-C-CH_3}}} \longrightarrow \underset{:N-NH_2}{\overset{\|}{CH_3-C-CH_3}} + H_2SO_4$$

$$\overset{\curvearrowleft}{\ominus:\ddot{O}-SO_2-OH}$$

This uses the liberated electron pair to make the π bond directly, giving the major contributing structure rather than a minor one. It is perfectly legitimate and is the more common way of writing it.

(b) The third and fourth steps in the third example remove $H^\oplus$ from N and add $H^\oplus$ to O using $H_2SO_4/HSO_4^\ominus$ as an $H^\oplus$ storage device. *In fact, we rarely know how the proton-transfer steps in a mechanism actually occur*, because they are so fast that ordinary methods yield no information on them. Consequently, we often write the proton-transfer steps just as equilibria, without curved arrows. If this is done, then successive proton-transfer steps can be collapsed to a single equilibrium that rearranges the $H^\oplus$ upon the basic sites so as to get them just where you need them. For instance, the third and fourth steps in the third example remove $H^\oplus$ from N and add $H^\oplus$ to O, using H_2SO_4 as the proton donor and $HSO_4^\ominus$ as the proton acceptor. These steps could just as well be written

$$\underset{\oplus NH_2-NH_2}{\overset{H-O}{\underset{|}{CH_3-\underset{|}{C}-CH_3}}} \rightleftharpoons \underset{NH-NH_2}{\overset{\overset{\oplus}{OH_2}}{\underset{|}{CH_3-\underset{|}{C}-CH_3}}}$$

The molecules in solution are in fact redistributing $H^\oplus$ among all basic sites at all times.

(c) Basic steps 1 to 3, above, appear over and over again in many mechanisms. In fact, the nine standard mechanisms of Table 10.1 are all just combinations of such basic steps. The total number of basic steps required to write all organic mechanisms is about 35.

(d) The mechanism of formation of the acetals and ketals, $\diagdown_C\diagup^{OR}_{OR}$, differs from the standard nucleophilic addition pattern because it is really a reaction of the hemiacetal or hemiketal, $\diagdown_C\diagup^{OH}_{OR}$, which is formed first. The $\diagdown_C\diagup^{OH}_{OR} \longrightarrow \diagdown_C\diagup^{OR}_{OR}$ reaction is just an S_N1 etherification with a resonance-stabilized carbonium ion intermediate:

$$\diagdown_C\diagup^{\oplus}{-}\ddot{O}{-}R \longleftrightarrow \diagdown_C\diagup{=}\overset{\oplus}{O}{-}R$$

(e) The Wittig reaction

$$\diagdown_C\diagup{=}O + \underset{|}{-\overset{\ominus}{C}-\overset{\oplus}{P}(C_6H_5)_3} \longrightarrow \diagdown_C\diagup{=}C\diagdown + O{-}P(C_6H_5)_3$$

is a genuine nucleophilic addition, but its mechanism deviates from the standard patterns after the addition step:

$$\underset{-\underset{|}{\overset{\oplus}{C}}\overset{}{-}P^\oplus(C_6H_5)_3}{\overset{\curvearrowright}{\diagdown C\overset{}{=}\ddot{O}:}} \longrightarrow \underset{-\underset{|}{C}-P^\oplus(C_6H_5)_3}{-\underset{|}{C}-\ddot{O}:^\ominus} \longrightarrow \underset{-\underset{|}{C}-P(C_6H_5)_3}{-\underset{|}{C}-O} \longrightarrow \underset{\underset{C}{\|}}{\overset{C}{\|}} + \underset{P(C_6H_5)_3}{\overset{O}{\uparrow}}$$

★ *Exercises*

Write a stepwise mechanism for each reaction.

1. 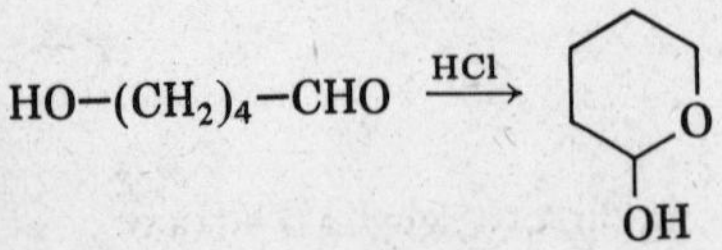 HO–(CH$_2$)$_4$–CHO $\xrightarrow{HCl}$ (tetrahydropyran-2-ol)

2. Ph–C(=O)–Ph + HCN $\xrightarrow{Na^{\oplus} \ CN^{\ominus}}$ Ph–C(OH)(CN)–Ph

3. cyclohexanone $\xrightarrow[\text{(2) H}_2\text{O/H}_3\text{O}^{\oplus}]{\text{(1) Ph–MgBr}}$ 1-phenylcyclohexanol

4. Ph–C(=O)–H + NH$_2$–OH $\xrightarrow{CH_3CO_2H}$ Ph–CH=N–OH

5. 2 CH$_3$–C(=O)–H $\xrightarrow{NaOH}$ CH$_3$–CH(OH)–CH$_2$–C(=O)–H

- - - - - - - - -

1. 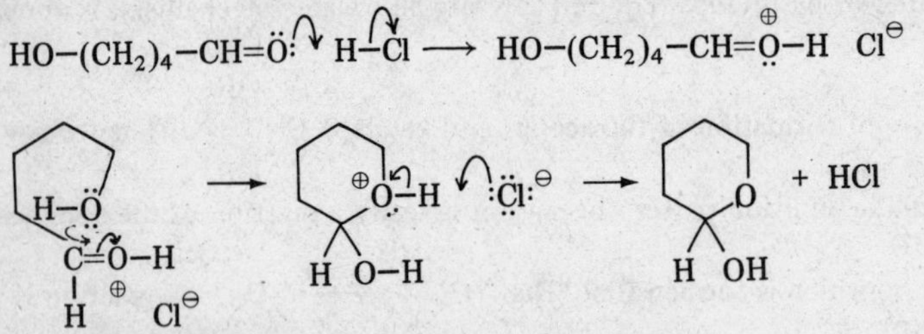 HO–(CH$_2$)$_4$–CH=Ö: ↶ H–Cl ⟶ HO–(CH$_2$)$_4$–CH=Ö–H Cl$^{\ominus}$

   H–Ö⋯
   C–O–H   ⟶   [ring with ⊕O–H, :Cl$^{\ominus}$, H O–H]   ⟶   [tetrahydropyran ring with O, H OH]   + HCl
   H Cl$^{\ominus}$

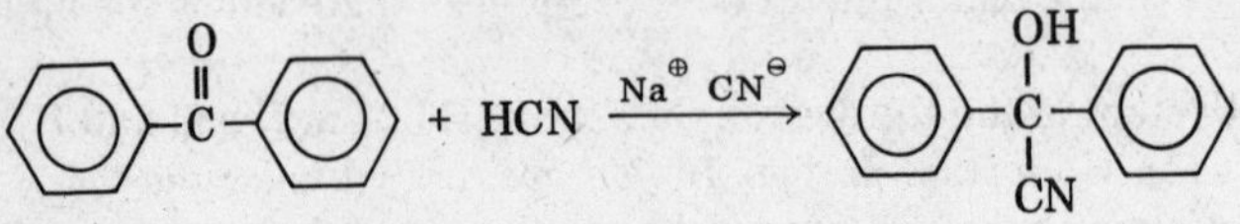

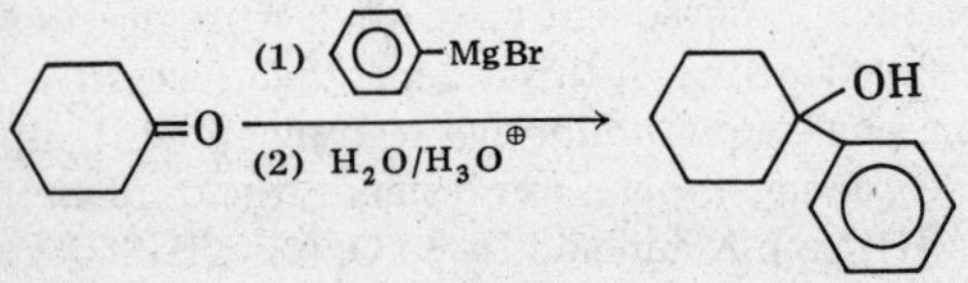

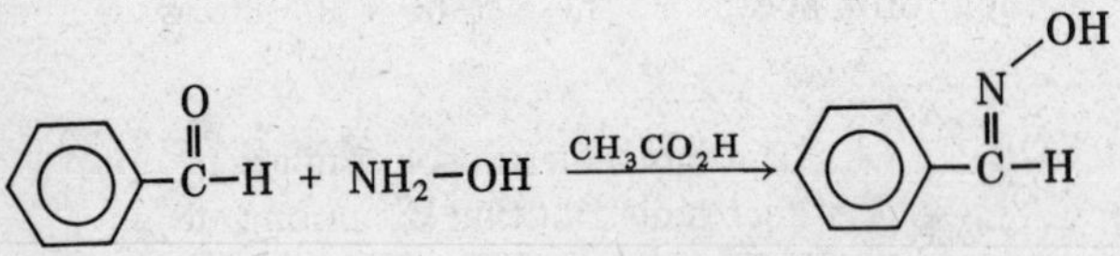

2. Ph–C(=O)–Ph with :C≡N:$^{\ominus}$ ⟶ Ph–C(:Ö:$^{\ominus}$)(C≡N:)–Ph

   Ph–C(:Ö:$^{\ominus}$)(CN)–Ph ↶ H–C≡N ⟶ Ph–C(O–H)(CN)–Ph + CN$^{\ominus}$

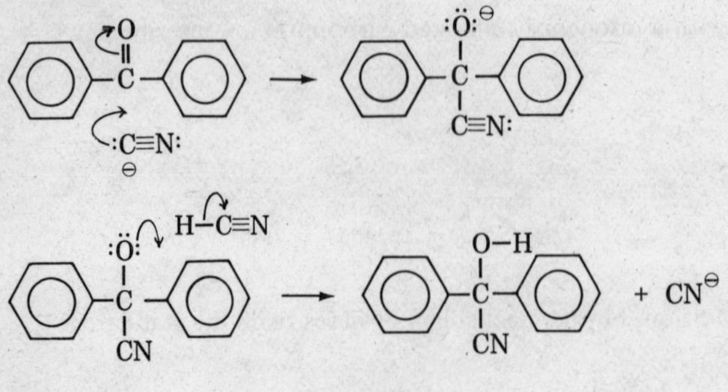

3. 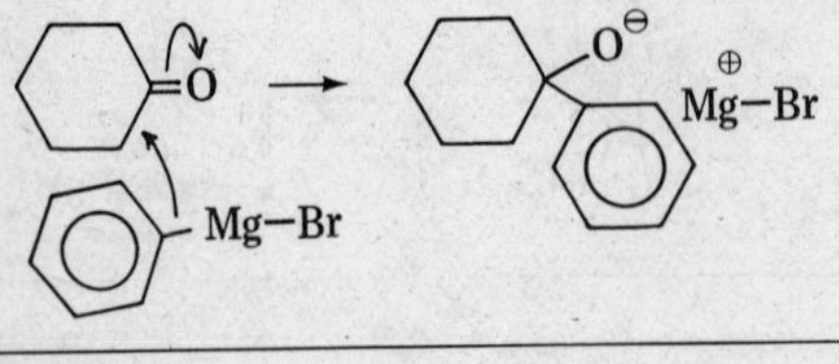 cyclohexanone + Ph–Mg–Br ⟶ 1-phenyl-cyclohexyl–O$^{\ominus}$ $^{\oplus}$Mg–Br

ALDEHYDES AND KETONES (CARBONYL COMPOUNDS) 427

$$\text{(Ph)(C}_6\text{H}_{11}\text{)C–}\overset{\ominus}{\text{O}}: \;\; H-\overset{\oplus}{\text{O}}H_2 \longrightarrow \text{(Ph)(C}_6\text{H}_{11}\text{)C–OH} + H_2O$$

4. $\text{Ph–C(=O)–H} \;\; :NH_2\text{–OH} \longrightarrow \text{Ph–C(–O}^{\ominus}\text{)–}\overset{\oplus}{NH_2}\text{–OH} \rightleftharpoons \text{Ph–CH(OH)–NH–OH}$

$\downarrow CH_3CO_2H$

$$\text{Ph–CH=}\overset{\oplus}{N}\text{(H)–OH} \longleftrightarrow \text{Ph–}\overset{\oplus}{CH}\text{–}\overset{..}{N}H\text{–OH} \longleftarrow \text{Ph–CH(–}\overset{\oplus}{O}H_2\text{)–NH–OH}$$

$\overset{\ominus}{:\underset{..}{O}}\text{–CO–CH}_3 \longrightarrow \text{Ph–CH=N–OH}$

5. $\underset{H}{\overset{\ominus}{\text{C}}}H_2\text{–}\overset{\text{O}}{\overset{||}{\text{C}}}\text{–H} \longrightarrow :\overset{\ominus}{CH_2}\text{–}\overset{\text{O}}{\overset{||}{\text{C}}}\text{–H} \longleftrightarrow CH_2\text{=}\overset{\overset{\ominus}{O}}{\overset{|}{\text{C}}}\text{–H}$

$:\overset{..}{\overset{..}{O}}H^{\ominus}$

$$CH_3\text{–}\overset{\text{O}}{\overset{||}{\text{C}}}\text{–H} \quad :\overset{\ominus}{CH_2}\text{–}\overset{\text{O}}{\overset{||}{\text{C}}}\text{–H} \longrightarrow CH_3\text{–}\overset{\overset{\ominus}{:\underset{..}{O}:}}{\underset{CH_2\text{–C(=O)–H}}{\overset{|}{\text{C}}}}\text{–H} \;\; H\text{–}\overset{..}{\text{O}}\text{–H} \longrightarrow CH_3\text{–CH(OH)–CH}_2\text{–CHO} + OH^{\ominus}$$

SPECIFIC PROBLEM TYPES

11. 1°, 2°, and 3° alcohols can all be made by appropriate reactions of carbonyl compounds with Grignard reagents. You should be able to look at any alcohol and see one or more combinations of these reactants that would produce the alcohol.

Examples

From what combinations of Grignard reagent and carbonyl compound are the following alcohols derived?

1. $CH_3\text{–}\underset{\underset{Ph}{|}}{\overset{\overset{OH}{|}}{C}}\text{–CH(CH}_3)_2$

Answer three ways.

2. (cyclohexyl)–CH(H)–CH$_2$–OH

428 HOW TO SUCCEED IN ORGANIC CHEMISTRY

Solutions:

1. (a) Identify the alcohol type by counting the H atoms attached to the carbon bearing the OH group. In this case: none, therefore, a 3° alcohol.

   (b) What kind of carbonyl compound gives 3° alcohols on treatment with Grignard reagents? Visualize Chart 17.2: R—CO—R′

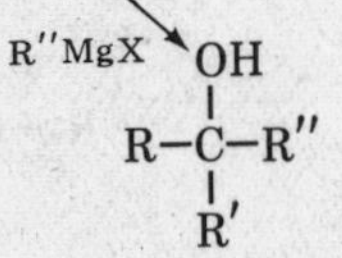

   $$R-CO-R' \xrightarrow{R''MgX} \underset{\underset{R'}{|}}{\overset{\overset{OH}{|}}{R-C-R''}}$$

   Answer: ketone

   (c) Identify the original C=O group as the present —C—OH part of the alcohol:

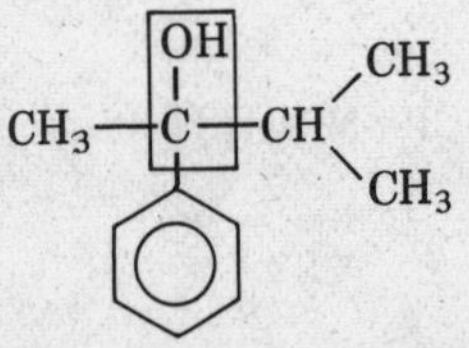

   $$CH_3-\boxed{\underset{|}{\overset{OH}{\overset{|}{C}}}}-CH\underset{\diagdown CH_3}{\diagup CH_3}$$
   (with phenyl attached to C)

   (d) There are three alkyl (or aryl) groups attached to the boxed —C—OH group; choose any one of them to play the role of R—Mg—X:

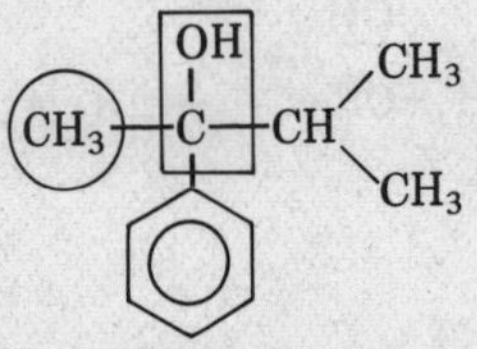

   $$(\widehat{CH_3})-\boxed{\underset{|}{\overset{OH}{\overset{|}{C}}}}-CH\underset{\diagdown CH_3}{\diagup CH_3}$$

   (e) Identify the remaining groups on —C—OH as belonging to the carbonyl compound:

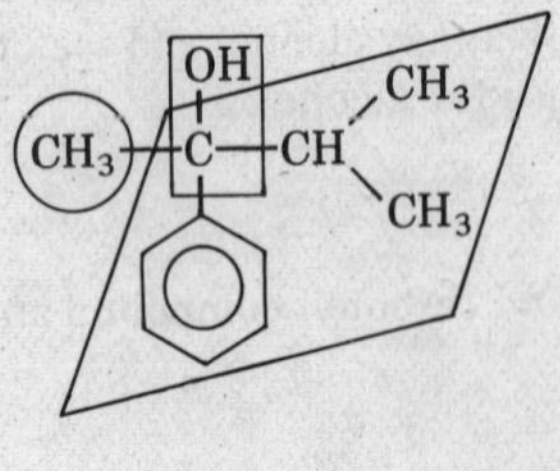

   $$(CH_3)-\boxed{\underset{|}{\overset{OH}{\overset{|}{C}}}}-CH\underset{\diagdown CH_3}{\diagup CH_3}$$
   (parallelogram encloses phenyl, C-OH, and CH(CH_3)_2)

   (f) Write the particular Grignard and carbonyl compound structures arrived at, and combine:

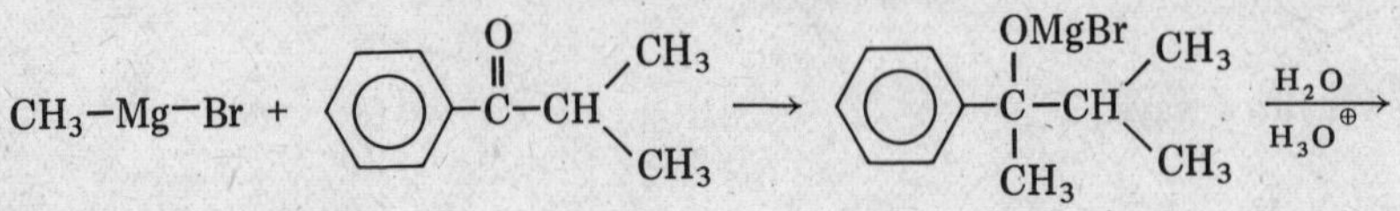

   $$CH_3-Mg-Br \; + \; Ph-\underset{\diagdown CH_3}{\overset{O}{\overset{\|}{C}}-CH\diagup CH_3} \longrightarrow Ph-\underset{\underset{CH_3}{|}}{\overset{\overset{OMgBr}{|}}{C}}-CH\underset{\diagdown CH_3}{\diagup CH_3} \xrightarrow[H_3O^\oplus]{H_2O}$$

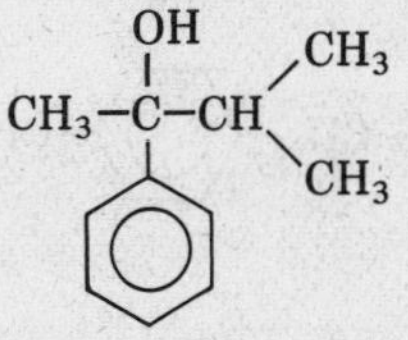

$$CH_3-\underset{\underset{Ph}{|}}{\overset{\overset{OH}{|}}{C}}-CH\overset{CH_3}{\underset{CH_3}{\diagdown}}$$

Repeat the process using alternative assignment of roles:

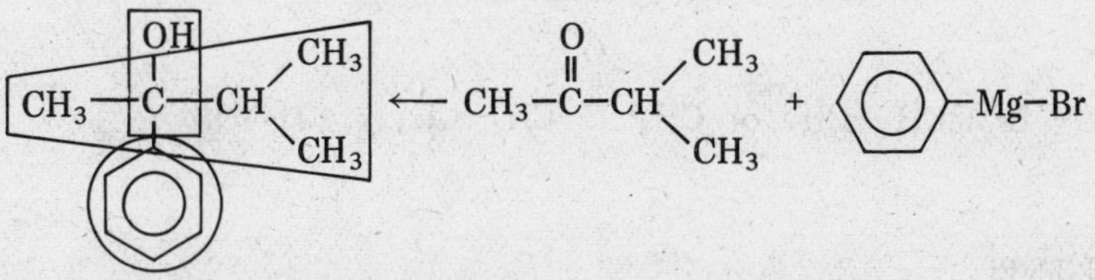

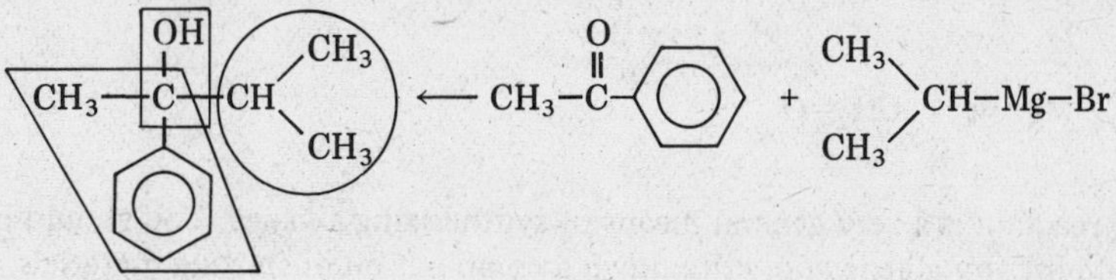

$$CH_3-\underset{\underset{Ph}{|}}{\overset{\overset{\boxed{OH}}{|}}{C}}-\boxed{CH\overset{CH_3}{\underset{CH_3}{\diagdown}}} \quad \longleftarrow \quad CH_3-\overset{\overset{O}{\|}}{C}-CH\overset{CH_3}{\underset{CH_3}{\diagdown}} \quad + \quad Ph-Mg-Br$$

$$CH_3-\underset{\underset{\boxed{Ph}}{|}}{\overset{\overset{OH}{|}}{C}}-CH\overset{CH_3}{\underset{CH_3}{\diagdown}} \quad \longleftarrow \quad CH_3-\overset{\overset{O}{\|}}{C}-Ph \quad + \quad \overset{CH_3}{\underset{CH_3}{\diagup}}CH-Mg-Br$$

2. (a) 1° ROH

(b) Grignard + $H_2C=O \longrightarrow$ 1° ROH

(c) 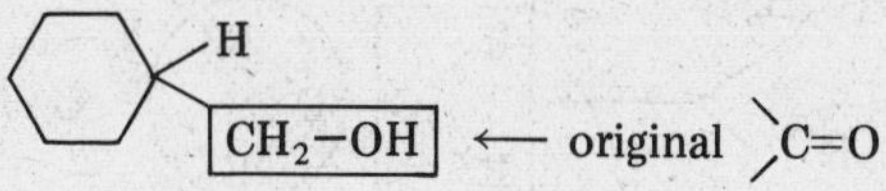 Ph—CH(H)—$\boxed{CH_2-OH}$ $\longleftarrow$ original $>C=O$

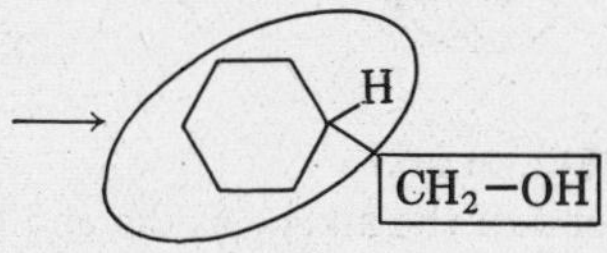

(d) and (e) Grignard $\longrightarrow$ Ph—CH(H)—$\boxed{CH_2-OH}$ (circled together)

(f) reaction: Ph—CH(H)—MgBr $\xrightarrow[(2)\ H_2O/H_3O^\oplus]{(1)\ H_2C=O}$ Ph—CH(H)—CH_2-OH

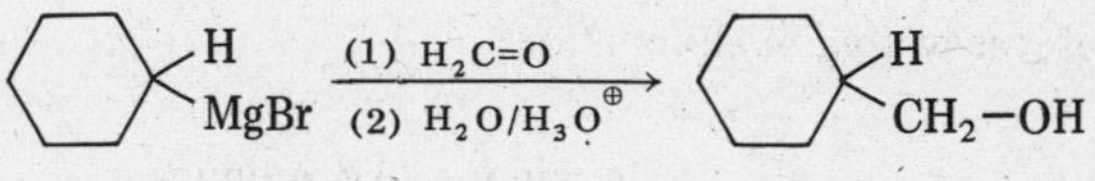

★ *Exercises*

Draw the structure of a Grignard reagent and a carbonyl compound that would react to give each alcohol.

1. 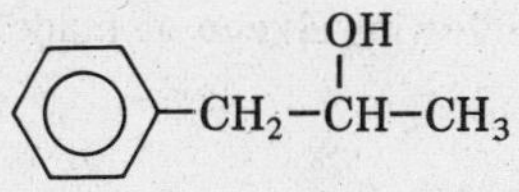 $Ph-CH_2-\underset{\underset{}{\overset{OH}{|}}}{CH}-CH_3$

2. $CH_3-\underset{\underset{CH_3}{|}}{\overset{\overset{CH_2-CH_3}{|}}{C}}-OH$

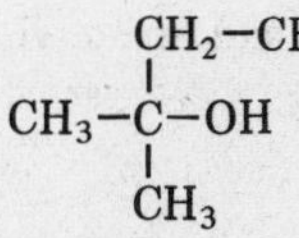

3. 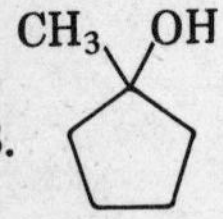 [1-methyl-1-hydroxycyclopentane: CH₃, OH on cyclopentane]

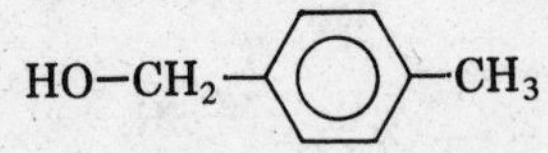

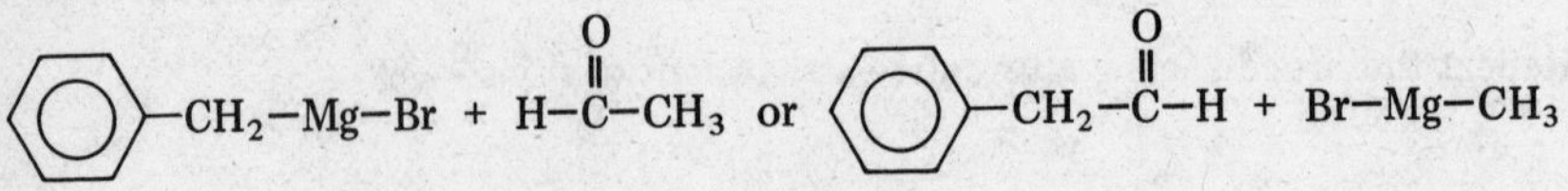

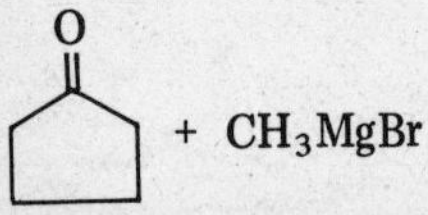

4. 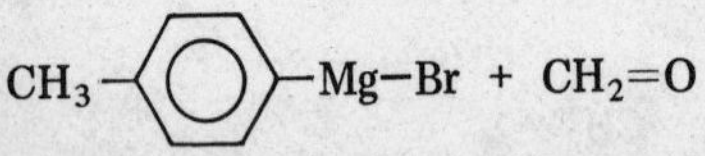 HO–CH₂–⟨C₆H₄⟩–CH₃

----

1. ⟨C₆H₅⟩–CH₂–Mg–Br + H–C(=O)–CH₃ or ⟨C₆H₅⟩–CH₂–C(=O)–H + Br–Mg–CH₃

2. CH₃–C(=O)–CH₃ + BrMgCH₂–CH₃ or CH₃–C(=O)–CH₂–CH₃ + CH₃MgBr

3. [cyclopentanone] + CH₃MgBr

4. CH₃–⟨C₆H₄⟩–Mg–Br + CH₂=O

12. The Wittig reaction is a very general means of synthesizing alkenes. You should be able to look at almost any alkene and write down a combination of starting materials whose Wittig reaction would produce the alkene in good yield. To do this, note that one end of the alkene is derived from an alkyl halide reactant; the other end is furnished by a carbonyl compound:

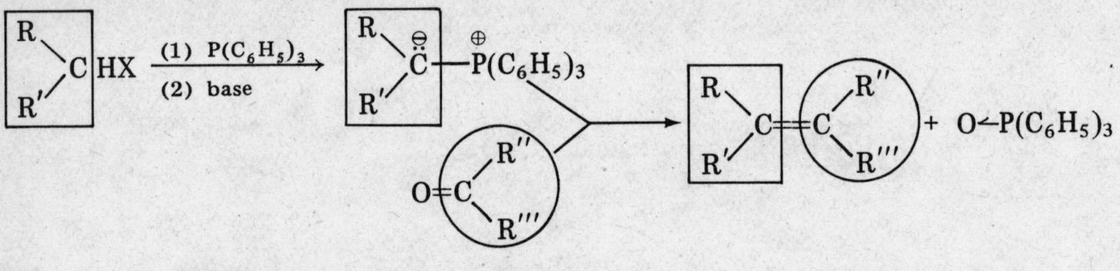

$$\underset{R'}{\overset{R}{>}}CHX \xrightarrow[(2)\ \text{base}]{(1)\ P(C_6H_5)_3} \underset{R'}{\overset{R}{>}}\overset{\ominus}{C}-\overset{\oplus}{P}(C_6H_5)_3 \quad \underset{R'''}{\overset{R''}{>}}C=O \longrightarrow \underset{R'}{\overset{R}{>}}C=C\underset{R'''}{\overset{R''}{<}} + O \leftarrow P(C_6H_5)_3$$

Here R, R′, R″, and R‴ can be H, alkyl, or aryl.

You choose the *least substituted* end of the alkene as the end to be derived from the *halide*. This choice guarantees the best results because the more highly substituted the halide is, the more of it is wasted in elimination:

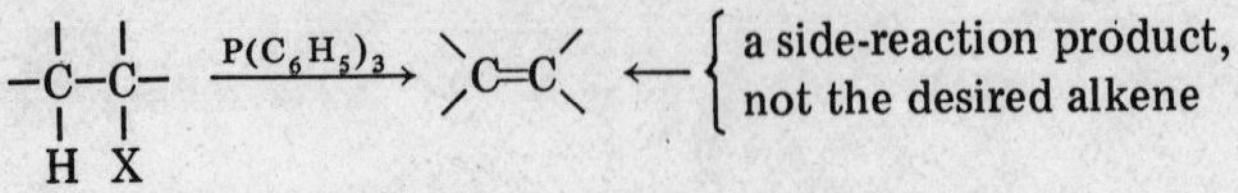

$$-\underset{\underset{H}{|}}{\overset{|}{C}}-\underset{\underset{X}{|}}{\overset{|}{C}}- \xrightarrow{P(C_6H_5)_3} >C=C< \longleftarrow \left\{\begin{array}{l}\text{a side-reaction product,}\\ \text{not the desired alkene}\end{array}\right.$$

Example

From which alkyl halide and carbonyl compound can the following alkene be made?

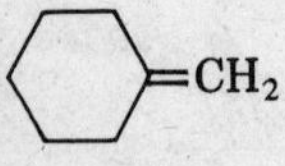

[cyclohexylidene]=CH₂

ALDEHYDES AND KETONES (CARBONYL COMPOUNDS)

Solution:

Label:

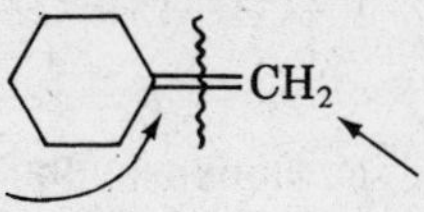

more substituted end—
derive it from carbonyl
compound:

less substituted end—derive it from
alkyl halide:

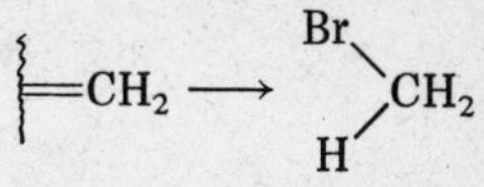

So

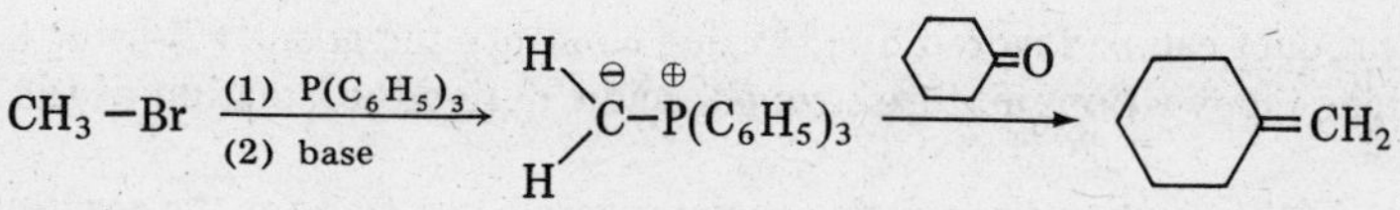

$$CH_3-Br \xrightarrow[(2)\ base]{(1)\ P(C_6H_5)_3} \underset{H}{\overset{H}{C}}-\overset{\oplus}{P}(C_6H_5)_3 \longrightarrow \text{(cyclohexanone)} \longrightarrow \text{(cyclohexylidene)=CH}_2$$

Exercises

From which alkyl halide and carbonyl compound should each alkene be made?

1. $CH_3-(CH_2)_{12}-CH=CH-(CH_2)_{10}-CH_3$ 2. 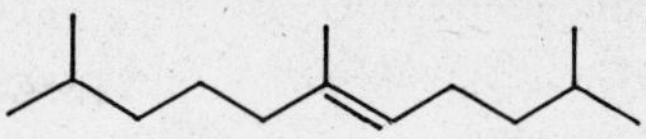 (branched alkene structure)

3. 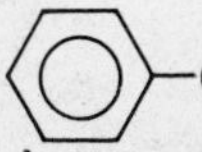 $\text{Ph}-CH_2-CH=CH-CH=CH_2$

— — — — — — — — —

1. $CH_3-(CH_2)_{12}-CH_2-Br\ +\ CH_3-(CH_2)_{10}-CHO$
   or $CH_3-(CH_2)_{10}-CH_2-Br\ +\ CH_3-(CH_2)_{12}-CHO$

2. 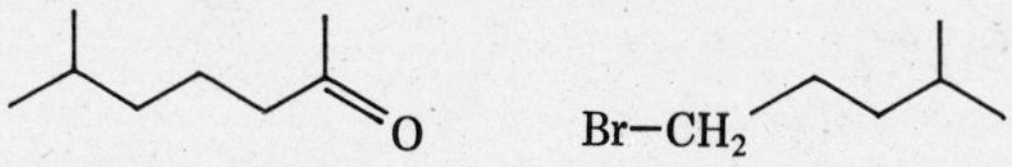 (branched ketone) + $Br-CH_2$-(branched)

3. 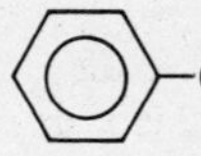 $\text{Ph}-CH_2-CH_2-Br\ +\ O=CH-CH=CH_2$ or $\text{Ph}-CH_2-CH=CH-CH=O$
   + CH_3Br

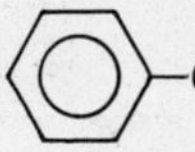

13. An aldol condensation is the nucleophilic addition of the enolate ion derived from an aldehyde or ketone to a fresh molecule of the aldehyde or ketone. A base (usually $OH^{\ominus}$ or $OR^{\ominus}$) is required to generate the enolate ion. The product is a β-hydroxy carbonyl compound, which, depending on the conditions, may or may not dehydrate to the α,β-unsaturated carbonyl compound:

$$2R-CH_2-\overset{O}{\underset{\|}{C}}-H \xrightarrow{:B} R-CH_2-\underset{R}{\underset{|}{CH}}-\overset{O-H}{\underset{|}{CH}}-\overset{O}{\underset{\|}{C}}-H \quad \text{an aldol}$$

or

$$R-CH_2-CH=C(R)-\overset{O}{\underset{\|}{C}}-H + H_2O$$

an aldol dehydration product

$$2R-CH_2-\overset{O}{\underset{\|}{C}}-R' \xrightarrow{:B} \text{ or }$$

$$R-CH_2-\overset{OH}{\underset{R}{C}}-\overset{}{\underset{R}{CH}}-\overset{O}{\underset{\|}{C}}-R'$$

$$R-CH_2-\overset{}{\underset{R'}{C}}=\overset{}{\underset{R}{C}}-\overset{O}{\underset{\|}{C}}-R' + H_2O$$

The dehydrated product can be favored by (1) using concentrated alkali, (2) using hot alkali, or (3) making the reaction mixture strongly acidic during the isolation of the product.

The three steps in the mechanism sum to the net, overall equation for the reaction:

$$R-CH_2-\overset{O}{\underset{\|}{C}}-H + \overset{\ominus}{O}-R' \rightleftharpoons \left\updownarrow \begin{array}{c} R-\overset{\ddot{O}:}{\underset{\|}{\overset{\ominus}{C}H}}-\overset{}{\underset{}{C}}-H \\ :\overset{\ominus}{\ddot{O}}: \\ R-CH=\overset{}{\underset{}{C}}-H \end{array}\right. + H-O-R'$$

$$R-CH_2-\overset{\ddot{O}:}{\underset{\|}{C}}-H \rightleftharpoons R-CH_2-\overset{:\ddot{O}:^\ominus}{\underset{|}{C}}-H$$
$$R-\overset{\ominus}{\underset{}{\ddot{C}H}}-\overset{O}{\underset{\|}{C}}-H \qquad R-CH-\overset{}{\underset{\|}{C}}-H$$
$$\qquad\qquad\qquad\qquad\qquad O$$

$$\overset{\ominus}{:\ddot{O}:}\curvearrowright H\overset{\curvearrowleft}{-\ddot{O}-R'} \qquad\qquad :\ddot{O}H$$
$$R-CH_2-CH \qquad\qquad R-CH_2-CH \quad\; O \qquad + \;\overset{\ominus}{O}-R'$$
$$R-CH-\overset{}{\underset{\|}{C}}-H \rightleftharpoons R-CH-\overset{}{\underset{\|}{C}}-H$$
$$\quad\;\; O$$

The molecule of base consumed in the first step is regenerated in the last step; it is a true catalyst. Each step is reversible.

You should be able to write the structure of the aldol condensation product for any aldehyde or ketone containing at least one α-hydrogen atom, and write the mechanism of that reaction. In addition, it is important to be able to recognize quickly if a compound has such a structure that it could either be the product of an aldol condensation, or be closely related to such a product. This skill is desirable because the aldol condensation, as a new C—C bond-forming reaction, has utility in building up a carbon skeleton from smaller fragments. As you can see, there are strict limitations on the kind of carbon skeleton that can result from the aldol condensation: it must be divisible into two identical halves, and the functional groups present in each half have particular positions.

First we practice recognizing genuine aldol products. There are two equally effective approaches: (1) work the mechanism through backwards, or (2) cut the molecule into

ALDEHYDES AND KETONES (CARBONYL COMPOUNDS) 433

halves and reconstruct either half into an aldehyde or ketone. The second method is faster; with practice it can be carried out at sight.

Examples

From what carbonyl compound were these aldol condensation products derived?

1. 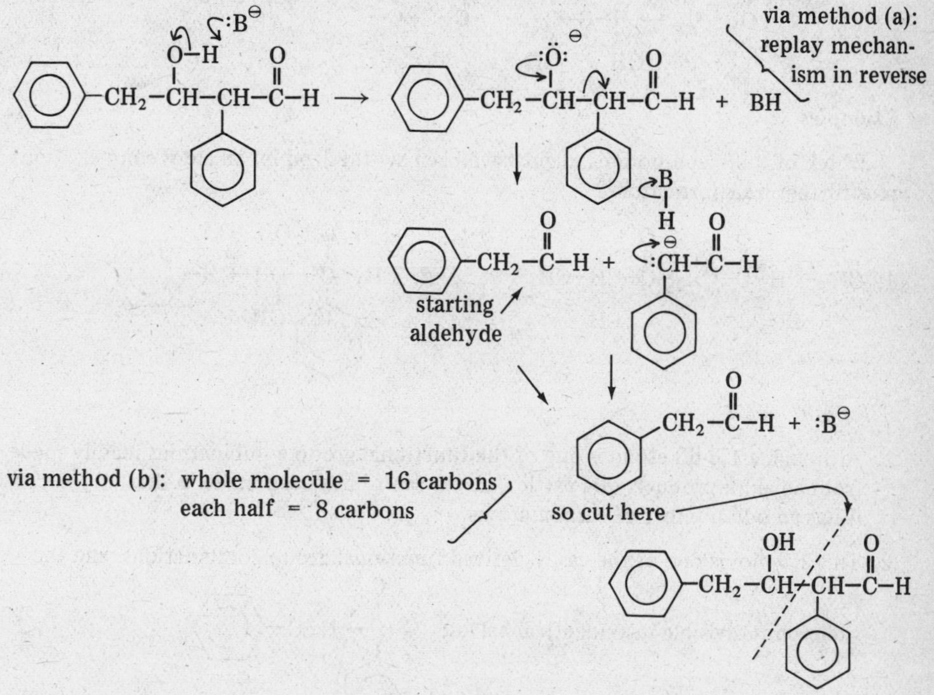 Ph—CH$_2$—CH(OH)—CH—C(=O)—H → Ph—CH$_2$—CH(O$^\ominus$)—CH(Ph)—C(=O)—H + BH   *via method (a): replay mechanism in reverse*

↓

Ph—CH$_2$—C(=O)—H + $^\ominus$:CH(Ph)—C(=O)—H
starting aldehyde

↓

Ph—CH$_2$—C(=O)—H + :B$^\ominus$

via method (b): whole molecule = 16 carbons, each half = 8 carbons, so cut here

Ph—CH$_2$—CH(OH)—CH(Ph)—C(=O)—H

Either half can be adjusted to reconstitute the starting carbonyl compound. The easiest half is the one containing the C=O group; replacing the missing α H in this fragment gives the original carbonyl compound:

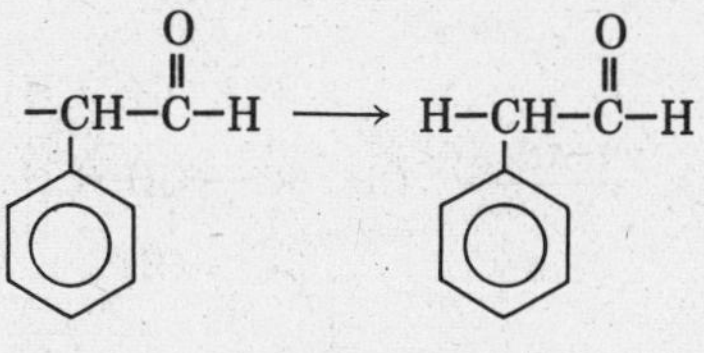

Ph—CH—C(=O)—H → Ph—CH$_2$—C(=O)—H

2. CH$_3$—C(=O)—CH=C(CH$_3$)(CH$_3$)   In reconstituting a dehydrated aldol product, you need to add two α hydrogens:

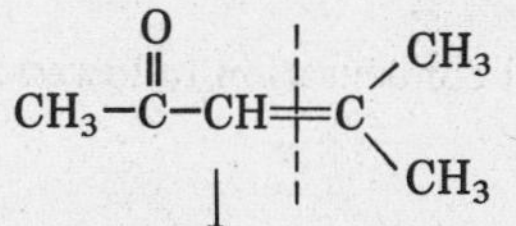

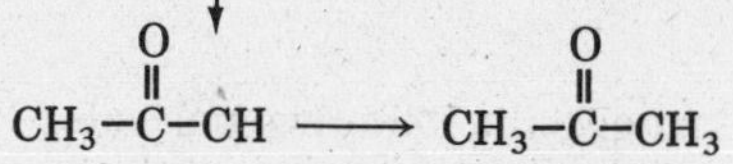

↓

CH$_3$—C(=O)—CH → CH$_3$—C(=O)—CH$_3$

To recognize structures that could have been derived from aldol condensation products by further reactions, focus on two aspects of the aldol product: (1) Think of the functional groups into which >C=O, —OH, and >C=C< (the aldol product's functions) can be readily transformed:

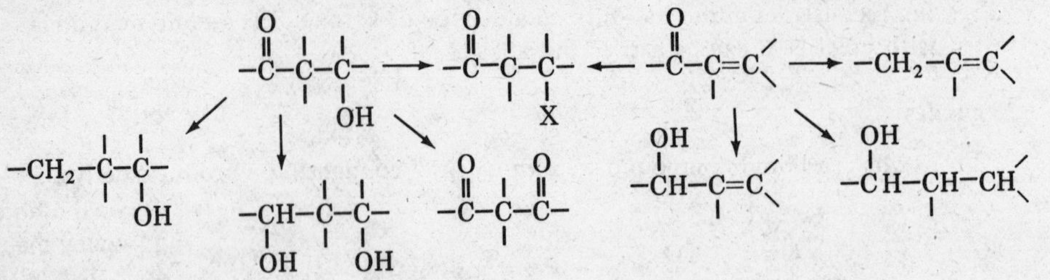

Examples

Which of these compounds might have been synthesized by an aldol condensation plus further transformations?

1. 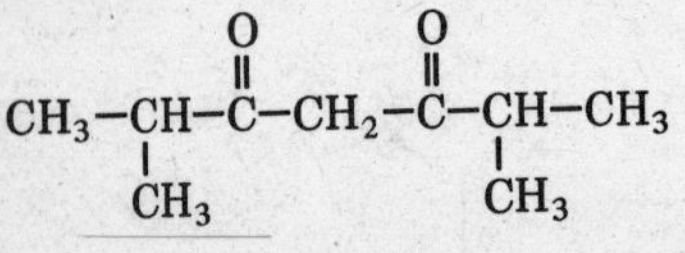 $CH_3-\underset{CH_3}{\underset{|}{CH}}-\overset{O}{\overset{||}{C}}-CH_2-\overset{O}{\overset{||}{C}}-\underset{CH_3}{\underset{|}{CH}}-CH_3$

2. $CH_3-\underset{\phi}{\overset{CH_2OH}{\overset{|}{C}}}-\underset{OH}{\underset{|}{CH}}-\overset{CH_3}{\overset{|}{CH}}-\phi$

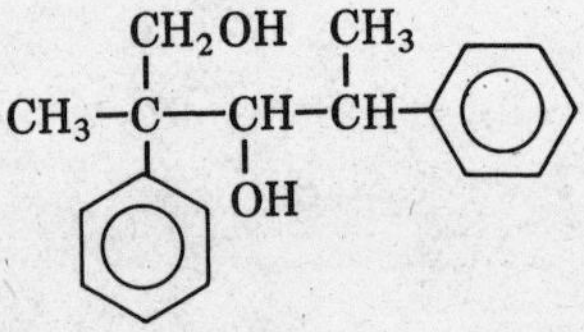

Solutions:

1. Although a 1,3-diketone is one of the functional group combinations readily made from an aldol product, this particular one can't have been made in that way because it has an odd number of carbon atoms.

2. This 1,3-diol is one of the easily derived functional group combinations, and the skeleton is divisible into identical halves:

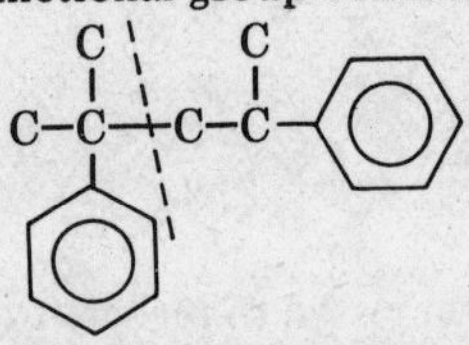

$C-\underset{\phi}{\overset{C}{\underset{|}{C}}} \,\vert\, C-\underset{\phi}{\overset{C}{\underset{|}{C}}}-\phi$

The position of the oxygen atom on the left half locates the original C=O group. The easiest thing to do is write and operate on what must have been the original carbonyl compound:

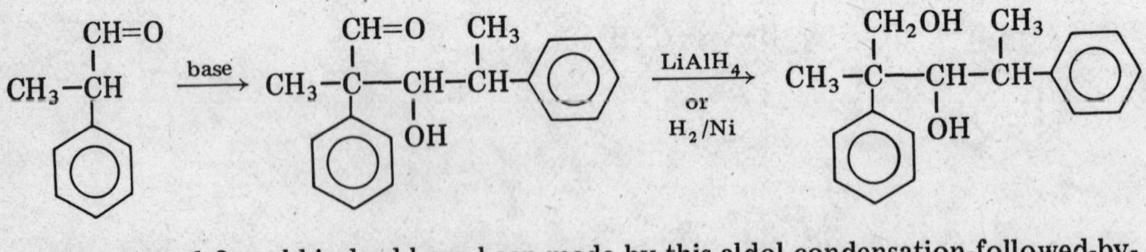

$CH_3-\underset{\phi}{\underset{|}{CH}}-CH=O \xrightarrow{\text{base}} CH_3-\underset{\phi}{\overset{CH=O}{\underset{|}{C}}}-\underset{OH}{\underset{|}{CH}}-\overset{CH_3}{\overset{|}{CH}}-\phi \xrightarrow[\text{or } H_2/Ni]{LiAlH_4} CH_3-\underset{\phi}{\overset{CH_2OH}{\underset{|}{C}}}-\underset{OH}{\underset{|}{CH}}-\overset{CH_3}{\overset{|}{CH}}-\phi$

So compound 2 could indeed have been made by this aldol condensation-followed-by-reduction sequence.

★ Exercises

1. Draw the structure of the carbonyl compounds that give the following molecules on aldol condensation.

(a) 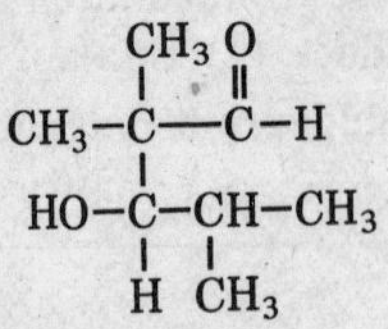 $\underset{\underset{H\;\;CH_3}{\underset{|\;\;\;\;|}{HO-C-CH-CH_3}}}{\overset{CH_3\;\;O}{\overset{|\;\;\;\;||}{CH_3-C-\!\!-C-H}}}$

(b) 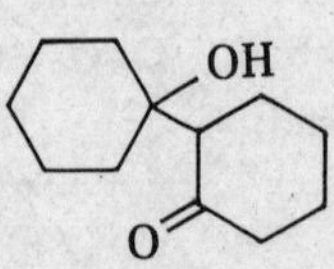 [cyclohexane ring with OH substituent attached to another cyclohexanone ring via shared carbon]

ALDEHYDES AND KETONES (CARBONYL COMPOUNDS)

(c) 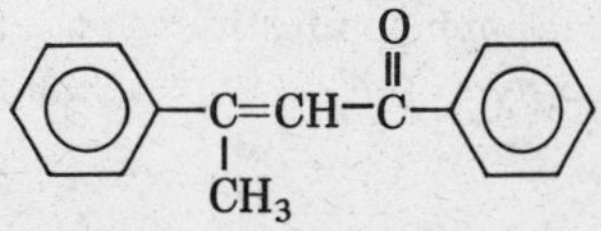 Ph−C(CH₃)=CH−C(=O)−Ph

(d) 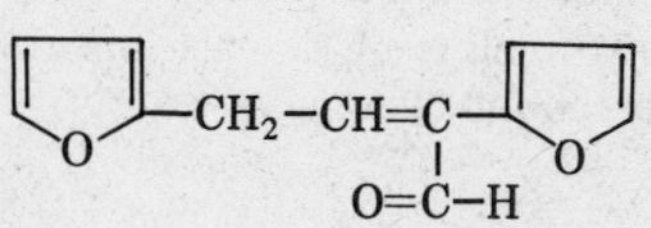 (furyl)−CH₂−CH=C(−CHO)−(furyl)

2. Which of the following compounds might have been synthesized by an aldol condensation plus further transformations?

(a) 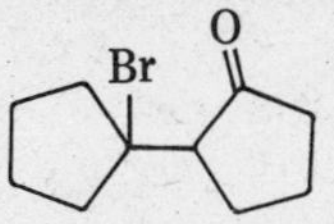 cyclopentyl(Br)−cyclohexanone

(b) 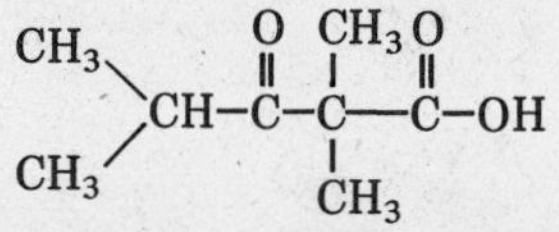 (CH₃)₂CH−C(=O)−C(CH₃)(OCH₃)−C(=O)−OH

(c) Ph−CH₂−CH₂−CH(CH₃)−CH₂−CH₂−CH₂−CH₂−Ph

- - - - - - - - - -

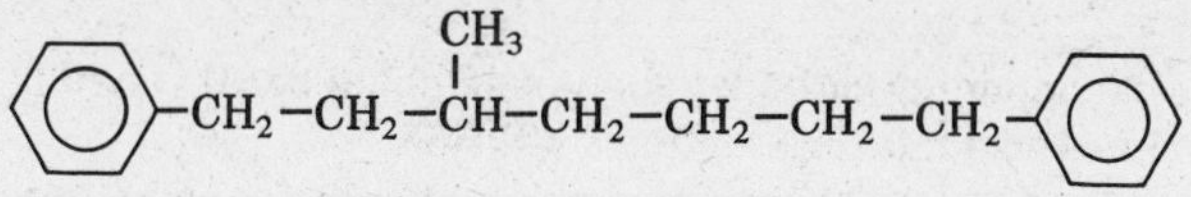

1. (a) 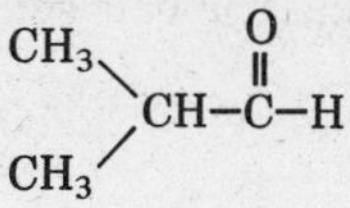 (CH₃)₂CH−CHO (b) 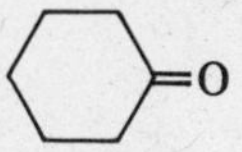 cyclohexanone
   (c) 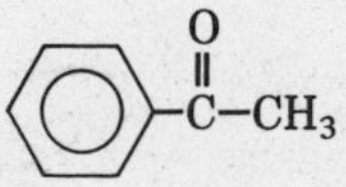 Ph−C(=O)−CH₃ (d) 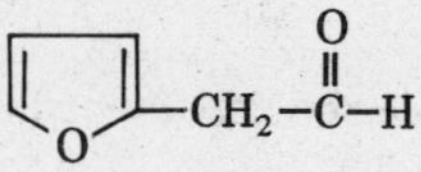 (furyl)−CH₂−C(=O)−H

2. (a) 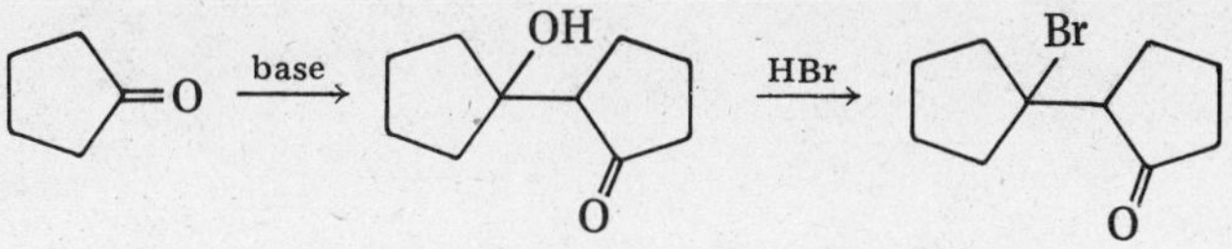 cyclopentanone $\xrightarrow{\text{base}}$ cyclopentyl-(OH)cyclopentanone $\xrightarrow{\text{HBr}}$ cyclopentyl-(Br)cyclopentanone

(b) 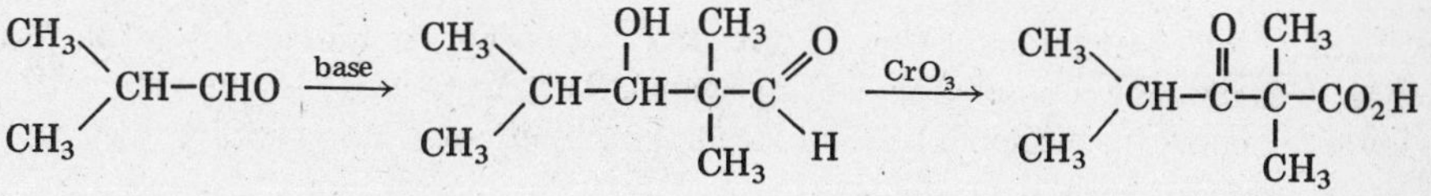 (CH₃)₂CH−CHO $\xrightarrow{\text{base}}$ (CH₃)₂CH−CH(OH)−C(CH₃)(CHO) $\xrightarrow{\text{CrO}_3}$ (CH₃)₂CH−C(=O)−C(CH₃)(CH₃)−CO₂H

(c) Ph−CH₂CH₂CH₂−C(=O)−H $\xrightarrow{\text{base}}$ Ph−CH₂CH₂CH₂−CH(OH)−CH(CHO)−CH₂CH₂−Ph
$\xrightarrow{\Delta}$ Ph−CH₂CH₂CH₂−CH=C(CHO)−CH₂CH₂−Ph
$\xrightarrow{\text{H}_2/\text{Pt}}$ Ph−(CH₂)₄−CH(CH₂OH)−(CH₂)₂−Ph
$\xrightarrow{\text{HBr}}$ Ph−(CH₂)₄−CH(CH₂Br)−(CH₂)₂−Ph
$\xrightarrow{\text{(1) Mg, (2) H}_2\text{O}}$ Ph−(CH₂)₄−CH(CH₃)−(CH₂)₂−Ph

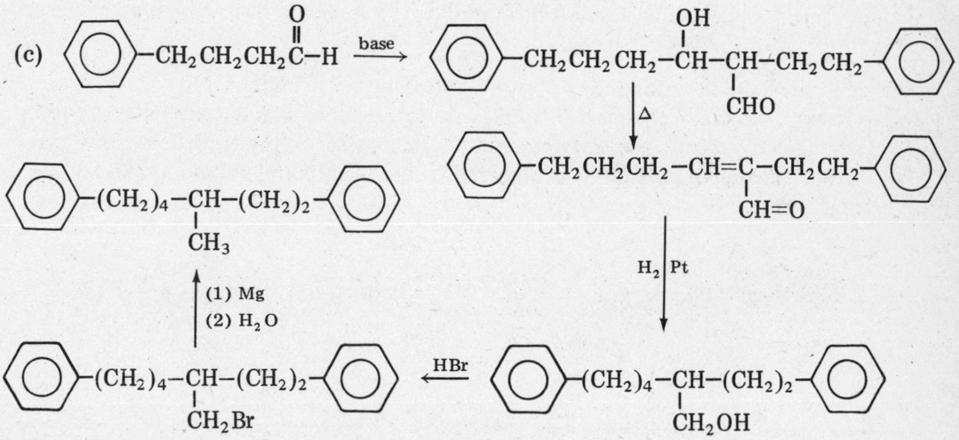

14. When an enolate ion acts as the nucleophile in a nucleophilic substitution (S_N2) reaction, the overall result is replacement of one of the α hydrogens in the original carbonyl compound by an alkyl group:

$$R'-CH_2-\overset{O}{\overset{\|}{C}}-R + :B^{\ominus} M^{\oplus} + R''-X \longrightarrow R'-\overset{R''}{\underset{|}{CH}}-\overset{O}{\overset{\|}{C}}-R + BH + M^{\oplus} X^{\ominus}$$

This process is called *alkylation* of the carbonyl compound. The mechanism is

$$R'-CH_2-\overset{O}{\overset{\|}{C}}-R + :B^{\ominus} \rightleftharpoons R'-\overset{\ddot{O}:}{\underset{|}{\overset{|}{C}H}}-\overset{}{\overset{|}{C}}-R \longleftrightarrow R'-CH=\overset{:\ddot{O}:^{\ominus}}{\underset{|}{C}}-R + BH$$

$$R'-CH=\overset{}{\underset{\overset{|}{:O:}^{\ominus}}{C}}-R \longleftrightarrow R'-\overset{}{\underset{\overset{|}{\overset{|}{O:}^{\ominus}}}{\overset{|}{C}H}}-\overset{R''\curvearrowleft X}{\overset{}{\underset{}{C}}}-R \longrightarrow R'-\overset{R''}{\underset{|}{CH}}-\overset{O}{\overset{\|}{C}}-R + X^{\ominus}$$

In principle, any α-hydrogen-containing carbonyl compound could be alkylated in this fashion. In practice, several factors limit the reaction's utility:

(a) If the carbonyl compound has more than one variety of α hydrogen, a mixture of isomeric alkylation products generally results.

(b) If the monoalkylated product has additional α hydrogens, the acid-base equilibrium,

$$R'-\overset{\ominus}{\underset{}{C}H}-\overset{O}{\overset{\|}{C}}-R + R'-\overset{}{\underset{\overset{|}{R''}}{CH}}-\overset{O}{\overset{\|}{C}}-R \rightleftharpoons R'-CH_2-\overset{O}{\overset{\|}{C}}-R + R'-\overset{\ominus}{\underset{\overset{|}{R''}}{C}}-\overset{O}{\overset{\|}{C}}-R$$

converts some of it to its enolate ion, which competes for the available R—X, leading to the dialkylated product

$$R-\overset{R''}{\underset{\overset{|}{R''}}{\overset{|}{C}}}-\overset{O}{\overset{\|}{C}}-R$$

(c) If the common bases $CH_3O^{\ominus}$ or $C_2H_5O^{\ominus}$ are used, the initial enolate ion formation does not go to completion, leaving some $CH_3O^{\ominus}$ or $C_2H_5O^{\ominus}$ to compete with the enolate ion for the available R''X and leading to by-product $R''-O-CH_3$ or $R''-O-C_2H_5$ and alkene from dehydrohalogenation (*E*2) of R''X.

(d) Depending on solvent and other variables, some of the alkylation may occur at the other nucleophilic site in the enolate ion, leading to $R'-CH=\overset{O-R''}{\underset{}{C}}-R$.

In many cases the problems associated with this reaction can be eliminated by using an *enamine* derived from the carbonyl compound in place of the enolate ion. An enamine is a nitrogen analog of an enol. It is prepared from the carbonyl compound and a secondary amine:

$$R'-CH_2-\overset{O}{\overset{\|}{C}}-R + \underset{\underset{H}{|}}{\overset{}{N}}\bigcirc \longrightarrow R'-CH_2-\overset{OH}{\underset{\overset{|}{N}\bigcirc}{\overset{|}{C}}}-R$$

ALDEHYDES AND KETONES (CARBONYL COMPOUNDS)

Since there are no remaining hydrogens on N, the addition product can dehydrate only by using an α hydrogen:

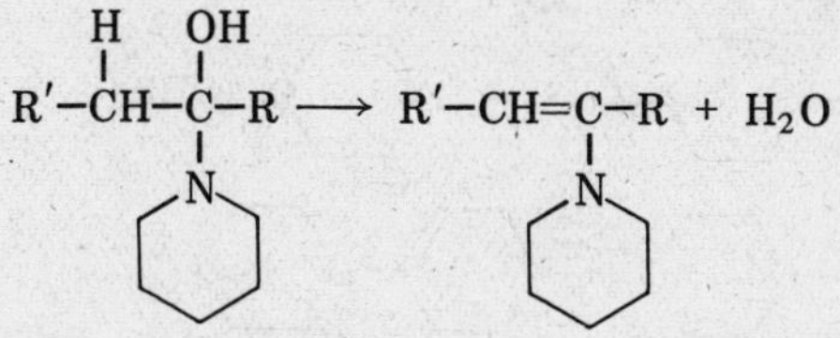

$$R'-\underset{\underset{N}{|}}{\underset{|}{C}H}-\underset{\underset{}{|}}{\overset{OH}{\underset{|}{C}}}-R \longrightarrow R'-CH=\underset{\underset{N}{|}}{C}-R + H_2O$$

The enamine is a resonance-hybrid molecule that possesses substantial carbanion character on the α carbon, and it gives reactions characteristic of enolate ions, for instance the alkylation reaction:

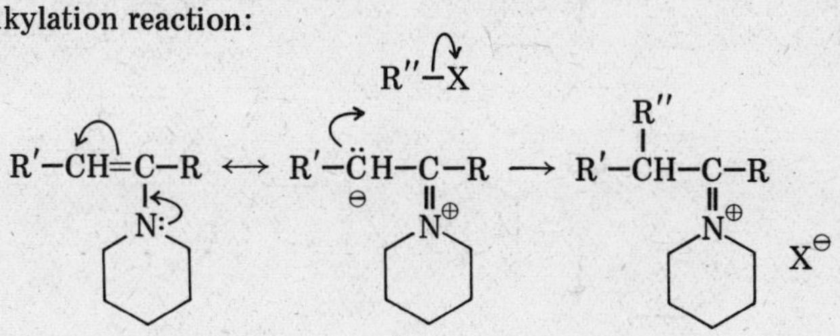

$$R'-CH=\underset{\underset{N:}{|}}{C}-R \longleftrightarrow R'-\overset{\ominus}{\underset{\underset{N^{\oplus}}{|}}{C}H}-\underset{\underset{}{||}}{C}-R \xrightarrow{R''-X} R'-\underset{\underset{N^{\oplus}}{|}}{\overset{R''}{\underset{|}{C}H}}-\underset{\underset{}{||}}{C}-R \quad X^{\ominus}$$

The product hydrolyzes to the alkylated carbonyl compound:

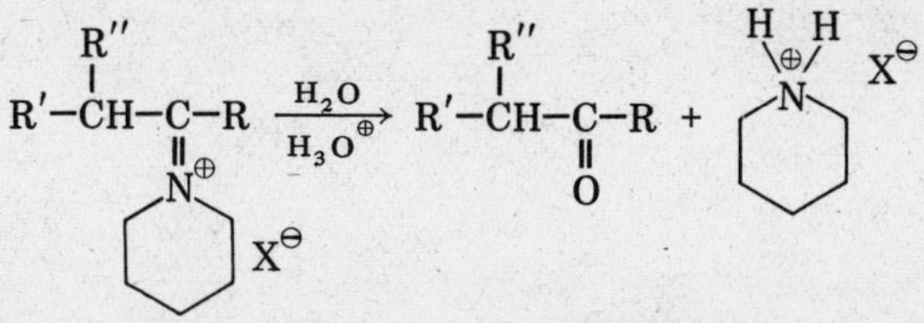

$$R'-\underset{\underset{N^{\oplus}}{|}}{\overset{R''}{\underset{|}{C}H}}-\underset{\underset{}{||}}{C}-R \quad X^{\ominus} \xrightarrow[H_3O^{\oplus}]{H_2O} R'-\underset{\underset{}{|}}{\overset{R''}{\underset{|}{C}H}}-\underset{\underset{O}{||}}{C}-R + \underset{\underset{N^{\oplus}}{\overset{H \ \ H}{\diagup}}}{} \quad X^{\ominus}$$

★ *Exercises*

1. Write the equations for alkylation via enamine using this choice of slot fillers:

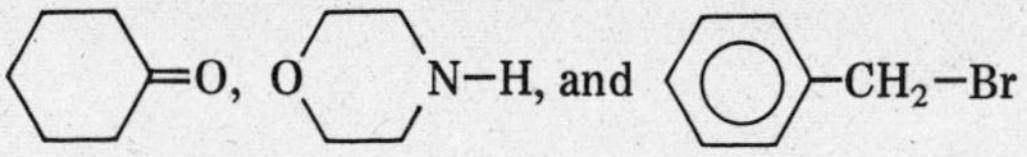

   ⬡=O, O⬡N—H, and ⌬—CH₂—Br

2. When CH₃-cyclohexanone=O is substituted for cyclohexanone=O in exercise 1, the final product is a mixture of two isomeric ketones. On the other hand, 2,6-dimethylcyclohexanone=O gives only one alkylation product. Write structures for the products in these cases and show why these results are observed.

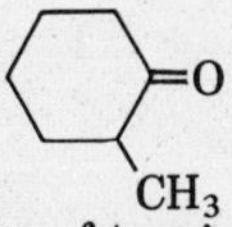

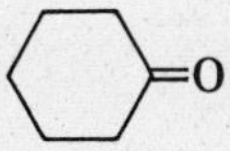

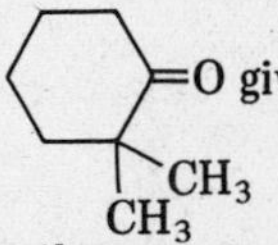

- - - - - - - - - - -

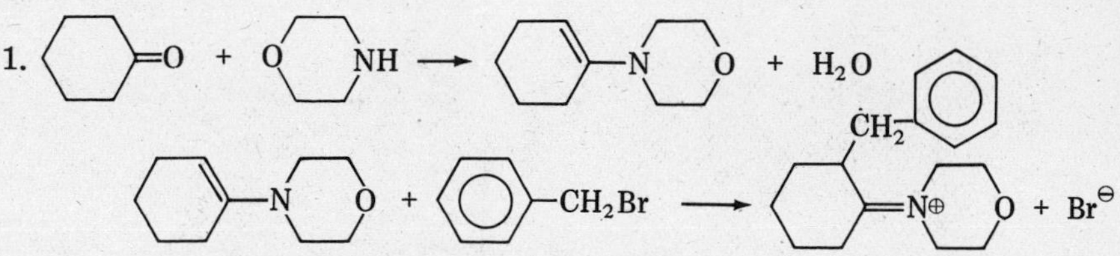

1. ⬡=O + O⬡NH ⟶ ⬡—N⬡O + H₂O

    ⬡=N⬡O + ⌬—CH₂Br ⟶ ⬡(CH₂—⌬)=N⊕⬡O + Br⁻

2.

UNIT EIGHTEEN
Carboxylic Acids and Their Derivatives

Here we are dealing for the first time with a set of closely related functional groups, each of which can be readily converted to the others. They are

$$R-\overset{\overset{O}{\|}}{C}-O-H \qquad R-\overset{\overset{O}{\|}}{C}-O-R' \qquad R-\overset{\overset{O}{\|}}{C}-Cl \qquad \begin{matrix} R-\overset{\overset{O}{\|}}{C} \\ \diagdown O \\ R'-\underset{\underset{O}{\|}}{C} \diagup \end{matrix}$$

carboxylic acid (the "parent function")   ester   acyl chloride   carboxylic acid anhydride

$$R-\overset{\overset{O}{\|}}{C}-NH_2 \qquad R-\overset{\overset{O}{\|}}{C}-NH-R' \qquad R-\overset{\overset{O}{\|}}{C}-N\diagup^{R'}_{\diagdown R''} \qquad R-C\equiv N$$

1° amide   2° amide   3° amide   nitrile

The conjugate acid and conjugate base of RCO_2H could also be included, but their reactions are closely related to those of RCO_2H, and in the interest of simplicity they are not listed separately in the charts that follow.

In all these compounds R, R', and R" can be alkyl or aryl groups.

Acyl fluorides, bromides, and iodides are known, but in practice the cheaper chlorides are always used for the reactions to be discussed here.

The two alkyl or aryl groups in the anhydride molecule are usually the same, R = R'. They can be different, but these mixed anhydrides, R ≠ R', are rare.

In some of their reactions, the 1°, 2°, and 3° amides behave quite differently. But in the basic suite of reactions discussed here, they are very similar, and I have simplified the charts by writing just $R-C\overset{\diagup O}{\diagdown NH_2}$.

The nitrile (R—C≡N), which lacks a carbonyl group, is the farthest removed from the others structurally. Yet it is closely related chemically; it is in the same oxidation state and is derived from RCO_2H by adding one NH_3 and removing 2 H_2O.

We will find it useful to symbolize the acids, esters, acyl chlorides, anhydrides, and amides with a single generic structure: $R-C\overset{\diagup O}{\diagdown G}$.

439

Frames 1 to 11 lay out the basic modes of reaction of these functions, the relationships of one to another and of all the R—CO—G to the functional groups in lower oxidation states. Frames 12 to 14 provide strategy and drills for learning the large number of reactions involved. Frames 15 and 16 consider specific problem types.

OBJECTIVES

When you have completed this unit, you should be able to:

- Draw and name the functional groups generally considered to be derivatives of the carboxylic acids, and write equations for the reactions that interconvert them (frame 1).
- Draw the structures of the conjugate base of RCO_2H and of the conjugate acids of RCO_2H and all its derivatives; indicate which of these species can be isolated as stable salts (frames 2, 3).
- Write equations for the important reactions connecting the carboxylic acid derivatives with compounds in other oxidation stages (frame 11).
- Supply the missing product or reagent in these equations if given the other species; predict products or reagents for many related reactions that you have not specifically studied (frames 12 to 14).
- Write all steps in the mechanisms of most of the reactions of the carboxylic acids and derivatives (frames 15, 16).
- Show how the acyl S_N2 mechanism differs from the S_N2 mechanism, and demonstrate how acid and base catalysis occur in the acyl S_N2 mechanism (frame 4).
- Propose syntheses of the carboxylic acids and derivatives from a variety of starting materials (frame 17).

If you believe you may already have achieved some or all of these objectives, take a self-test consisting of the exercises marked by a star in the margin. If you miss any of these exercises, reread the relevant frames and rework the exercises before proceeding. If you are not ready for a self-test, proceed to frame 1.

INTERCONVERSION OF CARBOXYLIC ACID DERIVATIVES

1. Chart 18.1 lays out the relationships between the various $R-C\underset{G}{\overset{O}{\diagup\!\!\!\diagdown}}$ in the most concise way possible. As before, absence of an arrow means that there is no good way to carry out that conversion.

Note the following points:

1. RCO_2H is the parent compound; each of the derivatives hydrolyzes back to RCO_2H.
2. The chart is very nearly symmetrical. This makes learning it easy. The deviations from symmetry are:
   (a) The conversion $R-CO-NH_2 \rightarrow R-CO-O-R'$ is missing (unknown or impractical).
   (b) $R-C\equiv N$ connects to the rest via $R-CO-NH_2$.
3. The symmetrically situated $R-CO-Cl$ and $(R-CO-)_2O$ do almost exactly the same things, and they play a central and important role in the interconversions. The preferred routes from RCO_2H to RCO_2R' or $R-CO-NH_2$ are not the direct ones but

CARBOXYLIC ACIDS AND THEIR DERIVATIVES 441

Chart 18.1

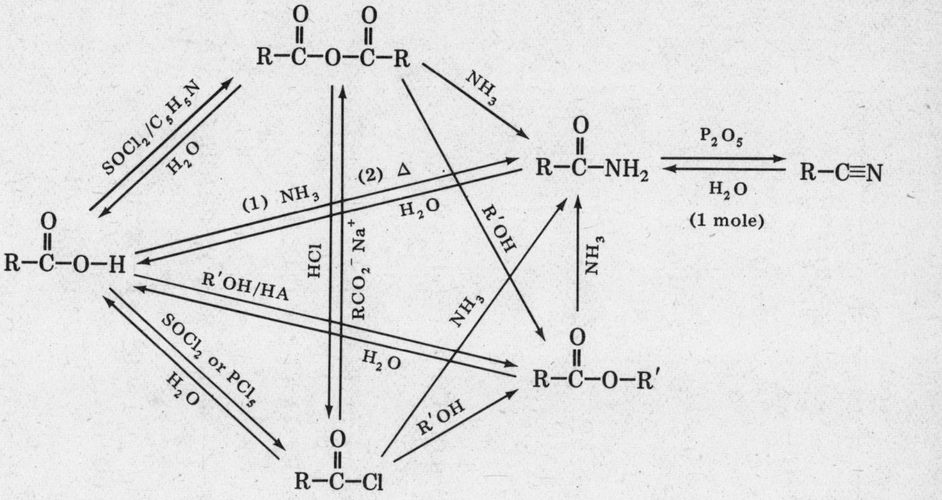

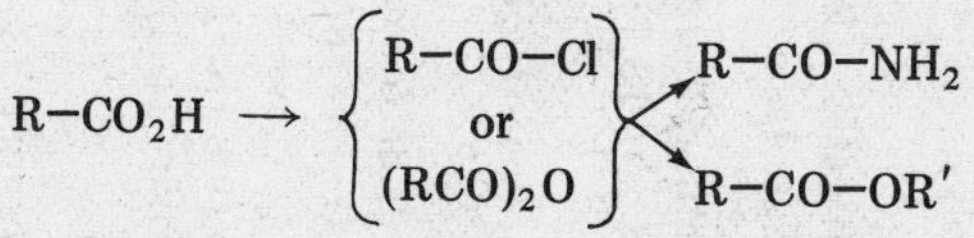

$$\begin{array}{c}
\text{O} \quad\quad \text{O} \\
\| \quad\quad \| \\
\text{R-C-O-C-R}
\end{array}$$

[Interconversion chart showing transformations among R-CO-OH, (RCO)$_2$O, R-CO-Cl, R-CO-NH$_2$, R-CO-OR′, and R-C≡N with reagents: SOCl$_2$/C$_5$H$_5$N, H$_2$O, (1) NH$_3$, (2) Δ, NH$_3$, P$_2$O$_5$, H$_2$O (1 mole), R′OH, R′OH/HA, HCl, RCO$_2^-$Na$^+$, SOCl$_2$ or PCl$_5$, H$_2$O.]

$$R\text{-}CO_2H \rightarrow \left\{\begin{array}{c} R\text{-}CO\text{-}Cl \\ \text{or} \\ (RCO)_2O \end{array}\right\} \begin{array}{c} \nearrow R\text{-}CO\text{-}NH_2 \\ \searrow R\text{-}CO\text{-}OR' \end{array}$$

4. Since R—CO—Cl is easier to make than (RCO)$_2$O, R—CO—Cl is more often used in practice [exception: (CH$_3$—CO—)$_2$O, acetic anhydride, which is readily available and cheap].
5. The hydrolysis of R—C≡N can be done stepwise if one wants to isolate R—CO—NH$_2$, but it can be carried directly to RCO$_2$H in one step if desired.

BASIC MODES OF REACTIVITY

2. Chart 18.2 analyzes the way in which the reactions displayed by the RCO$_2$H and derivatives flow from the structural features present. The next several frames take up in turn the basic modes of reactivity listed in Chart 18.2, beginning here with the acidity of RCO$_2$H.

Replacing R— in R—O—H with $\overset{\overset{\displaystyle O}{\|}}{R-C-}$ to make $\overset{\overset{\displaystyle O}{\|}}{R-C}-O-H$ increases the acidity. The quantitative increase in acidity is huge—a factor of about 10^{10} in K_a (Unit 11, frame 8). This results mainly from resonance delocalization of the negative charge in the conjugate base:

$R-\ddot{\underset{..}{O}}:^\ominus$ versus $R-C\overset{\displaystyle O}{\underset{\displaystyle \ddot{\underset{..}{O}}:^\ominus}{\diagup}} \updownarrow R-C\overset{\displaystyle \ddot{\underset{..}{O}}:^\ominus}{\underset{\displaystyle O}{\diagdown}}$

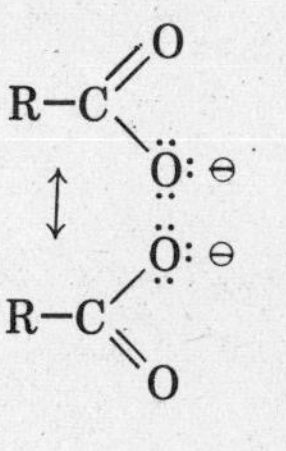

less stable,   more stable,
$\ominus$ localized on $\ominus$ delocalized
one O atom   over two O atoms

The more stable the product, the greater the equilibrium constant:

442 HOW TO SUCCEED IN ORGANIC CHEMISTRY

Chart 18.2

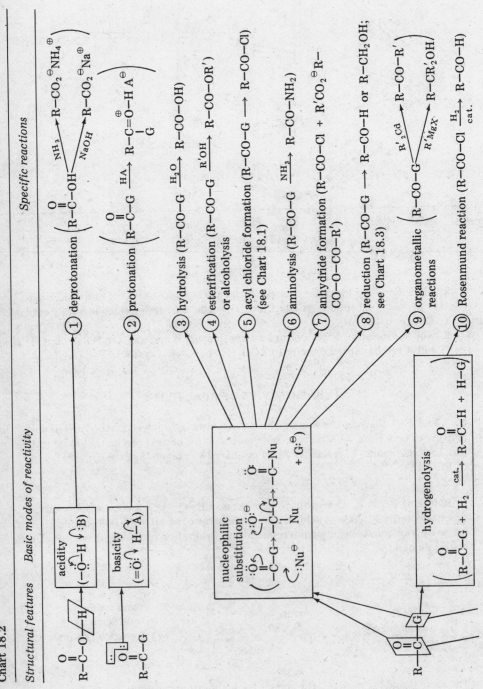

CARBOXYLIC ACIDS AND THEIR DERIVATIVES

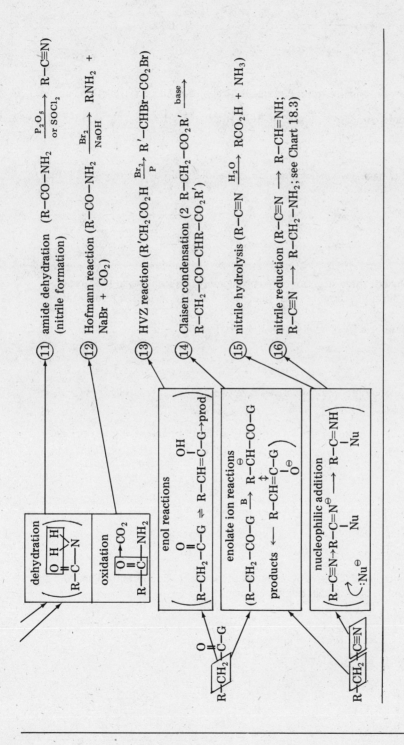

$$R-OH + H_2O \xrightleftharpoons{K_a \approx 10^{-15}} R-O^{\ominus} + H_3O^{\oplus}$$

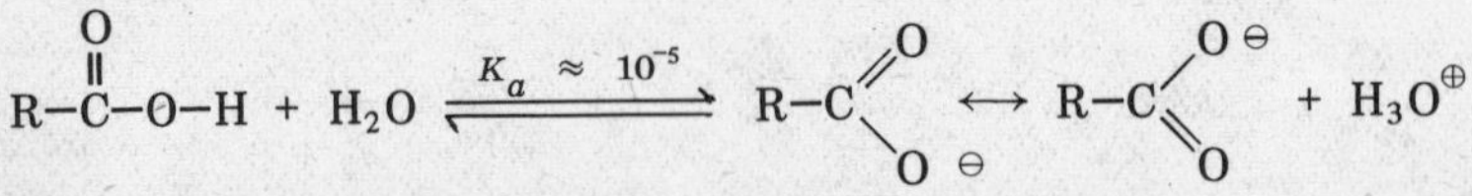

$$R-\overset{\overset{O}{\|}}{C}-O-H + H_2O \xrightleftharpoons{K_a \approx 10^{-5}} R-C\overset{\nearrow O}{\underset{\searrow O^{\ominus}}{}} \leftrightarrow R-C\overset{\nearrow O^{\ominus}}{\underset{\searrow O}{}} + H_3O^{\oplus}$$

Exercise

Use the theory of resonance to predict qualitatively how the acid strength of phenol, ⌬—OH, should compare with that of ROH and R—CO$_2$H.

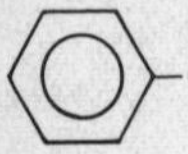

– – – – – – – – – –

Since the conjugate base is resonance-stabilized, phenol should be a stronger acid than ROH:

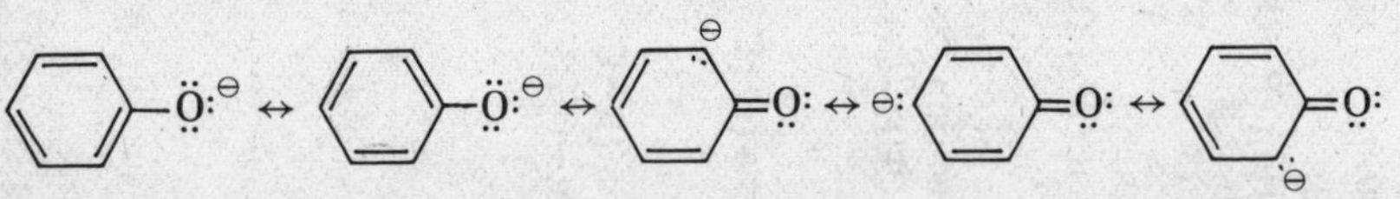

⌬—Ö:$^{\ominus}$ ↔ ⌬—Ö:$^{\ominus}$ ↔ ⌬=O: ↔ $^{\ominus}$:⌬=O: ↔ ⌬=O:

However, since the last three structures are considerably less stable than the first, the resonance stabilization should be less than that of $RCO_2^{\ominus}$, with the result that ⌬—OH should be a weaker acid than RCO_2H.

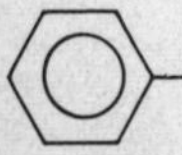

3. Some of the carboxylic acid derivatives have two basic sites $\left(\text{e.g., } R-\overset{\overset{\ddot{O}:}{\|}}{C}-\overset{..}{\overset{..}{O}}-H, \right.$

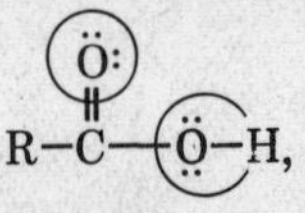

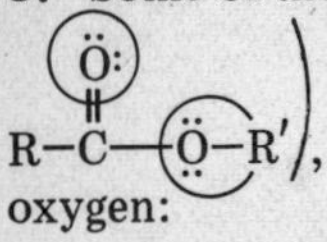

$\left. R-\overset{\overset{\ddot{O}:}{\|}}{C}-\overset{..}{\overset{..}{O}}-R' \right)$, but when they accept a proton it is bound preferentially to the C=O oxygen:

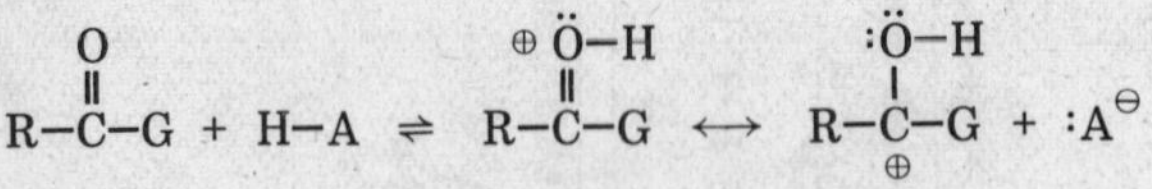

$$R-\overset{\overset{O}{\|}}{C}-G + H-A \rightleftharpoons R-\overset{\overset{\oplus \ddot{O}-H}{\|}}{C}-G \leftrightarrow R-\overset{\overset{:\ddot{O}-H}{|}}{\underset{\oplus}{C}}-G + :A^{\ominus}$$

Again the resonance delocalization of charge in the product is important. When G has unshared pairs, the ⊕ charge is delocalized further:

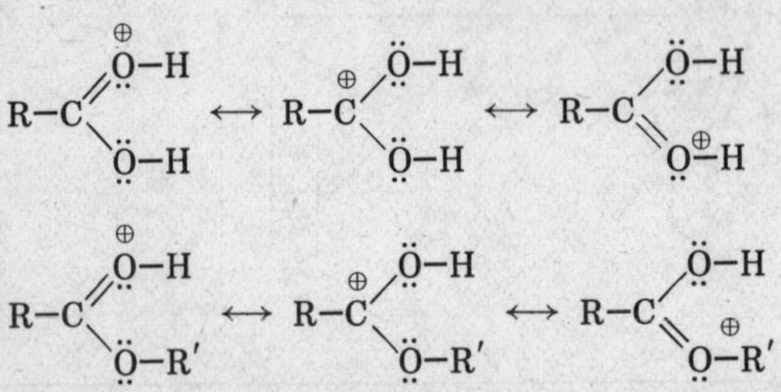

$$R-C\overset{\nearrow \overset{\oplus}{O}-H}{\underset{\searrow \ddot{O}-H}{}} \leftrightarrow R-C\overset{\nearrow \ddot{O}-H}{\underset{\searrow \ddot{O}-H}{}} \leftrightarrow R-C\overset{\nearrow \ddot{O}-H}{\underset{\searrow \overset{\oplus}{O}-H}{}}$$

$$R-C\overset{\nearrow \overset{\oplus}{O}-H}{\underset{\searrow \ddot{O}-R'}{}} \leftrightarrow R-C\overset{\nearrow \ddot{O}-H}{\underset{\searrow \ddot{O}-R'}{}} \leftrightarrow R-C\overset{\nearrow \ddot{O}-H}{\underset{\searrow \overset{\oplus}{O}-R'}{}}$$

As a result, these bases are 10- to 100-fold stronger than the $R-\overset{\overset{O}{\|}}{C}-R'$, in whose conjugate acids the $\oplus$ charge is restricted to two atoms: $R-\overset{\overset{\oplus O-H}{\|}}{C}-R' \leftrightarrow R-\overset{\overset{O-H}{|}}{\underset{\oplus}{C}}-R'$.

Nevertheless, we are dealing here with bases that are very much weaker than water. Consequently, in dilute aqueous solutions of mineral acid, the equilibrium

$$R-C\underset{O-H}{\overset{O}{\diagup\!\!\!\diagdown}} + H_3O^\oplus \rightleftharpoons R-C\underset{O-H}{\overset{O-H}{\diagup\!\!\!\diagdown}}{}^\oplus + H_2O$$

lies so far to the left that no more than one R—CO—G molecule in 10,000 exists in the form of the conjugate acid. Conversely, if one were to prepare pure $R-C\underset{O-H}{\overset{O-H}{\diagup\!\!\!\diagdown}}{}^\oplus \quad A^\ominus$ and drop it into water, it would be instantly and almost completely destroyed by the reversal of the reaction above.

Only in rather concentrated H_2SO_4 is $R-C\underset{G}{\overset{O-H}{\diagup\!\!\!\diagdown}}{}^\oplus$ formed in substantial quantity (e.g., CH_3CO_2H is half converted to $CH_3CO_2H_2{}^\oplus$ in 74% aqueous H_2SO_4). As in the case of the simple carbonyl compounds or the alcohols, however, even a small concentration of the conjugate acid can be important as an intermediate in the reactions of R—CO—G. If $R-C\underset{G}{\overset{OH}{\diagup\!\!\!\diagdown}}{}^\oplus$ reacts faster in some reaction than $R-\overset{\overset{O}{\|}}{C}-G$, that reaction is acid-catalyzed.

Exercise

Draw the conjugate acid of $R-C\equiv N$. Is it resonance-stabilized?

— — — — — — — — — —

$R-C\equiv\overset{\oplus}{N}-H \leftrightarrow R-\overset{\oplus}{C}=\ddot{N}-H$

4. Nucleophilic substitution occurs readily at the sp^2-hybridized carbonyl carbon atom of the carboxylic acid derivatives: $R-C\underset{G}{\overset{O}{\diagup\!\!\!\diagdown}} + Nu: \longrightarrow R-C\underset{Nu}{\overset{O}{\diagup\!\!\!\diagdown}} + G:$. These substitutions have several things in common with the S_N2 and S_N1 reactions that occur at a saturated (sp^3-hybridized) carbon atom in alkyl halides, alcohols, and alcohol derivatives. The two types of nucleophilic substitution differ in the species that can fill the nucleophile and leaving-group slots, and more important, they differ in mechanism. Nature *never* uses the S_N2 mechanism at an unsaturated carbon atom.

A two-step mechanism called the *acyl S_N2* mechanism (mechanism IXa, Table 10.1) applies for substitution at the $-C\underset{G}{\overset{O}{\diagup\!\!\!\diagdown}}$ function:

$$\underset{:Nu^\ominus}{\overset{\overset{O}{\|}}{R-C-G}} \longrightarrow \underset{Nu}{\overset{\overset{O^\ominus}{|}}{R-C-G}} \qquad \text{addition}$$

$$\underset{\underset{Nu}{|}}{\overset{\overset{:\ddot{O}:^{\ominus}}{|}}{R-C-G}} \longrightarrow \underset{}{\overset{\overset{\ddot{O}:}{\|}}{R-C-Nu}} + \cdot\cdot G:^{\ominus} \quad \text{elimination}$$

The first step is exactly the same as the first step in the nucleophilic addition mechanism (mechanism VIII). In both cases a tetrahedral (sp^3-hybridized at the carbonyl carbon) intermediate is formed:

$$\underset{:Nu^{\ominus}}{\overset{\overset{\ddot{O}:}{\|}}{R-C-R'}} \longrightarrow \underset{\underset{Nu}{|}}{\overset{\overset{:\ddot{O}:^{\ominus}}{|}}{R-C-R'}} \quad \text{mechanism VIII, first step}$$

$$\underset{:Nu^{\ominus}}{\overset{\overset{\ddot{O}:}{\|}}{R-C-G}} \longrightarrow \underset{\underset{Nu}{|}}{\overset{\overset{:\ddot{O}:^{\ominus}}{|}}{R-C-G}} \quad \text{mechanism IXa, first step}$$

The difference lies in what happens next, and this is determined by the fact that G is a leaving group ($-Cl, -O-C\overset{\overset{O}{\|}}{-}R', -OH, -OR', -NH_2$), whereas R' is not ($-R' \not\longrightarrow :R'^{\ominus}$; $:R^{\ominus}$ is an *unstable* anion in comparison with $Cl^{\ominus}, {}^{\ominus}O-CO-R$, etc.). As a result, the intermediate derived from R—CO—R' stabilizes itself by picking up a proton

$$\underset{\underset{Nu}{|}}{\overset{\overset{O^{\ominus}}{|}}{R-C-R'}} + H-A \longrightarrow \underset{\underset{Nu}{|}}{\overset{\overset{OH}{|}}{R-C-R'}} + A^{\ominus} \quad \text{protonation (mechanism VIII, second step)}$$

but the intermediate from R—CO—G spits out G and restores the $-\overset{\overset{O}{\|}}{C}-$ group, leading to a net, overall replacement of G by Nu—one nucleophile by another:

$$\underset{\underset{Nu}{|}}{\overset{\overset{:\ddot{O}:^{\ominus}}{|}}{R-C-G}} \longrightarrow \underset{}{\overset{\overset{\ddot{O}}{\|}}{R-C-Nu}} + :G^{\ominus} \quad \text{elimination of leaving group (mechanism IXa, second step)}$$

When G = $-OH, -OR'$, or $-NH_2$, acids catalyze the second step by making G into a viable leaving group (just as in the ordinary $S_N 2$ mechanism; see frame 1 of Unit 15), for instance

$$\underset{\underset{Nu}{|}}{\overset{\overset{O^{\ominus}}{|}}{R-C-OH}} \overset{HA}{\rightleftharpoons} \underset{\underset{Nu}{|}}{\overset{\overset{OH}{|}}{R-C-OH}} \overset{HA}{\rightleftharpoons} \underset{\underset{Nu}{|}}{\overset{\overset{:\ddot{O}H}{|}}{R-C-\overset{\oplus}{O}H_2}} A^{\ominus} \longrightarrow \underset{}{\overset{\overset{:O-H}{\|}}{R-C-Nu}} + H_2O$$

$$\downarrow A^{\ominus}$$

$$\underset{}{\overset{\overset{\ddot{O}:}{\|}}{R-C-Nu}} + HA$$

As in the case of nucleophilic addition, the *addition* step may also be acid-catalyzed, especially when the nucleophile Nu: is not very reactive. An example is ester hydrolysis:

$$\underset{}{\overset{\overset{O}{\|}}{R-C-O-R'}} + HA \rightleftharpoons \underset{}{\overset{\overset{\oplus OH}{\|}}{R-C-O-R'}} \longleftrightarrow \underset{\oplus}{\overset{\overset{OH}{|}}{R-C-O-R'}} + A^{\ominus}$$

CARBOXYLIC ACIDS AND THEIR DERIVATIVES

$$\overset{\oplus}{\text{:}\ddot{O}H} \qquad \text{:}\ddot{O}H \qquad \text{:}\ddot{O}H$$
$$R-\overset{\|}{C}-O-R' \longleftrightarrow R-\overset{|}{\underset{\oplus}{C}}-O-R' \longrightarrow R-\overset{|}{C}-O-R' \rightleftharpoons$$
$$\text{:}\ddot{O}H_2 \qquad\qquad\quad \underset{H}{\overset{:O}{\diagdown}}\overset{\oplus}{\underset{H}{\diagup}}$$

$$\overset{\text{:}\ddot{O}H}{\underset{\underset{OH}{|}}{R-C-\underset{H}{\overset{R'}{O}}}} \longrightarrow R-\overset{:O-H}{\underset{\|}{C}}-OH \;\;\overset{:A^{\ominus}}{\rightleftharpoons}\; R-\overset{O}{\underset{\|}{C}}-OH + HA$$

The $-\overset{O}{\underset{\|}{C}}-$ group in $R-\overset{O}{\underset{\|}{C}}-G$ is not as reactive toward nucleophiles as that in $R-CO-R'$. This can be visualized as resulting from decreased carbonium ion character at the carbonyl carbon due to electron donation of the unshared pair on G by resonance of the type

$$R-\overset{\text{:}\ddot{O}:}{\underset{\|}{C}}-\ddot{G} \longleftrightarrow R-\overset{\text{:}\ddot{O}:^{\ominus}}{\underset{\underset{\oplus}{|}}{C}}-\ddot{G} \longleftrightarrow R-\overset{\text{:}\ddot{O}:^{\ominus}}{\underset{|}{C}}=G^{\oplus}$$

Now exactly how does this acyl S_N2 mechanism relate to the ordinary S_N2 mechanism employed by the $R-X$, $R-OH_2^{\oplus}$, and so on? Acyl S_N2 is a two-step, addition-then-elimination mechanism as opposed to direct, or one-step, nucleophilic substitution:

$$\text{acyl } S_N2: \quad R-\overset{\text{:}\ddot{O}:}{\underset{\|}{C}}-G \longrightarrow R-\overset{\text{:}\ddot{O}:^{\ominus}}{\underset{\underset{Nu}{|}}{C}}-G \longrightarrow R-\overset{\text{:}\ddot{O}}{\underset{\|}{C}}-Nu + :G^{\ominus}$$
$$\quad\quad\quad\quad :Nu^{\ominus}$$

versus

$$S_N2: \quad Nu:^{\ominus} \; R-X \longrightarrow Nu-R + X^{\ominus}$$

Exercise

Compare the acyl S_N2 mechanism with the aryl S_N2 mechanism (mechanism IXb; see frame 3 of Unit 16).

— — — — — — — — —

The acyl S_N2 is, in fact, very closely analogous to the aryl S_N2 mechanism, which is also an addition + elimination:

$$O_2N-\underset{NO_2}{\underset{|}{\overset{}{\bigcirc}}}-G \longrightarrow O_2N-\underset{NO_2}{\underset{|}{\overset{\ominus}{\bigcirc}}}\overset{Nu}{\underset{G}{\diagdown}} \longleftrightarrow O_2N^{\ominus}-\underset{NO_2}{\underset{|}{\overset{}{\bigcirc}}}\overset{Nu}{\underset{G}{\diagdown}} \longleftrightarrow \underset{O^{\ominus}}{\overset{O^{\ominus}}{\diagdown}}N=\underset{NO_2}{\underset{|}{\overset{}{\bigcirc}}}\overset{Nu}{\underset{G}{\diagdown}}$$
$$\quad :Nu^{\ominus}$$

$$:G^{\ominus} + O_2N-\underset{NO_2}{\underset{|}{\overset{}{\bigcirc}}}-Nu \longleftarrow O_2N-\underset{NO_2}{\underset{|}{\overset{}{\bigcirc}}}\overset{Nu}{\underset{G}{\diagdown}} \longleftrightarrow O_2N-\underset{\underset{O^{\ominus}}{\overset{N}{\diagdown}O}}{\underset{}{\overset{}{\bigcirc}}}\overset{Nu}{\underset{G}{\diagdown}}$$

In both cases the intermediate anion is reasonably stable because the negative charge is lodged on electronegative O. The only difference is the number of electron shifts required to get $\ominus$ onto O in the aryl S_N2 case.

5. Hydrogenolysis is the word used to designate catalytic reactions of H_2 in which a σ rather than a π bond is broken, and the result is substitution rather than addition:

$$A{=}B + H{-}H \xrightarrow{\text{catalyst}} \underset{\underset{H\ H}{|\ \ |}}{A{-}B} \qquad \text{hydrogenation}$$

$$A{-}B + H{-}H \xrightarrow{\text{catalyst}} A{-}H + H{-}B \qquad \text{hydrogenolysis}$$

Hydrogenolysis reactions are not very general, and are useful mostly in special situations. Their mechanism is unknown. The hydrogenolysis of acyl chlorides to aldehydes (Rosenmund reaction) is important only because aldehydes are difficult to make, owing to the tendency to overoxidize or overreduce, destroying the product. In the Rosenmund reaction this is avoided by using a deliberately deactivated catalyst.

6. Dehydrations are not a class of mechanistically homogeneous reactions. They are brought about by reagents (e.g., H_2SO_4, P_2O_5, $SOCl_2$, PCl_5) that either tightly bind or destroy the water molecule, but the mechanisms of the dehydration reactions are diverse and rarely if ever involve simply removal of a water molecule from the product side of the reaction. Ordinarily, the dehydration reagent makes $-OH$ into a leaving group:

$$-OH \xrightarrow{H_2SO_4} -\overset{\oplus}{O}H_2$$

$$-OH \xrightarrow{P_2O_5} -O-\overset{O}{\underset{OH}{P}}-O-\overset{O}{\underset{OH}{P}}-OH$$

$$-OH \xrightarrow{SOCl_2} -O-\overset{O}{S}-Cl$$

This is followed by elimination or substitution, for example

$$\underset{H}{\overset{|}{\underset{|}{C}}}{-}\underset{H}{\overset{|}{\underset{|}{C}}}{-}O{-}H \xrightarrow{H_2SO_4} \underset{H}{\overset{|}{\underset{|}{C}}}{-}\underset{H}{\overset{|}{\underset{|}{C}}}{-}\overset{\oplus}{O}H_2 \longrightarrow \underset{H}{\overset{|}{\underset{\underset{\underset{:\ddot{O}{-}SO_3H}{\nwarrow}}{\ominus}}{C}}}{-}\overset{\oplus}{C}\overset{|}{\underset{|}{}} + H_2O$$

$$ \diagdown C{=}C \diagup + H_2SO_4$$

Exercise

Write a possible mechanism for the dehydration of $R{-}C\overset{\displaystyle O}{\underset{\displaystyle NH_2}{\diagdown}}$ to $R{-}C{\equiv}N$ by P_2O_5.

Use the tautomeric equilibrium $R{-}C\overset{\displaystyle O}{\underset{\displaystyle NH_2}{\diagdown}} \rightleftharpoons R{-}C\overset{\displaystyle OH}{\underset{\displaystyle NH}{\diagdown}}$ to get started. The easiest way to visualize esterification by P_2O_5 is via the partially hydrolyzed species $H_2P_2O_6$:

$$\underset{R-C\diagdown NH}{\overset{O}{\parallel}} \quad \underset{\ddot{O}-H}{\overset{O=P-O-P-OH}{\underset{O}{|}}} \overset{OH}{\underset{|}{|}}$$

- - - - - - - -

[Mechanism showing protonation/deprotonation steps leading to nitrilium ion and eventually nitrile $R-C\equiv N$:]

etc. $\longleftrightarrow R-C\equiv\overset{\oplus}{N}-H$ $R-C\equiv N:$

7. Only CO_2 is a higher oxidation state for carbon than the C in $R-C\diagup^{O}_{\diagdown OH}$. So oxidizing the $-CO_2H$ function requires breaking the $R-CO_2H$ bond. Several reactions are known that will accomplish this (Schmidt, Curtius, Lossen, and Hofmann reactions). In the Hofmann reaction the oxidizing agent is halogen:

$$R-C\diagup^{O}_{\diagdown NH_2} \underset{Br-Br}{\longrightarrow} R-C\diagup^{O}_{\diagdown \overset{\oplus}{NH_2} Br^{\ominus}} \xrightarrow{NaOH} R-C\diagup^{O}_{\diagdown NHBr} \xrightarrow{NaOH} R-C\diagup^{O}_{\diagdown N^{\ominus}-Br}$$

$$CO_2 + H_2NR \longleftarrow \underset{HO}{\overset{O}{\diagdown}}C-NH-R \xleftarrow{H_2O} O=C=N-R \longleftarrow R-C\diagup^{O}_{\diagdown \ddot{N}:} + Br^{\ominus}$$

8. Reactions of the enol form of $R-CH_2-CO_2H$ are rare, but one very useful reaction of $R-CH_2-CO_2H$ appears to be an electrophilic addition to the enol form of the acyl bromide:

$$R-CH_2-\overset{O}{\overset{\parallel}{C}}-OH \xrightarrow[PBr_3]{Br_2} R-CH_2-\overset{O}{\overset{\parallel}{C}}-Br \rightleftharpoons R-CH=\overset{OH}{\overset{|}{C}}-Br$$

$$\underset{Br}{\overset{}{R-CH}}-\overset{O}{\overset{\parallel}{C}}-Br \xleftarrow{-HBr} \underset{Br}{\overset{}{R-CH}}-\overset{OH}{\overset{|}{C}}-Br \xleftarrow{Br_2}$$
                                    $\qquad\qquad\qquad\qquad\;\; |$
                                    $\qquad\qquad\qquad\qquad\; Br$

Note that the addition of Br_2 followed by elimination of HBr amounts to substitution of Br for H overall. Note also that the halogenation occurs specifically at the α position since no other position is activated by enolization.

9. Like $-\overset{O}{\underset{\|}{C}}-R$, all the groups $-\overset{O}{\underset{\|}{C}}-G$ (and $-C\equiv N$) make adjacent α hydrogens somewhat acidic. Again this is due to resonance stabilization of the conjugate base, an enolate ion; for example:

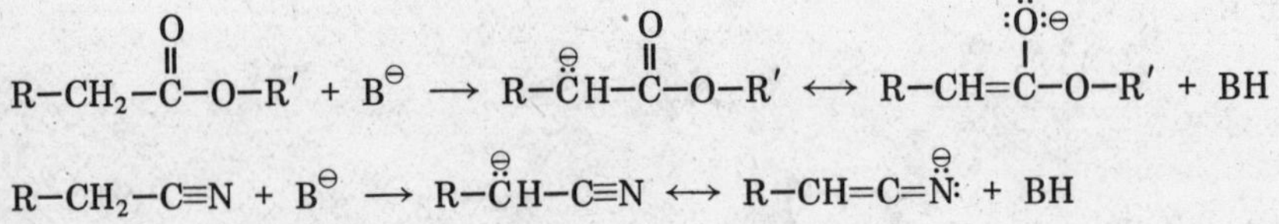

$$R-CH_2-\overset{O}{\underset{\|}{C}}-O-R' + B^{\ominus} \longrightarrow R-\overset{\ominus}{C}H-\overset{O}{\underset{\|}{C}}-O-R' \longleftrightarrow R-CH=\overset{:\overset{..}{O}:^{\ominus}}{\underset{|}{C}}-O-R' + BH$$

$$R-CH_2-C\equiv N + B^{\ominus} \longrightarrow R-\overset{\ominus}{C}H-C\equiv N \longleftrightarrow R-CH=C=\overset{\ominus}{\overset{..}{N}:} + BH$$

The α-H acidity in the carboxylic acid is less pronounced because of the presence of the moderately acidic $-OH$ group, which ionizes first. The α hydrogen must be removed from a species that already bears a full negative charge:

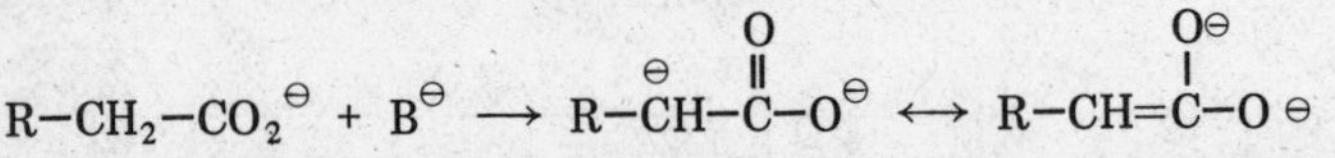

$$R-CH_2-CO_2{}^{\ominus} + B^{\ominus} \longrightarrow R-\overset{\ominus}{C}H-\overset{O}{\underset{\|}{C}}-O^{\ominus} \longleftrightarrow R-CH=\overset{O^{\ominus}}{\underset{|}{C}}-O^{\ominus}$$

Electrostatic repulsion of the two like charges destabilizes this enolate ion.

The $R-\overset{\ominus}{C}H-C\overset{\nearrow O}{\underset{\searrow G}{}} \longleftrightarrow R-CH=C\overset{\nearrow O^{\ominus}}{\underset{\searrow G}{}}$ enolate ions undergo the same reactions as those derived from aldehydes and ketones, namely nucleophilic additions and S_N2 reactions (alkylations). The alkylations suffer from the same limitations described for the $R-\overset{\ominus}{C}H-CO-R'$ ions (Unit 17, frame 14).

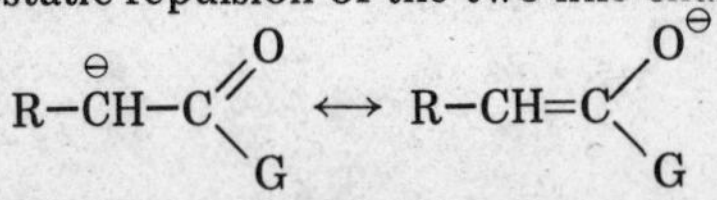

The enolate ions derived from carboxylic acid derivatives undergo an important new reaction type: acyl S_N2 reaction with its conjugate acid. In practice the reaction is carried out on the ester (Claisen condensation):

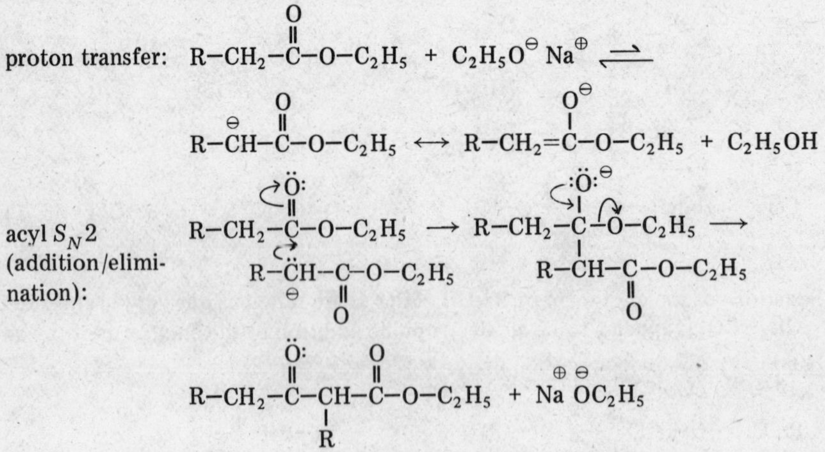

proton transfer: $R-CH_2-\overset{O}{\underset{\|}{C}}-O-C_2H_5 + C_2H_5O^{\ominus} Na^{\oplus} \rightleftharpoons$

$R-\overset{\ominus}{C}H-\overset{O}{\underset{\|}{C}}-O-C_2H_5 \longleftrightarrow R-CH_2=\overset{O^{\ominus}}{\underset{|}{C}}-O-C_2H_5 + C_2H_5OH$

acyl S_N2 (addition/elimination):

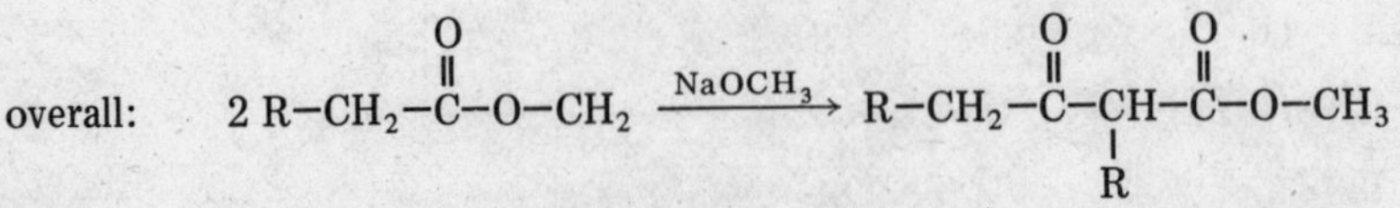

$R-CH_2-\overset{\overset{\curvearrowleft \overset{..}{O}:}{\|}}{\underset{\underset{\underset{O}{\|}}{R-\overset{\ominus}{C}H-C-O-C_2H_5}}{C}}-O-C_2H_5 \longrightarrow R-CH_2-\overset{:\overset{..}{O}:^{\ominus}}{\underset{\underset{\underset{O}{\|}}{R-\overset{}{C}H-C-O-C_2H_5}}{\underset{|}{C}\overset{\curvearrowright}{-}\overset{..}{O}-C_2H_5}} \longrightarrow$

$R-CH_2-\overset{\overset{..}{O}:}{\underset{\|}{C}}-\overset{}{\underset{\underset{R}{|}}{C}H}-\overset{O}{\underset{\|}{C}}-O-C_2H_5 + Na^{\oplus} OC_2H_5{}^{\ominus}$

overall: $2\ R-CH_2-\overset{O}{\underset{\|}{C}}-O-CH_2 \xrightarrow{NaOCH_3} R-CH_2-\overset{O}{\underset{\|}{C}}-\underset{\underset{R}{|}}{C}H-\overset{O}{\underset{\|}{C}}-O-CH_3$

This is an exact analog of the aldol condensation, only on a different oxidation level, and with nucleophilic substitution in place of nucleophilic addition. The Claisen reactant is one oxidation state above the aldol reactant, and the two functional groups of the product are each the next oxidation level up from the aldol product:

CARBOXYLIC ACIDS AND THEIR DERIVATIVES

aldol: $\quad 2\ R-CH_2-\overset{\overset{O}{\|}}{C}-H \longrightarrow R-CH_2-\boxed{CH}-CH-\boxed{\overset{\overset{O}{\|}}{C}-H}$
$\qquad\qquad\qquad\qquad\qquad\qquad\qquad\quad | $
$\qquad\qquad\qquad\qquad\qquad\qquad\qquad\quad R$
(with OH on the first boxed CH)

Claisen: $\quad 2\ R-CH_2-\overset{\overset{O}{\|}}{C}-O-CH_3 \longrightarrow R-CH_2-\boxed{\overset{\overset{O}{\|}}{C}}-CH-\boxed{\overset{\overset{O}{\|}}{C}-OCH_3}$
$\qquad\qquad\qquad\qquad\qquad\qquad\qquad\qquad\qquad\qquad\quad | $
$\qquad\qquad\qquad\qquad\qquad\qquad\qquad\qquad\qquad\quad\ R$

To write the Claisen condensation product for a given ester, just replace one of the α hydrogens of the ester with the acyl group of the ester, $R-CH_2-C\overset{\nearrow O}{\searrow OR'}$. For example,

$$CH_3-CH-C\overset{\nearrow O}{\searrow O-C_2H_5} \xrightarrow{base} CH_3-CH-C\overset{\nearrow O}{\searrow O-C_2H_5}$$
$\qquad\quad\ |\qquad\qquad\qquad\qquad\qquad\qquad\qquad\quad |$
$\qquad\ \boxed{H}\qquad\qquad\qquad\qquad\qquad\qquad\quad O=C$
$\qquad\qquad\qquad\qquad\qquad\qquad\qquad\qquad\qquad |$
$\qquad\qquad\qquad\qquad\qquad\qquad\qquad\qquad CH_2-CH_3$

★ *Exercises*

1. Draw the product of the Claisen condensation of each of these esters, or write "No reaction" if the ester is incapable of this reaction.

   (a) $\begin{array}{c}CH_3\\ \\ CH_3\end{array}\!\!\!\searrow\!\!CH-CH_2-C\overset{\nearrow O}{\searrow O-CH_3}$

   (b) $CH_3-CH_2-O-C\overset{\nearrow O}{\searrow \phi}$

   (c) $\phi-CH_2-C\overset{\nearrow O}{\searrow O-CH_2-CH_3}$

2. Write the mechanism for this *intramolecular* Claisen condensation:

$$CH_3-O-\overset{\overset{O}{\|}}{C}-(CH_2)_5-\overset{\overset{O}{\|}}{C}-O-CH_3 \xrightarrow{CH_3O^{\ominus}\ Na^{\oplus}} \text{(cyclohexanone with } CO_2CH_3 \text{ group)}$$

- - - - - - - - - - -

1. (a) $\begin{array}{c}CH_3\\ \\ CH_3\end{array}\!\!\!\searrow\!\!CH-CH-C\overset{\nearrow O}{\searrow O-CH_3}$
$\qquad\qquad\qquad\quad |$
$\qquad\qquad\qquad\ \ C$
$\qquad\qquad\quad O\!\!\nearrow\ \searrow\!\!CH_2-CH\!\!\!\begin{array}{c}\nearrow CH_3\\ \searrow CH_3\end{array}$

   (b) no reaction (no α hydrogens)

   (c) $\phi-CH-C\overset{\nearrow O}{\searrow O-CH_2-CH_3}$
$\qquad\ \ |$
$\qquad O=C$
$\qquad\quad |$
$\qquad\ CH_2-\phi$

2.
$$CH_3O-\overset{O}{\underset{\|}{C}}-(CH_2)_4-\underset{\underset{:\overset{\ominus}{O}CH_3}{H}}{CH}-\overset{O}{\underset{\|}{C}}-O-CH_3 \rightleftharpoons CH_3-O-\overset{O}{\underset{\|}{C}}-(CH_2)_4-\overset{\ominus}{\underset{|}{C}H}-\overset{O}{\underset{\|}{C}}-O-CH_3$$

$$+ HOCH_3$$

[cyclization scheme showing intermediates with $CH_3-O-\overset{O}{\underset{\|}{C}}$ groups and cyclohexanone-like product]

10. Unlike the other carboxylic acid derivatives, the nitriles, R—C≡N, have no leaving group and cannot undergo nucleophilic displacements. The cyano group, $\overset{+\delta\;-\delta}{-C\equiv N}$, behaves like carbonyl, $\overset{+\delta\;-\delta}{>C=O}$, and undergoes *nucleophilic addition* reactions, for instance

$$R-C\equiv N: \longrightarrow R-\underset{\underset{\oplus}{:\ddot{O}H_2}}{\overset{:\ddot{N}:^{\ominus}}{\underset{\|}{C}}} \rightleftharpoons R-\underset{\underset{}{:\ddot{O}H}}{\overset{\ddot{N}-H}{\underset{\|}{C}}} \rightleftharpoons R-\overset{\overset{\ddot{O}:}{\|}}{C}-\ddot{N}H_2$$

with :ÖH₂ as the incoming nucleophile,

and

$$R-C\equiv N: \longrightarrow R-\underset{\underset{\oplus}{Ar-NH_2}}{\overset{:\ddot{N}:^{\ominus}}{\underset{\|}{C}}} \rightleftharpoons R-\underset{\underset{}{:NH-Ar}}{\overset{\ddot{N}-H}{\underset{\|}{C}}}$$

with Ar—ṄH₂ as the incoming nucleophile.

These additions present two new facets not seen with carbonyl compounds: (a) The initial product may be the less stable of a pair of tautomers, and consequently may immediately tautomerize as in the example of R—C=N—H, above. (b) The product of the

$$\underset{OH}{|}$$

nucleophilic addition reaction may be an R—CO—G molecule that *does* have a leaving group and is capable of undergoing acyl nucleophilic *substitutions*. This is what happens in nitrile hydrolysis to R—CO₂H. The nucleophilic *addition* of water to give R—CO—NH₂ is followed by acyl nucleophilic *substitution* of NH₂ by —OH with water as nucleophile to give R—CO—OH (reaction ③, Chart 18.2).

Exercise

What reaction or reactions is the process

$$\text{Ph}-CH_2-C\equiv N \xrightarrow[\text{(strong base)}]{:B^{\ominus}} \text{Ph}-CH_2-\overset{NH}{\underset{\|}{C}}-\underset{\underset{Ph}{|}}{CH}-C\equiv N$$

most closely analogous to? Working by analogy, write a mechanism for this reaction.

— — — — — — — — —

Closely analogous to the aldol and (somewhat less closely) to the Claisen condensation:

$C_6H_5-CH-C\equiv N:$ → $C_6H_5-\overset{\ominus}{\underset{\curvearrowleft}{CH}}-C\equiv N:$ → $C_6H_5-CH-C\equiv N$
$|$ $\updownarrow$ $|$
H $C_6H_5-CH=C=\overset{\ominus}{\underset{..}{N}}:$ $:\overset{\ominus}{N}=C-CH_2-C_6H_5$
$\overset{\curvearrowleft}{:}B^{\ominus}$ $:N\equiv C-CH_2-C_6H_5$ $\downarrow BH^{\oplus}$
$C_6H_5-CH-C\equiv N$
$|$
$HN=C-CH_2-C_6H_5$

RELATIONSHIP TO OTHER FUNCTIONS

11. The first important group of reactions involving the $R-C\overset{\displaystyle O}{\underset{\displaystyle G}{\diagdown}}$ compounds consists of the reactions that convert these derivatives into one another. These were summarized in Chart 18.1. Practically all the remaining $R-C\overset{\displaystyle O}{\underset{\displaystyle G}{\diagdown}}$ chemistry involves conversions to and from the lower oxidation states. An oxidation state-alkylation state diagram such as Charts 15.2 and 17.2 is handy for this purpose, but now we are dealing with a whole packet of compounds in the slot formerly occupied by $R-CO_2H$ in oxidation stage III, Chart 18.3. Each of the $R-C\overset{\displaystyle O}{\underset{\displaystyle G}{\diagdown}}$ functions turns out to have a different relationship with

Chart 18.3

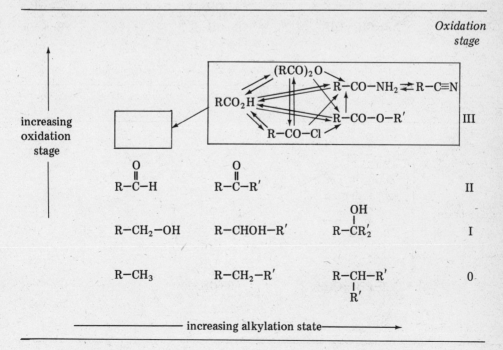

454 HOW TO SUCCEED IN ORGANIC CHEMISTRY

the lower oxidation states. Consequently, it is best to draw a separate chart for each carboxylic acid derivative. These are shown as Charts 18.3A to 18.3E. The

Chart 18.3A

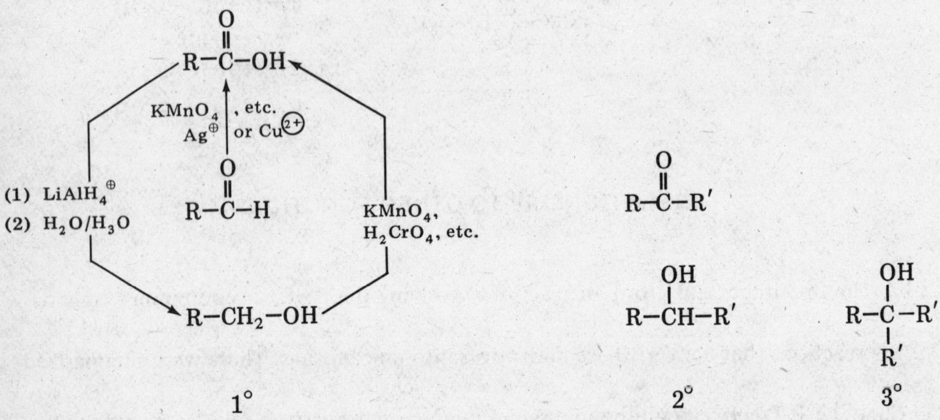

$$R-\overset{O}{\underset{\|}{C}}-OH$$

KMnO$_4$, etc.
Ag$^{\oplus}$ or Cu^{2+}

(1) LiAlH$_4^{\ominus}$
(2) H$_2$O/H$_3$O$^{\oplus}$

$$R-\overset{O}{\underset{\|}{C}}-H$$

KMnO$_4$, H$_2$CrO$_4$, etc.

$$R-\overset{O}{\underset{\|}{C}}-R'$$

R−CH$_2$−OH   R−CH−R'   R−C−R'
                      | |
                      OH   OH
                                        |
                                        R'
  1°   2°   3°

Chart 18.3B

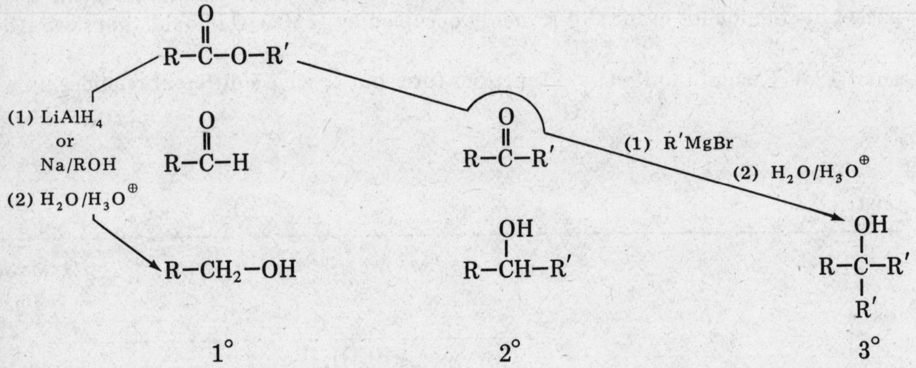

$$R-\overset{O}{\underset{\|}{C}}-O-R'$$

(1) LiAlH$_4$
   or
   Na/ROH
(2) H$_2$O/H$_3$O$^{\oplus}$

$$R-\overset{O}{\underset{\|}{C}}-H$$ $$R-\overset{O}{\underset{\|}{C}}-R'$$

(1) R'MgBr
(2) H$_2$O/H$_3$O$^{\oplus}$

R−CH$_2$−OH   R−CH−R'   R−C−R'
                      | |
                      OH   OH
                                        |
                                        R'
  1°   2°   3°

Chart 18.3C

$$R-\overset{O}{\underset{\|}{C}}-Cl$$

LiAlH(OR)$_3$ or   R'−Cd−R'
H$_2$/Pd on BaSO$_4$

(1) LiAlH$_4$
(2) H$_2$O/H$_3$O$^{\oplus}$

$$R-\overset{O}{\underset{\|}{C}}-H$$ $$R-\overset{O}{\underset{\|}{C}}-R'$$

R−CH$_2$−OH   R−CH−R'   R−C−R'
                      | |
                      OH   OH
                                        |
                                        R'
  1°   2°   3°

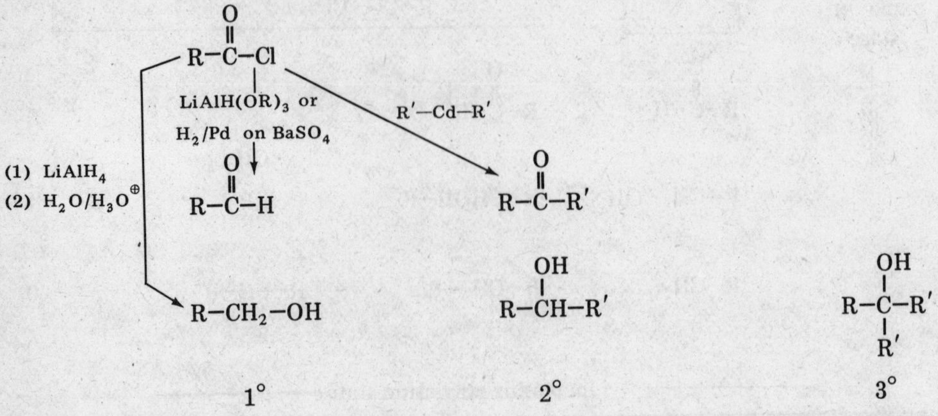

CARBOXYLIC ACIDS AND THEIR DERIVATIVES

Chart 18.3D

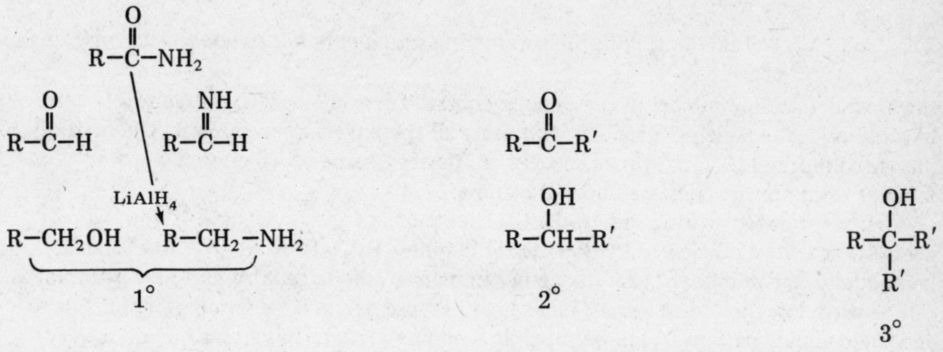

$$\begin{array}{c} O \\ \parallel \\ R-C-NH_2 \end{array}$$

$$\begin{array}{ccc} O & & NH \\ \parallel & & \parallel \\ R-C-H & & R-C-H \end{array} \qquad \begin{array}{c} O \\ \parallel \\ R-C-R' \end{array}$$

LiAlH$_4$

$$R-CH_2OH \qquad R-CH_2-NH_2 \qquad \begin{array}{c} OH \\ | \\ R-CH-R' \end{array} \qquad \begin{array}{c} OH \\ | \\ R-C-R' \\ | \\ R' \end{array}$$

$\underbrace{\qquad\qquad\qquad}_{1°} \qquad\qquad\qquad 2° \qquad\qquad\qquad 3°$

Chart 18.3E

$$R-C\equiv N$$

H$_2$/Ni or LiAlH$_4$ $\qquad$ LiAlH(OR)$_3$

$$\begin{array}{c} O \\ \parallel \\ R-C-H \end{array} \xleftarrow{H_2O/H_3O^{\oplus}} \begin{array}{c} NH \\ \parallel \\ R-C-H \end{array} \qquad \begin{array}{c} O \\ \parallel \\ R-C-R' \end{array}$$

$$R-CH_2OH \qquad R-CH_2-NH_2 \qquad \begin{array}{c} OH \\ | \\ R-CH-R' \end{array} \qquad \begin{array}{c} OH \\ | \\ R-C-R' \\ | \\ R' \end{array}$$

$\underbrace{\qquad\qquad\qquad}_{1°} \qquad\qquad\qquad 2° \qquad\qquad\qquad 3°$

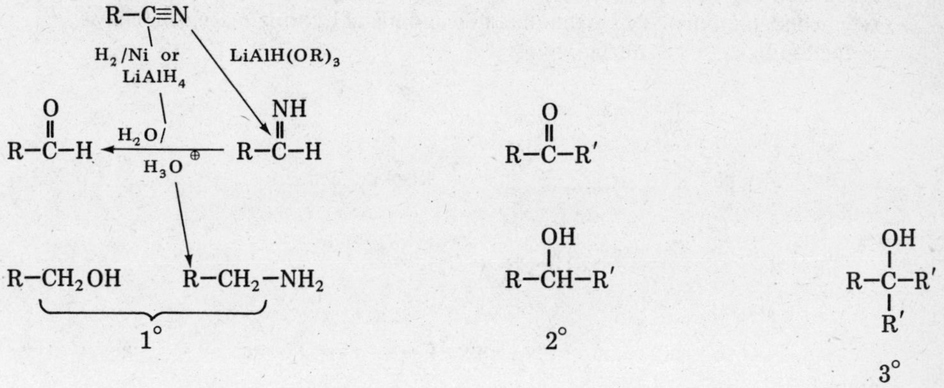

one reaction that each chart has in common is reduction of $R-\overset{\overset{O}{\parallel}}{C}-G$ to the lowest oxidation state via strong reducing agents such as LiAlH$_4$. Note, however, that the nitrogen-containing derivatives, $R-CO-NH_2$ and $R-C\equiv N$, reduce not to the alcohol, $R-CH_2-OH$, but to its nitrogen analog, the primary amine $R-CH_2-NH_2$. Similarly, the *partial* reduction of $R-C\equiv N$ gives $R-CH=NH$, which is the nitrogen analog of an aldehyde (Unit 19, frame 1). It is hydrolyzed to the aldehyde by aqueous acid.

Only two downward diagonals proceed from the $R-CO-G$ to more highly alkylated compounds via organometallic reagents. They are important reactions because they make new C—C bonds and are consequently valuable in synthetic problems. We do not develop that aspect here but in Unit 21. There we try to tie all these organometallic reactions together, place them in context with other synthetically useful reactions, and develop methods for solving synthetic problems with them.

Charts 18.1, 18.2, and 18.3 are useful in learning the $R-CO-G$ reactions. You will find these applications of the charts in the next three frames.

LEARNING THE REACTIONS INVOLVING THE R—CO—G

12. The $-C\overset{O}{\underset{G}{\diagdown}}$ functions, with their numerous and highly interrelated reactions, are the most demanding subject in the organic course. There are no difficult concepts here, but the task of organizing, recalling, and using all these facts is substantial. The method presented in frames 12 to 14 uses a variety of devices, some of which you will find familiar from previous units; some will be new.

First we work on writing the products of reactions of the R—CO—G. If you try to learn the reactions of each of the R—CO—G families (RCO_2H, $RCONH_2$, $RCOCl$, etc.) individually, the number of items becomes unmanageably large. Slot-and-filler techniques can be used, but they seem less efficient here because of the large number of slot fillers and the complex pattern of nonpermissible combinations. In this situation you are better off to do more figuring out and less remembering. The first scheme we will try has you taking your clues for the "figuring out" from a chemical classification of the reagent.

The basic reagent classification is shown in Table 18.1.

The reactions brought about by each class are partly obvious, partly not:

- Oxidizing agents oxidize, reducing agents reduce. But some of the reducing agents are also strong bases (e.g., R—Li, LiH) and may bring about base-catalyzed reactions (e.g., condensations). The organometallic and metal hydride reagents achieve reduction by acting as nucleophiles:

$$Li^{\oplus} \ H_3Al-H \quad \overset{R}{\underset{|}{CH_2}}-X \ \longrightarrow \ H_3Al \ + \ H-\overset{R}{\underset{|}{CH_2}} \ + \ X^{\ominus} \ Li^{\oplus}$$

   stage I   stage 0

$$Li-CH_3 \quad \overset{CH_3}{\underset{|}{CH}}=O \ \longrightarrow \ CH_3-\overset{CH_3}{\underset{|}{CH}}-O^{\ominus} \ Li^{\oplus}$$

   stage II   stage I

  These are reductions because they replace C—X or C—O bonds by C—H and C—C bonds.
- Acids form salts with basic reactants as do bases with acidic reactants. But in the presence of water, both can act as catalysts for hydrolysis reactions. Many acids also play the role of electrophile in electrophilic addition or substitution. Strong bases in anhydrous media also bring about condensations and other reactions of enolate ions.
- Nucleophiles are always bases also, although many of them are weak bases.
- Alkylating agents are combinations of skeleton + leaving group—compounds that give S_N reactions with nucleophiles.
- Dehydrating agents are those that make —OH into a leaving group by protonation or esterification.

Now the reactions of Chart 18.2 can be classified by reagent type, as shown in Table 18.2.

This is the best organization under which to memorize the R—CO—G reactions, but before you start drilling on them a few cautions and amplifications are in order:

- The reactions in Table 18.2 that are written with R—CO—G as reactant do not work for all five choices of G, except for the hydrolyses, R—CO—G + $H_2O \longrightarrow$ RCOOH. Remembering these restrictions—which choices of G work, which pairs of functions are connected by a workable reaction—is best done by other means taken up in frame 14.
- Some reactions appear in more than one place in the table, for example, hydrolysis

Table 18.1 `Classification of Reagents

Oxidizing agents	Reducing agents	Nucleophiles	Moderate-to-strong acids	Moderate-to-strong bases	Dehydrating agents	Alkylating agents
O_2	H_2	$ROH, RO^{\ominus}$	H_2SO_4	$M^{\oplus}OH^{\ominus}$	P_2O_5	$R-X$
H_2O_2	$LiAlH_4, NaBH_4$	$H_2O, OH^{\ominus}$	H_3PO_4	$M^{\oplus}NH_2^{\ominus}$	PCl_5	$R-O-SO_2-Ar$
O_3	$RLi, RMgX$	NH_3, RNH_2, etc.	$R-SO_3H$	$M^{\oplus}R^{\ominus} \longleftrightarrow M-R$	$POCl_3$	$R-\overset{\oplus}{N}\equiv N:$
$KMnO_4$	Metal + acid	$RCO_2^{\ominus}$	$HClO_4$	NH_3	$HO-\overset{\overset{O}{\|}}{P}-O-\overset{\overset{O}{\|}}{P}-OH$	$R-\overset{\oplus}{O}H_2$
MnO_2	N_2H_4	$X^{\ominus}$	HNO_3	RNH_2, etc.	$OHOH$	
HNO_3	Metal ions in their lower oxidation states	$ArOH, ArO^{\ominus}$	RCO_2H	$M^{\oplus}OR^{\ominus}$	H_2SO_4	
X_2		$H^{\ominus}$ (latent in $AlH_4^{\ominus}$, etc.)		$Na^{\oplus}H^{\ominus}$	$SOCl_2$	
$Fe^{③}, Cu^{②}$					$Ar-SO_2Cl$	
$Ag^{\oplus}$ (sometimes)					$Ar-SO_3H$	
					$R-N=C=N-R$	

Table 18.2 *Common Reactions of the R—CO—G Classified by Reagent Type*

Reagent type	Reactions, numbered as in Chart 18.2
Reducing agents	⑧ R—CO—G $\xrightarrow{a}$ R—CO—H ; $\xrightarrow{a}$ R—CH$_2$—OH
	⑨ R—CO—G $\xrightarrow{R'_2Cd}$ R—CO—R' ; $\xrightarrow{R'MgX}$ R—CR'$_2$—OH
	⑩ R—CO—Cl + H$_2$ $\xrightarrow[\text{catalyst}]{\text{poisoned}}$ R—CO—H
	⑯ R—C≡N $\xrightarrow{a}$ R—CH=NH ; $\xrightarrow{a}$ R—CH$_2$—NH$_2$
Oxidizing agents	⑫ R—CO—NH$_2$ + Br$_2$ + NaOH ⟶ R—NH$_2$ + CO$_2$ + Na$^{\oplus}$Br$^{\ominus}$
	⑬ R—CH$_2$—CO$_2$H + Br$_2$ $\xrightarrow{P}$ R—CHBr—CO—Br
Bases	① R—CO$_2$H $\xrightarrow{NaOH}$ R—CO$_2^{\ominus}$ Na$^{\oplus}$; $\xrightarrow{NH_3}$ R—CO$_2^{\ominus}$ NH$_4^{\oplus}$
	③ R—CO—G + H$_2$O $\xrightarrow{OH^{\ominus}}$ R—CO$_2^{\ominus}$ ⎫
	⑮ R—C≡N + H$_2$O $\xrightarrow{OH^{\ominus}}$ R—CO$_2^{\ominus}$ ⎬ (base-catalyzed hydrolyses)
	⑭ R—CH$_2$—COOR' $\xrightarrow{\text{base}}$ R—CH—COOR' | O=C—CH$_2$—R
Acids	② R—CO—G $\xrightarrow{HA}$ R—C(OH)G$^{\oplus}$ A$^{\ominus}$
	③ R—CO—G + H$_2$O $\xrightarrow{HA}$ R—COOH ⎫
	⑮ R—C≡N + H$_2$O $\xrightarrow{HA}$ R—COOH ⎬ (acid-catalyzed hydrolyses)
Nucleophiles	③ R—CO—G + H$_2$O ⟶ R—COOH
	④ R—CO—G + R'OH ⟶ R—CO—OR'
	⑤ R—CO—G + Cl$^{\ominus}$ ⟶ R—CO—Cl
	⑥ R—CO—G + NH$_3$ ⟶ R—CO—NH$_2$
	⑦ R—CO—G + R'CO$_2^{\ominus}$ ⟶ R—CO—O—CO—R'
	⑧ R—C≡N $\xrightarrow{\text{cold, conc. HCl}}$ R—CO—NH$_2$; $\xrightarrow{\text{hot } H_2O/H_3O^+}$ R—COOH
Dehydrating agents	⑪ R—CO—NH$_2$ $\xrightarrow[\text{or SOCl}_2]{P_2O_5}$ R—C≡N

^{a} See Chart 18.3.

can be brought about by hot water alone, but more commonly it is carried out with aqueous acid or alkali, which contains both reactant (H_2O) and catalyst ($H_3O^{\oplus}$ or $OH^{\ominus}$).

• All bases as strong or stronger than NH_3 (NH_3, NR_3, $HCO_3^{\ominus}$, $CO_3^{\ominus}$, $OH^{\ominus}$, $OR^{\ominus}$, $NH_2^{\ominus}$, $R^{\ominus}$) react with RCO_2H to give stable salts ($RCO_2^{\ominus}M^{\oplus}$, $RCO_2^{\ominus}NH_4^{\oplus}$, etc.) that can be isolated and stored. However, the reactions of the R—CO—G with *acids* do not yield isolable salts because the R—CO—G are such weak bases. Reasonable concentrations of the conjugate acids,
$R-C\overset{\oplus OH}{\underset{G}{\diagdown}} \leftrightarrow R-C\overset{OH}{\underset{G}{\diagdown{\oplus}}}$, are produced in solutions containing R—CO—G and a strong acid, but the salts are never isolated.

• NH_3 appears both among the nucleophiles and among the bases. It reacts as a nucleophile with all the R—CO—G; with R—CO—OH it shows both reactions:

$$R-CO_2H + NH_3 \rightleftharpoons R-CO_2^{\ominus}NH_4^{\oplus} \text{ (at room temperature)}$$

$$R-CO_2H + NH_3 \rightleftharpoons R-CO_2^{\ominus}NH_4^{\oplus} \longrightarrow R-C\overset{O}{\underset{NH_2}{\diagdown}} \text{ (at elevated temperatures—pyrolysis of dry } RCOO^{\ominus}NH_4^{\oplus}\text{)}$$

• The reagent $SOCl_2$ appears in two of the reactions. Its function is to convert —OH to $-O-S\overset{O}{\underset{Cl}{\diagdown}}$, a much more reactive leaving group:

$$\underset{O}{\overset{\|}{R-C}}-OH + SOCl_2 \longrightarrow \underset{O}{\overset{\|}{R-C}}-O-\underset{O}{\overset{\|}{S}}-Cl + HCl$$

If $SOCl_2$ is used alone, the nucleophile that displaces this leaving group is $Cl^{\ominus}$ (from the HCl formed in the first step). This gives R—CO—Cl as the final product:

$$\underset{O}{\overset{\|}{R-C}}-O-\underset{O}{\overset{\|}{S}}-Cl \xrightarrow{HCl} \underset{O}{\overset{\|}{R-C}}-Cl + SO_2$$

If pyridine, ⟨○⟩N, is also present, it generates the nucleophilic carboxylate ion via the equilibrium

$$\underset{O}{\overset{\|}{R-C}}-OH + \langle\bigcirc\rangle N \rightleftharpoons \underset{O}{\overset{\|}{R-C}}-O^{\ominus} + \langle\bigcirc\rangle\overset{\oplus}{N}-H$$

Then the final product is the anhydride:

$$\underset{O}{\overset{\|}{R-C}}-Cl + \underset{O}{\overset{\|}{R-C}}-O^{\ominus} \longrightarrow \underset{O}{\overset{\|}{R-C}}-O-\underset{O}{\overset{\|}{C}}-R + Cl^{\ominus}$$

• Strong bases like $OH^{\ominus}$ and $OR^{\ominus}$ bring about two types of reaction of the R—CH_2—CO—G. If water is present, hydrolysis occurs: R—CH_2—CO—G + H_2O $\xrightarrow{OH^{\ominus}}$ R—CH_2—$COO^{\ominus}$. Only if water is absent are the much more weakly acidic α hydrogens attacked, leading to the Claisen condensation:

$$2\ R-CH_2-C\overset{O}{\underset{OR'}{\diagdown}} \xrightarrow{Na^{\oplus}OR'^{\ominus}} R-CH_2-\overset{O}{\overset{\|}{C}}-\underset{R}{\overset{|}{CH}}-\overset{O}{\overset{\|}{C}}-OR' + R'OH$$

This reaction is usually carried out in R'OH as solvent.

• Oxidation by halogen (Cl_2 or Br_2) occurs at the carbonyl carbon in the case of amides (reaction ⑫):

$$R-\underset{\underset{III}{\uparrow}}{C}\overset{O}{\underset{NH_2}{\diagdown}} + Br_2 \xrightarrow{NaOH} R-NH_2 + \underset{\underset{IV}{\uparrow}}{CO_2} + Na^{\oplus} Br^{\ominus}$$

(The oxidation stages of the carboxylic carbon are marked.) The acids are oxidized (halogenated) at the α carbon (reaction ⑬):

$$R-\underset{\underset{0}{\uparrow}}{CH_2}-CO_2H \xrightarrow[PBr_3]{Br_2} R-\underset{\underset{I}{\uparrow}}{CHBr}-CO-Br$$

• Reduction of R—CO—G and R—C≡N by the strongest reducing agent, $LiAlH_4$, goes all the way down to oxidation stage I, giving $R-CH_2-OH$ or $R-CH_2-NH_2$ (see Charts 18.3A to E). Stopping at oxidation stage II (the aldehyde, or its nitrogen analog,

$$R-C\underset{H}{\overset{N-H}{\diagup}}\bigg)$$ requires milder reducing agents such as H_2/poisoned $Pd/BaSO_4$ or

$LiAlH(OR)_3$ (see Chart 18.3).

Exercises

1. Place each of the following reagents in one of the seven classes in Table 18.1 (oxidizing agents, reducing agents, bases, acids, nucleophiles, dehydrating agents, alkylating agents) without looking at the table.

   (a) H_2CrO_4
   (b) $Na^{\oplus}NH_2^{\ominus}$
   (c) CH_3-CH_2-OH
   (d) PCl_5
   (e) ⌬—SO_3H
   (f) ⌬—$MgBr$
   (g) ⌬—NH_2
   (h) H_3PO_4
   (i) CH_3Li
   (j) H_2/Pt
   (k) $SOCl_2$
   (l) Cl_2
   (m) ⌬—$CO_2^{\ominus}$
   (n) $NaBH_4$
   (o) Fe^{3+}
   (p) NaH
   (q) Zn/CH_3CO_2H
   (r) $CH_3CH_2-\overset{\oplus}{O}H_2$

2. Fix Table 18.2 in your mind by first memorizing the six reagent classes. Find some means of remembering the number of reactions in each class and subcategorizing the class; for example, the reactions with reducing agents break into vertical (↓) and diagonal (↘) reductions in the sense of Chart 18.3. The diagonal reducing agents are the organometallic compounds RMgX, R_2Cd, and RLi. The vertical reducing agents are either strong ($LiAlH_4$) or mild [$LiAlH(OR)_3$, H_2/poisoned $Pd/BaSO_4$], leading

CARBOXYLIC ACIDS AND THEIR DERIVATIVES

to stage I and stage II, respectively. After giving some thought of this sort to each class, try reconstructing the table from memory on blank paper. Check the result and repeat the drill.

★ 3. Draw the structural formula of the principal organic products of these reactions.

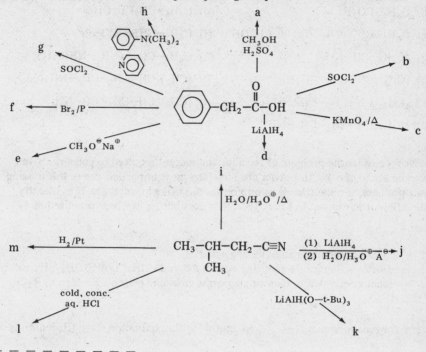

---

1. (a) oxidizing agent

   (b) strong base and nucleophile

   (c) nucleophile

   (d) dehydrating agent

   (e) strong acid

   (f) strong base, nucleophile and reducing agent

   (g) moderate base and nucleophile

   (h) strong acid

   (i) strong base, nucleophile and reducing agent

   (j) reducing agent

   (k) dehydrating agent

   (l) oxidizing agent

   (m) nucleophile

   (n) reducing agent and nucleophile

   (o) oxidizing agent

   (p) strong base, nucleophile, and reducing agent

(q) reducing agent

(r) alkylating agent

3. (a) $C_6H_5-CH_2-CO-O-CH_3$ (b) $C_6H_5-CH_2-CO-Cl$

(c) $C_6H_5-CO_2H$ (d) $C_6H_5-CH_2CH_2OH$

(e) $C_6H_5-CH_2-CO_2^{\ominus} Na^{\oplus} + CH_3OH$ (f) $C_6H_5-CHBr-CO-Br$

(g) $(C_6H_5-CO-)_2O$ (h) $C_6H_5-CO_2^{\ominus}\ C_6H_5-\overset{\oplus}{N}H(CH_3)_2$

(i) $(CH_3)_2CH-CH_2-CO_2H$ (j) $(CH_3)_2CH-CH_2-CH_2-NH_3^{\oplus} A^{\ominus}$

(k) $(CH_3)_2CH-CH_2-CH=NH$ (l) $(CH_3)_2CH-CH_2-CO-NH_2$

(m) $(CH_3)_2CH-CH_2-CH_2-NH_2$

13. Now consider the problem of recalling the reagent required to convert a given carboxylic acid derivative to a given product. The most important cue in this missing reagent problem, and the first decision you should make in solving it, is to identify the direction of movement in Chart 18.3. The possibilities can be symbolized as ↑, ↓, ↘, and ☐ :

↑ = oxidation, requiring an oxidizing agent
↓ = reduction, requiring a reducing agent of type $LiAlH_4$, $LiAlH(OR)_3$, H_2/catalyst
↘ = reductive alkylation, requiring an organometallic reagent, R—Mg—X, R_2Cd, R—Li

☐ = conversion of one $R-C{\overset{O}{\underset{G}{\diagdown}}}$ to another within oxidation stage III, requiring a nucleophile and perhaps a dehydrating agent.

Examples

Supply the missing reactant.

1. $CH_3-\underset{CH_3}{\underset{|}{CH}}-C{\overset{O}{\underset{O-CH_2CH_3}{\diagdown}}} \xrightarrow{?} CH_3-\underset{CH_3}{\underset{|}{CH}}-\underset{CH_3}{\underset{|}{\overset{OH}{\overset{|}{C}}}}-CH_3$

2. ⬡—C≡N $\xrightarrow{?}$ ⬡—CH_2-NH_2

3. ⬡—$\overset{O}{\overset{||}{C}}$—O—$\overset{O}{\overset{||}{C}}$—⬡ $\xrightarrow{?}$ ⬡—$\overset{O}{\overset{||}{C}}-NH_2$

Solutions:

1. The movement in Chart 18.3 is ↘: $\begin{matrix} R-CO-G \\ R-CO-H \\ R-CH_2OH \end{matrix} \begin{matrix} \\ R-CO-R' \\ R-CHOC-R' \end{matrix} \begin{matrix} \\ \\ R-\underset{OH}{\underset{|}{\overset{R}{\overset{|}{C}}}}-R'' \end{matrix}$

So an organometallic reagent is required, namely the reactive R—Mg—X (rather than the restrained R_2Cd, since we need to go all the way; see Charts 18.3B and C). The identity of the two new alkyl groups (CH_3—) fixes the identity of R—Mg—X:

$$CH_3-CH(CH_3)-C(=O)OCH_3 \xrightarrow[(2) H_2O/H_3O^\oplus]{(1) CH_3MgBr \text{ (2 moles)}} CH_3-CH(CH_3)-C(CH_3)(OH)-CH_3$$

2. ⬡—CH_2—NH_2 is a nitrogen analog of the 1° alcohol ⬡—CH_2—OH. The conversion is thus a reduction, ↓. Since the reduction goes all the way from oxidation stage III to stage I, the potent reducing agent $LiAlH_4$ is required:

$$\text{⬡}-C\equiv N \xrightarrow[(2) H_2O]{(1) LiAlH_4} \text{⬡}-CH_2-NH_2$$

3. The anhydride reactant and amide product are both derivatives of the same carboxylic acid, ⬡—CO_2H. The reaction occurs within the R—CO—G block, ☐. The nucleophile required for amide formation is NH_3 (see Chart 18.2):

$$\text{⬡}-\overset{O}{\underset{\|}{C}}-O-\overset{O}{\underset{\|}{C}}-\text{⬡} \xrightarrow{NH_3} \text{⬡}-\overset{O}{\underset{\|}{C}}-NH_2 + \text{⬡}-CO_2H$$

If you find you cannot visualize Chart 18.2 when faced with one of these missing reagent problems, you can proceed in another way. Analyze the change in the reactant's structural formula brought about by the reagent and try to fit the change into one of the following categories. Identification of the category is your cue to identification of the reagent required.

(a) Increase in O or X and/or decrease in H = oxidation.
(b) Decrease in O or X and/or increase in H = reduction.
(c) The changes $C\equiv N \to C=N$ or $C=O$, $C\equiv N \to C-N$, and $C=O \to C-O$ are always reductions.
(d) Replacement of C—X, C—O, or C—N bonds by C—N, C—O, or C—X bonds is usually nucleophilic substitution.

Examples

Identify the reaction type and decide what you can about the reagent required.

1. ⬡—CH_2—$\overset{O}{\underset{\|}{C}}$—Cl $\xrightarrow{?}$ ⬡—CH_2—$\overset{O}{\underset{\|}{C}}$—H

2. $CH_3=CH-C(=O)NH_2 \xrightarrow{?} CH_2=CH-C(=O)OH$

3. $CH_2-C\equiv N \xrightarrow{?} CH_3-\overset{O}{\underset{\|}{C}}-\text{⬡}$

Solutions:

1. C—Cl ⟶ C—H is reduction, requires a reducing agent. Since the product,

   Ph—CH$_2$—C(=O)H, is itself reducible (to Ph—CH$_2$—CH$_2$OH), a mild reagent is required: H$_2$/poisoned Pd/BaSO$_4$ or AlH(OR)$_3$Li.

2. —NH$_2$ ⟶ —OH is nucleophilic substitution. Nucleophile required = H$_2$O, generally used with acid or base catalysis.

3. C≡N ⟶ C=O is reduction. A new C—C bond is made at the carbon atom reduced. Therefore, suspect that the reducing agent is an organometallic compound. Although you have not seen this reaction before, you could make a good guess by writing

   Ph—MgBr or Ph—Li.

The same sort of analysis can be carried out using just the molecular formulas of reactant and product. Subtract the molecular formula of the reactant from the molecular formula of the product. The result is the "change" brought about by the reagent. Atoms that show up positive in this change are atoms gained in the reaction, negative atoms are atoms lost.

Examples

$$C_4H_9ON \longrightarrow C_4H_7N$$

$$\begin{array}{r} C_4H_7N \\ -C_4H_9NO \\ \hline \end{array}$$

$$\text{change} = -H_2 - O = -H_2O$$

$$C_5H_8O_2 \longrightarrow C_5H_7BrO$$

$$\begin{array}{r} C_5H_7BrO \\ -C_5H_8O_2 \\ \hline \end{array}$$

$$\text{change} = -H + Br - O$$

$$C_2H_6O \longrightarrow C_2H_4O_2$$

$$\begin{array}{r} C_2H_4O_2 \\ -C_2H_6O \\ \hline \end{array}$$

$$\text{change} = -2H + O$$

Next, check for obvious meanings in the changes. In the first example above, $-H_2O$ probably means dehydration. $-H + Br - O$ in the second example means in all likelihood that $-Br$ has replaced $-OH$ (i.e., nucleophilic substitution). The third result, $-2H + O$, means oxidation. Since the usual "missing reagent" problem gives you the reactant and product structures, you don't need to resort to the molecular formulas. However, this type of analysis is important in the "structure problems" discussed in Unit 20.

Exercises

1. Study the reagents required to bring about the interconversions of the R—C(=O)G shown in Chart 18.1. They are not reducing or oxidizing agents but nucleophiles

(or reagents that supply a nucleophile and also increase the reactivity of the leaving group—the dehydrating agents discussed in frame 12). Now put aside Chart 18.1 and fill in the missing reagents in this blank chart:

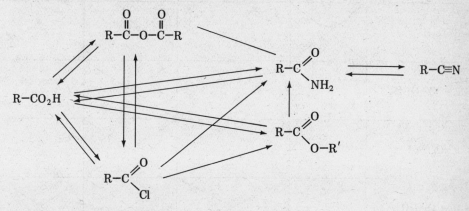

Check your answers and repeat the drill

2. Study the reagents required to produce the conversions in Charts 18.3A to E. These are all reducing or oxidizing agents. Now fill in the blank charts:

(a) RCO$_2$H
     ↑
     R—CHO
     ↓
     R—CH$_2$OH

(b) R—C(=O)OR'
    R—CHO   R—CO—R''
    R—CH$_2$OH   R—CH—R''   R—C(R'')(R'')—OH
                  |
                  OH

(c) R—CO—Cl
    ↙   ↘
    R—CO—H   R—CO—R'
    ↓   ↓
    R—CH$_2$OH   R—CH—R'
                 |
                 OH

(d)   R—CO—NH$_2$
      ↙   ↘
    R—CO—H   R—CNH—H   R—CO—R'
    ↓   ↓   ↓
    R—CH$_2$OH R—CH$_2$NH$_2$ R—CH—R'
                                 |
                                 OH

(e)   R—C≡N
           ↙   ↘
    R—CO—H   R—CNH—H
    ↓   ↓
    R—CH$_2$OH   R—CH$_2$—NH$_2$

Check your answers and repeat the drills.

3. Fill in each open box in the following table with the reagent required to convert the compound at left to the compound at the top.

466 HOW TO SUCCEED IN ORGANIC CHEMISTRY

Product Reactant	RCO_2H	$R-CO-OC_2H_5$	$R-CO-Cl$	$R-CO-NH_2$
RCO_2H				
$R-CO-OCH_3$				
$R-CO-Cl$				
$R-CO-NH_2$				

★ 4. Supply the missing reagent.

3.

	CH$_3$CH$_2$OH (HCl or H$_2$SO$_4$)	SOCl$_2$ or PCl$_5$	(1) NH$_3$ (2) heat
✗	CH$_3$CH$_2$OH (HCl or H$_2$SO$_4$)		(1) NH$_3$ (2) heat
H$_2$O (H$_3$O$^\oplus$ or OH$^\ominus$)	CH$_3$CH$_2$OH (HCl or H$_2$SO$_4$)	✗	NH$_3$
H$_2$O (H$_3$O$^\oplus$ or OH$^\ominus$)	CH$_3$CH$_2$OH	✗	NH$_3$
H$_2$O (H$_3$O$^\oplus$ or OH$^\ominus$)	✗	✗	✗

4. (a) H$_2$/Pd/BaSO$_4$/poison or LiAlH(OtBu)$_3$

   (b) HO—CH$_2$—CH$_2$—OH

   (c)  ⬡ /AlCl$_3$

   (d) 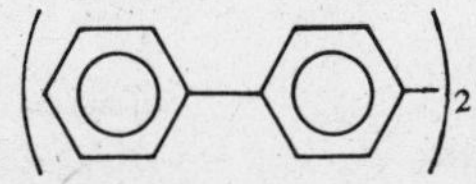 (⬡—⬡—)$_2$Cd

   (e) H$_2$O/NaOH

   (f) LiAlH$_4$, then H$_2$O/H$_3$O$^\oplus$

   (g) O⌒NH (morpholine)

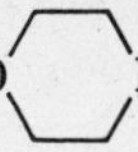

   (h) (CH$_3$)$_2$CH—CO$_2^\ominus$ Na$^\oplus$

   (i) LiAlH$_4$, then H$_2$O/H$_3$O$^\oplus$

   (j) CH$_3$CH$_2$O$^\ominus$ Na$^\oplus$, cold

   (k) 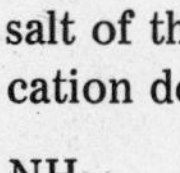 ⬡—O$^\ominus$ Na$^\oplus$/Δ (strong base to promote the Claisen condensation; choose Na salt of the same alcohol present in the ester so that simultaneous transesterification does not change it)

   (l) NH$_3$

   (m) H$_2$O/H$_3$O$^\oplus$/Δ

   (n) H$_2$O/NaOH/Δ

14. You need one more type of drill—recalling the entire set of reactions of the R—CO—G without cues and with particular attention to which conversions are possible and which ones are not.

You can cue yourself to recall the reactions of the R—CO—G by means of the classification scheme in Chart 18.2. Ask yourself what reactions take place at each part of the molecule in turn. Or you can break the list into classes defined by reaction type, for example proton transfers, substitutions, eliminations, additions, oxidations, and reductions. Such a scheme would look like this:

Proton transfers:

    ① deprotonation of RCO_2H ② protonation of R—CO—G

Substitutions:

    ③ hydrolysis of R—CO—G
    ④ esterification (alcoholysis) of R—CO—G
    ⑤ acyl chloride formation
    ⑥ amide formation (aminolysis)
    ⑦ anhydride formation
    ⑧ reduction of R—CO—G to R—CO—H
    ⑨ Cason's reaction, R—CO—Cl ⟶ R—CO—R′
    ⑩ Rosenmund reduction, R—CO—Cl ⟶ R—CO—H
    ⑬ HVZ reaction, R—CH_2—CO_2H ⟶ R—CHBr—CO—Br
    ⑭ Claisen condensation, 2 R—CH_2—CO—OR′ $\xrightarrow{base}$ R—CH—CO—OR′
                                                                                                 O=C—CH_2—R

Eliminations:

    ⑪ nitrile formation (amide dehydration)

Additions:

    ⑧ reduction of R—CO—G to R—CH_2—OH
    ⑯ reduction of R—C≡N to R—CH=NH and R—CH_2—NH_2
    ⑮ nitrile hydrolysis, R—C≡N ⟶ R—CO—NH_2

Oxidations:

    ⑫ Hofmann reaction, R—CO—NH_2 ⟶ R—NH_2 + CO_2

Note that in this case the reductions are handled under the addition category. This classification scheme is not ideal for purposes of recall because there are too many entries in the substitution class. One remedy would be to break the substitutions into subclasses on the basis of mechanism (electrophilic, nucleophilic, radical). Another way to subcategorize the substitutions is by the "incoming element," for example:

    R—CO—G ⟶ R—CO—ⓃH_2 incoming nitrogen
    R—CO—G ⟶ R—CO—R′   incoming carbon

Then you can use the periodic table as a source of cues, so that recall of the substitutions works like this:

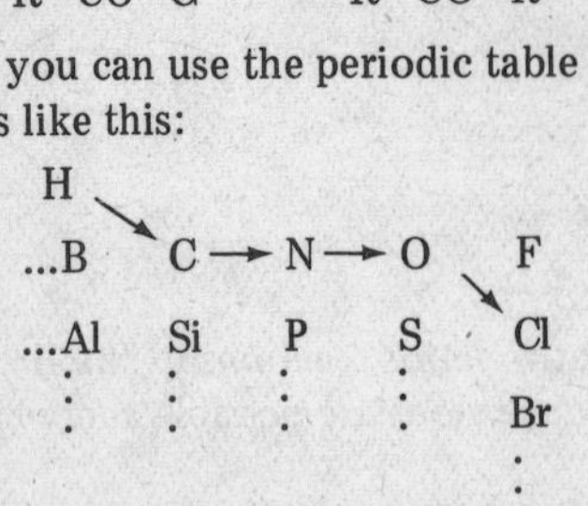

```
       H
        ↘
   ...B   C → N → O   F
                   ↘
   ...Al  Si  P   S   Cl
    :    :   :   :   Br
                     :
```

Incoming H:

⑧ reduction of R—CO—G to R—CO—H ⑩ Rosenmund reduction

Incoming C:

⑨ Cason's reaction ⑭ Claisen condensation

Incoming N:

⑥ aminolysis

Incoming O:

③ hydrolysis ④ esterification ⑦ anhydride formation

Incoming Cl:

⑤ acyl chloride formation

Incoming Br:

⑥ HVZ reaction

Whatever classification scheme you use, it helps if you can remember the number of items in each class. Then you are set up for a very rapid drill consisting of recalling the 16 standard reactions of the R—CO—G; "recall" in this drill means recalling the name of the reaction and visualizing the reactant, reagent, and product. This type of recall drill must be supplemented by practice in writing out each reaction recalled to be sure you can do so error-free.

Exercises

1. If you are in a classroom situation, edit the list of reactions in Chart 18.2 to conform to the coverage of your textbook or lecture notes.

2. Using whatever scheme you wish, write from memory all the generic reactions in your edited list from exercise 1. Check your answers and repeat the drill.

3. Here is Chart 18.1 minus the connecting arrows. Supply the missing arrows and reagents. (Omit any reactions that you have removed from Chart 18.2.)

$$R-CO-O-CO-R$$

$$R-CO-NH_2 \quad R-C\equiv N$$

$$R-COOH$$

$$R-CO-OR'$$

$$R-CO-Cl$$

Check your answers and repeat the drill.

4. Here are the various parts of Chart 18.3, minus the connecting arrows:

(a) R—COOH

R—CO—H   R—CO—R'

R—CH$_2$—OH   R—CH(OH)—R'   R—C(R'')(OH)—R'

(b) R—CO—OR'

    R—CO—H   R—CO—R''

                                              R''
                                              |
    R—CH$_2$—OH   R—CH—R''   R—C—R''
                                 | |
                                 OH   OH

(c) R—CO—Cl

    R—CO—H   R—CO—R'

                                            R''
                                          |
    R—CH$_2$—OH   R—CH—R'   R—C—R'
                             | |
                            OH   OH

(d)   R—CO—NH$_2$

    R—CO—H   R—CNH—H   R—CO—R'

                                                                                   R'
                                                                                    |
    R—CH$_2$—OH   R—CH$_2$—NH$_2$ R—CH—R'   R—C—R'
                                                                     | |
                                                                          OH   OH

(e)   R—C≡N

    R—CO—H   R—CNH—H   R—CO—R'

                                                                                    R'
    R—CH$_2$—OH   R—CH$_2$—NH$_2$ R—CH—R'   R—C—R'
                                                                       OH   OH

Supply the missing arrows and reagents. Check your answers and repeat the exercise.

SPECIFIC PROBLEM TYPES

15. The "Write the mechanism of these reactions" problem presents difficulties when it concerns the R—CO—G interconversion reactions. Because of the large number of combinations and variations on the acyl $S_N 2$ mechanism, it is impossible and undesirable to memorize all these processes. Instead, you need a method that will allow you to generate any specific acyl $S_N 2$ mechanism at will, given the reactant, reagent, and product. Before we develop such a method, let's study an example that illustrates all the possible steps that may be involved. This is the acid-catalyzed hydrolysis of simple esters:

$$R-\overset{\overset{\ddot{O}:}{\|}}{C}-\ddot{O}-R' + H_3O^\oplus \rightleftharpoons R-\overset{\overset{\overset{\oplus}{:O}-H}{\|}}{C}-\ddot{O}-R' \longleftrightarrow R-\overset{\overset{:\ddot{O}-H}{|}}{\underset{\oplus}{C}}-\ddot{O}-R' + H_2O$$

$$R-\overset{\overset{:\overset{\oplus}{O}-H}{\|}}{C}-\ddot{O}-R' \longleftrightarrow R-\overset{\overset{:\ddot{O}-H}{|}}{\underset{\oplus}{C}}-\ddot{O}-R' \longrightarrow R-\overset{\overset{:\ddot{O}-H}{|}}{\underset{\underset{H\ \overset{\oplus}{O}\ H}{|}}{C}}-\ddot{O}-R'$$
                                :OH$_2$

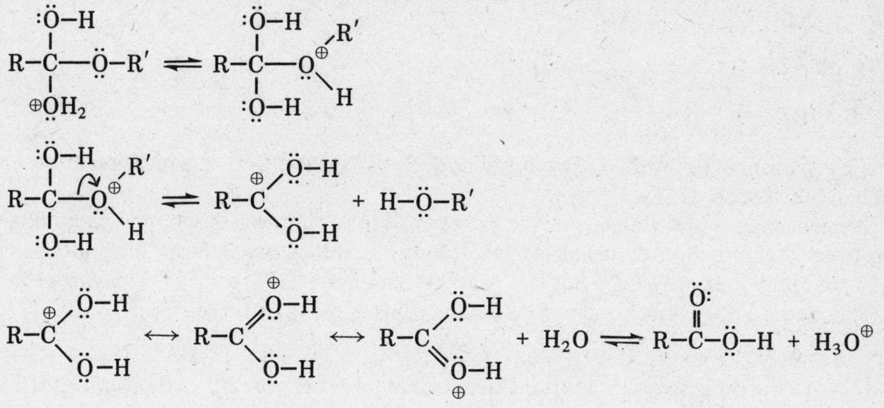

$$R-\underset{\overset{|}{\oplus}OH_2}{\overset{\overset{\ddot{O}-H}{|}}{C}}-\ddot{O}-R' \rightleftharpoons R-\underset{\overset{|}{:}\ddot{O}-H}{\overset{\overset{\ddot{O}-H}{|}}{C}}-\overset{\oplus}{O}\overset{R'}{\underset{H}{\diagdown}}$$

$$R-\underset{\overset{|}{:}\ddot{O}-H}{\overset{\overset{\ddot{O}-H}{|}}{C}}\overset{\curvearrowleft}{\underset{H}{\overset{\oplus}{O}}}\overset{R'}{\diagdown} \rightleftharpoons R-\underset{\overset{|}{\ddot{O}-H}}{\overset{\overset{\oplus}{\ddot{O}-H}}{C}} + H-\ddot{O}-R'$$

$$R-\underset{\overset{|}{\ddot{O}-H}}{\overset{\overset{\oplus}{\ddot{O}-H}}{C}} \leftrightarrow R-\underset{\overset{|}{\ddot{O}-H}}{\overset{\overset{O-H}{\|}}{C}} \leftrightarrow R-\underset{\overset{\|}{\underset{\oplus}{\ddot{O}-H}}}{\overset{\overset{\ddot{O}-H}{|}}{C}} + H_2O \rightleftharpoons R-\overset{\overset{\ddot{O}:}{\|}}{C}-\ddot{O}-H + H_3O^{\oplus}$$

The first step shows the function of the acid catalyst; it protonates the C=O group, which increases the fractional positive charge on the carbonyl carbon atom and consequently increases the rate of nucleophilic attack (second step). The third step turns —O—R into a good leaving group, $-\overset{\oplus}{O}\overset{R}{\underset{H}{\diagdown}}$, setting up the fourth step, departure of the leaving group:

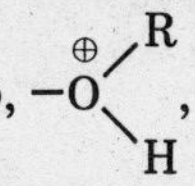

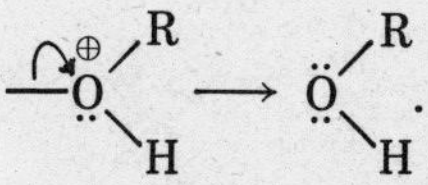

$-\overset{\curvearrowleft}{\underset{H}{\overset{\oplus}{O}}}\overset{R}{\diagdown} \longrightarrow \overset{R}{\underset{H}{\ddot{O}\diagdown}}$. In the last step, the protonated product gives up the extra proton, regenerating the catalyst.

Most of the variations on this mechanism require only some of these steps. Uncatalyzed reactions frequently consist of the middle three steps, for example ester hydrolysis in neutral solutions:

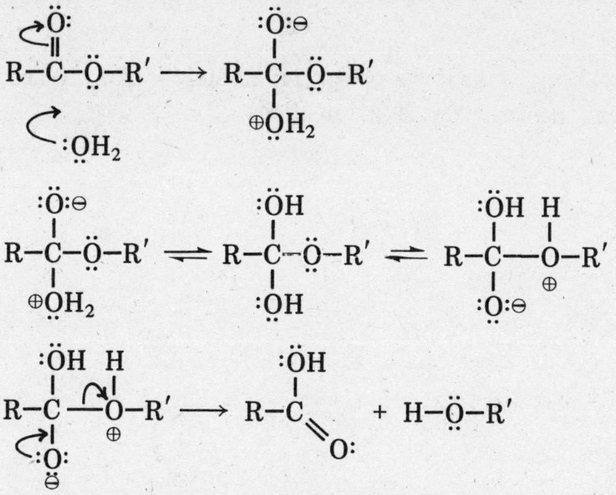

$$R-\underset{\underset{:OH_2}{\curvearrowleft}}{\overset{\overset{\ddot{O}:}{\|}}{C}}-\ddot{O}-R' \longrightarrow R-\underset{\overset{|}{\oplus}OH_2}{\overset{\overset{:\ddot{O}:^{\ominus}}{|}}{C}}-\ddot{O}-R'$$

$$R-\underset{\overset{|}{\oplus}OH_2}{\overset{\overset{:\ddot{O}:^{\ominus}}{|}}{C}}-\ddot{O}-R' \rightleftharpoons R-\underset{\overset{|}{:OH}}{\overset{\overset{:\ddot{O}H}{|}}{C}}-\ddot{O}-R' \rightleftharpoons R-\underset{\overset{|}{:\overset{\oplus}{O}:^{\ominus}}}{\overset{\overset{:\ddot{O}H}{|}}{C}}-\underset{}{\overset{H}{|}}O-R'$$

$$R-\underset{\underset{\ominus}{\overset{\curvearrowright}{:\ddot{O}:}}}{\overset{\overset{:\ddot{O}H}{|}}{C}}\overset{\curvearrowleft}{\underset{\oplus}{\overset{H}{|}}}O-R' \longrightarrow R-\underset{O:}{\overset{:\ddot{O}H}{\diagup\!\!\!\diagdown}} + H-\ddot{O}-R'$$

This process is slower than the acid-catalyzed hydrolysis because the unprotonated $\diagup\!\!\!\diagdown C=O$ group is less reactive than its conjugate acid, $\diagup\!\!\!\diagdown C=\overset{\oplus}{O}-H \leftrightarrow \diagup\!\!\!\diagdown\overset{\oplus}{C}-O-H$. The nucleophile is the same in both mechanisms.

Alkaline hydrolyses, aminolyses, and the reductions resulting from $H:^{\ominus}$ or $R:^{\ominus}$ acting as nucleophile are generally shorter still, because the middle step is missing:

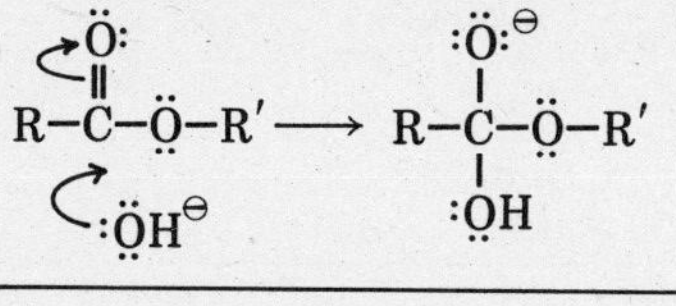

$$R-\underset{\underset{:\ddot{O}H^{\ominus}}{\curvearrowleft}}{\overset{\overset{\ddot{O}:}{\|}}{C}}-\ddot{O}-R' \longrightarrow R-\underset{\overset{|}{:OH}}{\overset{\overset{:\ddot{O}:^{\ominus}}{|}}{C}}-\ddot{O}-R'$$

$$R-\underset{\underset{:OH}{|}}{\overset{:\overset{\curvearrowleft}{\ddot{O}}:^{\ominus}}{C}}-\ddot{O}-R' \longrightarrow R-\overset{\overset{\ddot{O}:}{\|}}{C}-\ddot{O}-H + :\ddot{O}R'^{\ominus}$$

This mechanism is faster than the neutral hydrolysis because $OH^{\ominus}$ is a more reactive nucleophile than H_2O.

The mechanism just written employs a leaving group that was taboo in nucleophilic substitution via the S_N2 mechanism: $OR^{\ominus}$. In the acyl S_N2 mechanism this limitation does not apply; not only $OR^{\ominus}$ but $OH^{\ominus}$ and perhaps even $NH_2^{\ominus}$ are viable leaving groups.

The method of generating acyl S_N2 mechanisms consists of these steps:

1. Identify the nucleophile:
   - In the reductive substitutions the nucleophile is $H:^{\ominus}$ or $R:^{\ominus}$, latent in $H-AlH_3$ and $R-M$.
   - In basic solutions the nucleophile is $OH^{\ominus}$, $OR'^{\ominus}$, $^{\ominus}O-CO-R'$, NH_3, RNH_2, etc.
   - In neutral and acidic solutions the nucleophile is H_2O, $R'OH$.
   - $Cl^{\ominus}$ acts as nucleophile under all conditions.

2. (acidic solutions only) Write the protonation of $R-CO-G$:

$$R-\overset{\overset{:\overset{\curvearrowright}{\ddot{O}}:}{\|}}{C}-G \quad H\overset{\curvearrowleft}{-}A \longrightarrow R-\overset{\overset{:\overset{\oplus}{O}-H}{\|}}{C}-G + :A^{\ominus}$$

3. Write the addition of nucleophile to $R-CO-G$ or $R-\overset{\oplus}{C}OH-G$:

$$R-\overset{\overset{\overset{\curvearrowright}{O}}{\|}}{C}-G \longrightarrow R-\underset{\underset{Nu}{|}}{\overset{\overset{O^{\ominus}}{|}}{C}}-G \quad \text{or} \quad R-\overset{\overset{\oplus O-H}{\|}}{C}-G \longleftrightarrow R-\underset{\underset{\underset{:Nu}{\curvearrowleft}}{\oplus}}{\overset{\overset{O-H}{|}}{C}}-G \longrightarrow R-\underset{\underset{Nu}{|}}{\overset{\overset{OH}{|}}{C}}-G$$

$$\underset{:Nu}{\curvearrowleft}$$

4. (acidic and neutral solutions only) Transfer a proton, if available, from Nu to G (unless G = Cl, in which case protonation is not needed).

Examples

$$R-\underset{\underset{\oplus OH_2}{|}}{\overset{\overset{OH}{|}}{C}}-OR' \rightleftharpoons R-\underset{\underset{OH}{|}}{\overset{\overset{OH}{|}}{C}}-\overset{\oplus}{O}\overset{R'}{\underset{H}{\diagdown}} \quad \text{and} \quad R-\underset{\underset{\oplus NH_3}{|}}{\overset{\overset{OH}{|}}{C}}-O-\overset{\overset{O}{\|}}{C}-R \rightleftharpoons R-\underset{\underset{NH_2}{|}}{\overset{\overset{OH}{|}}{C}}-\overset{\oplus}{O}\overset{H}{\underset{CO-R}{\diagdown}}$$

5. Let G depart, and restore the C=O group.

Examples

$$R-\underset{\underset{O-C_2H_5}{|}}{\overset{\overset{:\overset{\curvearrowleft}{\ddot{O}}:^{\ominus}}{|}}{C}}-CH_3 \longrightarrow R-\overset{\overset{O}{\|}}{C}-O-C_2H_5 + \overset{\ominus}{O}-CH_3$$

and

$$R-\underset{\underset{OH}{|}}{\overset{\overset{:\overset{\curvearrowleft}{\ddot{O}}H}{\oplus|}}{C}}-\overset{R'}{\underset{H}{\overset{\diagup}{O}}} \longrightarrow R-\overset{\overset{\overset{\oplus}{O}-H}{\|}}{C}-OH + O\overset{R'}{\underset{H}{\diagdown}}$$

6. Protonate or deprotonate the product if necessary to put it in the state required by the environment. This is usually R—CO—G (not R—$\overset{\oplus}{\text{C}}$OH—G). If the product is carboxylic acid, it should end up as R—CO$_2$H in acidic solutions, as R—CO$_2^\ominus$ in basic solutions.

Example

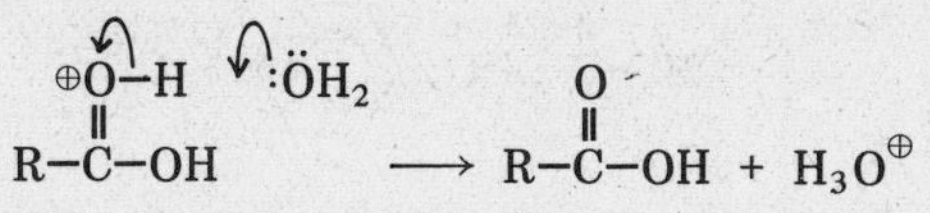

$$\underset{\text{R-C-OH}}{\overset{\overset{\displaystyle\oplus}{\text{O-H}}\;\;\;:\ddot{\text{O}}\text{H}_2}{\|}} \longrightarrow \underset{\text{R-C-OH}}{\overset{\text{O}}{\|}} + \text{H}_3\text{O}^\oplus$$

You will need some guidelines in judging when "acidic" or "basic" conditions are present:

- "Acidic conditions" means that a strong acid (HCl, H$_2$SO$_4$, ArSO$_3$H, etc.) is present. RCO$_2$H is not strong enough to protonate R—CO—G appreciably.
- The organometallic reagents RMgX, RLi, and R$_2$Cd can exist only in the absence of all acids, so their reactions with R—CO—G are always uncatalyzed.
- The most common "basic" conditions for these reactions involve the following combinations:

OH$^\ominus$ in H$_2$O

R'O$^\ominus$ in R'OH

NH$_3$, RNH$_2$, etc., in any solvent

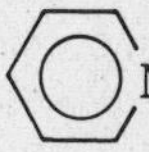

⟨pyridine⟩N as solvent; this is generally used to generate R'CO$_2^\ominus$ as nucleophile:

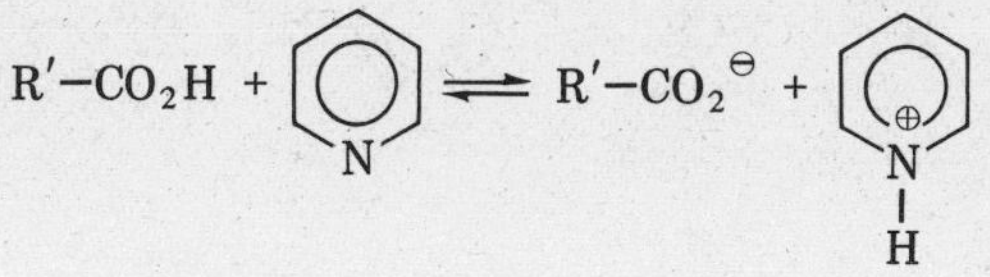

$$\text{R}'-\text{CO}_2\text{H} + \text{C}_5\text{H}_5\text{N} \rightleftharpoons \text{R}'-\text{CO}_2^\ominus + \text{C}_5\text{H}_5\overset{\oplus}{\text{N}}\text{H}$$

Examples

Write all the steps in the mechanism of each reaction.

1. 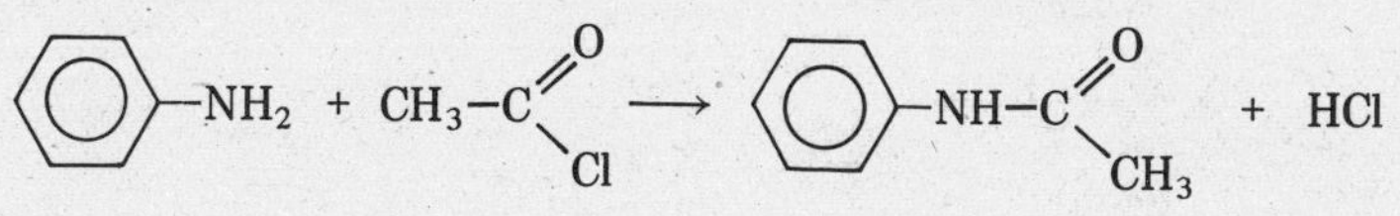 Ph—NH$_2$ + CH$_3$—COCl ⟶ Ph—NH—COCH$_3$ + HCl

2. Ph—NH—COCH$_3$ + Na$^\oplus$OH$^\ominus$ ⟶ Ph—NH$_2$ + CH$_3$—CO$_2^\ominus$ Na$^\oplus$

3. Ph—COOH + CH$_3$OH $\xrightarrow{\text{H}_2\text{SO}_4}$ Ph—CO—O—CH$_3$ + H$_2$O

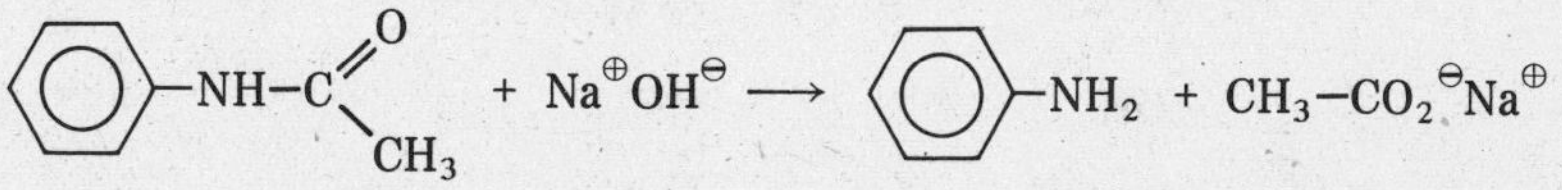

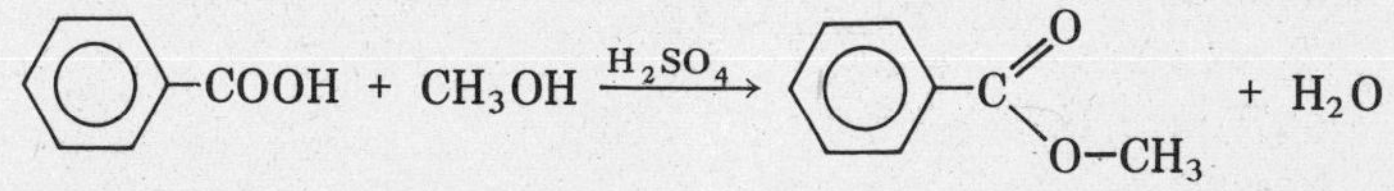

Solutions:

1. nucleophile = Ph—NH$_2$ conditions = basic   leaving group = —Cl ⟶ Cl$^\ominus$

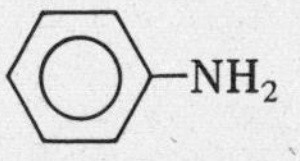

Use steps 3 and 5:

CH₃−C(=O)−Cl + H₂N−Ph → CH₃−C(O⁻)(Cl)−⁺NH₂−Ph

CH₃−C(O⁻)(Cl)(⁺NH₂−Ph) → CH₃−C(=O)−⁺NH₂−Ph + Cl⁻

Now see that step 6 is also needed:

CH₃−C(=O)−⁺NH(H)−Ph + Cl⁻ → CH₃−C(=O)−NH−Ph + HCl

2. Nucleophile = OH⁻ conditions = basic

Use steps 3 and 5:

Ph−NH−C(=O)−CH₃ + ⁻OH → Ph−NH−C(O⁻)(OH)−CH₃ →

Ph−NH⁻ + HO−C(=O)−CH₃

Need step 6:

Ph−NH⁻ + H−O−C(=O)−CH₃ → Ph−NH₂ + ⁻O−C(=O)−CH₃

3. Nucleophile = CH₃OH conditions = acidic

Use steps 1 to 6:

Ph−C(=O)−OH + H−O−SO₃H → Ph−C(⁺O−H)−OH + HSO₄⁻

Ph−C(=⁺O−H)−OH + CH₃−Ö−H → Ph−C(O−H)(OH)(⁺O(CH₃)(H))

Ph−C(OH)(O−CH₃/⁺H)−OH ⇌ Ph−C(OH)(O(H)(⁺H))−O−CH₃

Ph−C(⁺OH₂)(OH)−O−CH₃ → Ph−C(=⁺O−H)−O−CH₃ + H₂O

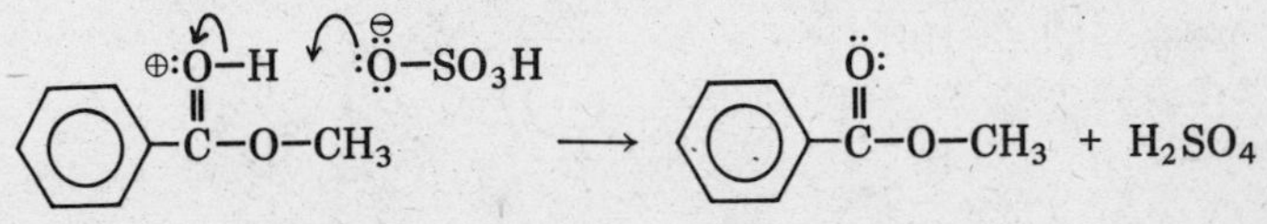

CARBOXYLIC ACIDS AND THEIR DERIVATIVES

Ph-C(=O)-O-CH₃ with ⊕:Ö-H protonated and :Ö-SO₃H⁻ → Ph-C(=O)-O-CH₃ + H₂SO₄

The following exercises give you practice on the straightforward cases that represent the large majority of possible combinations. In the next frame we look at the few special cases.

★ *Exercises*

Write all steps in the mechanism of each reaction.

1. Alkaline hydrolysis of CH₃–C(=O)–O–C(=O)–CH₃

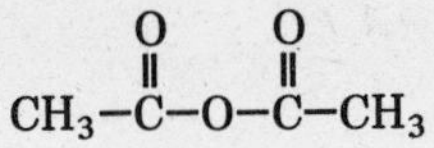

2. Neutral hydrolysis of Ph–C(=O)–Cl

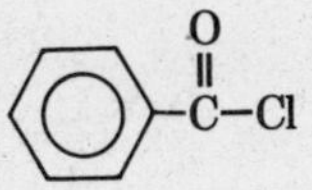

3. Acid hydrolysis of
$$CH_2\begin{matrix}CH_2-CH_2\\ CH_2-C\end{matrix}\begin{matrix}\\O\\ =O\end{matrix}$$
(γ-butyrolactone-like cyclic ester)

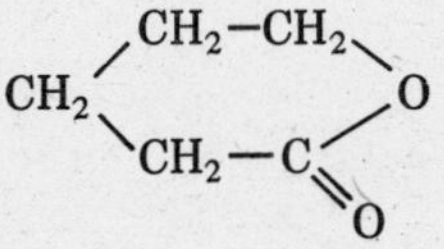

4. 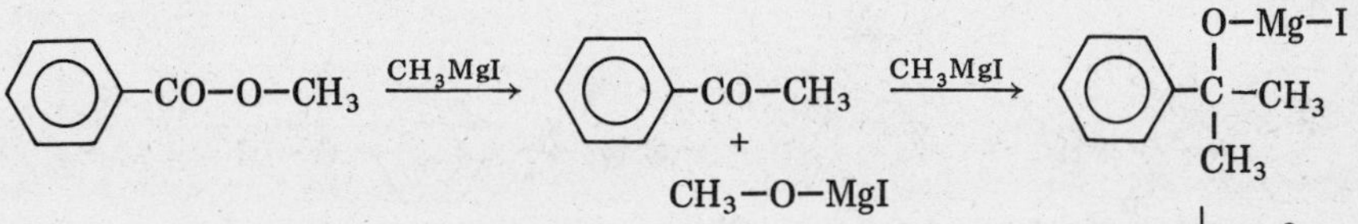 Ph–CO–O–CH₃ $\xrightarrow{CH_3MgI}$ Ph–CO–CH₃ + CH₃–O–MgI $\xrightarrow{CH_3MgI}$ Ph–C(O–Mg–I)(CH₃)₂

   $\downarrow$ H₂O | H₃O⊕

   Ph–C(OH)(CH₃)₂

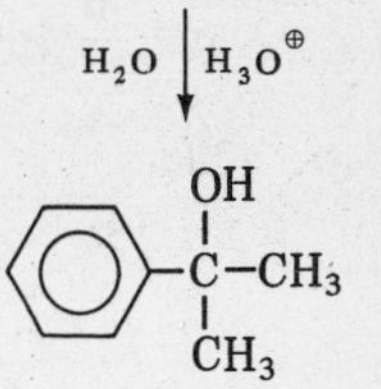

   (write the first stage, formation of Ph–CO–CH₃, only)

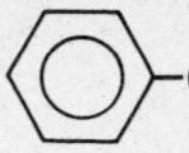

5. 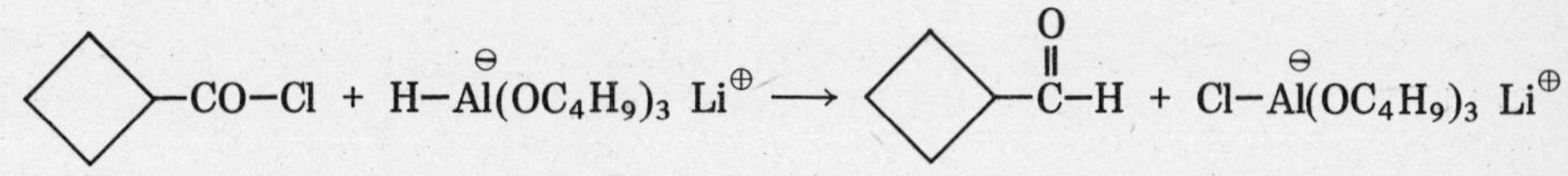 (cyclobutyl)–CO–Cl + H–Al(OC₄H₉)₃⁻ Li⊕ → (cyclobutyl)–C(=O)–H + Cl–Al(OC₄H₉)₃⁻ Li⊕

- - - - - - - - -

1. 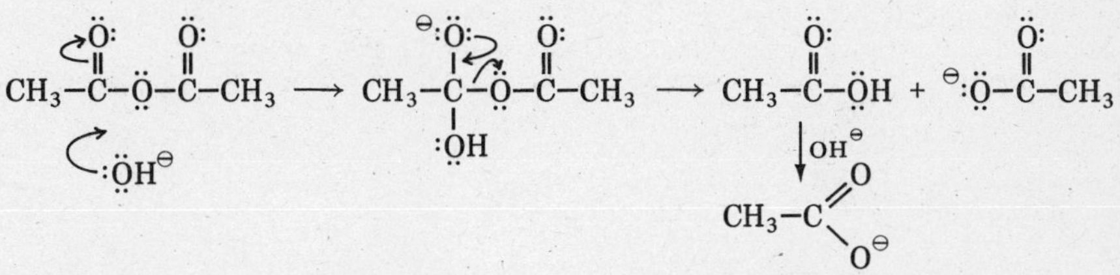 CH₃–C(=Ö:)–Ö–C(=Ö:)–CH₃ with :ÖH⁻ attacking → CH₃–C(–Ö:⁻)(OH)–Ö–C(=Ö:)–CH₃ → CH₃–C(=Ö:)–ÖH + ⁻:Ö–C(=Ö:)–CH₃

   ↓ OH⁻

   CH₃–C(=O)(O⁻)

2. 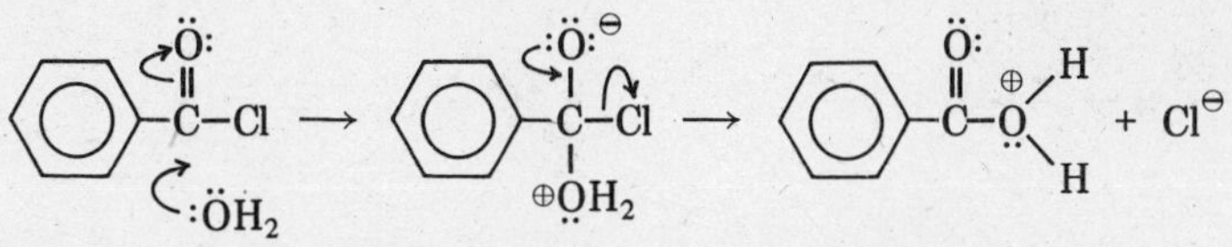 Ph–C(=Ö:)–Cl with :ÖH₂ → Ph–C(–Ö:⁻)(⊕ÖH₂)–Cl → Ph–C(=Ö:)–Ö(H)(H) + Cl⁻

$$\underset{H}{\overset{\overset{\displaystyle \overset{..}{O}:}{\|}}{\underset{}{\bigcirc}-C-\overset{\oplus}{\overset{..}{O}}-H}} + H_2\overset{..}{O}: \rightleftharpoons \bigcirc-\overset{\overset{\displaystyle \overset{..}{O}:}{\|}}{C}-\overset{..}{O}H + H_3\overset{\oplus}{\overset{..}{O}}$$

3. (cyclohexanone lactone) $+ H_3O^\oplus \rightleftharpoons$ (protonated) $\leftrightarrow$ (resonance) $+ H_2O$

[Series of cyclohexane ring structures showing protonation, water attack, tetrahedral intermediate formation, and proton transfers leading to ring-opened hydroxy acid]

:Ö—H + H₂O ⇌ O—H + H₃O⊕
C=Ö—H   C—Ö—H
:ÖH   ‖
               :O:

4. $\underset{CH_3-Mg-I}{\bigcirc-\overset{\overset{O}{\|}}{C}-O-CH_3} \rightarrow \underset{CH_3}{\bigcirc-\overset{\overset{:\overset{\ominus}{O}:}{|}}{C}-\overset{\oplus}{O}-CH_3}\;\;Mg\!-\!I \rightarrow \bigcirc-\overset{\overset{O}{\|}}{C}-CH_3 + Mg\!-\!I\;\;\overset{\ominus}{O}-CH_3$

5. $\underset{H-Al(OC_4H_9)_3\;Li^\oplus}{\square-\overset{\overset{O}{\|}}{C}-Cl} \rightarrow \underset{Al(OC_4H_9)_3}{\square-\overset{\overset{:\overset{\ominus}{O}:}{|}}{\underset{H}{C}}-Cl}\;\;Li^\oplus \rightarrow \square-\overset{\overset{:\overset{..}{O}:}{\|}}{C}-H + Cl^\ominus$

$:\overset{\ominus}{\underset{..}{Cl}}: \;\; Al(OC_4H_9)_3 \rightarrow :\overset{..}{\underset{..}{Cl}}-\overset{\ominus}{Al}(OC_4H_9)_3\;\;Li^\oplus$

16. A few of the acyl S_N2 reactions fit the standard pattern you have been practicing, only with a bit of modification.

Those reactions employing $SOCl_2$, PCl_5, $POCl_3$, etc., on the carboxylic acid, RCO_2H, are using an alternative way of making $-OH$ into a good leaving group:

The $R-CO_2H + PCl_5 \longrightarrow R-CO-Cl + POCl_3 + HCl$ reaction has this mechanism:

$$\underset{H}{R-\overset{\overset{O}{\|}}{C}-\overset{..}{O}:} \;\; PCl_4-Cl \longrightarrow \underset{H}{R-\overset{\overset{O}{\|}}{C}-\overset{\oplus}{O}-PCl_4} + Cl^\ominus$$

$$\underset{H}{R-\overset{\overset{O}{\|}}{C}-\overset{\oplus}{O}-PCl_4} \longrightarrow \underset{Cl\;\;H}{R-\overset{\overset{:\overset{\ominus}{O}:}{|}}{\underset{}{C}}-\overset{\oplus}{O}-PCl_4} \longrightarrow R-\overset{\overset{O}{\|}}{C}-Cl + HO-PCl_4 \rightleftharpoons HCl + POCl$$

$^\ominus\!:\!\overset{..}{\underset{..}{Cl}}:$

The R—COOH + $SOCl_2$ ⟶ R—CO—Cl + SO_2 + HCl reaction is similar:

$$R-\overset{O}{\underset{\|}{C}}-\overset{..}{\underset{..}{O}}-H \longrightarrow R-\overset{O}{\underset{\|}{C}}-\overset{\oplus}{O}-H \longrightarrow R-\overset{O}{\underset{\|}{C}}-O-\overset{O}{\underset{\|}{S}}-Cl + H-Cl$$

$$R-\overset{O}{\underset{\|}{C}}-O-\overset{O}{\underset{\|}{S}}-Cl + H-Cl \rightleftharpoons R-\overset{\overset{\oplus}{O}-H}{\underset{\|}{C}}-O-\overset{O}{\underset{\|}{S}}-Cl \longrightarrow R-\overset{:\overset{..}{O}-H}{\underset{Cl}{C}}-O-\overset{O}{\underset{\|}{S}}-Cl$$

$$R-\overset{O}{\underset{\|}{C}}-Cl + H-Cl \longleftarrow R-\overset{\overset{\oplus}{O}-H}{\underset{Cl}{C}} \longleftrightarrow R-\overset{\overset{..}{O}H}{\underset{Cl}{\overset{\oplus}{C}}} + SO_2 + Cl^{\ominus}$$

The combination $SOCl_2$ + ⟨pyridine⟩N, used on an excess of RCO_2H, generates a new nucleophile, $RCO_2^{\ominus}$, which wins out over $Cl^{\ominus}$, so that the product is the anhydride:

$$R-\overset{O}{\underset{\|}{C}}-OH + SOCl_2 \longrightarrow R-\overset{O}{\underset{\|}{C}}-O-\overset{O}{\underset{\|}{S}}-Cl$$

as before, but

$$R-C\overset{O}{\underset{OH}{\diagdown}} + \text{⟨pyr⟩}N \rightleftharpoons R-C\overset{O}{\underset{O^{\ominus}}{\diagdown}} + H-\overset{\oplus}{N}\text{⟨pyr⟩}$$

so

$$R-\overset{O}{\underset{\|}{C}}-O-\overset{O}{\underset{\|}{S}}-Cl \longrightarrow R-\overset{O}{\underset{\|}{C}}-O-\overset{O}{\underset{\|}{S}}-Cl \longrightarrow R-\overset{O}{\underset{\|}{C}}-O-\overset{O}{\underset{\|}{C}}-R + O=S=O + Cl^{\ominus}$$

The "acyl S_N2" reactions of the nitriles, R—C≡N, only occur after an initial nucleophilic addition reaction. For example, the acid hydrolysis proceeds in this way:

Nucleophilic addition of H_2O:
$$CH_3-C\equiv N: + H_3O^{\oplus} \rightleftharpoons CH_3-C\equiv \overset{\oplus}{N}-H \longleftrightarrow CH_3-\overset{\oplus}{C}=N-H + H_2O$$

$$CH_3-C\equiv \overset{\oplus}{N}-H \longleftrightarrow CH_3-\overset{\oplus}{C}=\overset{..}{N}-H \longrightarrow CH_3-\underset{\overset{\oplus}{O}H_2}{\overset{\|}{C}}=\overset{..}{N}-H$$

Proton transfer:
$$CH_3-\underset{\overset{\oplus}{O}H_2}{C}=\overset{..}{N}-H \rightleftharpoons CH_3-\underset{:\overset{..}{O}H}{C}=NH_2 \longleftrightarrow CH_3-\underset{\overset{\oplus}{O}H}{\overset{\|}{C}}-\overset{..}{N}H_2$$

etc. ⟷ $CH_3-\underset{\overset{\oplus}{O}-H}{\overset{\|}{C}}-\overset{..}{N}H_2 \longrightarrow CH_3-\underset{:\overset{..}{O}H}{\overset{:\overset{\oplus}{O}H_2}{C}}-\overset{..}{N}H_2 \rightleftharpoons CH_3-\underset{:\overset{..}{O}H}{\overset{:\overset{..}{O}H}{C}}-\overset{\oplus}{N}H_3$

478 HOW TO SUCCEED IN ORGANIC CHEMISTRY

Acyl $S_N 2$:

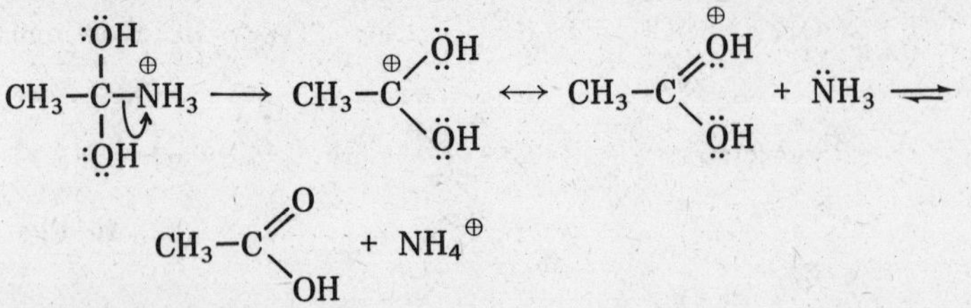

$$CH_3-\overset{:\ddot{O}H}{\underset{:\ddot{O}H}{\underset{|}{C}}}-\overset{\oplus}{N}H_3 \longrightarrow CH_3-\overset{\oplus\ddot{O}H}{\underset{\ddot{O}H}{C}} \longleftrightarrow CH_3-\overset{\overset{\oplus}{O}H}{\underset{\ddot{O}H}{C}} + \ddot{N}H_3 \rightleftharpoons$$

$$CH_3-\overset{O}{\underset{OH}{C}} + NH_4^{\oplus}$$

★ Exercises

Using only steps that are closely analogous to those we have used in this unit, write mechanisms for each reaction.

1. Ph−C≡N + CH₃−CH₂−O−H $\xrightarrow{HCl}$ Ph−C(=N⊕H₂)(O−CH₂CH₃) · Cl⁻ (with H on N)

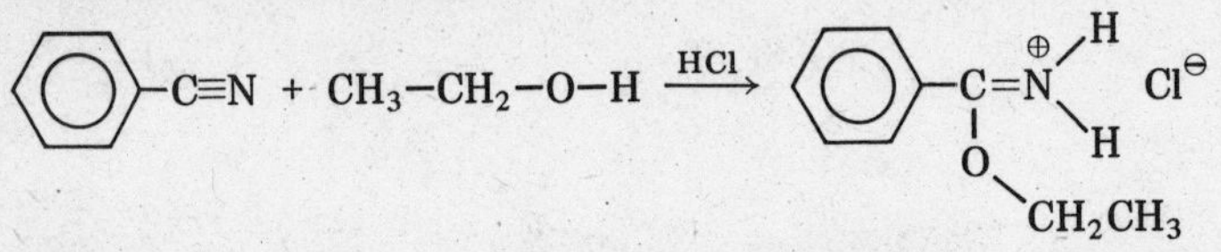

2. Ph−C(=N⊕H₂)(O−CH₂CH₃) Cl⁻ + 2 NH₃ ⟶ Ph−C(=NH)(NH₂) + NH₄⊕ Cl⁻ + CH₃CH₂OH

3. CH₃−C(=O)−O−CH₃ + NH₂−OH ⟶ CH₃−C(=O)−NH−OH + HOCH₃

4. CH₃CO₂H + Ph−C(=O)Cl $\xrightarrow[\text{pyridine}]{\text{trace}}$ CH₃−C(=O)Cl + Ph−CO₂H

5. o-C₆H₄(CO₂H)₂ $\xrightarrow[\text{pyridine}]{SOCl_2}$ phthalic anhydride

- - - - - - - - - - -

1. Ph−C≡N: ↷ H−Cl ⟶ Ph−C≡N−H Cl⁻

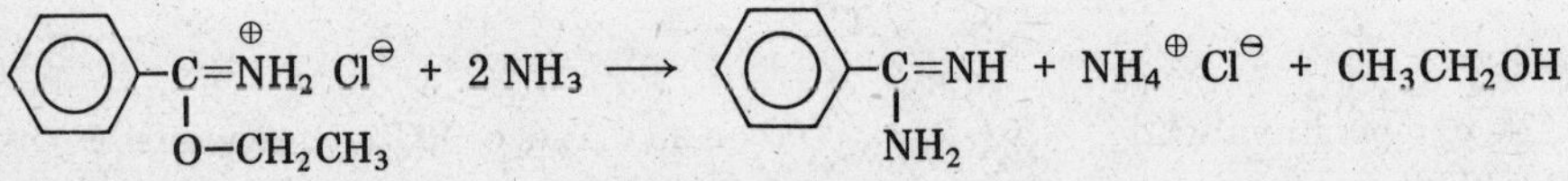

   Ph−C⊕≡N−H ⟶ Ph−C≡N̈−H ⇌ Ph−C(=N⊕H₂)(O−CH₂CH₃) Cl⁻
   H−Ö−CH₂−CH₃   H⊕O(CH₂CH₃)

2. Ph−C(=N⊕H···NH₃)(O−CH₂CH₃) Cl⁻ ⟶ Ph−C(⊕NH₃)(NH₂)(O−CH₂CH₃) ⇌ Ph−C(:NH₂)(−NH₂)(O⊕H−CH₂CH₃) Cl⁻

   ↓

   NH₄⊕ Cl⁻ + Ph−C(=NH)(−NH₂) $\xleftarrow{NH_3}$ Ph−C(⊕NH₂)(−NH₂) Cl⁻ + HOCH₂CH₃

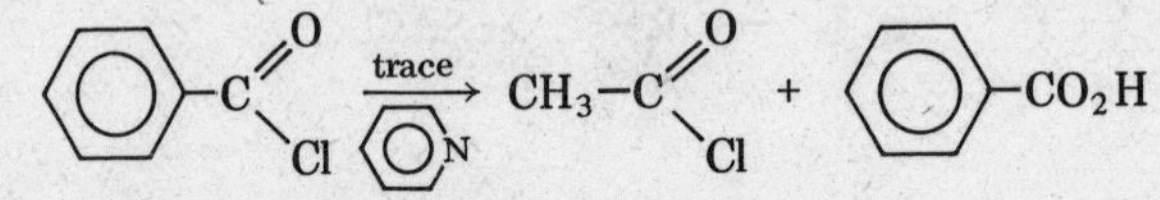

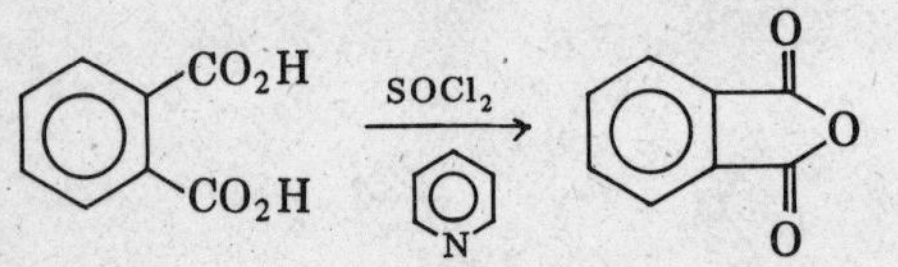

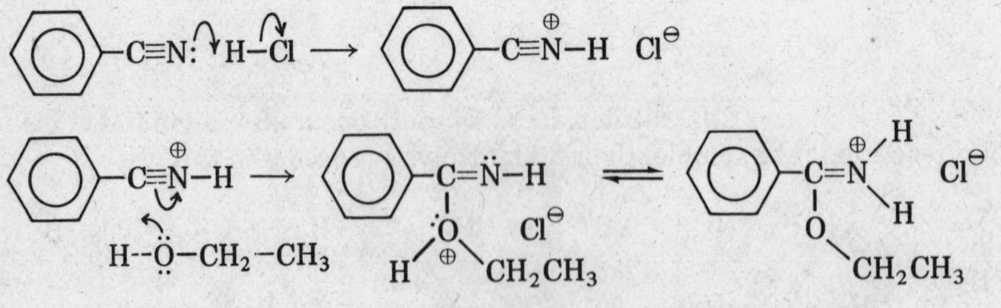

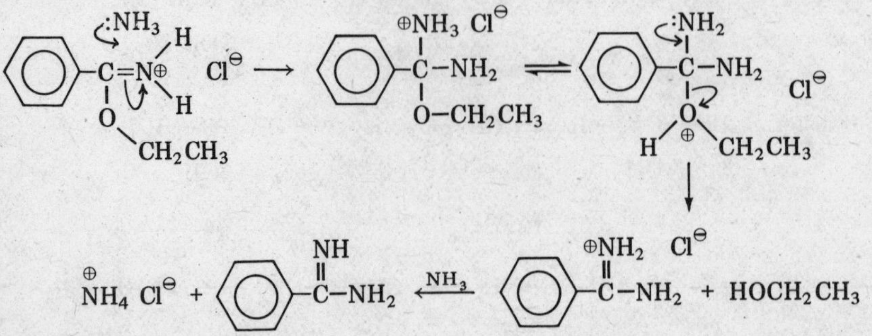

CARBOXYLIC ACIDS AND THEIR DERIVATIVES 479

3. $CH_3-\overset{\overset{\displaystyle :\ddot{O}:}{\|}}{C}-O-CH_3 \longrightarrow CH_3-\overset{\overset{\displaystyle :\ddot{O}:^{\ominus}}{|}}{\underset{\underset{\displaystyle \oplus NH_2-OH}{|}}{C}}-O-CH_3 \rightleftharpoons CH_3-\overset{\overset{\displaystyle :\ddot{O}:H}{|}}{\underset{\underset{\displaystyle :NH-OH}{|}}{C}}-\overset{\oplus}{O}-CH_3 \longrightarrow CH_3-\overset{\overset{\displaystyle :\ddot{O}}{\|}}{C}-NH-OH$
   $\overset{\curvearrowleft}{:NH_2-OH}$ + HOCH$_3$

4. $CH_3-CO_2H +$ ⌬N $\rightleftharpoons CH_3-COO^{\ominus} +$ ⌬$\overset{\oplus}{N}-H$

[Ph–CO–Cl + acetate → mixed anhydride + Cl⁻; then → PhCO₂⁻ + CH₃COCl schemes]

$\langle\text{Ph}\rangle-CO_2^{\ominus} +$ ⌬$\overset{\oplus}{N}-H$ $\rightleftharpoons$ $\langle\text{Ph}\rangle-CO_2H +$ ⌬N

5. [Salicylic acid type + SOCl₂ → acyl chlorosulfite intermediate → cyclic phthalic anhydride formation mechanism, leading to Cl⁻ + SO₂ + phthalic anhydride]

17. Since the R−CO−G are readily interconvertible, a synthesis of any one is easily turned into a synthesis of them all. Five approaches are available:

```
                              R-CO-O-CO-R
                                              R-CO-NH₂   R-C≡N
                H₂SO₄                         R-CO-OR'
   R-CCl₃  ─────────→  R-CO₂H                                    CN⁻
                H₂O        ↗ ↑      R-CO-Cl                         ↘
                  (1) CO₂  /                                         R-X
                  (2) H₂O/   H₂CrO₄    KMnO₄
                       H₃O⁺
          Mg
   R-X  ────→  R-Mg-X     R-CH₂-OH      R-CH=C
```

The two paths starting from R−X add one carbon atom (derived from CO$_2$ or CN$^{\ominus}$). The primary alcohol oxidation does not alter the carbon skeleton. The alkene oxidation cleaves the skeleton at the C=C. The hydrolysis of a 1,1,1-trihaloalkane looks like it ought to give a 1,1,1-trihydroxy compound:

$$R-CCl_3 \longrightarrow R-\underset{\underset{OH}{|}}{\overset{\overset{OH}{|}}{C}}-OH$$

According to Unit 17, such a structure should be unstable and lose H_2O, which it does, giving $R-C\overset{O}{\underset{OH}{\diagdown}}$ + H_2O.

★ *Exercises*

1. Show how a given straight-chain carboxylic acid, $R-CH_2-CH_2-CH_2-CO_2H$, could be converted, in four steps or less, to carboxylic acids with the following structures:

   (a) $R-CH_2-CH_2-CH_2-CH_2-CO_2H$

   (b) $R-CH_2-CH_2-CO_2H$

   (c) $R-CH_2-CH_2-CH_2-CH_2-CH_2-CO_2H$

   (d) $R-CH_2-CO_2H$

2. Show how the carboxylic acid $R-CH_2-CO_2H$ could be converted to the labeled acid $R-CH_2-^{14}CO_2H$ using $Ca^{14}CO_3$ as the source of labeled carbon.

— — — — — — — — —

1. (a) $R-(CH_2)_3-CO_2H \xrightarrow[(2)\ H_2O/H_3O^\oplus]{(1)\ LiAlH_4} R-(CH_2)_3-CH_2OH \xrightarrow{HBr} R-(CH_2)_3-CH_2Br$

   $\xrightarrow{Mg} R-(CH_2)_3-CH_2MgBr \xrightarrow[(2)\ H_3O^\oplus]{(1)\ CO_2} R-(CH_2)_3-CH_2-CO_2H$

   (b) as in (a), then $R-(CH_2)_2-CH_2-CH_2OH \xrightarrow[heat]{H_2SO_4} R-(CH_2)_2-CH=CH_2$

   $\xrightarrow[heat]{KMnO_4} R-(CH_2)_2-CO_2H$

   (c) as in (a), then $R-(CH_2)_3-CH_2-MgBr \xrightarrow[(2)\ H_2O/H_3O^\oplus]{(1)\ \triangle O} R-(CH_2)_3-CH_2-CH_2-CH_2OH$

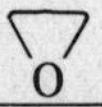

   $\xrightarrow{H_2CrO_4} R-(CH_2)_5-CO_2H$

   (d) $R-(CH_2)_2-CH_2-CO_2H \xrightarrow[(2)\ H_2O]{(1)\ P\ +\ Br_2} R-(CH_2)_2-\underset{\underset{Br}{|}}{CH}-CO_2H \xrightarrow{KOH}$

   $R-CH_2-CH=CH-CO_2H \xrightarrow[heat]{KMnO_4} R-CH_2-CO_2H$

2. $R-CH_2-CO_2H \longrightarrow R-CO_2H$ as in exercise 1(b), then $R-CO_2H \xrightarrow[(2)\ H_2O/H_3O^\oplus]{(1)\ LiAlH_4}$

   $R-CH_2OH \xrightarrow{HBr} R-CH_2Br \xrightarrow{Mg} R-CH_2-Mg-Br$

   $Ca^{14}CO_3 \xrightarrow{HCl} CaCl_2 + H_2O + {}^{14}CO_2$

   $R-CH_2-Mg-Br \xrightarrow[(2)\ H_3O^\oplus]{(1)\ {}^{14}CO_2} R-CH_2-{}^{14}CO_2H$

UNIT NINETEEN
Amines

The chemical behavior of the R—NH$_2$, $\begin{smallmatrix}R\\\\R\end{smallmatrix}$N—H, $\begin{smallmatrix}R\\\\R\end{smallmatrix}$N—R, and $\begin{smallmatrix}R\\\\R\end{smallmatrix}N^{\oplus}$$\begin{smallmatrix}R\\\\R\end{smallmatrix}$ functions is a direct extension of the chemistry of the R—OH, R—O—R, R—O$^{\oplus}\begin{smallmatrix}H\\\\H\end{smallmatrix}$, and $\begin{smallmatrix}R\\\\R\end{smallmatrix}O^{\oplus}$—H functions. Practically all the alcohol and ether reactions have close analogs among the amines, although the reagents and conditions are often different. The reaction mechanisms are the same S_N1, S_N2, acyl S_N2, $E1$, and $E2$ mechanisms that you learned in studying alcohol and alkyl halide chemistry. Consequently, it is most efficient to study the amines with the alcohols and ethers in the front of your mind. A few minutes spent refreshing your memory in Unit 15 will be helpful at this point.

After looking more closely at the analogy between alcohols/ethers and amines in frame 1, we take up the basic modes of reactivity in turn in frames 2 to 5. Frames 6 to 8 discuss learning strategies and provide general exercises.

OBJECTIVES

When you have completed this unit, you should be able to:

- Identify most of the reactions of the amines as simple extensions of one of the basic modes of reactivity of the alcohols or ethers, and predict their mechanisms by analogy with the alcohol/ether reactions (frame 1).
- Given the pH of the aqueous solution in which it is dissolved, predict the approximate percentage of ionization of an organic compound containing the functional group —NH$_2$, —NHR, —NR$_2$, —CO$_2$H, —SO$_3$H, or —OH (frame 2).
- Remember all the reactions involving amines by recalling the centers they attack as nucleophiles and the situations in which the amino group acts as a leaving group (frames 3, 4).
- Supply the missing reactant, reagent, or product for a reaction involving an amine or ammonium salt if given the other two species (frame 6).

If you believe you may be able to skip part or all of this unit, take a self-test consisting of the exercises marked with a star in the margin. If you miss any of these, review the frames involved and rework the exercises before proceeding.

If you are not ready for a self-test, proceed to frame 1.

1. Amines are the nitrogen analogs of the alcohols and ethers. If we mentally replace the oxygen atom in the important organic functional groups by a nitrogen atom, we get the stable, well-known compounds shown in Table 19.1. In some cases these O $\longrightarrow$ N exchange processes can actually be carried out: for example, RCO$_2$H $\xrightarrow[\Delta]{NH_3}$ R—CO—NH$_2$

+ H_2O and $R_2C=O$ + $ArNH_2$ ⟶ $R_2C=N-Ar$ + H_2O. The unsaturated nitrogen compounds in the table are readily converted to their oxygen analogs by acid hydrolysis:

Table 19.1 *Nitrogen Analogs of Oxygen Compounds*

O compound	N compound
R—OH	R—NH_2 and R_2N—H
R—O—R'	R—N(R')(R'')
R^1R^2C=O	R^1R^2C=N—H or R^1R^2C=N—R^3 or R^1R^2C=$\overset{\oplus}{N}$(R^3)(R^4)
R—C(=O)O—H	R—C(=O)NH_2 and R—C(=NH)NH_2

$$R-C(=O)NH_2 + H_2O \xrightarrow{H_3O^\oplus} R-C(=O)OH + \overset{\oplus}{NH_4}$$

$$R_1R_2C=NH + H_2O \xrightarrow{H_3O^\oplus} R_1R_2C=O + \overset{\oplus}{NH_4}$$

The amines (R—NH_2, R_2NH, R_3N) and their oxygen analogs (R—OH, R—O—R) are *not* directly interconvertible. Nevertheless, the reactions of the amines closely parallel the reactions of the alcohols, both in type and mechanism. The structural basis for the amine-alcohol analogy is as follows:

```
              C—Ö—H
              |
              H
   oxidizable   basic and     weakly      leaving group
                nucleophilic  acidic      when protonated
                    H   H
                    |
                    C—N̈
                       |
                       H
```

★ *Exercises*

1. Write the mechanism for these examples of interconversion of analogous N and O compounds:

(a) $CH_3-\overset{O}{\underset{\|}{C}}-CH_3$ + ⟨C₆H₅⟩-NH₂ $\xrightarrow{HA}$ $\overset{CH_3}{\underset{CH_3}{>}}C=N-$⟨C₆H₅⟩ + H_2O

(b) ⟨C₆H₅⟩-$\overset{O}{\underset{\|}{C}}$-NH₂ + H_2O $\xrightarrow{HA}$ ⟨C₆H₅⟩-$\overset{O}{\underset{\|}{C}}$-OH + NH_3

2. Both of the processes above can also be made to go in the reverse direction. What can you say about the mechanisms of these reverse reactions?

— — — — — — — — —

1. (a) $CH_3-\overset{O}{\underset{\|}{C}}-CH_3$ + HA ⇌ $CH_3-\overset{\overset{\oplus}{O}-H}{\underset{\|}{C}}-CH_3$ $A^\ominus$ ↔ $CH_3-\overset{O-H}{\underset{\oplus}{C}}-CH_3$ $A^\ominus$ ⟶

                                                                                        ↶:NH₂-⟨C₆H₅⟩

$CH_3-\overset{O-H}{\underset{\oplus NH_2-⟨C₆H₅⟩}{C}}-CH_3$ ⇌ $CH_3-\overset{\overset{\oplus}{OH_2}}{\underset{NH-⟨C₆H₅⟩}{C}}-CH_3$ $A^\ominus$ ⟶ $CH_3-\overset{\oplus}{\underset{N-H}{C}}-CH_3$ (with ⟨C₆H₅⟩ on N)

↔ $CH_3-\overset{\oplus N}{\underset{\|}{C}}-CH_3$ (with ⟨C₆H₅⟩ and H on N) ⟶ $CH_3-\overset{N}{\underset{\|}{C}}-CH_3$ (with ⟨C₆H₅⟩ on N) + HA^+
 $^\ominus A:↷H$

(b) ⟨C₆H₅⟩-$\overset{O}{\underset{\|}{C}}$-NH₂ + HA ⇌ ⟨C₆H₅⟩-$\overset{\overset{\oplus}{O}-H}{\underset{\|}{C}}$-NH₂ $A^\ominus$ ↔ ⟨C₆H₅⟩-$\overset{O-H}{\underset{\oplus}{C}}$-NH₂ $A^\ominus$ ⟶

                                                                                                    ↶:ÖH₂

⟨C₆H₅⟩-$\overset{O-H}{\underset{\oplus OH_2\ A^\ominus}{C}}$-NH₂ ⟶ ⟨C₆H₅⟩-$\overset{OH}{\underset{OH\ A^\ominus}{C}}$-$\overset{\oplus}{N}H_3$ ⟶ ⟨C₆H₅⟩-$\overset{OH}{\underset{OH}{C}}$⊕ $A^\ominus$ ↔

⟨C₆H₅⟩-$\underset{OH}{C}$=$\overset{\oplus}{O}-H$ $:A^\ominus$ ⟶ ⟨C₆H₅⟩-$\underset{OH}{C}$=O + HA

2. The reverse reaction has precisely the same mechanism with the steps taken in reverse order.

THE BASIC MODES OF AMINE REACTIVITY

2. Chart 19.1 shows how the reactions of the amines flow from five general chemical

Chart 19.1

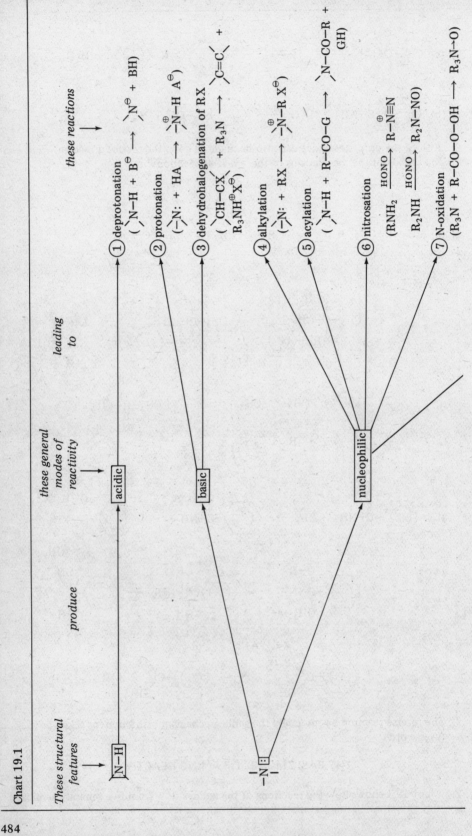

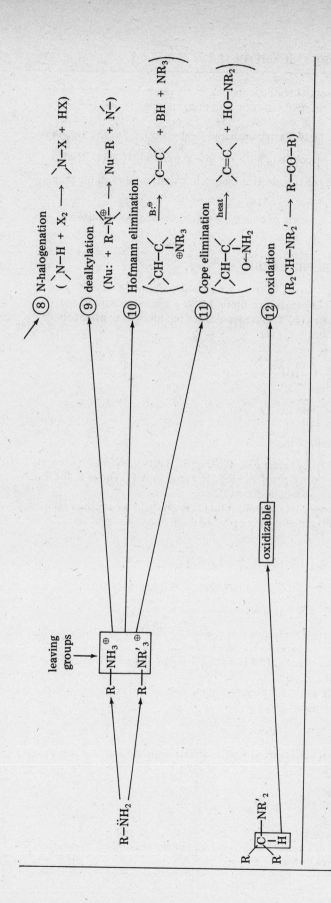

properties originating in the structural features present. In this and the next few frames we consider these five modes of reactivity in turn, beginning here with *basicity* and *acidity*.

Both alcohols and amines are basic, having unshared electron pairs, but the amines are much stronger bases than the alcohols. This is because the conjugate acid, $R\overset{\oplus}{N}H_3$, which has the positive charge on the less electronegative nitrogen, is more stable than $R-\overset{\oplus}{O}H_2$, which has the positive charge on oxygen.

For many years the standard measure of amine base strength was the equilibrium constant for the process

$$R-NH_2 + H_2O \overset{K_b'}{\rightleftharpoons} R-\overset{\oplus}{N}H_3 + OH^{\ominus}$$

which has values in the range 10^{-3} to 10^{-4} for simple aliphatic amines. You may still encounter these K_b values. Most chemists now prefer to use a single measure of the strength of all acids (AH) and bases (B); this measure is the equilibrium constant for deprotonation in aqueous solution:

$$\begin{array}{c} A-H \\ \text{or} \\ B-H^{\oplus} \end{array} + H_2O \overset{K_a}{\rightleftharpoons} \begin{array}{c} A^{\ominus} \\ \text{or} \\ B \end{array} + H_3O^{\oplus} \qquad (1)$$

For the amines this looks like

$$R-\overset{\oplus}{N}H_3 + H_2O \overset{K_a}{\rightleftharpoons} R-NH_2 + H_3O^{\oplus}$$

and K_a has values around 10^{-10} (pK_a values near 10; the pK_a bears the same relation to K_a as pH does to $[H^+]$: $pK_a = -\log K_a$, $pH = -\log [H^+]$; see Unit 11, frame 8, for the manner in which pK_a varies with acid-base strength).

Now, it is not difficult to prove that if equilibrium (1) is set up in an aqueous solution with some definite pH value, then the following relationships govern the position of the equilibrium:

when $pH = pK_a$: $[B] = [BH^{\oplus}]$, B is 50% ionized to $BH^{\oplus}$

when $pH < pK_a$: $[B] < [BH^{\oplus}]$, B is more than 50% ionized to $BH^{\oplus}$

when $pH > pK_a$: $[B] > [BH^{\oplus}]$, B is less than 50% ionized to $BH^{\oplus}$

Similar relations hold for the acids, A–H. Since in reasonably dilute aqueous solutions pH can vary only between about 1 and 13, this means that only those acids and bases that have pK_a values between about 1 and 13 can achieve equal concentrations of B and $BH^{\oplus}$ (or AH and $A^{\ominus}$) in aqueous solutions.

Acids with $pK_a < 1$ (strong acids) are almost completely ionized in water at all pH values, for instance the sulfonic acids:

$$R-SO_3H + H_2O \rightleftharpoons R-SO_3^{\ominus} + H_3O^{\oplus}$$

Bases with $pK_a < 1$ (very weak bases) are almost completely un-ionized in water at all pH values, for instance the ethers:

$$\underset{R}{\overset{R}{\diagdown}}\overset{\oplus}{O}-H + H_2O \rightleftharpoons \underset{R}{\overset{R}{\diagdown}}O + H_3O^{\oplus}$$

Acids with $pK_a > 13$ (very weak acids) are almost completely un-ionized in water at all pH values, for instance the alcohols:

$$R-O-H + H_2O \rightleftharpoons R-O^\ominus + H_3O^\oplus$$

Table 11.2 shows that only three functions have pK_a values between 1 and 13: RCO_2H, ArOH, and the amines. These are the compounds that have an interesting acid-base chemistry in aqueous media. Their physical, chemical, and physiological properties depend strongly on pH because the percentages of the two forms (e.g., $R-NH_2$ and $R-NH_3^\oplus$) shift as the pH changes. These percentages can be calculated; they are shown in Table 19.2.

Table 19.2 *Species Present in Aqueous Solutions at Various pH Values*

pH	Carboxylic acids		Aromatic amines		Phenols		Aliphatic amines	
	% RCO_2H	% $RCO_2^\ominus$	% $ArNH_2$	% $ArNH_3^\oplus$	% ArOH	% $ArO^\ominus$	% RNH_2	% $RNH_3^\oplus$
1	99.99	0.01	0.01	99.99	~100	10^{-7}	10^{-7}	~100
3	99	1	1	99	~100	10^{-5}	10^{-5}	~100
5	50	50	50	50	~100	10^{-3}	10^{-3}	~100
7	1	99	99	1	99.9	0.1	0.1	99.9
9	0.01	99.99	99.99	0.01	91	9	9	91
11	10^{-4}	~100	~100	10^{-4}	9	91	91	9
13	10^{-6}	~100	~100	10^{-4}	0.1	99.9	99.9	0.1

The following useful rules come from the table:

1. In physiological systems (pH near 7) the predominant species are $RCO_2^\ominus$, ArOH, $RNH_3^\oplus$, and $ArNH_2$.
2. To get carboxylic acids completely (say, ≥ 99.9%) in the nonelectrolyte form, RCO_2H—for example so that they crystallize out or can be extracted—requires dropping the pH to about 2.
3. To get aliphatic amines completely into the nonelectrolyte form, RNH_2, requires a pH of at least 13.

Table 19.2 also includes the aromatic amines ($ArNH_2$) and lists them at pK_a ~5, or about 100,000-fold weaker as bases than the alkylamines, RNH_2. This is because protonation destroys the resonance interaction of the $-NH_2$ function with the aromatic ring:

[Resonance structures of aniline showing delocalization of the nitrogen lone pair into the ring]

but

[Anilinium ion resonance structures — $-NH_3^\oplus$ only, no unshared pair]

The acidities/basicities of these four important families conveniently coincide on numbers that are easy to remember:

RCO_2H and $ArNH_3^\oplus$ have pK_a ~ 5

488 HOW TO SUCCEED IN ORGANIC CHEMISTRY

ArOH and $RNH_3^{\oplus}$ have $pK_a \sim 10$

The aliphatic amines are basic enough to bring about the $E2$ reaction of alkyl halides:

$$\underset{H}{\overset{X}{\underset{|}{\overset{|}{C-C}}}} \quad \rightarrow \quad \overset{X^{\ominus}}{\underset{\overset{\oplus}{N-H}}{C=C}}$$

The $S_N 2$ reaction with amine playing the role of nucleophile is a competing side reaction (see frame 3). Since the $S_N 2$ reaction is slowed down greatly by bulky groups near the nucleophilic atom, some highly hindered amines have been designed to do the $E2$ process without much competition from $S_N 2$ reactions. One favorite is pentamethylpiperidine:

[Structure of pentamethylpiperidine with CH₃ groups]

Finally, the primary and secondary amines are acids, but very weak ones—so much weaker than H_2O that their conjugate bases, $RNH^{\ominus}$ and $R_2N^{\ominus}$, cannot exist in aqueous or alcoholic solutions. They must be prepared from $R^{\ominus}$, $H^{\ominus}$, or alkali metals:

$$\text{Ph-}NH_2 + n\text{-}C_4H_9\text{-}Li \rightleftharpoons \text{Ph-}NH^{\ominus} Li^{\oplus} + n\text{-}C_4H_{10}$$

$$CH_3\text{-}NH_2 + Na \longrightarrow CH_3\text{-}NH^{\ominus} Na^{\oplus} + \tfrac{1}{2}H_2$$

★ Exercises

1. Use the ideas of Unit 11, frame 9, to predict toward which side the following equilibria will lie:

    (a) Ph-NH_2 + Ph-$\overset{\oplus}{NH_3}$ ⇌ Ph-$\overset{\oplus}{NH_3}$ + Ph-NH_2

    (b) Ph-OH + Ph-$CO_2^{\ominus}$ ⇌ Ph-$O^{\ominus}$ + Ph-CO_2H

    (c) $CH_3\text{-}CH_2\text{-}CH_2\text{-}NH_2$ + $CH_3\text{-}CO_2H$ ⇌ $CH_3\text{-}CH_2\text{-}CH_2\text{-}\overset{\oplus}{NH_3}$ + $CH_3\text{-}CO_2^{\ominus}$

    (d) Ph-OH + Ph-NH_2 ⇌ Ph-$O^{\ominus}$ + Ph-$\overset{\oplus}{NH_3}$

2. Will CO_2 probably be evolved from a solution of $CH_3\text{-}CH_2\text{-}\overset{\oplus}{NH_2}\text{-}CH_3$ $Cl^{\ominus}$ and $Na^{\oplus}$ $HCO_3^{\ominus}$ in H_2O?

3. Will the odor of $CH_3\text{-}NH_2$ probably be apparent above an aqueous solution made up by dissolving $CH_3\text{-}\overset{\oplus}{NH_3}$ $Cl^{\ominus}$ and Na_3PO_4 in water? ($HPO_4^{\,2\ominus}$ + H_2O $\overset{K}{\rightleftharpoons}$ $H_3O^{\oplus}$ + $PO_4^{\,3\ominus}$ has K near 10^{-12}.)

4. Considering the reasons for the relative base strengths of Ph-NH_2 and $CH_3\text{-}NH_2$,

would you predict ⟨O⟩-CO₂H to be stronger, weaker, or about the same in acid strength compared with CH_3-CO_2H?

5. Explain why $R-NH_2$ is a weaker acid than $R-OH$.

6. Considering possible polar and resonance effects, would you predict the N—H bonds in R-C(=O)-NH₂ to be more strongly or more weakly acidic than those in RNH_2?

7. The reaction $R-OH + NH_3 \longrightarrow R-NH_2 + H_2O$ gives practical yields of the amine only under rather extreme conditions (temperature, pressure, pH just right) that are useful only in industrial, not laboratory, chemistry. Considering what you know about the mechanism of nucleophilic substitution reactions, propose an explanation for the sluggishness of this reaction.

- - - - - - - - -

1. (a) Ph–NH₂ + Ph–NH₃⁺ ⇌ Ph–NH₃⁺ + Ph–NH₂
        wb   wa   sa   sb

   (b) Ph–OH + Ph–CO₂⁻ ⇌ Ph–O⁻ + Ph–CO₂H
        wa   wb   sb   sa

   (c) $CH_3-CH_2-CH_2-NH_2 + CH_3-CO_2H \rightleftharpoons CH_3-CH_2-CH_2-NH_3^{\oplus} + CH_3-CO_2^{\ominus}$
            sb   sa   wa   wb

   (d) Ph–OH + Ph–NH₂ ⇌ Ph–O⁻ + Ph–NH₃⁺
        wa   wb   sb   sa

2. $CH_3-CH_2-\overset{\oplus}{NH_2}-CH_3 + HCO_3^{\ominus} \rightleftharpoons CH_3-CH_2-NH-CH_3 + H_2CO_3$
           wa   wb   sb   sa
                                                                                                      ⇕
                                                                                                $H_2O + CO_2$

   The acid-base equilibrium lies to the left; it does not furnish substantial H_2CO_3 concentrations and probably not enough CO_2 to exceed its H_2O solubility.

3. $CH_3-NH_3^{\oplus} + PO_4^{3-} \rightleftharpoons CH_3-NH_2 + HPO_4^{2-}$
      sa   sb   wb   wa

   The equilibrium lies to the right, furnishing substantial concentrations of CH_3-NH_2, which should be smellable.

4. Search for resonance interaction between ring and function in the acid and conjugate base:

[Resonance structures of benzoic acid and benzoate anion shown, with "no interaction" labels]

In both cases resonance is present only within the ring and within the functional groups. Thus, as far as resonance is concerned, $C_6H_5-CO_2H$ and CH_3-CO_2H have equal stabilization, and so do $C_6H_5-CO_2^\ominus$ and $CH_3-CO_2^\ominus$. So $C_6H_5-CO_2H$ and CH_3-CO_2H should have about the same K_a (except for the difference in polar effect of CH_3- versus C_6H_5-, which is not large).

5. The conjugate base $R-O^\ominus$ puts the negative charge on oxygen, $R-NH^\ominus$ puts it on nitrogen. Since oxygen is more electronegative than nitrogen, $R-O^\ominus$ is more stable than $R-NH^\ominus$ (relative to the parent acids, ROH and RNH_2). So $R-OH + B^- \rightleftharpoons RO^\ominus + BH^\oplus$ lies farther to the right than $R-NH_2 + B \rightleftharpoons R-NH^\ominus + BH^\oplus$, which means that ROH is more acidic.

6. Polar effect:

$$\underbrace{R-\overset{O}{\overset{\|}{C}}\leftarrow \overset{\ominus}{\ddot{N}}H}_{\text{electron-withdrawing}} \quad \text{versus} \quad \underbrace{R\rightarrow \overset{\ominus}{\ddot{N}}H}_{\text{electron-donating}}$$

Therefore, charge is delocalized in $R-CO-NH^\ominus$, stabilizing it, while charge is localized in $R-NH^\ominus$, destabilizing it.

Resonance effect: $R-\overset{\ddot{O}}{\underset{\ddot{N}H^\ominus}{C}} \leftrightarrow R-\overset{\ddot{O}{:}^\ominus}{\underset{\ddot{N}H}{C}}$ vs. $R-\ddot{N}H^\ominus$

resonance-stabilized
conjugate base

The two effects agree in predicting greater stability for $R-CO-NH^\ominus$ and consequently greater acidity for $R-CO-NH_2$.

7. OH can only be a leaving group in the S_N2 reaction if it is protonated:

$$H_3N:\curvearrowright R\overset{\curvearrowright}{-}\overset{\oplus}{O}H_2 \longrightarrow \overset{\oplus}{H_3N}-R + H_2O$$

But since NH_3 is a much stronger base than ROH, the acid-base equilibrium

$$H_3N + R-\overset{\oplus}{O}H_2 \rightleftharpoons \overset{\oplus}{NH_4} + ROH$$

lies far to the right. Since $NH_4^{\oplus}$ no longer has an unshared pair, it is not nucleophilic, and both reactants required for the S_N2 process, $ROH_2^{\oplus}$ and $:NH_3$, have been destroyed.

3. Amines are nucleophiles for the same reason they are bases. They can play the role of nucleophile in the S_N2, acyl S_N2, and aryl S_N2 mechanisms.

The S_N2 reactions of amines introduce alkyl groups on nitrogen (*alkylate* the nitrogen). This process applies equally well to ammonia (an unalkylated amine):

$$H_3N:\curvearrowright R\overset{\curvearrowright}{-}X \longrightarrow \underset{1°}{\overset{\oplus}{H_3N}-R} + X^{\ominus}$$

Since the acid-base equilibrium

$$\underset{1°}{\overset{\oplus}{H_3N}-R} + NH_3 \rightleftharpoons \underset{1°}{H_2N-R} + \overset{\oplus}{NH_4}$$

is set up as soon as the first $RNH_3^{\oplus}$ forms, NH_3 and RNH_2 compete for the available RX; this results in formation of some $R_2NH_2^{\oplus}$:

$$\underset{1°}{R-\ddot{N}H_2}\curvearrowright R\overset{\curvearrowright}{-}X \longrightarrow \underset{2°}{R-\overset{\oplus}{N}H_2-R}\ X^{\ominus}$$

and so on:

$$\underset{2°}{R-\overset{\oplus}{N}H_2-R} + NH_3 \rightleftharpoons \underset{2°}{R-NH-R} + NH_4^{\oplus}$$

$$\underset{2°}{R-\ddot{N}H-R}\curvearrowright R\overset{\curvearrowright}{-}X \longrightarrow \underset{3°}{R-\overset{\oplus}{N}H\underset{R}{\overset{R}{\diagdown}}}\ X^{\ominus}$$

$$\underset{3°}{R-\overset{\oplus}{N}H\underset{R}{\overset{R}{\diagdown}}} + NH_3 \rightleftharpoons \underset{3°}{R-N\underset{R}{\overset{R}{\diagdown}}} + NH_4^{\oplus}$$

$$\underset{3°}{R_3N:}\curvearrowright R\overset{\curvearrowright}{-}X \longrightarrow \underset{4°}{\underset{R}{\overset{R}{\diagup}}\overset{\oplus}{N}\underset{R}{\overset{R}{\diagdown}}}\ X^{\ominus}$$

The stages of alkylation are labeled in these equations; the primary (1°), secondary (2°), and tertiary (3°) compounds exist as amines and their conjugate acids, the 1°, 2°, or 3° ammonium cations. In the quaternary (4°) stage, only the cation is possible. Note that to end up with a single product some special arrangement is required. One possibility is to provide some other base to absorb the acidic hydrogens produced, thus converting everything to the quaternary ammonium salt:

$$NH_3 + 4\,RX + 3\,NaHCO_3 \longrightarrow R_4N^{\oplus}\,X^{\ominus} + 3\,CO_2 + 3\,H_2O + 3\,Na^{\oplus}X^{\ominus}$$

or

$$R_2NH + 2\,R'X + NaHCO_3 \longrightarrow R_2\overset{\oplus}{N}R'_2\,X^{\ominus} + CO_2 + H_2O + Na^{\oplus}X^{\ominus}$$

These S_N2 reactions work well on 1° RX, but a 3° RX leads mainly to the elimination reaction discussed in frame 2: $\overset{}{-}N: + R-CH_2-CXR'_2 \longrightarrow \overset{\oplus}{-}NH + R-CH=CR'_2 + X^-$.

Just as the acyl S_N2 reactions of NH_3 give amides (Unit 18, frame 1), so 1° and 2° amines give substituted amides:

$$\underset{}{R-\overset{\overset{O}{\|}}{C}-O-CH_3} + R'-NH_2 \longrightarrow \underset{}{R-\overset{\overset{O}{\|}}{C}-NH-R'} + CH_3OH$$

$$\underset{}{R-\overset{\overset{O}{\|}}{C}-Cl} + R'-NH_2 \longrightarrow \underset{}{R-\overset{\overset{O}{\|}}{C}-NH-R'} + HCl$$

$$\underset{}{R-\overset{\overset{O}{\|}}{C}-O-\overset{\overset{O}{\|}}{C}-R} + R'-NH_2 \longrightarrow \underset{}{R-\overset{\overset{O}{\|}}{C}-NH-R'} + HO-\overset{\overset{O}{\|}}{C}-R$$

When the by-product is a moderately strong acid (HX, RCO_2H), it competes for the available amine, and a second mole of amine is required unless some other base (e.g., NaOH) is available to neutralize the acid produced. The mechanism should be familiar from Unit 18:

$$R-\overset{\overset{:\ddot{O}:}{\|}}{C}-Y \longrightarrow R-\overset{\overset{:\ddot{O}:^{\ominus}}{|}}{\underset{\underset{}{_{\oplus}NH_2-R}}{C}}-Y \rightleftharpoons R-\overset{\overset{:\ddot{O}-H}{|}}{\underset{\underset{}{NHR}}{C}}-Y \longrightarrow R-\overset{\overset{:O^{\oplus}H}{\|}}{C}-NHR + Y^{\ominus}$$

$$R-\overset{\overset{:\ddot{O}:^{\ominus}}{|}}{\underset{\underset{}{NHR}}{C}}-Y-H \longrightarrow R-\overset{\overset{:O:}{\|}}{C}-NHR + YH$$

The reactions of amines with three other types of reagent (nitrosating agents, peroxides, and halogenating agents) are S_N2 reactions in which the amine again plays the role of nucleophile, but attacks a nitrogen, an oxygen, or a halogen atom. The nitrosating agents are derivatives of nitrous acid, O=N-OH:

$$\overset{}{-}N: \quad \overset{\overset{O}{\|}}{N}-\overset{\oplus}{O}H_2 \longrightarrow \overset{}{-}\overset{\oplus}{N}-\overset{\overset{O}{\|}}{N} + OH_2$$

$$\overset{}{-}N: \quad \overset{\overset{O}{\|}}{N}-Cl \longrightarrow \overset{}{-}\overset{\oplus}{N}-\overset{\overset{O}{\|}}{N} + Cl^{\ominus}$$

What happens next depends on the structure of the amine.

The initially formed $\overset{}{-}\overset{\oplus}{N}-N=O$ species deprotonates to give a stable N-nitroso compound if the amine is secondary (aliphatic or aromatic). If derived from a primary amine, the intermediate dehydrates to the diazonium cation, $R-\overset{\oplus}{N}\equiv N$ or $Ar-\overset{\oplus}{N}\equiv N$.

$$\begin{array}{c}\text{R-NH}_2\\ \text{or}\\ \text{Ar-NH}_2\end{array} + \text{HONO} + \text{HA} \longrightarrow [\overset{\oplus}{\text{-NH}_2}\text{-N=O A}^{\ominus}] \longrightarrow \begin{array}{c}\text{R-}\overset{\oplus}{\text{N}}\equiv\text{N A}^{\ominus}\\ \text{or}\\ \text{Ar-}\overset{\oplus}{\text{N}}\equiv\text{N A}^{\ominus}\end{array} + \text{H}_2\text{O}$$

diazonium salt

$$\begin{array}{c}\text{R}\\ \diagdown\\ \text{NH}\\ \diagup\\ \text{R}\end{array} + \text{HONO} \longrightarrow \begin{array}{c}\text{R}\\ \diagdown\\ \text{N-N=O}\\ \diagup\\ \text{R}\end{array} + \text{H}_2\text{O}$$

N-nitrosamine

The $\text{Ar}-\overset{\oplus}{\text{N}}\equiv\text{N}$ and $\text{R}-\overset{\oplus}{\text{N}}\equiv\text{N}$ are very different in stability. $\text{Ar}-\overset{\oplus}{\text{N}}\equiv\text{N}$ is fairly stable at 0°, and a variety of nucleophiles can be introduced to react with it:

$$\text{Ar}-\overset{\oplus}{\text{N}}\equiv\text{N} + \text{Nu:}^{\ominus} \longrightarrow \text{Ar-Nu} + \text{:N}\equiv\text{N:}$$

These synthetically useful reactions are summarized in Unit 21.

The $\text{R}-\overset{\oplus}{\text{N}}\equiv\text{N}$, on the other hand, decompose immediately via the $S_N 1$ path, and the products are those derived from the carbonium ion $\text{R}^{\oplus}$—in aqueous solution the alcohol and alkene:

$$:\text{N}\equiv\overset{\oplus}{\text{N}}-\overset{|}{\underset{\text{H}}{\text{C}}}-\overset{|}{\text{C}}\diagup \longrightarrow :\text{N}\equiv\text{N}: + \underset{\text{H}_2\overset{..}{\text{O}}:}{\overset{\oplus}{\text{C}}-\overset{|}{\underset{\text{H}}{\text{C}}}\diagup} \longrightarrow \text{HO}-\overset{|}{\underset{\text{H}}{\text{C}}}-\overset{|}{\text{C}}\diagup + \diagup\!\!\text{C=C}\!\!\diagdown$$

$$\text{and } \overset{..}{\text{C}}\!:\!\overset{..}{\text{O}}\text{H}_2$$

This reaction is rarely synthetically useful, but it can be used to remove the amino group, and the evolution of N_2 is used to assay primary amines.

The reactions with peroxides (compounds containing an —O—O— group) look like this:

$$\text{R}_3\text{N:} \downarrow \overset{\frown}{\text{O}}-\overset{\frown}{\underset{\text{H}}{\text{O}}}-\text{H}$$

or

$$\text{R}_3\text{N:} \downarrow \overset{\frown}{\text{O}}-\overset{\oplus}{\underset{\text{H}}{\text{O}}}\text{H}_2 \longrightarrow \begin{array}{c}\text{R}\\ \diagdown\\ \overset{\oplus}{\text{R-N-O}}\\ \diagup |\\ \text{R} \text{H}\end{array} + \text{L}^{\ominus} \longrightarrow \begin{array}{c}\text{R}\\ |\\ \overset{\oplus}{\text{R-N}}\!\!-\!\!\overset{\ominus}{\text{O}}\\ |\\ \text{R}\end{array}$$

3° amine oxide

or

$$\text{R}_3\text{N:} \downarrow \overset{\frown}{\text{O}}-\overset{\frown}{\underset{\text{H}}{\text{O}}}-\text{SO}_3\text{H}$$

The 3° amine oxides are stable, strongly basic solids. The nitrogen atom is in the -1 oxidation state, up from -3 in the amine. Primary and secondary amines are oxidized, but often to a variety of products.

Compounds containing the groups X—X, X—$\overset{\oplus}{\text{OH}_2}$, X—NH—$\overset{\overset{\text{O}}{||}}{\text{C}}$—R, X—N$\diagup^{}_{\diagdown}$, and X—$\overset{\oplus}{\underset{|}{\text{N}}}\!\!\diagdown^{|}_{}$ are capable of halogenating strong nucleophiles like the amines via $S_N 2$ reactions:

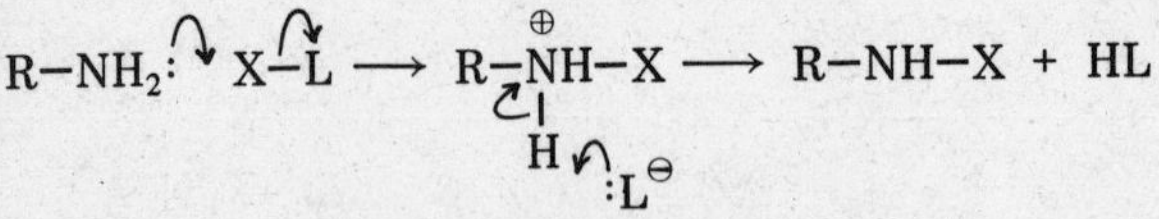

494 HOW TO SUCCEED IN ORGANIC CHEMISTRY

$$R-NH_2: \; X-L \longrightarrow R-\overset{\oplus}{NH}-X \longrightarrow R-NH-X + HL$$
$$\qquad\qquad\qquad\qquad \underset{H}{|} \; :L^{\ominus}$$

The product, an *N*-haloamine, is itself a halogenating agent capable of passing the halogen atom on to some new nucleophile:

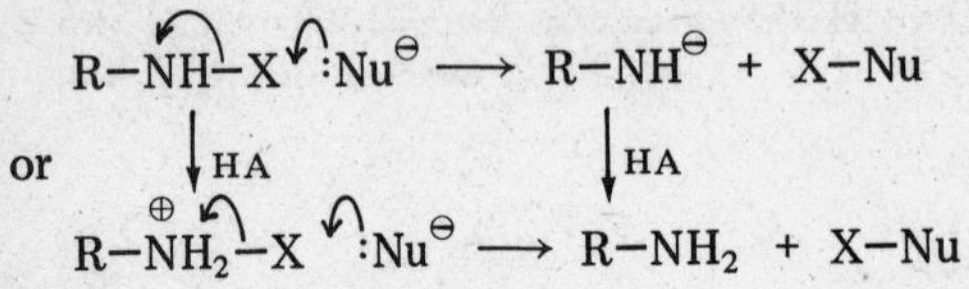

$$R-NH-X \; :Nu^{\ominus} \longrightarrow R-NH^{\ominus} + X-Nu$$

or $\quad \downarrow HA \qquad\qquad\qquad \downarrow HA$

$$R-\overset{\oplus}{NH_2}-X \; :Nu^{\ominus} \longrightarrow R-NH_2 + X-Nu$$

★ *Exercises*

1. Write the structures of all products formed in the reaction of $CH_3CH_2CH_2NH_2$ with CH_3I.

2. How many pairs of starting materials can you think of that react to give the salt

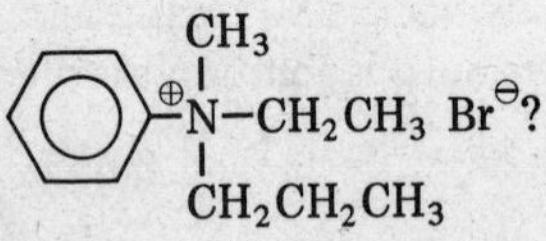

   Ph−$\overset{\oplus}{N}$(CH$_3$)(CH$_2$CH$_3$)(CH$_2$CH$_2$CH$_3$) $Br^{\ominus}$?

3. The 3° amine (cyclohexyl)−N:(piperidyl) is alkylated by CH_3I, but the product is (cyclohexyl with =N$^{\oplus}$−CH$_3$) $I^{\ominus}$, not (cyclohexyl)−$\overset{\oplus}{N}$(CH$_3$)(piperidyl) $I^{\ominus}$. Write a mechanism that accounts for formation of the observed product.

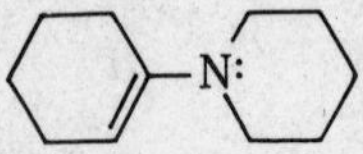

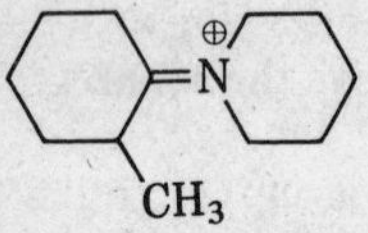

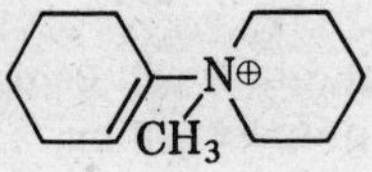

- - - - - - - - -

1. $CH_3CH_2CH_2-\overset{\oplus}{NH_2}-CH_3 \; I^{\ominus}$ $\qquad\qquad CH_3CH_2CH_2-NH-CH_3$

   $CH_3CH_2CH_2-\overset{\oplus}{NH}(CH_3)_2 \; I^{\ominus}$ $\qquad\qquad CH_3CH_2CH_2-N(CH_3)_2$

   $CH_3CH_2CH_2-\overset{\oplus}{N}(CH_3)_3 \; I^{\ominus}$ $\qquad\qquad CH_3CH_2CH_2-\overset{\oplus}{NH_3} \; I^{\ominus}$

2. 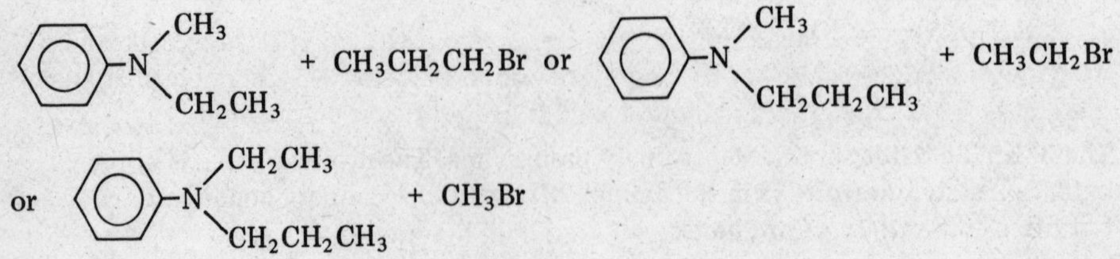 Ph−N(CH$_3$)(CH$_2$CH$_3$) + $CH_3CH_2CH_2Br$ or Ph−N(CH$_3$)(CH$_2$CH$_2$CH$_3$) + CH_3CH_2Br

   or Ph−N(CH$_2$CH$_3$)(CH$_2$CH$_2$CH$_3$) + CH_3Br

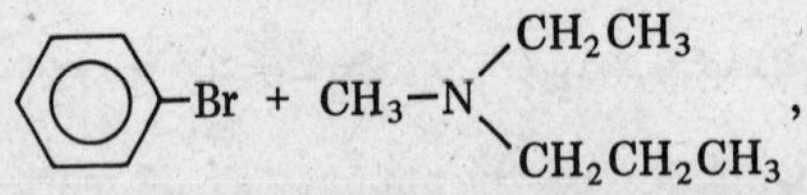

   but not Ph−Br + CH_3−N(CH$_2$CH$_3$)(CH$_2$CH$_2$CH$_3$) , because unactivated aryl halides don't undergo nucleophilic substitution reactions.

3. 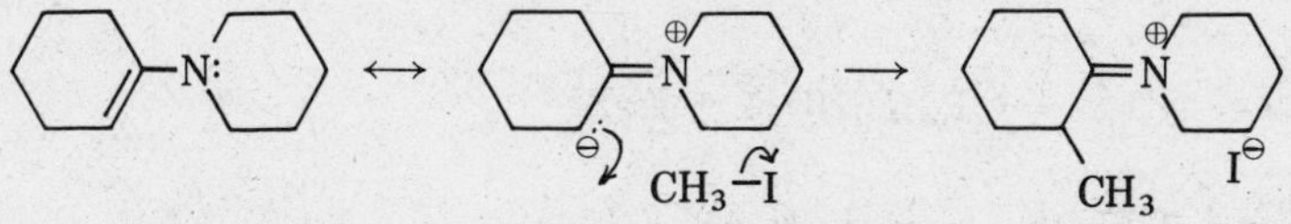 [Ph–N: ↔ Ph=N⁺ (with CH₃–I) → Ph=N⁺–CH₃ I⁻]

4. Under what circumstances can amino groups act as leaving groups? Like —OH and —OR groups, they do not leave in standard S_N2 reactions in their native condition:

$$\text{Nu}:^{\ominus} \;\; R-OH \;\not\to\; \text{Nu}-R + OH^{\ominus}$$

$$\text{Nu}:^{\ominus} \;\; R-NH_2 \;\not\to\; \text{Nu}-R + NH_2^{\ominus}$$

(We saw in Unit 18, however, that these are viable leaving groups in the *acyl* S_N2 mechanism.) Protonation makes —NH_2, —NHR, and —NR_2 groups into fair leaving groups for the S_N2 mechanism just as it does in the case of —OH. This is demonstrated in the S_N2 thermal decomposition of substituted ammonium halides:

$$:X:^{\ominus} \;\; R-\overset{\oplus}{N}H_3 \longrightarrow X-R + NH_3$$

If the nucleophile gets much more basic than halide ion, it deprotonates the ammonium cation and again no substitution occurs:

$$\text{Nu}:^{\ominus} + R-\overset{\oplus}{N}H_3 \rightleftharpoons R-NH_2 + \text{Nu}-H$$

However in the *quaternary* ammonium salts the amino group is locked into the cationic condition and is a fairly good leaving group under all conditions:

$$\text{Nu}:^{\ominus} \;\; R-\overset{\oplus}{N}(R)_2-R \longrightarrow \text{Nu}-R + NR_3$$

This reaction occurs for $\text{Nu}:^{\ominus}$ = $F^{\ominus}$, $Cl^{\ominus}$, $Br^{\ominus}$, $I^{\ominus}$, $SCN^{\ominus}$, $CN^{\ominus}$, $ArO^{\ominus}$, $R-CO_2^{\ominus}$, $NO_3^{\ominus}$, and $Ar-S^{\ominus}$. When the nucleophile becomes more basic (e.g., $OH^{\ominus}$ or $OR^{\ominus}$), the predominant reaction shifts from substitution to elimination via the $E2$ path, provided that a β H is available:

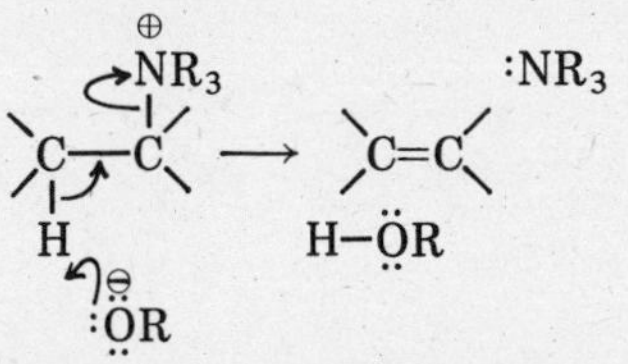

[Diagram showing β-H elimination: C(H)–C(NR₃⁺) with ⁻OR → C=C + :NR₃ + H–OR]

This is the Hofmann elimination reaction; it produces the least substituted alkene in cases where a choice between different β H atoms is possible. An intramolecular analog takes place when the tertiary amine N-oxide is heated (Cope elimination):

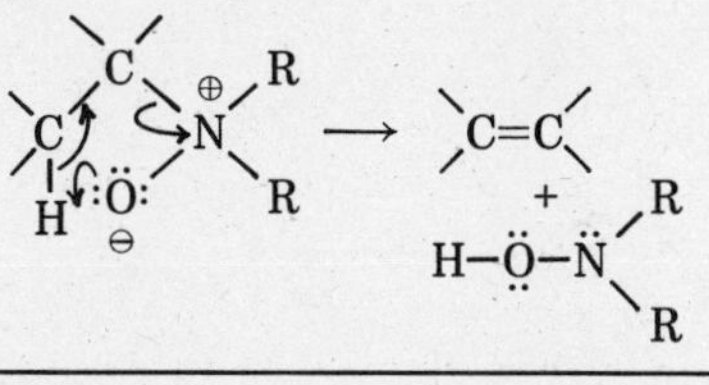

[Diagram showing Cope elimination: C(H)–C–N⁺(R)(R)–O⁻ → C=C + H–Ö–N(R)(R)]

To carry out these reactions of the quaternary ammonium cations, we first exchange the original halide ion for the desired anion by such simple ion reactions as

$$R_4\overset{\oplus}{N}\ Br^{\ominus} + Ag^{\oplus}\ NO_3^{\ominus} \longrightarrow R_4\overset{\oplus}{N}\ NO_3^{\ominus} + AgBr$$

$$R_4\overset{\oplus}{N}\ Br^{\ominus} \xrightarrow{Ag_2O} R_4\overset{\oplus}{N}\ OH^{\ominus} \xrightarrow{HA} R_4\overset{\oplus}{N}\ A^{\ominus} + H_2O$$
$$+$$
$$AgBr$$

★ Exercises

1. Predict the structure of the major organic products of these reactions:

(a) $CH_3-\overset{O}{\underset{\|}{C}}-O^{\ominus}\ (CH_3)_4\overset{\oplus}{N} \xrightarrow{heat}$

(b) $CH_3-\overset{O}{\underset{\|}{C}}-\overset{\overset{CH_3}{|}}{\underset{\underset{CH_3}{|}}{\overset{\oplus}{N}}}-CH_3\ HSO_4^{\ominus} \xrightarrow[heat]{CH_3OH}$

(c) $CH_3-\overset{\overset{CH_2CH_2CH_3}{|}}{\underset{\underset{\underset{CH_3}{|}}{CH_2-CH-CH_3}}{\overset{\oplus}{N}}}-CH_3 \quad OH^{\ominus} \xrightarrow{heat}$

(d) 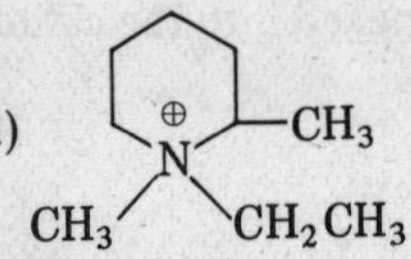 [piperidinium ring with N⁺ bearing CH₃ and CH₂CH₃, with a CH₃ substituent] $OH^{\ominus} \xrightarrow{heat}$

2. Reactions (a) and (b) in exercise 1 give the same product but by two different mechanisms. Write the mechanisms.

— — — — — — — —

1. (a) $CH_3-\overset{O}{\underset{\|}{C}}-O-CH_3 + N(CH_3)_3$

(b) $CH_3-\overset{O}{\underset{\|}{C}}-O-CH_3 + H\overset{\oplus}{N}(CH_3)_3\ HSO_4^{\ominus}$

(c) $CH_3-\underset{\underset{\underset{CH_3}{|}}{CH_2-CH-CH_3}}{N}-CH_3 \quad + CH_2=CH-CH_3 + H_2O$

(d) 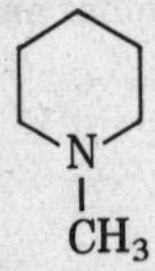 [piperidine with N-CH₃] $+ CH_2=CH_2 + H_2O$

2. (a) 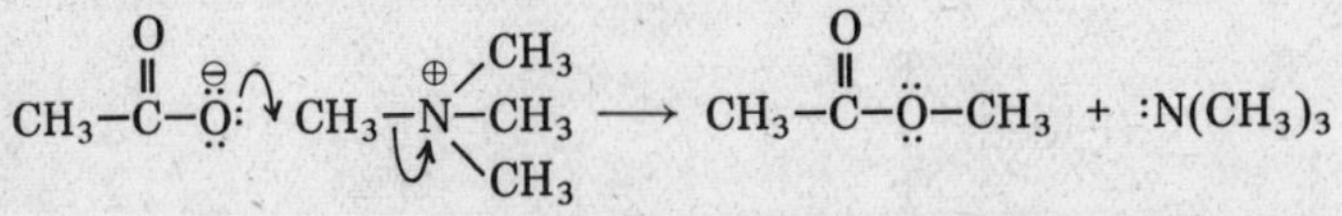 $CH_3-\overset{O}{\underset{\|}{C}}-\overset{..}{\underset{..}{O}}{:}^{\ominus} \curvearrowright CH_3\overset{\overset{CH_3}{\diagup}}{\underset{\underset{CH_3}{\diagdown}}{-\overset{\oplus}{N}-CH_3}} \longrightarrow CH_3-\overset{O}{\underset{\|}{C}}-\overset{..}{\underset{..}{O}}-CH_3 + :N(CH_3)_3$

(b) $CH_3-\overset{O}{\underset{\|}{C}}\overset{\oplus}{=}\underset{\underset{H-\ddot{O}-CH_3}{|}}{\underset{CH_3}{N}}-CH_3 \longrightarrow CH_3-\overset{:\ddot{O}:^{\ominus}}{\underset{\underset{H}{\overset{|}{\underset{\oplus}{O}}}}{\underset{|}{C}}}\overset{CH_3}{\underset{CH_3}{N}}-CH_3 \longrightarrow CH_3-\overset{O}{\underset{\|}{C}}-\overset{\oplus}{\underset{H}{O}}\overset{CH_3}{\underset{:N(CH_3)_3}{\diagup}} \quad HSO_4^{\ominus}$

$$\downarrow$$

$$CH_3-\overset{O}{\underset{\|}{C}}-\ddot{O}-CH_3 + H\overset{\oplus}{N}(CH_3)_3 HSO_4^{\ominus}$$

5. Oxidizing agents can attack amines at two different sites. Those containing O—O, N—O, O—X, or X—X bonds attack nitrogen; we discussed these reactions in frame 3. $KMnO_4$, MnO_2, $Ag^{\oplus}$, and others attack α hydrogens. This reaction is strictly analogous to the oxidation of a secondary alcohol to the ketone:

$$\underset{R}{\overset{R}{\diagdown}}CH-NR'_2 \xrightarrow[\text{etc.}]{KMnO_4} \underset{R}{\overset{R}{\diagdown}}C=\overset{\oplus}{N}\underset{R'}{\overset{R'}{\diagup}} \quad \text{an iminium salt}$$

The iminium salt is a nitrogen analog of the ketone and is hydrolyzable to the ketone:

$$\underset{R}{\overset{R}{\diagdown}}\overset{\oplus}{C}=N\underset{R'}{\overset{R'}{\diagup}} \xrightarrow{H_2O} \underset{R}{\overset{R}{\diagdown}}C=O + R'_2NH$$

These oxidations are less clean cut for 1° and 2° amines, and they may be less distinct mechanistically from oxidation on N. The oxidation of 3° amines by one reagent, nitrous acid, begins with attack at N but finally produces an imine:

$$R_2CH-\ddot{N}R'_2 \curvearrowright \overset{O}{\underset{\|}{N}}-\overset{\oplus}{O}H_2 \longrightarrow R_2CH-\overset{\overset{N=O}{|}R'}{\underset{\oplus}{N}}\underset{R'}{\diagup} + H_2O$$

$$\underset{\underset{H}{\overset{|}{R_2C}}\overset{\curvearrowleft}{\underset{O}{\diagup}}}{\overset{R'}{\overset{\oplus}{N}}\overset{R'}{\diagup}}\overset{\curvearrowright}{\underset{N}{\diagdown}} \longrightarrow \underset{R}{\overset{R}{\diagdown}}C=\overset{\oplus}{N}\underset{R'}{\overset{R'}{\diagup}} + [HON]$$
$$\downarrow$$
$$\text{unknown products}$$

LEARNING THE REACTIONS AND MECHANISMS

6. Next, we apply some of the standard drills developed in previous units to missing-product and missing-reagent problems involving the amines.

Exercises

1. Edit the list of reactions in Chart 19.1 to conform to your textbook or lecture coverage.

2. Lay out the edited list in three-column [reactant, reagent, product(s)] format as in the exercises in Unit 12, frame 16. Cover the product column with scratch paper and fill in the missing products for each reaction in turn. Check, correct, and repeat the drill.

3. Cover the middle column and supply the missing reagents. Check, correct, and repeat.

4. Cover the left column and supply the missing reactants. Check, correct, and repeat.

★ 5. Supply the missing reactants, intermediates, and products.

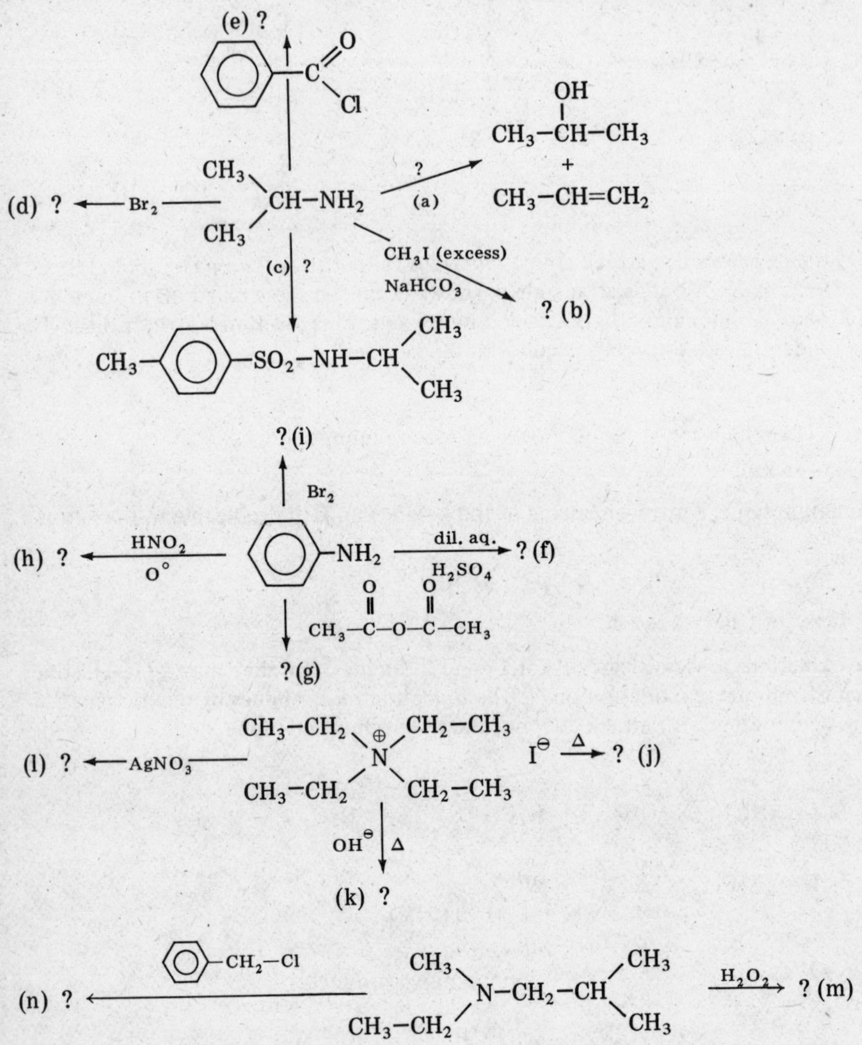

498 HOW TO SUCCEED IN ORGANIC CHEMISTRY

[Reaction scheme showing transformations of (CH$_3$)$_2$CH-NH$_2$ (isopropylamine) with various reagents, labeled (a) through (n).]

Central compound: (CH$_3$)$_2$CH-NH$_2$

- (e) ? ← C$_6$H$_5$-COCl
- (d) ? ← Br$_2$
- (a) ? → CH$_3$-CH(OH)-CH$_3$ + CH$_3$-CH=CH$_2$ (via HNO$_2$)
- CH$_3$I (excess), NaHCO$_3$ → ? (b)
- (c) ? → CH$_3$-C$_6$H$_4$-SO$_2$-NH-CH(CH$_3$)$_2$

From C$_6$H$_5$-NH$_2$ (aniline):
- ? (i) ← Br$_2$
- (h) ? ← HNO$_2$ / 0°
- dil. aq. H$_2$SO$_4$ → ? (f)
- CH$_3$-C(O)-O-C(O)-CH$_3$ → ? (g)

From (CH$_3$-CH$_2$)$_4$N$^⊕$ I$^⊖$:
- (l) ? ← AgNO$_3$
- Δ → ? (j)
- OH$^⊖$, Δ → (k) ?

From (CH$_3$)$_2$CH-N(CH$_2$-CH$_3$)$_2$
- C$_6$H$_5$-CH$_2$-Cl + CH$_3$-CH$_2$-N(CH$_3$)-CH$_2$-CH(CH$_3$)$_2$ ... (n) ? ←
- H$_2$O$_2$ → ? (m)

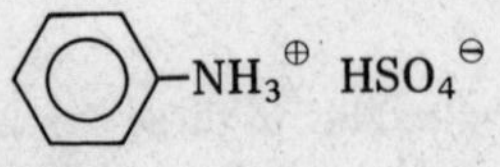

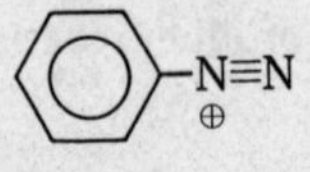

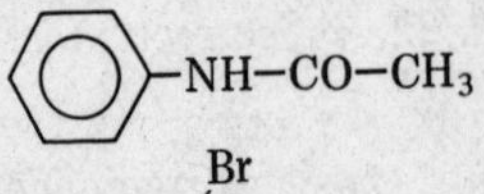

---

5. (a) HNO$_2$
   (b) (CH$_3$)$_2$CH$\overset{\oplus}{N}$(CH$_3$)$_3$ I$^\ominus$
   (c) CH$_3$-C$_6$H$_4$-SO$_2$-Cl
   (d) (CH$_3$)$_2$CH-NHBr
   (e) 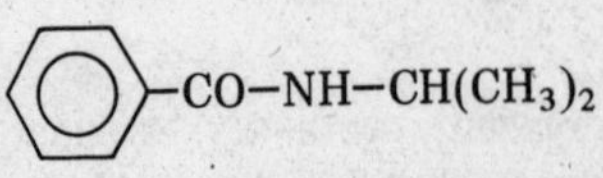 C$_6$H$_5$-CO-NH-CH(CH$_3$)$_2$
   (f) C$_6$H$_5$-$\overset{\oplus}{N}$H$_3$ HSO$_4^\ominus$
   (g) C$_6$H$_5$-NH-CO-CH$_3$
   (h) C$_6$H$_5$-N≡$\overset{\oplus}{N}$
   (i) 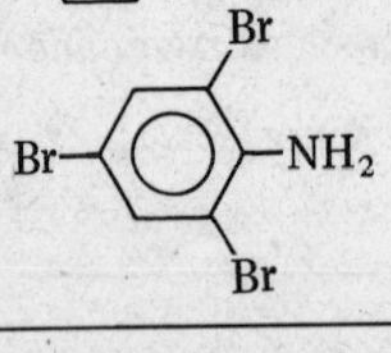 2,4,6-tribromoaniline (Br$_3$C$_6$H$_2$-NH$_2$)
   (j) CH$_3$-CH$_2$-I + (CH$_3$-CH$_2$)$_3$N

(k) $CH_2=CH_2 + (CH_3CH_2)_3N$

(l) $AgI + (CH_3CH_2)_4N^{\oplus} NO_3^{\ominus}$

(m) $CH_3CH_2-\underset{\underset{CH_3}{|}}{\overset{\overset{O^{\ominus}}{|}}{N^{\oplus}}}-CH_2CH(CH_3)_2$

(n) $\text{C}_6\text{H}_5-CH_2-\underset{\underset{CH_2CH_3}{|}}{\overset{\overset{CH_3}{|}}{N^{\oplus}}}-CH_2-CH(CH_3)_2$

7. Except for reaction ⑫, all the reactions in Chart 19.1 are either simple proton transfers (B:↷ H–A) or else follow the S_N2 or $E2$ mechanisms, sometimes preceded or followed by proton transfer.

★ Exercises

1. The following are familiar reactions of various oxygen-containing functional groups. Write the equation for the reaction of a nitrogen compound that you think is most closely analogous to each of them.

   (a) ether cleavage
   (b) ether formation
   (c) alcohol dehydration
   (d) alcohol esterification
   (e) ketone reduction to 2° alcohol

2. Write all steps in the mechanism of each reaction written in exercise 1.

3. Propose a mechanism for the reaction

$$\text{C}_6\text{H}_5-\overset{\oplus}{N}H_2-CH_2-CH_2-CH_3 \;\; Br^{\ominus} \xrightarrow{300°C} (CH_3)_2CH-\text{C}_6\text{H}_4-NH_2 + CH_3-CH=CH_2 + CH_3-CH_2-CH_2-Br$$

- - - - - - - - -

1. (a) $R_4N^{\oplus} I^{\ominus} \xrightarrow{\Delta} R-I + R_3N$

(b) $R-NH_2 + R'Br \longrightarrow R-\overset{\oplus}{N}H_2-R' \; Br^{\ominus} \xrightarrow{base} R-NH-R'$

(c) $R-CH_2-CH_2-\underset{\underset{CH_3}{|}}{\overset{\overset{CH_3}{|}}{N^{\oplus}}}-CH_3 \; OH^{\ominus} \xrightarrow{\Delta} R-CH=CH_2 + CH_3-N(CH_3)_2 + H_2O$

(d) $R-NH_2 + R'-\overset{O}{\overset{\|}{C}}-Cl \longrightarrow R'-\overset{O}{\overset{\|}{C}}-NH-R$

(e) $\underset{R}{\overset{R}{\diagdown}}C=NH \xrightarrow[(2) \; H_2O]{(1) \; LiAlH_4} \underset{R}{\overset{R}{\diagdown}}CH-NH_2$

2. (a) $R_3\overset{\oplus}{N}-R \;\; :\overset{..}{I}:^{\ominus} \longrightarrow R_3N: + R-I$

(b) $R-NH_2: \;\; R'-Br \longrightarrow R-\underset{H}{\overset{\oplus}{N}}H-R' \; Br^{\ominus} \;\; :B \longrightarrow R-NH-R' + BH^{\oplus} \; Br^{\ominus}$

(c) $R-CH_2-CH_2-\underset{\underset{CH_3}{|}}{\overset{\overset{CH_3}{|}}{N^{\oplus}}}-CH_3 \longrightarrow R-CH=CH_2 + :N(CH_3)_2-CH_3$
    $\;\; H \;\; :\overset{..}{O}H^{\ominus}$
    $\qquad\qquad\qquad\qquad\qquad\qquad\qquad + H-O-H$

(d)
$$R'-\overset{O}{\underset{\underset{R-NH_2}{\uparrow}}{C}}-Cl \longrightarrow R'-\underset{\underset{R-NH_2}{|}}{\overset{:\overset{\ominus}{O}:}{\underset{|}{C}}}-Cl \longrightarrow R'-\underset{\underset{H}{|}}{\overset{O}{\underset{|}{C}}}-\overset{\oplus}{N}H-R \longrightarrow R'-\overset{O}{\underset{|}{C}}-\ddot{N}H-R + H-Cl$$
$$:\overset{\ominus}{\ddot{C}l}:$$

(e)
$$\overset{\overset{\curvearrowright}{\ddot{N}-H}}{R-\underset{\underset{H-AlH_3^{\ominus}}{\uparrow}}{\overset{||}{C}}-R} \longrightarrow \overset{:\overset{\oplus}{N}H\quad AlH_3}{R-\underset{\underset{H}{|}}{C}-R} \longrightarrow \overset{:NH-AlH_3}{R-\underset{\underset{H}{|}}{C}-R} \longrightarrow (R_2CH-NH)_4Al^{\ominus} \xrightarrow{H_2O}$$

$$R_2CH-NH_2$$

3. The N—C bond is broken, which could be accomplished by Hofmann elimination or by S_N2 displacement of ⟨O⟩—NH_2 by $Br^{\ominus}$. The $CH_3-CH=CH_2$ could come from the Hofmann elimination, but there is no strong base present to bring that about. So try the S_N2:

$$\text{⟨O⟩}-\overset{\oplus}{N}H_2-CH_2-CH_2-CH_3 \longrightarrow \text{⟨O⟩}-NH_2 + CH_3-CH_2-CH_2-Br$$
$$\ominus :\ddot{B}r:$$

This accounts for the alkyl halide formed. The aromatic product has an alkyl group on the ring, which could only have gotten there by electrophilic aromatic substitution. What is the electrophile required? $R^{\oplus}$. Where could it come from? Only from R—Br:

$$CH_3-CH_2-CH_2\overset{\curvearrowright}{-}Br \longrightarrow CH_3-CH_2-CH_2^{\oplus} + Br^{\ominus}$$

But the group introduced is *iso*propyl. So:

$$CH_3-\underset{\underset{H}{|}}{CH}-CH_2^{\oplus}(1°) \longrightarrow CH_3-\overset{\oplus}{CH}-CH_3 \quad (2°)$$

$$\overset{CH_3}{\underset{CH_3}{\diagdown}}\overset{\oplus}{CH}\quad H\quad\text{⟨O⟩}-NH_2 \longrightarrow \overset{CH_3}{\underset{CH_3}{\diagdown}}CH\overset{\oplus}{\text{⟨O⟩}}-NH_2 \longleftrightarrow \text{etc.}$$

$$\text{etc.} \longleftrightarrow \overset{CH_3}{\underset{CH_3}{\diagdown}}\overset{H}{\underset{\oplus}{\text{⟨O⟩}}}CH\text{⟨O⟩}-NH_2 \longrightarrow \overset{CH_3}{\underset{CH_3}{\diagdown}}CH-\text{⟨O⟩}-NH_2$$
$$:\overset{\ominus}{\ddot{B}r}:$$

The alkene product is also now explained—some alkene is expected in all reactions involving $R^{\oplus}$:

$$CH_3-\overset{\oplus}{CH}\overset{\curvearrowleft}{-}CH_2 \longrightarrow CH_3-CH=CH_2 + HBr$$
$$\underset{\underset{:\ddot{B}r:^{\ominus}}{\curvearrowright}}{H}$$

8. Perhaps the easiest way to remember the amine-ammonium salt reactions is in two groups:
   1. Amine as electron donor (= base or nucleophile) reacting with electron acceptors (= acid or electrophile):

$$\ce{>N:} \begin{cases} \text{H--L} & \text{(reaction ②)} \\ \text{H--C--C--L} & \text{(reaction ③)} \\ \text{>C--L} & \text{(reaction ④)} \\ \text{>C=O} \text{ (L)} & \text{(reaction ⑤)} \\ \text{NO--L} & \text{(reaction ⑥)} \\ \text{OH--L} & \text{(reaction ⑦)} \\ \text{X--L} & \text{(reaction ⑧)} \end{cases}$$

The electron acceptors are easy to recall by thinking of the periodic table: H ↘ C → N → O → X, which gives you the identity of the atoms actually attacked by the amine. In reconstructing the full equations from these cues you must supply a specific leaving group to fill the L slot. Chart 19.1 and frames 2 to 5 used L = Cl, Br, I in almost all cases, but L = $O-SO_2-R$, $O-SO_2-Ar$, $O-PO(OR)_2$, and $O-NO_2$ also work in most cases. Learning the reactions in this highly condensed form actually gives you the power to predict the existence of a number of reactions beyond those ordinarily studied in the organic course, just by plugging in groups that you have learned to be good leaving groups in other contexts (e.g., Unit 14, frame 1).

   2. Electron donors (= bases, nucleophiles) reacting with amines as electron acceptors (acids, electrophiles):

$$\text{Nu:} \begin{cases} \text{H--N<} & \text{(reaction ①)} \\ \text{H--C--C--N}^{\oplus} & \text{(reactions ⑩, ⑪)} \\ \text{>C--N}^{\oplus}\text{<} & \text{(reaction ⑨)} \end{cases}$$

Exercises

Practice reconstructing your edited list of amine-ammonium salt reactions from memory, on paper. Cue yourself by thinking of either the classification developed in this frame or the classification based on structural features used in Chart 19.1.

UNIT TWENTY
Structure Problems

A "structure problem" poses this question: Given certail chemical and physical properties for a compound, what can you deduce about its structural formula? The amount and usefulness of the data are variable, and they determine the completeness of solution that is possible. Given enough information, you can deduce a structural formula complete with all stereochemical details. Given insufficient data, you can deduce some structural features but not others. We will treat all possible cases as a single problem whose goal is as much structural detail as can be deduced. The best way to express what is known and what is not known about a structure is by drawing for the known part a *part structure* of the type used in previous units (e.g., $-\overset{|}{\underset{|}{C}}-\overset{|}{C}H-C\equiv N$, in which the open bonds indicate that the groups attached at those sites are unknown or unspecified). So the solutions to structure problems will be either part structures or complete structures. These problems have always been considered an important part of learning organic chemistry because of their pedagogical value and because they occur continually in real life. Every structural formula that we write so confidently in the organic chemistry course was once a real-life structure problem. Some of these problems were solved completely by a single chemist using experimental data gathered with his own hands. Others were carried only to the part-structure stage originally, then picked up by later chemists, who, with more data, refined the part structures further.

OBJECTIVES

When you have completed this unit, you should be able to:
- Combine information such as the molecular formula, results of diagnostic reactions (frames 4 to 7), structures of the products of degradative reactions (frames 11 to 15), and infrared and proton magnetic resonance spectra to deduce the structural formulas of simple compounds.
- Express incomplete chemical and physical information in the form of part structures (frames 1, 2).

If you think you may be able to skip all or part of this unit, take a self-test consisting of the exercises marked with a star in the margin. If you miss some problems, review the frames in question and rework the problems. If you are not ready for a self-test, proceed to frame 1.

1. In attacking a structure problem we sift through the given data a piece at a time, putting aside irrelevant data and deducing what we can from each relevant piece. As in a jigsaw puzzle, we can't put everything together at once. To keep track of what you have found out about the structure at each stage, you need something to build onto, piece by piece. That something is a part structure in which the known and unknown structural features are summarized in juxtaposition. Let's first get into your vocabulary some possible ways of drawing such part structures.

Examples

1. If you know that Cl and NO_2 are attached to a benzene ring, but you don't know their relative locations, you can write

   [benzene ring]—Cl, —NO_2 or   [benzene ring] {—Cl, —NO_2}

2. If you know that a compound is 2-methyl-4-*tert*-butylcyclohexanol, and you know that the —OH and $(CH_3)_3C$— groups are cis, but you don't know whether the CH_3— group is cis or trans to these, you can write a squiggle bond, which means "configuration unknown":

   [cyclohexane structure with OH, CH_3 (squiggle bond), and C(CH_3)_3 substituents]

3. Sometimes you can express known and unknown parts by writing the molecular formula in place of the structural formula for an unknown group, as in

   HOOC—[benzene ring]—C_6H_{13}

   A hexyl group of some kind is in the *p*-position, but it could be 1-, 2-, or 3-hexyl,

   $$-CH_2-CH-CH_2-CH_2-CH_3, \quad -\underset{\underset{CH_3}{|}}{\overset{\overset{CH_3}{|}}{C}}-CH_2-CH_2-CH_3, \text{ etc.}$$
   with CH_3 branch

4. A certain compound has molecular formula $C_{16}H_{17}Cl$. What is

   $\left.\begin{array}{l} Cl-[C_6H_4]- \\ \;\;\;\;[C_6H_5]- \\ \;\;\;\;CH_3- \end{array}\right\} C_3H_5,$

   the part structure, saying about the compound? This is a format we will use repeatedly: the structural features known to be present are listed on the left; the remainder, of unknown structure, is on the right. This particular part structure says that the presence of a methyl, a phenyl, and a *p*-chlorophenyl group in the molecule has been deduced, but nothing more. These groups account for

   CH_3
   C_6H_5
   $\underline{C_6H_4Cl}$
   $C_{13}H_{12}Cl$

   out of a total of $C_{16}H_{17}Cl$. The difference is C_3H_5, about whose structure nothing is known except that it holds the other three groups together.

5. A certain compound has molecular formula $C_8H_{10}O$. Its nmr spectrum (frame 8) suggests the presence of $CH_3-\underset{\underset{H}{|}}{\overset{\overset{|}{}}{C}}-$ and [benzene ring]— groupings. We use the format of example 4 as follows:

list the known groups: CH₃–CH⟨ (phenyl)   subtract from the molecular formula: combine: CH₃–CH⟨(phenyl)–OH

add: C₈H₉

$C_8H_{10}O - C_8H_9 = HO$

★ *Exercises*

1. Write part structures embodying all the structural information given.

    (a) Compound X, C_6H_{12}, gives evidence for the presence of a double bond [reaction with Br_2(dark) and $KMnO_4$]. It contains only one H on doubly bonded carbon.

    (b) Compound Y, $C_{11}H_{12}$, is a terminal alkyne and has an isopropyl group.

2. Draw a part structure embodying the following information. Compound X, $C_5H_{11}OCl$, shows chemical reactivity and nmr spectra characteristic of a secondary alcohol. It also behaves like a primary alkyl chloride.

— — — — — — — — — —

1. (a) $\rangle C=C\langle_H$ } C_4H_{11} (b) $CH_3\rangle CH-$, CH_3, $-C\equiv C-H$ } C_6H_4

2. $HO-CH\langle$, $Cl-CH_2-$ } C_3H_7 If you wrote $HO-$, $Cl-$ } C_5H_{10} instead, you failed to include all the information that was available, namely that the alcohol is 2° and the chloride 1°.

2. As we learn more about the structure of a compound by sifting through more and more data, the number of atoms listed on the left-hand (known) side of the part structure grows and the number on the right (unknown) side shrinks. When we know only the molecular formula, a tremendous number of complete structural formulas are usually possible (as your practice in Unit 6 showed). But as the right side of the developing part structure shrinks, we are weeding out more and more of the isomeric structures as not consistent with the data. If enough data are available, all but one can be weeded out. When has the part structure been developed to the point where it is feasible to write out all the possible complete structures? Usually, when the number of carbon atoms in the unknown side has been whittled down to 3 or fewer.

One of the advantages of writing and revising part structures is this: it helps you avoid starting to write specific, complete structures while there are still too many of them in the running. Use of part structures also reminds you to keep an open mind with respect to *all structural possibilities* when working structure problems. One major source of student errors in structure problems is settling too soon on a single structure for the molecule or one of its constituent groups when other structures are equally probable. We want to develop a grand strategy of *staying general* as long as possible; when the facts *force* us to adopt a single, specific structure, the problem is solved.

Example

Draw all the complete structures implied by this part structure for C_3H_4OCl:

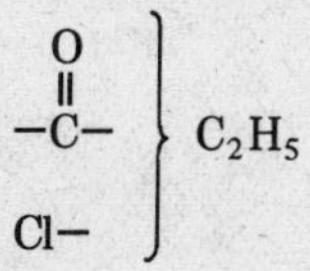

$$\left.\begin{array}{c} \overset{O}{\underset{\|}{-C-}} \\ Cl- \end{array}\right\} C_2H_5$$

Solution:

The unknown part must be $-CH_2-CH_3$, right? Wrong. There is no reason why these two carbons must be bonded to one another, and we should avoid jumping to this conclusion. As usual, you should use a systematic approach to avoid missing possible structures:

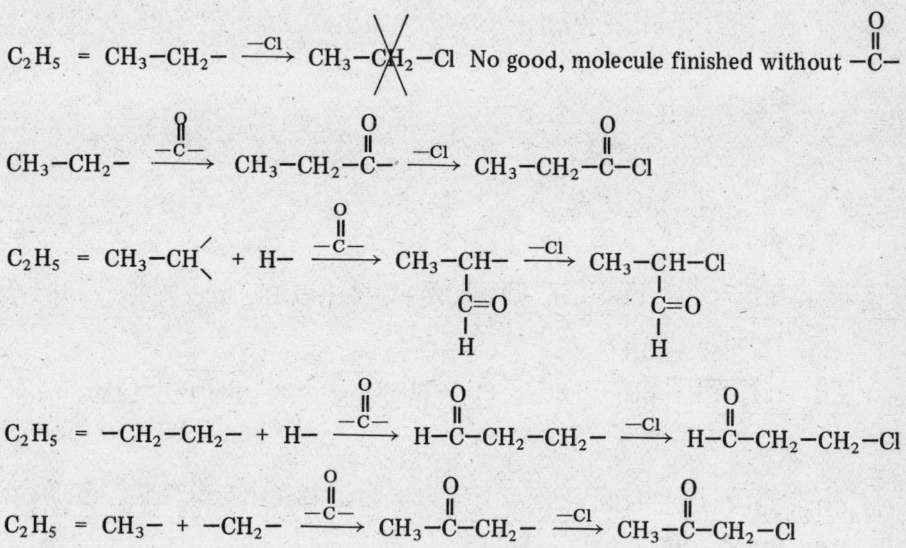

$C_2H_5 = CH_3-CH_2- \xrightarrow{-Cl} CH_3-\cancel{CH}_2-Cl$ No good, molecule finished without $-\overset{O}{\underset{\|}{C}}-$

$CH_3-CH_2- \xrightarrow{-\overset{O}{\underset{\|}{C}}-} CH_3-CH_2-\overset{O}{\underset{\|}{C}}- \xrightarrow{-Cl} CH_3-CH_2-\overset{O}{\underset{\|}{C}}-Cl$

$C_2H_5 = CH_3-CH\backslash + H- \xrightarrow{-\overset{O}{\underset{\|}{C}}-} CH_3-\underset{\underset{H}{|}}{\underset{C=O}{\overset{|}{CH}}}- \xrightarrow{-Cl} CH_3-\underset{\underset{H}{|}}{\underset{C=O}{\overset{|}{CH}}}-Cl$

$C_2H_5 = -CH_2-CH_2- + H- \xrightarrow{-\overset{O}{\underset{\|}{C}}-} H-\overset{O}{\underset{\|}{C}}-CH_2-CH_2- \xrightarrow{-Cl} H-\overset{O}{\underset{\|}{C}}-CH_2-CH_2-Cl$

$C_2H_5 = CH_3- + -CH_2- \xrightarrow{-\overset{O}{\underset{\|}{C}}-} CH_3-\overset{O}{\underset{\|}{C}}-CH_2- \xrightarrow{-Cl} CH_3-\overset{O}{\underset{\|}{C}}-CH_2-Cl$

★ *Exercises*

Write all the structural formulas implied by each part structure.

1. C_9H_{12}: $\left.\begin{array}{c} \bigcirc- \\ CH_3- \end{array}\right\} C_2H_4$

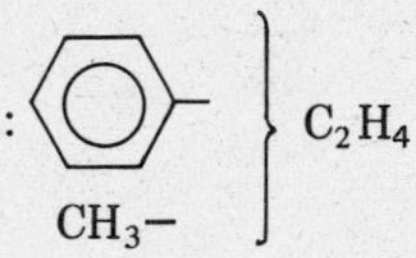

2. C_9H_{10}: $\left.\begin{array}{c} \bigcirc- \\ CH_3- \end{array}\right\} C_2H_2$

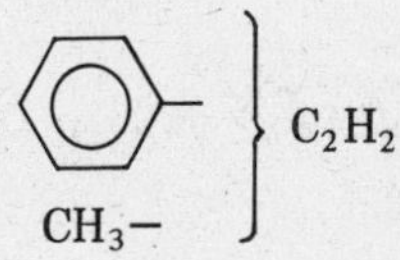

3. $C_5H_{12}O_2$: $\left.\begin{array}{c} -\overset{|}{\underset{|}{C}}-O-\overset{|}{\underset{|}{C}}- \\ HO- \end{array}\right\} C_3H_{11}$

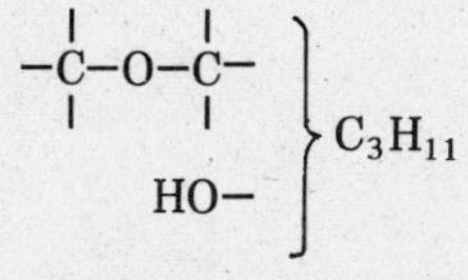

4. $C_{13}H_{12}$: $\left.\begin{array}{c} \bigcirc-\overset{|}{\underset{|}{C}}- \\ \bigcirc-\overset{|}{\underset{|}{C}}- \end{array}\right\}$

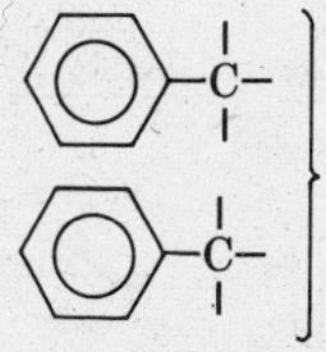

5. C_3H_9NO: $\left.\begin{array}{c} HO- \\ CH_3- \end{array}\right\} C_2H_5N$

- - - - - - - -

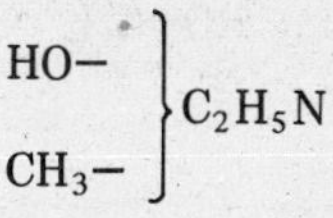

1.  $-CH_2-CH_2- \rightarrow \bigcirc-[CH_2-CH_2]-CH_3$

    and

$-\text{CH}-\text{CH}_3 \longrightarrow \text{Ph}[\text{CH}-\text{CH}_3]$
                                              $\quad\quad\quad\quad\quad \text{CH}_3$

2. $\text{C}=\text{C}\begin{smallmatrix}\text{H}\\\text{H}\end{smallmatrix}\begin{smallmatrix}\text{H}\\\text{H}\end{smallmatrix} \longrightarrow \begin{smallmatrix}\text{Ph}\\\text{CH}_3\end{smallmatrix}\text{C}=\text{C}\begin{smallmatrix}\text{H}\\\text{H}\end{smallmatrix}$

$\text{H}\begin{smallmatrix}\\\end{smallmatrix}\text{C}=\text{C}\begin{smallmatrix}\text{H}\\\text{H}\end{smallmatrix} \longrightarrow \begin{smallmatrix}\text{H}\\\text{Ph}\end{smallmatrix}\text{C}=\text{C}\begin{smallmatrix}\text{H}\\\text{CH}_3\end{smallmatrix}$

$\text{H}\begin{smallmatrix}\\\end{smallmatrix}\text{C}=\text{C}\begin{smallmatrix}\text{H}\\\text{H}\end{smallmatrix} \longrightarrow \begin{smallmatrix}\text{H}\\\text{Ph}\end{smallmatrix}\text{C}=\text{C}\begin{smallmatrix}\text{CH}_3\\\text{H}\end{smallmatrix}$

3. $[\text{CH}_3-\text{CH}_2-\text{CH}_2\!\!-\!\!] \longrightarrow [\text{CH}_3-\text{CH}_2-\text{CH}_2\!\!-\!\!]\text{CH}_2-\text{O}-\text{CH}_2-\text{OH}$

and $[\text{CH}_3-\text{CH}_2-\text{CH}_2\!\!-\!\!]\overset{\text{OH}}{\text{CH}}-\text{O}-\text{CH}_3 \quad [\text{CH}_3-\text{CH}-\text{CH}_3] \longrightarrow [\text{CH}_3-\text{CH}-\text{CH}_3]$
$\quad\quad\quad\quad\quad\quad\quad\quad\quad\quad\quad\quad\quad\quad\quad\quad\quad\quad\quad\quad\quad\quad\quad \text{CH}_2-\text{O}-\text{CH}_2-\text{OH}$

and $[\text{CH}_3-\text{CH}-\text{CH}_3] \quad [-\text{CH}_2-\text{CH}_2-\text{CH}_2-] \longrightarrow \text{CH}_3-\text{O}-\text{CH}_2[-\text{CH}_2-\text{CH}_2-\text{CH}_2-]\text{OH}$
$\quad \text{HO}-\text{CH}-\text{O}-\text{CH}_3$

$[-\text{CH}_2-\text{CH}-\text{CH}_3] \longrightarrow \text{CH}_3-\text{O}-\text{CH}_2[-\text{CH}_2-\text{CH}-\text{CH}_3] \text{ and } \text{HO}[-\text{CH}_2-\text{CH}-\text{CH}_3]$
$\quad\quad\quad\quad\quad\quad\quad\quad\quad\quad\quad\quad\quad\quad \text{OH} \quad\quad\quad\quad\quad\quad\quad\quad\quad\quad \text{CH}_2-\text{O}-\text{CH}_3$

$[-\text{CH}-\text{CH}_2-\text{CH}_3] \longrightarrow \text{CH}_3-\text{O}-\text{CH}_2[-\text{CH}-\text{CH}_2-\text{CH}_3]$
$\quad\quad\quad\quad\quad\quad\quad\quad\quad\quad\quad\quad\quad\quad \text{OH}$

$[\text{CH}_3-\overset{|}{\text{C}}-\text{CH}_3] \longrightarrow [\text{CH}_3-\overset{\text{OH}}{\underset{|}{\text{C}}}-\text{CH}_3]$
$\quad\quad\quad\quad\quad\quad\quad\quad\quad \text{CH}_2-\text{O}-\text{CH}_3$

$[\text{CH}_3-\text{CH}_2-] + [\text{CH}_3-] \longrightarrow \text{CH}_3-\text{O}-\overset{\text{OH}}{\underset{|}{\text{C}}}[-\text{CH}_2-\text{CH}_3] \text{ and } \text{HO}-\text{CH}_2-\text{O}-\text{CH}[-\text{CH}_2-\text{CH}_3]$
$\quad\quad\quad\quad\quad\quad\quad\quad\quad\quad\quad\quad\quad\quad [\text{CH}_3] \quad\quad\quad\quad\quad\quad\quad\quad\quad\quad\quad\quad [\text{CH}_3]$

and $[\text{CH}_3\!\!-\!\!]\text{CH}_2-\text{O}-\text{CH}[-\text{CH}_2-\text{CH}_3] \text{ and } [\text{CH}_3-\text{CH}_2\!\!-\!\!]\text{CH}_2-\text{O}-\text{CH}[-\text{CH}_3]$
$\quad\quad\quad\quad\quad\quad\quad\quad\quad\quad \text{OH} \quad\quad\quad\quad\quad\quad\quad\quad\quad\quad\quad\quad\quad\quad \text{OH}$

$[-\text{CH}_2-\text{CH}_2-] + [-\text{CH}_3] \longrightarrow \text{HO}[-\text{CH}_2-\text{CH}_2-]\text{CH}_2-\text{O}-\text{CH}_2[-\text{CH}_3]$

$\boxed{CH_3-CH_2}\!-\! + \!-\!\boxed{CH_2}\!-\! \longrightarrow OH\!-\!\boxed{CH_2}\!-\!CH_2-O-CH_2\!-\!\boxed{CH_2-CH_3}$

$\boxed{CH_3-\overset{|}{CH}}\!-\! + \!-\!\boxed{CH_3} \longrightarrow HO\!-\!\boxed{\overset{CH_3}{\underset{|}{CH}}}\!-\!CH_2-O-CH_2\!-\!\boxed{CH_3}$

$\boxed{CH_3-\overset{|}{CH}}\!-\! + \!-\!\boxed{CH_2}\!-\! \longrightarrow H\!-\!\boxed{\overset{CH_3}{\underset{|}{CH}}}\!-\!CH_2-O-CH_2\!-\!\boxed{CH_2}\!-\!OH$ (duplicate)

and $HO\!-\!\boxed{\overset{CH_3}{\underset{|}{CH}}}\!-\!CH_2-O-CH_2\!-\!\boxed{CH_2}\!-\!H$ (duplicate)

$\boxed{CH_3}\!-\! + \!-\!\boxed{CH_3} + \boxed{CH_3}\!-\! \longrightarrow \boxed{CH_3}\!-\!\overset{CH_3}{\underset{CH_3}{C}}\!-\!O-CH_2-OH$

and $\boxed{CH_3}\!-\!\overset{H}{\underset{CH_3}{C}}\!-\!O-\overset{CH_3}{\underset{H}{C}}\!-\!OH$

and $\boxed{CH_3}\!-\!CH_2-O-\overset{CH_3}{\underset{CH_3}{C}}\!-\!OH$

$\boxed{CH_3}\!-\! + \!-\!\boxed{CH_3} + \!-\!\boxed{CH_2}\!-\! \longrightarrow \overset{\boxed{CH_3}}{\underset{\boxed{CH_3}}{}}\!\!\!>CH-O-CH_2\!-\!\boxed{CH_2}\!-\!OH$

and $\boxed{CH_3}\!-\!CH_2-O-\overset{CH_3}{CH}\!-\!\boxed{CH_2}\!-\!OH$

and $CH_3-O-\overset{CH_3}{\underset{CH_3}{C}}\!-\!\boxed{CH_2}\!-\!OH$

$\boxed{CH_3}\!-\! + \!-\!\boxed{CH_2}\!-\! + \!-\!\boxed{CH_2}\!-\! \longrightarrow H\!-\!\boxed{CH_2}\!-\!\overset{CH_3}{CH}\!-\!O-CH_2\!-\!\boxed{CH_2}\!-\!OH$ (duplicate)

and $HO\!-\!\boxed{CH_2}\!-\!\overset{CH_3}{CH}\!-\!O-CH_2\!-\!\boxed{CH_2}\!-\!H$ (duplicate)

$\!-\!\boxed{CH_2}\!-\! + \!-\!\boxed{CH_2}\!-\! + \!-\!\boxed{CH_2}\!-\! \longrightarrow$ none

4. ⌬—CH₂—⌬

5.

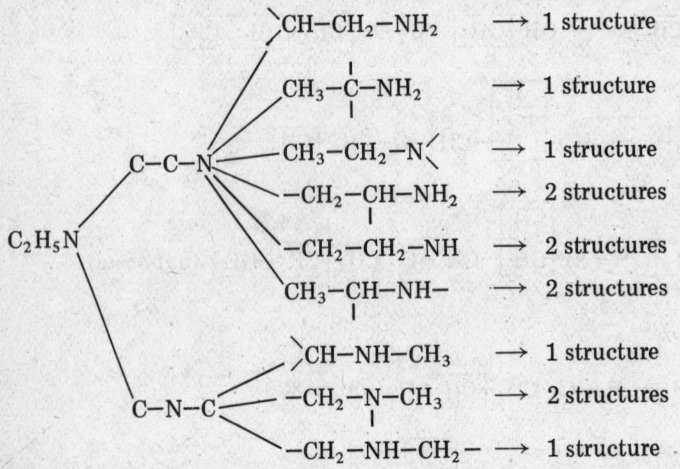

C_2H_5N
- C–C–N
  - $>$CH–CH$_2$–NH$_2$ → 1 structure
  - CH$_3$–C(–NH$_2$)– → 1 structure
  - CH$_3$–CH$_2$–N$<$ → 1 structure
  - –CH$_2$–CH–NH$_2$ → 2 structures
  - –CH$_2$–CH$_2$–NH– → 2 structures
  - CH$_3$–CH–NH– → 2 structures
- C–N–C
  - $>$CH–NH–CH$_3$ → 1 structure
  - –CH$_2$–N(–CH$_3$)– → 2 structures
  - –CH$_2$–NH–CH$_2$– → 1 structure

3. Occasionally, the available data include a standard reaction (one of those reproducible and predictable ones learned in Units 14 to 19) that gives a compound of known structure as the product. This reduces the structure problem to a simple "missing starting material problem": ? $\xrightarrow{\text{reagent}}$ product. In many cases the reactant structure can be written immediately.

Example

Compound X gives (cyclopentane with Br and CH$_2$Br) –CH$_2$Br on treatment with Br$_2$ in the dark. What is the structure of X?

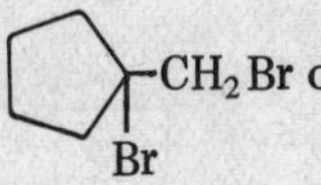

Solution:

1. Recognize that the problem is just

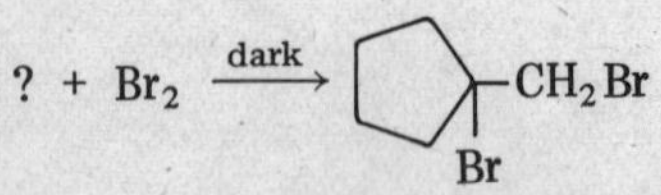

   ? + Br$_2$ $\xrightarrow{\text{dark}}$ (cyclopentane)–CH$_2$Br with Br

2. Identify the reaction: the only common reaction of Br$_2$ in the absence of light or catalyst is addition to C=C.

3. Identify location of original C=C by location of Br atoms in product:

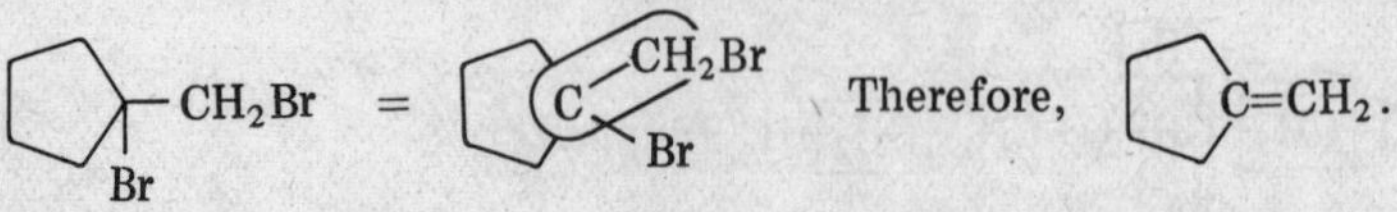

   (cyclopentane)–CH$_2$Br with Br = (cyclopentane)C(–CH$_2$Br)(Br)   Therefore, (cyclopentane)C=CH$_2$.

★ *Exercise*

Compound X, a hydrocarbon, gives *n*-butane on treatment with excess H$_2$ and a Pt catalyst. Write the possible structure(s) for X.

$$CH_2=CH-CH_2-CH_3 \quad \underset{H}{\overset{CH_3}{\diagdown}}C=C\underset{CH_3}{\overset{H}{\diagup}} \quad \underset{H}{\overset{CH_3}{\diagdown}}C=C\underset{H}{\overset{CH_3}{\diagup}}$$

$$CH_2=CH-CH=CH_2 \quad H-C\equiv C-CH_2-CH_3 \quad CH_3-C\equiv C-CH_3 \quad H-C\equiv C-C\equiv C-H$$

$$H-C\equiv C-CH=CH_2 \quad H_2C=C=C-CH_3 \quad H_2C=C=C=CH_2$$

(Compounds with cumulated double bonds, C=C=C, are rare, but they exist.) If you wrote only $CH_2=CH-CH_2-CH_3$, $\underset{H}{\overset{CH_3}{\diagdown}}C=C\underset{CH_3}{\overset{H}{\diagup}}$, and $\underset{H}{\overset{CH_3}{\diagdown}}C=C\underset{H}{\overset{CH_3}{\diagup}}$, you overlooked the possibility that more than 1 mole of H_2 was available when the reagent was given as "excess H_2."

DIAGNOSTIC REACTIONS

4. One of the richest sources of input data for the structure problem is a suite of chemical reactions selected to reveal the presence of specific functional groups. We will refer to these as *diagnostic* reactions. A diagnostic reaction is most useful if its information is delivered visually, in the form of a color change, a gas evolved, or a precipitate formed. Table 20.1 lists the most common cases; most of these can be evaluated visually.

In the most common form of the structure problem you are not given conclusions from data (e.g., "a R–C(=O)NH₂ function is present in compound X"), but the raw data themselves (e.g., compound X gives the odor of ammonia on heating with 1 M aqueous NaOH). Each positive diagnostic result given allows you to add one functional group to the known side of the developing part structure for the unknown. Each negative result given is less useful, but it may weed out functional groups that are otherwise possibilities.

As soon as you see one or more oxygen atoms in the molecular formula, you should look for ways to narrow down the possible functional groups present ($-CO_2H$, $-CO_2R$, $\diagdown C=O$, $-OH$, $-O-$). If only one O is present, you can rule out $-CO_2H$ and $-CO_2R$. Note that calculation of the unsaturation number Δ (Unit 6, frame 4) is as good as a diagnostic reaction for this purpose. If $\Delta = 0$, $-CO_2H$ or $\diagdown C=O$ cannot be present. The ether function, $-O-$, is so inert that its presence often must be established by negative evidence: O is present, but diagnostic reactions show $-CO_2H$, $-CO_2R$, $\diagdown C=O$, and $-OH$ to be absent; therefore, the oxygen must be in an ether function.

Example

Compound X, $C_6H_{12}O_2$, does not react with aqueous $NaHCO_3$, but it reduces $Ag^\oplus$ to Ag and reacts with $CH_3-CO-Cl$ to give compound Y, $C_8H_{14}O_3$. Draw a part structure for X that incorporates all the available information.

Solution:

1. Calculate $\Delta = 1$. With 2 oxygens present, the possibilities are *(cont'd on p. 511)*

Table 20.1 *Diagnostic Reactions*

Diagnostic for:	Reagent	Positive test =	Tells the number of functional groups present?
$C\equiv C-H$	$AgNO_3$	White precipitate	No
$C=C$ or $C\equiv C$	H_2	Gas disappears	Yes, by gas-volume measurement
	Br_2	Decolorized	Yes, by titration
	Cold $KMnO_4$	Decolorized	No
3° R—X	$AgNO_3$ in ethanol	AgX precipitates	No
1° R—Cl	NaI	NaCl precipitates	No
1° R—OH	Conc. aq. HCl + $ZnCl_2$	Clouds only on heating	No
2° R—OH		Clouds within 5 min. at room temp.	
3° R—OH		Clouds immediately at room temp.	
R—OH	CH_3MgCl	CH_4 bubbles off	Yes, by gas volume measurement
R—OH	Excess $(CH_3CO)_2O$ or $CH_3-CO-Cl$	Not visually detectable	Yes, if product (R—O—CO—CH_3) is isolated and its molecular formula determined
$R-NH_2$, R_2NH	Excess $(CH_3CO)_2O$ or $CH_3-CO-Cl$	Not visually detectable	Yes, if product (R—NH—CO—CH_3 or R_2NH—CO—CH_3) is isolated
$R-NH_2$, R_2NH, R_3N	Standard aqueous HCl	Titrate to end point	Yes
$R-NH_2$	HNO_2	N_2 evolved	Yes
R—CHO	$Ag^\oplus$ or Cu^{2+}	Precipitate of Ag or a $Cu^\oplus$ salt	No

Table 20.1 Continued

Diagnostic for:	Reagent	Positive test =	Tells the number of functional groups present?
$\begin{array}{c}R\\ \diagdown\\ C=O, R-CHO\\ \diagup\\ R\end{array}$	Ar—NH—NH$_2$	Ar—NH—N=C$\diagdown$ precipitated	No
	NH$_2$OH	HO—N=C$\diagdown$ precipitated	No
	NaHSO$_3$	Na$^{\oplus\ominus}$O$_3$S—C— $\mid$ OH precipitated	No
R—CO$_2$H	NaHCO$_3$	CO$_2$ evolved	No
	Standard NaOH	Titrate to end point	Yes
ArOH	Na	H$_2$	No
	NaHCO$_3$	No CO$_2$ evolved	No
R—COOR	Standard KOH/EtOH	Titrate to end point	Yes
R—CO—NH$_2$	NaOH/heat	NH$_3$ evolved	No

$$1\ -C\diagup\!\!\!\!\!\!{}^O_{\diagdown OH}$$

or 1 $-\overset{O}{\overset{\|}{C}}-$ or $-\overset{O}{\overset{\|}{C}}-H$ + 1 —OH

or 1 $-\overset{O}{\overset{\|}{C}}-$ or $-\overset{O}{\overset{\|}{C}}-H$ + 1 —O—

or 1 C=C + 2 —OH
or 1 C=C + 2 —O—

or 1 C=C + 1 —OH + 1 —O—

2. X $\xrightarrow{\text{NaHCO}_3}$ no reaction; ∴ no —CO$_2$H group present

3. X $\xrightarrow{\text{Ag}^{\oplus}}$ Ag; ∴ $-C\diagup\!\!\!\!\!\!{}^O_{\diagdown H}$ present, ∴ the second O is in an —OH or a —O— function.

So X = $\left.-\overset{O}{\overset{\|}{C}}-H\right\}$ C$_5$H$_{11}$O.

4. X (C$_6$H$_{12}$O$_2$) $\xrightarrow{\text{CH}_3-\overset{O}{\overset{\|}{C}}-\text{Cl}}$ Y(C$_8$H$_{14}$O$_3$)   change in molecular formula = $\begin{array}{c}C_8H_{16}O_3\\ \underline{C_6H_{12}O_2}\\ +C_2H_2O\end{array}$

- Recognize the reaction with $CH_3-CO-Cl$ as diagnostic for $-OH$.
- Recognize the gain of C_2H_2O as the result with one $-OH$ group present:

$$R-OH + CH_3-CO-Cl \longrightarrow R-O-CO-CH_3 \qquad \begin{array}{r} RC_2H_3O_2 \\ -RHO \\ \hline +\ C_2H_2O \end{array}$$

∴ the second O is in an $-OH$ function, so $X = \left.\begin{array}{l} -\overset{\overset{O}{\|}}{C}-H \\ -OH \end{array}\right\} C_5H_{10}$

Experience shows that students often misread the given data in structure problems. It is not uncommon to find people translating the information "X gives Y on treatment with excess CH_3COCl" as $Y \xrightarrow{CH_3COCl} X$ (backwards!) rather than $X \xrightarrow{CH_3COCl} Y$ (correct). Probably it is best to immediately translate all the information into equations:

$$Y(C_8H_{14}O_3) \xleftarrow[CH_3COCl]{excess} X(C_6H_{12}O_2) \xrightarrow{aq.\ NaHCO_3} \text{no reaction}$$
$$\Delta = 1$$
$$\Big\downarrow Ag^\oplus$$
$$Ag$$

This allows all the information to be taken in at a glance.

A second point of difficulty is associated with the use of the labels X, Y, A, B, etc., in these problems. Students often ask "what do X and Y stand for?" They are just names for the compounds involved. The systematic name can't be used (it gives the structure away), and some label is needed.

★ *Exercise*

Compound A ($C_5H_{10}O_2$) decolorizes cold aqueous $KMnO_4$, but fails to react with ⟨O⟩—NH—NH_2. Treatment of A with excess Br_2 in CCl_4 gives B ($C_5H_{10}Br_2O_2$). With excess $CH_3-CO-Cl$ A gives C ($C_9H_{14}O_4$). Deduce what you can about the structure of A and express this information in a part structure.

— — — — — — — —

decolorized $\xleftarrow{\text{cold } KMnO_4}$ A ($C_5H_{10}O_2$, $\Delta = 5 + 1 - 10/2 = 1$)

∴ ① $\diagdown C=C \diagup$ ⟨O⟩—NH—NH_2 $\Big\downarrow Br_2$ $\xrightarrow[CH_3COCl]{excess}$ C ($C_9H_{14}O_4$)

or $-CHO$ present

no reaction   B($C_5H_{10}Br_2O_2$)

∴ ② no $\diagdown C=O$ gain of Br_2

groups present   ∴ ③ $\diagdown C=C \diagup$ present

$\begin{array}{r} C_9H_{14}O_4 \\ -C_5H_{10}O_2 \\ \hline \end{array}$

gain of $C_4H_4O_2 = 2 \times C_2H_2O$

two $-CO-CH_3$ groups incorporated ∴ ④ two $-OH$, $-NH_2$, or $-NH-$ groups present in A

Put the parts together:

conclusion ① says A = $\diagup\!\!\!\text{C}=\text{C}\!\!\!\diagdown$ } $C_3H_{10}O_2$

or $-\overset{\overset{O}{\|}}{C}-H$ } C_4H_9O

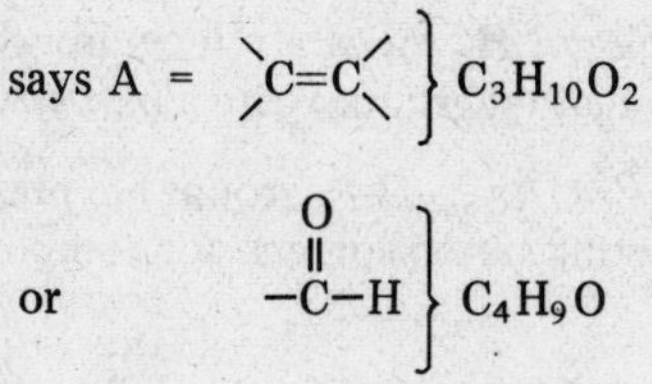

conclusion ② eliminates the second structure, so

A = $\diagup\!\!\!\text{C}=\text{C}\!\!\!\diagdown$ } $C_3H_{14}O_2$
accounts ∴ this portion saturated
for Δ = 1

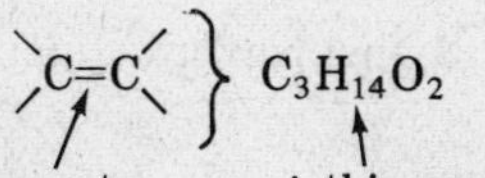

conclusion ③ confirms the above, but adds nothing new.
conclusion ④ says that two —OH groups are present in A (since N is absent from
the molecular formula); ∴ A = $\begin{matrix}\diagup\!\!\!\text{C}=\text{C}\!\!\!\diagdown\\-OH\\-OH\end{matrix}$ } C_3H_8

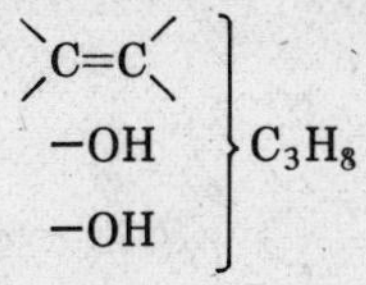

QUANTITATIVE DIAGNOSTIC REACTIONS

5. Several of the diagnostic reagents in Table 20.1 can be made to reveal not only that a certain function is present, but also how many of these functional groups are present per molecule. These reagents are identified in Table 20.1. Next we illustrate three of the most useful ones.

You obtain the total number of multiple bonds (C=C counting as one, C≡C counting as two) plus total number of rings directly from the molecular formula by calculating Δ. If the given data tell you the number of multiple bonds present, you can calculate the number of rings by subtraction. The number of multiple bonds is given by the number of moles of H_2 gas absorbed by 1 mole of compound X.

Example

Compound X, $C_4H_3F_3$, takes up 2 moles of H_2 for each mole of X treated with excess H_2/Pt. Write a part structure or structures.

Solution:

1. Δ = 4 + 1 − 3/2 − 3/2 = 2

2. 2 H_2 consumed = 2 π bonds per molecule, must be —C≡C— or 2 $\diagup\!\!\!\text{C}=\text{C}\!\!\!\diagdown$

3. Number of rings = 2 − 2 = 0

4. Part structure—we need two of them, since the functional group information is indefinite:

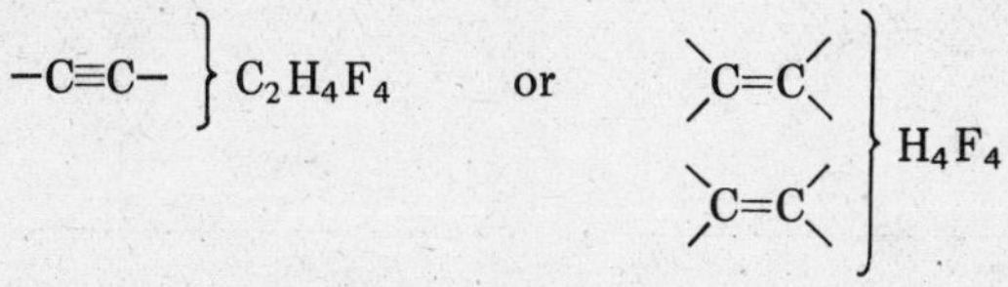

—C≡C— } $C_2H_4F_4$ or $\begin{matrix}\diagup\!\!\!\text{C}=\text{C}\!\!\!\diagdown\\\diagup\!\!\!\text{C}=\text{C}\!\!\!\diagdown\end{matrix}$ } H_4F_4

★ Exercises

1. Compound X, C_7H_8, takes up 3 moles of H_2 on catalytic hydrogenation. Spectroscopy shows that it has at least one methyl group. Write a part structure or structures.

2. If spectroscopic data show that no C≡C or C=C=C groups are present, draw all the possible complete structures (neglecting stereoisomers) for compound X in exercise 1.

– – – – – – – – –

1. (a) $\Delta = 7 + 1 - 8/2 = 4$

    (b) Number of multiple bonds/molecule = 3; first part structure:

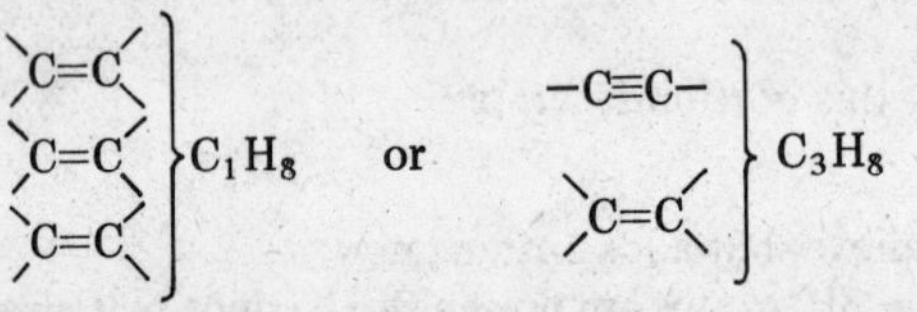

    $\left.\begin{array}{c}\text{C=C}\\\text{C=C}\\\text{C=C}\end{array}\right\} C_1H_8$ or $\left.\begin{array}{c}-\text{C}\equiv\text{C}-\\\text{C=C}\end{array}\right\} C_3H_8$

    (c) Number of rings = 4 − 3 = 1

    Third part structure: $\left.\begin{array}{c}\text{C=C}\\\text{C=C}\\\text{C=C}\\\text{1 ring}\end{array}\right\} C_1H_8$ or $\left.\begin{array}{c}-\text{C}\equiv\text{C}-\\\text{C=C}\\\text{1 ring}\end{array}\right\} C_3H_8$

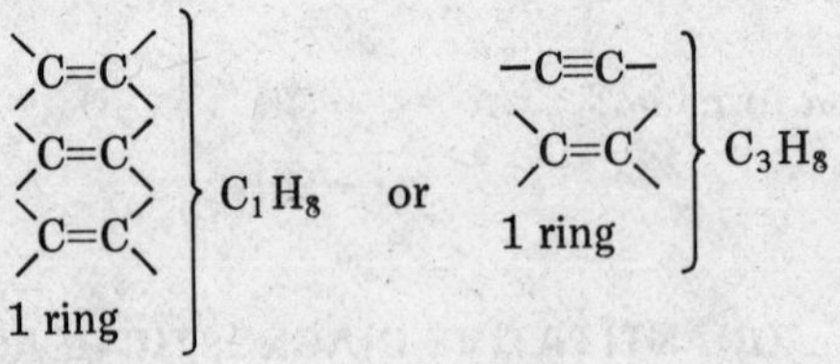

    (d) 1 or more CH_3 present

    Third part structure: $\left.\begin{array}{c}\text{C=C}\\\text{C=C}\\\text{C=C}\\\text{1 ring}\\-CH_3\end{array}\right\} H_5$ or $\left.\begin{array}{c}-\text{C}\equiv\text{C}-\\\text{C=C}\\\text{1 ring}\\-CH_3\end{array}\right\} C_2H_5$

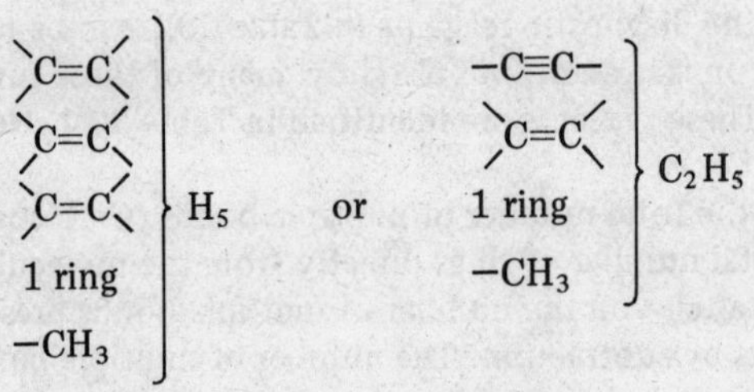

2. [structures of C_7H_8 isomers]

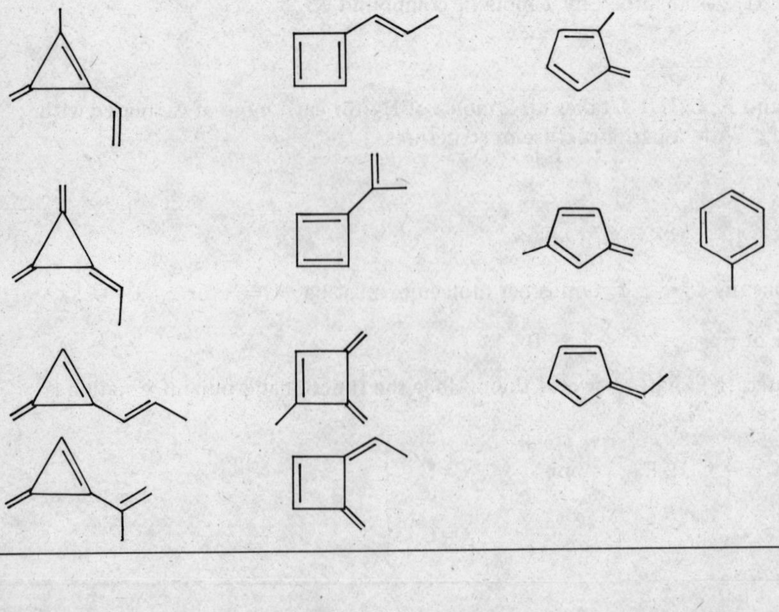

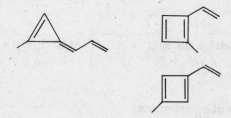

6. The second quantitative diagnostic reaction is the titration of acids and bases. We will illustrate this using RCO_2H, but everything said applies to RNH_2, R_2NH, and R_3N as well. The titration results are expressed in terms of the neutralization equivalent, NE, which is *calculated* from the equation

$$NE = \frac{\text{no. g } RCO_2H \text{ titrated}}{\text{no. moles NaOH required}} = \frac{\text{no. mg } RCO_2H \text{ titrated}}{\underset{\underset{\text{(moles/liter)}}{\text{molarity}}}{M_{NaOH}} \times \underset{\underset{\text{(ml)}}{\text{volume}}}{V_{NaOH}}} \quad (1)$$

NE is *interpreted* by means of the equation

$$NE = \frac{\text{molecular weight of } RCO_2H}{\text{no. } CO_2H \text{ groups per molecule}} \quad (2)$$

If the molecular weight is known, from the molecular formula or from other given data, equation (2) immediately gives the number of CO_2H groups in the structure. If the molecular weight is not known, you can use NE as a tentative value of the molecular weight under the assumption that there is one CO_2H group per molecule (the most common case). Of course, you must be ready to abandon the 1 CO_2H hypothesis if other data require it.

Example

Compound X contains C, H, and O. It decolorizes Br_2 in CCl_4 and liberates CO_2 from $NaHCO_3$. Titration gives NE = 100 ± 1.* Summarize what you can deduce about X in the form of a part structure and draw all the complete structures implied by your part structure.

Solution:

$$\text{decolorized} \xleftarrow{Br_2} X(C_?H_?O_?) \xrightarrow{NaHCO_3} CO_2 \therefore -CO_2H \text{ present}; \therefore C=C \text{ present}$$

Without any knowledge of the molecular weight (MW) we can only proceed by hypothesis:

if there is 1 CO_2H/molecule, then $NE = \frac{MW}{1}$, so MW = 100 ± 1

if there are 2 CO_2H/molecule, then $NE = \frac{MW}{2}$, so MW = 200 ± 2

if there are 3 CO_2H/molecule, then $NE = \frac{MW}{3}$, so MW = 300 ± 3

*The ±1 is the experimental error. The experiment says that NE can be 99, 100, or 101.

Try the simplest hypothesis as a part structure:

$\left.\begin{array}{r}C=C \\ CO_2H\end{array}\right\}$ C,H skeleton ⟵ the skeleton's MW must be $100 - $ MW of $C_3O_2H = 100 \pm 1 - (36 + 32 + 1) = 31 \pm 1$, that is, 30, 31, or 32.
total = C_3O_2H

Does a skeleton with MW 31 ± 1 have a rational interpretation?

$C_1 = 12$ ⟵ too small
$C_2 = 24$
$C_3 = 36$ ⟵ too big; ∴ probably 2 C present in skeleton

so 30, 31, or 32
    − 24
    ─────
    6, 7, or 8 MW units left for H,

⟶ 5 open bonds (or 3 if $\diagup\!\!\!\!C=C\diagdown$ and $-CO_2H$ are bonded together as $-C=C-COOH$)

$\left.\begin{array}{c}\diagup\!\!\!\!C=C\diagdown \\ -CO_2H\end{array}\right\} C_2H_{6-8}$

Now C_2H_6 has 0 or 2 open bonds (CH_3-CH_3 or CH_3- and CH_3-), which can't give a finished structure with $\left.\begin{array}{c}HOOC- \\ \diagup\!\!\!\!C=C\diagdown\end{array}\right\}$, which requires 3 or 5. C_2H_7 has 3 or 5 open bonds ($CH_3- + CH_3- + H-$ or $-CH_2-CH_2- + 3 H-$, etc.), which does match with $\left.\begin{array}{c}HOOC- \\ \diagup\!\!\!\!C=C\diagdown\end{array}\right\}$.

C_2H_8 has 4 or 6 open bonds ($CH_3- + CH_3- + 2H-$ or $-CH_2-CH_2- + 4H-$, etc.), which doesn't match. ∴ X must be

$\left.\begin{array}{c}C=C \\ -CO_2H\end{array}\right\} C_2H_7$

Try structures:

$\diagup\!\!\!\!C=C\diagdown + -CO_2H + \underbrace{CH_3-CH_2- + 2 H-}_{C_2H_7} \longrightarrow CH_3-CH_2-CH=CH-CO_2H$ (E and Z)

$\diagup\!\!\!\!C=C\diagdown + -CO_2H + \underbrace{-CH_2-CH_2- + 3 H-}_{C_2H_7} \longrightarrow CH_2=CH-CH_2-CH_2-CO_2H$ and $CH_2=C\diagup^{CO_2H}_{\diagdown CH_2-CH_3}$

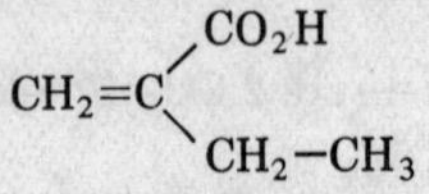

$\diagup\!\!\!\!C=C\diagdown + -CO_2H + \underbrace{CH_3- + CH_3- + H-}_{C_2H_7} \longrightarrow \begin{array}{c}CH_3\diagdown \\ CH_3\diagup\end{array}C=C\begin{array}{c}\diagup CO_2H \\ \diagdown H\end{array}$ and

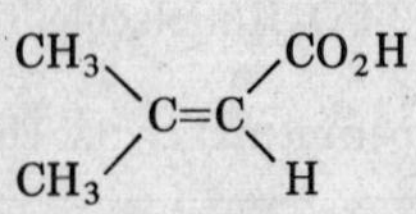

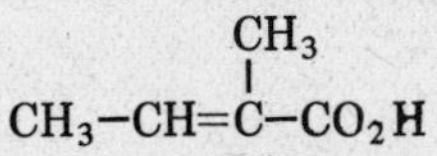

STRUCTURE PROBLEMS 517

$$\underset{\text{(E and Z)}}{CH_3-CH=\underset{\underset{CH_3}{|}}{C}-CO_2H}$$

$$\text{\Large$>$}C=C\text{\Large$<$} + -CO_2H + \underbrace{-CH_2- + CH_3-}_{C_2H_7} + 2\,H- \longrightarrow CH_3-CH=CH-CH_2-CO_2H \text{ and}$$

(E and Z)

$$CH_2=\underset{\underset{CH_3}{|}}{C}-CH_2-CO_2H$$

$$\text{\Large$>$}C=C\text{\Large$<$} + -CO_2H + -CH_2- + -CH_2- + 3\,H- \longrightarrow \text{nothing new}$$

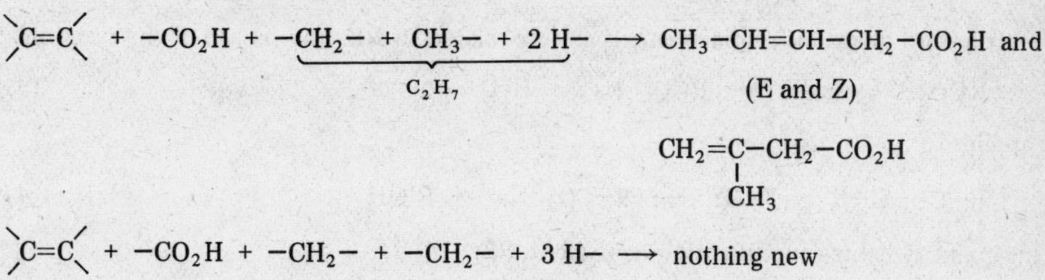

★ *Exercises*

1. A 125-mg sample of compound L requires 16.4 ml of 0.1 N NaOH for titration. Calculate NE.

2. Compound L in exercise 1 has MW = 152 and contains only C, H, O, and Cl. L does not react with H_2/Pt, Br_2 in CCl_4, or cold $KMnO_4$, but it does give CO_2 on treatment with $NaHCO_3$. Deduce what you can about the structure of L.

— — — — — — — — —

1. $NE = \dfrac{125}{0.1 \times 16.4} = 76$

2. The reactions show C=C not present, CO_2H present. Since both NE and MW are available, calculate the number of $-CO_2H$ groups $= \dfrac{MW}{NE} = \dfrac{156}{76} = 2$.

$\left.\begin{array}{l}-CO_2H\\-CO_2H\end{array}\right\}$ rest with MW = 152 - 90 = 62

$C_2O_4H_2 \longleftarrow$ MW = 90

Now "rest" must contain Cl. Since 2 Cl has MW 71, only one can be present.

$L = \left.\begin{array}{l}-CO_2H\\-CO_2H\\-Cl\end{array}\right\}$ rest with MW = 62 - 35 = 27 $\longleftarrow$ could be $\begin{cases} C_1H_{15} & \longleftarrow \text{unlikely}\\ C_2H_3 & \longleftarrow \text{try this}\end{cases}$

C_2H_3 can only be $CH_3-\underset{|}{\overset{|}{C}}-$ or $-CH_2-CH\text{\Large$<$}$

Assemble the fragments:

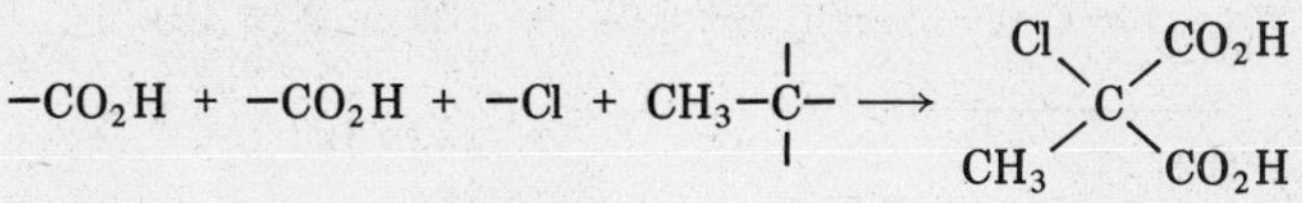

$-CO_2H + -CO_2H + -Cl + CH_3-\underset{|}{\overset{|}{C}}- \longrightarrow \underset{CH_3}{\overset{Cl}{\diagdown}}C\underset{CO_2H}{\overset{CO_2H}{\diagup}}$

and

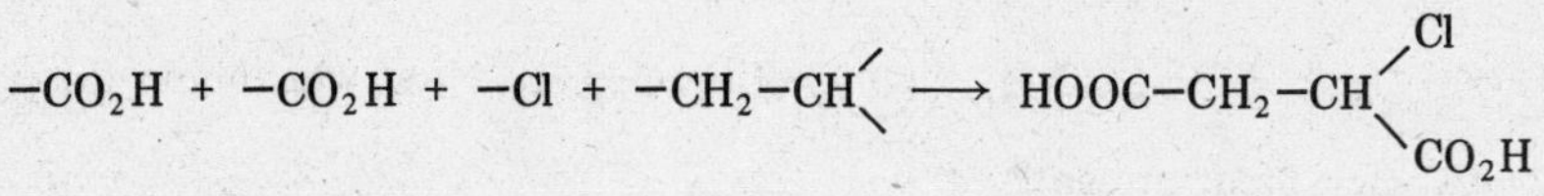

$-CO_2H + -CO_2H + -Cl + -CH_2-CH\text{\Large$<$} \longrightarrow HOOC-CH_2-CH\underset{CO_2H}{\overset{Cl}{\diagup}}$ or

518 HOW TO SUCCEED IN ORGANIC CHEMISTRY

$$Cl-CH_2-CH\begin{array}{c}CO_2H\\CO_2H\end{array}$$

7. The third quantitative diagnostic is closely related to NE, but instead of the reaction

$$RCO_2H + NaOH \longrightarrow RCO_2^{\ominus}Na^{\oplus} + H_2O \qquad (3)$$

it applies to

$$R-CO-O-R' + NaOH \longrightarrow R-CO_2^{\ominus}Na^{\oplus} + R'OH \qquad (4)$$

This quantity is the saponification equivalent, SE:

$$SE = \frac{\text{mg of ester}}{V_{NaOH} \times M_{NaOH}} = \frac{\text{MW of compound}}{\text{no. } -\underset{\underset{O}{\|}}{C}-O- \text{ groups/molecule}} \qquad (5)$$

★ *Exercise*

Compound A has SE = 135 ± 2. After reaction (4) has been carried out, the solution is acidified and benzoic acid, ⟨◯⟩-CO_2H, precipitates. What is the structure of A?

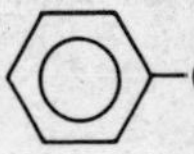

— — — — — — — —

With SE = 135 but no knowledge of MW, try several hypotheses by using eq. (5):

Number of $-CO-O-$ groups present in A	MW must be
1	1 × (135 ± 2) = 135 ± 2
2	2 × (135 ± 2) = 270 ± 4
3	3 × (135 ± 2) = 405 ± 6

Since ⟨◯⟩-CO_2H is isolated, A must be ⟨◯⟩-CO-O-R.

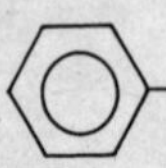

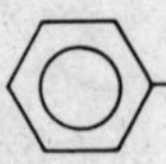

If 1 -CO-O- is present, then A =

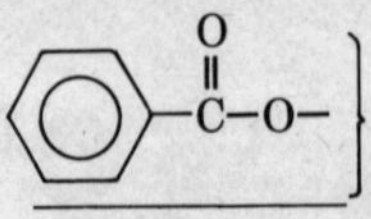

⟨◯⟩-$\overset{\overset{O}{\|}}{C}$-O- } rest ⟵ MW = 135 ± 2 - 121
= 14 ± 2
= 12 to 16

MW = 121 ⟶ $C_7H_5O_2$

Does MW 12 to 16 ring a bell? Yes, it is the MW of $-CH_2-$ or CH_3-. So A could be

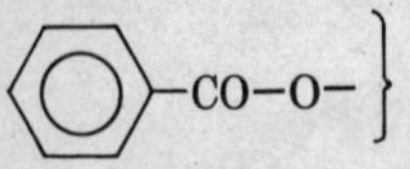

⟨◯⟩-CO-O-} -CH_3, that is, ⟨◯⟩-CO-O-CH_3. Now, what about the possibility of 2 -CO-O- being present?

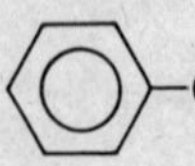

$$A = \left.\begin{array}{c}\langle\bigcirc\rangle\text{-CO-O-} \\ \langle\bigcirc\rangle\text{-CO-O-}\end{array}\right\} \text{rest} \longleftarrow MW = 272 \pm 4 - 242 = 30 \pm 4, \text{ that is,}$$

$$C_{14}H_{10}O_4 \longleftarrow MW = 242$$

Can one make a skeleton of MW 26 to 34 and two open bonds? $-CH_2-$, MW = 14, doesn't fit. But $-CH_2-CH_2-$ or $CH_3-CH\diagdown$ has MW = 28, which fits. So A could also

be $\langle\bigcirc\rangle\text{-CO-O-CH}_2\text{-CH}_2\text{-O-CO-}\langle\bigcirc\rangle$ or $CH_3-CH\diagup\begin{array}{c}\text{O-CO-}\langle\bigcirc\rangle \\ \text{O-CO-}\langle\bigcirc\rangle\end{array}$.

There are probably other structures.

INFRARED AND NUCLEAR MAGNETIC RESONANCE SPECTRA

8. A compound's infrared (ir) spectrum is a record of the absorption of light by the compound in the infrared wavelength region. The absorption occurs only at a collection of specific wavelengths, creating a pattern of bands characteristic of the compound's structure. The absorption of light is selective because it occurs only at those wavelengths where the energy of a quantum of the light is precisely the amount required to excite some group of atoms in the molecule to perform a characteristic mechanical vibration. Usually, each group present in a compound causes one or more bands to appear in the ir spectrum of the compound. Since the frequency of the band produced by, say, a C=O group does not depend much on the compound in which the C=O group is situated, there is an approximately 1:1 correspondence between the types of groups present in the compound and the frequencies of the bands in the ir spectrum. This allows us to prepare a correlation chart that identifies the common groups corresponding to a given band frequency.

To apply the data in an infrared spectrum to deducing the structure of an unknown compound, you don't need to know anything about the theory of the origin of spectra or why the characteristic bands fall at their particular frequencies. The method is a correlative method: since many compounds containing the group $\diagup C=C-C=O$ have a

band at frequency x, a band at frequency x in an unknown compound's spectrum is probably due to the presence of this same group in the unknown.

The correlation method has two limitations:

1. It is not foolproof because there are reasons why expected bands sometimes do not appear, and because other nearby groups in the molecule may cause the frequency of a group to shift out of its normal position. Actually, these shifts due to local environment are also beneficial; they often allow us to deduce not only the type of group present, but also the type of site it is attached to.

2. The presence of a band in the spectrum tells you that a certain group is probably present, but it tells you nothing about how many of them are present in each molecule. So the ir spectrum usually lets you identify the type of skeleton (aromatic, aliphatic, saturated, unsaturated) and which functional groups are present, but it tells you very little about the size of the molecule or how the different parts are put together. This means that the ir spectrum is best used in combination with other data, usually chemical behavior and especially the nuclear magnetic resonance spectrum.

As the name implies, nuclear magnetic resonance (nmr) spectra originate from the atomic nuclei in organic molecules. Although in recent years it has become practical to measure nmr spectra of quite a number of elements, including carbon, the spectra measured on ^{1}H nuclei—proton magnetic resonance (pmr) spectra—are still the most commonly available.

Measurement of the pmr spectrum is effectively a sensitive measurement of the actual, local magnetic field prevailing at each H nucleus in a sample of an organic compound placed in a uniform magnetic field (the "applied field"). The field varies from locale to locale within the molecule because (1) the electrons in the molecule partly shield out the applied magnetic field, and the electron density varies from atom to atom in the molecule and from one compound to the next; (2) the ^{1}H nuclei are themselves small magnets, so the local field depends on how many hydrogen atoms are in the neighborhood and how near they are (near being reckoned along the path of the covalent bonds rather than directly through space); and (3) the π electrons in certain groups, especially aromatic rings, produce internal magnetic fields with particular spatial orientations that add to or subtract from the field measured at nearby locations.

Thus, interpreting a pmr spectrum is (in part) a process of inferring groups present from the observed magnetic fields using a correlation chart, as with the ir spectrum. In the pmr spectrum, however, important structural information is also provided by the shape of the bands (usually called signals or lines in the nmr spectrum).

The pmr spectrum is extremely useful in structure problems because the information it provides complements that from other sources so beautifully. The IR spectrum speaks most directly of the functional groups present in the molecule. The pmr spectrum, on the other hand, originates in the H atoms of the skeleton and reveals the details of the C,H skeleton best, so we can infer something about the functional groups present from their effects on neighboring C—H bonds. Furthermore, whereas the ir spectrum can identify various groups of the skeleton but is unable to tell anything about the number of each present, the pmr spectrum usually furnishes the number of H atoms of each type unambiguously.

We will not practice the interpretation of ir and nmr spectra in this book, but we will use typical results of such interpretations as part of the data in the structure problems in the remainder of this unit.

INFORMATION ON CHIRALITY AND SYMMETRY

9. We cannot get information on the symmetry properties of the molecules of an unknown compound in any direct way, but we can sometimes infer something useful indirectly from the ir spectrum and from any available data on the ability or inability of the compound to rotate polarized light. The nmr spectrum also reflects molecular symmetry.

If a compound is optically active, the molecule must be dissymmetric (chiral), and nine times out of ten this means that an asymmetric carbon is present. If a compound is optically inactive but it can be resolved into optically active components, asymmetric carbon(s) is present, but the sample is a racemic mixture of enantiomers. If the compound is inactive and cannot be resolved, it has a plane of symmetry and usually lacks $\diagdown_{\diagup}\!C^{*}\!\diagdown_{\diagup}$.

★ *Example*

Compound X, $C_{10}H_{14}$, contains a phenyl group as shown by the ir spectrum, and it is optically active. Deduce its structure.

STRUCTURE PROBLEMS

Solution:

1. Part structure: Ph—}C$_4$H$_9$. It must be a butylbenzene.

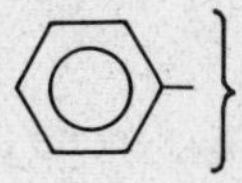

2. 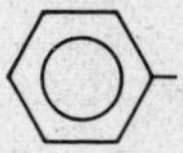 Ph— contains no >C*< ; therefore, the butyl group must contain >C*<.

3. Which of the isomeric butyl groups contain >C*< ?

   CH$_3$—CH$_2$—CH$_2$—CH$_2$—   no

   CH$_3$—CH(CH$_3$)—CH$_2$—   no

   (CH$_3$)$_3$C—   no

   CH$_3$—C*H(—)—CH$_2$—CH$_3$ yes

4. X is Ph—CH*(CH$_3$)—CH$_2$—CH$_3$

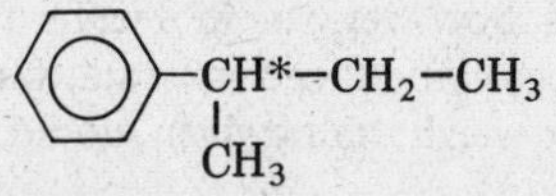

The ir bands due to the C=C and C≡C stretching vibrations disappear when the plane bisecting them is a plane of symmetry for the whole molecule. This is true of other vibrations as well. One consequence of this effect of molecular symmetry on the ir spectrum is sometimes a useful clue in structure problems. An ir spectrum that contains very few bands is usually due to a molecule that possesses several planes of symmetry.

★ *Exercise*

Compound X, C$_9$H$_{12}$, reacts with a deficiency of bromine under illumination to give *only two* products, Y and Z, both with molecular formula C$_9$H$_{11}$Br. Y is optically inactive and cannot be resolved. Z is also optically inactive, but it *can* be resolved into a pair of enantiomers. The ir spectrum of X contains bands characteristic of Ph— groups. Draw structures for X, Y, and Z.

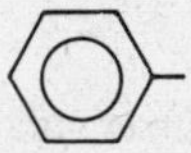

- - - - - - - - -

For X, Δ = 4. Since one Ph— group accounts for 4 units of unsaturation, the remainder of the molecule must be saturated. So X = Ph—}C$_3$H$_7$, apparently a propyl benzene. Of the two propyl groups, —^{a}CH$_2$—^{b}CH$_2$—^{c}CH$_3$ and —^{a}CH(^{b}CH$_3$)(^{b}CH$_3$), the first should give three monobromo compounds, but the second should give only two, as observed. X is therefore Ph—CH(CH$_3$)$_2$. Of the two bromides, Ph—C(CH$_2$Br)(H)(CH$_3$)

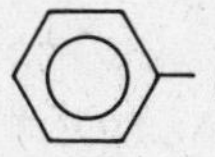

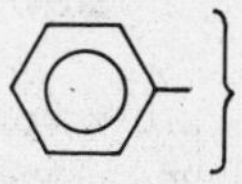

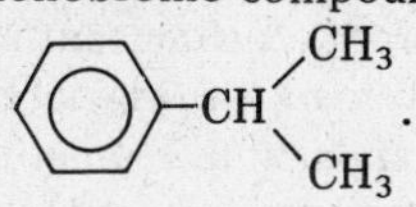

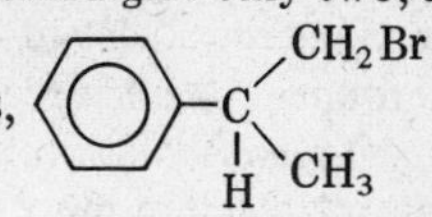

and Ph-C(CH₃)(Br)(CH₃) , neither should be optically active, but the first contains an asymmetric carbon and should be resolvable; it is thus Z. The second is achiral and hence unresolvable; it is Y.

MASS SPECTRA

10. A mass spectrometer is a device for removing an electron from an organic molecule in the gas phase, accelerating the resulting positive ions in a given direction and making measurements on their flight path in an electric field. Since deflection by the field depends upon the particle's mass, this mass can be calculated from the measured deflection. Since the mass of the missing electron is negligible, the measured mass of the positive ion (in atomic mass units per molecule = g/mole) is the molecular weight of the original compound. The molecular weight is a useful piece of data in solving structure problems, although not as useful as the molecular formula. In real-life structure determination we ordinarily have both. In practice problems, however, the molecular formula is sometimes withheld to force greater reliance on indirectly deduced information. Exercise 2 in frame 6 was of that variety; the molecular weight value given would have come from mass spectrometry.

Since considerable energy can be put into the gas-phase molecules in the process of ionizing them, some of the positive ions that result (parent ions) can be made to decompose into fragment ions whose molecular weights are then measured, all as part of the original experiment. It turns out that these gas-phase fragmentations are often understandable and predictable reactions associated with the presence of certain functional (or skeletal) groups in the molecule. Thus a certain sequence of mass losses corresponding to the sloughing of certain atoms or groups by the parent ion can be recognized as diagnostic for the presence of a certain functional group. Many textbooks now list these characteristic "fragmentation patterns" for each chemical family studied. Since mass spectra (MS) fragmentation data come to you in very familiar quantities (mass in molecular weight units) that can be directly interpreted in structural terms (disappearance of 44 mass units is very likely to mean loss of O=C=O), they are easier to deal with than ir or nmr spectra. Again, we will not practice interpretation of MS fragmentation patterns here. If your textbook does treat mass spectra, the simplest way to study them is to include the characteristic MS fragmentation as one of the standard set of reactions of each function, and drill on it just as you do on the reactions that involve a reagent and take place in solution.

DEGRADATIVE REACTIONS

11. Degradation is the process of breaking a molecule of unknown structure into smaller molecules by chemical reactions. One generally degrades the unknown because it has proven difficult to deduce the structure from diagnostic reactions and spectra alone, either because it is especially large or especially complex. The ideal degradation breaks the molecule into large fragments by means of reactions that are selective, well understood, and not likely to cause rearrangement of the carbon skeleton. In that case, once the structures of the degradation products (fragment molecules) have been determined, one can reason back to the structure of the original molecule. This conceptual reconstruction may or may not be unambiguous—sometimes the fragments can be reassembled in more than one way, as we shall see. Beyond mild conditions and selectivity for a certain type of function, a degradative reaction should give good yields and be easy to run on small quantities of unknown, since in many real-life structure problems on naturally occurring unknowns, only small amounts are available. The reactions assembled in Table 20.2 possess most of these qualifications. Their use in solving structure problems is discussed and practiced in the next few frames.

12. The ozonolysis reaction is discussed in detail in Unit 12, frame 13, where the focus is on writing the structures of the products of ozonolysis of a given alkene. Now we do the reverse, which proves to be easier: given the ozonolysis product structures, the problem is to write the structures of all the alkenes that could give this set of products.

To do this, bring the two carbonyl groups close to one another, remove all the oxygen atoms, and connect the two fragments with a double bond between the carbon atoms.

Example

What alkene gives only $CH_3-\overset{O}{\underset{\parallel}{C}}-H + H-\overset{O}{\underset{\parallel}{C}}-H$ on ozonolysis?

Solution:

$$CH_3-\underset{H}{\overset{H}{C}}\underset{O}{\overset{OO}{}}\overset{H}{\underset{}{C}}-H \longrightarrow \underset{H}{\overset{CH_3}{>}}CC\overset{H}{\underset{H}{<}} \longrightarrow \underset{H}{\overset{CH_3}{>}}C=C\overset{H}{\underset{H}{<}}$$

This technique works also for the products of ozonolysis with oxidative workup (O_3, then H_2O_2).

Example

Draw the structure of the alkene that gives ⬡=O + ⬡–$\overset{O}{\underset{\parallel}{C}}$–OH on treatment with O_3 followed by H_2O_2.

Solution:

⬡–C(H)(O O)–⬡ ⟶ ⬡ + C(H)–⬡ ⟶ ⬡=C(H)–⬡

Now notice what happens if in each carbonyl compound the two groups attached to $\overset{\backslash}{/}C=O$ are different from one another, for instance in

$\underset{CH_3}{\overset{H}{>}}C=O$, $\underset{CH_3-CH_2}{\overset{CH_3}{>}}C=O$, or $\underset{HO}{\overset{⬡}{>}}C=O$, but not in

$\underset{H}{\overset{H}{>}}C=O$ or ⬡–$\overset{O}{\underset{\parallel}{C}}$–⬡ . If this is the case for *both* of the

Table 20.2 *Common Degradative Reactions*

Frame	Name	Equation	Comments
12	Ozonolysis	$R-CH=CH-R' \xrightarrow[(2) \ Zn/H_2O]{(1) \ O_3} R-CHO + R'-CHO$	(for other degrees of alkylation see Unit 12, frame 13)
13	Aromatic side-chain oxidation	$Ar-G \xrightarrow{hot \ KMnO_4} Ar-CO_2H$	G = any group attached to Ar by a C atom
14	Hofmann degradation	$R-CH_2-CH_2-\overset{\oplus}{N}R'_3 \ OH^\ominus \xrightarrow{heat} R-CH=CH_2 + R_3N$	
14	Cope degradation	$R-CH_2-CH_2-NH_2 \to O \xrightarrow{heat} R-CH=CH_2 + R_2N-OH$	
14	Ketone oxidation	$R-CH_2-CO-CH_2-R' \xrightarrow{CF_3COOOH} RCH_2-CO-O-CH_2R' + RCH_2-O-CO-CH_2R'$ $\downarrow H_2O \ | \ NaOH \qquad \qquad \downarrow H_2O \ | \ NaOH$ $R-CH_2-COOH + \qquad \qquad R-CH_2-OH +$ $R'-CH_2-OH \qquad \qquad R'-CH_2-COOH$	

Barbier-Wieland degradation

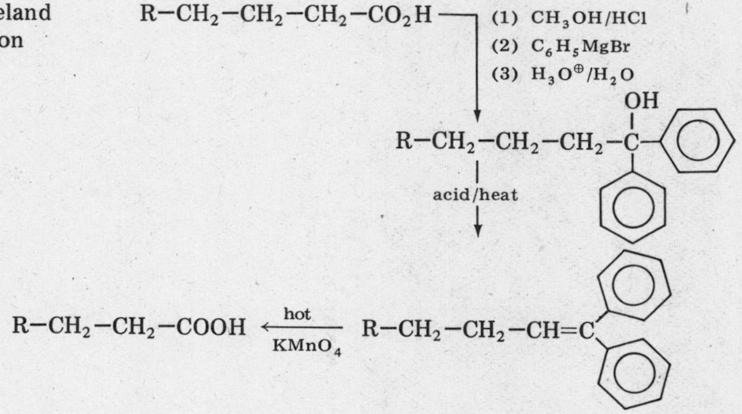

$R-CH_2-CH_2-CH_2-CO_2H \xrightarrow[\text{(3) } H_3O^\oplus/H_2O]{\text{(1) } CH_3OH/HCl \\ \text{(2) } C_6H_5MgBr}$

$R-CH_2-CH_2-CH_2-\underset{OH}{\underset{|}{C}}(C_6H_5)_2 \xrightarrow{\text{acid/heat}} R-CH_2-CH_2-CH=C(C_6H_5)_2$

$\xrightarrow[KMnO_4]{\text{hot}} R-CH_2-CH_2-COOH$

Each cycle removes one carbon; repeat *ad lib.* A branched chain → ketone.

Hydrolysis

$R-\underset{\underset{O}{\|}}{C}-O-R' \xrightarrow[\text{(2) } H_3O^\oplus]{\text{(1) hot, aq. NaOH}} R-CO_2H + R'OH$

$R-\underset{\underset{O}{\|}}{C}-NHR' \xrightarrow[\text{(2) } H_3O^\oplus]{\text{(1) hot, aq. NaOH}} R-CO_2H + R'\overset{\oplus}{N}H_3 \xrightarrow{NaOH} R'NH_2$

The easiest and most reliable, where applicable.

Ether cleavage

$R-O-R' \xrightarrow[\text{or } N\overset{\oplus}{H}_4Cl^\ominus/\text{heat}]{\text{hot, conc., aq. HI}} R-X + R'-X$

$Ar-O-R \xrightarrow{\text{hot, conc., aq. HI}} Ar-OH + R-I$

ozonolysis products, then *either of two isomeric alkenes* could have produced these products. Without further information one cannot decide which of these two structures the alkene really has.

Example

Draw the structure of the alkene(s) whose oxidative cleavage products are

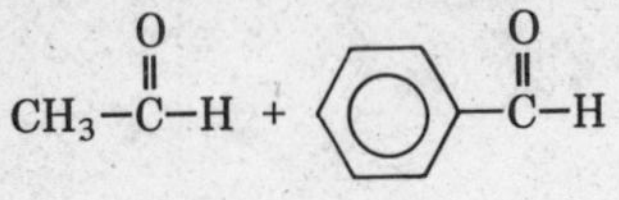

$$CH_3-\overset{O}{\underset{\|}{C}}-H \quad + \quad \text{(Ph)}-\overset{O}{\underset{\|}{C}}-H$$

Solution:

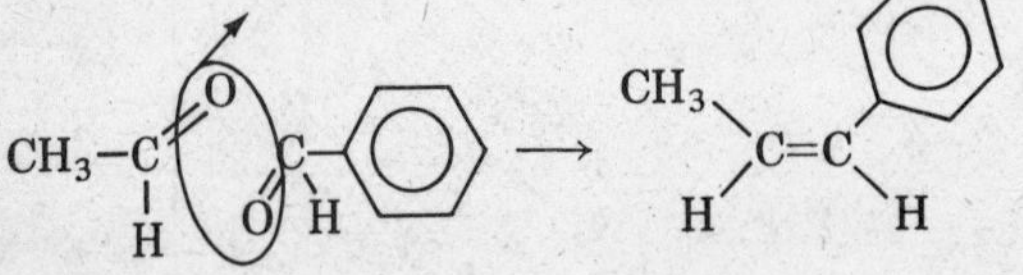

[Structure showing ozonide intermediate leading to cis alkene: CH₃ and Ph on same side, with H's on other side] cis

or

[Structure showing ozonide intermediate leading to trans alkene: CH₃ and H on one side, H and Ph on other] trans

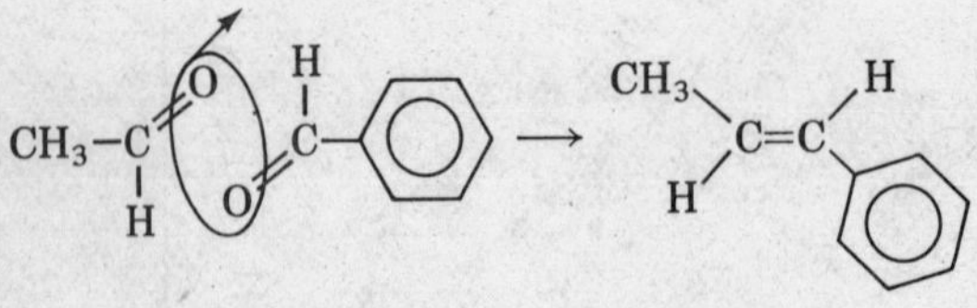

A cyclic alkene is cleaved into *one* fragment:

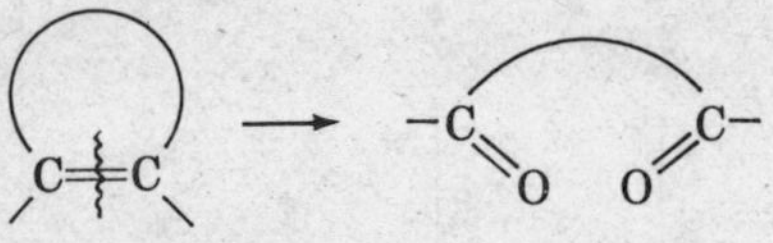

[Cyclic alkene cleaving to give a single dicarbonyl fragment]

Example

Draw the structure of the alkene whose oxidative cleavage product is

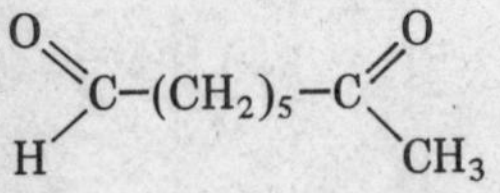

$$\overset{O}{\underset{H}{\succ}}C-(CH_2)_5-C\overset{O}{\underset{CH_3}{\nwarrow}}$$

Solution:

[Cyclic structure with CH₂ groups and two carbonyl groups connecting back to form methylcycloheptene]

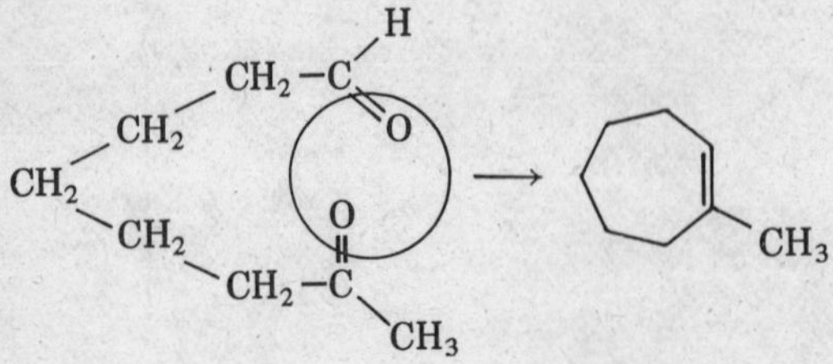

An alkene containing more than one double bond cleaves to give more than two carbonyl fragments. The number of fragments is always one greater than the number of $\overset{}{\underset{}{>}}C=C\overset{}{\underset{}{<}}$. A simple analog of this problem is cutting a string:

One cut gives two pieces:

—————{————— → ————— + —————

STRUCTURE PROBLEMS

$$\underset{C}{\overset{C}{\diagdown}}\underset{C}{\overset{H}{C=C}}\underset{C}{\overset{C}{\diagup}}\underset{C}{\overset{}{\diagdown}}\xrightarrow{\text{hot}}\underset{\text{KMnO}_4,\text{ etc.}} \underset{C}{\overset{C}{\diagdown}}\underset{C}{\overset{CO_2H}{\diagup}} + \text{HOOC}\underset{C}{\overset{C}{\diagdown}}\underset{C}{\overset{C}{\diagup}}$$

Two cuts give three pieces:

——{——{—— → —— + —— + ——

$$\underset{C}{\overset{C}{\diagdown}}\underset{C=C}{\overset{}{\diagup}}\underset{C}{\overset{C}{\diagdown}}\underset{C}{\overset{C=C}{\diagup}}\underset{C}{\overset{}{\diagdown}}\underset{C}{\overset{C}{\diagup}}\xrightarrow{\text{hot}}_{\text{KMnO}_4,\text{ etc.}} \underset{C}{\overset{C}{\diagdown}}\underset{CO_2H}{\overset{}{\diagup}} + \text{HOOC}\underset{C}{\overset{C}{\diagdown}}\underset{C}{\overset{CO_2H}{\diagup}}$$

$$+\;\underset{\text{HOOC}}{\overset{}{\diagdown}}\underset{C}{\overset{C}{\diagup}}$$

An ozonolysis that gives more than two fragments usually has some ambiguity about it: the fragments can be mentally reassembled in more than one way. Note in the example of a diene oxidation above that one of the fragment molecules is unique—it has a carbonyl group at *each end*. This marks it as the *center* fragment. The products with a single carbonyl group are always *end* fragments. The ambiguity arises when the center fragment is unsymmetrical, so that its ends are not identical. In this case it is impossible to tell which end fragment goes on which end of the center fragment:

⁓⁓Ⓐ ⎫
⁓⁓Ⓑ ⎬ end fragments

Ⓒ⁓⁓Ⓒ′ center fragment

⁓⁓Ⓐ Ⓒ⁓⁓Ⓒ′ Ⓑ⁓⁓   ⁓⁓Ⓑ Ⓒ⁓⁓Ⓒ′ Ⓐ⁓⁓

↓ reassemble   ↓ reassemble

one diene   another diene

Examples

What can you say about the structure of a diene whose oxidative cleavage products are the following?

1. $CH_3-\overset{\overset{O}{\|}}{C}-CH_3 + H-\overset{\overset{O}{\|}}{C}-CH_2-CH_2-\overset{\overset{O}{\|}}{C}-H + H_2CO$

2. $CH_3-\overset{\overset{O}{\|}}{C}-CH_3 + CH_3-\overset{\overset{O}{\|}}{C}-CH_2-CH_2-\overset{\overset{O}{\|}}{C}-H + H_2CO$

3. $2\;\bigcirc\!\!=\!O\; + \;\underset{HO}{\overset{O}{\diagdown}}C-CH_2-\underset{\underset{CH_3}{|}}{\overset{\overset{CH_3}{|}}{C}}-\underset{OH}{\overset{O}{\diagup}}C$

Solutions:

1. $\underset{CH_3}{\overset{CH_3}{\diagdown}}C\underset{O}{\overset{O}{\bigcirc}}C\underset{H}{\overset{CH_2-CH_2}{\diagup}}\underset{H}{\overset{}{\diagdown}}C\underset{O}{\overset{O}{\bigcirc}}C\underset{H}{\overset{H}{\diagup}} \rightarrow \underset{CH_3}{\overset{CH_3}{\diagdown}}C=C\underset{H}{\overset{CH_2-CH_2}{\diagup}}\underset{H}{\overset{}{\diagdown}}C=C\underset{H}{\overset{H}{\diagup}}$

This is the only possible alkene. Since the center fragment is symmetrical

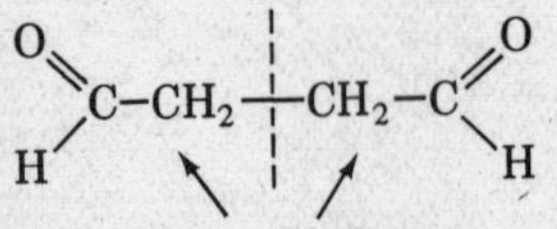

$$\underset{H}{\overset{O}{\diagdown}}C-CH_2 \mid CH_2-C\underset{\diagdown H}{\overset{O}{\diagup}}$$

identical halves

we get the same alkene with $\underset{CH_3}{\overset{CH_3}{\diagdown}}C=O$ left and $O=C\underset{\diagdown H}{\overset{H}{\diagup}}$ right, or vice versa.

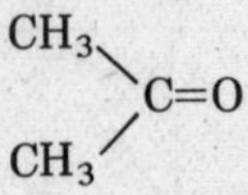

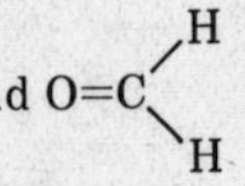

2. Expect two solutions because the center fragment is unsymmetrical; so

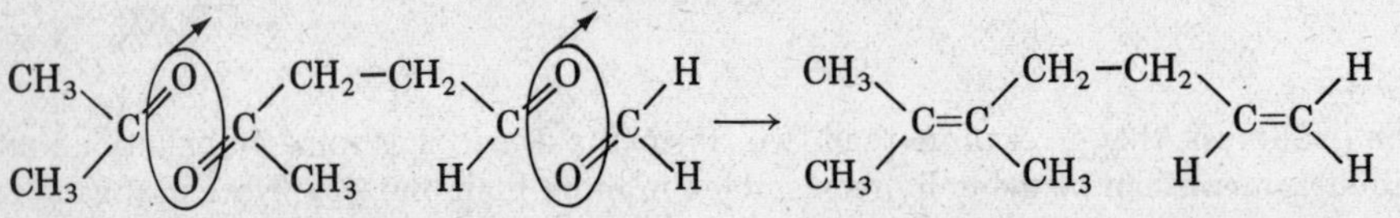

$$\underset{CH_3}{\overset{CH_3}{\diagdown}}C=C\underset{CH_3}{\overset{CH_2-CH_2}{\diagdown}}C=C\underset{H}{\overset{H}{\diagup}} \quad A$$

and with end fragments reversed:

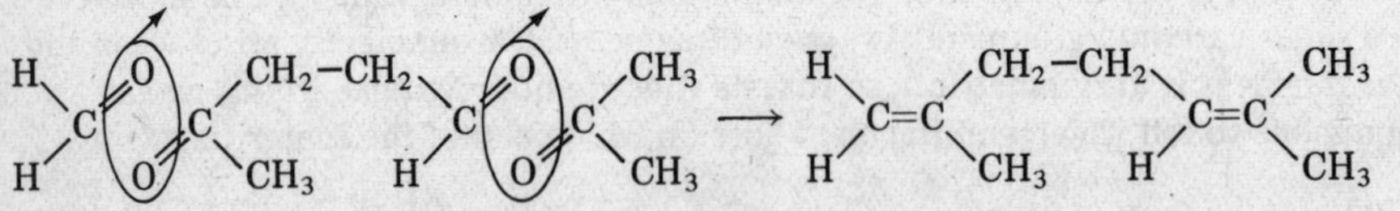

$$\underset{H}{\overset{H}{\diagdown}}C=C\underset{CH_3}{\overset{CH_2-CH_2}{\diagdown}}C=C\underset{CH_3}{\overset{CH_3}{\diagup}} \quad B$$

The original diene must have either structure A or structure B; we can't tell which from the cleavage data alone.

3. 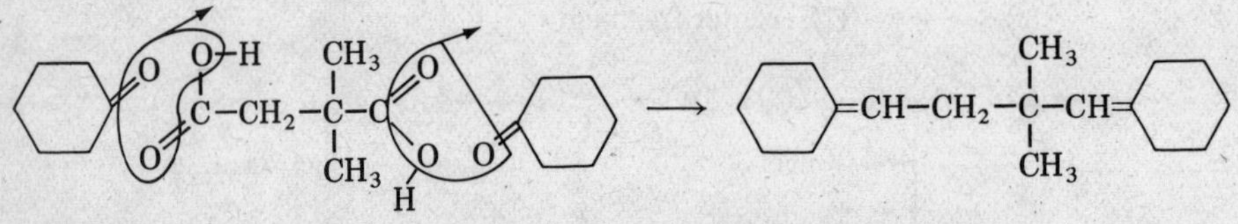 [structure] → [alkene product with cyclohexyl=CH-CH₂-C(CH₃)₂-CH=cyclohexyl]

The alkene must have this structure. When the end fragments are identical, there is no ambiguity, even if the center fragment is unsymmetrical.

★ *Exercises*

1. Draw the structure(s) of the alkene or alkenes whose ozonolysis products are

(a) 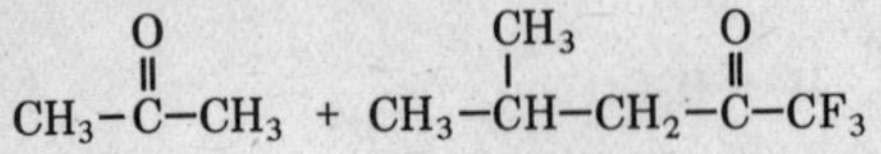 $CH_3-\overset{O}{\underset{\|}{C}}-CH_3 + CH_3-\overset{CH_3}{\underset{|}{CH}}-CH_2-\overset{O}{\underset{\|}{C}}-CF_3$

(b) $\text{Ph}-\overset{O}{\underset{\|}{C}}-H + CH_3-O-CH_2-CH_2-CH_2-\overset{O}{\underset{\|}{C}}-H$

(c) cyclopentanone $=O + H_2CO$

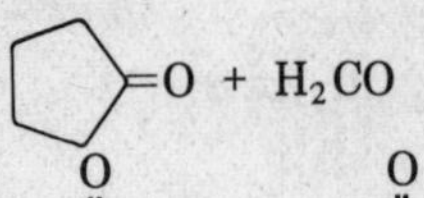

(d) 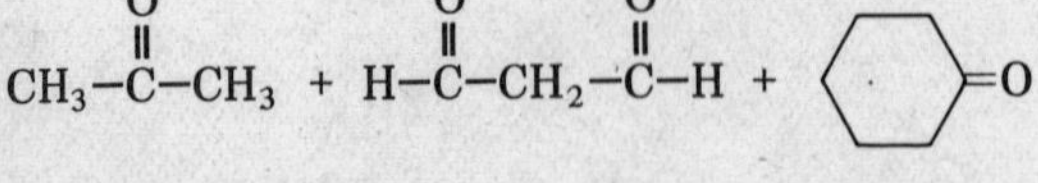 $CH_3-\overset{O}{\underset{\|}{C}}-CH_3 + H-\overset{O}{\underset{\|}{C}}-CH_2-\overset{O}{\underset{\|}{C}}-H + \text{cyclohexyl}=O$

(e) 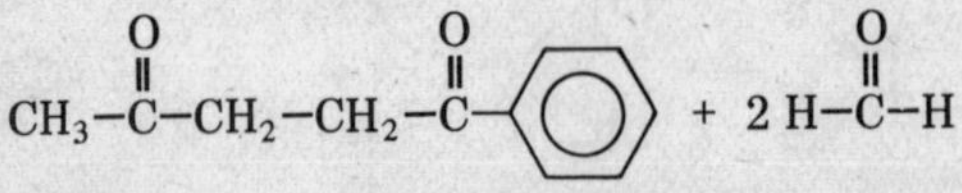 $CH_3-\overset{O}{\underset{\|}{C}}-CH_2-CH_2-\overset{O}{\underset{\|}{C}}-\text{Ph} + 2\,H-\overset{O}{\underset{\|}{C}}-H$

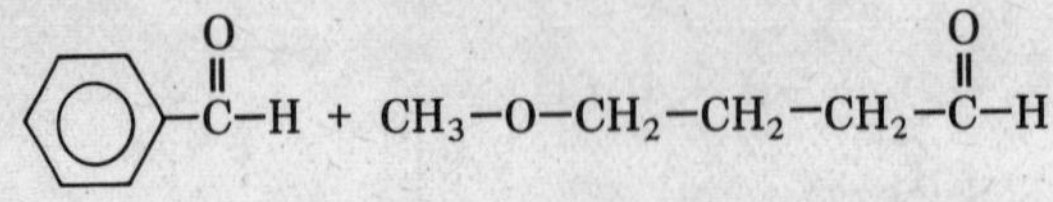

(f) $CF_3-\overset{O}{\underset{\|}{C}}-CF_2-CF_2-\overset{O}{\underset{\|}{C}}-H$ + C$_6$H$_5$-$\overset{O}{\underset{\|}{C}}$-C$_6H_5$ + $CF_3-\overset{O}{\underset{\|}{C}}-CF_3$

(g) $CF_3-\overset{O}{\underset{\|}{C}}-CF_2-CF_2-\overset{O}{\underset{\|}{C}}-CF_3$ + C$_6$H$_5$-$\overset{O}{\underset{\|}{C}}$-CF$_3$ + $CF_3-\overset{O}{\underset{\|}{C}}-H$

(h) o-C$_6$H$_4$(CHO)(CH$_2$-CHO)

2. Compounds A, B, C, and D are isomers (C_6H_{12}). A gives 1 mole each of CH_3-CH_2-CHO and $CH_3-CO-CH_3$ on ozonolysis, B and C fail to react with ozone or H_2/Pt. The pmr spectrum of C implies that all 12 H's are equivalent; that of B suggests the presence of $>CH-CH_3$. D gives a single compound on ozonolysis. What are the structures of A to D?

----

1. (a) $(CH_3)_2C=C(CF_3)(CH_2-CH(CH_3)_2)$

(b) $\underset{H}{\overset{C_6H_5}{>}}C=C\underset{CH_2-CH_2-CH_2-O-CH_3}{\overset{H}{<}}$ or $\underset{C_6H_5}{\overset{H}{>}}C=C\underset{CH_2-CH_2-CH_2-O-CH_3}{\overset{H}{<}}$

(c) cyclopentylidene=CH$_2$

(d) $(CH_3)_2C=CH-CH_2-CH=$cyclohexylidene

(e) $CH_2=C(CH_3)-CH_2-CH_2-C=CH_2$ with phenyl

(f) $C_6H_5-\underset{C_6H_5}{\overset{CF_3}{C}}=C-CF_2-CF_2-CH=\underset{CF_3}{\overset{}{C}}-CF_3$ or $CF_3-\underset{CF_3}{\overset{CF_3}{C}}=C-CF_2-CF_2-CH=\underset{C_6H_5}{\overset{}{C}}-C_6H_5$

(g) $\underset{CF_3}{\overset{C_6H_5}{>}}C=C\underset{CF_2-CF_2}{\overset{CF_3}{<}}\underset{}{\overset{CF_3}{>}}C=C\underset{CF_3}{\overset{H}{<}}$, $\underset{C_6H_5}{\overset{CF_3}{>}}C=C\underset{CF_2-CF_2}{\overset{CF_3}{<}}\underset{}{\overset{CF_3}{>}}C=C\underset{CF_3}{\overset{H}{<}}$,

$\underset{C_6H_5}{\overset{CF_3}{>}}C=C\underset{CF_2-CF_2}{\overset{CF_3}{<}}\underset{}{\overset{CF_3}{>}}C=C\underset{H}{\overset{CF_3}{<}}$, or $\underset{CF_3}{\overset{C_6H_5}{>}}C=C\underset{CF_2-CF_2}{\overset{CF_3}{<}}\underset{}{\overset{CF_3}{>}}C=C\underset{H}{\overset{CF_3}{<}}$

(h) 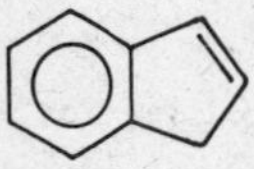 [indene structure]

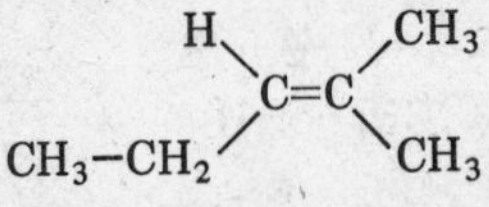

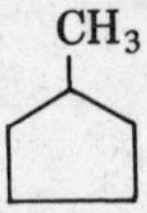

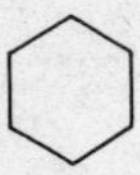

2. A = $\text{CH}_3\text{-CH}_2\text{-CH=C(CH}_3\text{)-CH}_3$ (with H on one carbon)

B = methylcyclopentane

C = cyclohexane

D = $(\text{CH}_3)_2\text{C=C}(\text{CH}_3)_2 \longrightarrow 2\,\text{CH}_3\text{COCH}_3$

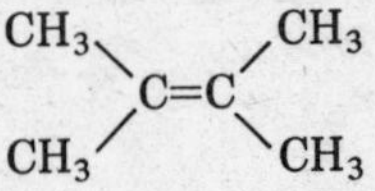

13. Hot, alkaline, aqueous $KMnO_4$ degrades Ar–R to Ar–COOH + H_2O + CO_2. The side chain, R, can be long, short, straight, or branched, and it can contain any functional group. The only requirement is that the first atom out, that is, the one directly attached to the aromatic ring, must be an aliphatic carbon atom. If the starting molecule contains two side chains, two $-CO_2H$ groups appear in the degradation product and their location shows where the original chains were attached. Since all of the mono, di, tri, etc., benzene carboxylic acids are well-characterized solids, the acid obtained in the degradative experiment can be identified by simple physical means, for example by melting point.

Example

Compound X, $C_{11}H_{13}Cl$, displays ir bands characteristic of an aromatic ring. X rapidly decolorizes cold $KMnO_4$. Treatment of X with hot, alkaline $KMnO_4$ followed by acidification produces ⟨Ph⟩–COOH. Draw a part structure summarizing what is known about X.

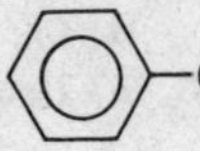

Solution:

Oxidation to ⟨Ph⟩–COOH proves (a) the presence of a benzene ring with one alkyl group only on the ring, and (b) that the chlorine atom is in the side chain (if it were on the ring, the degradation product would be one of the chlorobenzoic acids, Cl–⟨Ph⟩–COOH). Therefore, X = ⟨Ph⟩–C_5H_8Cl. $\Delta = 11 + 1 - 13/2 - 1/2 = 5$, of which ⟨Ph⟩– accounts for 4 units; ∴ there is a ring or double bond in the side chain. Oxidation by cold $KMnO_4$ shows that this fifth unit of Δ is a double bond. So

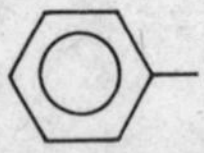

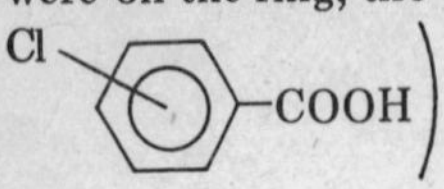

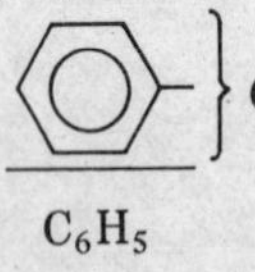

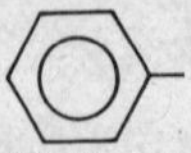

X = ⟨Ph⟩–C(=C)–C_3H_8 with Cl–

C_8H_5Cl

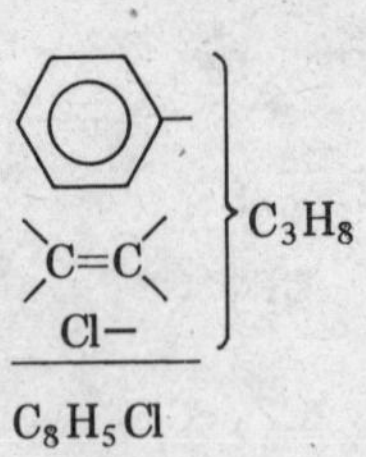

STRUCTURE PROBLEMS

★ *Exercises*

1. Compound Y, $C_9H_{11}NO_2$, gives HOOC—C$_6$H$_3$(NO$_2$)—COOH on treatment with hot, alkaline $KMnO_4$ followed by acidification. What can you say about the structure of Y?

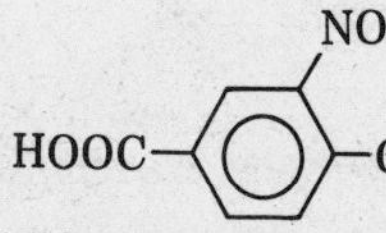

2. Suppose that one more experiment, ozonolysis, is done on compound X, $C_{11}H_{13}Cl$, in the example above. If some $CH_3-CH_2-\overset{\overset{O}{\|}}{C}-H$ can be isolated from the ozonolysis products, revise the part structure and write all the possible full structures for X.

---

1. Y = [—C—C$_6$H$_3$(NO$_2$)—C—] CH_8, which can only be CH_3-CH_2—C$_6$H$_3$(NO$_2$)—CH_3 or

   $C_8H_3NO_2$

   CH_3—C$_6$H$_3$(NO$_2$)—CH_2-CH_3

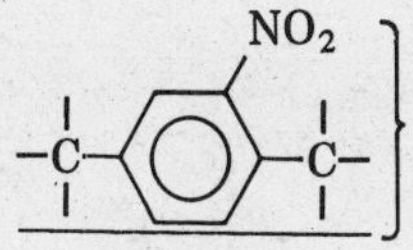

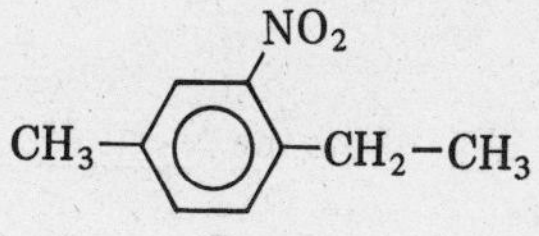

2. X 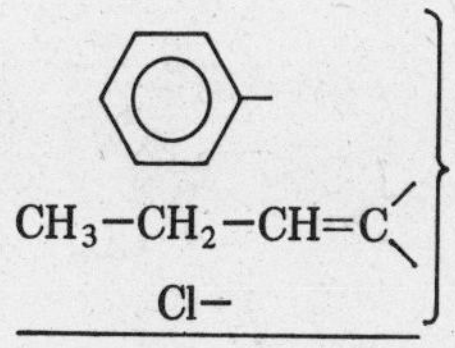 = [C$_6$H$_5$—] CH_2, which can only be one of four compounds:

   $CH_3-CH_2-CH=C\underset{Cl-}{\diagdown}$

   $C_{10}H_{11}Cl$

   $\underset{H}{\overset{CH_3-CH_2}{\diagdown}}C=C\underset{CH_2Cl}{\overset{C_6H_5}{\diagup}}$ $\underset{H}{\overset{CH_3-CH_2}{\diagdown}}C=C\underset{C_6H_5}{\overset{CH_2Cl}{\diagup}}$

   $\underset{H}{\overset{CH_3-CH_2}{\diagdown}}C=C\underset{Cl}{\overset{CH_2-C_6H_5}{\diagup}}$ $\underset{H}{\overset{CH_3-CH_2}{\diagdown}}C=C\underset{CH_2-C_6H_5}{\overset{Cl}{\diagup}}$

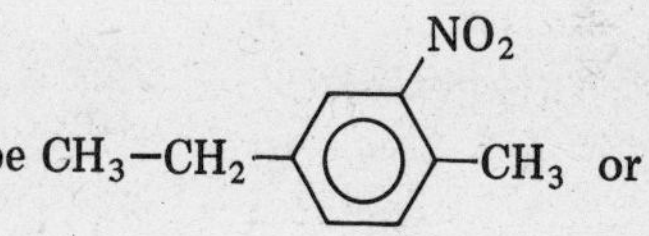

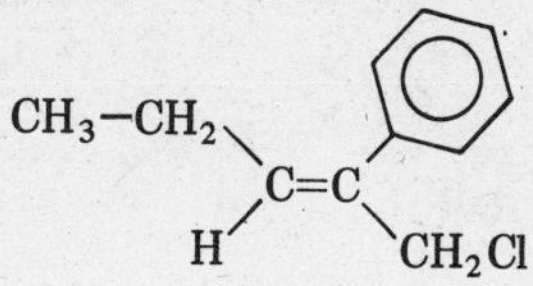

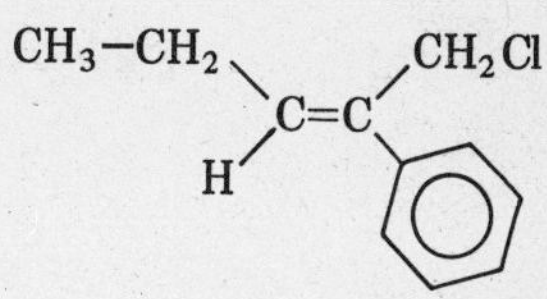

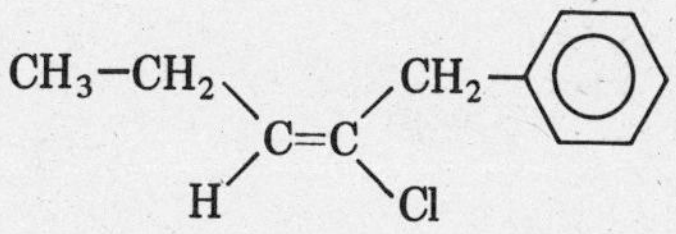

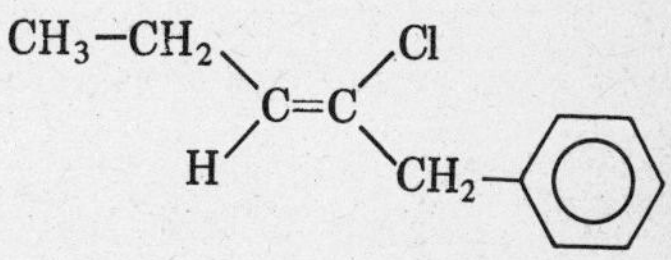

14. The Hofmann and Cope degradations provide the same structural information and are handled the same way in structure problems. They are often applied repeatedly, with the overall result that the nitrogen atom is removed from the molecule. The degradation reactions proper operate on the quaternary ammonium salt or the amine oxide, but one generally uses them to study the structure of *amines* since an amine can be easily converted to the quaternary ammonium salt or amine oxide. The complete sequence looks like this for a 1° amine:

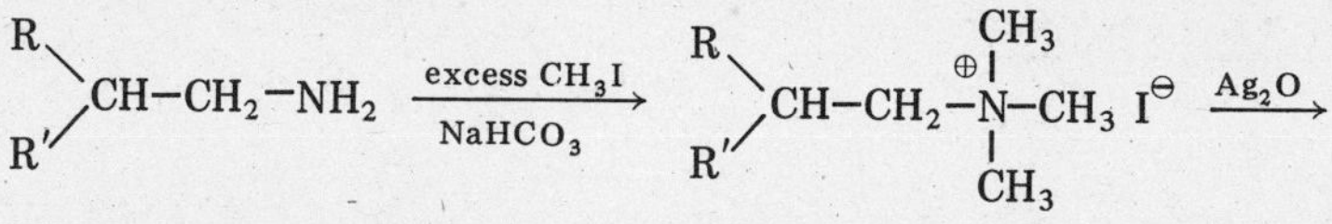

$\underset{R'}{\overset{R}{\diagdown}}CH-CH_2-NH_2 \xrightarrow[NaHCO_3]{excess\ CH_3I} \underset{R'}{\overset{R}{\diagdown}}CH-CH_2-\overset{\overset{CH_3}{|}}{\underset{\underset{CH_3}{|}}{\overset{\oplus}{N}}}-CH_3\ I^{\ominus} \xrightarrow{Ag_2O}$

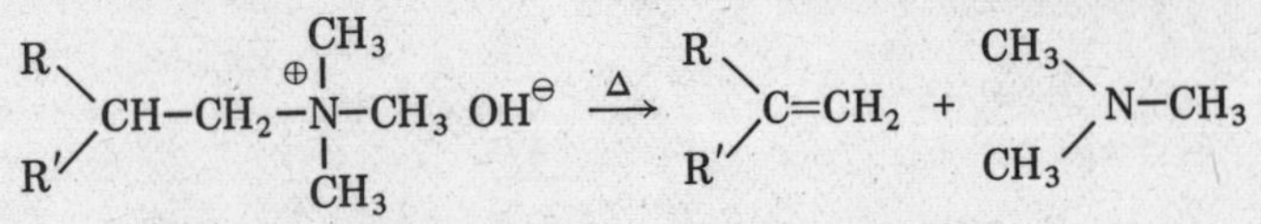

$$\underset{R'}{\overset{R}{>}}CH-CH_2-\overset{\overset{CH_3}{\underset{|}{\oplus}}}{\underset{\underset{CH_3}{|}}{N}}-CH_3 \ OH^\ominus \xrightarrow{\Delta} \underset{R'}{\overset{R}{>}}C=CH_2 + \underset{CH_3}{\overset{CH_3}{>}}N-CH_3$$

For a 2° or 3° amine one or two alkyl groups are removed with the N:

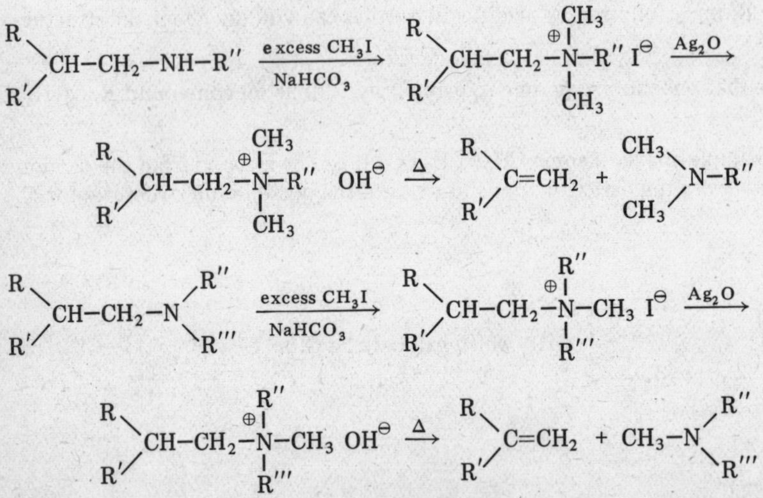

$$\underset{R'}{\overset{R}{>}}CH-CH_2-NH-R'' \xrightarrow[NaHCO_3]{excess\ CH_3I} \underset{R'}{\overset{R}{>}}CH-CH_2-\overset{\overset{CH_3}{\underset{|}{\oplus}}}{\underset{\underset{CH_3}{|}}{N}}-R''\ I^\ominus \xrightarrow{Ag_2O}$$

$$\underset{R'}{\overset{R}{>}}CH-CH_2-\overset{\overset{CH_3}{\underset{|}{\oplus}}}{\underset{\underset{CH_3}{|}}{N}}-R''\ OH^\ominus \xrightarrow{\Delta} \underset{R'}{\overset{R}{>}}C=CH_2 + \underset{CH_3}{\overset{CH_3}{>}}N-R''$$

$$\underset{R'}{\overset{R}{>}}CH-CH_2-N\underset{R'''}{\overset{R''}{<}} \xrightarrow[NaHCO_3]{excess\ CH_3I} \underset{R'}{\overset{R}{>}}CH-CH_2-\overset{\overset{R''}{\underset{|}{\oplus}}}{\underset{\underset{R'''}{|}}{N}}-CH_3\ I^\ominus \xrightarrow{Ag_2O}$$

$$\underset{R'}{\overset{R}{>}}CH-CH_2-\overset{\overset{R''}{\underset{|}{\oplus}}}{\underset{\underset{R'''}{|}}{N}}-CH_3\ OH^\ominus \xrightarrow{\Delta} \underset{R'}{\overset{R}{>}}C=CH_2 + CH_3-N\underset{R'''}{\overset{R''}{<}}$$

In all the cases above, where N is not a part of any ring, one "cycle" gets the N out of the original molecule (now the alkene). When N is part of one ring, two cycles are required to remove N:

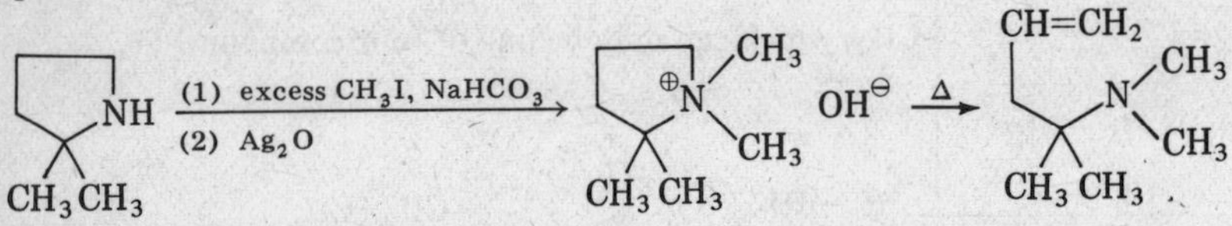

1st cycle

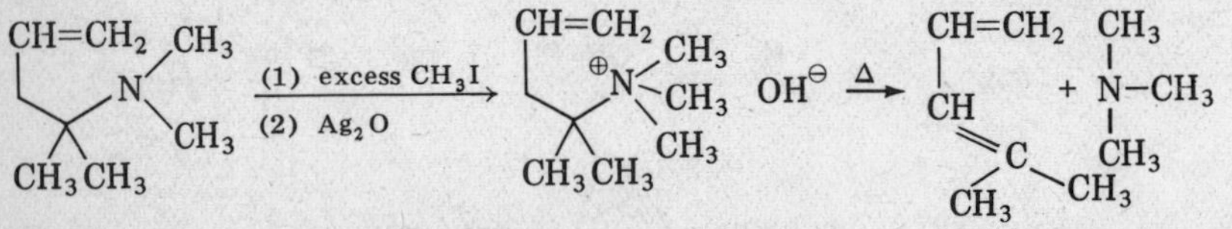

2nd cycle

When N is part of two rings, three cycles are required:

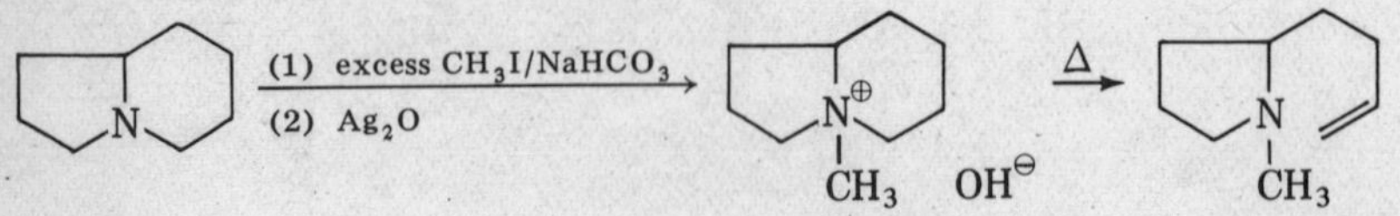

1st cycle

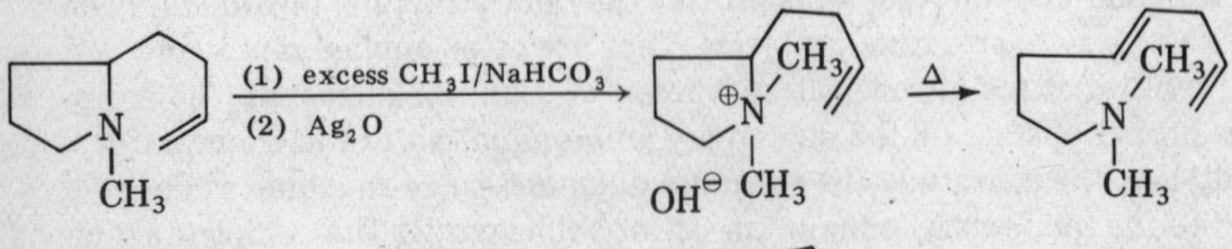

2nd cycle

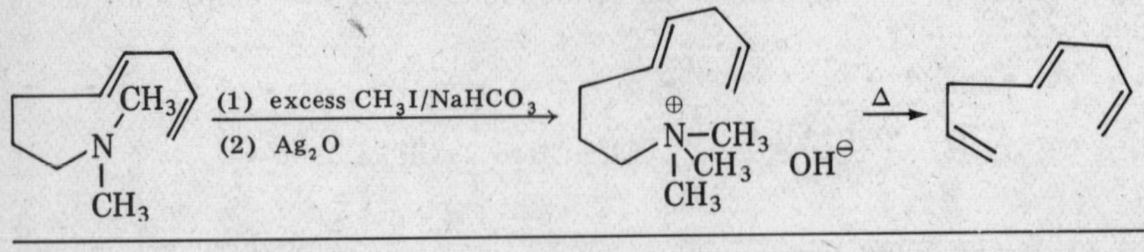

3rd cycle

Note that if the degradation is carried through enough cycles, the nitrogen always ends up as trimethylamine; for this reason the whole process has been called *exhaustive methylation*. In the illustrations above only one product has been written for each cycle. However, it should be clear that the elimination can usually go in more than one direction; for example, the two-cycle example above could have been written

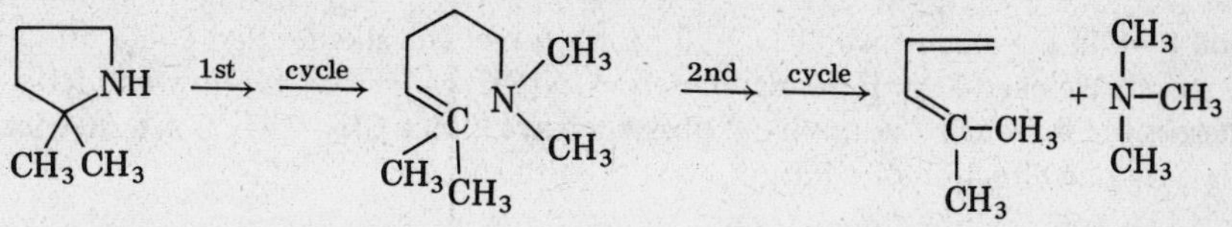

$$\underset{CH_3\ CH_3}{\bigcirc\!\!-\!NH} \xrightarrow{\text{1st cycle}} \underset{\underset{CH_3}{\overset{|}{C}}\!CH_3}{\bigcirc\!\!=\!\!C\!-\!N\!\!\begin{array}{c}CH_3\\CH_3\end{array}} \xrightarrow{\text{2nd cycle}} \underset{\underset{CH_3}{|}}{C\!-\!CH_3} + \underset{\underset{CH_3}{|}}{N\!-\!CH_3}$$

In actual fact, both paths occur simultaneously although one usually predominates. The fact that more than one path is followed can be ignored in solving structure problems (unless the structures of the intermediates are specifically desired).

To interpret these degradative reactions you just mentally reverse them by adding the amine product back onto the alkene product(s) in as many ways as possible. Since the methylation steps have introduced some extraneous CH_3 groups on N, you must discard these. Discard as many as necessary (replacing them with H) to make the number of N—H bonds in the amine equal to the number of double bonds in the alkene.

Example

Compound X, $C_7H_{17}N$, is treated with excess $CH_3I/NaHCO_3$ followed by Ag_2O, then heat (Hofmann degradation). From this reaction $CH_3-CH_2-CH_2-N(CH_3)_2$ and $CH_2=CH-CH_3$ can be isolated. What structures are possible for X?

Solution:

The alkene product has one C=C, so we need one N—H bond on the product amine. Therefore, replace one CH_3 by H: $CH_3-CH_2-CH_2-N\begin{array}{c}CH_3\\CH_3\end{array} \longrightarrow$

$CH_3-CH_2-CH_2-N\begin{array}{c}H\\CH_3\end{array}$. Now add this N—H bond across C=C in both of the possible directions:

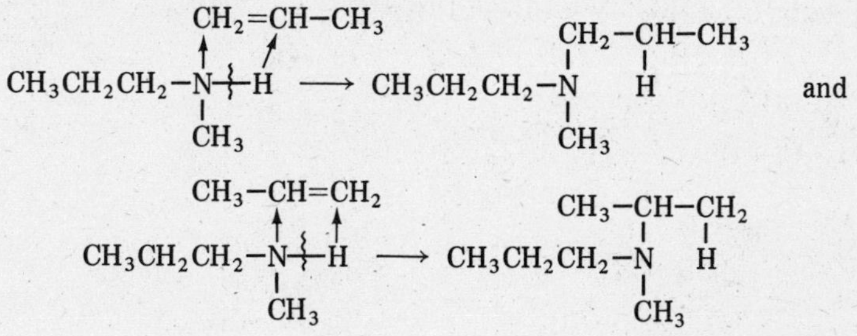

$$\underset{\underset{CH_3}{|}}{CH_3CH_2CH_2-N\!+\!H} \overset{CH_2=CH-CH_3}{\underset{\uparrow\quad\nearrow}{}} \longrightarrow \underset{\underset{CH_3}{|}}{CH_3CH_2CH_2-N} \overset{CH_2-CH-CH_3}{\underset{|\quad\ |}{H}} \quad \text{and}$$

$$\underset{\underset{CH_3}{|}}{CH_3CH_2CH_2-N\!+\!H} \overset{CH_3-CH=CH_2}{\underset{\uparrow\quad\uparrow}{}} \longrightarrow \underset{\underset{CH_3}{|}}{CH_3CH_2CH_2-N} \overset{CH_3-CH-CH_2}{\underset{|\quad\ |}{H}}$$

Therefore, X must be either $CH_3CH_2CH_2-\underset{\underset{CH_3}{|}}{N}-CH_2CH_2CH_3$ or $CH_3CH_2CH_2-\underset{\underset{CH_3}{|}}{N}-CH\begin{array}{c}CH_3\\CH_3\end{array}$.

Example

Compound A, $C_{11}H_{23}N$, on Hofmann degradation gives $CH_2=CH_2$ plus B, $C_{10}H_{21}N$. Hofmann degradation of B gives C, $C_{11}H_{23}N$. Hofmann degradation of C gives $(CH_3)_3N$

534 HOW TO SUCCEED IN ORGANIC CHEMISTRY

plus a diene that proves to be $CH_2=C(CH_3)-CH_2-CH=CH-CH(CH_3)-CH_3$. What are the possible structures for A?

Solution:

Total C=C in the alkene products = 3 (don't forget the alkenes that come off in the earlier steps). We need 3 N−H bonds; therefore, $N(CH_3)_3 \longrightarrow NH_3$. Now add NH_3 back across the three C=C bonds in all possible ways. Since $CH_2=CH_2$ is symmetrical, there is only one direction of addition, so start with this one:

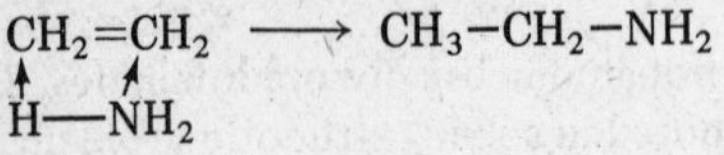

$$CH_2=CH_2 + H-NH_2 \longrightarrow CH_3-CH_2-NH_2$$

Next,

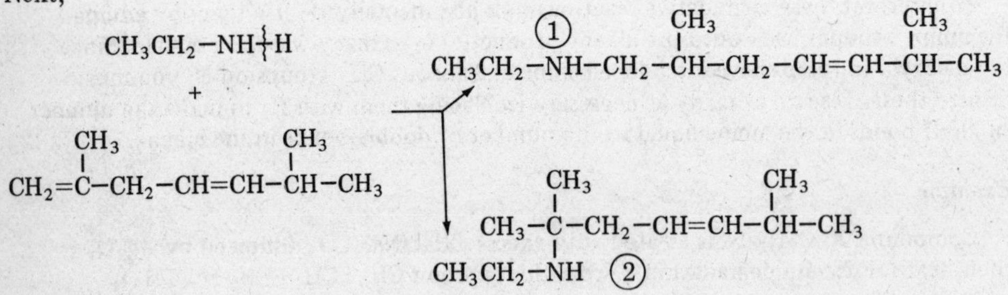

$$CH_3CH_2-NH-H \; + \; CH_2=C(CH_3)-CH_2-CH=CH-CH(CH_3)-CH_3$$

gives:

① $CH_3CH_2-NH-CH_2-CH(CH_3)-CH_2-CH=CH-CH(CH_3)-CH_3$

② $CH_3-C(CH_3)(NHCH_2CH_3)-CH_2-CH=CH-CH(CH_3)-CH_3$

Next,

① = intramolecular cyclization forming a pyrrolidine ring with ethyl and isopropyl substituents:

CH_3-CH_2-N with $CH_2-CH(CH_3)-CH_2$ and $H \rightarrow CH=CH-CH(CH_3)-CH_3$ gives an N-ethyl-2-isopropyl-pyrrolidine derivative.

① = alternative cyclization forming a piperidine ring:

CH_3-CH_2-N with $CH_2-CH(CH_3)-CH_2-CH=CH-CH(CH_3)-CH_3$ gives an N-ethyl-isopropyl-piperidine.

② = cyclization of the branched intermediate:

$CH_3-C(CH_3)-CH_2 \cdots CH=CH-CH(CH_3)-CH_3$ with CH_3-CH_2-N-H gives a bicyclic / substituted pyrrolidine:

$$\begin{array}{c} CH_3 \\ | \\ CH_3-C-CH_2 \\ | \quad\quad | \\ CH_2-N-CH \\ | \quad\quad | \\ CH_3 \quad CH_2 \\ \quad\quad | \\ CH_3-CH-CH_3 \end{array}$$

and an azetidine-type ring structure with N-isopropyl substitution.

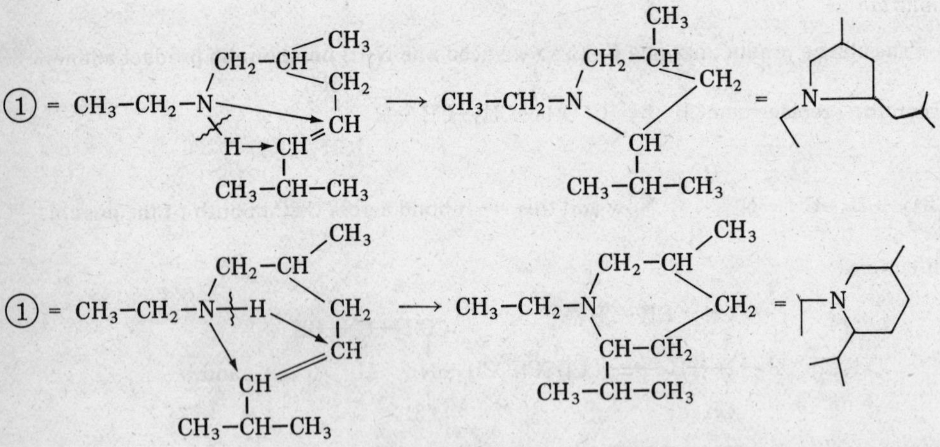

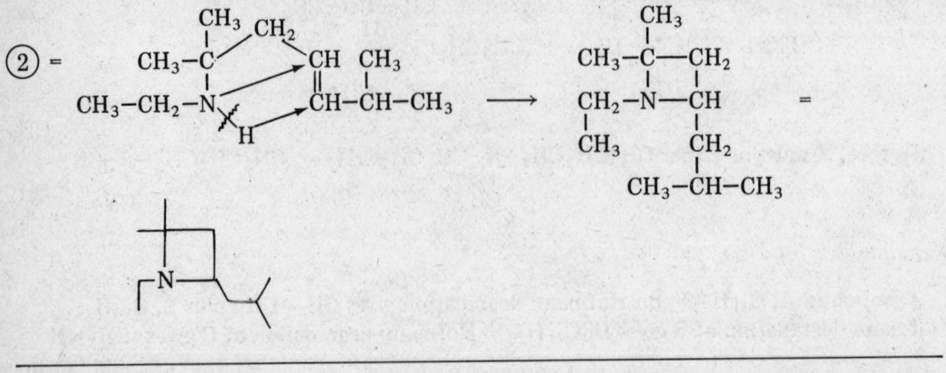

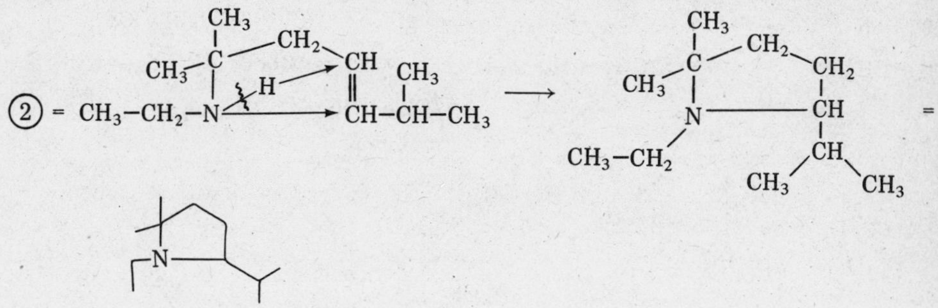

STRUCTURE PROBLEMS

[Structures at top showing cyclization reactions]

★ *Exercises*

1. Hofmann degradation of compound J, $C_{12}H_{17}N$, gives K, $C_{14}H_{21}N$. Hofmann degradation of K gives $(CH_3)_3N$ and $CH_2=CH-CH(CH_3)-CH=CH-C_6H_5$. Draw the possible structures for J.

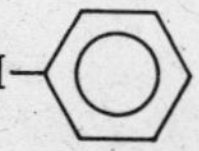

2. On Hofmann degradation compound L, $C_{12}H_{15}F_{10}N$ gives M ($C_{13}H_{17}F_{10}N$). Hofmann degradation of M gives $CH_3-CH_2-N(CH_3)_2$ plus N ($C_{10}H_8F_{10}$). Ozonolysis of N produces 1 mole of $CF_3-\overset{O}{\underset{\|}{C}}-CF_2-CF_2-\overset{O}{\underset{\|}{C}}-CF_3$ and 2 moles of CH_3-CHO. Deduce all the structures possible for L, M, and N.

— — — — — — — — —

1. $(CH_3)_3N \longrightarrow CH_3-NH_2 \xrightarrow{diene} CH_3-NH-CH_2-CH_2-CH(CH_3)-CH=CH-C_6H_5$

[Four possible cyclic structures shown]
$CH_3-CH-CH-CH=CH-C_6H_5$ with CH_3-NH and CH_3

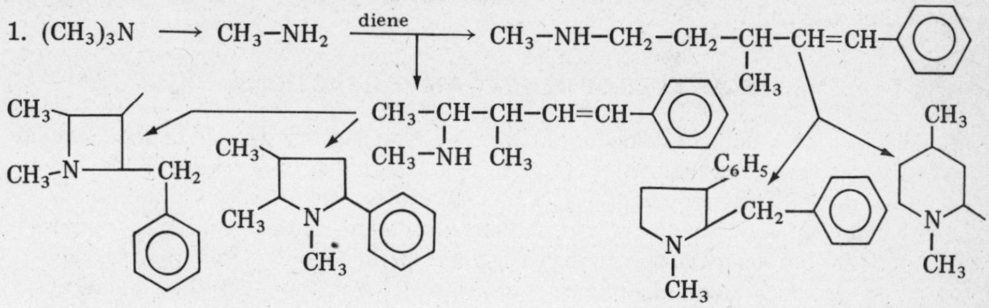

Since these are all $C_{13}H_{19}N$ instead of $C_{12}H_{17}N$, the N–CH$_3$ groups are derived from CH_3I, not compound J.

J = [four possible structures shown: azetidine with phenyl, pyrrolidine with phenyl (2 isomers), piperidine with phenyl]

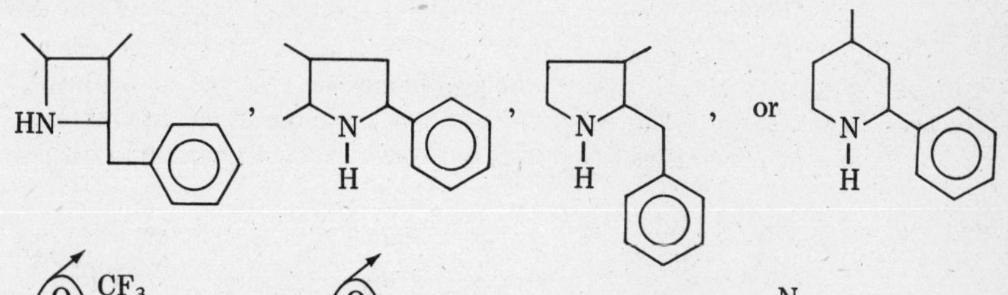

2. $CH_3-CH\overset{O}{\underset{O}{\diagdown\diagup}}\overset{CF_3}{\underset{CF_3}{C}}-CF_2-CF_2-\overset{CF_3}{\underset{O}{C}}\overset{O}{\diagdown\diagup}CH-CH_3 \longrightarrow CH_3-CH=\overset{CF_3}{C}-CF_2-CF_2-\overset{CF_3}{C}=CH-CH_3$

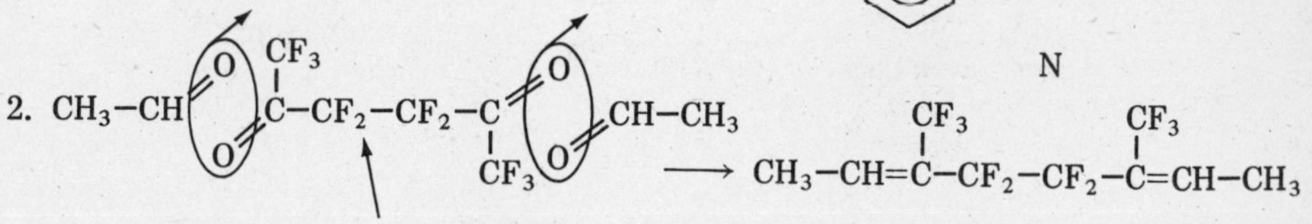

symmetrical, only one diene structure possible

536 HOW TO SUCCEED IN ORGANIC CHEMISTRY

$$CH_3-CH_2-N\begin{smallmatrix}CH_3\\CH_3\end{smallmatrix} \longrightarrow CH_3-CH_2-N\begin{smallmatrix}H\\H\end{smallmatrix} \longrightarrow \underset{CH_3-CH_2-NH}{CH_3-\overset{CF_3}{\underset{}{CH}}-\overset{(A)}{\underset{}{CH}}CF_2CF_2\overset{CF_3}{\underset{}{C}}=CHCH_3}$$

$$\underset{CH_3-CH_2-NH}{CH_3-CH_2-\overset{CF_3}{\underset{}{C}}-CF_2-CF_2-\overset{CF_3}{\underset{}{C}}=CH-CH_3} \quad (B)$$

(with ring structures labeled ①, ②, ③, ④)

④ CH_3CH_2–$\overset{CF_2-CF_2}{\underset{CF_3\,CH_2\,CF_3}{N}}$–$CH_2CH_3$ with CH_3

③ (ring with CF_2, CF_2, CF_3, N, CH_2–CH_3, CH_3CH_2, CF_3)

② (ring with CF_3, CF_2, CF_2, CH_3, N, CF_3, CH_2CH_3, CH_2CH_3)

① (ring with CF_2–CF_2, CF_3, CF_3, CH_3, N, CH_3, CH_2–CH_3)

L must be ①, ②, ③, or ④.

M is related to (A) or (B)—a half-degraded L. Since the Hofmann product is always a tertiary amine, must have an N—CH$_3$ in place of N—H:

$$CH_3CH_2-\underset{CH_2CH_3}{\overset{CH_3}{\underset{}{N}}}-\overset{CF_3}{\underset{}{C}}-CF_2CF_2-\overset{CF_3}{\underset{}{C}}=CH-CH_3 \quad \text{or} \quad CH_3CH_2-\overset{CH_3}{\underset{}{N}}-\overset{CH_3}{\underset{}{CH}}-\overset{CF_3}{\underset{}{CH}}-CF_2CF_2-\overset{CF_3}{\underset{}{C}}=CH-CH_3$$

Since we could just as well have operated on the right-hand C=C of N first in the very first step above, there would ordinarily be two more possible structures for M; in this case, however, compound N is symmetrical and the two additional structures are identical to A and B.

SEQUENCES OF DEGRADATIVE REACTIONS

15. When the technique of deducing structure by degradation is applied to very complex molecules, the fragments themselves are often so complex that they cannot be identified. That is, they are themselves compounds whose structures have never been determined. So we cannot just isolate and purify such fragments, determine their molecular formulas by analysis for the elements, measure melting point, boiling point, refractive index, and various spectra, then look in the literature to find which compound has these properties, because the compounds are not yet in the literature.

When this happens, the original structure problem turns into new structure problems that must be solved before the original problem can be solved. However, the new problems should be simpler than the original one because the fragment molecules are smaller and presumably simpler than the original. Such problems can often go on to third-, fourth-, fifth-, or sixth-generation fragments before known compounds are encountered. These complex structure problems can be linear, but they are more commonly branched:

$$A \xrightarrow{\text{expt.}\ a} B \xrightarrow{\text{expt.}\ b} C \xrightarrow{\text{expt.}\ c} D\ \text{(identifiable)} \quad \text{linear structure problem}$$

$$\underbrace{}_{\text{reason back}}\ \underbrace{}_{\text{reason back}}\ \underbrace{}_{\text{reason back}}$$

$$A \longrightarrow B + C \longrightarrow G$$
$$\quad\quad\quad \downarrow \quad\quad \downarrow$$
$$\quad\quad\quad D \quad E + F \longrightarrow H$$
$$\quad\quad\quad\quad\quad\quad \downarrow$$
$$\quad\quad\quad\quad\quad I + J$$

branched structure problem

(D, G, H, I, and J identifiable)

STRUCTURE PROBLEMS 537

Most of the Hofmann degradation problems we worked on in frame 14 were linear structure problems. Each individual step in a many-stage problem can be attacked by the most promising physical or chemical means. You will find all combinations of diagnostic, spectroscopic, and degradative data in these problems. The appearance of these multi-step structure problems, when laid out graphically as above, has produced the name "roadmap problems."

The first step in solving such a problem is to translate the data (usually given in words) into graphic (roadmap) form.

The second step is to locate the "way into" the problem. This is the point or points where fragment compound(s) have been identified and either their structural formulas or names given.

The third step is to reason back to the structure of the immediate precursor of the known fragment. You then repeat this process, reasoning backward a step at a time until you come to the original compound.

★ *Exercise*

Compound A, $C_{19}H_{28}O$, reacts with CF_3CO_3H to give B, $C_{19}H_{28}O_2$. B is easily hydrolyzed with aqueous NaOH to give C, $C_9H_{18}O$, and (after acidification) D, $C_{10}H_{12}O_2$. Treatment of C with concentrated H_2SO_4 gives E, C_9H_{16}. Ozonolysis of E gives acetone and cyclohexanone in roughly equal amounts. D is degraded by the Barbier-Wieland sequence to F, $C_9H_{10}O_2$. One more Barbier-Wieland cycle gives G, $C_8H_8O_2$. G fails to undergo a Barbier-Wieland cycle, but hot alkaline $KMnO_4$ converts G to terephthalic acid, HOOC—⌬—CO_2H. Deduce the possible structure(s) of A to G.

— — — — — — — — —

Diagram:

$$A\ (C_{19}H_{28}O)$$
$$\downarrow CF_3COOH$$
$$B\ (C_{19}H_{28}O_2)$$
$$\downarrow H_2O$$

$E(C_9H_{16}) \xleftarrow{H_2SO_4} C\ (C_9H_{18}O) + D\ (C_{10}H_{12}O_2) \xrightarrow{B-W} F\ (C_9H_{10}O_2) \xrightarrow{B-W} G\ (C_8H_8O_2)$

$O_3 \downarrow \longrightarrow CH_3-\underset{\underset{O}{\|}}{C}-CH_3\ +\ $ cyclohexanone

Attack problem at E and G. E: cyclohexanone + $O=C(CH_3)_2$ → cyclohexylidene=$C(CH_3)_2$ = E

Now C → E is $C_9H_{18}O$ → C_9H_{16}, change = $C_9H_{18}O - C_9H_{16}$

H_2O lost, together with H_2SO_4 reagent and alkene product means alcohol dehydration. The alcohols that dehydrate to

cyclohexylidene=$C(CH_3)_2$ are cyclohexyl–CH(OH)–CH(CH_3)CH_3 and 1-(HO)(CH_3)-cyclohexyl with –CH(H)CH_3. C is one or the other.

Now start with G ($C_8H_8O_2$). As a Barbier-Wieland degradation product, G should be a carboxylic acid, and indeed it has two oxygens. Δ = 8 + 1 − 8/2 = 5, of which

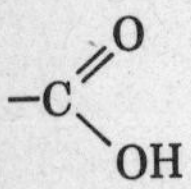

$-C{\stackrel{\displaystyle O}{\diagdown_{OH}}}$ accounts for one unit; the remaining 4 units suggest ⬡. Oxidation of G to HOOC–⬡–CO₂H means C–⬡–C } H₄O₂. Since one –CO₂H is known to be present, G must be C–⬡–CO₂H } H₃ or H₃C–⬡–CO₂H. Since the Barbier-Wieland degradation removes one –CH₂– per cycle, we reason back to F and D by adding CH₂ adjacent to –COOH:

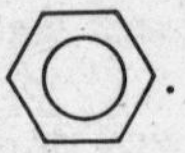

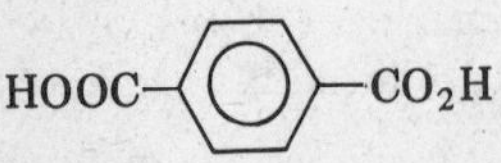

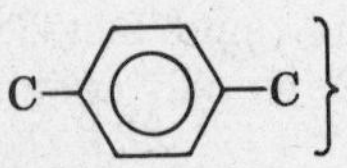

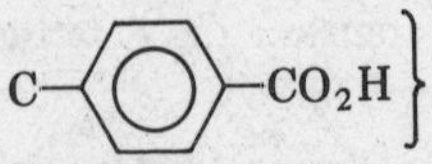

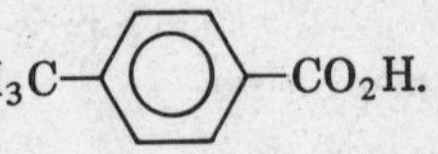

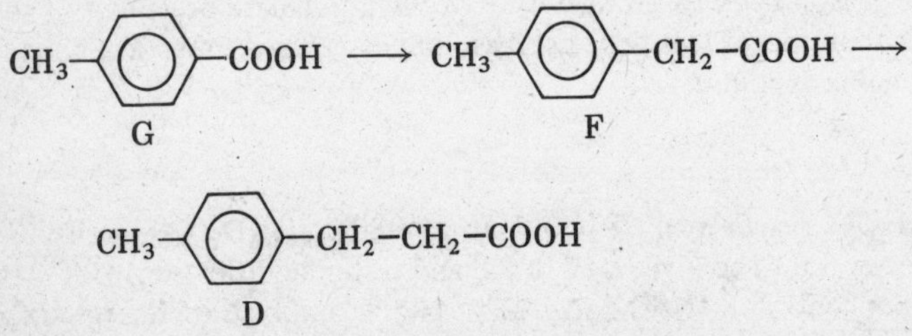

CH₃–⬡–COOH → CH₃–⬡–CH₂–COOH →
    G   F

CH₃–⬡–CH₂–CH₂–COOH
        D

Now for B. B($C_{19}H_{28}O_2$) → C($C_9H_{18}O$) + D($C_{10}H_{12}O_2$) is gain of H_2O, as a hydrolysis reaction should be. Since the products are alcohol and acid, C is the corresponding ester:

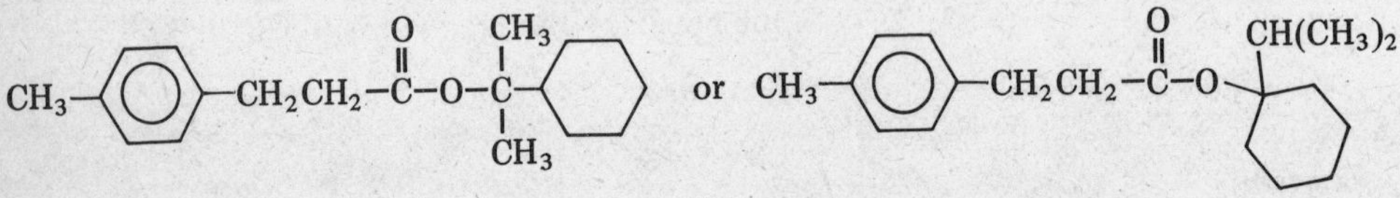

CH_3–⬡–CH_2CH_2–C(=O)–O–C(CH_3)(CH_3)(cyclohexyl)   or CH_3–⬡–CH_2CH_2–C(=O)–O–(cyclohexyl with CH(CH_3)$_2$)

ESTABLISHING STRUCTURE BY SYNTHESIS

16. A structure deduced from degradative, spectroscopic, and diagnostic evidence should be considered probably correct, but subject to confirmation or correction if more or better data become available. Some links in the argument are always weaker than others. And it is always possible that an unsuspected rearrangement of the carbon skeleton occurred during degradation. One way to confirm a structure deduced by the means discussed in this unit is to synthesize the compound with that structure, using only reliable synthetic reactions to put the molecule together from simpler starting materials of well-established structure. Then, if the synthetic sample has properties identical with the original material used for the degradative work, the evidence for the correctness of the deduced structure is very strong.

Synthesis can also play a simpler role in structure problems. Sometimes the structure of one of the degradation products can only be narrowed down to two or three possibilities. In this case one then looks over the additional experiments that one could do to choose between the alternative structures. Often this is more spectroscopic or degradative work, but sometimes it is simpler to just synthesize both structures and see which material is identical with the degradation product. The following exercise uses this approach.

★ *Exercise*

Compounds A and B both have the molecular formula $C_{14}H_{20}$. Both A and B give the same products on treatment with O_3 followed by H_2O/Zn, namely a mixture of compounds C ($C_9H_{10}O$) and D ($C_5H_{10}O$). C can be synthesized by treating benzene with $CH_3-CH_2-CO-Cl$ in the presence of $AlCl_3$. D does not give diagnostic chemical reactions characteristic of an aldehyde. D can be hydrogenated with H_2 on a nickel catalyst to E, $C_5H_{12}O$. E reacts with warm, concentrated H_2SO_4 to give F (C_5H_{10}).

F can be synthesized by a Wittig reaction of CH_3COCH_3 with $CH_3-\overset{\ominus}{CH}-\overset{\oplus}{P}(C_6H_5)_3$. Write possible structures for compounds A to F.

— — — — — — — — —

$$\text{A and B} = \begin{array}{c} C_6H_5 \\ CH_3-CH_2 \end{array}\!\!\!C=C\!\!\!\begin{array}{c} CH_3 \\ CH \\ \diagdown CH_3 \\ CH_3 \end{array} \quad \text{and} \quad \begin{array}{c} CH_3-CH_2 \\ C_6H_5 \end{array}\!\!\!C=C\!\!\!\begin{array}{c} CH_3 \\ CH \\ \diagdown CH_3 \\ CH_3 \end{array}$$

$$C = \underset{}{\bigcirc}\!-\!\overset{O}{\underset{\|}{C}}\!-\!CH_2\!-\!CH_3 \qquad D = CH_3\!-\!\overset{O}{\underset{\|}{C}}\!-\!CH\!\!\begin{array}{c} CH_3 \\ \diagdown CH_3 \end{array}$$

$$E = CH_3\!-\!\underset{\underset{OH}{|}}{CH}\!-\!CH\!\!\begin{array}{c} CH_3 \\ \diagdown CH_3 \end{array} \qquad F = CH_3\!-\!CH\!=\!C\!\!\begin{array}{c} CH_3 \\ \diagdown CH_3 \end{array}$$

THE FINAL ARBITER

17. Complete certainty in the assignment of the structural formula for any compound can only be achieved by the most direct physical methods, which in practice means X-ray crystallography. Interpretation of X-ray diffraction data gives not only the gross structural formula; it provides all the actual bond distances and angles—the complete three-dimensional geometry (including, possibly, configuration in the case of chiral compounds). Crystallographic examination has usually confirmed the correctness of structures deduced from chemical and spectroscopic data, but in a small number of cases it has made subtle or important corrections. In the past few years the art and science of X-ray crystallography have been refined and facilitated to the point where, for some compounds, it is now the most economical procedure to determine their structures by diffraction directly, without using chemical or spectroscopic data at all.

UNIT TWENTY-ONE
Synthetic Problems

Synthesis is the conversion of available compounds into desired but unavailable ones. Usually, this means conversion of smaller, simpler molecules into larger, more complex ones. Synthesis of a desired compound (the "target molecule") may require one or many steps (sequential reactions). The process can be written

$$A \xrightarrow{B} C \xrightarrow{D} E \xrightarrow{F} \cdots \longrightarrow TM$$

where A, B, D, F, etc., are starting materials, C, E, etc., are intermediates, and TM is the target molecule. The starting materials either exist in nature—petroleum, coal tar, plant and animal materials—or they are commercially available compounds that someone else has previously synthesized from natural materials.

In addition to synthesizing compounds for resale, there are several motives for synthetic work. First, synthesis provides a means of confirming the structure deduced for an unknown compound by chemical and spectroscopic means. To do this we synthesize the compound represented by using starting materials of thoroughly established structure and dependable, well-known reactions (an *unambiguous synthesis*). If the compound so prepared proves to be identical in all its properties with the original unknown, the deduced structure must be correct. This and the other means of structure confirmation are discussed in Unit 20.

Second, new compounds are often synthesized for use in testing theoretical predictions of their behavior. The theory of organic chemistry has often developed according to the following pattern:

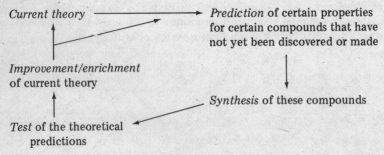

Third, we synthesize compounds that we expect will have practical uses. This motive for synthetic work has led to the discovery and applications of dyes, drugs, flavorants/odorants, pheromones, building materials (polymers), pesticides, photographic materials, food additives, and innumerable specialty items, some good, some bad. Discovery of useful synthetic compounds has often followed the principle of analogy with naturally occurring compounds: deduce the structure of a natural product with useful properties, synthesize various *analogs* of that structure (higher and lower homologs, 6-rings in place of 5-rings, heteroatoms in place of carbons or vice versa, different stereoisomers, compounds with major parts of the original molecule lopped off, etc.).

Fourth, synthetic chemists, like mountain climbers, sometimes solve hard problems just to show they can do it.

In frame 1 we discuss problem types and general strategy. Frames 3 to 10 treat progressively more complex syntheses of aromatic target molecules. Frames 11 to 13 do the same for aliphatic target molecules.

OBJECTIVES

When you have completed this unit, you should be able to plan synthetic sequences leading from simple starting materials to reasonably complex aromatic and aliphatic compounds.

If you think you may be able to skip all or part of this unit, take a self-test consisting of the exercises marked with a star in the margin. If you miss any problems, review the frames in question and rework the exercise before going on. If you are not ready for a self-test, proceed to frame 1.

NATURE OF SYNTHETIC PROBLEMS AND BASIC APPROACH

1. In real-life synthetic work the starting materials are determined by two very simple considerations: the physical availability of the compounds in the quantities required and their cost. To minimize labor costs and time, the starting materials should approach the target molecule's structure as closely as possible and thus require a minimum of transformation. However, to minimize materials costs, since more complex starting materials are more expensive, the simplest starting materials should be used. The compromise reached between these two inclinations depends on the identity of the compounds; availability and prices of chemicals; money, labor, and time available; the quantities of product required and its intended use. So the starting materials must be decided in each case individually. Chemists interested in naturally occurring compounds use the phrase *total synthesis* for synthesis of a complex natural product starting from simple compounds available from petroleum, coal tar, and so on. A synthesis in which one complex natural product is made from another or is remade from some of its own degradation products is called a *partial synthesis*.

In the case of practice synthetic problems for student use, the allowed starting materials and other restrictions imposed (e.g., permissible number of steps) make up a more or less arbitrary set of rules. These can be chosen to simulate real-life problems or for pedagogical purposes. Some choices of allowed starting materials used in student problems are:

- Any (that is, no restrictions).
- Alcohols of four or fewer carbon atoms, benzene, toluene, and any desired inorganic compounds.
- Any commercially available compound.

In both real-life and practice syntheses, the aromatic ring is not ordinarily built as part of the synthesis. Instead, starting materials with preformed aromatic rings are chosen. This means that aromatic syntheses usually consist mainly of introducing and manipulating functional groups on a largely preformed carbon skeleton. The main exceptions to this statement are the problems in which an aliphatic side chain is attached to or gradually built up on an aromatic ring and those in which a second ring is closed onto the first.

Syntheses of aliphatic target molecules usually consist of two parts:

- Building the carbon skeleton.
- Introducing the functional group or groups in their proper places.

542　HOW TO SUCCEED IN ORGANIC CHEMISTRY

The two parts of the task are interdependent. On the one hand, you can make the skeleton in such a way that it is then practically impossible to get the functions right—for example, by constructing the skeleton in the form of an alkane, intending to then introduce the functional groups. This generally is impossible because of the paucity and poor selectivity of reactions undergone by alkanes. On the other hand, with sufficient forethought, you can put most of the functions in their proper places as you construct the carbon skeleton. Even if you cannot put the *correct* function at each site as you build the skeleton, it is usually best to put *some* function there to mark the spot.

Rule: It is usually easier to *change* a function than to *introduce* one.
Corollary: Alkanes are not useful intermediates in syntheses.

★ *Exercises*

1. Here is a general outline for a synthesis. Can you see any basic flaw in this plan?

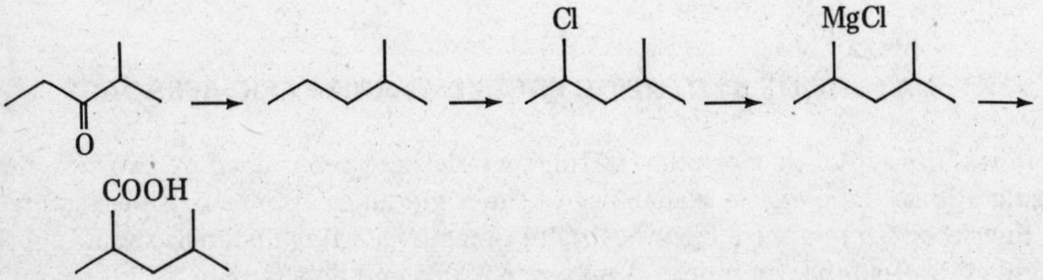

2. Does this sequence have the same flaw?

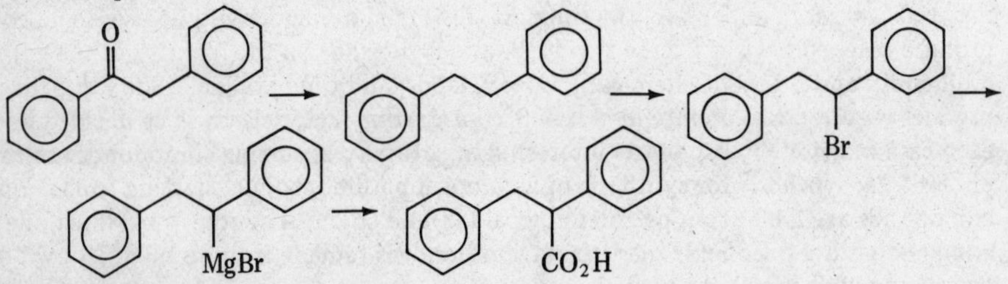

3. Is there a basic flaw in this sequence?

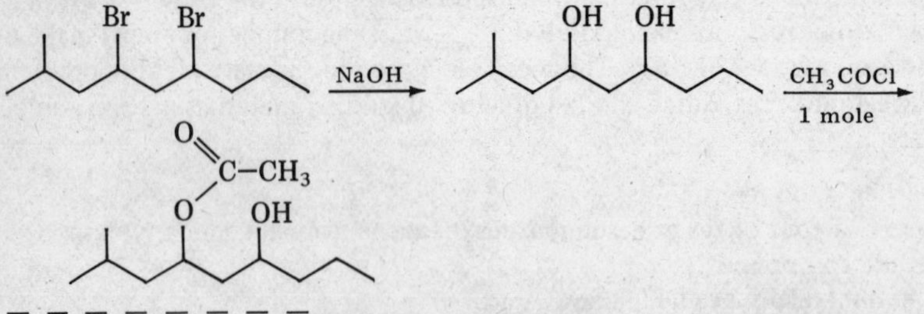

---

1. Yes, the chlorination step will not be specific for one position as written, but will give a mixture of six isomeric monochloro compounds. The yield of the desired isomer will be poor and the product will be difficult to isolate in pure condition.

2. This sequence is OK; the two saturated carbon atoms are equivalent, and monobromination will give a single compound.

3. The last step will actually give about equal amounts of two isomeric monoesters because there is little difference in the properties of the two OH groups in the first intermediate. This diol intermediate is a mistake; once you have it, you cannot operate on one OH selectively.

ONE-STEP SYNTHESES

2. Now, let's consider the simplest of all synthetic problems, the one-step conversion. In earlier units we have often practiced the missing-reagent and missing-product problems:

$$\text{reactant} \xrightarrow{?} \text{product}$$
$$\text{reactant} \xrightarrow{\text{reagent}} ?$$

The missing-reactant problem

$$? \xrightarrow{\text{reagent}} \text{product}$$

is, in fact, an introductory synthetic problem in which product = target molecule and the number of variables has been reduced to one—the missing starting material.

A more valuable starting point for the study of synthesis is the problem

$$? \xrightarrow{?} \text{product}$$

which asks "given your choice of materials, how could you make this compound in one step?" In solving synthetic problems you must ask yourself this question repeatedly. Obviously, there are usually many answers. Since one step does not allow for much skeleton building, these one-step syntheses are mainly conversions of functional groups. The problem boils down to this: Given functional group A (the one in the target molecule), from what other functional groups can we make it in one step (and with what reagent)? Facility in synthetic problems requires that you be able to visualize all the basic reactions that produce a given functional group and sift through them quickly to pick those most suitable to the problem at hand. Practice on individual $? \xrightarrow{?}$ TM problems helps you mobilize previously learned functional-group reactions for this kind of use in the full-scale synthetic problems to come.

Example

From what other functional groups can the functional group $-\overset{\overset{\displaystyle O}{\|}}{C}-NH_2$ be made in one step?

Solution:

$$\begin{array}{c} -C(=O)-O-C(=O)- \\ -C(=O)-Cl \\ -C(=O)-OR \\ -C\equiv N \end{array} \xrightarrow{NH_3} -C(=O)-NH_2 \xleftarrow{H_2O \text{ (cold, conc. HCl)}} -C\equiv N$$

with $-CO_2^{\ominus}NH_4^{\oplus} \xrightarrow{\Delta} -C(=O)-NH_2$

These routes to $-CONH_2$, and those to any other function, are most easily recalled by bringing to mind the maplike schemes in Charts 14.2, 15.2, 17.2, 18.1, and 18.3. If you have difficulty with the following exercises, you may want to review these charts briefly. (Remember that not all the arrows were drawn in on some of these charts.) One

544 HOW TO SUCCEED IN ORGANIC CHEMISTRY

alternative approach is to generate lists of the reactions that produce a given function and drill on these.

★ *Exercises*

For each functional group, identify all the other functions from which it can be made in one step, and write the equation for each of those conversions.

1. $\diagup_{\diagdown}\!C\!-\!CH\!-\!C\!\diagup^{\diagdown}$
   $\quad\;\;\;|$
   $\quad\;\;\,Br$

2. $-CH_2OH$

3. $CH_3-\overset{\overset{O}{\|}}{C}-$

4. $-COOH$

5. $-C\equiv N$

6. $\diagup^{\diagdown}\!C\!=\!C\!\diagup^{\diagdown}$

7. $-CHO$

- - - - - - - - -

1.

$\diagup_{\diagdown}\!C\!-\!CH\!-\!C\!\diagup^{\diagdown}$ with OH
$\quad\quad\;\;\downarrow$ HBr or PBr$_3$

$\diagup_{\diagdown}\!C\!-\!CH\!-\!C\!\diagup^{\diagdown}$ with OR $\xrightarrow{HBr}$ $\diagup_{\diagdown}\!C\!-\!CH\!-\!C\!\diagup^{\diagdown}$ with Br $\xleftarrow{HBr}$ $\diagup^{\diagdown}\!C\!=\!C\!\diagup^{\diagdown}$

$\quad\quad\quad\quad\quad\quad\quad\;\uparrow$ Br$_2$/light

$\diagup_{\diagdown}\!C\!-\!CH_2\!-\!C\!\diagup^{\diagdown}$ (unselective in presence of other C—H bonds)

2. $RMgBr \xrightarrow[(2)\;H_3O^\oplus/H_2O]{(1)\;H_2CO} R-CH_2-OH$

$RMgBr \xrightarrow[(2)\;H_3O^\oplus/H_2O]{(1)\;\triangle\text{(epoxide)}} R-CH_2-CH_2-OH$

$\left.\begin{array}{l}-COOH\\ \text{or}\\ -COOR\\ \text{or}\\ -CHO\end{array}\right\} \xrightarrow[(2)\;H_3O^\oplus/H_2O]{(1)\;LiAlH_4} -CH_2OH$

$-CH_2OH \xleftarrow[H_2O]{NaOH} -CH_2-X$

$-CH_2OH \xleftarrow[(2)\;H_2O_2/NaOH]{(1)\;B_2H_6} \diagup^{\diagdown}\!C\!=\!CH_2$

3. $\diagup^{\diagdown}\!C\!=\!C\!\diagdown^{CH_3}$ $\xrightarrow{KMnO_4/\Delta \text{ or } O_3}$ $CH_3-\overset{\overset{O}{\|}}{C}-$ $\xleftarrow{H_2O/H_3O^\oplus,\;Hg^{2+}}$ $H-C\equiv C-$

$\overset{\overset{O}{\|}}{-C}-Cl$ $\xrightarrow{(CH_3)_2Cd}$ $CH_3-\overset{\overset{O}{\|}}{C}-$ $\xleftarrow{H_2O}$ $CH_3-\overset{\overset{X}{|}}{\underset{\underset{X}{|}}{C}}-$

$CH_3-\overset{\overset{O}{\|}}{C}-$ $\xleftarrow{CrO_3, \text{ etc.}}$ CH_3-CH-
$\quad\quad\quad\quad\quad\quad\quad\quad\quad\quad\;\;|$
$\quad\quad\quad\quad\quad\quad\quad\quad\quad\;\;OH$

4. −CHO or −CH$_2$OH

$$\begin{array}{c} \text{−CH=C} \xrightarrow[\text{or}]{\text{KMnO}_4/\Delta} \\ \text{(1) O}_3 \text{ (2) H}_2\text{O}_2 \end{array} \xrightarrow{\text{CrO}_3, \text{ etc.}} \text{−COOH} \xleftarrow[\text{NaOH}]{\text{H}_2\text{O}} \left\{\begin{array}{l} \text{−COOR} \\ \text{−CO−X} \\ \text{−CONH}_2 \\ \text{−C≡N} \\ \text{−C−O−C−R} \\ \quad \parallel \quad \parallel \\ \quad \text{O} \quad \text{O} \end{array}\right.$$

$$\text{−MgX} \xrightarrow{\text{(1) CO}_2}_{\text{(2) H}_2\text{O}} \text{−COOH}$$

Ar−R $\xrightarrow[\Delta]{\text{KMnO}_4}$ Ar−COOH

5. −CONH$_2$ $\xrightarrow{\text{P}_2\text{O}_5}$ −C≡N $\xleftarrow{\text{NaCN}}$ −Br

6. CH−C
   |
   OH

$$\xrightarrow{\text{H}_2\text{SO}_4} \text{C=C} \xleftarrow{} \text{C−P}\phi_3^{\ominus\oplus} + \text{O=C}$$

−C≡C− $\xrightarrow[\text{or Na/NH}_3(\ell)]{\text{H}_2/\text{cat.}}$ C=C $\xleftarrow{\text{KOH}}$ CH−C with X or O−SO$_2$−R
                                                            |
                                                           CH−C

7. −C−Cl
   ‖
   O

$$\text{−C−Cl} \xrightarrow[\text{poison}]{\text{H}_2/\text{Pd/BaSO}_4} \text{−C−H} \xleftarrow{\text{(RO)}_3\text{AlH}^{\ominus} \text{Li}^{\oplus}} \text{−C−OR}$$
                                          ‖   ‖
                                          O   O

−Mg−X $\xrightarrow{\text{HC(OR)}_3}$ −C−H $\xleftarrow[\text{(2) Zn/H}_2\text{O}]{\text{(1) O}_3}$ C=C with H

−CH$_2$OH $\xrightarrow{\text{Cu}/\Delta}$

SYNTHESIS OF AROMATIC TARGET MOLECULES

3. Now let's discuss the synthesis of the simplest aromatic compounds, those with one group on a benzene ring. The most common starting material is benzene itself. The question to be answered in this problem is: What is the most efficient sequence of reactions for attaching group G to the ring? This is also the first step in solving more complex aromatic synthesis problems, and we consider it in detail here.

The "synthetic trees" in Tables 21.1 to 21.3 give you a simple way of choosing these sequences. All or practically all of the reactions in these tables should be familiar to you, although stringing them together in sequences may be new. You can use drill techniques from earlier units to fix these sequences in your mind. Note that some groups can be obtained in more than one way. This doesn't mean that you need know only one of the paths; in later, more complex problems, we need the flexibility of alternative routes to the same group. Ordinarily, the shortest route is the best.

Table 21.1 *Groups Introduced via Electrophilic Aromatic Substitution*

Groups introduced directly	Derived groups		
	1st generation	2nd generation	3rd generation

−H → {
−Tl(OCOCF$_3$)$_2$ → −I, −OH
−OH → −O−CO−R, −O−R
−NO$_2$ → −NH$_2$ → −NH−CO−R, −N$_2^{\oplus}$ → see Table 21.2
−SO$_3$H → −SO$_3$R
−Cl, −Br, −D → −MgX → −CO−R, −CHOH−R, −C(R)(R′)OH, −CH$_2$OH, −COOH
−MgX → −C(R)(R′)OH
−CO−R → −CHOH−R
−R → −COOH → −COOR, −CONH$_2$, −CO−X, −CH$_2$OH, −C≡N
−CO−X → −CO−R, −CHO
−COOH → −CH$_2$OH
−CH$_2$Cl → −CH$_2$OH
}

Table 21.2 Groups Introduced via Nucleophilic Aromatic Substitution on Diazonium Salts

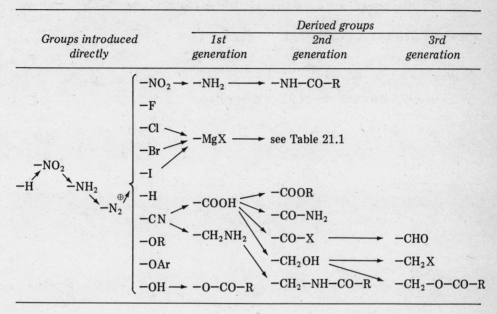

Table 21.3 Groups Introduced via Nucleophilic Aromatic Substitution on Activated Aryl Halides

	Groups introduced directly	Derived groups
Ar—H ⟶ Ar—X ⟶ (activating groups[a] must be present)	—NH$_2$	⟶ —NH—CO—R
	—NHR	⟶ —NH—CO—R′
	—NR$_2$	
	—SH	⟶ —S—CO—R
	—SR	
	—OH	
	—OR	
	—O—CO—R	

[a] At least one, and preferably two, of the following groups ortho or para to —X: —NO$_2$, —C≡N, —SO$_2$R, —CO—R. See Unit 16, frames 3, 6.

Example

Starting from benzene and any desired aliphatic or inorganic reagents, propose an efficient synthesis of Ph–CO_2H.

Solution:

The trees of Tables 21.1 and 21.2 give three paths:

$$\text{benzene} \longrightarrow \begin{cases} \text{Ph–X} \longrightarrow \text{Ph–MgX} \longrightarrow \text{Ph–COOH} \\ \text{Ph–R} \longrightarrow \text{Ph–COOH} \\ \text{Ph–NO}_2 \longrightarrow \text{Ph–NH}_2 \longrightarrow \text{Ph–N}_2^{\oplus} \longrightarrow \text{Ph–C}{\equiv}\text{N} \longrightarrow \text{Ph–COOH} \end{cases}$$

The middle path is shortest.

★ Exercises

Starting from benzene and any desired aliphatic and inorganic reagents, propose efficient syntheses of the following compounds:

1. Ph–NH–C(=O)–CH_2–CH_3
2. Ph–C(=O)–O–CH_2–CH(CH_3)(CH_3)
3. Ph–C(OH)(CH_3)–CH_2–CH_3
4. Ph–C(CH_3)=cyclopentane

- - - - - - - - - - - - - - -

1. –H $\xrightarrow{\text{HNO}_3 / \text{H}_2\text{SO}_4}$ –NO_2 $\xrightarrow[\text{aq. HCl}]{\text{Fe}}$ –NH_2 $\xrightarrow{\text{CH}_2\text{CH}_2\text{COCl}}$ –NH–CO–C_2H_5

2. –H $\xrightarrow[\text{AlCl}_3]{\text{RCl}}$ –R $\xrightarrow{\text{KMnO}_4}$ –COOH $\xrightarrow[\text{(2) }i\text{-BuOH}]{\text{(1) SOCl}_2}$ –COO–i-Bu

3. –H $\xrightarrow[\text{AlCl}_3]{\text{CH}_3\text{COCl}}$ –CO–CH_3 $\xrightarrow[\text{(2) H}_2\text{O}]{\text{(1) CH}_3\text{CH}_2\text{MgX}}$ –C(OH)(CH_3)–CH_2CH_3

4. 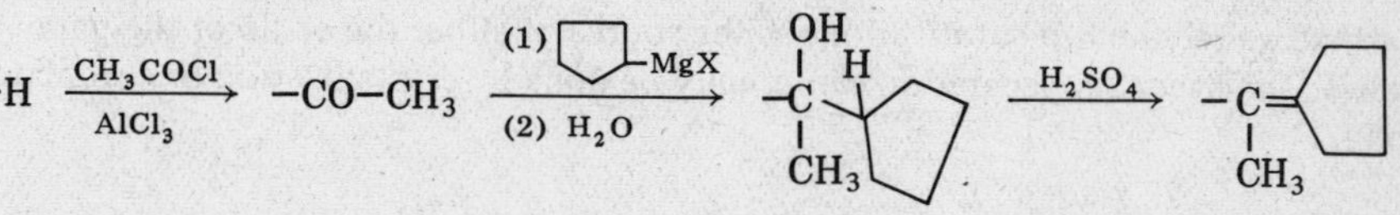 $-H \xrightarrow{CH_3COCl, AlCl_3} -CO-CH_3 \xrightarrow[(2)\ H_2O]{(1)\ \text{cyclopentyl-MgX}} \underset{CH_3}{\overset{OH\ H}{-C-}}\text{cyclopentyl} \xrightarrow{H_2SO_4} -\underset{CH_3}{C}=\text{cyclopentene}$

4. Planning the synthesis of a TM that has two groups on a benzene ring starts out the same way. First select the most efficient sequence available for putting in each group. The only remaining problem then is whether the required relationship of the two groups on the ring (*o*, *m*, or *p*) can be achieved.

If the two can be put on the ring by electrophilic substitution (alone or followed by chemical transformation), then the problem of position can always be solved simply in the following two cases:

- If the two groups are ortho or para to one another and at least one of them is an *o/p* director, or
- If the two groups are meta to one another and at least one of them is an *m* director.

In these cases choice of the correct order of operations solves the problem.

Examples

Starting from benzene and any desired aliphatic or inorganic reagents, propose an efficient synthesis of each compound.

1. 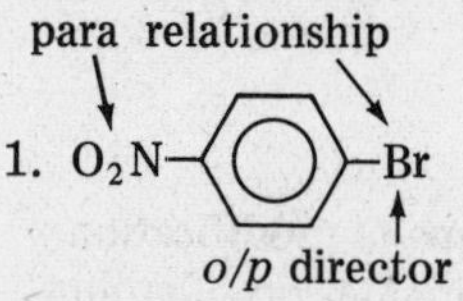 para relationship: $O_2N-\text{C}_6H_4-Br$ (*o/p* director)

2. 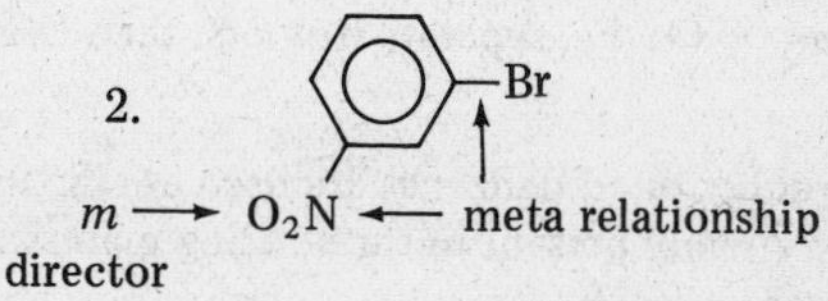 $O_2N-\text{C}_6H_4-Br$ meta relationship (*m* director)

Solutions:

In both examples the groups are directly introducible by electrophilic aromatic substitution. This can be done in two ways:

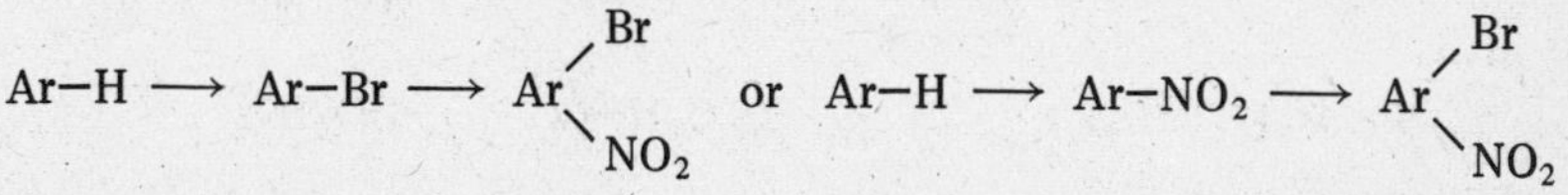

$Ar-H \longrightarrow Ar-Br \longrightarrow Ar\begin{smallmatrix}Br\\NO_2\end{smallmatrix}$ or $Ar-H \longrightarrow Ar-NO_2 \longrightarrow Ar\begin{smallmatrix}Br\\NO_2\end{smallmatrix}$

Since $-Br$ orients *o/p* and $-NO_2$ orients *m*, the two orders of operations give different products:

1. 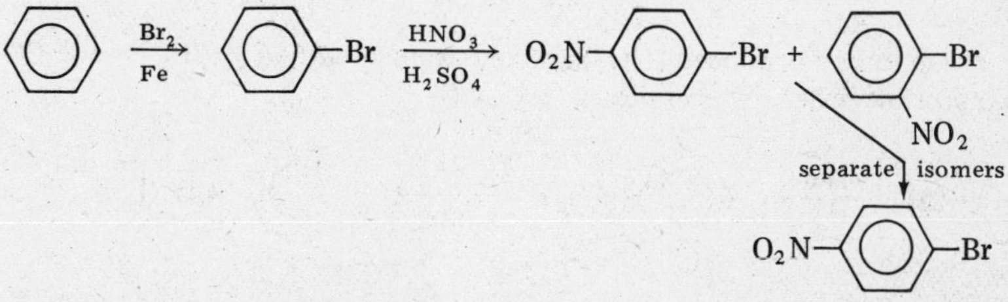 $C_6H_6 \xrightarrow{Br_2, Fe} C_6H_5-Br \xrightarrow{HNO_3, H_2SO_4} O_2N-C_6H_4-Br + C_6H_4(Br)(NO_2)$

separate isomers → $O_2N-C_6H_4-Br$

2. $C_6H_6 \xrightarrow{HNO_3, H_2SO_4} C_6H_5-NO_2 \xrightarrow{Fe, Br_2} C_6H_4(NO_2)-Br$

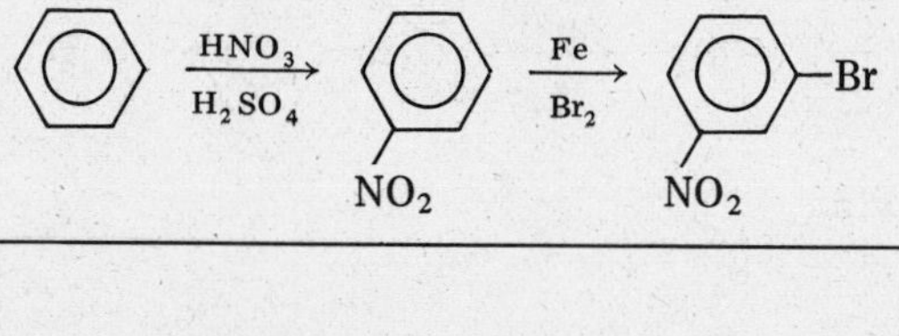

Note that we rely on *o/p* substitutions as the source of either the ortho or the para product. The necessary separation can usually be done by crystallization or chromatography.

★ *Exercises*

Starting from benzene, phenol, and any desired aliphatic and inorganic compounds, propose an efficient synthesis of the following compounds.

1. 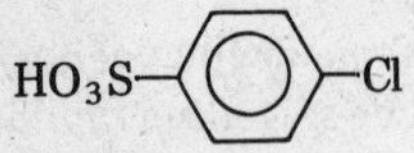 $HO_3S-\bigcirc-Cl$

2. $(CH_3)_2CH-\bigcirc-NO_2$

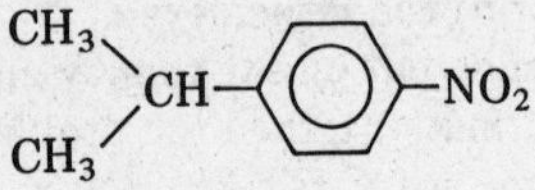

3. 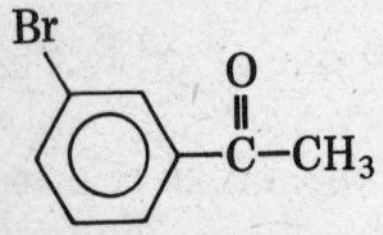 $Br-\bigcirc-C(=O)-CH_3$ (Br ortho to acetyl)

4. 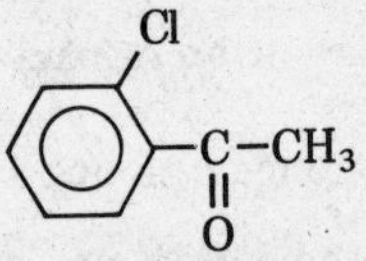 $Cl-\bigcirc-C(=O)-CH_3$ (Cl ortho to acetyl)

- - - - - - - - - -

1. benzene + Cl_2/Fe, separate isomers, then H_2SO_4
2. benzene + $(CH_3)_2CHCl/AlCl_3$, separate isomers, then HNO_3/H_2SO_4
3. benzene + $CH_3COCl/AlCl_3$, then Br_2/Fe
4. benzene + Cl_2/Fe, separate isomers, then $CH_3COCl/AlCl_3$

5. When substituted benzenes are used as starting materials, the time of modification of a group already present in the starting material may govern the orientation of groups in the target molecule.

Example

Synthesize the following compounds from toluene plus any desired aliphatic and inorganic reagents.

1. 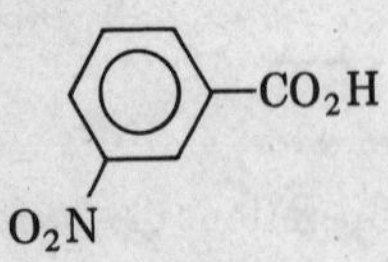 $\bigcirc-CO_2H$ with O_2N meta

2. 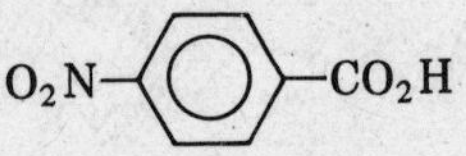 $O_2N-\bigcirc-CO_2H$

Solutions:

With one group to be introduced (NO_2) and one to be modified ($-CH_3 \longrightarrow -CO_2H$), there are two possible orders of operations:

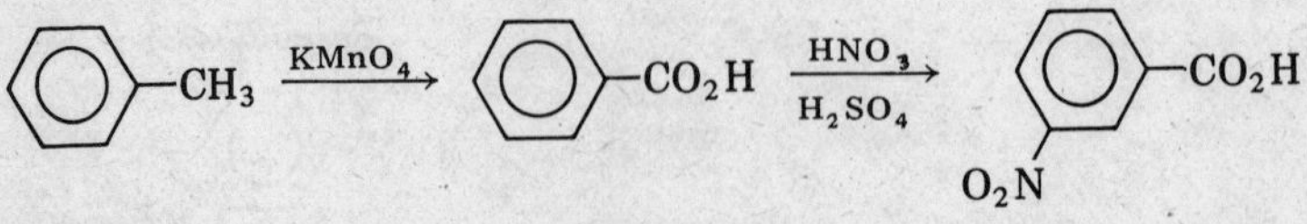

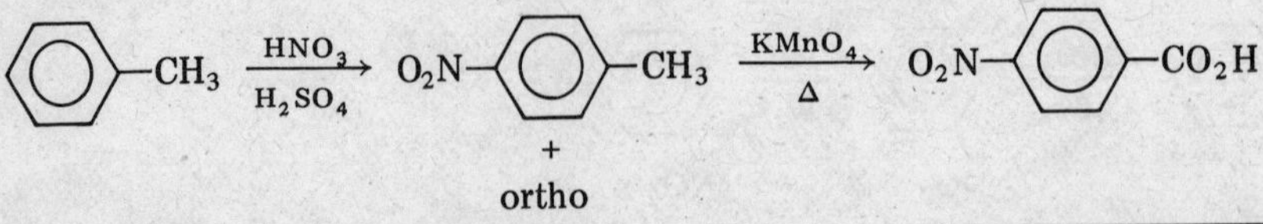

$\bigcirc-CH_3 \xrightarrow{KMnO_4} \bigcirc-CO_2H \xrightarrow[H_2SO_4]{HNO_3} O_2N\text{-}m\text{-}\bigcirc-CO_2H$

$\bigcirc-CH_3 \xrightarrow[H_2SO_4]{HNO_3} O_2N-\bigcirc-CH_3 \xrightarrow[\Delta]{KMnO_4} O_2N-\bigcirc-CO_2H$
                                               + ortho

The order of operations determines whether the substitution is done while the substituent present directs *o/p*, or whether it is done after the substituent present has been turned into a *m* director. The top path solves example 1. The bottom path solves example 2.

★ *Exercises*

Using any desired aliphatic and inorganic reagents, carry out the indicated conversions.

1. 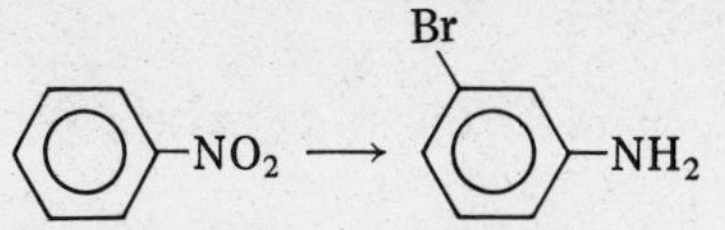 Ph–NO_2 → (4-Br)Ph–NH_2

2. Ph–NO_2 → Br–Ph–NH_2 (para)

3. Ph–CO–CH_3 → HO_3S–Ph–CH_2–CH_3 (para)

4. Ph–CO–CH_3 → Ph–CH_2CH_3 with HO_3S (meta)

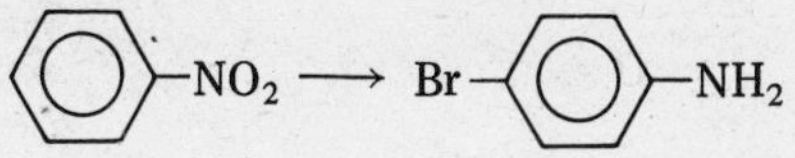

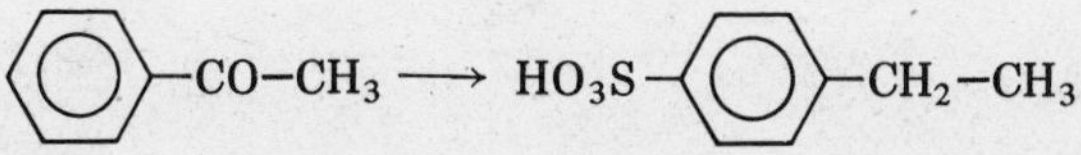

― ― ― ― ― ― ―

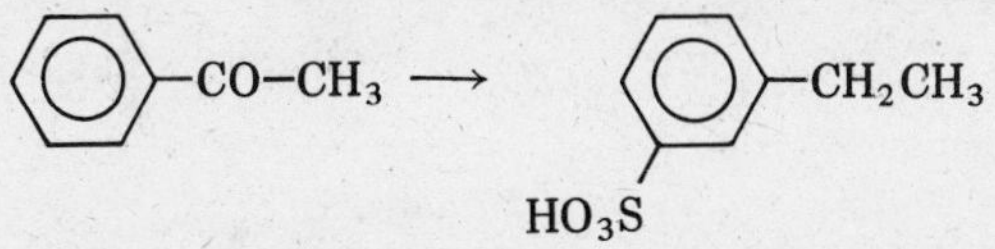

1. Br_2/Fe, then Fe/aq. HCl

2. Fe/aq. HCl, then Br_2/Fe, separate isomers, or better: Fe/aq. HCl, then $(CH_3CO)_2O$, then Br_2/Fe, then aq. NaOH/Δ, separate isomers (this avoids polybromination)

3. $N_2H_4/OH^\ominus/\Delta$ or Zn·Hg/conc. HCl, then H_2SO_4, separate isomers

4. H_2SO_4, then $N_2H_4/OH^\ominus/\Delta$

6. Before we go on to the next, slightly more complex case, you should note that there are two additional considerations that may decide the order of introduction of two groups on a benzene ring.

First, if you have a choice, put in the most strongly activating group (or the most weakly deactivating group) first. This improves yields by avoiding running reactions on less active benzene rings.

Example

Propose an efficient synthesis of this compound starting from benzene and using any desired reagents: Br–Ph–CH_2–CH_3.

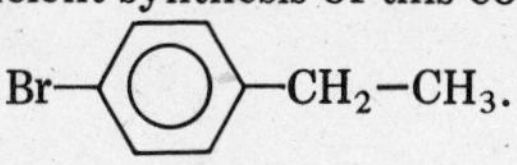

Solution:

benzene →[CH_3CH_2Cl, $AlCl_3$] Ph–CH_2CH_3 →[Br_2/Fe] Br–Ph–CH_2CH_3

benzene →[Br_2/Fe] Br–Ph →[CH_3CH_2Cl, $AlCl_3$] Br–Ph–CH_2CH_3

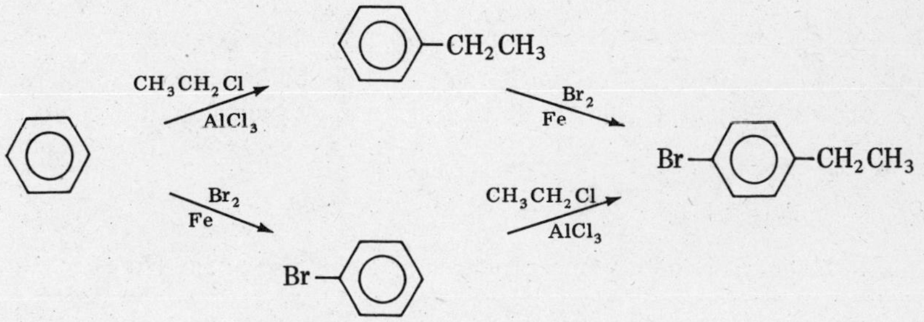

The top path is the better solution because the second substitution is performed on an activated rather than a deactivated ring.

Second, you must be alert for cases in which the order of steps is dictated by restrictions inherent in the reactions used. The main restrictions are summarized in Tables 16.1 to 16.4.

Examples

Propose efficient syntheses of the following compounds starting from benzene, toluene, phenol, and any desired reagents.

1. 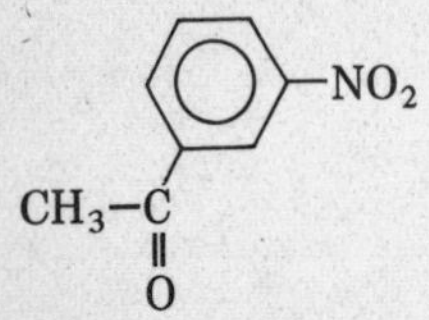 $CH_3-C(=O)-C_6H_4-NO_2$ (meta)

2. $(CH_3)_2N-C_6H_4-N=N-C_6H_4-NO_2$

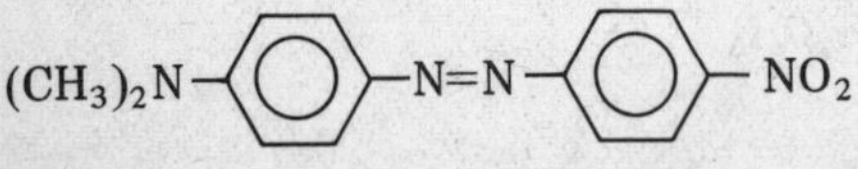

Solutions:

1. 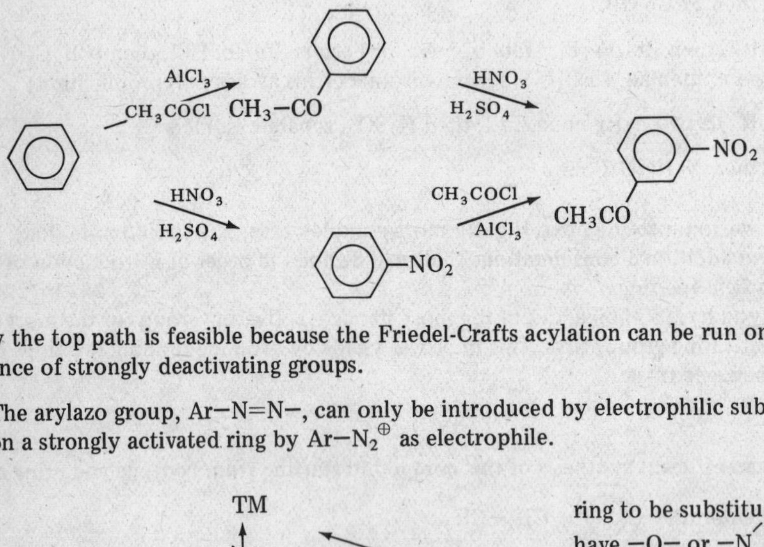 Benzene → (AlCl$_3$, CH$_3$COCl) → CH$_3$–CO–C$_6$H$_5$ → (HNO$_3$, H$_2$SO$_4$) → CH$_3$CO–C$_6$H$_4$–NO$_2$

   Benzene → (HNO$_3$, H$_2$SO$_4$) → C$_6$H$_5$–NO$_2$ → (CH$_3$COCl, AlCl$_3$) → CH$_3$CO–...

Only the top path is feasible because the Friedel-Crafts acylation can be run only in the absence of strongly deactivating groups.

2. The arylazo group, Ar—N=N—, can only be introduced by electrophilic substitution on a strongly activated ring by Ar—N$_2^{\oplus}$ as electrophile.

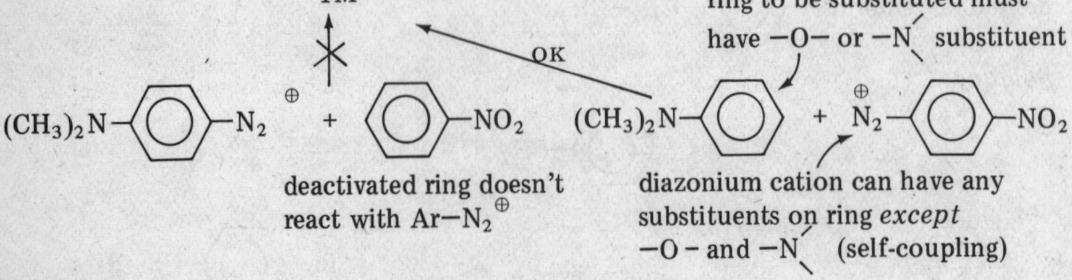

   TM ✗ $(CH_3)_2N-C_6H_4-N_2^{\oplus}$ + $C_6H_5-NO_2$ — deactivated ring doesn't react with Ar—N$_2^{\oplus}$

   OK $(CH_3)_2N-C_6H_5$ + $N_2^{\oplus}-C_6H_4-NO_2$ — ring to be substituted must have —O— or —N substituent; diazonium cation can have any substituents on ring *except* —O— and —N (self-coupling)

★ *Exercises*

1. Propose efficient reaction sequences for accomplishing these conversions:

(a) 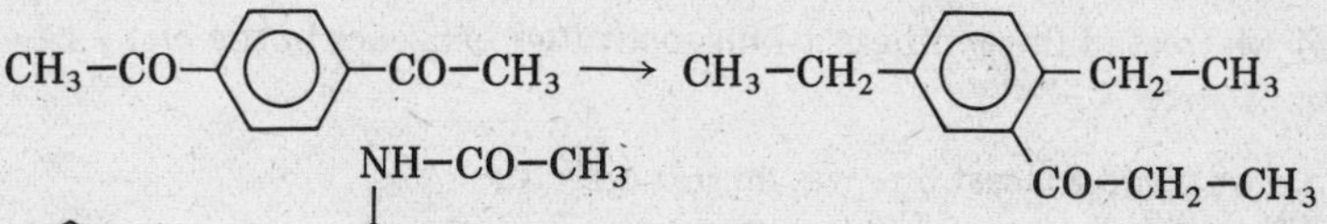 CH₃—CO—⟨C₆H₄⟩—CO—CH₃ → CH₃—CH₂—⟨C₆H₃(CO—CH₂—CH₃)⟩—CH₂—CH₃

(b) 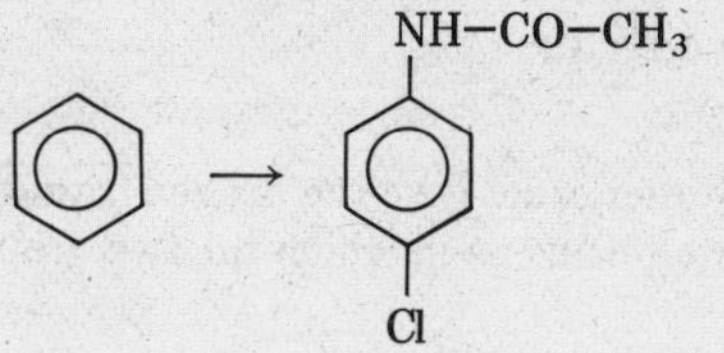 ⟨C₆H₆⟩ → ⟨C₆H₄(NH—CO—CH₃)(Cl)⟩

2. What pair of aromatic reactants (diazonium salt + substituted benzene) reacts to give each of these azo dyes?

(a) 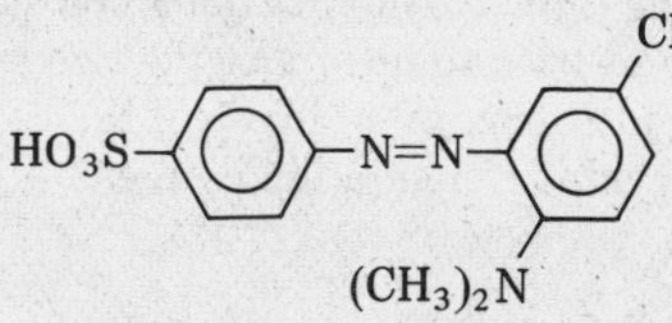 HO₃S—⟨C₆H₄⟩—N=N—⟨C₆H₃(CH₃)⟩—N(CH₃)₂

(b) 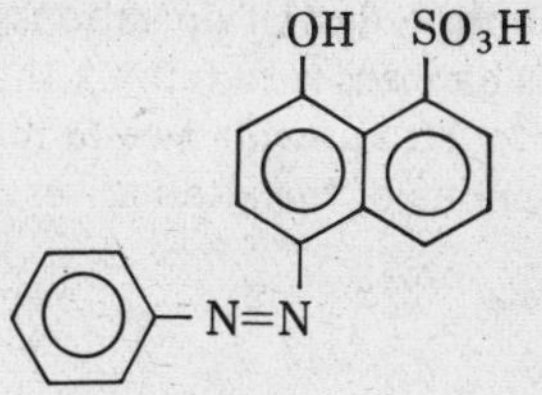 ⟨C₆H₅⟩—N=N—⟨naphthalene(OH)(SO₃H)⟩

- - - - - - - - - -

1. (a) Try to put —CO—CH₂CH₃ on by Friedel-Crafts acylation. Then it must go on *after* CH₃—CO— → CH₃—CH₂— because the Friedel-Crafts acylation does not work if deactivating groups are present. So:

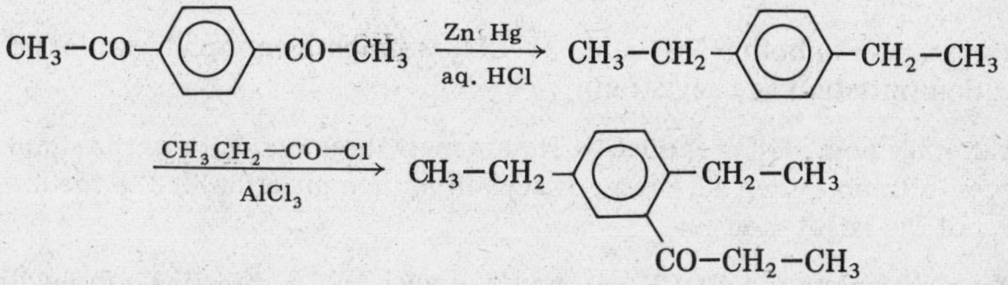

CH₃—CO—⟨C₆H₄⟩—CO—CH₃ $\xrightarrow{\text{Zn Hg, aq. HCl}}$ CH₃—CH₂—⟨C₆H₄⟩—CH₂—CH₃

$\xrightarrow{\text{CH}_3\text{CH}_2\text{—CO—Cl, AlCl}_3}$ CH₃—CH₂—⟨C₆H₃(CO—CH₂—CH₃)⟩—CH₂—CH₃

(b) Since both —Cl and —NH—CO—CH₃ orient *o/p*, either can be introduced first as far as orientation is concerned. However, the yield will be best if the activator (—NH—CO—CH₃) is introduced before the deactivator (—Cl). So:

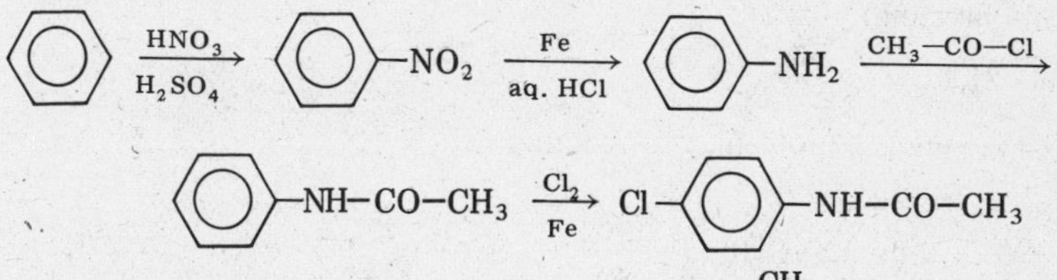

⟨C₆H₆⟩ $\xrightarrow{\text{HNO}_3, \text{H}_2\text{SO}_4}$ ⟨C₆H₅⟩—NO₂ $\xrightarrow{\text{Fe, aq. HCl}}$ ⟨C₆H₅⟩—NH₂ $\xrightarrow{\text{CH}_3\text{—CO—Cl}}$

⟨C₆H₅⟩—NH—CO—CH₃ $\xrightarrow{\text{Cl}_2, \text{Fe}}$ Cl—⟨C₆H₄⟩—NH—CO—CH₃

2. (a) from HO₃S—⟨C₆H₄⟩—N≡N⁺ HSO₄⁻ + ⟨C₆H₄(CH₃)⟩—N(CH₃)₂

required activator

(b) from ⟨C₆H₅⟩—N≡N⁺ HSO₄⁻ + ⟨naphthalene(OH)(SO₃H)⟩

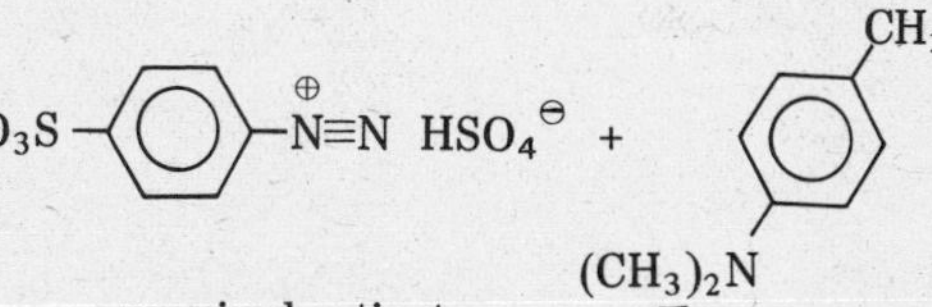

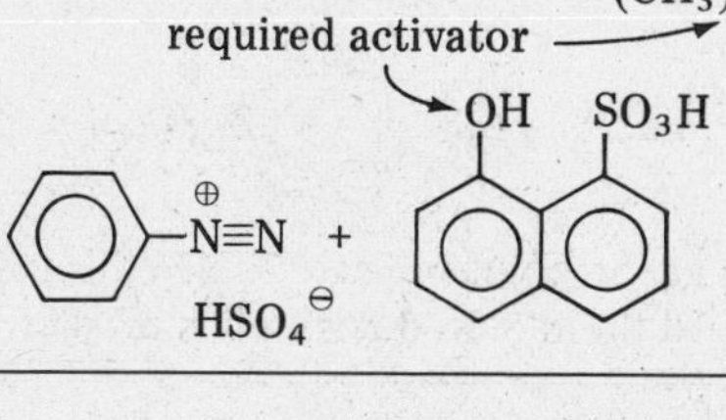

7. In frame 4 we treated the synthesis of disubstituted benzenes in the lucky cases where the groups were:

- Situated *o* or *p* and at least one was an *o/p* director,
- Situated *m* and at least one was a *m* director,

so that introducing the groups in the properly chosen order gave the desired isomer.

Now suppose that we must make a disubstituted benzene in which the two groups are:

- Situated *o* or *p* and neither is an *o/p* director.
- Situated *m* and neither is a *m* director.

In this case both orders of operations give the wrong isomer. The first thing you should do when this happens is to rethink the reactions to be used to introduce the two groups. Tables 21.1 to 21.3 provide two or more paths for introduction of many of the common groups. Often one of the alternatives allows you to get the orientation correct.

Example

Synthesize $O_2N-\langle\bigcirc\rangle-CO-CH_3$ from benzene plus aliphatic and inorganic reagents.

Solution:

1. The shortest path to both $-NO_2$ and $-COCH_3$ is simple electrophilic aromatic substitution (nitration and acylation).

2. However, since both $-NO_2$ and $-COCH_3$ are meta directors, both of the simple sequences (nitration, then acylation; or acylation, then nitration) led to the meta isomer, not the target molecule.

3. Look for a new route to $-COCH_3$. It should involve an *o/p*-directing intermediate at some stage. One possibility is the route

$$-H \longrightarrow -R \longrightarrow -CO_2H \longrightarrow -COCl \longrightarrow -CO-R$$
$$\uparrow$$
$$(o/p \text{ directing})$$

from Table 21.1.

4. Applied to the present problem, this becomes

$\bigcirc \xrightarrow[AlCl_3]{C_2H_5Cl} \bigcirc-C_2H_5 \xrightarrow[H_2SO_4]{HNO_3} O_2N-\bigcirc-C_2H_5 \xrightarrow[\Delta]{KMnO_4}$

$O_2N-\bigcirc-CO_2H$

$O_2N-\bigcirc-\overset{O}{\underset{\|}{C}}-CH_3 \xleftarrow{(CH_3)_2Cd} O_2N-\bigcirc-CO-Cl \xleftarrow{SOCl_2}$

The nitration can only be carried out at the point indicated since this is the only point at which a group directing toward the required position is present.

SYNTHETIC PROBLEMS

★ *Exercises*

Synthesize the following compounds from benzene + aliphatics and inorganics.

1. 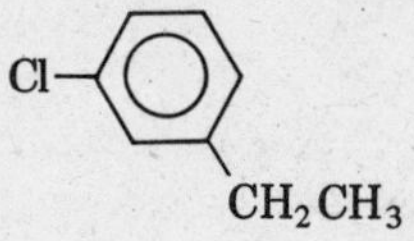 Cl–⟨⟩–CH$_2$CH$_3$

2. O$_2$N–⟨⟩–NO$_2$

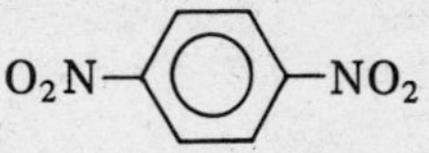

- - - - - - - - - -

1. 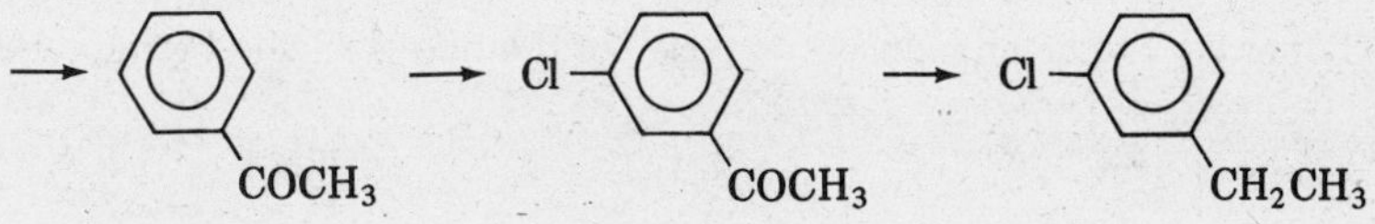 → ⟨⟩–COCH$_3$ → Cl–⟨⟩–COCH$_3$ → Cl–⟨⟩–CH$_2$CH$_3$

2. → ⟨⟩–NO$_2$ → ⟨⟩–NH$_2$ → ⟨⟩–NH–COCH$_3$ →

O$_2$N–⟨⟩–NH–COCH$_3$ → O$_2$N–⟨⟩–NH$_2$ → O$_2$N–⟨⟩–N$_2^{\oplus}$ → TM

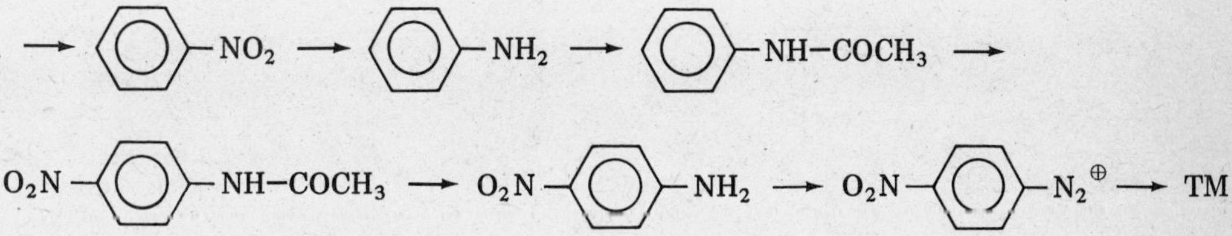

8. In many cases the method of frame 7 fails—no reaction sequence can be found that gives the desired isomer. However, the problems can still be solved. This is accomplished by introducing a removable, orientation-controlling or blocking group. The method is described in this and the next frame.

To introduce a group G_2 meta to some o/p directing group, G_1, we proceed as follows:
- If either G_1 or G_2 can be introduced via replacement of $-N_2^{\oplus}$ (reactions of Table 21.2) and if neither G_1 nor G_2 is an alkyl group, use the following scheme:

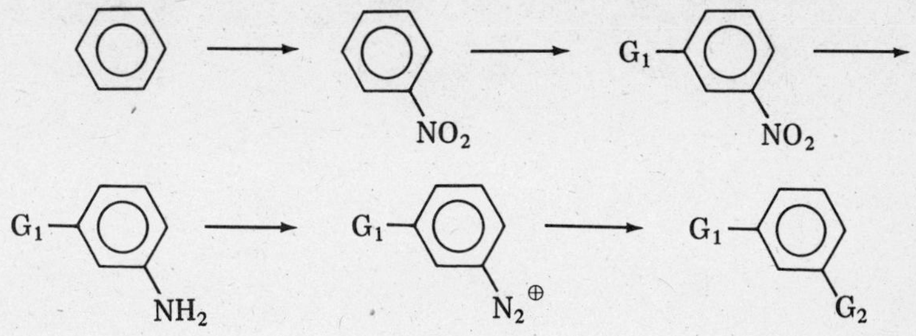

⟨⟩ → ⟨⟩–NO$_2$ → G_1–⟨⟩–NO$_2$ →

G_1–⟨⟩–NH$_2$ → G_1–⟨⟩–N$_2^{\oplus}$ → G_1–⟨⟩–G_2

Example

Synthesize the following compound from benzene or toluene and inorganic reagents.

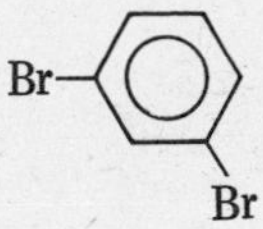

Br–⟨⟩–Br

Solution:

(a) Two o/p directors situated meta.

(b) Both introducible by $-N_2^{\oplus}$ and neither = alkyl.

So:

- If neither G_1 nor G_2 can be introduced via $-N_2^{\oplus}$ or if either is an alkyl group, use this scheme:

temporary $-NH_2$ or $-NH-CO-CH_3$ forces substitution ortho to itself [diagram: benzene with G₁ and two G₂ groups (meta variation) ← benzene with G₁, G₂, NH₂ ← benzene with G₁, G₂, N₂⁺ → benzene with G₁, G₂ (removal of temporary group)]

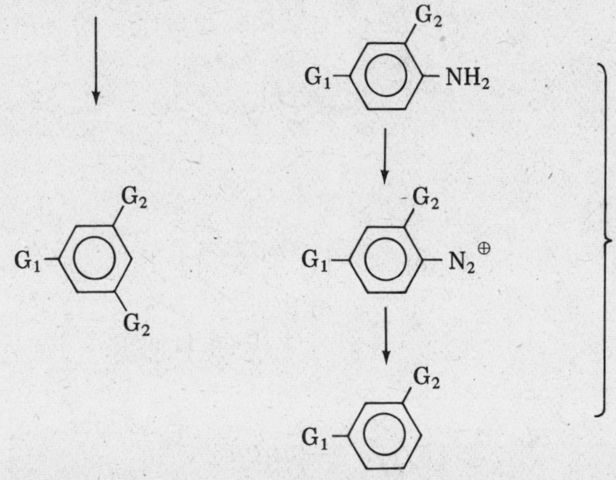

Notice that the left-hand variation allows you to put in *two* substituents meta to G₁.

Example

Synthesize this compound from benzene or toluene and inorganic reagents.

[structure: benzene with Br and CH₃ meta]

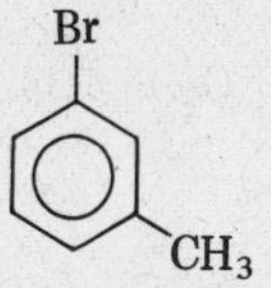

Solution:

(a) Two o/p directors, situated meta.

(b) One = alkyl.

So:

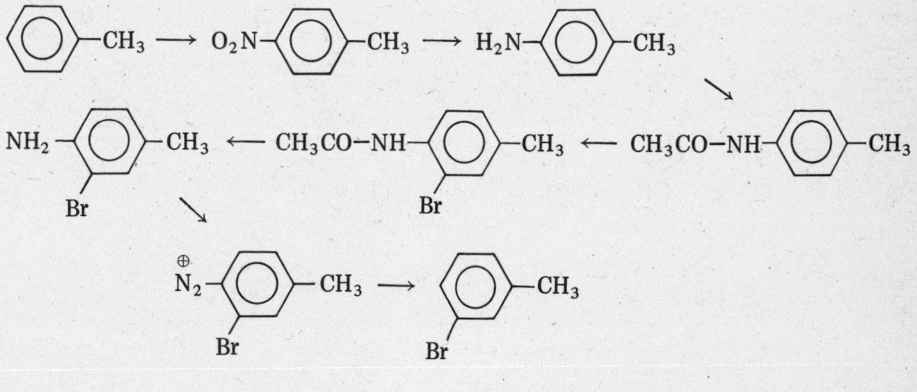

PhCH₃ → O₂N–C₆H₄–CH₃ → H₂N–C₆H₄–CH₃

NH₂–C₆H₃(Br)–CH₃ ← CH₃CO–NH–C₆H₃(Br)–CH₃ ← CH₃CO–NH–C₆H₄–CH₃

↘

N₂⁺–C₆H₃(Br)–CH₃ → C₆H₄(Br)–CH₃

★ *Exercises*

1. Supply the reagents required to accomplish the conversion in the examples of this frame.

2. Synthesize the following compounds from benzene or toluene, aliphatic and inorganic reagents.

(a) 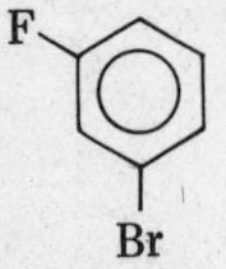 [3-bromofluorobenzene: F and Br meta]

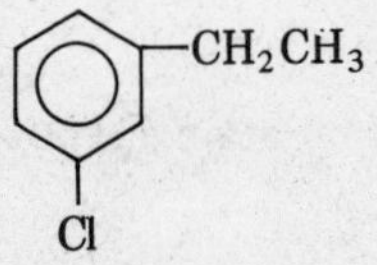

(b) 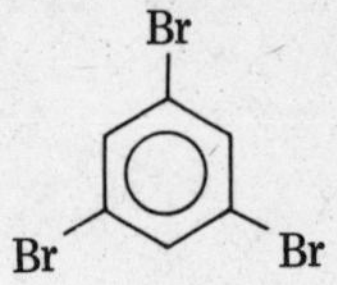 [1,3,5-tribromobenzene]

(c) 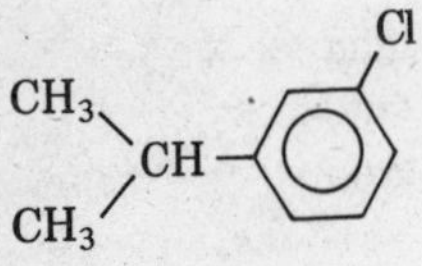 [1-chloro-3-isopropylbenzene]

(d) 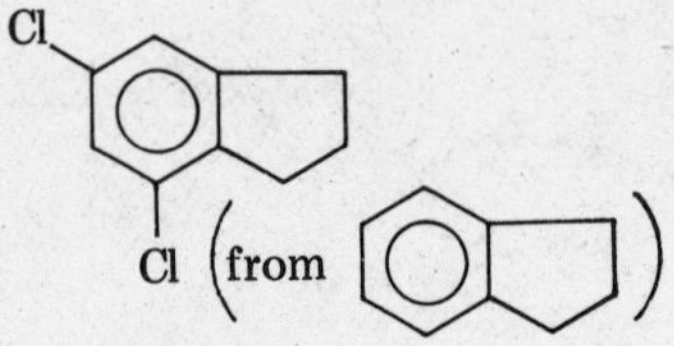 [4,7-dichloroindane (from indane)]

(e) 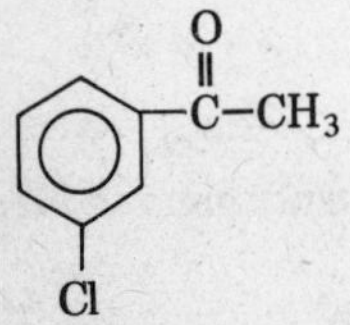 [3'-chloroacetophenone]

(f) [1-chloro-3-ethylbenzene]

- - - - - - - - - -

1. Br—⟨⟩—Br (1,3-dibromobenzene) : HNO_3/H_2SO_4; then Br_2/Fe; then Fe/aq. HCl; then HNO_2 in cold, dil. aq. H_2SO_4; then CuBr.

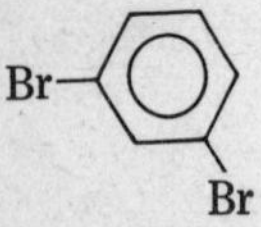

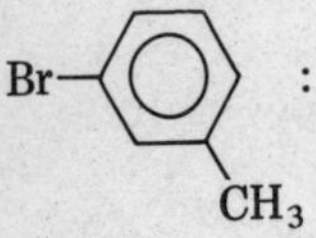

   Br—⟨⟩—CH₃ (3-bromotoluene) : HNO_3/H_2SO_4; then Fe/aq. HCl; then CH_3—CO—Cl; then Br_2/Fe; then $H_2O/NaOH$/heat; then HNO_2 in cold, dil. aq. H_2SO_4; then H_3PO_2.

2. (a) 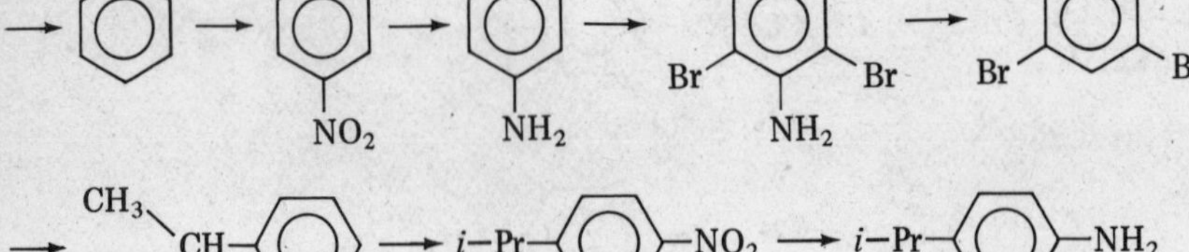 → [NO₂-benzene] → [NO₂, Br (meta)] → [NH₂, Br] → [N₂⁺, Br] → [F, Br]

   (b) 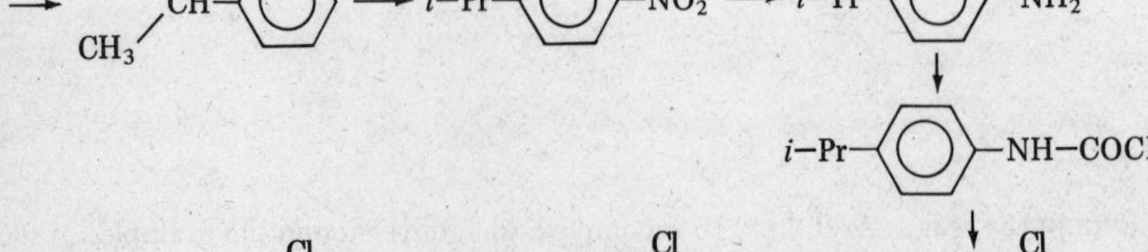 → [Br-benzene] → [Br, NO₂ (para)] → [Br, NH₂ (para)] → [Br, NH₂ with 2 Br ortho to NH₂] → [1,3,5-tribromobenzene]

   (c) 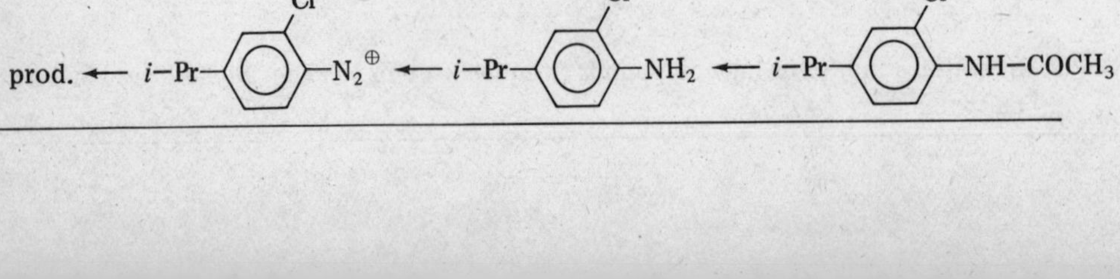 → [isopropylbenzene, $(CH_3)_2CH$—⟨⟩] → i-Pr—⟨⟩—NO₂ → i-Pr—⟨⟩—NH₂

   ↓

   i-Pr—⟨⟩—NH—COCH₃

   ↓

   i-Pr—⟨⟩(Cl)—NH—COCH₃ → i-Pr—⟨⟩(Cl)—NH₂ → i-Pr—⟨⟩(Cl)—N₂⁺ → prod.

(d) 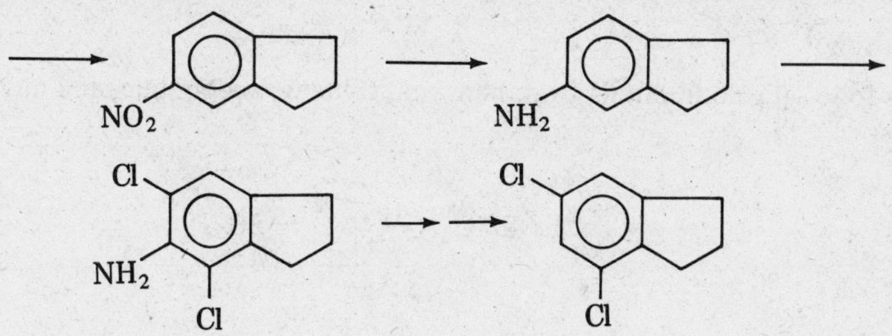 → [benzene ring fused to cyclopentane with NO₂] → [same with NH₂] →

[Cl, NH₂, Cl substituted indane] → → [Cl, Cl substituted indane]

(e) 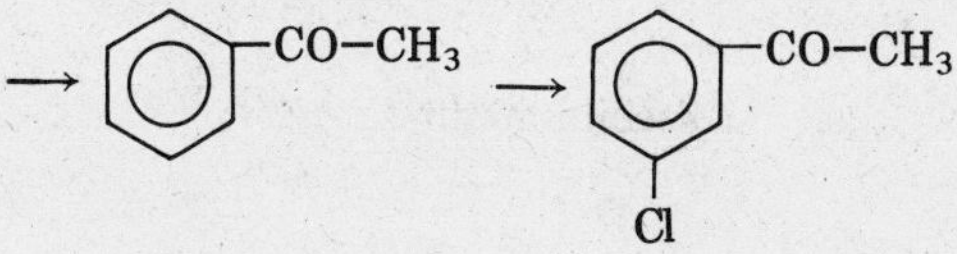 → [C₆H₅–CO–CH₃] → [Cl-substituted C₆H₄–CO–CH₃]   one meta director present, ∴ simple two-step

(f) 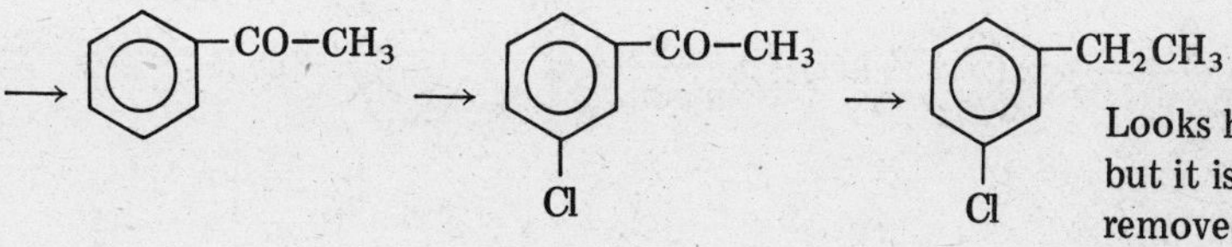 → [C₆H₅–CO–CH₃] → [Cl-C₆H₄–CO–CH₃] → [Cl-C₆H₄–CH₂CH₃]   Looks hard, like (c), but it is just once removed from (e), so you need not introduce –NH₂.

9. To introduce a group ortho or para to a meta-directing group, use one of the following schemes:

- If one of the groups (G_2) can be introduced via $-N_2^\oplus$, do it this way:

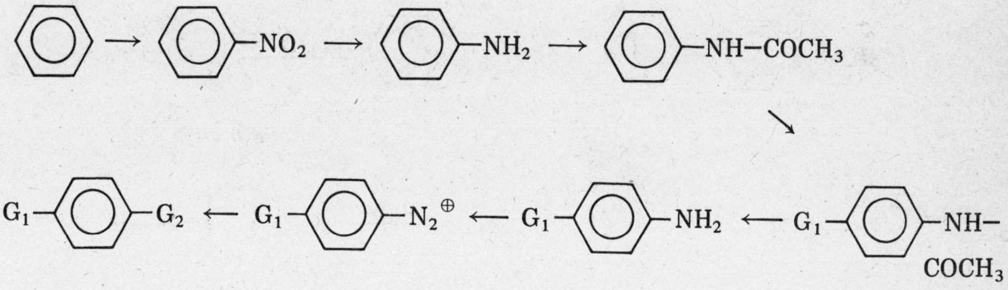

[benzene] → [Ph–NO₂] → [Ph–NH₂] → [Ph–NH–COCH₃]

[G_1–Ph–G_2] ← [G_1–Ph–$N_2^\oplus$] ← [G_1–Ph–NH₂] ← [G_1–Ph–NH–COCH₃]

- If both G_1 and G_2 can only be put in via electrophilic substitution, do it this way:

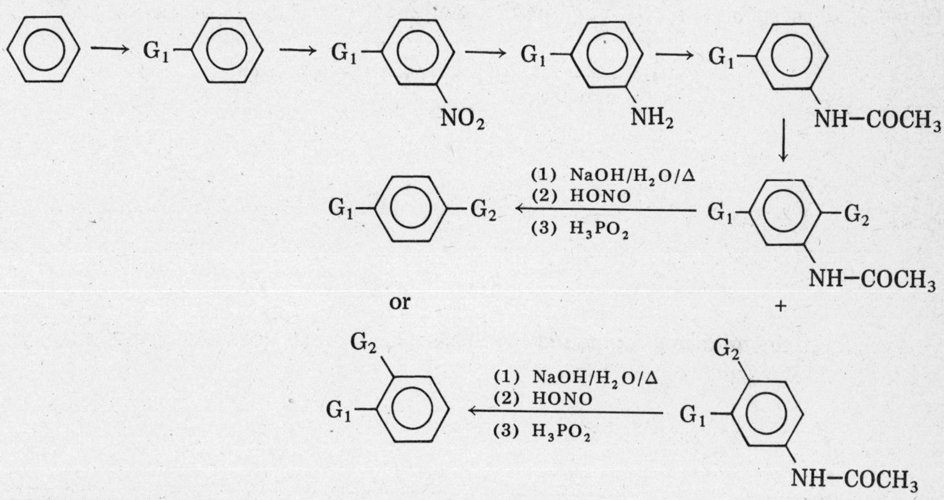

[benzene] → [G_1–Ph] → [G_1–Ph–NO₂] → [G_1–Ph–NH₂] → [G_1–Ph–NH–COCH₃]

↓

[G_1–Ph–G_2] ←(1) NaOH/H₂O/Δ (2) HONO (3) H₃PO₂— [G_1–Ph(G_2)–NH–COCH₃]

or   +

[G_2, G_1 on Ph] ←(1) NaOH/H₂O/Δ (2) HONO (3) H₃PO₂— [G_2, G_1 on Ph with NH–COCH₃]

Examples

Synthesize the following compounds from benzene, toluene, acetanilide, and any inorganic compounds.

1. benzene-1,2-disulfonic acid (ortho -SO$_3$H, -SO$_3$H)

2. 4-nitrobenzenesulfonic acid (O$_2$N— —SO$_3$H)

Solutions:

1. (a) Neither group accessible via -N$_2^\oplus$; use the second method.

(b) Benzene $\xrightarrow{H_2SO_4}$ C$_6$H$_5$-SO$_3$H $\xrightarrow{HNO_3 / H_2SO_4}$ m-O$_2$N-C$_6$H$_4$-SO$_3$H $\xrightarrow{HCl / Fe}$ m-H$_2$N-C$_6$H$_4$-SO$_3$H $\xleftarrow{(CH_3CO)_2O}$ m-CH$_3$CO-NH-C$_6$H$_4$-SO$_3$H

$\xrightarrow{SO_3/\text{pyridine}}$ CH$_3$CO-NH-C$_6$H$_3$(HO$_3$S)-SO$_3$H + CH$_3$CO-NH-C$_6$H$_3$(SO$_3$H)-SO$_3$H

$\xrightarrow{\text{aq. NaOH}/\Delta}$ NH$_2$-C$_6$H$_3$(SO$_3$H)-SO$_3$H $\xrightarrow{HNO_2}$ N$_2^\oplus$-C$_6$H$_3$(SO$_3$H)-SO$_3$H $\xrightarrow{H_3PO_2}$ C$_6$H$_4$(SO$_3$H)-SO$_3$H

2. (a) NO$_2$ accessible via -N$_2^\oplus$; use the first method.

(b) C$_6$H$_5$-NH-COCH$_3$ $\xrightarrow{SO_3/\text{pyridine}}$ HO$_3$S-C$_6$H$_4$-NH-COCH$_3$ $\xrightarrow{H_3O^\oplus / H_2O/\Delta}$ $^\ominus$O$_3$S-C$_6$H$_4$-NH$_3^\oplus$ $\xrightarrow{HNO_2}$ $^\ominus$O$_3$S-C$_6$H$_4$-N$_2^\oplus$ $\xrightarrow{NO_2^\ominus / Cu}$ $^\ominus$O$_3$S-C$_6$H$_4$-NO$_2$ → HSO$_3$-C$_6$H$_4$-NO$_2$

★ Exercises

Synthesize the following compounds from benzene, toluene, acetanilide, and any aliphatic and inorganic compounds.

1. N≡C— —C(CH$_3$)$_3$

2. 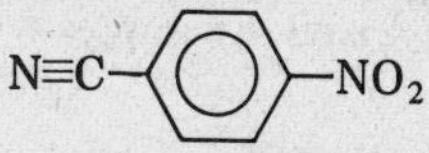 N≡C—⟨C6H4⟩—NO2

3. 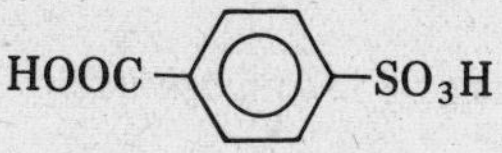 HOOC—⟨C6H4⟩—SO3H

4. 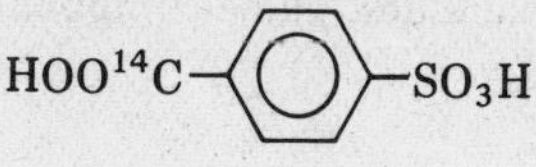 HOO¹⁴C—⟨C6H4⟩—SO3H (from labeled benzoic acid, ⟨C6H5⟩—¹⁴COOH)

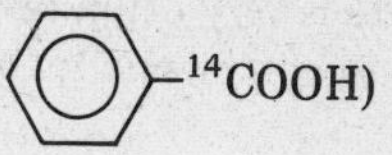

5. 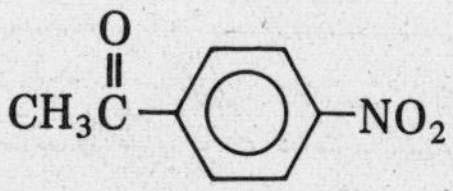 CH3C(=O)—⟨C6H4⟩—NO2

─ ─ ─ ─ ─ ─ ─ ─ ─ ─ ─ ─ ─ ─ ─ ─ ─

1. 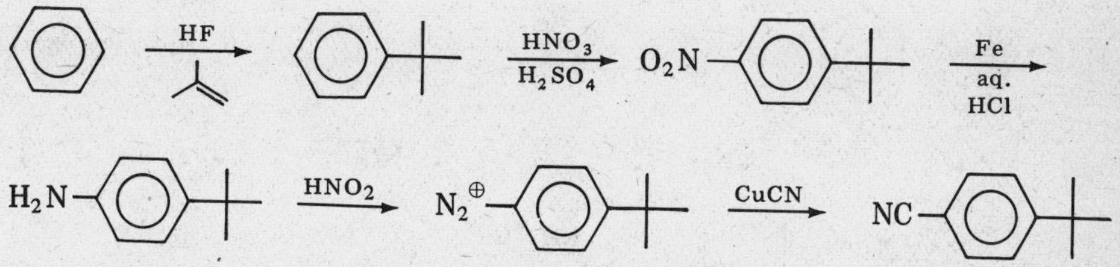 ⟨C6H6⟩ $\xrightarrow{\text{HF, }CH_2=C(CH_3)_2}$ ⟨C6H5⟩—C(CH3)3 $\xrightarrow{HNO_3 / H_2SO_4}$ O2N—⟨C6H4⟩—C(CH3)3 $\xrightarrow[\text{aq. HCl}]{Fe}$

   H2N—⟨C6H4⟩—C(CH3)3 $\xrightarrow{HNO_2}$ N2⊕—⟨C6H4⟩—C(CH3)3 $\xrightarrow{CuCN}$ NC—⟨C6H4⟩—C(CH3)3

2. ⟨C6H5⟩—NH—COCH3 $\xrightarrow{HNO_3 / H_2SO_4}$ NO2—⟨C6H4⟩—NH—COCH3 $\xrightarrow{NaOH}$

   O2N—⟨C6H4⟩—C≡N $\xleftarrow{CuCN}$ O2N—⟨C6H4⟩—N2⊕ $\xleftarrow{HNO_2}$ O2N—⟨C6H4⟩—NH2

3. CH3—⟨C6H5⟩ $\xrightarrow{H_2SO_4}$ CH3—⟨C6H4⟩—SO3H $\xrightarrow{KMnO_4}$ HOOC—⟨C6H4⟩—SO3H

4. HOO¹⁴C—⟨C6H5⟩ $\xrightarrow{HNO_3 / H_2SO_4}$ HOO¹⁴C—⟨C6H4⟩—NO2 (meta) $\xrightarrow{Fe / HCl}$ HOO¹⁴C—⟨C6H4⟩—NH2 (meta)

   $\downarrow (CH_3CO)_2O$

   HOO¹⁴C—⟨C6H3(SO3H)⟩—NHCOCH3 + HOO¹⁴C—⟨C6H3⟩(SO3H)(NHCOCH3) $\xleftarrow{H_2SO_4}$ HOO¹⁴C—⟨C6H4⟩—NHCOCH3

   $\xrightarrow{NaOH} \xrightarrow{HNO_2} \xrightarrow{H_3PO_2}$ HOO¹⁴C—⟨C6H4⟩—SO3H

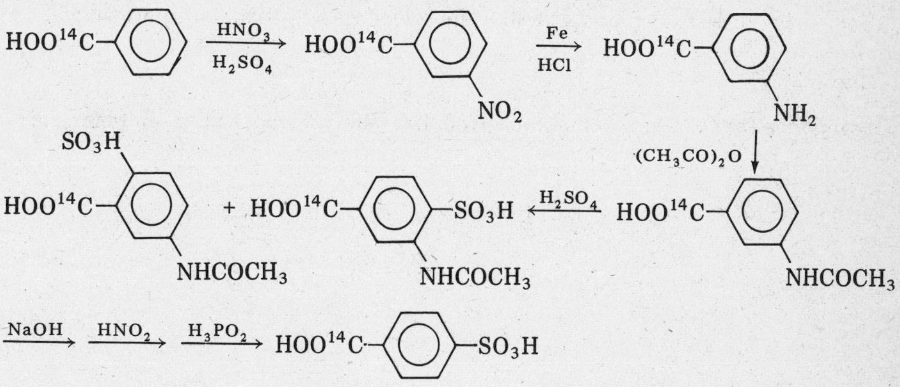

5. 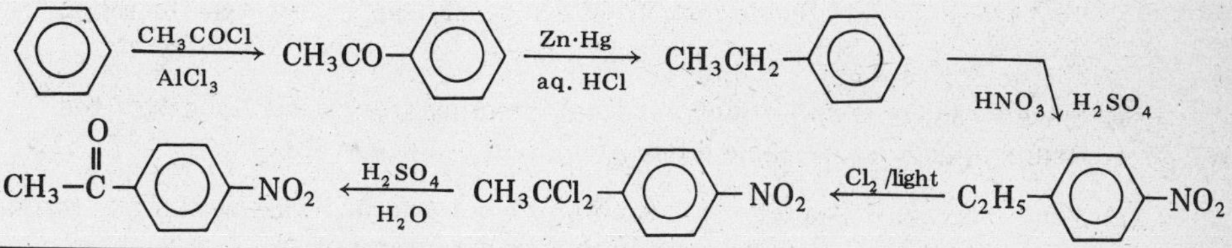 ⟨C6H6⟩ $\xrightarrow[AlCl_3]{CH_3COCl}$ CH3CO—⟨C6H5⟩ $\xrightarrow[\text{aq. HCl}]{Zn·Hg}$ CH3CH2—⟨C6H5⟩ $\xrightarrow{HNO_3 / H_2SO_4}$

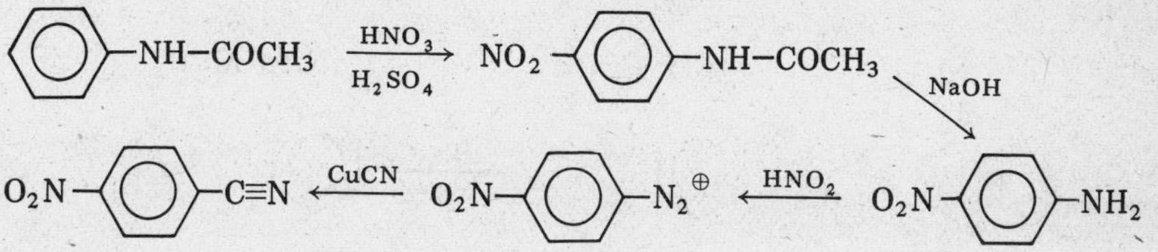

   CH3—C(=O)—⟨C6H4⟩—NO2 $\xleftarrow[H_2O]{H_2SO_4}$ CH3CCl2—⟨C6H4⟩—NO2 $\xleftarrow{Cl_2 \text{/light}}$ C2H5—⟨C6H4⟩—NO2

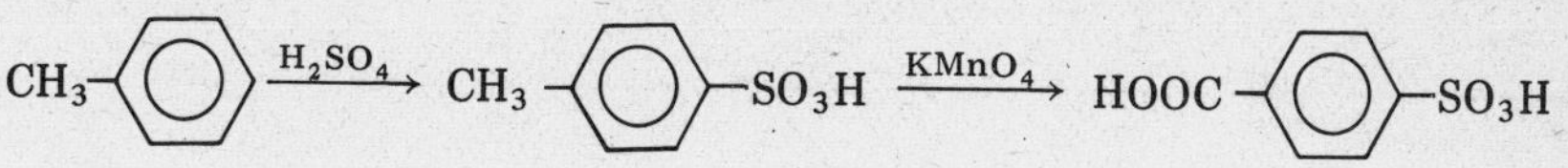

562 HOW TO SUCCEED IN ORGANIC CHEMISTRY

10. A few molecules cannot be made by the techniques of frames 4 to 9. Examples are R-C₆H₃(R)(G₁) and G₁-C₆H₃(R)(G₂) when G_1 and G_2 must be introduced via $-N_2^{\oplus}$ displacement. In these and all other cases where the set methods fail, the rule is to *work backward from the target molecule*.

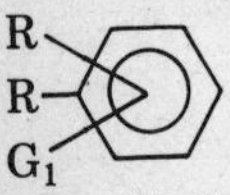

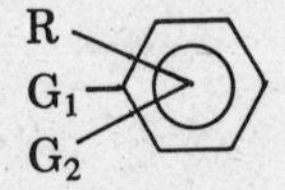

Such solutions very often revolve around the isomeric nitroanilines. These compounds are readily (and commercially) available:

Industrial source   *Laboratory source*

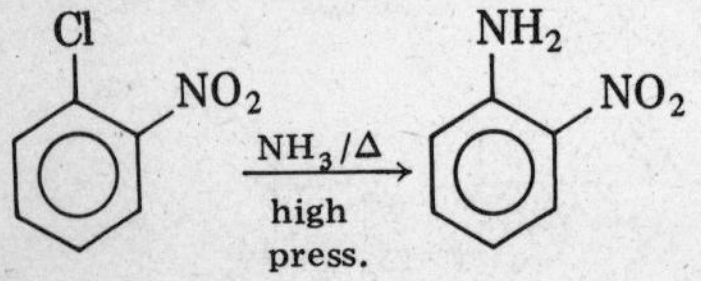

o-chloronitrobenzene →(NH₃/Δ, high press.) o-nitroaniline   (no easy way)

m-dinitrobenzene →(Fe (1 mole), HCl) m-nitroaniline ←(NH₄⁺ HS⁻) 1,3-dinitrobenzene

p-chloronitrobenzene →(NH₃/Δ, high press.) p-nitroaniline ←(NaOH, H₂O/Δ) p-nitroacetanilide ←(HNO₃, H₂SO₄) acetanilide

---

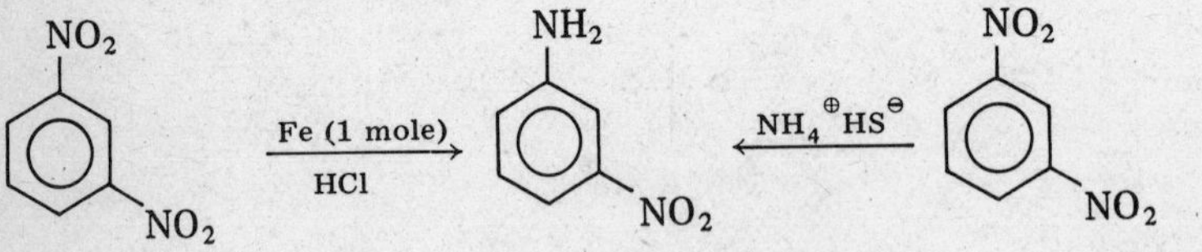

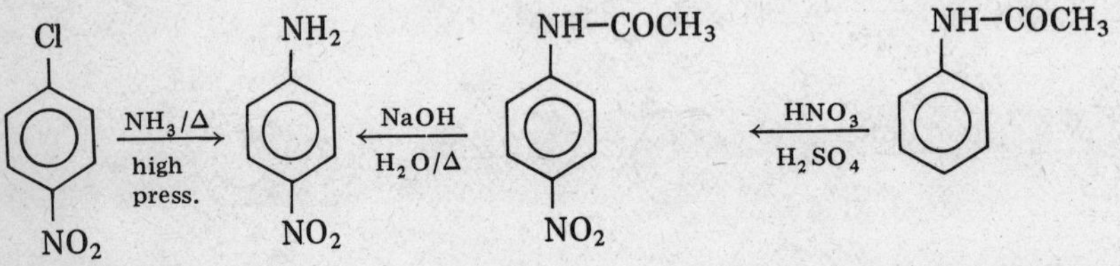

Two of the three examples that follow are solved via a nitroaniline intermediate.

Examples

Synthesize the following compounds from benzene, toluene, aniline, and inorganic reagents.

1. 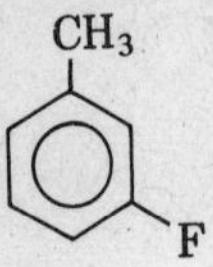 m-fluorotoluene   2. 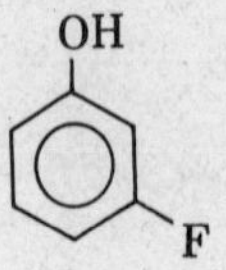 m-fluorophenol   3. 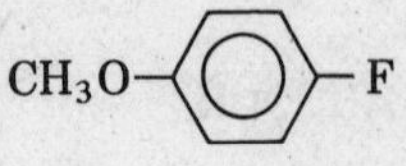 p-fluoroanisole (CH₃O–C₆H₄–F)

Solutions:

1. (a) Two *o/p* directors situated meta to one another, which is the case treated in frame 8.

   (b) Since F can come only from $-N_2^{\oplus}$ and the other group is alkyl, both of the standard methods of frame 8 fail.

   (c) Work backward. The following equations are drawn such that working or reading from left to right takes the steps in the desired reverse order:

[Scheme showing two synthetic paths for preparing 3-fluorotoluene and related meta-substituted toluenes]

Path from m-nitrotoluene: m-nitrotoluene →(Fe/HCl)→ m-toluidine →(HNO₂/HBF₄)→ ArN₂⁺BF₄⁻ →(Δ)→ m-fluorotoluene.

Two possible strategies at this point: (1) build CH₃ out of an *m*-directing group (—COCH₃, —CN, —COOH, etc.) or (2) start with toluene, then block the para position with a removable *o/p* director that will force nitration *m* to CH₃. Write out both and compare:

two variations on path (1):

Upper variation: m-aminotoluene ← (LiAlH₄) ← m-(CH₂OSO₂Ar)-aniline ← (ArSO₂Cl) ← m-aminobenzyl alcohol (CH₂OH / NH₂) ← (LiAlH₄) ← m-aminobenzoic acid. TM ← (as above) ← m-toluidine.

Lower variation: m-nitrobenzaldehyde ← (Rosenmund) ← m-nitrobenzoyl chloride ← (SOCl₂) ← m-nitrobenzoic acid ← (HNO₃/H₂SO₄) ← benzoic acid ← (KMnO₄) ← toluene. Then m-nitrobenzaldehyde →(HCl, Zn·Hg)→ m-nitrotoluene → TM.

path (2):

toluene →(HNO₃/H₂SO₄)→ p-nitrotoluene →(Fe, aq. HCl)→ p-toluidine →((CH₃CO)₂O)→ p-NHCOCH₃-toluene →(HNO₃/H₂SO₄)→ 2-nitro-4-acetamidotoluene →(H₂O/Δ, NaOH)→ 2-nitro-4-aminotoluene →(HNO₂, NO₂⁻ ... actually HNO₂)→ diazonium →(H₃PO₂)→ m-nitrotoluene →(as above)→ TM.

Path (2) is the longest. The lower variation of path (1) is probably cheapest.

2. (a) Two *o/p* directors situated meta as in example 1.

   (b) Both OH and F must come from —N₂⁺; ∴ the set methods fail.

564 HOW TO SUCCEED IN ORGANIC CHEMISTRY

(c) Work backward:

[Retrosynthesis scheme showing: m-fluorophenol ← m-fluorobenzenediazonium ← m-fluoroaniline ← m-fluoronitrobenzene ← fluorobenzene; and below: 3-nitrobenzenediazonium ← m-nitroaniline ← m-dinitrobenzene ← benzene]

3. (a) Two *o/p* directors situated para; should be easy.

   (b) But neither group is introducible via electrophilic aromatic substitution.

   (c) The set methods fail; work backward.

[Scheme showing synthesis of p-methoxyfluorobenzene:
CH₃O–C₆H₄–F ←(CH₃OH)— N₂⁺–C₆H₄–F ←(HNO₂)— NH₂–C₆H₄–F ↑(Fe/HCl) O₂N–C₆H₄–F ←(HNO₃/H₂SO₄)— C₆H₅–F ←(Δ)— C₆H₅–N₂⁺ BF₄⁻ ←(HNO₂/HBF₄)— C₆H₅–NH₂

And alternative route:
C₆H₅–NH–COCH₃ →(HNO₃/H₂SO₄)→ O₂N–C₆H₄–NH–COCH₃ →(H₂O/Δ, NaOH)→ O₂N–C₆H₄–NH₂ →(HBF₄/HNO₂)→ O₂N–C₆H₄–N₂⁺ BF₄⁻ →(Δ)→ O₂N–C₆H₄–F]

★ Exercises

Synthesize the following compounds from benzene, toluene, aniline, and inorganic and aliphatic reagents. This is a composite set covering all the aromatic target-molecule types and using all the solution methods discussed in frames 4 to 10.

1. p-chlorobenzonitrile (CN and Cl para)

2. m-chlorobenzonitrile (CN and Cl meta)

3. 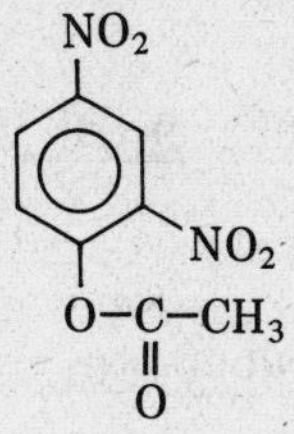 [2,4-dinitrophenyl acetate]

4. $O_2N-C_6H_4-N=N-C_6H_4-N(CH_3)_2$

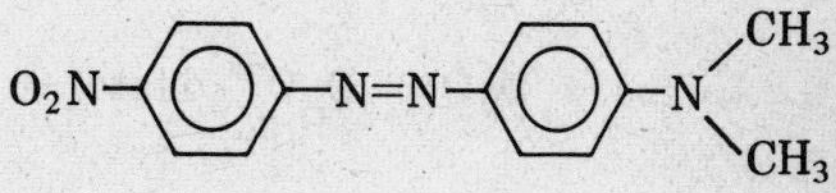

5. 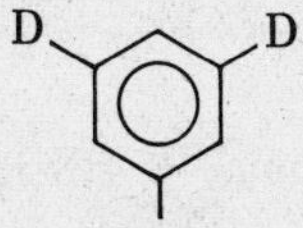 1,3-dideutero-5-methylbenzene

6. 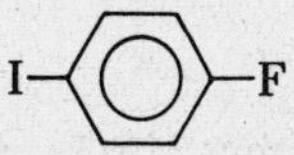 $I-C_6H_4-F$ (para)

7. 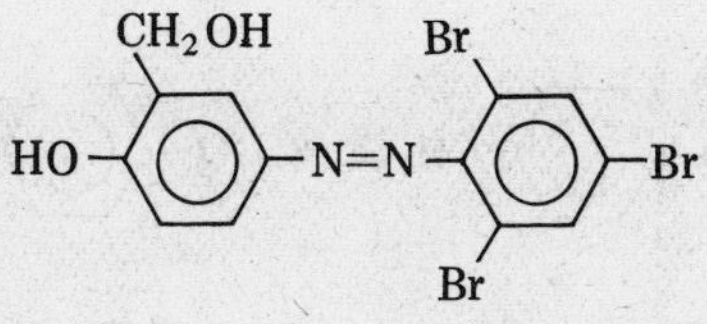 $HO-C_6H_3(CH_2OH)-N=N-C_6H_2(Br)_3$ (2,4,6-tribromo on right ring; OH and CH$_2$OH on left ring)

8. 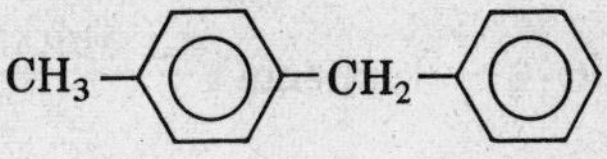 $CH_3-C_6H_4-CH_2-C_6H_5$

9. 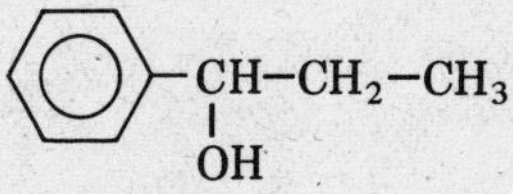 $C_6H_5-CH(OH)-CH_2-CH_3$

10. 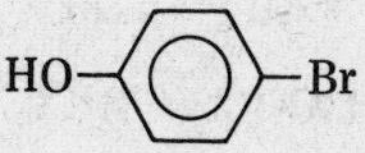 $HO-C_6H_4-Br$ (para)

11. 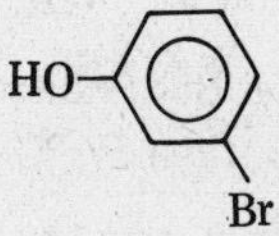 $HO-C_6H_4-Br$ (meta)

12. 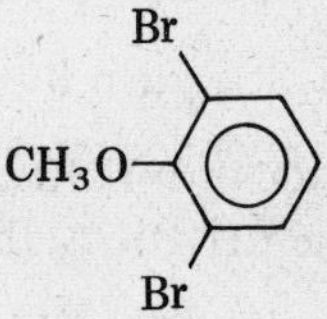 $CH_3O-C_6H_3(Br)_2$ (2,6-dibromo)

13. 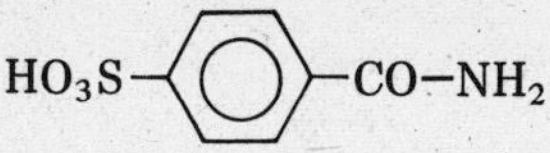 $HO_3S-C_6H_4-CO-NH_2$ (para)

14. 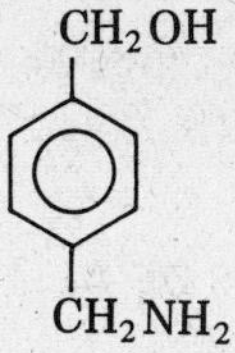 para-$HOCH_2-C_6H_4-CH_2NH_2$

— — — — — — — — —

1. Toluene + Cl$_2$/Fe, separate isomers, then hot KMnO$_4$, then SOCl$_2$, then NH$_3$, then P$_2$O$_5$/Δ.

2. Toluene + hot KMnO$_4$, then SOCl$_2$, then NH$_3$, then P$_2$O$_5$/Δ, then Cl$_2$/Fe.

3. Benzene + Cl$_2$/Fe, then HNO$_3$/H$_2$SO$_4$/Δ, then Na$^{\oplus}$CH$_3$CO$_2^{\ominus}$.

4. Aniline + (CH$_3$CO)$_2$O, then HNO$_3$/H$_2$SO$_4$, then aq. NaOH/Δ, then HNO$_2$/cold

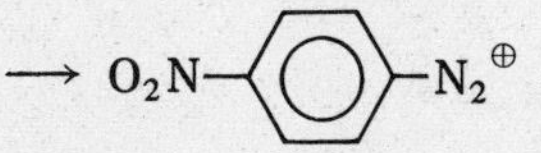

→ $O_2N-C_6H_4-N_2^{\oplus}$ = A. Aniline + CH$_3$I (2 moles) + NaHCO$_3$ →

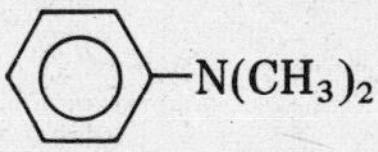

$C_6H_5-N(CH_3)_2$ = B. A + B ⟶ product.

5. Aniline + D_2O + $D_3O^\oplus$ (dilute) + time ⟶ [2,4,6-trideuterioaniline: H_2N-C$_6$H$_2$D$_3$] $\xrightarrow{HNO_2}$ $\xrightarrow{H_3PO_2}$ prod.

6. Aniline + HNO_2 + HBF_4 ⟶ $C_6H_5-N_2^\oplus BF^\ominus$ $\xrightarrow{\Delta}$ C_6H_5-F $\xrightarrow{Tl(OCOCF_3)_3}$ $\xrightarrow{KI}$ prod.

7. Aniline + excess Br_2, then HNO_2 ⟶ [2,4,6-tribromo-$N_2^\oplus$] = A. Aniline + HNO_2,

then H_2O/Δ ⟶ phenol $\xrightarrow{H_2CO/HCl/ZnCl_2}$ HO-C$_6$H$_4$-CH$_2$Cl + HO-C$_6$H$_4$-CH$_2$Cl (ortho) = B.

B $\xrightarrow[(2)\ NaOH]{(1)\ A}$ product.

8. Benzene + H_2CO + HCl + $ZnCl_2$, then toluene/$AlCl_3$.

9. Benzene + CH_3CH_2COCl + $AlCl_3$, then $NaBH_4$.

10. Aniline + $(CH_3CO)_2O$, then Br_2/Fe, then NaOH/Δ, then HNO_2, then H_2O/Δ.

11. Benzene + HNO_3/H_2SO_4, then Br_2Fe, then Fe/HCl, then HNO_2/cold, then H_2O/Δ.

12. Aniline + $(CH_3CO)_2O$, then HNO_3/H_2SO_4, then hot aq. NaOH, then Br_2 (excess), then HNO_2/cold, then CH_3OH/Δ, then Fe/aq. HCl, then HNO_2/cold, then H_3PO_2.

13. Toluene + H_2SO_4/Δ, then hot $KMnO_4$, then $SOCl_2$, then NH_3, then $H_3O^\oplus$/cold.

14. Toluene + HNO_3/H_2SO_4, then Fe/HCl, then HNO_2/cold, then CuCN, then hot $KMnO_4$, then excess $LiAlH_4$.

SYNTHESIS OF ALIPHATIC TARGET MOLECULES

11. When we shift from aromatic to aliphatic target molecules, the nature of the problem changes considerably. We become much more preoccupied with constructing the carbon skeleton of the target molecule, and we have much more limited possibilities of introducing functional groups on a hydrocarbon skeleton once we have built it. This means that the strategy adopted for solving an aliphatic synthesis problem must consider skeleton building and functional group placement simultaneously. In the method we are going to use, the strategic high point in the synthesis is a step in which the carbon skeleton is formed by connecting two or three preformed pieces in such a way that a functional group is established in the proper position. We do this key step first, although it comes near the middle of the synthetic sequence. Then in two subsequent stages we connect the key intermediate forward to TM on the one hand and connect it back to allowed, available starting materials on the other:

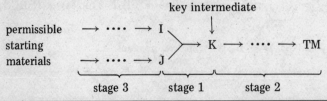

Table 21.4 *Carbon-Carbon Bond-Forming Reactions via Nucleophilic Addition*

A. Via Organometallic Reagents

$$CO_2 \xrightarrow{R_1 MgX} [R_1CO_2H,\ R_1COCl,\ R_1COOR'] \xrightarrow{R_2CdR_2} R_1\!-\!\underset{\underset{}{\|}}{\overset{O}{C}}\!-\!R_2 \xrightarrow{R_3 MgX} \begin{bmatrix} R_1\!-\!\underset{\underset{R_2}{|}}{\overset{OH}{\underset{|}{C}}}\!-\!R_3,\ R_1\!-\!\underset{\underset{R_2}{|}}{\overset{OH}{\underset{|}{C}}}\!-\!R_2 \end{bmatrix}$$

$$H\!-\!C(OR)_3 \xrightarrow{R_1 MgX} H\!-\!\overset{O}{\underset{\|}{C}}\!-\!H \xrightarrow{R_1 MgX} R_1\!-\!\overset{O}{\underset{\|}{C}}\!-\!H \xrightarrow{R_2 MgX} R_1\!-\!\overset{OH}{\underset{|}{C}H}\!-\!R_2 \xrightarrow{R_2 MgX}$$

$$CH_3OH \xrightarrow{R_1 MgX} R_1\!-\!CH_2OH$$

B. Via Wittig Reaction

$$\underset{R_2}{\overset{R_1}{{>}}}\!C\!=\!O\ +\ \phi_3\overset{\oplus}{P}\!-\!\overset{\ominus}{\underset{R_4}{\overset{R_3}{{<}}}}C \longrightarrow \underset{R_2}{\overset{R_1}{{>}}}\!C\!=\!C\underset{R_4}{\overset{R_3}{{<}}} \qquad \phi = C_6H_5$$

C. Via Enolate Ion Condensations

$$R_1\!-\!CH_2\!-\!\overset{O}{\underset{\|}{C}}\!-\!R_2 \xrightarrow{base} R_1\!-\!CH_2\!-\!\underset{\underset{R_1}{\underset{|}{R_2}}}{\overset{OH}{\underset{|}{C}}}\!-\!CH\!-\!\overset{O}{\underset{\|}{C}}\!-\!R_2 \qquad \text{Aldol}$$

$$R_1\!-\!CH_2\!-\!\overset{O}{\underset{\|}{C}}\!-\!R_2\ +\ R_3\!-\!\overset{O}{\underset{\|}{C}}\!-\!R_4 \xrightarrow{base} R_3\!-\!\underset{\underset{R_1}{\underset{|}{R_4}}}{\overset{OH}{\underset{|}{C}}}\!-\!CH\!-\!\overset{O}{\underset{\|}{C}}\!-\!R_2 \qquad \text{Crossed aldol}$$

$$R\!-\!CH_2\!-\!\overset{O}{\underset{\|}{C}}\!-\!O\!-\!C_2H_5 \xrightarrow{base} R\!-\!CH_2\!-\!\overset{O}{\underset{\|}{C}}\!-\!\underset{\underset{R}{|}}{CH}\!-\!COOC_2H_5 \qquad \text{Claisen}$$

$$\left. \begin{array}{l} -\overset{O}{\underset{\|}{C}}\!-\!CH\!-\!\overset{O}{\underset{\|}{C}}\!-\! \\[4pt] -\overset{O}{\underset{\|}{C}}\!-\!\underset{|}{CH}\!-\!\overset{O}{\underset{\|}{C}}\!-\!O\!-\!C_2H_5 \\[4pt] C_2H_5O\!-\!\overset{O}{\underset{\|}{C}}\!-\!\underset{|}{CH}\!-\!\overset{O}{\underset{\|}{C}}\!-\!OC_2H_5 \end{array} \right\}\ +\ R_1\!-\!\overset{O}{\underset{\|}{C}}\!-\!R_2 \xrightarrow{base} R_1\!-\!\underset{\underset{R_2}{|}}{\overset{OH}{\underset{|}{C}}}\!-\!\underset{\underset{\|}{\underset{O}{C}}\!-}{\overset{\overset{O}{\|}}{\underset{}{C}}\!-}{C} \qquad \begin{array}{l}\text{Knoevenagel,}\\ \text{etc.}\end{array}$$

We treat stage 1 in this frame, stage 2 in frame 12, and stage 3 in frame 13. We will limit consideration to compounds with no more than two functional groups (difunctional compounds).

568 HOW TO SUCCEED IN ORGANIC CHEMISTRY

Stage 1, the I + J ⟶ K reaction, can be any reaction that makes a new carbon-carbon bond, and as you gain experience in synthetic problems you will want to exercise progressively more freedom in the choice of this reaction. However, it is best to start with a more rigid scheme; this has the advantage that you will always be able to proceed, because you have a definite method to apply. In this method K will usually be an alcohol or an alkene, and the I + J ⟶ K reaction will usually be an organometallic type of alcohol preparation or a Wittig alkene synthesis. But before we study this method in

Table 21.5 *Carbon-Carbon Bond-Forming Reactions via Nucleophilic Substitution*

A. Enamine Alkylation

$$\underset{\overset{\|}{-\text{C}-\text{CH}}}{\overset{\text{O}}{}} \xrightarrow{R_2NH} \underset{\underset{/\backslash}{\overset{|}{\text{N:}}}}{-\text{C}=\text{C}} \leftrightarrow \underset{\underset{/\backslash}{\overset{\|}{\text{N}^\oplus}}}{-\text{C}-\overset{\ominus}{\text{C}}} \xrightarrow{R-X} \underset{\underset{/\backslash}{\overset{|}{\text{N}^\oplus\ \ R}}}{-\text{C}-\text{C}} \ \ X^\ominus \xrightarrow{\underset{H_3O^\oplus}{H_2O}} \underset{\overset{\|}{-\text{C}-\text{C}}\ R}{\overset{\text{O}}{}}$$

B. Terminal Alkyne Alkylation

$$R-C\equiv C-H \xrightarrow{NaNH_2} R-C\equiv C^\ominus_\oplus \ \ Na \xrightarrow{R'X} R-C\equiv C-R'$$

C. Superenolate Ion Alkylations

1. malonic ester synthesis

$$\underset{\underset{\text{CO}_2\text{C}_2\text{H}_5}{\overset{|}{\text{CH}_2}}}{\overset{\text{CO}_2\text{C}_2\text{H}_5}{|}} \xrightarrow{NaOC_2H_5} \underset{\underset{\text{CO}_2\text{C}_2\text{H}_5}{\overset{|}{\text{CH:}^\ominus \text{ Na}^\oplus}}}{\overset{\text{CO}_2\text{C}_2\text{H}_5}{|}} \xrightarrow{R-X} \underset{\underset{\text{CO}_2\text{C}_2\text{H}_5}{\overset{|}{\text{CH}-R}}}{\overset{\text{CO}_2\text{C}_2\text{H}_5}{|}} \xrightarrow[H_3O^\oplus]{H_2O/heat} \underset{\underset{\text{COOH}}{\overset{|}{\text{CH}-R}}}{\overset{\text{COOH}}{|}} \xrightarrow{heat} R-\text{CH}_2-\text{COOH}$$

↓ repeat

$$\underset{\underset{\text{CO}_2\text{C}_2\text{H}_5}{\overset{|}{R-\text{C}-R'}}}{\overset{\text{CO}_2\text{C}_2\text{H}_5}{|}} \xrightarrow[H_3O^\oplus]{H_2O/heat} \underset{\underset{\text{COOH}}{\overset{|}{\text{C}}}}{\overset{\text{COOH}}{\underset{R'}{\overset{R}{|}}}} \xrightarrow{heat} \underset{R'}{\overset{R}{\diagdown}}\text{CH}-\text{COOH}$$

2. acetoacetic ester synthesis

$$\text{CH}_3-\overset{\overset{\text{O}}{\|}}{\text{C}}-\text{CH}_2-\text{CO}_2\text{C}_2\text{H}_5 \xrightarrow{NaOC_2H_5} \text{CH}_3-\overset{\overset{\text{O}}{\|}}{\text{C}}-\overset{\ominus}{\text{CH}}-\text{CO}_2\text{C}_2\text{H}_5 \ \ Na^\oplus \xrightarrow{R-X} \text{CH}_3-\overset{\overset{\text{O}}{\|}}{\text{C}}-\underset{R}{\overset{|}{\text{CH}}}-\text{CO}_2\text{C}_2\text{H}_2$$

↓ $H_2O/heat\ H_3O^\oplus$

$$\text{CH}_3-\overset{\overset{\text{O}}{\|}}{\text{C}}-\text{CH}_2-R \xleftarrow{heat} \text{CH}_3-\overset{\overset{\text{O}}{\|}}{\text{C}}-\underset{R}{\overset{|}{\text{CH}}}-\text{COOH}$$

detail, let us collect in one place all the carbon-carbon bond-forming reactions that can be used in the I + J ⟶ K step.

Tables 21.4 and 21.5 contain the reactions that join two preformed chunks of carbon skeleton and leave one functional group at or near the new C—C bond. These reactions are arranged by chemical type. Those in Table 21.4 are what we have previously called reductive alkylations. They all start with addition of a carbanionlike species (organometallic compound, ylide, enolate ion) to a C=O group. Those in Table 21.5 are alkylations; they leave the functional group of the reactant unchanged, and just replace a weakly acidic H with R.

Table 21.6 contains the reactions that join preformed pieces of skeleton and leave *two* functions near the junction. Chemically they are a grabbag. The spacing of functional groups that they produce is their only common feature.

Table 21.6 *Reactions Forming Difunctional Compounds*

A. 1,1 Compounds

$$\text{>C=O} + \text{HCN} \longrightarrow \text{>C(OH)(CN)}$$

$$\text{>C=O} + \text{ROH} \longrightarrow \text{>C(OH)(OR)}$$

$$\text{>C=O} + PX_5 \longrightarrow \text{>C(X)(X)}$$

$$*\text{>C(H)(C=O-)} + X_2 \xrightarrow{H^\oplus \text{ or } OH^\ominus} \text{>C(X)(C=O-)}$$

$$*-CH_2-COOH \xrightarrow[(2)\ H_2O]{(1)\ Br_2/P} -CH(Br)-COOH$$

B. 1,2 Compounds

$$\text{>C=C<} + X_2 \longrightarrow \text{>C(X)-C(X)<}$$

$$\text{>C=C<} \xrightarrow[\text{cold}]{KMnO_4} \text{>C(OH)-C(OH)<}$$

$$\text{>C-C<}\ \text{(epoxide)} + HA \longrightarrow \text{>C(OH)-C(A)<}$$

(see footnote on page 570)

Table 21.6 *Continued*

$$* \;\; \mathrm{\underset{|}{C}=\underset{|}{C}-\underset{|}{\overset{|}{C}}-H} \xrightarrow{NBS} \mathrm{\underset{|}{C}=\underset{|}{C}-\underset{|}{\overset{|}{C}}-X}$$

$$* \;\; \mathrm{-\overset{O}{\overset{\|}{C}}-CH_2-} \xrightarrow{SeO_2} \mathrm{-\overset{O}{\overset{\|}{C}}-\overset{O}{\overset{\|}{C}}-}$$

C. 1,3 Compounds

$$\mathrm{\underset{C}{C{-}{-}C}} \xrightarrow{Br_2} \mathrm{Br-\underset{|}{\overset{|}{C}}-\underset{|}{\overset{|}{C}}-\underset{|}{\overset{|}{C}}-Br}$$

$$* \;\; 2\text{-}\overset{O}{\overset{\|}{C}}\text{-}CH\diagup \xrightarrow{base} \mathrm{-\overset{O}{\overset{\|}{C}}-\underset{|}{\overset{|}{C}}-\overset{OH}{\overset{|}{C}}-CH}\diagup$$

$$* \;\; 2 \;\; \mathrm{\diagdown CH-\overset{O}{\overset{\|}{C}}-OR} \xrightarrow{base} \mathrm{\diagdown CH-\overset{O}{\overset{\|}{C}}-\underset{|}{\overset{|}{C}}-\overset{O}{\overset{\|}{C}}-OR}$$

$$* \;\; \square \xrightarrow[heat]{KMnO_4} \mathrm{HOOC-CH_2-CH_2-CH_2-COOH}$$

D. 1,4 Compounds

$$\mathrm{\diagdown C=\underset{|}{C}-\underset{|}{C}=C\diagup} \begin{array}{c} \xrightarrow{AB} \\ \\ \xrightarrow{X_2} \end{array} \begin{array}{c} \mathrm{\diagdown\underset{|}{C}-\underset{B}{\overset{|}{C}}=\underset{|}{C}-\underset{A}{\overset{|}{C}}\diagup} \\ \\ \mathrm{\diagdown\underset{X}{\overset{|}{C}}-\underset{|}{C}=\underset{|}{C}-\underset{X}{\overset{|}{C}}\diagup} \end{array}$$

$$\underset{O}{\square} \xrightarrow{HX} \mathrm{X-CH_2-CH_2-CH_2-CH_2-X}$$

$$* \;\; \bigcirc \xrightarrow[heat]{KMnO_4} \mathrm{HOOC-CH_2-CH_2-CH_2-CH_2-COOH}$$

---

*There are various ways to classify some of these combinations, depending, for example, on whether one takes $\mathrm{-\underset{\overset{\|}{O}}{C}-\underset{|}{\overset{|}{C}}-\underset{\overset{\|}{O}}{C}-}$ as 1,3 functionalization $\left(\mathrm{-\underset{\boxed{O}}{C}-\underset{|}{\overset{|}{C}}-\underset{\boxed{O}}{C}-}\right)$ or as 1,1 functionalization $\left(\mathrm{-\boxed{\underset{\overset{\|}{O}}{C}}-\underset{|}{\overset{|}{C}}-\boxed{\underset{\overset{\|}{O}}{C}}-}\right)$.

SYNTHETIC PROBLEMS

Here is the set method of carrying out stage 1, described first in words, then in flow-chart form.

First, you identify the compound that is to play the role of key intermediate K. You do this by writing the structure of that alcohol or alkene that has the same carbon skeleton as TM and has the closest functional-group relationship to TM. The most "closely related" functional group is the one that can be converted into the functional group of TM in the smallest number of steps. For example, if TM is Ph–CH(CH$_3$)–C(O)–CH$_2$–CH(CH$_3$)$_2$, you choose K = Ph–CH(CH$_3$)–CH(OH)–CH$_2$–CH(CH$_3$)$_2$ since the –CH(OH)– function can be oxidized to >C=O in one step. Similarly, if TM = Ph–CH(CH$_3$)–CH$_2$–CH$_2$–CH(CH$_3$)$_2$, you choose K = Ph–CH(CH$_3$)–CH=CH–CH(CH$_3$)$_2$ or Ph–C(CH$_3$)=CH–CH$_2$–CH(CH$_3$)$_2$, each of which is just one step (catalytic hydrogenation) removed from TM.

Second, you write the equation for a reductive alkylation reaction or a Wittig reaction (Table 21.4A or B) that forms K from smaller compounds.

If TM has two functional groups, you would like to find a reaction I + J ⟶ K that makes the complete carbon skeleton, places *two* functional groups in the correct positions, and makes both of these functions as closely related to the functions in TM as possible. In fact, you can usually find a reaction giving a product with the correct positions of both functional groups and the correct skeleton, but you cannot usually form that skeleton from smaller fragments in the same step that puts the two functions in place. For example, suppose that TM = Ph–CHCl–CH(OC(O)CH$_3$)–Ph. You could choose K = Ph–CHCl–CH(OH)–Ph and make this K from Ph–CH(–O–)CH–Ph (epoxide) + HCl (Table 21.6B). This puts two well-chosen functions in the right places on the right skeleton and is a good stage 1 solution, even though it doesn't put the skeleton together from smaller fragments. Assembling the skeleton can be done in stage 3.

Chart 21.1 lays out stage 1 as a flowchart. The flowchart accomplishes the same thing as the written procedure, but it makes it possible for you to see the whole operation at once.

In the following examples the solutions follow the scheme of Chart 21.1. Solutions are carried only through stage 1. We take up the same examples again in frames 12 and 13 and carry the solutions through stages 2 and 3.

Chart 21.1 *Stage 1*

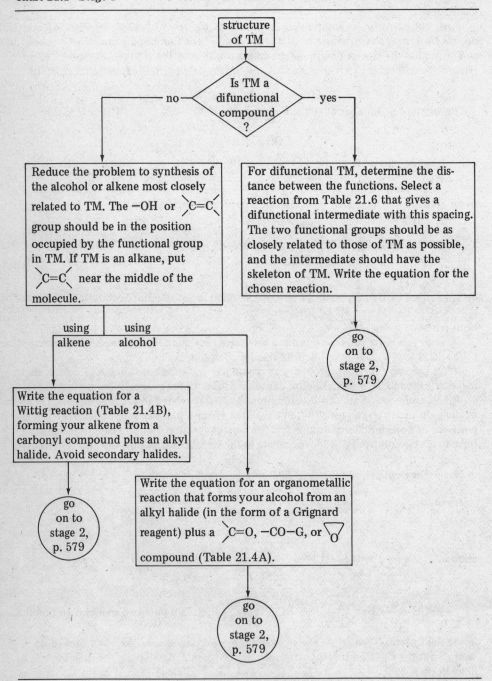

Examples

Devise a workable synthesis (stage 1 only) of each of the following compounds starting from alcohols with no more than four carbon atoms, benzene, toluene, and any inorganic reagents.

SYNTHETIC PROBLEMS 573

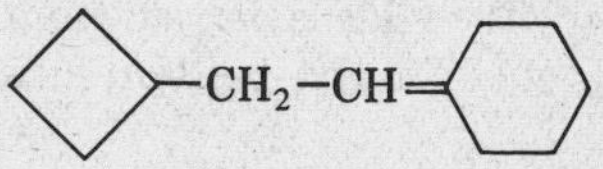

1. 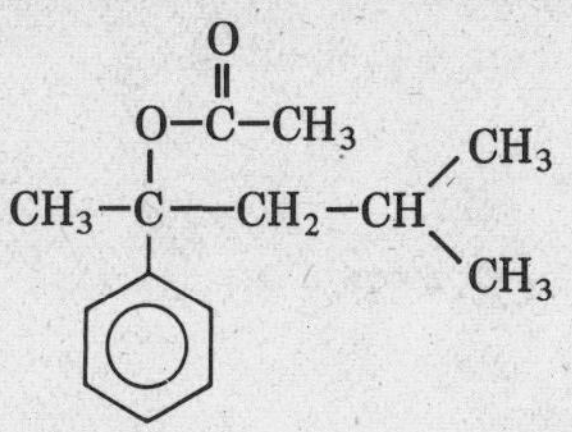 $\text{CH}_3-\underset{\underset{\phi}{|}}{\text{C}}-\text{CH}_2-\text{CH}\underset{\text{CH}_3}{\overset{\text{CH}_3}{\diagdown}}$ with O-C(=O)-CH$_3$ on the central C

2. ◻—CH$_2$—CH=⬡

3. ⬡—CH$_2$—C(OH)(CN)(φ)—CH(φ)—C(=O)—NH—CH$_2$—CH$_2$—CH$_3$

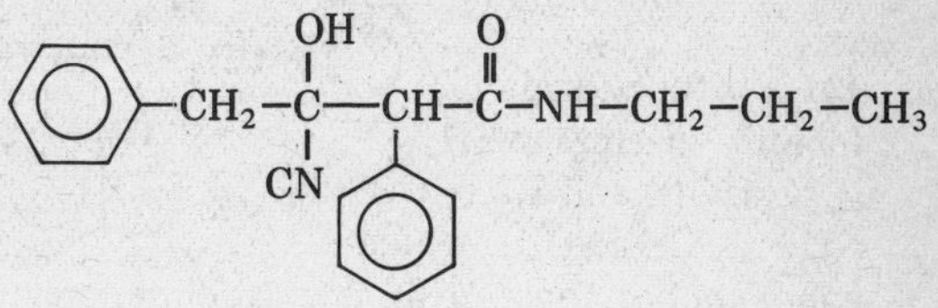

Solutions:

1. TM = $\text{CH}_3-\underset{\underset{\phi}{|}}{\text{C}}(\text{O-C(=O)-CH}_3)-\text{CH}_2-\text{CH}(\text{CH}_3)_2$ = monofunctional

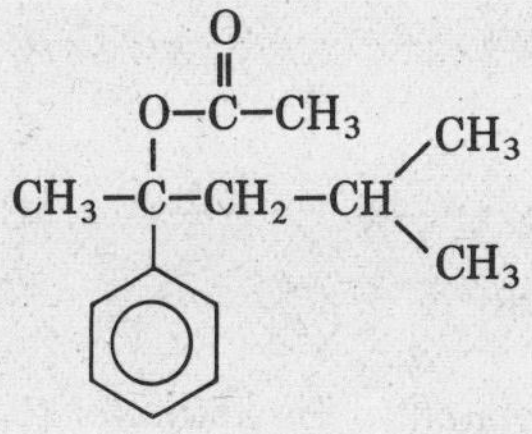

   Is TM closer to an alcohol or to an alkene? It is an ester = derivative of an alcohol. So make that alcohol the key intermediate. What kind of alcohol? 3° with no two groups alike. What kind of Grignard reaction makes such an alcohol? Ketone + Grignard (Table 21.4A). What would be the possible combinations?

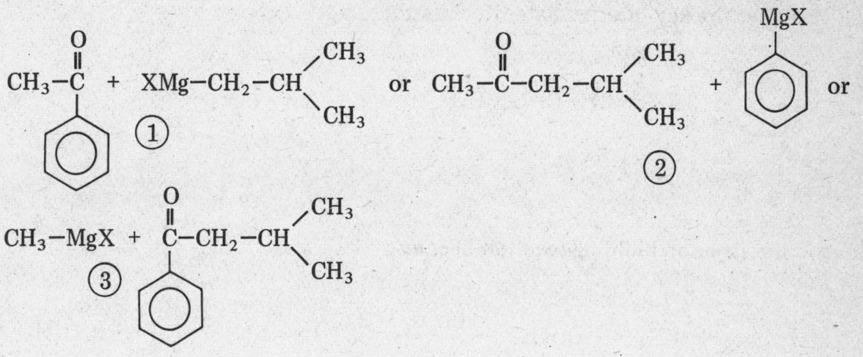

   $\text{CH}_3-\underset{\overset{\parallel}{\text{O}}}{\text{C}}-\phi$ + XMg—CH$_2$—CH(CH$_3$)$_2$ ① or CH$_3$—C(=O)—CH$_2$—CH(CH$_3$)$_2$ + φ—MgX ② or

   CH$_3$—MgX + φ—C(=O)—CH$_2$—CH(CH$_3$)$_2$ ③

   Are these all feasible reactions? Yes. Which will match up to allowed starting materials best? ① looks most logical because it uses C$_4$ (allowed) + C$_8$ [= benzene (allowed) + C$_2$ (allowed)].

   So write for stage 1: CH$_3$—C(=O)—φ $\xrightarrow[\text{(2) H}_2\text{O}]{\text{(1) BrMg—CH}_2\text{—CH(CH}_3\text{)}_2}$ CH$_3$—C(OH)(φ)—CH$_2$—CH(CH$_3$)$_2$

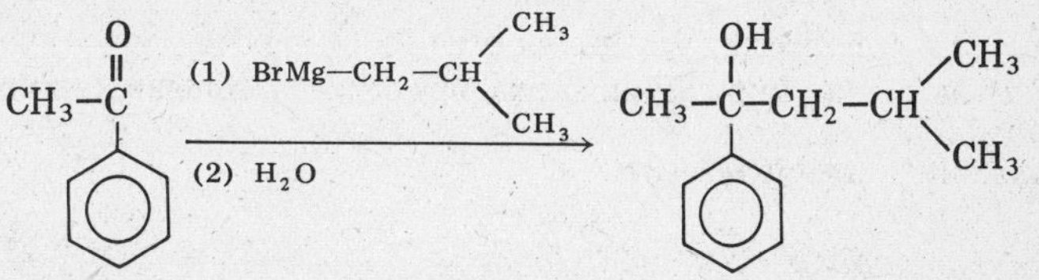

2. TM = alkene; ∴ write the Wittig reaction for that alkene = ◻—CH$_2$—CH=⬡

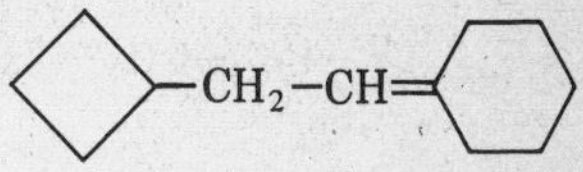

   Either ① ◻—CH$_2$—CHO + φ$_3$P$^\oplus$—$^\ominus$⬡ → ◻—CH$_2$—CH=⬡

   or ② ◻—CH$_2$—$\overset{\ominus}{\text{C}}$H—P$^\oplus$φ$_3$ + O=⬡

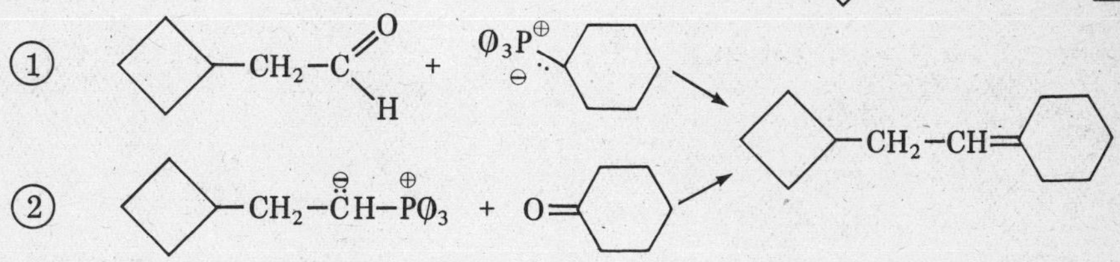

Which is better? The ylide is derived from a 2° halide in ①, from a 1° halide in ②; therefore, expect ② to give better results.

3. TM is trifunctional: Identify spacings as 1,1 (CN vs. OH) and 1,3 (–CO–NH– vs. CN or OH). Which

Ph–CH$_2$–C(OH)(CN)–CH(Ph)–C(=O)–NH–CH$_2$–CH$_2$–CH$_3$

compound of Table 21.6C is most closely related to these functions? Evidently either the β-hydroxyketone or the β-ketoester:

$$\begin{array}{c} -\overset{|}{\underset{|}{C}}-OH \\ -\overset{|}{\underset{|}{C}}- \\ \overset{|}{\underset{|}{C}}=O \\ \text{①} \end{array} \xrightarrow{?} \cdots \rightarrow \begin{array}{c} NC-\overset{|}{\underset{|}{C}}-OH \\ -\overset{|}{\underset{|}{C}}- \\ \overset{|}{\underset{|}{C}}=O \\ \overset{|}{N}\diagdown \end{array} \quad \text{or} \quad \begin{array}{c} \overset{|}{\underset{|}{C}}=O \\ -\overset{|}{\underset{|}{C}}- \\ \overset{|}{\underset{|}{C}}=O \\ \text{OR ②} \end{array} \xrightarrow{?} \cdots \rightarrow \begin{array}{c} NC-\overset{|}{\underset{|}{C}}-OH \\ -\overset{|}{\underset{|}{C}}- \\ \overset{|}{\underset{|}{C}}=O \\ \overset{|}{N}\diagdown \end{array}$$

Close examination reveals that TM is very closely related to the ketoester (path ②): carbonyl goes to its cyanohydrin, ester goes to amide, without any changes in oxidation state. Path ① requires two oxidations and further modifications. Choose ②. This gives the key intermediate, K, the structure

Ph–CH$_2$–C(=O)–CH(Ph)–C(=O)–OC$_2$H$_5$

and the skeleton-building reaction becomes

2 Ph–CH$_2$–C(=O)–OC$_2$H$_5$ $\xrightarrow{\text{NaOC}_2\text{H}_5}$ Ph–CH$_2$–C(=O)–CH(Ph)–C(=O)–OC$_2$H$_5$

★ *Exercises*

Carry out stage 1 in devising syntheses of the following compounds starting from alcohols of no more than four carbon atoms plus benzene, toluene, tetrahydrofuran (furan), and any inorganic reagents.

1. CH$_3$–CH$_2$–CH$_2$–CH(CH$_3$)–CH$_2$–CH$_2$–CH$_3$

2. Ph–CH$_2$–CH(OH)–CH$_2$–CH$_2$–CH$_3$

3. (cyclobutyl)=CH–CH$_2$–Ph

4. 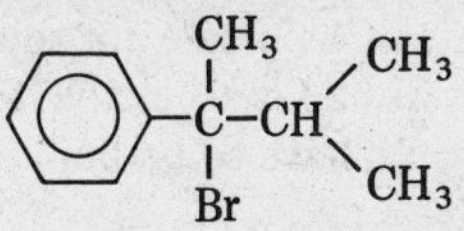 C$_6$H$_5$—C(CH$_3$)(Br)—CH(CH$_3$)CH$_3$

   (phenyl)—C(CH$_3$)(Br)—CH(CH$_3$)$_2$

5. CH$_3$—CH$_2$—CH$_2$—CH(CHO)—CH$_2$—CH$_3$

6. 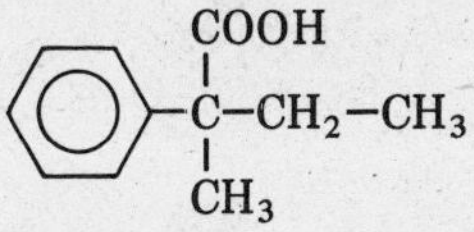 (phenyl)—C(COOH)(CH$_3$)—CH$_2$—CH$_3$

7. CH$_3$—CH$_2$—O—CH$_2$—CH$_2$—CH$_2$—CH$_2$—O—CH$_2$—CH$_3$

8. 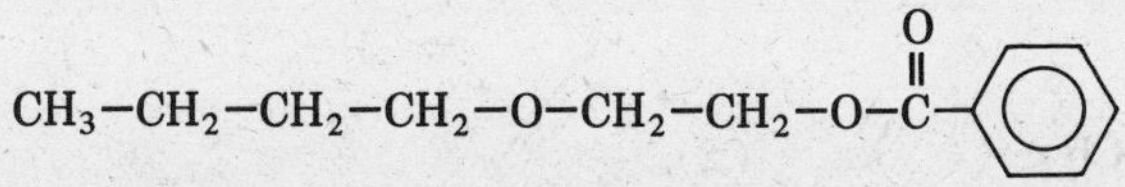 CH$_3$—CH$_2$—CH$_2$—CH$_2$—O—CH$_2$—CH$_2$—O—C(=O)—(phenyl)

9. 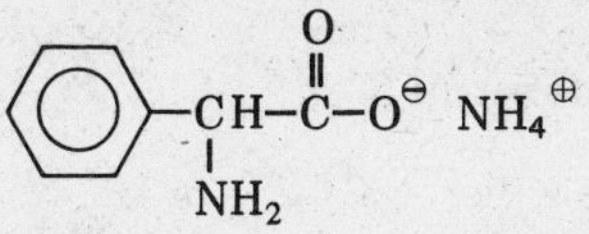 (phenyl)—CH(NH$_2$)—C(=O)—O$^\ominus$ NH$_4^\oplus$

10. CH$_3$—CH$_2$—CH(Cl)—CH(CH$_3$)—CH(O—CH$_3$)(O—CH$_3$)

---

1. Alkane ("nonfunctional"); therefore, choose an alkene intermediate with central or near-central C=C:

$$\underset{①}{CH_3-CH_2-CH_2-\underset{\|}{\overset{CH_2}{C}}-CH_2-CH_2-CH_3} \qquad \underset{②}{CH_3-CH_2-CH=\underset{|}{\overset{CH_3}{C}}-CH_2-CH_2-CH_3}$$

Both can be made via the Wittig reaction using a 1° halide and both use one fragment ⩽ C$_4$ (allowed) and one fragment ⩾ C$_4$ (not allowed, will have to be built up eventually). In such cases it is usually best to choose the path combining the more nearly equal-sized fragments, in this case ②. Write it:

$$CH_3-CH_2-\underset{\overset{|}{\oplus P\phi_3}}{CH\!:^\ominus} \; + \; \underset{\overset{|}{CH_3}}{\overset{\overset{O}{\|}}{C}}-CH_2-CH_2-CH_3 \longrightarrow CH_3-CH_2-CH=\underset{\overset{|}{CH_3}}{C}-CH_2-CH_2-CH_3$$

2. TM = an alcohol, choose it as the key intermediate built in stage 1. It is 2°; ∴ make it via Grignard + aldehyde:

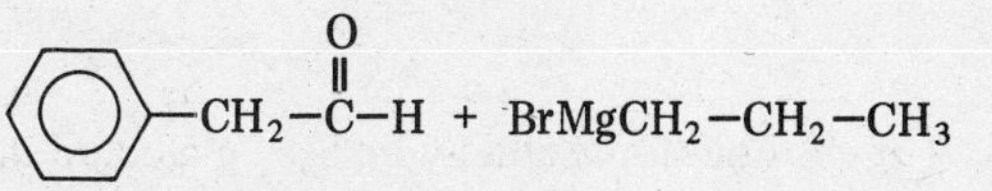

$$(phenyl)-CH_2-\overset{\overset{O}{\|}}{C}-H \; + \; BrMgCH_2-CH_2-CH_3 \qquad ①$$

or

$$(phenyl)-CH_2-MgBr \; + \; H-\overset{\overset{O}{\|}}{C}-CH_2-CH_2-CH_3 \qquad ②$$

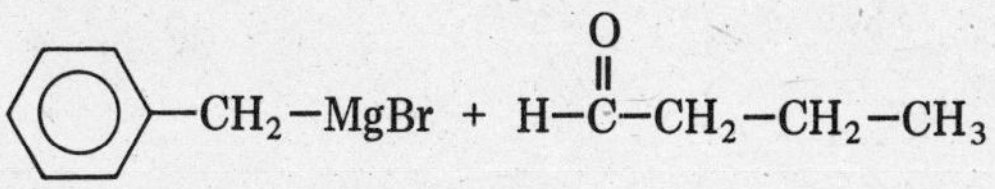

They look equally good chemically. The right-hand fragments in each case are equally easily accessible from ≤ C₄ alcohols. The left-hand fragment in ② is accessible from toluene; the C₈ skeleton of the left-hand fragment in ① would have to be built up in stage 3. Therefore, choose ②:

$$\text{Ph-CH}_2\text{-MgBr} \xrightarrow[\text{(2) H}_2\text{O}]{\text{(1) H-C(=O)-CH}_2\text{CH}_2\text{CH}_3} \text{Ph-CH}_2\text{-CH(OH)-CH}_2\text{-CH}_2\text{-CH}_3$$

3. TM = alkene = stage 1 product. Choice of Wittig reactions:

$$\text{cyclobutyl-}\bar{\text{C}}\text{-P}\phi_3^+ \;+\; \text{H-C(=O)-CH}_2\text{-Ph} \xrightarrow{①}$$

$$\text{cyclobutyl=CH-CH}_2\text{-Ph}$$

$$\text{cyclobutyl=O} \;+\; \phi_3\text{P}^+\text{-}\bar{\text{C}}\text{H-CH}_2\text{-Ph} \xrightarrow{②}$$

Choose ②, where the ylide is derived from a 1° halide versus 2° in ①.

4. TM = 2° alkyl halide. It could come in one step from either >C=C< or $-\underset{\underset{\text{OH}}{|}}{\text{CH}}-$. Addition of HBr to >C=C< has an orientation problem: >C=C< + HBr

→ $-\underset{\underset{\text{Br}}{|}}{\text{C}}-\underset{\underset{\text{H}}{|}}{\text{C}}-$ or $-\underset{\underset{\text{H}}{|}}{\text{C}}-\underset{\underset{\text{Br}}{|}}{\text{C}}-$. Therefore, it is simpler to use the alcohol as the key

intermediate. Therefore, write K = $\text{Ph-}\underset{\underset{\text{OH}}{|}}{\overset{\overset{\text{CH}_3}{|}}{\text{C}}}\text{-CH(CH}_3\text{)}_2$. This alcohol is 3°, no two

groups alike; ∴ via Grignard plus ketone. Three possible places to join the skeleton:

②—from C₁ (available) + C₁₀ (must build up)

$$\text{Ph}\underset{①}{\overset{\text{CH}_3}{-\text{C}-}}\underset{③}{\overset{}{\underset{\text{OH}}{|}}}\text{CH(CH}_3\text{)}_2$$

from C₆ (benzene, allowed) + C₅ (must build up)

from C₈ (benzene + C₂, available) + C₃ (available)

③ looks easiest: $\text{Ph-C(=O)-CH}_3 \;+\; \text{BrMg-CH(CH}_3\text{)}_2 \longrightarrow \text{Ph-}\underset{\underset{\text{OH}}{|}}{\overset{\overset{\text{CH}_3}{|}}{\text{C}}}\text{-CH(CH}_3\text{)}_2$

5. $\text{CH}_3\text{-CH}_2\text{-CH}_2\text{-}\underset{\underset{\text{CHO}}{|}}{\text{CH}}\text{-CH}_2\text{-CH}_3$ = monofunctional aldehyde; use an alcohol intermediate = $\text{CH}_3\text{-CH}_2\text{-CH}_2\text{-}\underset{\underset{\text{CH}_2\text{OH}}{|}}{\text{CH}}\text{-CH}_2\text{-CH}_3$ = 1°, accessible from Grignard + H₂CO:

$$CH_3-CH_2-CH_2-\underset{\underset{MgBr}{|}}{CH}-CH_2-CH_3 \xrightarrow[\text{(2) } H_2O]{\text{(1) } \underset{H}{\overset{H}{>}}C=O} CH_3-CH_2-CH_2-\underset{\underset{|}{CH_2OH}}{CH}-CH_2-CH_3$$

6. $\phi-\underset{\underset{CH_3}{|}}{\overset{\overset{CO_2H}{|}}{C}}-CH_2CH_3$ Since carboxylic acids are most readily made via R—Mg—X + $CO_2 \rightarrow R-COOH$ or $R-X \xrightarrow{NaCN} R-CN \xrightarrow[H_3O^{\oplus}]{H_2O/\Delta}$

R—COOH, it is best to choose as intermediate the alcohol with one carbon less than the TM, in this case $\phi-\underset{\underset{CH_3}{|}}{\overset{\overset{OH}{|}}{C}}-CH_2-CH_3$. This is 3°, no two groups alike; ∴ make it

via Grignard + ketone. The logical joining point is $\phi\underset{\underbrace{\hspace{1em}}_{C_6 \text{ from benzene}}}{\overset{\overset{OH}{|}}{\vdash}}\underset{\underbrace{\hspace{2em}}_{C_4 \text{ from allowed alcohol}}}{\overset{|}{\underset{CH_3}{|}}C-CH_2-CH_3}$. So:

$$\phi-MgBr \xrightarrow[\text{(2) } H_2O]{\text{(1) } CH_3-\overset{\overset{O}{\|}}{C}-CH_2-CH_3} \phi-\underset{\underset{CH_3}{|}}{\overset{\overset{OH}{|}}{C}}-CH_2CH_3$$

7. $CH_3CH_2-O-CH_2CH_2-CH_2CH_2-O-CH_3$ Note immediately the plane of symmetry; symmetry usually means a synthetic shortcut. TM is difunctional with 1,4 spacing. This 1,4 diether is closely related to which 1,4 compound of Table 21.6D? To $X-(CH_2)_4-X$. Therefore, write $\underset{\text{(allowed)}}{\square_O} \xrightarrow[\text{excess}]{HBr} Br-CH_2CH_2CH_2CH_2-Br$.

8. $CH_3-CH_2-CH_2-CH_2-O-CH_2-CH_2-O-\underset{\underset{O}{\|}}{C}-\phi$ Difunctional, 1,2 compound, ether + ester functions. Best precursor from Table 21.6B = $\underset{\underset{OH\ \ A}{|\ \ \ |}}{CH_2-CH_2}$ with —A = —O—CH_2CH_2CH_2CH_3, so write

$$CH_3CH_2CH_2CH_2-O^{\ominus}Na^{\oplus} + \triangle_O \longrightarrow CH_3CH_2CH_2CH_2-O-CH_2CH_2-OH = K$$

9. $\phi-\underset{\underset{NH_2}{|}}{CH}-C\underset{O}{\overset{O}{\lessgtr}}\ \ominus NH_4^{\oplus}$ Difunctional, 1,1-compound, amino + carboxylic acid (salt) functions. Check Table 21.6A. Closest pair of functions is $\underset{\underset{X}{|}}{>}C-COOH$, so write

$$\phi-CH_2-COOH + Br_2 \xrightarrow{P} \phi-\underset{\underset{Br}{|}}{CH}-COOH$$

10. $CH_3-CH_2-CHCl-CH(CH_3)-CH(O-CH_3)_2$ Difunctional, 1,2 (or 1,3, depending on how one counts) compound, halo + acetal functions. Check Table 21.6B and C. Closest match = alcohol + $>C=O$, in this case $-CH(OH)-CH-C(=O)H$. These β-hydroxyaldehydes are easy to make only when one can use the aldol condensation, as in Table 21.6C. When is this? When the molecule can be divided into halves with identical skeletons (see examples in Unit 17, frame 13). Does the present intermediate qualify? Write it and see:

$\boxed{CH_3-CH_2-CH(OH)}\boxed{CH(CH_3)-CHO}$ Yes: $2 CH_3-CH_2-CHO \xrightarrow{base}$

$$CH_3-CH_2-CH(OH)-CH(CH_3)-CHO$$

12. Stage 1 produced a key intermediate, K, with the skeleton of TM and with some functional groups (not necessarily the correct ones) in the same positions as TM. In stage 2 we adjust the functional groups of K to match exactly those desired in TM:

$$K \xrightarrow{L} M \longrightarrow \cdots \longrightarrow W \xrightarrow{X} Y \xrightarrow{Z} TM$$

At this point you don't know how many steps will be required. You figure this out as you go, using the scheme laid out in flowchart form in Chart 21.2.

You may already have visualized the reaction or sequence of reactions that will do this when you chose the main skeleton-forming reaction in stage 1. How easy it is to convert K to TM depends on the difficulty of the problem, your previous experience, and your ability to scan rapidly your knowledge of the basic reactions. You may find this slow going at first, but the method developed here will give you a systematic way to proceed. This should solve the common problem of getting stuck in synthetic problems due to "not seeing" possible reaction sequences connecting one compound with another.

In words, you first determine if the functional group(s) in K can be turned into those in TM by a single reaction. To do this you use the methods practiced in frame 2. If you do solve the K $\longrightarrow$ TM problem as a one-step conversion, write it out and go on to stage 3. If not, try to solve it as a two-step conversion as follows. Ask yourself what compound you *could* convert into the target molecule in one step. When you have found such a compound (Y), compare it with K and again look for a one-step conversion, this time of K $\longrightarrow$ Y. If you find it, you have completed a two-step conversion of K to TM: K $\longrightarrow$ Y $\longrightarrow$ TM. If you don't, you then find a compound W that *can* be converted to Y and look for a three-step conversion: K $\longrightarrow$ W $\longrightarrow$ Y $\longrightarrow$ TM. And so on until K has been connected with TM.

If you are forced into a K $\longrightarrow$ TM conversion sequence that is more than three steps long, you should consider two alternatives that may be more profitable than plugging away further on this path. First, you can decide that perhaps the intermediate Y chosen is not a good choice and replace it with another. Second, you can decide that the stage 1 solution (I + J $\longrightarrow$ K) arrived at was an unfortunate choice, scrap it, and substitute another K that may be more easily connected with TM.

The following examples illustrate the method. They carry the examples in frame 11 through stage 2.

SYNTHETIC PROBLEMS 579

Chart 21.2 *Stage 2*

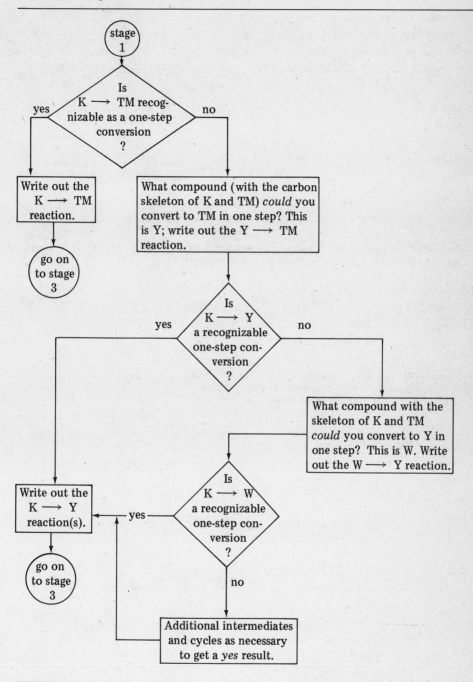

Examples

1. The stage 1 solution was

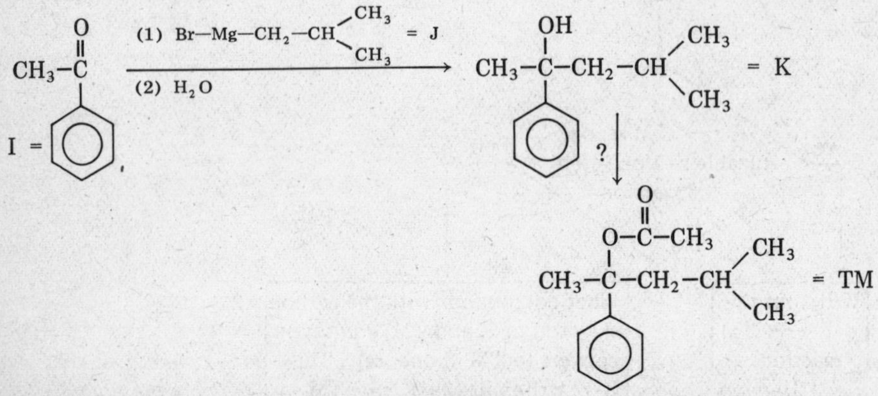

$$CH_3-\underset{I\,=\,\text{Ph}}{\underset{\|}{C}}-\text{Ph} \quad \xrightarrow[(2)\ H_2O]{(1)\ Br-Mg-CH_2-CH(CH_3)_2 \,=\, J} \quad CH_3-\underset{\text{Ph}}{\underset{|}{C}}(OH)-CH_2-CH(CH_3)_2 \,=\, K$$

$$\xrightarrow{?} \quad CH_3-\underset{\text{Ph}}{\underset{|}{C}}(O-\underset{\|}{C}(=O)-CH_3)-CH_2-CH(CH_3)_2 \,=\, TM$$

The K ⟶ TM conversion is recognized as esterification of an alcohol. This is a one-step conversion that completes stage 2 of this problem:

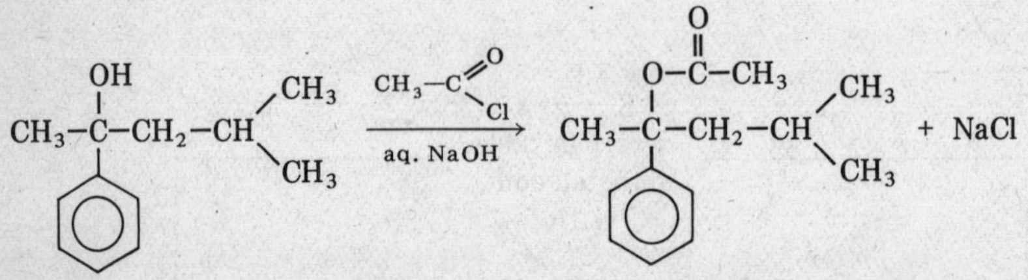

$$CH_3-\underset{\text{Ph}}{\underset{|}{C}}(OH)-CH_2-CH(CH_3)_2 \xrightarrow[\text{aq. NaOH}]{CH_3-C(=O)Cl} CH_3-\underset{\text{Ph}}{\underset{|}{C}}(O-C(=O)-CH_3)-CH_2-CH(CH_3)_2 + NaCl$$

2. K already = TM, no stage 2 required.

3. 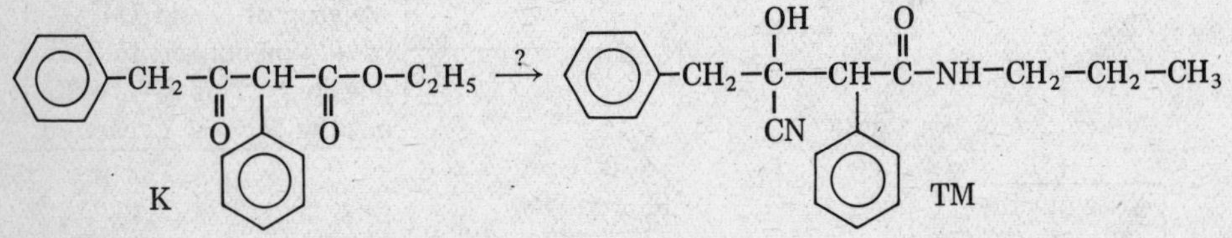 $\text{Ph}-CH_2-\underset{\|}{C}(=O)-CH-\underset{\|}{C}(=O)-O-C_2H_5 \xrightarrow{?} \text{Ph}-CH_2-\underset{CN}{\underset{|}{C}}(OH)(\text{Ph})-CH-C(=O)-NH-CH_2-CH_2-CH_3$

K   TM

With two functions to be converted, more than one step will be required. What compound Y could ⟶ TM in one step?

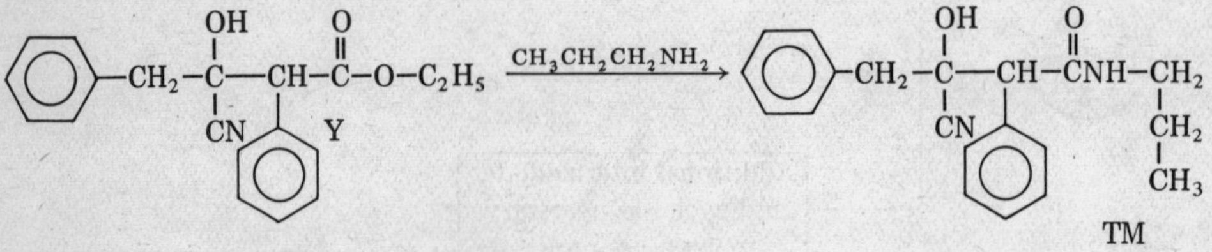

$$\text{Ph}-CH_2-\underset{CN}{\underset{|}{C}}(OH)(\text{Ph})-CH-\underset{\|}{C}(=O)-O-C_2H_5 \xrightarrow{CH_3CH_2CH_2NH_2} \text{Ph}-CH_2-\underset{CN}{\underset{|}{C}}(OH)(\text{Ph})-CH-C(=O)NH-CH_2-CH_2-CH_3$$

Y   TM

Now can K ⟶ this Y in one step? Yes:

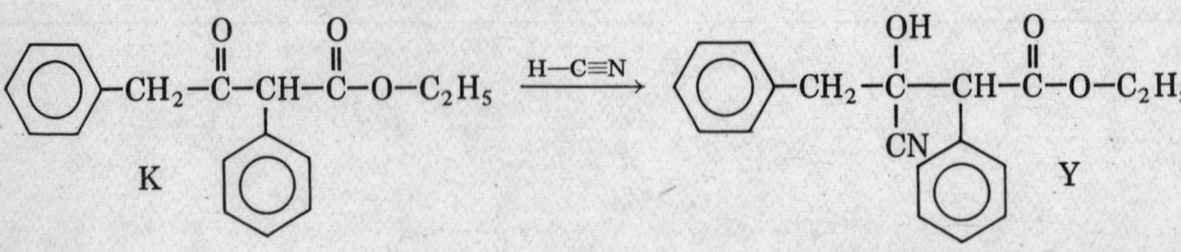

$$\text{Ph}-CH_2-\underset{\|}{C}(=O)-CH(\text{Ph})-\underset{\|}{C}(=O)-O-C_2H_5 \xrightarrow{H-C\equiv N} \text{Ph}-CH_2-\underset{CN}{\underset{|}{C}}(OH)(\text{Ph})-CH-\underset{\|}{C}(=O)-O-C_2H_5$$

K   Y

SYNTHETIC PROBLEMS 581

The sequence K ⟶ Y ⟶ TM completes stage 2 of this problem.

★ *Exercises*

Carry out stage 2 in solving the synthetic problems in the exercises in frame 11.

---

1. Stage 1 solution:

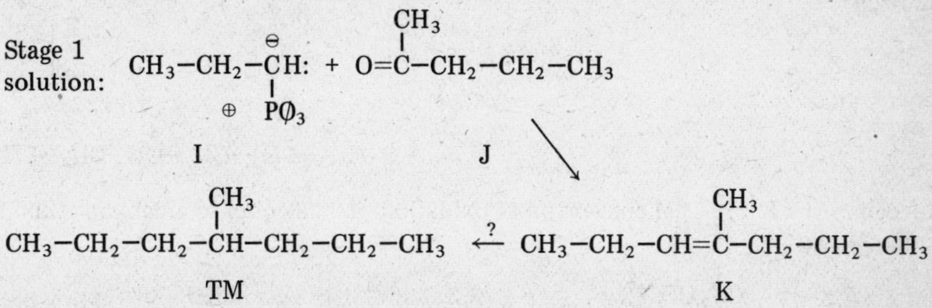

$$CH_3-CH_2-\overset{\ominus}{\underset{\oplus P\varnothing_3}{CH:}} + O=\overset{CH_3}{\underset{|}{C}}-CH_2-CH_2-CH_3$$

   I J

$$CH_3-CH_2-CH_2-\overset{CH_3}{\underset{|}{CH}}-CH_2-CH_2-CH_3 \overset{?}{\leftarrow} CH_3-CH_2-CH=\overset{CH_3}{\underset{|}{C}}-CH_2-CH_2-CH_3$$

   TM   K

Can K ⟶ TM be written as a one-step conversion? Yes:

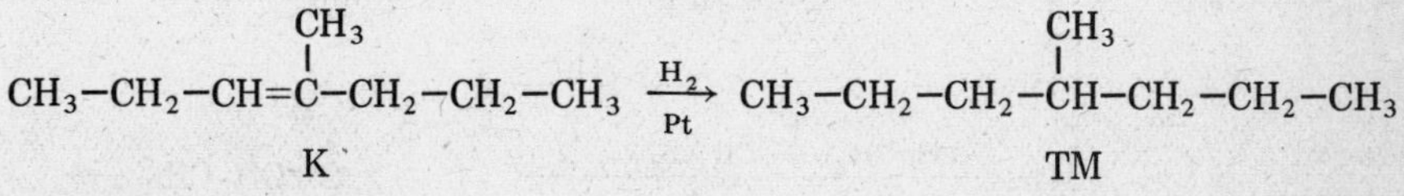

$$CH_3-CH_2-CH=\overset{CH_3}{\underset{|}{C}}-CH_2-CH_2-CH_3 \xrightarrow{H_2 \atop Pt} CH_3-CH_2-CH_2-\overset{CH_3}{\underset{|}{CH}}-CH_2-CH_2-CH_3$$

   K   TM

This completes stage 2.

2. Stage 1 solution:

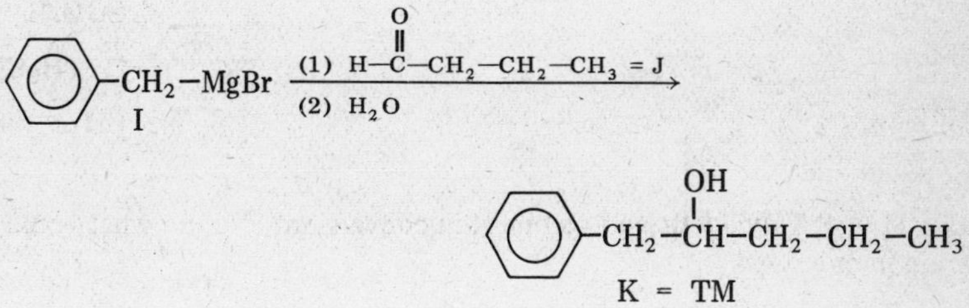

$$\bigcirc\!\!-CH_2-MgBr \xrightarrow[(2)\ H_2O]{(1)\ H-\overset{O}{\underset{\|}{C}}-CH_2-CH_2-CH_3\ =\ J}$$
   I

$$\bigcirc\!\!-CH_2-\overset{OH}{\underset{|}{CH}}-CH_2-CH_2-CH_3$$
   K = TM

No stage 2.

3. Stage 1 solution:

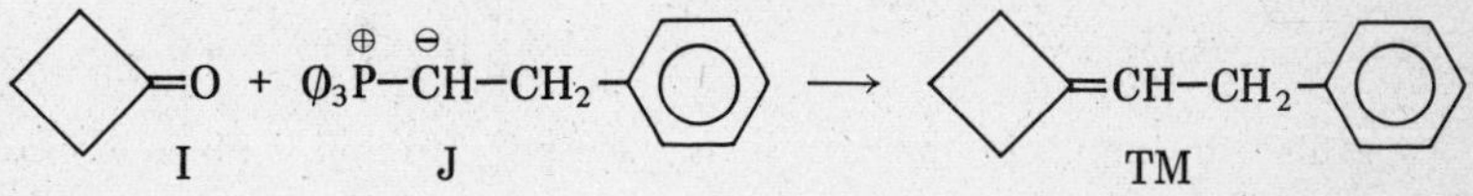

$$\diamondsuit\!\!=O + \varnothing_3\overset{\oplus}{P}-\overset{\ominus}{CH}-CH_2-\!\!\bigcirc \longrightarrow \diamondsuit\!\!=CH-CH_2-\!\!\bigcirc$$
   I   J   TM

No stage 2.

4. Stage 1 solution:

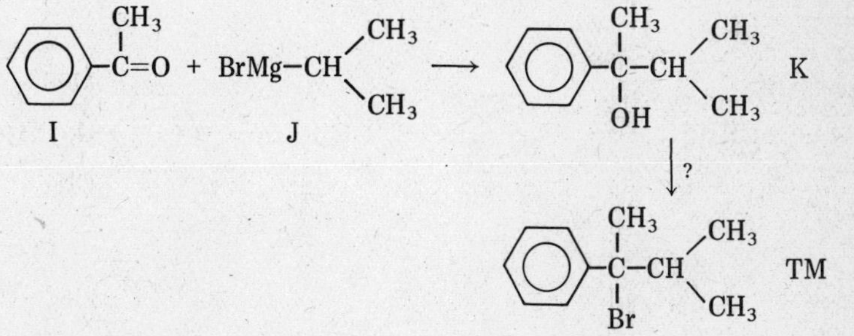

$$\bigcirc\!\!-\overset{CH_3}{\underset{|}{C}}=O + BrMg-CH\!\!\begin{array}{c}CH_3\\ \\ CH_3\end{array} \longrightarrow \bigcirc\!\!-\overset{CH_3}{\underset{\underset{OH}{|}}{C}}-CH\!\!\begin{array}{c}CH_3\\ \\ CH_3\end{array}\quad K$$
   I   J

$$\downarrow ?$$

$$\bigcirc\!\!-\overset{CH_3}{\underset{\underset{Br}{|}}{C}}-CH\!\!\begin{array}{c}CH_3\\ \\ CH_3\end{array}\quad TM$$

Now K ⟶ TM can be recognized as a one-step conversion:

582 HOW TO SUCCEED IN ORGANIC CHEMISTRY

$$C_6H_5-\underset{\underset{OH}{|}}{\overset{\overset{CH_3}{|}}{C}}-CH\underset{CH_3}{\overset{CH_3}{\diagup}} \xrightarrow{HBr} C_6H_5-\underset{\underset{Br}{|}}{\overset{\overset{CH_3}{|}}{C}}-CH\underset{CH_3}{\overset{CH_3}{\diagup}}$$

5. Stage 1 solution:

$$CH_3-CH_2-CH_2-\underset{\underset{I}{\overset{|}{MgBr}}}{CH}-CH_2-CH_3 \xrightarrow{H_2CO = J} CH_3-CH_2-CH_2-\underset{\overset{|}{CH_2OH}}{CH}-CH_2-CH_3 \underset{?}{\overset{K}{\downarrow}}$$

$$CH_3-CH_2-CH_2-\underset{\overset{|}{CHO}}{CH}-CH_2-CH_3 \quad TM$$

Recognize the K ⟶ TM conversion as oxidation of 1° alcohol to aldehyde (Unit 15, frame 2). So:

$$CH_3CH_2CH_2-\underset{\overset{|}{CH_2OH}}{CH}-CH_2CH_3 \xrightarrow[\text{in } C_5H_5N \text{ CHCl}_3]{CrO_3} CH_3CH_2CH_2-\underset{\overset{|}{CHO}}{CH}-CH_2CH_3 \quad TM$$

6. Stage 1 solution:

$$C_6H_5-MgBr \xrightarrow[(2) H_2O]{(1) CH_3-\overset{O}{\overset{\|}{C}}-CH_2CH_3 = J} C_6H_5-\underset{\underset{CH_3}{|}}{\overset{\overset{OH}{|}}{C}}-CH_2CH_3 \xrightarrow{?} K$$

$$C_6H_5-\underset{\underset{CH_3}{|}}{\overset{\overset{COOH}{|}}{C}}-CH_2CH_3 \quad TM$$

Can K ⟶ TM be written as a one-step conversion? No. So what could one make TM from in one step?

$$C_6H_5-\underset{\underset{CH_3}{|}}{\overset{\overset{MgBr}{|}}{C}}-CH_2CH_3 \xrightarrow[(2) H_3O^\oplus/H_2O]{(1) CO_2} C_6H_5-\underset{\underset{CH_3}{|}}{\overset{\overset{COOH}{|}}{C}}-CH_2CH_3 \quad TM$$
$$Y$$

Can this Y be connected back to K in one step? No, but it obviously comes from

$$C_6H_5-\underset{\underset{CH_3}{|}}{\overset{\overset{Br}{|}}{C}}-CH_2CH_3 \text{, which we can recognize as one step away from K:}$$

$$C_6H_5-\underset{\underset{CH_3}{|}}{\overset{\overset{OH}{|}}{C}}-CH_2CH_3 \xrightarrow{HBr} C_6H_5-\underset{\underset{CH_3}{|}}{\overset{\overset{Br}{|}}{C}}-CH_2CH_3 \xrightarrow{Mg} C_6H_5-\underset{\underset{CH_3}{|}}{\overset{\overset{MgBr}{|}}{C}}-CH_2CH_3 \xrightarrow[(2) H_2O/H_3O^\oplus]{(1) CO_2}$$
$$K W Y$$

$$C_6H_5-\underset{\underset{CH_3}{|}}{\overset{\overset{COOH}{|}}{C}}-CH_2CH_3 \quad TM$$

7. Stage 1 solution:

$$\text{(tetrahydrofuran)} \xrightarrow[\text{excess}]{\text{HBr}} \underset{K}{\text{Br-CH}_2\text{CH}_2\text{CH}_2\text{CH}_2\text{-Br}} \xrightarrow{?} \underset{TM}{\text{CH}_3\text{CH}_2\text{-O-CH}_2\text{CH}_2\text{CH}_2\text{CH}_2\text{-O-CH}_2\text{CH}_3}$$

Though two functions must be converted, they are identical and can be converted in one step:

$$\underset{K}{\text{Br-(CH}_2)_4\text{-Br}} \xrightarrow{2\ \text{C}_2\text{H}_5\text{-O}^\ominus\ \text{Na}^\oplus} \text{CH}_3\text{CH}_2\text{-O-(CH}_2)_4\text{-O-CH}_2\text{CH}_3 = \text{TM}$$

8. Stage 1 solution:

$$\underset{I}{\text{CH}_3(\text{CH}_2)_3\text{-O}^\ominus\ \text{Na}^\oplus} + \underset{J}{\text{(ethylene oxide)}} \longrightarrow \underset{K}{\text{CH}_3(\text{CH}_2)_3\text{-O-CH}_2\text{CH}_2\text{-OH}}$$

$$\downarrow ?$$

$$\text{CH}_3\text{CH}_2\text{CH}_2\text{CH}_2\text{-O-CH}_2\text{-CH}_2\text{-O-}\overset{\text{O}}{\underset{\|}{\text{C}}}\text{-(phenyl)} = \text{TM}$$

Recognize this K ⟶ TM as a one-step esterification:

$$\underset{K}{\text{CH}_3(\text{CH}_2)_3\text{-O-CH}_2\text{CH}_2\text{-OH}} \xrightarrow{\text{(phenyl)-C(=O)-Cl}} \underset{TM}{\text{CH}_3(\text{CH}_2)_3\text{-O-CH}_2\text{CH}_2\text{-O-C(=O)-(phenyl)}}$$

9. Stage 1 solution:

$$\underset{I}{\text{(phenyl)-CH}_2\text{-COOH}} + \underset{J}{\text{Br}_2} \xrightarrow{P} \underset{K}{\text{(phenyl)-CH(Br)-COOH}} \xrightarrow{?}$$

$$\underset{TM}{\text{(phenyl)-CH(NH}_2)\text{-CO}_2^\ominus\ \text{NH}_4^\oplus}$$

Recognize the one-step conversion:

$$\underset{K}{\text{(phenyl)-CH(Br)-COOH}} \xrightarrow[\text{NH}_3]{\text{excess}} \underset{TM}{\text{(phenyl)-CH(NH}_2)\text{-CO}_2^\ominus\ \text{NH}_4^\oplus}$$

10. Stage 1 solution:

$$2\ \text{CH}_3\text{CH}_2\text{CHO} \xrightarrow{\text{base}} \underset{K}{\text{CH}_3\text{CH}_2\text{-CH(OH)-CH(CH}_3)\text{-CHO}} \xrightarrow{?}$$

$$\underset{TM}{\text{CH}_3\text{CH}_2\text{-CH(Cl)-CH(CH}_3)\text{-CH(OCH}_3)_2}$$

Two things must be done: alcohol ⟶ halide and aldehyde ⟶ acetal. So either of two orders of operations, ① or ②:

$$K = CH_3-CH_2-\underset{\underset{CH_3}{|}}{\overset{\overset{OH}{|}}{CH}}-CH-CHO \xrightarrow[\text{①}]{HCl} CH_3-CH_2-\underset{\underset{CH_3}{|}}{\overset{\overset{Cl}{|}}{CH}}-CH-CHO = Y$$

$$\downarrow H^{\oplus} \mid CH_3OH \qquad\qquad ② \qquad\qquad \downarrow H^{\oplus} \mid CH_3OH$$

$$CH_3-CH_2-\underset{\underset{CH_3}{|}}{\overset{\overset{OH}{|}}{CH}}-CH-CH\underset{OCH_3}{\overset{OCH_3}{<}} \xrightarrow{HCl} CH_3-CH_2-\underset{\underset{CH_3}{|}}{\overset{\overset{Cl}{|}}{CH}}-CH-CH\underset{OCH_3}{\overset{OCH_3}{<}}$$
$$\qquad\qquad Y \qquad\qquad\qquad\qquad\qquad\qquad TM$$

$H^{\oplus}/CH_3OH$ probably will not interfere with the $-\underset{\underset{Cl}{|}}{CH}-$ function, but since water is

formed in the $-\underset{\underset{OH}{|}}{CH}- \rightarrow -\underset{\underset{Cl}{|}}{CH}-$ reaction, this reaction could conceivably hydrolyze

the $-CH\underset{OCH_3}{\overset{OCH_3}{<}}$ function back to $-CHO$. Therefore, ① is probably the better

order of operations.

13. At this point you have the hardest parts of the synthesis behind you. The problem remaining to be solved in stage 3 is synthesis of the intermediates I and J from the starting materials allowed in the problem. The gap separating them may be large or small:

permissible → ··· → I
starting   ⟩→ K → ··· → TM
materials → ··· → J

If I and J both fall within the list of permissible starting materials, the problem has already been finished in stage 2. Exercise 7 from frames 11 and 12 was of this type.

Or in some cases the carbon skeletons of I and J may fit the description of allowed starting materials, but their functional groups may not. When this happens you need to find a reaction or reaction sequence that turns some allowed function into the function needed in I or J. You do this by exactly the same method used in stage 2 (frame 12) to make the function(s) of K into those of TM.

If the skeleton of I or J is larger or fancier than allowed for starting materials, you must build it up from allowed starting materials. This is just an easier version of stage 1. It is easier because, if you used an alcohol or an alkene for K and made it via a Grignard or a Wittig reaction, then I and J are an alkyl halide (or its Grignard reagent) and a carbonyl compound. Both are easily made. This would usually be done via the corresponding alcohol. In that case the scheme would become

permissible → $\rangle C-OH$ → $\begin{matrix}\rangle C-X \text{ or} \\ \rangle C-Mg-X \\ I\end{matrix}$
starting   ⟩→ K → etc.
materials → $\rangle C-OH$ → $\rangle C=O$
                                                              J

These alcohols are made from the same Grignard + carbonyl compound reactions used previously for the larger alcohol, I or J. If both I and J must be built up from smaller compounds, we have

$$\left.\begin{array}{l}\text{permissible}\\ \text{starting}\\ \text{materials,}\\ \text{e.g., short-}\\ \text{chain ROH}\end{array}\right\} \begin{array}{l}\longrightarrow \geq\!\!\text{C-Mg-X}\\ \longrightarrow \geq\!\!\text{C=O}\\ \longrightarrow \geq\!\!\text{C-Mg-X}\\ \longrightarrow \geq\!\!\text{C=O}\end{array} \longrightarrow \geq\!\!\text{C-OH} \longrightarrow \begin{array}{l}\geq\!\!\text{C-X or}\\ \geq\!\!\text{C-Mg-X}\\ \text{I}\end{array} \longrightarrow \text{K} \longrightarrow \text{etc.}$$

Suppose that alcohols with no more than four carbon atoms are the allowed starting materials. Then the alcohols from which I and J are made can have up to 8 carbons, and K can have up to 16 carbons. This will solve practically all student problems. But notice that we have in this solution a repeating cycle:

$$\begin{array}{l}\text{alcohol A} \longrightarrow \text{Grignard}\\ \qquad\qquad\qquad\qquad\qquad \searrow\\ \qquad\qquad\qquad\qquad\qquad\quad \text{larger}\\ \qquad\qquad\qquad\qquad\qquad\quad \text{alcohol}\\ \text{alcohol B} \longrightarrow \geq\!\!\text{C=O} \qquad\quad \text{C}\end{array}$$

The product alcohol from the first Grignard reaction can be used as a starting alcohol for a second cycle. By repetition an alcohol with a skeleton of any desired complexity could be built up. Synthetic chemists use such cycles, although usually in versatile combination with other reactions.

In addition to this scheme for building up I and J, it is useful for you to focus on a set of reactions that make small adjustments in a carbon skeleton, adding or removing one or two carbons. These are sometimes handy for connecting up to given starting materials. The most common reactions of this type are listed in Table 21.7; you have studied most of them already.

The following examples continue those from frames 11 and 12, completing them by carrying out stage 3.

Examples

1. I = CH$_3$–C(=O)–C$_6$H$_5$ and J = (CH$_3$)$_2$CH–CH$_2$–MgBr to be obtained from alcohols of $\leq$ C$_4$, benzene, and toluene. I (C$_8$) could be made from toluene (C$_7$) plus a C$_1$ fragment, or from benzene (C$_6$) plus a C$_2$ fragment. Try to visualize reactions of these types leading directly to I:

① C$_6$H$_6$ $\xrightarrow[\text{AlCl}_3]{\text{CH}_3\text{–COCl}}$ CH$_3$–C(=O)–C$_6$H$_5$

② C$_6$H$_5$–C(=O)–Cl $\xrightarrow{\text{(CH}_3\text{)}_2\text{Cd}}$

In ① we need CH$_3$–C(=O)–Cl, which comes from CH$_3$COOH, which can be made from the C$_3$ alcohol:

Table 21.7 *Reactions for Lengthening and Shortening Carbon Chains*

A. Lengthening Reactions

Number of carbons added	Equation
1	$R-MgBr + CO_2 \longrightarrow R-COOH$
1	$R-MgBr + HC(OCH_3)_3 \longrightarrow R-CHO$
1	$R-MgBr + H_2CO \longrightarrow R-CH_2-OH$
1	$R-\overset{O}{\underset{\|}{C}}-Cl \xrightarrow[(2)\ H_2O]{(1)\ CH_2N_2/Ag_2O} R-CH_2-CO_2H$
2	$R-MgBr + \underset{O}{\triangle} \longrightarrow R-CH_2-CH_2-OH$
2	$R-Br \xrightarrow[(2)\ H_2O/H_3O^{\oplus}/\Delta]{(1)\ Na^{\oplus\ominus}CH(COOC_2H_5)_2} R-CH_2-CO_2H$

B. Shortening Reactions

Number of carbons removed	Equation
1	$R-CH=CH_2 \begin{array}{c} \xrightarrow{KMnO_4/\Delta} R-COOH \\ \xrightarrow[(2)\ Zn/H_2O]{(1)\ O_3} R-CHO \end{array}$
1	$R-COOH \xrightarrow{(1)\ AgNO_3,\ (2)\ Br_2} R-Br + CO_2$

$CH_3CH_2OH \xrightarrow{Na_2Cr_2O_7} CH_3COOH \xrightarrow{SOCl_2} CH_3-CO-Cl$

In ② we need ⌬—CO—Cl and $(CH_3)_2Cd$. ⌬—CO—Cl comes from

⌬—COOH, which can be gotten from toluene:

⌬—$CH_3 \xrightarrow{KMnO_4}$ ⌬—COOH $\xrightarrow{SOCl_2}$ ⌬—CO—Cl

The $(CH_3)_2Cd$ comes from CH_3-MgX, which comes from CH_3X, which can be made from CH_3OH:

$CH_3-OH \xrightarrow{HX} CH_3-X \xrightarrow{Mg} CH_3-MgX \xrightarrow{CdCl_2} (CH_3)_2Cd$

So altogether ① requires three steps and ② requires six steps to bridge the gap to allowed starting materials. Choose ①. J is made from $(CH_3)_2CH-CH_2-Br$, which can come from an allowed alcohol:

$$(CH_3)_2CH-CH_2-OH \xrightarrow{HBr} (CH_3)_2CH-CH_2-Br$$

The completed solution is:

$$CH_3CH_2OH \xrightarrow{Na_2Cr_2O_7} CH_3COOH \xrightarrow{SOCl_2} CH_3COCl \xrightarrow[AlCl_3]{\phi} CH_3CO-\phi$$

$$(CH_3)_2CH-CH_2OH \xrightarrow{HBr} (CH_3)_2CH-CH_2Br \xrightarrow{Mg} (CH_3)_2CH-CH_2MgBr$$

$$(CH_3)_2CH-CH_2-\underset{CH_3}{\underset{|}{\overset{O-CO-CH_3}{\overset{|}{C}}}}-\phi \xleftarrow{CH_3COCl} (CH_3)_2CH-CH_2-\underset{CH_3}{\underset{|}{\overset{OH}{\overset{|}{C}}}}-\phi$$

Examination shows that in stage 2 we used CH_3COCl (last step) without preparing it from an alcohol; but since this has now been done in steps 1 and 2, it need not be repeated.

2. $I = \phi-CH_2-\overset{\ominus}{C}H-\overset{\oplus}{P\phi_3}$ $J = O=\bigcirc$

The immediate precursor of Wittig reagent I is $\phi-CH_2-CH_2-Br$. According to our discussion above, this would come from the alcohol $\phi-CH_2-CH_2-OH$, which might be built up by an alcohol cycle from alcohols with $\leq C_4$. Starting with a C_4 alcohol, a C_2 extension is needed. Since $\phi-CH_2-CH_2-OH$ is a primary alcohol, the reaction $\phi-Mg-Br + CH_2\overset{O}{-\!\!-\!\!-}CH_2 \longrightarrow \phi-CH_2-CH_2OH$ (Table 21.7A) suggests itself. The completed link between allowed starting materials and I then looks like this:

$$\phi-OH \xrightarrow{HBr} \phi-Br \xrightarrow{Mg} \phi-MgBr \xrightarrow[(2)\ H_2O]{(1)\ \triangledown\!O} \phi-CH_2-CH_2-OH$$

$$\downarrow HBr$$

$$\phi-CH_2-\overset{\ominus}{C}H-\overset{\oplus}{P\phi_3} \xleftarrow[(2)\ base]{(1)\ \phi_3P} \phi-CH_2-CH_2-Br$$

You can see one simple alcohol ⟶ complex alcohol cycle in this scheme.

Prepare J: C_6 and cyclic, not easy from alcohols $\leqslant C_4$, which cannot contain the six-ring preformed. Benzene is a better source, and the first step must be

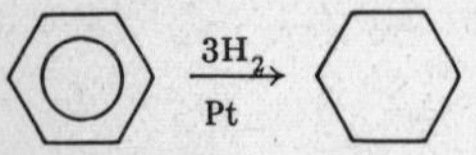

Ph $\xrightarrow{3H_2, Pt}$ Cy. The skeleton is correct, the functional group transformation required is $\begin{array}{c}\diagdown\\\diagup\end{array}C\begin{array}{c}H\\H\end{array} \longrightarrow \begin{array}{c}\diagdown\\\diagup\end{array}C=O$. Search memory for such a reaction, find none, ∴ cannot be done in one step. For a two-step path, the alcohol is the likely intermediate:

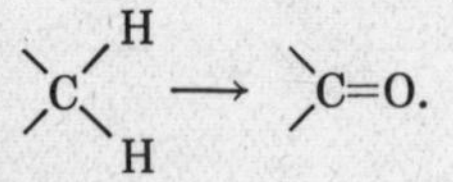

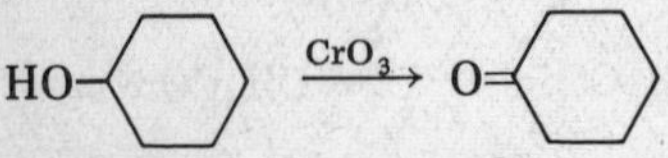

HO–Cy $\xrightarrow{CrO_3}$ O=Cy. Now can HO–Cy be made from Cy in one step? Memory contains no such reaction. So repeat, write a possible precursor of the alcohol: Br–Cy ⟶ HO–Cy. Now can Br–Cy be made from Cy in one step? Yes, recognize the gas-phase photobromination of the alkane:

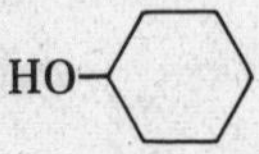

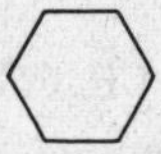

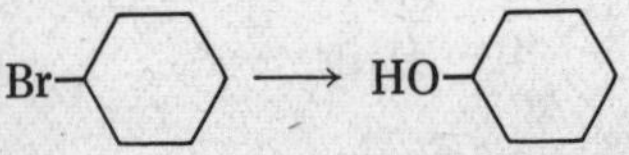

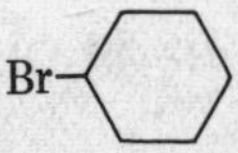

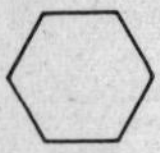

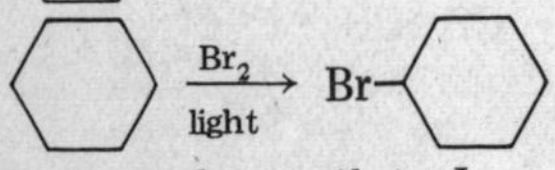

Cy $\xrightarrow[\text{light}]{Br_2}$ Br–Cy

The complete path to J:

Ph $\xrightarrow{3\,H_2, Pt}$ Cy $\xrightarrow[\text{light}]{Br_2}$ Cy–Br $\xrightarrow{NaOH}$ Cy–OH $\xrightarrow{CrO_3}$ Cy=O

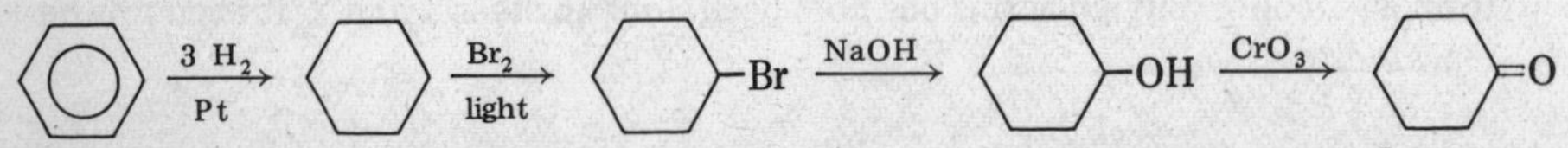

J

This is a successful exception to the "no alkane intermediate" rule; why does it succeed? Because all C—H bonds in this alkane are equivalent.

Combined with the above synthesis of I and the synthesis of TM from I + J, this sequence makes a completed solution to problem (2).

3. I = J = Ph–CH_2–$\overset{\overset{O}{\|}}{C}$–O–$C_2H_5$

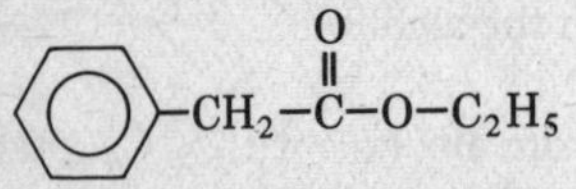

Presumably the ester will be made from the acid:

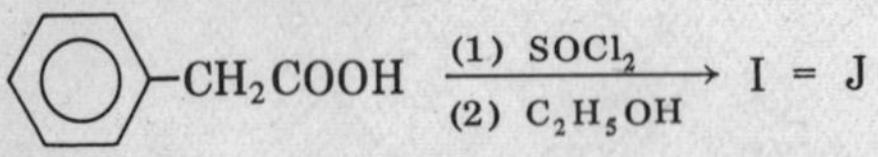

Ph–CH_2COOH $\xrightarrow[\text{(2) }C_2H_5OH]{\text{(1) }SOCl_2}$ I = J

The acid (C_8) could be gotten via C_6 (benzene) + C_2 or C_7 (toluene) + C_1. Table 21.7 lists both a C_1 and a C_2 chain extension leading to the CO_2H function. Applied to this case:

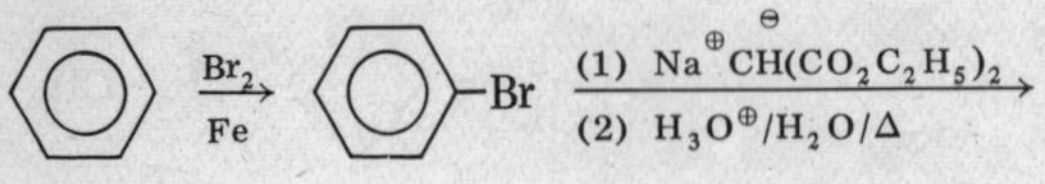

Ph $\xrightarrow[Fe]{Br_2}$ Ph–Br $\xrightarrow[\text{(2) }H_3O^\oplus/H_2O/\Delta]{\text{(1) }Na^\oplus\,{}^\ominus CH(CO_2C_2H_5)_2}$ fails, no S_N2 displacements on unactivated aryl halides

But

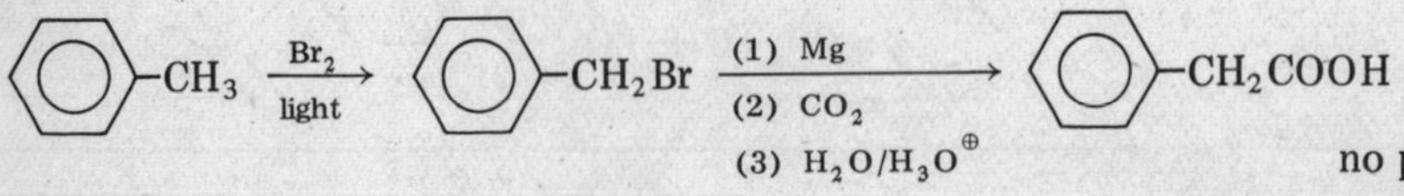

Ph–CH_3 $\xrightarrow[\text{light}]{Br_2}$ Ph–CH_2Br $\xrightarrow[\text{(3) }H_2O/H_3O^\oplus]{\text{(1) Mg} \atop \text{(2) }CO_2}$ Ph–CH_2COOH

no problem

SYNTHETIC PROBLEMS 589

The completed synthesis:

$$Ph\text{-}CH_3 \xrightarrow[\text{light}]{Br_2} Ph\text{-}CH_2Br \xrightarrow[\text{(2) } CO_2]{\text{(1) Mg}}_{\text{(3) } H_3O^\oplus} Ph\text{-}CH_2COOH \xrightarrow[\text{HCl}]{\text{excess } C_2H_5OH}$$

$$Ph\text{-}CH_2\text{-}\underset{\underset{O}{\|}}{C}\text{-}OC_2H_5 \xrightarrow{NaOC_2H_5} Ph\text{-}CH_2\text{-}\underset{\underset{O}{\|}}{C}\text{-}CH(Ph)\text{-}CO_2C_2H_5 \xrightarrow{HCN}$$

$$Ph\text{-}CH_2\text{-}\underset{\underset{CN}{|}}{\overset{\overset{OH}{|}}{C}}\text{-}CH(Ph)\text{-}\underset{\underset{O}{\|}}{C}\text{-}OC_2H_5 \xrightarrow{CH_3CH_2CH_2NH_2} Ph\text{-}CH_2\text{-}\underset{\underset{CN}{|}}{\overset{\overset{OH}{|}}{C}}\text{-}CH(Ph)\text{-}\underset{\underset{O}{\|}}{C}\text{-}NH(CH_2)_2CH_3$$

Have any unallowed materials been used? Yes, $CH_3CH_2CH_2\text{-}NH_2$. Make it from 1-propanol: $CH_3CH_2CH_2OH \xrightarrow{HBr} CH_3CH_2CH_2Br \xrightarrow[NH_3]{\text{excess}} CH_3CH_2CH_2NH_2$

★ *Exercises*

Carry out stage 3 in solving the synthetic problems in the exercises of frames 11 and 12. Put the stages together and write out the whole synthesis.

— — — — — — — — — —

1. $CH_3CH_2CH_2OH \xrightarrow{HBr} CH_3CH_2CH_2Br \xrightarrow[\text{(2) base}]{\text{(1) } \phi_3P} CH_3CH_2\text{-}\overset{\ominus}{C}H\text{-}\overset{\oplus}{P}\phi_3$

$CH_3CH_2CH_2CH_2OH \longrightarrow CH_3CH_2CH_2COOH \longrightarrow CH_3CH_2CH_2COCl$

$CH_3OH \longrightarrow CH_3Br \longrightarrow CH_3MgBr \longrightarrow (CH_3)_2Cd \longrightarrow CH_3\underset{\underset{O}{\|}}{C}CH_2CH_2CH_3 \longrightarrow$ TM

2. $Ph\text{-}CH_3 \xrightarrow[\text{light}]{Br_2} Ph\text{-}CH_2Br \longrightarrow Ph\text{-}CH_2MgBr$

$CH_3CH_2CH_2CH_2OH \longrightarrow CH_3CH_2CH_2CHO \longrightarrow$ TM

3. $\text{cyclobutyl-OH} \xrightarrow{KMnO_4} \text{cyclobutanone} \longrightarrow$ TM

$Ph\text{-}CH_2\overset{\ominus}{C}H\text{-}\overset{\oplus}{P}\phi_3 \uparrow$

$Ph\text{-}CH_2CH_2Br \uparrow$

$Ph\text{-}CH_2CH_2OH \uparrow$

$Ph \longrightarrow Ph\text{-}Br \longrightarrow Ph\text{-}MgBr$

4. $CH_3CH_2OH \rightarrow CH_3COOH \rightarrow CH_3COCl \rightarrow CH_3CO-C_6H_5 \rightarrow$ TM

$\begin{array}{c}CH_3\\ CH_3\end{array}\!\!>\!\!CH-OH \rightarrow \begin{array}{c}CH_3\\ CH_3\end{array}\!\!>\!\!CH-Br \rightarrow \begin{array}{c}CH_3\\ CH_3\end{array}\!\!>\!\!CH-MgBr$

5. $CH_3CH_2CH_2OH \rightarrow CH_3CH_2CHO$
   $\hspace{5em}\searrow$
   $\hspace{7em} CH_3CH_2\overset{OH}{\underset{|}{C}}HCH_2CH_3$

   $CH_3CH_2OH \rightarrow CH_3CH_2Br \rightarrow CH_3CH_2MgBr$
   $\hspace{19em}\downarrow$
   $\hspace{16em}CH_3CH_2\overset{Br}{\underset{|}{C}}HCH_2CH_3$
   $\hspace{19em}\downarrow$
   $\hspace{16em}CH_3CH_2\overset{MgBr}{\underset{|}{C}}HCH_2CH_3$

   $CH_3OH \rightarrow HCHO \hspace{2em}$
   $\hspace{8em}\downarrow$
   $\hspace{7em}$TM

6. $C_6H_5-Br \rightarrow C_6H_5-MgBr \rightarrow C_6H_5-\overset{OH}{\underset{CH_3}{\overset{|}{\underset{|}{C}}}}-CH_2-CH_3$

   $CH_3-CH_2-CH(OH)-CH_3 \rightarrow CH_3-CH_2-\overset{O}{\overset{||}{C}}-CH_3$

   TM $\leftarrow C_6H_5-\overset{MgBr}{\underset{CH_3}{\overset{|}{\underset{|}{C}}}}-CH_2-CH_3 \leftarrow C_6H_5-\overset{Br}{\underset{CH_3}{\overset{|}{\underset{|}{C}}}}-CH_2-CH_3$

7. (tetrahydrofuran) $\xrightarrow[\text{excess}]{HBr}$ Br-(CH$_2$)$_4$-Br
   $\hspace{15em}\searrow$
   $\hspace{17em}$TM
   $CH_3-CH_2-OH \xrightarrow{Na} CH_3-CH_2-O^\ominus Na^\oplus$
   excess $\hspace{5em}$ excess

8. $CH_3CH_2CH_2CH_2OH \xrightarrow{Na} CH_3CH_2CH_2CH_2O^\ominus Na^\oplus$
   $\hspace{20em}\searrow$
   $\hspace{20em}CH_3(CH_2)_3-O-CH_2CH_2-O^\ominus$
   $CH_3CH_2OH \xrightarrow[\text{heat}]{H_2SO_4} CH_2=CH_2 \xrightarrow[Ag]{O_2} CH_2\overset{O}{\diagdown\!\!\diagup}CH_2 \hspace{5em} Na^\oplus$

   $C_6H_5-CH_3 \rightarrow C_6H_5-CO_2H \rightarrow C_6H_5-\overset{O}{\overset{||}{C}}-Cl$
   $\hspace{20em}\searrow$
   $\hspace{20em}$TM

9. Ph–CH₃ → Ph–CH₂Br → Ph–CH₂MgBr → Ph–CH₂CO₂H
   ↓
   Ph–CH(NH₂)–C(=O)–O⁻ NH₄⁺ ← Ph–CH(Br)–CO₂H

10. CH₃–CH₂–CH₂OH → CH₃–CH₂–CHO → CH₃–CH₂–CH(OH)–CH(CH₃)–CHO
    ↓
    CH₃–CH₂–CH(Cl)–CH(CH₃)–CH(OCH₃)₂ ← CH₃–CH₂–CH(Cl)–CH(CH₃)–CHO

You have now completed *How to Succeed in Organic Chemistry*. Saving your solutions to the exercises and the summaries that you prepared may prove worthwhile if you review the subject in connection with later courses or for the MCAT, etc. Remember also that the techniques used in this book to learn organic chemistry can be applied successfully and with little change to many other subjects. Good learning!

Index

Acetal, 395
Acetoacetic ester synthesis, 568
Achiral, 260, 286
Acid–base equilibria, 192, 317
  of alcohols, 335
  of carbonyl compounds, 393, 398
  of carboxylic acids, 441
  position as function of pH, 487
Acids (Lewis and Brønsted), 178, 187
  catalysis by, 396, 446, 470
  strength, 191
Activating groups, 378
Acyl S_N2 mechanism, 445, 470
Addition reactions, 202
Aldol condensation, 431
Aliphatic/aromatic structures, 39
Aliphatic syntheses, 566
Alkylation, 436
Alkynes, alkylation of, 568
Allylic position (site), 43, 71
Alpha hydrogens, 398
Ammonium salts, 491
Amphiprotic compounds, 189
Analogies, 236
  N for O in functional groups, 481
Anti addition, 300
Anti conformer, 57
Aromatic syntheses, 545
Aryl S_N2 mechanism, 371, 383
Asymmetric carbon atom, 254, 262, 280

Bases, 178, 187
  in elimination reactions, 309, 320, 321
  strength, 191
Benzylic position (site), 43, 71
Benzyne, 372
Blocking group, 555
Bonding units, 13
Branching, 72

Cannizzaro reaction, 408
Carbonium ions, 139, 212, 306, 320, 322
  destruction of, 326
  rearrangement of, 349

Carbonyl group, 393
Carboxylic acids, strength, 441
Chain-extending/shortening reactions, 586
Chiral(ity), 260, 295, 298
  in structure problems, 520
Cis/trans isomers, 252, 266
Claisen condensation, 450
Clockwise/counterclockwise orders, 253, 269, 271
Condensed structural formula, 12, 28, 250
Configuration, 256, 266, 272, 274, 278
Conformation, conformer, 57, 248, 258
Conjugate acid (base), 182
Conjugation, 44, 157
Contributing structure, 142
Cope elimination (degradation), 531

Deactivating group, 378
Degradative reactions, 522, 536
Dehydration, 353, 448
Dehydrogenation, 343
Delocalization of charge, 307
Diagnostic reactions, 509, 513
Diastereomer, 265, 271, 285, 287, 296
Difunctional compounds
  naming, 125
  synthesis, 569
Dissymmetry, 260

Eclipsed conformation, 57
Electron-acceptor functions, 154
Electron-donor functions, 154
Electrophiles, 212, 381
Electrophilic addition mechanism, 212
  stereochemistry, 298
Electrophilic substitution
  aliphatic, 309
  aromatic, 363, 369
Elimination reactions, 202, 320, 321
Enamine alkylation, synthetic
  applications, 568
Enantiomer, 265, 271, 285, 287, 296

Energy of molecules, 133, 150
Enol, 401, 449
Enolate ion, 398
   condensation reactions, 567
   as nucleophile, 406, 432, 450
Equivalent sites, 78
$E2$ reaction (mechanism), 321
Error checks, 238, 243
Exhaustive methylation, 533
E/Z isomers (configurations), 274

Flowcharts, 127
Formal charges, 24
Function (functional group), 32

Gauche conformer, 57
Generic structures, 26
Geometry of molecules, 47
   and resonance, 151
Grignard syntheses, 427, 567, 585
Group names, 111

Handedness, 252, 264, 272
Hemiacetal/hemiketal, 395
Hoffman degradation (elimination), 531
Hydride (ion) transfer, 408
Hydrogenation, 221, 448
Hydrogenolysis, 448
Hydrolysis, 316, 440, 471

Infrared spectra, 519
Intermediates, 138, 199
Ionic bond, 91
IR spectra, 519
Isomers, 77, 248
IUPAC names, 96

Ketal, 395

Language of organic chemistry, 2
Learning reactions, 199, 222, 314, 375,
   409, 456, 497
Leaving groups, 305, 471
   esterified, 448, 476
   protonated, 335, 495
   quaternized-N, 495
Lewis structures, 26
Limiting structure, 142
Linear carbon, 50, 52, 55
Line-segment structures, 40

Malonic ester synthesis, 568
Mass spectra, 522

Mechanisms, 165, 239
   standard, 175
   stereochemical results of, 289, 332
Memorization, 315, 341, 376
Meso compounds, 287
Meta director, 378
Mirror images, 260
Models, 60
Molecular formula, 17

Naming compounds, 96
   amines, 115
   aromatics, 116
   ethers, 114
   groups, 111
Neutralization equivalent, 515
Newman projections, 57
Nitrogen analogs of oxygen functions,
   481
NMR spectra, 519
Notation, student difficulties with, 3, 257
Nuclear magnetic resonance, 519
Nucleophiles, 308, 335, 416
Nucleophilic addition, 394, 415
   mechanism, 396, 418
   to nitriles, 452
Nucleophilic substitution, 308, 319, 337
   by amines, 491
Numbering (in naming), 104, 118

Open bond (open valence), 27
Order of groups, 253, 256, 269
Organometallic compounds, 309, 316
   reaction mechanism, 326
Orientation, 377, 385
Ortho/para director, 378
Oxidation reactions, 202
   of alcohols, 341
   of amines, 492, 497
Oxidation stages, 220, 317, 342, 410,
   453
Ozonolysis, 218
   in structure problems, 526

Parent compound, 116
Part structures, 503
Perspective structures, 67
pH, 486
Plane of symmetry, 260
Polarized light, 271
Predominant product, 232, 234
Primary carbon or position (site), 42, 71
Principal group, 96, 106

Problem-solving strategies, 5, 223, 329, 344

Racemic mixtures, 272, 332
Racemization, 290, 293
Radical addition mechanism, 214
   stereochemistry, 303
Radical substitution mechanism, 211
   stereochemistry, 291
Reaction rates (reactivity), 327, 375, 379
Reagents, classification, 456
Reduction reactions, 203, 394, 408, 455
Resonance, 142
   effects on acidity and basicity, 490
   energy, 150
   stabilization, 150, 356
R/S isomers (configurations), 272, 282

Saponification equivalent, 518
Saturated carbons and skeletons, 39
Sawhorse structures, 58
Side-chain oxidation, 530
Skeleton, 32
Slot filler, 208, 212, 308, 338, 344, 363, 416
$S_N1/E1$ reaction (mechanism), 320
   stereochemistry, 332
S_N2 reaction, 319
   stereochemistry, 319, 332
Solvent polarity, 322
Space perception, 7
Specification of configuration, 272, 278
Specific rotation, 264, 271, 279
Stability of molecules, 133
   effect of resonance, 150
   prediction of, 136

Staggered conformation, 57
Starting material, 199
Stereochemistry of reactions, 289
Stereoisomers, 248, 265, 287
Strength of acids and bases, 191
Structural formula, 12, 17
   errors in, 22
   three-dimensional, 257
Structural isomers, 77
Study methods, 8, 10
Subsidiary group, 97, 107
Substituent, 97, 109, 121
Substitution, 37
   reactions, 202
Superimposability, 61, 253, 268
Syn addition, 300
Syntheses, one-step, 543
Synthesis, purposes, 540
Systematic names, 96

Tautomers, 402
Terminal alkyne, 37
Tetrahedral, 49
Trigonal-planar (carbon or geometry), 50, 52, 55

Unsaturation, 39
Unsaturation number, 86
   use in structure problems, 509

Vinylic position (site), 43, 71

Wittig reaction, 430, 567
Wurtz reaction, 311

Yield, 200